Surgical Robotics

Jacob Rosen • Blake Hannaford
Richard M. Satava
Editors

Surgical Robotics

Systems Applications and Visions

Springer

Editors
Jacob Rosen
Department of Computer Engineering
Jack Baskin School of Engineering
University of California Santa Cruz
1156 High Street, Santa Cruz
CA 95064, USA
rosen@ucsc.edu

Blake Hannaford
Department of Electrical Engineering
University of Washington
Box 325500, Seattle
Washington 98195-2500
USA
hannaford@ee.washington.edu

Richard M. Satava
Department of Surgery
University of Washington Medical Center
Box 356410
1959 Pacific Street NE, Seattle
Washington 98195, USA
rsatava@u.washington.edu

DOI 10.1007/978-1-4419-1126-1
Springer New York Dordrecht Heidelberg London

Printed on acid-free paper

Springer is part of Springer Science+Business Media (www.springer.com/mycopy)

Preface

The dictum *Primum non nocere* (First, do no harm) and the dictum "Primum Succurrere" (First, hasten to help) as the prime directives of ethics in medicine may dictate two orthogonal approaches of practicing medicine, both of which are aimed to provide the best health care to the patient. The conservative approach relies on decades of evidence-based practice and clinical experience with a specific medical or surgical approach. However, every now and then, a scientific, technological, or clinical breakthrough occurs (alone or in combination) which leads to a paradigm shift along with disruptive new approach to health care. To some extent, this progressive approach is regulated by rigorous clinical trials as dictated by the Federal and Drug Administration (FDA) aimed at demonstration of safety and effectiveness. Although the progressive treatment approach results in a relativity high risk, there is a concomitant high reward in terms of healing and regaining a high quality of life.

Surgical robotics is a recent and very significant breakthrough in surgery. The introduction of a surgical robot into the operating room (OR) combines a technological breakthrough with a clinical breakthrough in developing new surgical techniques and approaches to improve the quality and outcome of surgery. As significant as these breakthroughs are, it is not surprising that they occurred because they are based on more than a decade of innovation in field of robotics in both academia and industry. The promise of surgical robotics is to deliver high levels of dexterity and vision to anatomical structures that cannot be approached by the surgeon's fingers and viewed directly by the surgeon's eyes. Making this technology available to surgeons has led to new surgical techniques that could not be accomplished previously. It is likely that clinical knowledge accumulated using these new systems or even by simply realizing their capabilities will lead to the development of new surgical robotic systems in the future. The surgical robot and various imaging modalities may be viewed as mediators between the surgeon's hands and eyes and the surgical site, respectively; however, these two elements are part of a larger information system that will continue to evolve and affect every aspect of surgery and healthcare in general. Archived medical history, preoperative scans, preplanning, quantitative recording of the surgical execution, follow-up and outcome assessment are all part of feed forward and feedback mechanisms that will improve the quality of healthcare.

As product of a rapidly evolving research field, this assembly of monographs aimed to capture a wide spectrum of topics spanning from ambitious visions for the future down to today's clinical practice. The book is divided into four sections:

1. *The vision and overviews section* reviews the field from the civilian and military perspectives. It includes chapters discussing the Trauma Pod concept – a vision of an OR without humans. The importance of the trauma pod project was that it demonstrated the capability of automating all the services in the OR – services that are currently provided today by a scrub nurse and a circulation nurse that have been demonstrated to be translates to services by a robotic cell – robotic arms and information systems. Whether this concept of automation will be extended into clinical practice and thereby emphasizing even more the role of a surgeon as a decision maker while the operation is executed by the surgical robot automatically is yet to be seen.
2. *The systems section* is divided into two subsections including chapters describing key efforts in systems development and integration of macro- (first section) and micro- (second section) surgical robots in both academia and industry. Developing a macro-surgical robotic system is challenging in part due to the difficulties in translating qualitative clinical requirements into quantitative engineering specifications. Moreover, a successful system development as a whole is often a result of multidisciplinary and interdisciplinary efforts including all the subdisciplines of engineering and surgery – efforts that should not be taken lightly. In addition to challenges of macro-systems development, developing surgical robotics on a micro-system level introduces a significant reduction in scale. Forces, torques, pressures, and stresses do not scale down linearly with the geometrical dimensions. These interesting scaling properties challenge many engineering and surgical concepts. Inspired by the film "Fantastic Voyage," the promise of a micro-robotic system is the capability to travel in the human body and provide local treatment. This concept is still in its infancy, and the academic research currently conducted in this field is focused on fundamental aspects of the system such as propulsion, navigation, energy source, manipulation, and control.
3. *The engineering developments section* covers technologies, algorithms, and experimental data to enhance and improve the current capabilities of surgical robotics. Topics of chapters in this section include tactile and force feedback, motion tracking, needle steering, soft tissue biomechanics of internal organs, and objective assessment of surgical skill. All of these will be incorporated into different layers of the surgical robotic systems in the future and will eventually put a superior robotic system in the hands of the surgeon for improving the outcome.
4. *The clinical applications section* includes chapters authored by surgeons who use surgical robotic systems clinically and describe the current clinical

applications of surgical robotics in several subdisciplines of surgery including urology, cardiology, neurosurgery, pediatric surgery gynecology, and general surgery as well as telesurgery. Most of these chapters also provide some thoughts about future applications of surgical robots in surgery. The generic nature of the surgical robotic system allows the surgeon to explore many surgical procedures that were not targeted by the robot's original developers. Moreover, today's growing vast array of clinical applications of surgical robotics demonstrates that the clinical community can adopt new surgical approaches once a capable tool such as a robot is made available.

Jacob Rosen
Blake Hannaford
Richard M. Satava

Contents

Part III Engineering Developments

Part IV Clinical Applications/Overviews

Contributors

Jake J. Abbott
Department of Mechanical Engineering, University of Utah, 50 S. Central Campus Dr., Salt Lake City, UT 84112, USA
jake.abbott@utah.edu

Rony Abovitz
MAKO Surgical Corp., 2555 Davie Road, Ft. Lauderdale, FL 33317, USA

Ron Alterovitz
Department of Computer Science, University of North Carolina, Chapel Hill, NC 27599, USA
ron@es.unc.edu

Mehran Anvari
Department of Surgery, McMaster Institute for Surgical Innovation, Invention and Education, Faculty of Health Sciences, McMaster University, Hamilton, ON, Canada;
St. Joseph's Healthcare, 50 Charlton Ave, East Room 805, Hamilton, ON L8N 4C6, Canada
anvari@mcmaster.ca

Özkan Bebek
Department of Electrical Engineering and Computer Sciences, Case Western Reserve University, Cleveland, OH, USA
ozkan.bebek@case.edu

Brian C. Becker
Robotics Institute, Carnegie Mellon University, 5000 Forbes Ave, Pittsburgh, PA 15213, USA

Christos Bergeles
Institute of Robotics and Intelligent Systems, ETH Zurich, Tannenstr. 3, CLA H 17.1, 8092, Zurich, Switzerland
cbergeles@ethz.ch

Peter Berkelman
Department of Mechanical Engineering, University of Hawaii-Manoa,
2540 Dole St, Honolulu, HI 96822, USA
peterb@hawaii.edu

James W. Bisley
Center for Advanced Surgical and Interventional Technology (CASIT),
University of California Los Angoles, Los Angeles, CA 90095, USA;
Department of Neurobiology, UCLA, Los Angeles, CA 90095, USA;
Department of Psychology, UCLA, Los Angeles, CA 90095, USA

Antônio P.L. Bó
LIRMM, Montpellier France

John F. Boggess
The Division of Gynecologic Oncology, Department of Obstetrics
and Gynecology, University of North Carolina, Campus Box 7572,
Chapel Hill, NC 27599-7572, USA
jboggess@med.unc.edu

Timothy J. Broderick
US Army Medical Research and Materiel Command Telemedicine,
Advanced Technology Research Center MCMR-TT, 504 Scott
St Fort Detrick, Frederick, MD 21702, USA
timothy.j.broderick@us.army.mil

Jeff Brown
Intuitive Surgical Inc., 1266 Kifer Road, Sunnyvale, CA, USA

M. Cenk Çavuşoğlu
Department of Electrical Engineering and Computer Sciences,
Case Western Reserve University, 308 Glennan Building, Cleveland, OH, USA
cavusoglu@case.edu

Gregory S. Chirikjian
Department of Mechanical Engineering, Johns Hopkins University,
Baltimore, MD 21218, USA

W. Rodolph Chitwood
East Carolina Heart Institute, Department of Cardiovascular Sciences,
East Carolina University, Greenville, NC 27834, USA
chitwoodw@ecu.edu

Howie Choset
The Robotics Institute, Carnegie Mellon University, Pittsburgh, PA 15213, USA
choset@cs.cmu.edu

Michael Conditt
MAKO Surgical Corp., 2555 Davie Road, Fort Lauderdale, FL 33317, USA
mconditt@makosurgical.com

Noah J. Cowan
Department of Mechanical Engineering, Johns Hopkins University,
Baltimore, MD 21218, USA
ncowan@jhu.edu

Martin O. Culjat
Center for Advanced Surgical and Interventional Technology (CASIT),
University of California, Los Angeles, Los Angeles, CA 90095, USA;
Department of Surgery, UCLA, Los Angeles, CA 90095, USA;
Department of Bioengineering, UCLA, Los Angeles, CA 90095, USA
mculjat@mednet.ucla.edu

Paolo Dario
Scuola Superiore Sant'Anna, Pisa, Italy

Smita De
Department of Electrical Engineering, University of Washington,
Seattle, WA, USA

Amir Degani
The Robotics Institute, Carnegie Mellon University, Pittsburgh, PA 15213, USA

Simon DiMaio
Intuitive Surgical Inc., 1266 Kifer Road, Sunnyvale, CA, USA
simon.dimaio@intusurg.com

Charles R. Doarn
Departments of Surgery and Biomedical Engineering, University of Cincinnati,
2901 Campus Drive, Cincinnati, OH 45221, USA
charles.doarn@uc.edu

Erik P. Dutson
Center for Advanced Surgical and Interventional Technology (CASIT),
UCLA, Los Angeles, CA 90095, USA;
Department of Surgery, UCLA, Los Angeles, CA 90095, USA

Olgaç Ergeneman
Institute of Robotics and Intelligent Systems, ETH Zurich, Tannenstr. 3,
CLA H 17.1, 8092 Zurich, Switzerland
oergeneman@ethz.ch

Richard E. Fan
Center for Advanced Surgical and Interventional Technology (CASIT), UCLA,
Los Angeles, CA 90095, USA;
Department of Bioengineering, UCLA, Los Angeles, CA 90095, USA

Shane M. Farritor
Department of Mechanical Engineering, University of Nebraska-Lincoln,
N104 SEC, Lincoln, NE 68588-0656, USA
sfarritor@unl.edu

Gabor Fichtinger
Queen's University, Kingston, ON, Canada K7L 3N6

D. Friedman
Department of Electrical Engineering, University of Washington, Seattle, WA, USA

Pablo Garcia
SRI International, 333 Ravenswood Avenue, Menlo Park, CA 94025, USA
pablo.garcia@sri.com

Gary R. Gilbert
Georgetown University Imaging Science and Information Systems (ISIS) Center, US Army Medical Research and Materiel Command Telemedicine;
Advanced Technology Research Center MCMR-TT, 504 Scott, St Fort Detrick, Frederick, MD 21702, USA
gary.gilbert@tatrc.org

D. Glozman
Department of Computer Engineering, Baskin School of Engineering SOE-3, University of California Santa Cruz, 1156 High Street, Santa Cruz, CA 95064, USA

Ken Goldberg
University of California, Berkeley, CA 94720, USA

Paul G. Griffiths
Johns Hopkins University, Baltimore, MD 21218, USA

Warren S. Grundfest
Center for Advanced Surgical and Interventional Technology (CASIT), UCLA, Los Angeles, CA 90095, USA;
Department of Surgery, UCLA, Los Angeles, CA 90095, USA;
Department of Bioengineering, UCLA, Los Angeles, CA 90095, USA;
Department of Electrical Engineering, UCLA, Los Angeles, CA 90095, USA

James C. Gwilliam
Johns Hopkins University, Baltimore, MD 21218, USA

Benny Hagag
MAKO Surgical Corp., 2555 Davie Road, Ft. Lauderdale, FL 33317, USA
bhagag@makosurgical.com

Gregory D. Hager
Computer Science Department, Johns Hopkins University, 3400 N. Charles Street, Baltimore, MD 21218, USA

Rabbie K. Hanna
The Division of Gynecologic Oncology, Department of Obstetrics and Gynecology, University of North Carolina, Campus Box 7572, Chapel Hill, NC 27599-7572, USA

Blake Hannaford
Department of Electrical Engineering, University of Washington, Seattle, WA, USA

Mike Hanuschik
Intuitive Surgical Inc., 1266 Kifer Road, Sunnyvale, CA, USA

Kanako Harada
Scuola Superiore Sant'Anna, Pisa, Italy

Santiago Horgan
Department of Surgery, University of California San Diego,
San Diago CA 92403
shorgan@ucsd.edu

Ryan S. Hsi
Seattle Children's Hospital, 4800 Sand Point Way Northeast, Seattle, WA, USA

Vinutha Kallem
University of Pennsylvania, Philadelphia, PA 19104, USA

Hyosig Kang
MAKO Surgical Corp., 2555 Davie Road, Ft. Lauderdale, FL 33317, USA

Louis J. Kim
Department of Neurological Surgery, University of Washington,
325, 9th Avenue, Seattle, WA 98104, USA
ljkim1@u.washington.edu

Chih-Hung King
Center for Advanced Surgical and Interventional Technology (CASIT), UCLA, Los Angeles, CA 90095, USA;
Department of Bioengineering, UCLA, Los Angeles, CA 90095, USA

Gábor Kósa
Computer Vision Laboratory, Department of Information Technology and Electrical Engineering, ETH Zurich, Switzerland
kosa@vision.ee.ethz.ch

Usha Kreaden
Intuitive Surgical Inc., 1266 Kifer Road, Sunnyvale, CA, USA

Michael P. Kummer
Institute of Robotics and Intelligent Systems, ETH Zurich, Tannenstr. 3,
CLA H 17.1, 8092, Zurich, Switzerland
kummerm@ethz.ch

Amy C. Lehman
Department of Mechanical Engineering, University of Nebraska-Lincoln,
N104 SEC, Lincoln, NE 68588-0656, USA
alehman3@gmail.com

E.J. Lehr
Department of Cardiovascular Sciences, East Carolina Heart Institute, East Carolina University, Greenville, NC 27834, USA
ericjlehr@gmail.com

Thomas S. Lendvay
Seattle Children's Hospital, 4800 Sand Point Way Northeast, Seattle, WA, USA
thomas.lendvay@seattlechildrens.org

Mitchell Lum
4801 24th Ave, NE #505, Seattle, WA 98105, USA
mitchlum@u.washington.edu

Ji Ma
Department of Mechanical Engineering, University of Hawaii-Manoa, 2540 Dole St, Honolulu, HI 96822, USA
jima@hawaii.edu

Robert A. MacLachlan
Robotics Institute, Carnegie Mellon University, 5000 Forbes Ave, Pittsburgh, PA 15213, USA

Richard A. Malthaner
800 Commissioners Road EastSuite E2-124 London, ON N6A 5W9, Canada
richard.malthaner@lhsc.on.ca

Sylvain Martel
NanoRobotics Laboratory, Department of Computer and Software Engineering, Institute of Biomedical Engineering École Polytechnique, de Montréal (EPM), Station Centre-Ville, Montréal, QC, Canada
sylvain.martel@polymtl.ca

John Meehan
Department of Surgery, Seattle Children's Hospital, University of Washington, Seattle, Washington, USA
john.meehan@seattlechildrens.org

Ozanan Meireles
Department of Surgery, University of California, San Diago, San Diago, CA 92103
omeireles@ucsd.edu

Arianna Menciassi
Scuola Superiore Sant'Anna, Viale Rinaldo Piaggio 34-5602, Pisa, Italy
arianna.menciassi@sssup.it

Sarthak Misra
University of Twente, 7500 AE Enschede, The Netherlands

K. Moe
Department of Neurological Surgery, University of Washington, 325, 9th Avenue, Seattle, WA 98104, USA

Gerald R. Moses
University of Maryland, Baltimore, MD, USA
gmoses@smail.umaryland.edu

Michael D. Naish
Department of Mechanical & Materials Engineering,
Department of Electrical & Computer Engineering, The University
of Western Ontario, London, Ontario, Canada N6A 5B9;
Lawson Health Research Institute (LHRI), Canadian Surgical Technologies &
Advanced Robotics (CSTAR), 339 Windermere Road London, Ontario,
Canada N6A 5A5
mnaish@uwo.ca; michael.naish@lawsonresearch.com

Bradley J. Nelson
Institute of Robotics and Intelligent Systems, ETH Zurich, Tannenstr. 3,
CLA H 17.1, 8092, Zurich, Switzerland
bnelson@ethz.ch

Allison M. Okamura
Department of Mechanical Engineering, Johns Hopkins University,
Baltimore, MD 21218, USA
aokamura@jhu.edu

Dmitry Oleynikov
Department of Surgery, University of Nebraska Medical Center,
983280 Nebraska Medical Center, Omaha, NE 68198-3280, USA
doleynik@unmc.edu

Takeyoshi Ota
Division of Cardiothoracic Surgery, University of Pittsburgh,
Pittsburgh, PA 15213, USA

Wooram Park
Johns Hopkins University, Baltimore, MD 21218, USA

Rajni V. Patel
Department of Mechanical & Materials Engineering,
Department of Electrical & Computer Engineering, The University
of Western Ontario, London, Ontario, Canada N6A 5B9;
Lawson Health Research Institute (LHRI), Canadian Surgical Technologies
& Advanced Robotics (CSTAR), 339 Windermere Road London, Ontario,
Canada N6A 5A5

Nicholas A. Patronik
Robotics Institute, Carnegie Mellon University, Pittsburgh, PA, USA

Melissa T. Perri
Department of Mechanical & Materials Engineering,
Department of Electrical & Computer Engineering, The University of Western Ontario, London, Ontario, Canada N6A 5B9;
Lawson Health Research Institute (LHRI), Canadian Surgical Technologies & Advanced Robotics (CSTAR), 339 Windermere Road London, Ontario, Canada N6A 5A5

Philippe Poignet
LIRMM, Montpellier, France
philippe.poignet@lirmm.fr

D. Ramanathan
Department of Neurological Surgery, University of Washington, 325, 9th Avenue, Seattle, WA 98104, USA

Kyle B. Reed
University of South Florida, Tampa, FL 33620, USA

Rogério Richa
LIRMM 161 Rue Add, 34392 Montpellier Cedex 5, France
philippe.poignet@lirmm.rf

Cameron N. Riviere
Robotics Institute, Carnegie Mellon University, 5000 Forbes Ave, Pittsburgh, PA 15213, USA
camr@ri.cmu.edu

E. Rodriguez
Department of Cardiovascular Sciences, East Carolina Heart Institute, East Carolina University, Greenville, NC 27834, USA
rodrigueze@ecu.edu

Jacob Rosen
Department of Computer Engineering, Jack Baskin School of Engineering, University of California Santa Cruz, 1156 High Street, Santa Cruz, CA 95064, USA
rosen@ucsc.edu

Richard M. Satava
Department of Surgery, University of Washington Medical Center, Box 356410, 1959 Pacific Street NE, Seattle, Washington 98195, USA;
US Army Medical Research and Material Command, Fort Detrick, Frederick, MD, USA
rsatava@u.washington.edu

Brian Schmitz
MAKO Surgical Corp., 2555 Davie Road, Ft. Lauderdale, FL 33317, USA

David Schwartzman
Cardiovascular Institute, University of Pittsburgh, Pittsburgh, PA 15213, USA

L.N. Sekhar
Department of Neurological Surgery, University of Washington,
325, 9th Avenue, Seattle, WA 98104, USA
lsekhar@u.washington.edu

Moshe Shoham
Robotics Laboratory, Department of Mechanical Engineering,
Technion – Israel Institute of Technology, Haifa, Israel;
Mazor Surgical Technologies, Cesarea, Israel
shoham@technion.ac.il

Mika Sinanan
Department of Surgery, University of Washington Medical Center,
1959 Pacific Street NE, Seattle, WA 98195, USA

Ana Luisa Trejos
Department of Mechanical & Materials Engineering,
Department of Electrical & Computer Engineering, The University
of Western Ontario, London, Ontario, Canada N6A 5B9;
Lawson Health Research Institute (LHRI), Canadian Surgical Technologies
& Advanced Robotics (CSTAR), 339 Windermere Road London, Ontario,
Canada N6A 5A5

Pietro Valdastri
Scuola Superiore Sant'Anna, Pisa, Italy

Lawton N. Verner
Johns Hopkins University, Baltimore, MD 21218, USA

Sandrine Voros
Computer Science Department, Johns Hopkins University, 3400 N. Charles Street,
Baltimore, MD 21218, USA

Christopher Wottawa
Center for Advanced Surgical and Interventional Technology (CASIT), UCLA,
Los Angeles, CA 90095, USA;
Department of Bioengineering, UCLA, Los Angeles, CA 90095, USA

Cornell Wright
The Robotics Institute, Carnegie Mellon University, Pittsburgh, PA 15213, USA

Tomonori Yamamoto
Johns Hopkins University, Baltimore, MD 21218, USA

Andrew C. Yoo
University of Cincinnati Department of Surgery, 231 Albert Sabin Way,
P.O. Box 670558, Cincinnati, OH 45267-0558,
yooaw@ucmail.uc.edu

Marco A. Zenati
Division of Cardiothoracic Surgery, University of Pittsburgh, Pittsburgh, PA 15213, USA

Brett Zubiate
Bioengineering Department, University of Pittsburgh, Pittsburgh, PA 15213, USA

Part I
Visions and Overviews

Chapter 1
Future Directions in Robotic Surgery

Richard M. Satava

Abstract Robotic surgery has become an established part of clinical surgery. The advantages of using a robot have been enumerated by many clinicians, however the true potential has yet to be realized. In addition, the systems available today are extraordinarily simple and cumbersome relative to the more sophisticated robotic systems used in other industries. However more important is the fact that the fundamental principles underlying robotics have yet to be exploited, such as systems integration, feedback control, automatic performance, simulation and rehearsal and integration into healthcare enterprise. By looking at robotic implementation in other industries, and exploring the new robotic technologies in the laboratories, it is possible to speculate on the future directions which would be possible in surgical robotics.

1.1 Introduction

A robot is not a machine – it is an information system. Perhaps it has arms, legs, image capture devices (eyes) or various chemical or biologic sensors. However the primary functions are threefold – to acquire information about the world, to 'process' that information and to perform an action in the world. Simply put, robotics can be reduced to input, analysis and output. Some robotic systems interpose a human (instead of a computer) between the input and output – these are tele-manipulation (or for surgery, tele-surgical) systems. The complexity (and benefits) arise as each

R.M. Satava (✉)
Department of Surgery, University of Washington Medical Center, Box 356410, 1959 Pacific Street NE, Seattle, Washington, 98195, USA;
US Army Medical Research and Material Command, Fort Detrick, Frederick, MD, USA
e-mail: rsatava@u.washington.edu

J. Rosen et al. (eds.), *Surgical Robotics: Systems Applications and Visions*,
DOI 10.1007/978-1-4419-1126-1_1, © Springer Science+Business Media, LLC 2011

component is developed. On the input side, there are an enormous number of devices, from mechanical, chemical and biologic sensors to imagers of all portions of the electromagnetic spectrum. The 'processor' or analyzer of the information from the various sensors and/or imagers can be either a human, or a computer system, the former for human control while the latter is for 'autonomous' or semi-autonomous control, depending upon the level of sophistication of 'artificial intelligence' which is incorporated. Finally, on the output side there is likewise a wide variety of devices to interact with the world, including manipulators (instruments) and directed energy devices (electrocoagulation, lasers, etc.), all of which can be on a macro-scale of organs and tissues, or micro- and nano-scale for cells and intracellular structures.

However the most important concept is that robotic systems are nothing more than tools – admittedly very sophisticated tools – but tools nevertheless. The species *Homo sapiens* began with only teeth and fingernails to manipulate the world, progressing to sticks and stones, metal and finally energy. Over hundreds of thousands of years (though recently, only a few thousand years), the ability to interact and shape our world has provided the opportunity to free us from the vagaries of nature and to actually control our environment to a greater extent than ever before. Healthcare has always been a straggler, rarely inventing a new technology, but rather succeeding by adopting technologies from other disciplines and industries. Robotics is but one of the many areas where success has been achieved – to the greater benefit of our patients.

There is a new opportunity for medical and surgical robotics, one in which healthcare (or biomedical science) can take the lead – and that is in bio-inspired (or bio-mimicry) devices, whereby observing living systems, new robotic devices and/or systems can be developed. The fertile creativity of the physical and engineering sciences will continue to provide remarkable new ideas and systems, and together with biologic systems, will take robotics well beyond any of the possible projections of today. However, it must be kept in mind that the fundamental purpose is to extend human performance beyond the limitations of the human body, just as stone ax, metal scissor or microscope extended human capabilities in the past, with the stated intent to improve the surgeon's ability to provide higher quality and safer patient care.

1.2 Systems Integration

A capability that is unique to robotic surgery systems (as opposed to open, flexible endoscopy, laparoscopy and Natural Orifice Transluminal Endoscopic Surgery (NOTES)) is systems integration, a characteristic which is emphasized in engineering science. One of the principle advantages of the robotic surgical system is the ability to integrate the many aspects of the surgical care of a patient into a single place (the surgical console) and at a single time (just before or during performing surgery) (Fig. 1.1). At the console the surgeon can perform open or minimally invasive surgery, remote tele-surgery, pre-operative planning or surgical rehearsal, pre-operative

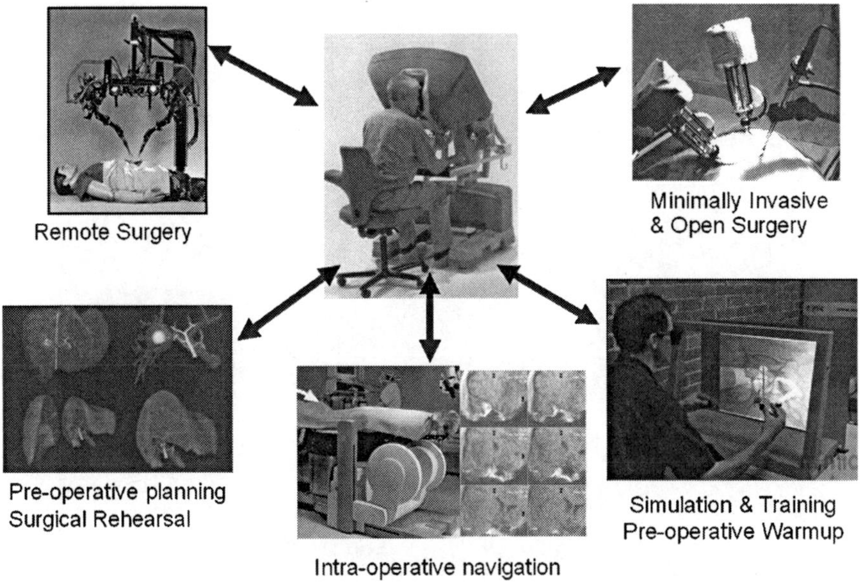

Fig. 1.1 Integration of operative procedures using the surgical work station (Courtesy of the author)

warm-up, intra-operative navigation and tele-mentoring (if a dual-console is used). In addition, training can be performed "off line" in a simulation laboratory or on the actual console.

Today's robotic surgical systems are stand alone, usually moved into the operating room (or for some image-guided systems, mounted on a boom or stationed in a part of the room with a CT or MRI scanner). Then surgeons, radiologist, cardiologists, etc. must operate together with their team of nurses, technicians, etc. When an instrument or catheter needs to be replaced, a scrub nurse is needed; when a new supply such as suture or gauze is needed, the circulation nurse is needed. This is not the case in industry – robotic systems incorporate multiple robots into a single 'robotic cell'. When a different tool is needed, the robotic tool changer performs the function; when a new supply (like a nut, bolt, etc.) needs to be inserted, this is provided by the robotic supply dispenser. The military has developed the 'Trauma Pod' surgical system [1], a prototype system of an "operating room without people' in which the scrub nurse is replaced by a robotic tool changer, and the circulation nurse is replaced with an automatic supply dispenser – modified from a standard pharmacy medication dispenser (Fig. 1.2). When the surgeon needs to change an instrument, the voice command is given (for example, scalpel for right hand) and the robotic tool changer automatically performs the function. When a supply is needed, a voice command (for example, 2–0 chromic catgut on a GI needle) is given and one of the 120 different sterile trays with supplies is chosen and 'handed' to the surgeon (robotic manipulator) to remove the supply and use it. The accuracy

Fig. 1.2 Prototype 'surgical robotic cell' – the 'operating room without people' (Courtesy of Pablo Garcia, SRI International, Menlo Park, CA 2008)

is 99% and the speed is approximately the same as the corresponding scrub or circulating nurse, which is about 17 s. The advantage is that this frees up the nurses to perform more intellectually demanding tasks, rather than standing around for hours, simply handing instruments or supplies to the surgeon.

As indicated above, because the robot is truly an information system, it can be incorporated into the entire hospital information enterprise. The information encoded into the robotic instruments or the supply trays can be collected, analyzed and distributed (in real-time) beyond the operating room to the other hospital support functions. When a disposable instrument or used supply is discarded, that information can be instantly sent to the Central Supply department, where a replacement can be automatically ordered and the inventory adjusted. This allows the hospital to not only accurately track all the instruments and supplies, but can also decrease the amount of inventory which is stored or goes out of date because of tracking and immediate re-ordering. This is standard practice in most industries, and referred to as asset-tracking and supply-chain management. Efficiency and cost savings are realized by decreased supplies on the shelf and decreased personnel needed to inventory and order all the supplies. Incorporating these capabilities directly into the robotic system functioning simply extends the efficiency and cost saving all the way into the operating room.

Another unique aspect of the robotic systems is the ability to store the video of the procedure and track hand motions [2]. These data can be stored in a 'black box' like the aircraft flight recorder and can lead to automatically generating the operative note (from analysis of video and hand motions) as well as mining the data for errors. As in inventory control, this data could automatically be sent to the quality improvement and risk management systems, greatly reducing the amount of time and effort to collect and analyze the data to improve patient care and safety while decreasing the time required by the surgeon to dictate operative reports, review quality assurance reports, etc. Such documentation could also be used by the hospital credentialing and privileging committee when the surgeon requests annual review of hospital operating procedures. Whether such an implementation of the robotic systems will occur is a different matter – it is no longer a technical issue but rather one of policy, privacy, cost or practicality. Thus, using the perspective that the surgical robot is just one more node of the hospital information enterprise demonstrates the value added of robotic systems beyond their mechanical and operative value.

For direct patient care, the importance of integrating the entire 'process' of surgery into operative procedures can be facilitated by a surgical robotic system. The current practice of surgery includes the pre-operative evaluation f the patient, with the result decision to operate and a plan for the surgical procedure. However, the 'plan' is in the surgeon's head, based upon the diagnostic information which has been gathered, and must be executed in real time during the surgical procedure, without complete information about the anatomy, anatomical variations due to the disease process, congenital anomalies, or other variations from the 'normal' and expected anatomy. The result is that the surgeon will encounter unexpected variations, hopefully recognize them in time to modify the procedure for a successful completion. All other industries use a 3-D model of their 'products' (Computer Aided Design/Computer Aided Manufacturing or CAD/CAM models) to rehearse a procedure through simulation before performing the procedure. In a non-pejorative way, the patient is the 'product' for healthcare, so when surgical procedures are performed without previous planning or rehearsal on a model, there frequently are resultant errors. There is the beginning of computer-based pre-operative planning and surgical rehearsal on patient-specific 3-D models, derived from the patient's own CT or MRI scan. Marescaux et al. [3] have reported pre-operative planning and surgical rehearsal for complex liver resections for hepatic cancer, with a result of a significant decrease in operating time, blood loss and errors. In the future for difficult surgical procedures, it will become commonplace for a surgeon to import the patient-specific 3-D image from the patient's CT or MRI scan, plan and rehearse the operation directly on the surgical console, repeat the difficult parts of the operation until no mistakes are made, and thereafter conduct a near perfect performance during the procedure. In the more distant future, the operation will be recorded while being rehearsed and errors will be 'edited out' of the stored performance of the procedure; when the surgeon is satisfied with the edited operation, it will be sent to the robot to perform under 'supervisory control' of the surgeon, with many times the precision and speed, and virtually error free.

One final component that will be integrated into the surgical console will be specific exercises for pre-operative warm-up. It is a priori that all professionals (soccer, basketball, symphony, dance, etc.) improve their performance by warming up before performing their professional skill, yet surgeons have not accepted this obvious advantage. Initial data has demonstrated that performing 15 min of pre-op warm-up exercises on a virtual reality simulator is able to decrease operative time and errors [4]. Soon these exercises will be incorporated into the surgical workstation and become a required preliminary part of every operation. This is yet one more way of incorporating simulation into daily clinical practice.

1.3 Automatic and Autonomous Surgery

Surgeons pride themselves on being completely in control of a surgical procedure, being able to deal with unexpected anatomy or events during a surgical procedure in order to complete a safe operation. Yet other industries use automatic (i.e. specifically executed pre-programmed 'steps' or tasks) or autonomous (i.e., perform a task in an unstructured environment rather than according to a pre-programmed sequence) robotic systems to perform procedures. With the exception of the LASIK procedure in ophthalmology [5], there are no automatic or autonomous systems in surgery. The closest analogy would be the surgical stapling devices, which can clamp, seal (staple) and cut bowel or other structures with a single application – but these are hand held and have no sensors to detect proper position, strength of application, etc. Yet looking at the clothing industry, an automatically sewn seam is far superior to a hand-sewn garment. Likewise, autonomous sorting robotic systems (pick and place robots) far exceed human performance both in accuracy and speed in identifying objects and moving them to a specific position, such as sorting different candies into box. The basic principles behind these actions are very well known and well proven, the challenge is to be able to adapt such systems or tasks to an unstructured environment in living systems for surgical procedures. While this is very hard, due to the large variability from patient to patient, continuous motion due to heart beat, breathing, etc., the problem is not intractable. It is computationally intense and requires micro-second adaptation to the dynamic situations, including such tasks as image recognition, analysis, registration (without fiducials), adaptive control, etc., however it theoretically could be achieved with known technology. It is likely that the first steps will be automatic tasks, such as an anastomosis, in which the surgeon performs a resection and then sets up the severed ends, and issues a "connect" command for the robotic system to automatically sew the ends together. Beyond this initial task, multiple other automatic tasks could be sequenced in such a fashion to have a simple autonomous procedure. Combined with a previously 'rehearsed' and 'saved' surgical procedure, there will eventually be the option to rehearse the operative procedure, edit out the errors, and then send the completed procedure to the robotic system to complete faster and with greater accuracy.

1.4 Intelligent Instruments

Today's surgical instruments are very simple mechanical devices, controlled directly by the surgeon's unaided hand. The surgeon proceeds to dissect, transect and other maneuvers with the instruments, unaware of what may lie just below the surface and depends upon the subjective 'sense of touch' to assist when visualization of the structures is not possible. Various haptic sensors and displays have been investigated, however there are no mechanical sensors for the sense of touch that are integrated into current surgical instruments.

Using 'information science', instruments can become intelligent. Within the area of 'information' the use of energy (rather than mechanical) systems should be included, since most energy systems are reliant upon some form of information (computer) control. Both information and energy are 'intangible' and thus are complimentary parts of the Information Age, and the combination of the two is creating the next generation of intelligent surgical (robotic) instruments.

There are a number of prototype laparoscopic instruments with various sensors, such as force reflecting graspers [5]. But future instruments will go beyond sensing and they will be energy directed rather than mechanical instruments. The advantage of using intelligent surgical instruments is that they can provide both diagnosis and therapy, in real-time, in a single instrument. One current example is combining diagnostic ultrasound with High Intensity Focused Ultrasound (HIFU) [6], in which both Doppler imaging for diagnosis and HIFU for therapy are combined to both detect internal hemorrhage with the Doppler, and instantly stop the bleeding with the HIFU, and then recheck with the Doppler to insure hemostasis is complete. This is performed transcutaneously, without any incisions, rather than the standard method of a full surgical procedure with large (or even laparoscopic) incisions and direct control of the bleeding. By moving into the energy spectrum (rather than mechanical instruments) it is possible to move from minimally invasive to non-invasive therapies. There are many such opportunities by incorporating biophotonics, ultrasonics and other energy-based systems into the instruments of a robotic surgery system, since they can be controlled by the surgical console. Such integration goes well beyond the scope of human performance, not only physical but cognitive. Using closed loop feedback, the therapeutic modality (laser, ultrasound, etc.) can be monitored and when a specific quantitative threshold has been reached, the instrument can be shut off in milliseconds, even before the threshold is perceived by the surgeon. Healthcare has just begun to exploit the potential of the energy spectrum; in spite of having a number of different energy-based systems, such as X-ray, ultrasound, lasers, ultraviolet, near-infrared, etc., less than 5% of the electromagnetic spectrum has been investigated. The utilization of energy to diagnose, treat, and monitor with closed-loop feedback will lead to the next generation of surgical and robotic devices and systems.

1.5 Molecular Surgery (Biosurgery) with Micro-Systems and Nano-Systems

The era of molecular biology, genetic engineering and other micro/nano scale procedures has been in the laboratory for decades and is finally emerging into clinical practice. Instruments and devices have been developed to both sense/diagnose as well as manipulate/treat cellular and intracellular structures. By working at the molecular level, the results are changing the biology of the patient, but not necessarily the anatomy – changing function, not structure. Cellular biologists and other basic science researchers are now using new tools, such as femtosecond lasers, optical tweezers, micro-electro-mechanical systems (MEMS), atomic force microscopes, etc. to make incisions into individual cells, and manipulate the mitochondria, Golgi apparatus and even to into the nucleus and 'operate' upon individual chromosomes. In the future, such systems will begin performing 'genetic engineering' by directly removing specific defective genes and replacing them with normal functioning genes. Yet, manipulation at the micro and nano-scale is not possible with human hands – it requires a sophisticated tele-operated work station, which is not different from the current surgical workstation. But what the surgeon 'sees' on the monitor at the cellular level is radically different from looking at organs or tissues. In viewing cells, the structures are caused to fluoresce – auto-fluorescence, induced fluorescence, or with molecular marker fluorescent 'probes' – in order to follow the progress of the procedure. For a cellular surgical procedure, the surgeon will be looking at the colors of the individual proteins within the cell which will change as they are treated. Working on the surface of the cell membrane will present a view similar to looking at a mountain range, where large craters (ion channels) will be the entry ports for various proteins that need to be inserted into the cell. Given such powerful new tools and an unfamiliar 'landscape', the surgeon will need to retrain not only their psychomotor skills, but also need to develop new cognitive skills.

1.6 From Soldiers to Generals

All of the surgical tools and systems, including robotic systems, are designed for the surgeon to directly operate upon the organ or tissues, with an assistant or two to help retract. It is quite literally the surgeon battling the disease one-on-one. There are interesting new robotic systems being developed at the micro level. Beginning with the endoscopic capsule [7], which is a video system in a pill that is swallowed with sequential photos taken as it passively is propelled by peristalsis through the GI tract, new concepts in surgical robotics are being investigated. Micro-robots, which go beyond simple visualization systems, are attempting to add locomotion to the capsules, and include various types of micro-manipulators to perform surgery. There are a number of very difficult challenges, nevertheless progress has been made and tiny robots are in the experimental laboratory. However at this small scale, it is very difficult for a single robot to perform all of the necessary functions of light source, visualization, locomotion, manipulating, etc. The concept has been raised that this level of micro-robots may require that each robot perform a single function, such as camera, light source, grasper, scissor, etc.; therefore it will be necessary to use a group of function-specific micro-robots in order to perform an operation. After inserting many micro-robots into the abdomen (through a tiny umbilical incision), the surgeon will then need to control a dozen or so of these at a time – much like a commander controlling a squad of soldiers. The surgeon will need to radically change perspective and behavior, and will need to begin acting like a general in charge of a squad of soldiers, rather than behaving like an individual soldier attacking a specific target (disease). This would be an extremely radical way of conducting a surgical procedure, but may well provide a revolutionary new way of performing "surgery".

1.7 Conclusion

Robotic and computer aided systems have finally brought surgery into the Information Age. Current instruments will continue to evolve and new systems, especially energy based and those systems on a much smaller scale, will be added to the surgical armamentarium. In general, the instruments and systems will become more intelligent and integrated, not only in the operating room but throughout the entire hospital information enterprise. The purpose for the speculative nature of this chapter is to create a vision of what is possible, and to offer a challenge to the biomedical engineering community as a whole. Surely many of these will not be realized, others will materialize even beyond these modest predictions, and then there will be the outliers, the "unknown unknowns", that will be the game-changers to disrupt the predictable progress and take surgery into a completely new direction. However, there is one thing which is certain: The future is not what it used to be.

References

1. Garcia, P., Rosen, J., Kapoor, C., Noakes, M., Elbert, G., Treat, M., Ganous, T., Hanson, M., Manak, J., Hasser, C., Rohler, D., Satava, R.: Trauma pod: a semi-automated telerobotic surgical system. Int. J. Med. Robot. **5**(2), 136–146 (2009)
2. Rosen, J., Brown, J.D., Chang, L., Sinanan, M., Hannaford, B.: Generalized approach for modeling minimally invasive surgery as a stochastic process using a discrete markov model. IEEE Trans. Biomed. Eng. **53**(3), 399–413 (2006) [JP 9]
3. Mutter, D., Dallemagne, B., Bailey, C., Soler, L., Marescaux, J.: 3-D virtual reality and selective vascular control for laparoscopic left hepatic lobectomy. Surg. Endos. **23**, 432–435 (2009)
4. Kahol, K., Satava, R.M., Ferrara, J., Smith, M.L.: Effect of short-term pretrial practice on surgical proficiency in simulated environments: a randomized trial of the "preoperative warm-up" effect. J. Am. Coll. Surg. **208**(2), 255–268 (2009)
5. De, S., Rosen, J., Dagan, A.,Swanson, P., Sinanan, M.,Hannaford, B.: Assessment of tissue damage due to mechanical stresses. Int. J. Robot. Res. **26**(11–12), 1159–1171 (2007)
6. Vaezy, S., Martin, R., Keilman, G., Kaczkowski, P., Chi, E., Yazaji, E., Caps, M., Poliachik, S., Carter, S., Sharar, S., Cornejo, C., Crum, L.: Control of splenic bleeding by using high intensity ultrasound. J. Trauma. **47**(3), 521–525 (1999)
7. Knorz, M.C., Jendritza, B.: Topographically-guided laser in situ keratomileusis to treat corneal irregularities. Ophthalmology **107**(6), 1138–1143 (2000)
8. Joseph J.V,., Oleynikov, D., Rentschler, M., Dumpert, J., Patel, H.R.: Microrobot assisted laparoscopic urological surgery in a canine model. J. Urol. **180**(5), 2202–2205 (2008)
9. Iddan, G., Meron, G., Glukhovsky, A., Swain, P.: Wireless capsule endoscopy. Nature. **405** (6785), 417 (2000)

Chapter 2
Military Robotic Combat Casualty Extraction and Care

Andrew C. Yoo, Gary R. Gilbert, and Timothy J. Broderick

Abstract Buddy treatment, first responder combat casualty care, and patient evacuation under hostile fire have compounded combat losses throughout history. Force protection of military first responders is complicated by current international and coalition troop deployments for peacekeeping operations, counter terrorism, and humanitarian assistance missions that involve highly visible, politically sensitive, low intensity combat in urban terrain. The United States Department of Defense (DoD) has significantly invested in autonomous vehicles, and other robots to support its Future Force. The US Army Telemedicine and Advanced Technology Research Center (TATRC) has leveraged this DoD investment with augmented funding to broadly focus on implementing technology in each phase of combat casualty care. This ranges from casualty extraction, physiologic real-time monitoring, and life saving interventions during the "golden hour" while greatly reducing the risk to first responders.

The TATRC portfolio of projects aims to develop, integrate, and adapt robotic technology for unmanned ground and air battlefield casualty extraction systems that operate in hostile environments that include enemy fire. Work continues on multiple ground extraction systems including a prototype dynamically balanced bipedal Battlefield Extraction Assist Robot (BEAR) capable of extracting a 300–500 pound casualty from a variety of rugged terrains that include urban areas and traversing stairs. The TATRC and the Defense Advanced Research Projects Agency (DARPA) are collaborating to investigate the use of Unmanned Aircraft Systems (UAS) to conduct casualty evacuation (CASEVAC) missions. TATRC has also sponsored research in robotic implementation of Raman and Laser-Induced Breakdown Spectroscopy (LIBS) to detect and identify potential chemical and biological warfare agents and explosive hazards to casualties and first responders during the extraction

G.R. Gilbert, (✉)
Georgetown University Imaging Science and Information Systems (ISIS) Center,
US Army Medical Research and Materiel Command Telemedicine;
Advanced Technology Research Center MCMR-TT, 504 Scott,
St Fort Detrick, Frederick, MD 21702, USA
e-mail: gary.gilbert@tatrc.org

J. Rosen et al. (eds.), *Surgical Robotics: Systems Applications and Visions*,
DOI 10.1007/978-1-4419-1126-1_2,

process, and patient monitoring equipment with sophisticated telemedicine and patient monitoring equipment such as "smart stretchers" that allow for real-time physiologic monitoring throughout the combat casualty care process, from extraction to definitive care. Other projects are intended to build upon these monitoring systems and incorporate telerobotic and near autonomous casualty assessment and life saving treatment to the battlefield. These have included the DARPA Trauma Pod and several TATRC efforts to integrate robotic arms with the Life Support for Trauma and Transport (LSTAT) litter for robotic implementation of non-invasive technologies such as acoustic cauterization of hemorrhage via High Intensity Focused Ultrasound (HIFU). Several projects have explored the essential telecommunication link needed to implement telesurgery and telemedicine in extreme environments. UAS were leveraged to establish a telecommunication network link for telemedicine and telesurgery applications in extreme situations. Another collaborative telesurgery research project at the NASA Extreme Environment Mission Operations (NEEMO) included performing telesurgery in an undersea location.

Research into identification and solutions of the limitations of telecommunication and robotics that prevent robust casualty interventions will allow future medical robots to provide robust casualty extraction and care that will save the lives and limbs of our deployed warfighters.

Keywords Surgical robotics · Military robotics · da Vinci · Zeus · BEAR · Battlefield Extraction Assist Robot · LSTAT · Life Support for Trauma and Transport · TAGS-CX · UAS · Unmanned Aircraft Systems · Trauma Pod · M7 · RAVEN · HIFU · High Intensity Focused Ultrasound · Tissue Welding · RAMAN Spectroscopy · Golden Hour · Hemorrhage · Telesurgery · Telemedicine · Teleoperator · Combat Casualty Care · Casualty Extraction · Trauma · DoD · Department of Defense · DARPA · Defense Advanced Research Projects Agency · TATRC · Telemedicine and Advanced Technology Research Center · NASA · NEEMO · MRMC · Medical Resarch and Material Command · Army · Military · Computer Motion · Intuitive Surgical

2.1 Introduction

Advancement in telecommunication and robotics continue to shift the paradigm of health care delivery. During the 1980s, the nascent field of telemedicine developed and allowed for increasing distribution of medical knowledge to large populations with limited local medical infrastructure and capabilities. Despite technologic strides, telemedicine has been primarily used in diagnostic applications such as radiology and pathology. However, telemedicine continues to evolve and will soon incorporate the full spectrum of medicine from diagnosis to treatment.

The United States military has provided significant impetus, focus and funding for telemedicine and medical robotics. The U.S. Army Medical Research and Materiel Command (MRMC), Telemedicine and Advanced Technology Research Center

(TATRC), and the Defense Advanced Research Projects Agency (DARPA) have spurred innovation in areas such as surgical robotics and the emerging field of "telesurgery." Telecommunication and robotic limitations that prevent robust intervention at a distance are areas of continued military research and development. Medical robots are force multipliers that can distribute expert trauma and subspecialty surgical care across echelons of care. This chapter provides a historical context of and future opportunities in military robotic casualty extraction and care that will save the lives and limbs of our deployed warfighters.

Military robotic combat casualty care has three primary goals: safely extracting patients from harm's way; rapidly diagnosing life threatening injuries such as non-compressible hemorrhage, tension pneumothorax and loss of airway; and delivering life-saving interventions. For optimum effect, medical robots must robustly operate in extreme environments and provide effective combat casualty care as close as possible to the point and time of injury. Robotic tactical combat casualty care begins with the extraction of casualties from the battlefield. In the short term, extraction robots will decrease the risk to the soldier and combat medic by safely moving wounded warfighters out of the line of fire. In the longer term, teleoperated and autonomous surgical robots will deliver expert surgical care within the "golden hour" on the battlefield as well as during transport to military treatment facilities.

DARPA and MRMC/TATRC partnered to develop the Digital Human Robotic Casualty Treatment and Evacuation Vision with robotic systems targeted on these priorities:

1. Mobility
2. Plan/execute search in unmapped interior environments, find and identify wounded soldiers
3. Track, record, transmit and act upon real-time physiological information
4. Conduct both remote and real-time diagnosis using heuristic algorithms integrated with pattern recognition imaging systems and physiological sensors
5. Perform semi-autonomous and autonomous medical procedures and interventions
6. Evacuate casualties from the battlefield using semi-autonomous and autonomous evacuation platforms and patient support systems like LSTAT

2.2 Assessment of Current State and Future Potential for Robotic Combat Casualty Care Within the Army

The Training and Doctrine Command (TRADOC) is the Army's organization for developing new doctrine on how the Army will fight in the future and what capabilities will be needed to support that operational doctrine. In 2009, TATRC contributed to TRADOC's assessment of the state of medical robotics and their potential application to combat casualty care. Currently only a few Warfighter Outcomes are involved with robotics use in medical and surgical tasks. The U.S. Department of Defense Uniformed Joint Task List suggests several topics for improvements in the areas of field medical care and force health protection through robotics. Areas of focus in combat casualty care and surgery are faster casualty recovery and evacuation by fewer personnel, faster and more certain recognition of injuries, and communications supporting remote telemedicine. TRADOC's desired future "Force Operating Capabilities" publication states that: "Future Soldiers will utilize unmanned vehicles, robotics, and advanced standoff equipment to recover wounded and injured soldiers from high-risk areas, with minimal exposure. These systems will facilitate immediate evacuation and transport under even the harshest combat or environmental hazard conditions; medical evacuation platforms must provide en route care," and TRADOC's "Capability Plan for Army Aviation Operations 2015–2024," states that "unmanned cargo aircraft will conduct autonomous..... extraction of wounded." Following was are assessment of the current state and future potential for the combat casualty care robotic applications cited by TRADOC:

1. *Perform battlefield first aid (tourniquets, splints, shots, IV drips, etc.)*: Self-assistance by the soldier or the availability of buddy care cannot be assumed in all combat situations; likewise, there are never enough combat medics or combat life savers (combat arms soldier with additional medical training) to treat and extract all casualties, especially during intense close combat or in contaminated or otherwise hostile environments. The Army Institute for Soldier Nanotechnology at MIT has ongoing basic research in uniform-based diagnostics and emergency injections. Further, sewn-in tourniquet loops on uniforms are under consideration for fielding, with Soldier-actuation required. Autonomous and robotic first aid treatment may dovetail well with robotic recovery and evacuation tasks. Slow progress is being made in the development of sophisticated sensors, autonomous analysis of sensory input, and autonomous application of intervention and treatment procedures, but deployment of such robots is years away. Likewise, local cultural concerns or confusion among the wounded may complicate acceptance of close contact by a first aid robot.
2. *Recover battlefield casualties*: As with battlefield first aid, universal availability of combat medics, combat life savers, or other soldiers assigned to perform extraction and recovery of casualties under fire or in otherwise hostile environments cannot be assumed. Therefore, a means to autonomously find, assess, stabilize, then extract casualties from danger for further evacuation is needed. This may be complicated by the unknown nature of injuries, which may complicate or confound a rote mechanical means of body movement. For example, a compound fracture or severed limb might not be gripped or gripping may increase injury. As part of several ongoing research and development efforts in both unmanned ground and air systems for casualty evacuation (CASEVAC), the MRMC is actively addressing the potential complications of robotic casualty extraction. Discussed further below, the tele-operated semi-autonomous Battlefield Extraction Assist Robot (BEAR) represents a developing casualty extraction capability which can carry a 300–500 pound load while traversing rough and urban terrain with dismounted soldiers. A fully autonomous version of the BEAR would need significant additional artificial intelligence programming and a transparent hands-free soldier-robot interface to integrate and perform this

mission in combat while keeping the soldier-operator focused on their primary mission. Research in autonomous flight control and navigation technologies needed for CASEVAC via Unmanned Air Systems (UAS) is ongoing (described below) but actual employment of operational systems is probably years away because of the current immaturity of autonomous en route casualty care systems.

3. *Robotic detection and identification of force health protection threats*: Detection and identification of chemical and biological threats to which combat casualty patients may have been exposed, along with segregation and containment of contaminated casualties prior to receiving casualties in forward medical and surgical treatment facilities are critical capability needs. The MRMC has several completed and ongoing research projects in robotic detection and identification of chemical and biological agents and chemical contaminants. The goal is to produce modular threat detection and identification systems that can be implemented on robots performing other missions, such as casualty extraction. These efforts utilize robotic enabled Raman spectroscopy, florescence, and Laser Induced Breakdown Spectroscopy (LIBS) as well as antigen-based technologies. One of these projects is discussed below.
4. *Perform telemedicine/surgery*: Remote tele-operated medicine is feasible, but with limitations. Visual examination information is planar and may lack depth and full five-sense information (e.g. tactile feedback). As a human assistant will likely be required, a question arises as to the feasibility of doing better than having a trained human assistant, local to the patient, relaying information back to the remotely located surgeon. However, vital signs (e.g. skin temperature, pulse, blood pressure) may be available via biomonitors contained on a simple robotic platform arm. Proof of concept projects have demonstrated the feasibility of remote robotic diagnosis and treatment of patients. The DARPA 'Trauma Pod' project discussed below was an attempt to leverage emerging advanced imaging technologies and robotics to enable autonomous casualty scan, diagnosis and intervention, MRMC also has several physiological sensor and image-based robotic casualty assessment and triage research projects underway. However, these capabilities are currently only experimental and are non-ruggedized, teleoperated component capabilities at best. The idea of far forward combat telesurgery in combat is compelling; a surgeon controlling a robot's movements in a distant location to treat an injured soldier could serve as a force multiplier and reduce combat exposure to highly trained medical personnel. At first glance, remote tele-operated surgery capability appears to already exist since minimally invasive operations have been remotely performed using dedicated fiber optic networks, the Zeus and da Vinci surgical robots have been and are currently used in civilian hospitals and many other telesurgery demonstrations and experiments have been conducted around the world. Military funded research as discussed below has demonstrated that surgical robotic systems can be successfully deployed to extreme environments and wirelessly operated via microwave and satellite platforms. However, significant additional research is required to develop supervisory controlled autonomous robots that can overcome the operational communication challenges of limited bandwidth, latency, and loss of signal in the deployed combat environment. Addressing acute and life threatening injuries such as major non-compressible

vascular injury requires development of new surgical robots that move beyond stereoscopic, bimanual telemanipulators and leverage advances such as autonomous imaging analysis and application of directed energy technologies already used in non-medical military robotic systems.

2.3 Robotic Casualty Extraction, Evaluation and Evacuation

The US military has funded multiple robotic projects focused on casualty extraction, evaluation and evacuation. Robotic casualty extraction research is focused on the development of semi-autonomous systems that will safely extract the casualty from the line of fire, deliver the casualty to medical care, and limit risk to care providers.

Representative systems are briefly described below.

TAGS CX (Tactical Amphibious Ground Support system – Common Experimental).

The Army Medical Robotics Research through the Army's SBIR (Small Business Innovation Research) Program through TATRC contracted Applied Perceptions Inc. (Cranberry Township, PA) as the primary research entity for an extraction and evacuation vehicle. A tele-operated semi-autonomous control system capable of maneuvering a marsupial robotic vehicle was developed with a three module concept. The initial novel dual design prototype vehicle consisted of a small, mobile manipulator Robotic Extraction (REX) robot for short-range extraction from the site of injury to the first responder and a larger faster Robotic Extraction Vehicle (REV), which would deliver the wounded soldier to a forward medical facility. The smaller vehicle resided within the larger REV, which was equipped with two L-STAT stretchers and other life support systems. The TAGS platform provides a modular and interoperable ground robot system that could be modified for multiple purposes. The Joint Architecture for Unmanned Systems (JAUS) control platform was used to enable a standardized C2 interface for the OCU (Operational Control Unit) along with standardized mechanical, electrical, and messaging interfaces capable of supporting multiple unique "plug and play" payloads. This prototype robotic extraction vehicle also integrated other control technologies. These include GPS-based autonomous navigation, search and rescue sensing, multi-robot collaboration, obstacle detection, vehicle safe guard systems, autonomic vehicle docking and telemedicine systems (Fig. 2.1).

Subsequent to completion of the initial REV and REX prototypes, the US Army's TARDEC (Tank-Automotive Research, Development, and Engineering Center) developed a ground mobility, robotics systems integration and evaluation laboratory, TARDEC's Robotic SkunkWorks facility. This laboratory's goal is to assess and integrate novel unmanned systems technologies to support efficient conversion of these technologies to PM/PEO (program managers/program executive officer) and ATO (Advanced Technology Office) programs. The first unmanned system evaluated was the TAGS-CX, an enhanced version of the original TAGS designed to support multiple modular mission payloads. The most

Fig. 2.1 Robotic extraction (REX) and Robotic evacuation vehicle (REV) prototypes (*left*); REX towing casualty on litter in snow (*right*)

Fig. 2.2 Tactical amphibious ground system – common experimental (*left*); CX with patient transport & attendant modules (*Right*)

significant identified issue during the trials of the original REV vehicle was that the REV was designed to be completely unmanned and as a dedicated MEDEVAC vehicle. Currently and for the foreseeable future the US Army would not allow wounded soldiers to travel without a human medic or attendant. Based on this feedback the TAGS-CX concept was redesigned to incorporate a removable center module for an on-board medic and would allow for manual operation of the vehicle. Additionally the patient transport bays were designed and constructed as modular "patient pods" which would enable the TAGS-CX to be used for multiple combat support missions, CASEVAC being just one (Fig. 2.2).

2.4 BEAR: Battlefield Extraction Assist Robot

The BEAR (Vecna Technologies Cambridge Research Laboratory, Cambridge, MA) prototype was initially started with a TATRC grant in 2007 with the objective of creating a powerful mobile robot, which was also highly agile. It would have the capability to find and then lift and carry a combat casualty from a hazardous area in varying terrain. Vecna Technologies Inc. initially produced a proof of concept prototype (BEAR Version 6), which was featured in Time Magazine's Best Inventions of 2006. This machine was intended to be capable of negotiating any general hazardous terrain and not be limited only to the battlefield. The BEAR robot is extremely strong and agile approximately the size of an adult male. The original prototype was composed of an upper torso with two arm actuators and a lower body built around the Segway RMP base with additional tank tracks on its analogous thighs and calves. It is designed to lift 300–500 lbs (the approximate weight of a fully equipped soldier) and move at $\sim$10 miles/h. It utilizes gyroscopic balance that enables it to traverse rough and uneven terrain (Fig. 2.3).

The latest iteration of the BEAR (version 7) has several redesigned components. These include a sleeker, stronger, and more humanoid appearing upper torso, integration of NASA's Actin software for coordinated control of limbs and upper torso, and a lower body with separately articulating tracking leg subsystems, a novel connection and integration of the lower body and upper torso components, completion of the "finger-like" end effectors, and a Laser Induced Breakdown Spectroscopy (LIBS) detector for chemical, biological, and explosive agents. The system will incorporate a variety of input devices including multiple cameras and audio input. The initial control of the BEAR is via a remote human operator but work is underway for more complicated semi-autonomous behaviors in which the robot understands and carries out increasingly higher-level commands. Other planned inputs include pressure sensors that will allow it to have sensitivity to a human cargo. Another milestone is the completion of the first phase of continuing BEAR characterization and operational simulation and assessment at the Army Infantry Center Maneuver Battle Lab (MBL). The humanoid form enables the robot to access most places that a human would, including stairs. The versatility of this robot includes applications within hospitals and nursing homes where infirmed patients with limit mobility could be easily moved.

Fig. 2.3 Battlefield extraction assist robot (BEAR) prototype (*left*); BEAR extracting casualty with foldable litter (*right*)

2.5 Combat Medic UAS for Resupply and Evacuation

TATRC has also provided support for aerial robotic systems. This project focused on autonomous UAS (Unmanned Aircraft Systems) takeoff, landing, navigation in urban and wooded environments and the coordination and collaboration between UAS ground controllers and human combat medics so that proper care and evacuation can be performed during the golden hour. Five Phase I SBIR grants were given out to identify notional concepts of operation as well as develop technical models that recognize requirements in implementable UAS system designs. Phase II grants went to Dragon Fly Pictures Inc. and Piasecki Aircraft both of Essington, PA. Phase II focuses on navigation through urban/wooded terrain to combat site of injury, selection of a suitable autonomous landing and takeoff site with minimal human input, autonomous safe landing and takeoffs, communication with a human medical team, and carrying a payload of medical supplies including a Life Support for Trauma and Transport (L-STAT) system. Phase II concludes with live demonstrations of these capabilities using real aircraft.

2.6 Raman Chem/Bio/IED Identification Detectors

Research interest exists in providing these unmanned ground vehicles (UGV) extraction platforms with chemical, biological, and explosive (CBE) detection systems based on Raman spectroscopy so that they have the operational ability to identify environmental toxins and provide force protection. Currently UGVs are unable to provide any early information as to the possible toxic hazards in the environment. TATRC along with MRMC and other governmental agencies have funded development of several JAUS compliant robotic CBE identification systems that could be placed on unmanned extraction vehicles.

The Raman detection technological advantages are that it is reagentless, which simplifies deployability and can detect a broad range of CBE threats in a single measurement cycle. Reagent based detection methods must start with some assumption as to the possible threat. The Raman Effect has been used for years and depends on the phenomenon that when a photon encounters a molecule it imparts vibrational bond energy to this molecule. This exchange creates a slight disturbance in the frequency in a small amount of scattered light. Each chemical bond has its own unique frequency shift, which allows for creation of the Raman spectrum and the identification of chemicals. Further research sponsored by the Army Research Laboratory (ARL) has shown that concurrent deployment of both Raman and LIBS systems results in a significant improvement in sensitivity and accuracy of agent detection when the results are merged through a fusion algorithm developed by ChemImage Corporation, designer of the proximity RAMAN detector shown in Fig. 2.4a.

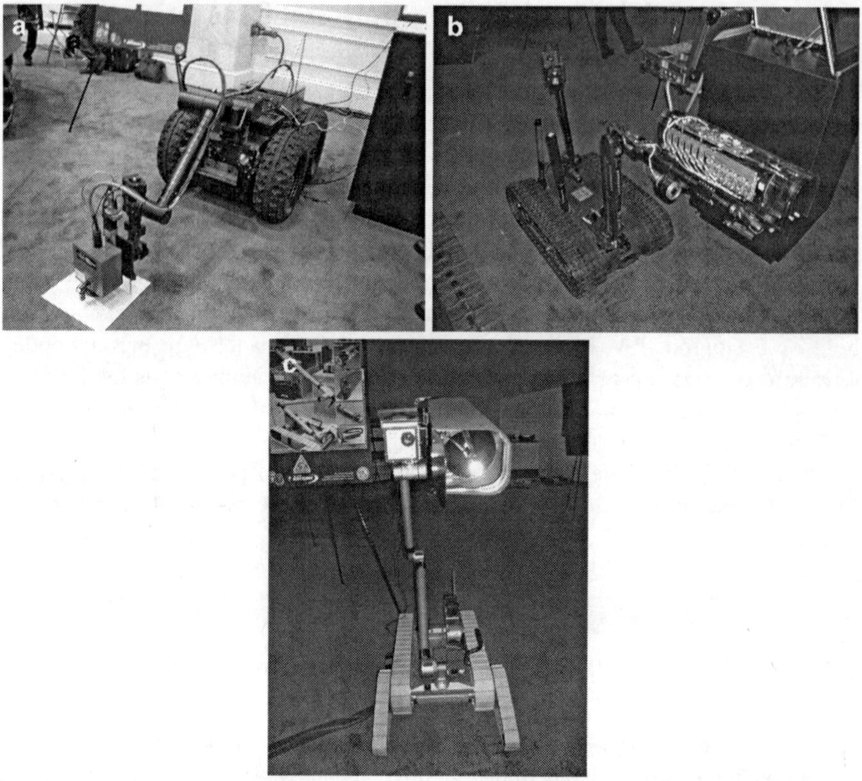

Fig. 2.4 (**a**) Chemimage proximity robotic Raman spectroscopy chem/bio/explosive detector on ARES robot; (**b**) Transparent model (enlarged) of photon systems stand-off robotic Raman florescence chem/bio/explosive detector on Talon robot; (**c**) Photon systems stand-off robotic Raman florescence chem/bio/explosive detector on packbot

The overall concept of this technology is to integrate a Raman sensor head onto a manipulator arm on the UGV, which is then coupled to an onboard or self contained spectrometer analyzer. When integrated with a robot, Raman spectroscopy detectors contain a video camera and fine positioning system that will allow for targeting of the head, laser illumination of the sample to induce the Raman Effect, optics to collect and focus the scattered light, and a fiber optic bundle to transport the scattered light to a spectral analyzer. In proximity applications, the Raman detector needs to be close but not necessarily touching the object; in stand-off applications the laser, spectroscope, and analysis computer can operate from a distance. Once the materials unique Raman effect has been detected it can then be compared to a spectra library of known materials to provide robust identification of whether the chemical is a threat. Several TATRC funded proximity and stand-off prototypes have been developed and integrated with robots.

2.7 L-STAT

The L-STAT system (Life Support for Trauma and Transport, Integrated Medical Systems Inc., Signal Hill, CA) was developed by a DARPA funded grant in 1999 in conjunction with the United States Marines. This system has seen deployed to the 28th and 31st Combat Support Hospitals (CSH) in Iraq and Afghanistan, Navy amphibious assault ships, national guard units in Alaska and Hawaii, special operations teams in the Philippines and Cambodia, and also domestically at select United States trauma centers (University of Southern California and the Navy Trauma Training Center both in Los Angeles, CA). L-STAT could be used to integrate components of intensive care monitoring and life support functions during unmanned CASEVAC (Fig. 2.5). This platform acts as a force multiplier and allows for patients to be cared for with less direct attention by medical personnel during transport to Combat Support Hospitals or Battalion Stations. As stated before focus on the golden hour of trauma is due to the fact that 86% of all battlefield mortality occurs within the first 30 mins. The majority of which are due to hemorrhage (~50%) followed by head trauma which leads to seizures and ischemic reperfusion injuries and these are the focus of L-STAT. The original version of L-STAT was

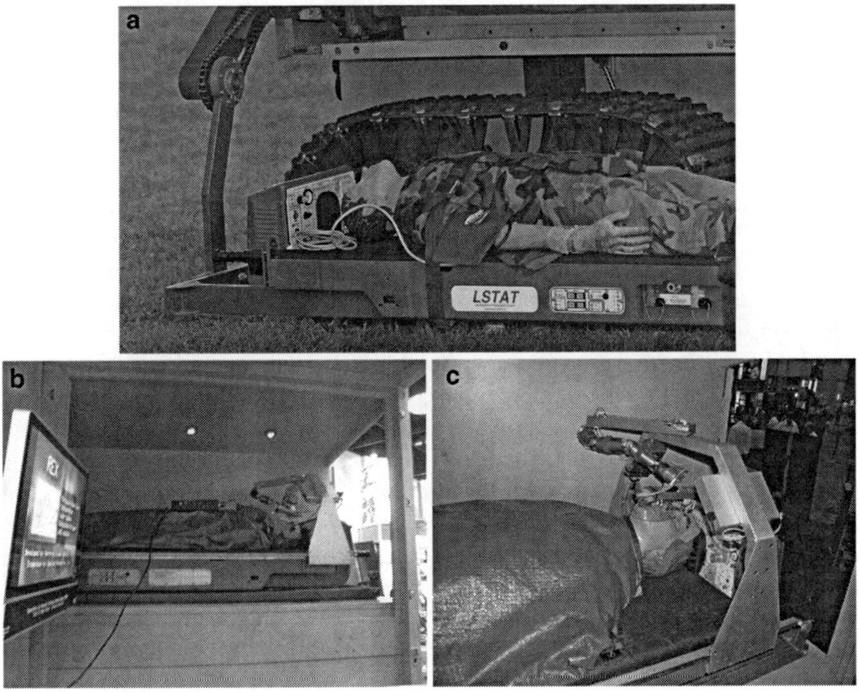

Fig. 2.5 Life support for trauma and transport (L-STAT): (**a**) integrated with REV; (**b**) L-STAT mounted in TAGS-CX patient transport pod;(**c**) L-STAT with Carnegie Mellon University serpentine robotic arm casualty assessment prototype

extremely cumbersome and weighed 200 lbs which severely limited its utility. Some more recent systems are much more mobile and include the L-STAT Lite, MOVES, and Lightweight Trauma Module.

A review of L-STAT identified possible future technologies for the next generation L-STAT (NG-LSTAT) and concluded there were multiple areas of potential improvement in diagnostic and therapeutic capabilities. The possible diagnostic additions included digital X-ray, portable ultrasound, medical image display and telediagnosis via remote controlled camera. Prospective therapeutic additions included the utilization of serpentine robotic manipulators for performing intubation, ultrasound catheterization for intravenous access and assisting in the application of HIFU (High Intensity Focused Ultrasound) for treating hemorrhage. The addition of bioinformatics, wireless data communication, additional imaging capabilities, robotic manipulators, and increased mobility would move the NG-LSTAT further toward the goal of an autonomous field deployable surgical platform. A lightweight version of the LSTAT called the MedEx-1000 which weighs less than 40 lbs and can be used independently of a litter was developed and released for sale in 2009.

2.8 Robotic Combat Casualty Care

The definition of telesurgery varies but in general practice it is the "use of telecommunication technology to aid in the practice of surgery." This commonly used broad based definition of telesurgery encompasses the entire gamut of surgical practice from ancillary guidance or evaluation to direct patient interventions. The initial roll of telesurgery focused on the supplementary or instructive components include: pre-, intra, and postoperative teleconsultation and teleevaluation, to intraoperative telementoring and teleproctoring. Recently a more limited view of telesurgery focuses on telecommunication and a distributed surgeon performing direct patient interventions through robotic telemanipulation and telepresent robotic surgery. This revolutionary idea was borne from the combination of advances in communication and robotic technology and the explosion in minimally invasive surgery in the 1990s. There are two components of telesurgery, first is the "teleoperator" which encompasses the insertion of technology between the surgeon and the patients so that the surgeon never directly touches the patient, and the second is the use of telecommunication technology to allow for the geographic distribution of surgical care. The idea of teleoperators is not a new phenomenon, but built upon ideas developed much earlier in the twentieth century. Ray Goertz of Argonne National Labs in the late 1940s and early 1950s coupled mechanical effectors with a mechanical system of cables and pulleys, which allowed for manipulation of objects at a distance. Though extremely effective and still in contemporary use this technology was fundamentally limited to short distances and similar size scales. The modern surgical teleoperator arose from the technologic advances that established the potential platform for technical feasibility while the laparoscopic context provided the surgeon the skill set needed to manipulate and master this new potential surgical schema. These fields developed synergistically with surgical demand driving a critical mass of technology in the form of optical cameras and manipulators which allowed the surgeon better visualization and dexterous manipulation of tissue in minimally invasive surgery. The physical disconnection of the surgeon from the patient created surgical telemanipulators, the first component of telesurgery.

In the early 1990s, Stanford Research Institute International (SRI, Menlo Park, CA) developed a two-handed teleoperated surgery unit through a DARPA (Defense Advanced Research Projects Agency) funded project. This provided the direct progenitor for the development of surgical robots currently in use. Two start-up companies were created to address the civilian surgical market: Computer Motion Inc. (Goletta, CA) and Intuitive Surgical Inc. (ISI), which was spun off of SRI in 1995. Both of these companies subsequently produced FDA approved surgical robot platforms, Computer Motion's Zeus and ISI's da Vinci system. These companies merged in 2003 effectively eliminating private surgical robotic competition. The da Vinci system is used around the world with more than 1,000 systems used currently.

2.9 Telesurgery

Strides toward the realization of the widespread application of telesurgery have been made with several historic procedures. The seminal event in telesurgery was Project Lindberg. On September 7, 2001, Jacques Marescaux in conjunction with his team at the European Institute of Telesurgery (EITS)/Universite Louis Pasteur in Strasbourg, France established the feasibility of telesurgery by performing the first transatlantic telerobotic laparoscopic cholecystectomy. Marescaux performed this successful operation with the Zeus robot (Computer Motion, Inc., Goleta, CA now operated by Intuitive Surgical, Inc., Sunnyvale, CA) in New York City on a patient located in Strasburg, France. Mehran Anvari has since extended viable telesurgery by bringing surgical therapy to underserved populations in rural Canada. He utilized a modified Zeus surgical system (Zeus TS) with a private network to perform advanced laparoscopic surgery from the Centre for Minimal Access Surgery (CMAS)/McMaster University, Hamilton, Ontario. He has performed 25 telesurgeries including laparoscopic fundoplications, colectomies, and inguinal hernias with outcomes comparable to traditional laparoscopic surgery.

In 2005, The US military funded the first transcontinental telesurgery utilizing the da Vinci robot. This collaborative project included Intuitive Surgery, Inc., Walter Reed Army Medical Center, Johns Hopkins University, and the University of Cincinnati. The experimental setup had the porcine subject in Sunnyvale, CA while the remote surgeon performed a nephrectomy and was located in Cincinnati, OH (March 21–23) or Denver, CO (April 17–19). A da Vinci console was located at both the remote and local sites and control of the three manipulator arms were shared by the two surgeons with the local surgeon controlling the electrocautery. This novel experiment performed several telesurgery firsts including: utilization of a non-dedicated public Internet connection, the first stereoscopic telesurgical procedure, and collaborative telesurgery with two separate consoles controlling different parts of the same robot.

The replication of patient side activity from a distance represents the fundamental goal of telesurgery. Recent telesurgery experiments have focused on the fidelity of replication without incidence, which would allow for confidence in the overall safety. A multidisciplinary, multi-institutional team approach has been undertaken because of the need to incorporate diverse, substantial expertise including robotics, surgery, and telecommunications. The general approach of these experiments in telesurgery has utilized a surgical robot in combination with high bandwidth, low latency, and high-Quality of Service (QoS) telecommunications. The Zeus (no longer commercially available) and the da Vinci robotic systems represent the past and current versions of commercially available robotic telesurgery platforms. During initial experimental and clinical trials, common problems and themes arose and generated a common vocabulary of technical terms specific to this burgeoning field. The most important definitions refer to the time delay inherent to telesurgery and include: control latency, visual discrepancy, round trip delay, and the CODEC. Control latency represents the time delay between the remote telesurgeon's controller manipulation and when the surgical manipulator moves within the patient. Simply it is the flow of information from the surgeon to the patient. Visual discrepancy is the time delay between an operative field action and when the surgeon appreciates this action at the remote controller site and represents the duration of time that visual information egresses from patient to surgeon. Round trip delay is the additive time increments of control latency and visual discrepancy and is the time it takes for a telesurgeon to manipulate a tool at the remote site and then be able to acknowledge the effect in the patient's surgical environment. An important software technology is the coder-decoder (CODEC) which through compression reduces the bandwidth required for video transmission. TATRC has funded multiple research projects to mitigate the effect of operationally relevant telecommunication limitations.

2.10 Extreme Environment Surgical Robotics

While the embryonic field of telesurgery has primarily utilized robots designed for minimally invasive surgery, the military goal is to develop battlefield telerobotic surgery for use in trauma. The current minimally invasive robotic surgery system cannot be used in an operationally and clinically relevant manner for battlefield or en route combat casualty care as trauma surgery currently requires open exposure to identify and manually treat injuries (e.g. abdominal packing, tissue mobilization, retraction, etc). The commercially available current surgical platform, da Vinci, is large, bulky and has a time consuming and complicated setup and is generally ill suited for trauma. Future battlefield robotic surgical system will have to provide life saving trauma care and will incorporate novel technologies that permit distributed and automated performance of simple "damage control" procedures. Battlefield interventions need to focus on the idea of the "golden hour" where the majority of trauma casualty deaths occur and where immediate and definitive care can prove life saving. The principle injuries that would be amenable to expeditious intervention include control of: airway, tension pneumothorax and non-compressible bleeding, Telesurgery is a potential force multiplier that could protects surgical assets and deliver immediate and definitive care to wounded soldiers. Due to the extreme nature of battlefield environments, the next generation mobile surgical robots will be smaller, robust trauma focused systems that leverage non-medical military telecommunication, computing, imaging, and mechanical resources.

A couple of robotic surgical platforms have been routinely used in the research and development of surgical robotics for use in extreme environments: the University of Washington RAVEN and the SRI International M7. The RAVEN is a small deployable surgical robot being developed at the University of Washington BioRobotics Laboratory with support from multiple government agencies including the US Army. The system consists of a slave component that resides with the patient and a master controller permitting remote control of the slave by the surgeon. The master site has a surgeon console that currently employs dual PHANToM Omni devices to control two surgical manipulators/instruments, a foot pedal, and a video screen displaying images from the surgical site. The video and robot control are transmitted using standard Internet communication protocols. The user interface uses open source commercial off the shelf technology and therefore it is remarkably low cost, portable, and interoperable (i.e. can readily control other systems with limited modifications). SRI's M7 surgical robot was initially developed in 1998 with funding from the US Army. The M7 leveraged military funded development of SRI's original telepresence surgical system. The features of this robot include a large workspace accessible via two anthropomorphic robotic arms with seven force-reflective degrees of freedom. These robotic arms manipulate conventional "open" surgical instruments allowing for complex surgical tasks to be performed. The system was recently upgraded with high definition stereoscopic vision, ergonomic hand controllers, and limited automation. Both of these surgical robotic systems have been utilized in extreme environments to evaluate feasibility as well as guide future research and development.

2.11 NASA Extreme Environment Mission Operations (Neemo)

Collaborative telesurgery research was conducted within the NASA Extreme Environment Mission Operations (NEEMO) program. US Army TATRC telesurgery research within NEEMO missions has been conducted in collaboration with the National Aeronautics and Space Administration (NASA), the National Oceanographic and Atmospheric Administration (NOAA), the Centre for Minimal Access

Surgery (CMAS), and the Canadian Space Agency. The NEEMO missions occur within the NOAA National Undersea Research Center Aquarius habitat located at 19 m depth within the Florida Keys.

In 2006, NEEMO 9 explored the use of telementoring, and telerobotic surgery to provide emergency diagnostic and surgical capabilities in an extreme environment. Mission accomplishments included: the first successful deployment and use of a surgical robot (SRI's M7) in an extreme environment and the use of microwave wireless telecommunications in support of telesurgery. Simulated surgical procedures were performed to evaluate the effect of increasing latency on surgeon performance. Latency of over 500 ms was found to greatly impact performance. While the remote surgeon was able to suture simulated tissue despite 2 s latency, placing and tying a single suture in 10 mins is not clinically relevant. These experiments demonstrated that latency compensation up to approximately 500 ms was possible by modifying surgical technique to include slow, one handed movements. Several technologic solutions were successfully used to overcome sub-second latency such as motion scaling. These M7 telesurgical experiments suggested further research in automation was necessary. Astronauts on NEEMO 9 also evaluated the University of Nebraska – Lincoln (UNL) in vivo robots. These novel miniature mobile robots were deployed inside a laparoscopic simulator and found to improve visualization of the surgical field.

In 2007, NEEMO 12 primarily focused on evaluation of image guided, supervisory controlled autonomous function to overcome latency. A modified M7 was used to perform an ultrasound guided, semi-autonomous needle insertion into a simulated blood vessel. The RAVEN surgical robot was also deployed and used to objectively assess telesurgical performance of SAGES' (Society of American Gastrointestinal and Endoscopic Surgeons) Fundamentals of Laparoscopic Surgery (FLS) tasks.

2.12 Mobile Robotic Telesurgery

Battlefield operations are dynamic and challenging, and do not permit routine operational use of traditional "wired" telecommunications. As geosynchronous orbit is 35,900 km above the earth's surface, satellite-based communications has a minimum round trip communication latency between surgeon and patient above 500 ms. Unmanned airborne vehicles (UAV) represent a readily available battlefield asset that could provide an extremely low latency "last mile" solution for telesurgery. TATRC funded the High Altitude Platform/Mobile Robotic Telesurgery (HAPs/MRT) project to evaluate the feasibility of deploying a mobile surgical robotic system to the high desert and operating this system using of a UAV based telecommunication link.

In 2006, a collaborative research team including the University of Cincinnati, University of Washington, AeroVironment Inc. (Monrovia, CA), and HaiVision Inc. (Montreal, Canada) conducted this research in the high desert of southern California.

A high bandwidth and low latency network was created using AeroVironment's PUMA (Point Upgraded Mission Ability) small UAV. The PUMA is hand-launched and currently in use in Iraq and Afghanistan primarily for local reconnaissance. The radio link onboard the PUMA provided a digital link between the RAVEN slave and master controller which were located in separate tents within Simi Valley. The UAV based network provided over 1 Mbps bandwidth with transmission times of less than 10 ms. The remote surgeon successfully used the UAV – RAVEN mobile robotic system to perform simulated surgical tasks. This experiment demonstrated that a readily deployed surgical robot and a routinely used small UAV could potentially deliver surgical capabilities to the battlefield. Challenges encountered during this research emphasized the need for continued development of telecommunications hardware and software to facilitate operationally and clinically relevant telesurgery.

U.S. Army TATRC also funded University of Cincinnati and SRI to explore distributed, automated surgical robotics as a means to augment en route care of injured warfighters. In 2007, the M7 was modified to overcome acceleration and movement routinely encountered during vehicle transport. Three-axis acceleration compensation was developed to dampen turbulence and apply a neutralizing force during periods of more constant acceleration. Multiple military personnel, including a U.S. Air Force Critical Care Air Transport (CCAT) surgeon, demonstrated robust performance of the acceleration compensating M7 during parabolic flight aboard NASA's C-9 aircraft.

2.13 Next Generation Technologies

The military continues to develop diagnostic and therapeutic modalities to improve care of injured warfighters. Two of the more promising technologies that could readily be incorporated into medical robotic systems are HIFU and laser tissue welding.

2.13.1 High Intensity Focused Ultrasound (HIFU)

HIFU continues to be evaluated as a non-invasive method of controlling bleeding. Military funded research has demonstrated that HIFU can seal vascular injuries of up to 3 mm in diameter. Recently, a DARPA funded project, "Deep Bleeder Acoustic Coagulation" (DBAC) has begun to develop a prototype HIFU device capable of limiting blood loss from non-compressible vessels. DBAC would be applied in a combat situation by minimally trained operators, automatically detect the location and severity of bleeding, and use HIFU to coagulate the bleeding vessel. The project includes Doppler based automated hemorrhage detection algorithms coupled with volumetric data to localize the bleeding source. HIFU delivery and dosing for safe acoustic hemostasis has been proven to raise the tissue temperature to a range of 70–95 C in an operationally relevant 30-s timeframe.

2.13.2 *Robotic Laser Tissue Welding*

TATRC funded SRI to investigate robotic assisted laser tissue welding as a means to circumvent the need for suturing. As previously mentioned, telerobotic suturing is especially challenging at longer latencies which would be encountered during robotic combat casualty care. These experiments used a robot to uniformly deliver laser energy as well as tissue pressure and apposition. Two methods were demonstrated for direct tissue welding: bovine serum albumin/hyaluronate acid solders and chitosan films. Robot controlled tissue welding of lacerations in explanted pig eyes decreased the total time of tissue apposition from a manual suturing from approximately 8 to 3 min. Laser welded tissue had similar burst pressure as manually sutured tissue. These experiments demonstrated that robotic laser tissue welding has great potential value and further research is indicated.

2.14 Trauma Pod: Distributed, Automated Robotic Combat Casualty Care

Trauma Pod (TP) is a DARPA program to develop automated robotic combat casualty care. Trauma Pod represents a semi-autonomous telerobotic surgical system that can be rapidly deployed and provide critical diagnostic and acute life-saving interventions in the field.

The Phase I proof of concept platform was comprised of a da Vinci Classic surgical robot supported by an automated suite of commercially available and custom designed robots. The surgeon remotely controlled the robotic suite to perform representative tasks that included placing a shunt in a simulated blood vessel and performing a bowel anastamosis.

TP footprint was 8 × 18 ft to fit within an International Standards Organization (ISO) shipment container for ready deployability. The Phase I system is comprised of 13 subsystems that include: the Surgical Robot (SRS), the Scrub Nurse (SNS), Tool Rack (TRS), Supervisory Controller (SCS), Patient Imaging (PIS), and the User Interface (UIS). The Scrub Nurse Subsystem system was developed by Oak Ridge National Laboratory and automatically delivered instruments and supplies to the surgical robot within 10 s (typically faster than a human). The University of Washington developed the Tool Rack System which held, accepted, and dispensed each of 14 surgical tools. The University of Texas developed the Supervisory Controller System which provided high-level control of all automated subsystems involved in supply dispensing/tool changing and coordinated these functions with the SRS. The Patient Imaging (GE Research) utilized the L-STAT platform to embed CT like capabilities as well as 2-D fluoroscopic data. The User Interface System developed by SRI International provided a visual, verbal, aural, and gesture based interface between the surgeon and TP system. The visual display consisted of a stereoscopic view of the surgical site augmented by physiologic data, icons and other supporting information.

In 2007, the phase I demonstration included:

1. Automatic storing and dispensing of surgical tools by the TRS with 100% accuracy
2. Automatic storing, de-packaging dispensing and counting supplies by the SDS (Supply Dispenser)
3. Automatic change of surgical tools and delivery and removal of supplies by SNS
4. Speech-based interface between a tele-operating surgeon and the TP system through the UIS
5. Automatic coordination and interaction between SRS and SNS
6. Performing iliac shunt and bowel anastamosis procedures by a tele-operated SRS on a phantom patient

Phase 1 proved that a single operator can effectively tele-operate a surgical robot and integrated suite of automated support robots to perform relevant surgical procedures on a simulated patient. Phase II of this project will integrate TP subsystems into a single robot designed to rapidly diagnose and innovatively treat life threatening battlefield.

2.15 Summary

The technological revolution of the past three decades is catalyzing a paradigm shift in the care of battlefield casualties. Telecommunications and robotic technology can revolutionize battlefield care by safely extracting patients from harm's way, rapidly diagnosing life threatening injuries, and delivering life-saving interventions. Telecommunication and robotic limitations that prevent robust intervention at a distance are areas of continued military research and development. As these limitations are overcome, medical robots will provide robust casualty extraction and care that will save the lives and limbs of our deployed warfighters.

Disclaimer The views expressed in this chapter are those of the authors and do not reflect official policy or position of the Department of the Army, Department of Defense or the U.S. Government.

Bibliography

1. Anvari, M., Broderick, T., Stein, H., Chapman, T., Ghodoussi, M., Birch, D., et al.: The impact of latency on surgical precision and task completion during robotic-assisted remote telepresence surgery. Comput. Aided Surg. **10**(2), 93–99 (2005)
2. Curley, K., Broderick, T., Marchessault, R., Moses, G., Taylor, R., et al.: Surgical robotics – the next steps. Integrated research team final report. Telemedicine and Advanced Technology Research Center. U.S. Army Medical Research and Materiel Command, Fort Detrick, MD (2005)
3. Defense Advanced Research Projects Agency: Arlington, VA. http://www.darpa.gov (Internet)

4. Doarn, C.R., Anvari, M., Low, T., Broderick, T.J.: Evaluation of teleoperated surgical robots in an enclosed undersea environment. Telemed. e-Health **15**(4), 325–335 (2009)
5. FY2009–2034 Unmanned systems integration roadmap. 2nd Edition. US Department of Defense (2009)
6. Garcia, P., Mines, M.J., Bower, K.S., Hill, J., Menon, J., Tremblay, E., Smith, B.: Robotic laser tissue welding of sclera using chitosan films. Lasers Surg. Med. **41**(1), 60–67 (2009)
7. Garcia, P., Rosen, J., Kapoor, C., Noakes, M., Elbert, G., Treat, M., Ganous, T., Hanson, M., Manak, J., Hasser, C., Rohler, D., Satava, R.: Trauma pod: a semi-automated telerobotic surgical system. Int. J. Med. Robot. Comput. Assist. Surg. **5**(2), 136–146 (2009)
8. Gardner, C.W., Treado, P.J., Jochem, T.M., Gilbert, G.R.: Demonstration of a robot-based Raman spectroscopic detector for the identification of CBE threat agents. In: Proceedings of 25th Army Science Conference (2006)
9. Hanly, E.J., Marohn, M.R., Schenkman, N.S., Miller, B.E., Moses, G.R., Marchessault, R., Broderick, T.J.: Dynamics and organizations of telesurgery. Eur. Surg. Acta Chir. Austriaca **37**(5), 274–278 (2005)
10. Hanly, E.J., Broderick, T.J.: Telerobotic surgery. Oper. Tech. Gen. Surg. **7**(4), 170–181 (2005)
11. Harnett, B.M., Doarn, C.R., Rosen, J., Hannaford, B., Broderick, T.J.: Evaluation of unmanned airborne vehicles and mobile robotic telesurgery in an extreme environment. Telemed. e-Health **14**(6), 539–544 (2008)
12. King, H.H., Low, T., Hufford, K., Broderick, T.: Acceleration compensation for vehicle based telesurgery on earth or in space. In: IEEE/RSJ International Conference on Intelligent Robots and Systems, IROS, pp. 1459–1464 (2008)
13. Kirkpatrick, A.W., Doarn, C.R., Campbell, M.R., Barnes, S.L., Broderick, T.J.: Manual suturing quality at acceleration levels equivalent to spaceflight and a lunar base. Aviat. Space Environ. Med. **79**(11), 1065–1066 (2008)
14. Lum, M.J., Friedmanm D.C., Kingm H.H., Donlinm R., Sankaranarayanan, G., Broderick, T. J., Sinanan, M.N., Rosen, J., Hannaford, B.: Teleoperation of a surgical robot via airborne wireless radio and transatlantic internet links. Springer Tracts Adv. Robot. **42**, 305–314 (2008)
15. Lum, M.J., Rosen, J., King, H., Friedman, D.C., Donlin, G., Sankaranarayanan, G., Harnett, B., Huffman, L., Doarn, C., Broderick, T., Hannaford, B.: Telesurgery via Unmanned Aerial Vehicle (UAV) with a field deployable surgical robot. Stud. Health. Technol. Inform. **125**, 313–315 (2007)
16. Osborn, J., Rocca, M.: (Carnegie-Mellon University Pittsburgh, PA.) Conceptual study of LSTAT integration to robotics and other advanced medical technologies. Final Report. U.S. Army Medical Research and Materiel Command, Fort Detrick, MD (2004)
17. Rytherford, M.: Ultrasound cuff to stop internal bleeding on the battlefield. CNET. http://news.cnet.com/8301-13639_3-10071564-42.html (2008)
18. Seip, R., Katny, A., Chen, W., Sanghvi, N.T., Dines, K.A., Wheeler, J.: Remotely operated robotic high intensity focused ultrasound (HIFU) manipulator system for critical systems for trauma and transport (CSTAT). Proc IEEE Ultrason. Symp. **1**, 200–203 (2006)
19. Telemedicine and Advanced Technology Research Center. Fort Detrick, MD. http://www.tatrc.gov (Internet)
20. Thirsk, R., Williams, D., Anvari, M.: NEEMO 7 undersea mission. Acta Astronaut. **60**, 512–517 (2007)
21. US Army TRADOC Pamphlet 525–66, Force Operating Capabilities, para 4–69b(5), p. 130. (2008)
22. US Army TRADOC Pamphlet 527–7–15, The United States Army Capability Plan for Aviation Operations 2015–2024. para 211b(3)(e), p43 (2008)
23. US Army TRADOC Robotics Strategy White Paper, pp. 16–17 (2009)
24. Vecna Technologies, Inc. Cambridge, MA. URL:http://www.vecna.com (Internet)

Chapter 3
Telemedicine for the Battlefield: Present and Future Technologies

Pablo Garcia

Providing medical care in the field is very challenging because of the limited availability of medical resources. The current practice in military operations is to stabilize patients far forward and evacuate them to better equipped medical facilities, such as combat field hospitals. This strategy has proven very successful in recent conflicts. However, it is possible to save more lives and ameliorate the consequences of long term injuries by providing an accurate diagnosis earlier, specialized surgical care faster and more sophisticated intensive care during transport. This chapter focuses on technologies that allow augmenting the diagnostics and treatment capabilities of medical teams in the field.

The decisions made by the first responders are critical for the patient chances of survival and they are done with very little information on the real status of the patient. For example, recognizing an internal head injury or abdominal bleeding is critical to determine where the patient should be evacuated but this information is currently not available. Once a patient is evacuated to a medical facility it is impossible to provide all specialties required to treat complex injuries (it is not uncommon for general trauma surgeons to conduct head procedures because they do not have a neurosurgeon in the staff). Finally, once patients are stable enough to transport them outside the country they are taken on long transcontinental flights with a small team of nurses and doctors who have limited means to react to adverse conditions in patients.

Technology can be used to force multiply the capabilities of the medical personnel in the field by allowing them to perform more sophisticated diagnosis and treatments closed to the point of injury. Figure 3.1 illustrates the different layers of technology which can be used to treat patients in the field. Each additional layer adds a level of sophistication to the one below but is applicable to a smaller pool of patients.

P. Garcia (✉)
SRI International, 333 Ravenswood Avenue, Menlo Park, CA 94025, USA
e-mail: pablo.garcia@sri.com

J. Rosen et al. (eds.), *Surgical Robotics: Systems Applications and Visions*,
DOI 10.1007/978-1-4419-1126-1_3, © Springer Science+Business Media, LLC 2011

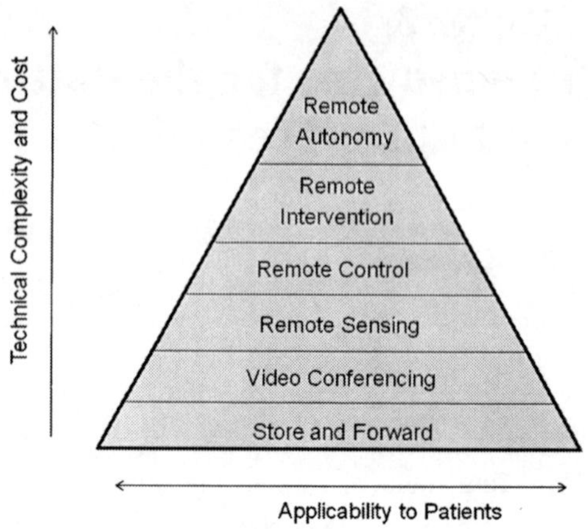

Fig. 3.1 Pyramid of telemedicine technologies for the battlefield

3.1 Store and Forward Communication

Clinicians in the field are able to get advice by using email or web applications, forwarding files or pictures about the patients they are treating.

The Army Knowledge Online (AKO) critical care tele-consultation system is a web-based service used to perform consultation on patients being treated in the field. The clinician in the field sends an email consultation request to a central address and the request is routed to specialists around the world who provide an answer with recommendations within a window of a few hours. The system has been used successfully in thousands of consultations.

MedWeb (San Francisco, CA) provides a web-based tele-radiology service which allows clinicians to transmit X-ray, CT MRI and Ultrasound data sets from patients in the field to be analyzed by radiologists around the world. This capability allows specialists to provide input remotely which might be critical in deciding the next steps for the patient. The Medweb system has been very successful and is routinely used to provide assistance diagnosing patients (Fig. 3.2).

3.2 Video-Conferencing Communication

Real time video-conferencing communication requires a minimum bandwidth (200–2,000 kbs) and maximum latency (up to 2 s) to conduct a real time consultation. The recent addition of the Joint Telemedicine Network (JTMN) has allowed an expanded bandwidth to conduct medical consultations with Iraq and the same network

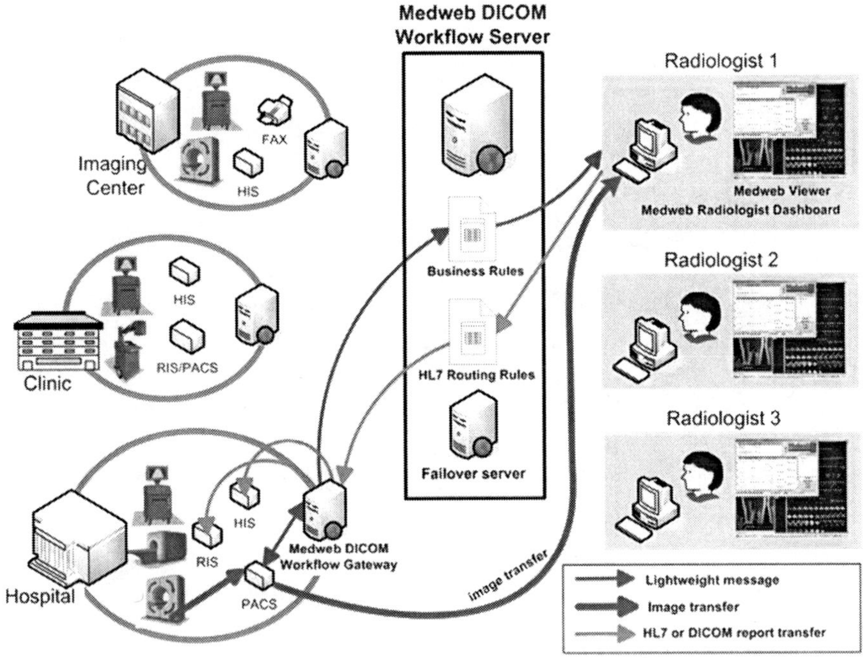

Fig. 3.2 Medweb network for teleradiology

will be transferred to Afghanistan. A network of conventional video-conferencing systems is currently available to provide consultation services between the US and combat field hospitals in Iraq although it is not widely used yet.

A telementoring system recently developed by SRI was successfully used to mentor a live chest surgery in Mosul (Iraq) from Brooke Army Medical Center (BAMC) in Texas. The system uses a pan-tilt-zoom camera attached to the operating room lights and a camera on the surgeon's head. The specialist providing the consultation has full control of the cameras and can manipulate the images to guide the surgeon working on the patient. X-rays or CT images can be shared before or during the surgery and surgical manuals can be discussed live to illustrate the next steps required in the case. The system is software based and can be downloaded in a few seconds to any machine within the military network, allowing the consulting clinician to participate in a consultation from wherever they happen to be. The system provides the surgeon in the field to get assistance from specialists in the US for cases in which they may not have as much experience (Fig. 3.3).

The real challenge with real-time consultation is the acceptance by the clinical community. Clinicians in the field have to be able to reach specialists they trust when they need them, and those specialists have to be willing to participate in the consultation which represents one more responsibility added to clinical duties.

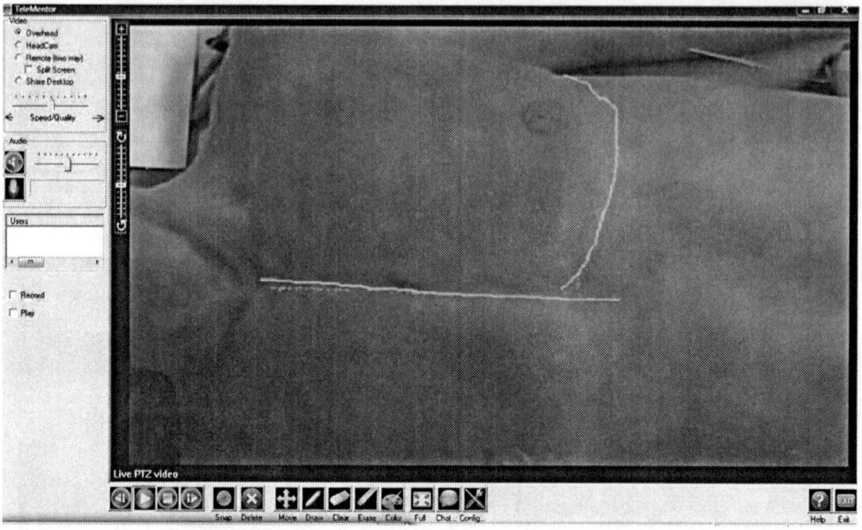

Fig. 3.3 SRI's surgical telementoring system

3.3 Remote Monitoring

There are various technologies available to monitor the vitals of patients in critical condition remotely. This is not currently used in the battlefield, but it is used in civilian applications and is the next logical step in military telemedicine network.

One of them is pioneered by TeleVital (San Jose, CA) web-based application which can monitor vital signs, EEG, ECGs in real time and store them in a data base for review by the clinicians. The application allows a clinician or intensivist to monitor a patient from anywhere the world, and provide real-time advice on the therapy required by that patient.

InTouch Health (Santa Barbara, CA) has taken the idea one step further by designing a mobile platform which can be controlled remotely around the hospital to check on the status of patients and provide consultations. The platform was successfully tested from the US to Landstuhl (Germany) to providing over 300 consultations on patients returning from Iraq and Afghanistan over a 15 month period.

3.4 Remote Tele-Operated Interventions

SRI International pioneered the use of tele-operated robots to perform surgical procedures in the 1990s. The technology was commercialized through Intuitive Surgical to augment the dexterity of laparoscopic surgeons in hospitals. Computer Motion developed a similar system and demonstrated the first time live transatlantic procedure from New York to Strasbough in 2001 [Ref].

In 2006 SRI demonstrated the feasibility of conducting remote procedures with portable robotic arms in an extreme environment in collaboration with the NASA Extreme Environment Mission Operations (NEEMO) 12 [Ref] and Fig. 3.4. The NEEMO 12 mission took place on board the National Oceanic and Atmospheric Administration (NOAA) Aquarius Underwater Laboratory, located more than 900 miles from Nashville and 60 feet underwater off the coast of Key Largo, Florida. The underwater habitat is used to train astronauts for 1 month to simulate the conditions in the international space station. The test was repeated in 2008 during NEEMO 13 to demonstrate the feasibility of conducting a closed loop procedure, an ultrasound guided intravenous insertion, using robotic manipulators (Fig. 3.5).

In 2006 the University of Washington and the University of Cincinnati conducted experiments in the southern California desert to prove the feasibility of controlling a mobile robotic telesurgery system through use of an unmanned aerial vehicle-based (UAV) communication system in an the extreme environment. The primary objective was to develop and validate a High Altitude Platforms (HAPs) Mobile Robotic

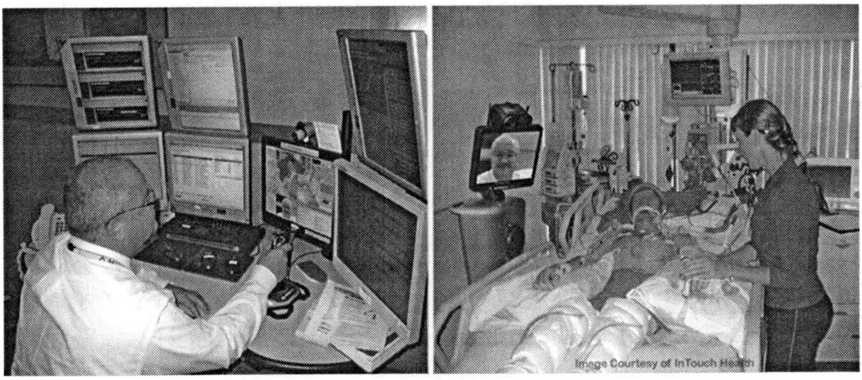

Fig. 3.4 In-touch remote control monitoring system

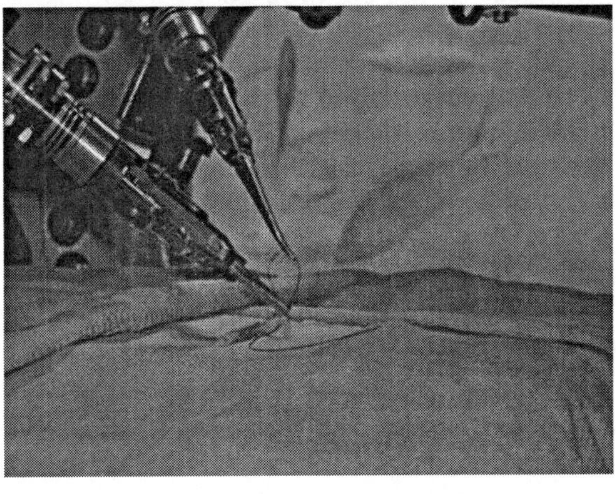

Fig. 3.5 SRI's surgical robot tested in aquarius underwater habitat during NEEMO 12

Telesurgery (MRT) system that allows a remote surgeon to effectively operate on an injured soldier regardless of the soldier's location or environment [Ref].

Although the feasibility of conducting surgical interventions remotely has been demonstrated, there has not been market or clinical drivers strong enough to justify its implementation. In the case of trauma surgery for the battlefield one of the challenges is the wide variety of possible interventions required and the large forces and workspace required by most procedures. Tele-operated procedures with current robotic tools have excelled at targeted delicate procedures such as heart or prostate procedures. To perform trauma relevant procedures remotely a new generation of surgical robots and tools needs to be developed more targeted towards those particular procedures.

3.5 Remote Autonomy

Guaranteeing a communication link to the battlefield is not always possible and coordinating with remote medical personnel can be very challenging. Furthermore some monitoring functions and therapies are better suited to closed loop computer control than to direct human control.

When injured soldiers are transported over the ocean hooked up to ventilators and vital sign monitors they have to rely solely on the judgment and resources of the medical personnel on board of the aircraft. Closed loop ventilation has been demonstrated to be more precise and require less oxygen than ventilation controlled by medical staff [Ref]. Similar studies have been performed for fluid administration and anesthetics {Ref]. A closed loop life support system can monitor the patient second by second and perform small adjustments based on the information collected. This frees medical personnel to dedicate more attention to those patients that might be having serious problems.

Life-saving procedures such as securing an airway or placing a chest tube require a significant level of training and skill. Having devices or robots that can perform those procedures would allow lower skilled medical personnel or even other soldier buddies to perform life saving interventions when there is no other alternative.

The Trauma Pod program described below was intended to integrate all the technologies described in the previous sections into a single platform which could be eventually deployed in small vehicle and controlled remotely. Below is a summary of the first phase of the program which focused on demonstrating the feasibility of such a concept.

3.6 Trauma Pod System Development

For decades, surgery and robotics have progressed separately. Only in the last decade have surgery and robotics matured to a point that allows safe merger of the two, thus creating a new kind of operating room (OR) [1–9]. The ultimate goal of the trauma Pod (TP) system is to produce a rapidly deployable telerobotic platform, capable of performing critical diagnostics and acute life-saving interventions in the field for an injured person who might otherwise die of airway loss or hemorrhage. The TP damage-control interventions will be much more invasive and effective than currently available first-aid treatments, and are envisioned only as necessary measures before the wounded reach a site where conventional medical care is provided. Controlled remotely by a surgeon or intensivist, these interventions will involve procedures that preserve life and limb, such as the ability to obtain an airway, insert an intravenous or intraosseous line, perform hemostasis, manipulate damaged tissues, and perhaps place monitoring devices. The TP will be used when the timely deployment of proper medical personnel is not possible or too risky and the patient cannot be evacuated quickly enough to an appropriate medical facility.

The vision is to provide a system that can perform critical life-saving procedures with the assistance of a corpsman who lacks a high level of medical training. Tasks that require a high level of medical skill will be teleoperated and made as autonomous as permitted by the specific condition of the patient and the complexity of the procedure involved.

Phase I of the TP research program, which is described in this paper, was aimed at exploring what human functions of an operating procedure can be adopted by a machine. Analysis of the functions performed in the OR of today indicates that, in many cases, a surgeon can engage a patient through a surgical robot teleoperational mode [9]. Surgical tool and supply functions (currently handled by a scrub nurse) and OR information (currently managed by a circulation nurse) can be automated. In the OR of the future hands-on functions, currently performed by surgeons, nurses, and imaging technicians, will be done by computer-controlled surgical robots, automated tool-change and supply-delivery systems, and automated intraoperative imaging devices. Software systems will automatically record, document, assist, and assess the surgical operation, and will also provide inventory management.

Subsequent phases of the program will address critical functionality such as closed loop anesthesia, sterilization or the design of custom modules to perform life saving procedures. Miniaturization of systems, deployment in the field and operation during transport will are also important problems which will be addressed in future work.

3.6.1 Overall Approach

We have developed the Phase-I version of a TP system capable of performing portions of specified surgical procedures teleoperated by a surgeon. These procedures include placing a shunt in a major abdominal vessel (e.g., iliac) and performing a bowel anastomosis; both procedures entail surgical skills required in trauma surgery. Eventually, the TP is expected to perform many other surgical procedures entailed in preserving life and limb.

The first objective for this phase of the program was to demonstrate these procedures can be performed by a surgeon teleoperating the surgical robot and supported by automated nursing functions. Ideally, the procedures would be performed as effectively and smoothly as with a human team. The second objective was to demonstrate the feasibility of performing full body CT scans in the field during the operation. All the procedures were demonstrated on phantoms in a controlled environment.

The system was prototyped in a fixed room using a combination of off-the-shelf and custom systems. The project focused on the process flow and architecture required to accomplish the surgical procedures, rather than on developing hardware perfectly tailored for trauma procedures. For example, the surgical robot chosen, the daVinci robot from Intuitive Surgical (ISI) is not designed for trauma surgery, but it is a good placeholder for a surgical robot to demonstrate the flow of tasks during a trauma procedure.

The system is able to detect and recover from errors through the remote intervention of an administrator. The philosophy was to design a system that operates automatically during normal operating conditions but switches to a manual mode during an error condition. All systems have safe states they automatically revert to when a system wide error is encountered.

The TP system consists of 13 subsystems. The abbreviation and description of each TP subsystem, as well as its developer, are listed in Table 3.1 for the convenience of reading this paper. The SRS is capable of performing basic surgical functions (e.g., cutting, dissecting, and suturing) through teleoperation. Except for the PRS, the remaining subsystems are capable of autonomously serving the SRS by changing tools and dispensing supplies, as required by the surgeon, and by recording every TP activity. Figure 3.6 shows the layout of the main physical TP components (those of the SRS, PRS, SNS, SDS (including a Fast Cache), and TRS).

The TP system architecture, shown in Fig. 3.7 is hierarchical. System tasks are initiated by the surgeon and interpreted by the UIS, which issues commands to the SCS, and coordinates all the system tasks. The surgeon has direct control over the SRS through voice commands and a teleoperated joystick interface. The AMS

Table 3.1 Trauma pod subsytems

Subsystems		Developer
SRS	Surgical robot subsystem	Intuitive Surgical, Inc.
AMS	Administrator and monitoring subsystem	SRI International
SNS	Scrub nurse subsystem	Oak Ridge National Lab.
TAS	Tool autoloader subsystem	Oak Ridge National Lab.
SDS	Supply dispenser subsystem	General dynamics robotics
TRS	Tool rack subsystem	University of Washington
SCS	Supervisory controller subsystem	University of Texas
PRS	Patient registration subsystem	Integrated medical systems
PIS	Patient imaging subsystem	GE Research
MVS	Machine vision subsystem	Robotic Surgical Tech, Inc.
RMS	Resource monitoring subsystem	University of Maryland
SIM	Simulator subsystem	University of Texas
UIS	User interface subsystem	SRI International

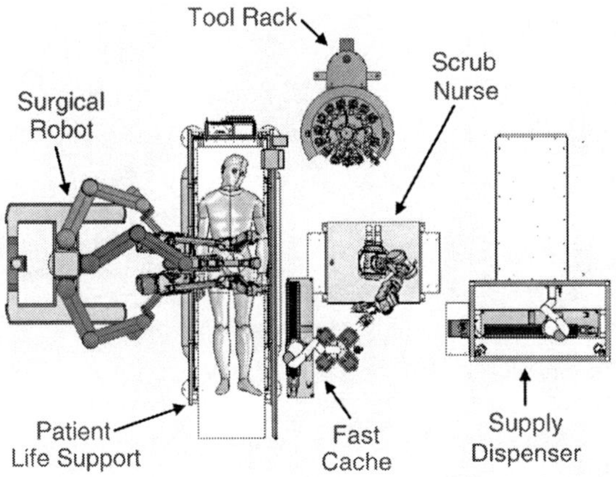

Fig. 3.6 Layout of trauma pod main components

monitors the system status and error conditions, and relays that information to a human administrator, who can manually control any subsystem and correct any error. The subsystems are loosely coupled, and communicate with each other using XML messages through a gigabit Ethernet network. Each subsystem uses its own operating system with a layer of well-defined interfaces that can accept and send XML messages using the appropriate format. Subsystem communication includes a command-and-response protocol along with an exchange of high-speed sensory data. Sensory data are shared in an open-loop fashion and are used by the SCS and SNS to generate collision-free SNS trajectories.

The TP system has been designed, assembled, tested, and debugged by SRI International. We have demonstrated the following TP system capabilities:

- Autonomous change of surgical tools and delivery of supplies by the SNS robot
- Storing, de-packaging, and dispensing of supplies
- Automatic counting of supplies
- User interface between a teleoperating surgeon and the TP system
- Coordination and interaction between the teleoperated surgical robot and the autonomous nursing robots
- Performing iliac shunt and bowel anastomosis procedures on a patient phantom.

The physical TP system, shown in Fig. 3.8, consists of two major cells: a *control cell*, where the surgery is monitored and controlled, and a *surgical cell*, where the surgery is performed. In a real application, the two cells will be far apart: the surgical cell will be deployed in the battlefield while the control cell will be located in a safer place behind; the two cells will telecommunicate via a wideband wireless link. In the current development (Phase I of this project), the two cells are located adjacent to each other and separated by a glass wall.

The control cell includes control stations for the surgeon and the system administrator. It also contains multiple displays of video and sensory information that

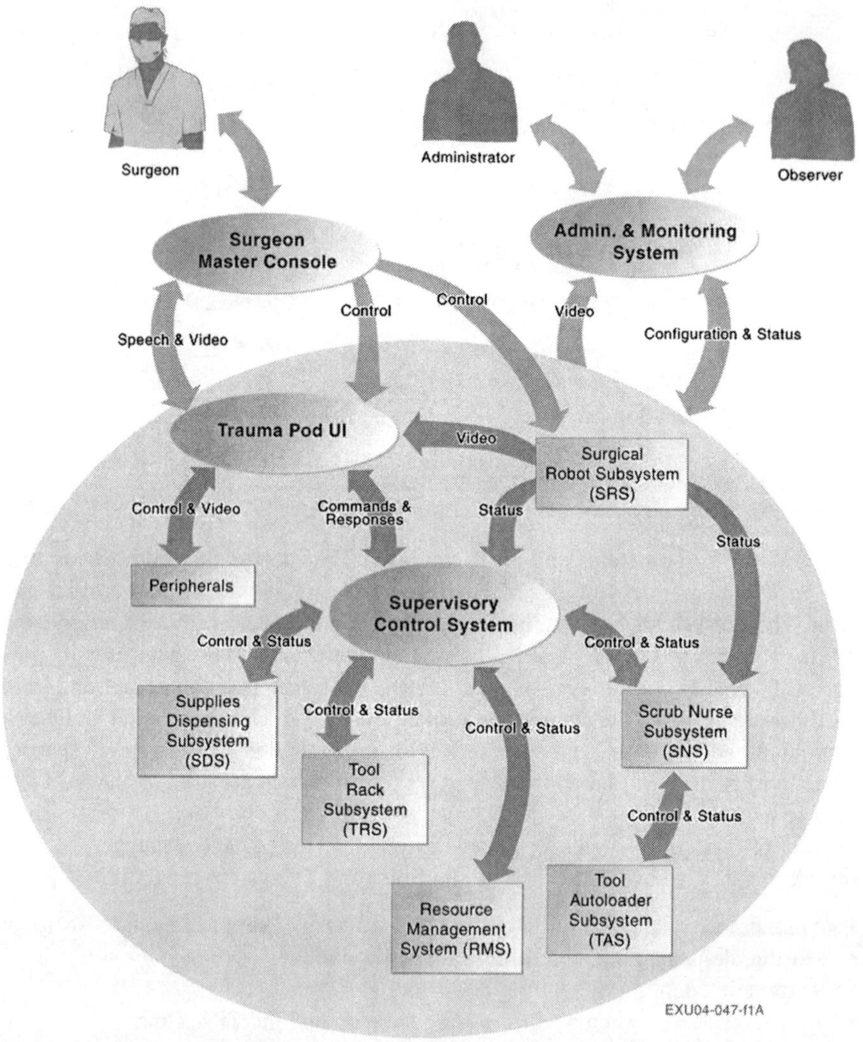

Fig. 3.7 Trauma pod architecture

assist in monitoring and controlling the surgical cell. The surgical cell contains all the TP subsystems required to perform a surgical procedure. The surgical cell footprint (8 × 18 ft) can fit within an International Standards Organization (ISO) container for shipment as cargo.

3.6.2 TP Subsystems

The Phase I TP subsystems are described below.

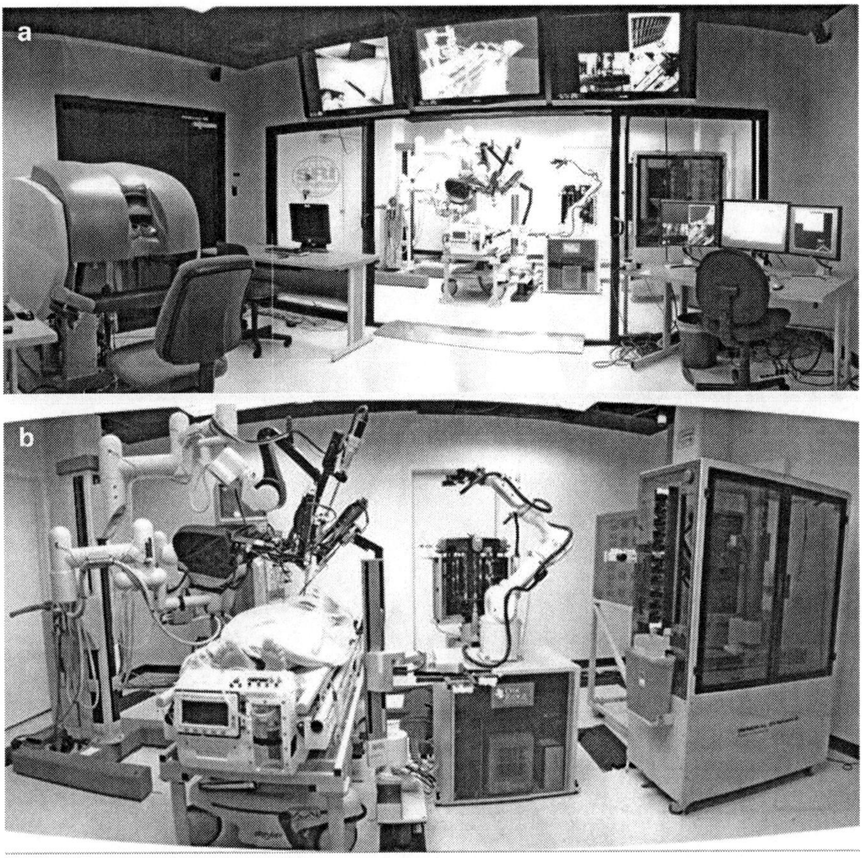

Fig. 3.8 The TP system (a) control cell (b) surgical cell

3.6.2.1 Surgical Robot Subsystem (SRS)

The SRS is capable of performing surgical operations on the patient by teleoperation.

SRS Hardware

The SRS hardware consists of two major parts: a surgical master console and a slave surgical robot. The robot is a first-generation da Vinci Surgical System (see Fig. 3.9), made by Intuitive Surgical, Inc. The surgical master console consists of a user interface for the surgeon and a video cart for acquiring and routing video. The slave surgical robot consists of three arms: one arm with a light source and a stereo endoscopic camera, and two surgical arms, each able to hold and manipulate one of sixteen different types of surgical tools.

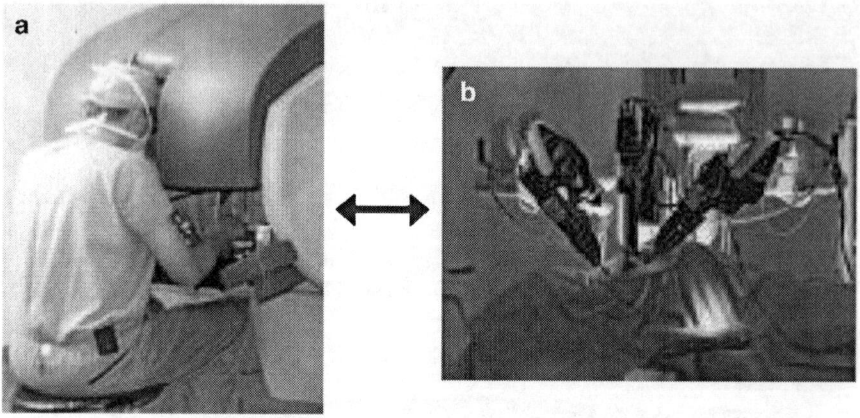

Fig. 3.9 The da Vinci surgical system. (**a**) surgical master console (**b**) slave surgical robot

Because the da Vinci system was leased during Phase I, it has been used without hardware modifications, except for the addition of devices for accepting and latching tools. Although not ideally suited for the TP because of its small workspace, weak forces, and large size, the da Vinci system was able to perform the tasks (dissection and suturing) selected to demonstrate the general TP principles.

The da Vinci system is controlled entirely by the surgical console. Video from the stereo endoscopic camera is sent to the console view port and, using a pair of haptic (force-sensing) master arms, the surgeon controls movement of each slave arm over the patient and its surgical manipulation. Automatically changing surgical tools (which cannot and should not be performed by the surgeon) requires precise positioning of the slave arms. A standard PC, using a proprietary fiber optic connection, was assigned to enable the surgeon to control the da Vinci slave arms.

SRS Software

Intuitive Surgical, Inc. provides an Application Program Interface (API) to collect data from a running robot for research purposes. For this project, the API was enhanced to enable two-way communication, including setting various robot modes and robotic arm control. Since the enhanced API (Version 5) runs only on the Linux operating system, the external SRS computer also runs on Linux.

A second API was developed to provide the surgeon with *positional* haptic feedback. Up to six parallel planes, defined relative to the surgical site, are specified for each arm. When "haptic feedback" is enabled, if a slave arm penetrates any of these planes, the API generates a force, felt by the surgeon, on the corresponding master arm. This feature is used mainly to establish a "haptic region" – a "keep out" zone to prevent the surgeon from accidentally moving the slave arm into a position that might result in a collision with the SNS during a tool change.

Although all the real-time robot control is handled by the two APIs, most of that control is executed by the software embedded in the da Vinci system. Like every other TP subsystem, the SRS has several message-based external software interfaces, defined as XML schemas that allow the SCS and the AMS to control the SRS. Application software, developed by SRI International, implements these interfaces and uses the APIs to directly control the da Vinci robot. Four primary SRS interface functions are: (1) parking and unparking arms during a tool change, (2) monitoring interlock regions for gestures (see the UIS description in Section 3.6.2.11 below), (3) enabling and disabling haptic regions, and (4) providing real-time joint-position data for each slave arm in messages over the sensor network. To facilitate debugging and low-level error handling by an operator, the application also includes a graphical user interface that can be started or stopped as necessary.

3.6.2.2 Scrub Nurse Subsystem (SNS)

The function of the SNS, shown in Fig. 3.10a, is to autonomously serve the SRS by providing supplies and exchanging surgical tools. Specifically, the SNS delivers supplies from the SDS to the SRS and to the fast cache; exchanges tools between the TRS and the SRS by means of a dual gripper end-effector (Fig. 3.10b); and performs geometric calibration and registration of the TP subsystems. Handling of supplies and tools must be done quickly and reliably upon verbal command from the surgeon. The target goal for each handling task is 10 s or less, which is faster than typical human performance of each of these tasks.

SNS Hardware

The SNS employs a 7 degrees-of-freedom (DOF) Mitsubishi PA10 electric arm (Fig. 3.10a). Seven degrees of freedom were deemed necessary to meet various obstacle-avoidance constraints. Other pertinent specifications include a maximum payload of 10 kg, a wrist-reach of 1 m with ±0.1 mm repeatability, and a maximum Cartesian speed of 1 m/s.

The da Vinci tools were designed for tool exchange by a human hand, not by robot manipulation. Thus, for the TP application, a dual-gripper end-effector and an alternative tool cover were developed to fit the original da Vinci tools with a robotic capture mechanism. The end-effector (Fig. 3.10b) includes a force/torque sensor to allow force control in 6-DOF and employs two Schunk pneumatic grippers, each consisting of two jaws, at an angled V-interface. This design permits tool grasping and placement without contacting any part of the TRS or the SRS as well as supply handling. Pneumatic actuation is used to achieve high speed and light weight. The jaws can handle a da Vinci tool or the lugs of a supply tray (see SDS description below). Proximity sensors are used to indicate the open/closed state of each gripper.

In addition, a tool autoloader subsystem (TAS), developed by the Oak Ridge National Laboratory, is mounted on a da Vinci arm as an end-effector for capturing

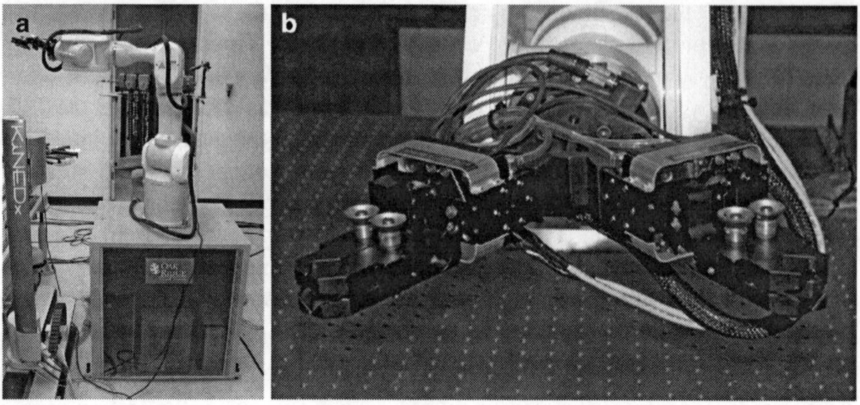

Fig. 3.10 Scrub nurse subsystem. (**a**) Mitsubishi PA10 robot on a pedestal base; (**b**) dual gripper end-effector

and holding a tool (see Fig. 3.11). The TAS is controlled by the SNS as part of a coordinated robotic tool-exchange activity.

SNS Software

Two key requirements for the SNS control software are (1) perform high speed motions smoothly and without collisions, and (2) perform contact tasks without exceeding a specified force threshold. These requirements are crucial because the SNS is essential to the TP performance and is expected to operate over the full TP workspace. For example, a supply dispense operation requires moving the SNS fast and smoothly between two extremes of the SNS workspace while holding the supply tray horizontally. A simple joint interpolation is not adequate for this operation because tray orientation cannot be maintained, and a straight-line move is not possible because of the distance. As another example, programmed compliance and chamfers in the contact interactions obviate the need for force control during pick-and-place operations. However, force control is used during geometric calibration of the TP subsystems because each subsystem is designed with flexible calibration lugs. During the calibration process of a subsystem, the SNS grabs each of these lugs and uses force control to zero the forces. The resulting joint angles render an accurate 3-D pose (position and orientation) of the subsystem. This information is then relayed to the SCS for the overall TP calibration and updating of the world model.

The SNS control software consists of two major components: a Motion Planner (MP), developed by the University of Texas at Austin, and a Motion Executor (ME), developed by the Oakridge National Laboratory.

The SNS MP performs trajectory generation, kinematics with redundancy resolution, obstacle avoidance, and detection (and alert) of imminent collision, as shown in Fig. 3.12. A previously developed software framework, Operational

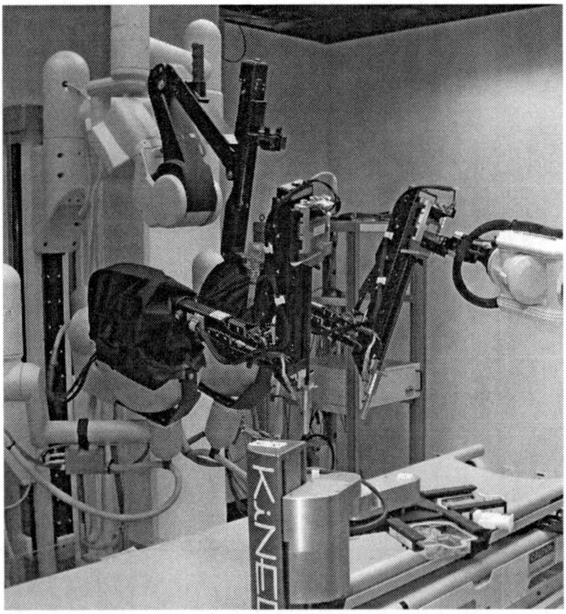

Fig. 3.11 Tool autoloader subsystem (TAS) mounted on each da Vinci arm

Software Components for Advanced Robotics [10–14], was used in the MP development.

Obstacle avoidance is based on a primitive model of the SNS and models of the other TP subsystems. The purpose of obstacle avoidance is to guide the SNS motions around obstacles. Detection of an imminent collision, on the other hand, makes sure that the SNS can be stopped prior to a *potential* collision. Collisions can occur as a result of failure in obstacle avoidance or an error in a geometric model caused by approximations. Therefore, the collision-detection software module for the TP uses a high fidelity computer-aided-design (CAD) model of all the TP subsystems.

The ME is responsible for reaching the SNS joints corresponding to the poses specified by the MP, and for performing low-level force-control operations during contact-based tasks. Low-level software includes a real-time kernel that manages the interaction with the SNS, the coordination between the real-time layers and the asynchronous higher layers, force limiting, calibration software, and trajectory compensation. The SNS controller is a proprietary black box that permits minimal parameter adjustment for control optimization. The added low-level real-time control was implemented under RTLinux. For most moves, the velocity of the SNS end-effector is at or near its maximum value. Hence, the end-effector's 6-axis force/torque sensor is used for force limiting rather than conventional force control, which is too slow for dynamic interaction with tools and supply trays.

While the SNS arm is optimized for velocity, its position tracking is not fast enough for handling tools and supply trays or for avoiding obstacles because a lag in

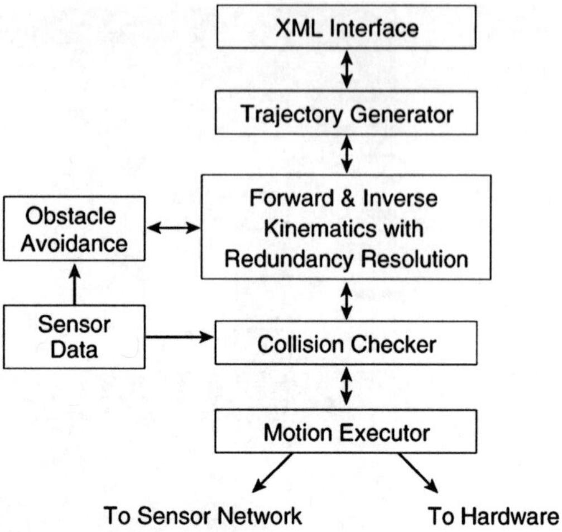

Fig. 3.12 Scrub nurse subsystem motion planner

tracking the arm's position causes diversion from a desired timely path. To correct for such a tracking lag, a trajectory-compensation filter was developed. The compensation filter acts on the MP output to minimize the risk of interaction with the SNS internal controller. Figure 3.13 illustrates the effect of the compensation filter on tracking the trajectory of Joint 2 (shoulder pitch) of the SNS arm as a function of time. Note that the improved accuracy of the compensated trajectory also shortens the time required to complete the move by 16.6% (from 1.2 to less than 1 s).

3.6.2.3 Supply Dispensing Subsystem (SDS)

The primary role of the SDS is to store, de-package, and dispense sterile consumable medical supplies. The SDS, shown in Fig. 3.14a, was custom designed and constructed by General Dynamics Robotic Systems (GDRS). The supplies of each type are stored in small trays that are, in turn, stored in a horizontal restocking cartridge; each cartridge can hold up to seven trays. The SDS includes up to 30 cartridges in a 3×10 array and an xyz gantry arm positioned in front of that array. Upon a request for a supply type, the gantry arm moves to the appropriate cartridge and de-packages a tray containing the desired supply. The tray is then placed in a tray-capture mechanism, which passively holds the tray using a series of compliant springs. Every tray is molded with two unique grasping lugs: one grasped by the SDS gantry arm during de-packaging, and the other grasped by the SNS while transporting the tray to the desired location in the TP surgical cell.

Each supply cartridge includes a passive read/write radio frequency identification (RFID) tag. The tag can be activated only when a read/write antenna is within a range of about 10–12 mm. With the antenna mounted on the end-effector of the gantry arm, the RFID tags are used to: (1) provide an inventory of existing supplies and their locations when the SDS boots up, and (2) update the information stored in

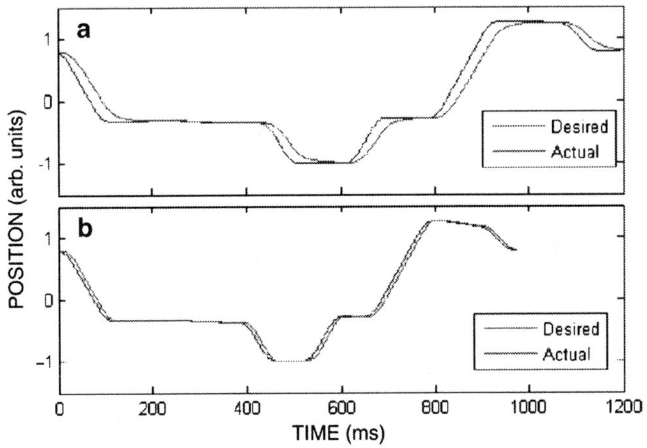

Fig. 3.13 SNS Joint-2 position versus time. (**a**) without a compensation filter; (**b**) with a compensation filter

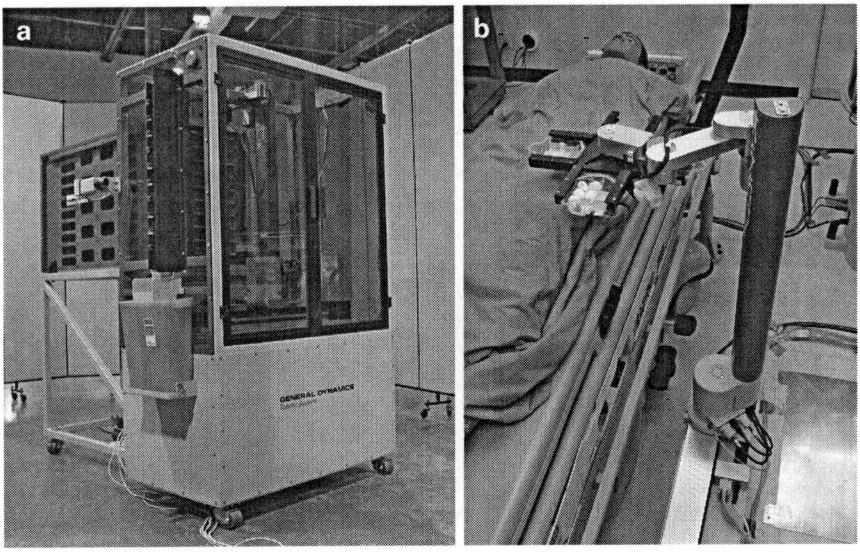

Fig. 3.14 Supply dispensing subsystem. (**a**) main unit; (**b**) fast cache (near patient)

the tags as supply trays are used so that inventory is always current. This information is used to generate a supply-inventory data base and to monitor usage during a surgical procedure.

Partially used trays are stored in either the fast cache next to the patient bed (Fig. 3.14b) or a slow cache, which is part of the 10-slot transfer array on the side of the SDS main unit. The SDS designates which supplies are stored in the fast cache and which in the slow cache. The function of the fast cache is to allow frequently used items to arrive at the work site quickly. While it may take 10 s to deliver a new supply to the surgeon from the SDS main unit, the fast cache can be used to deliver supplies in about 5 s. Empty supply trays and full waste trays are identified by machine vision and placed in a medical waste container by the SNS.

During Phase I, the shape and outside dimensions of a tray were the same for all medical supplies. However, different tray inserts were used for each supply type to position supplies in the tray and restrain their motion. This arrangement is also advantageous for determining the number of supplies remaining in each tray by machine vision (see Section H below). Figure 3.15 shows partially populated supply trays.

3.6.2.4 Tool Rack Subsystem (TRS)

The TRS, developed by the University of Washington, is a fully automated tool changer, whose overall dimensions are 0.18 × 0.68 × 1.5 m, that is capable of holding, accepting, dispensing, and maintaining the sterility of 14 surgical tools for the da Vinci robot.

TRS Hardware

The TRS (Fig. 3.16a) contains a round magazine, of 0.45 m diameter and 0.687 kg m^2 inertia, with 15 tool positions. The magazine is located in the TRS stationary base. One tool position is occupied by a compliant calibration lug used to locate the TRS with respect to all the other TP subsystems. The magazine can be

Fig. 3.15 Supply trays. Clockwise from top-left: shunt kit, sutures and ties, waste tray, spherical sponges, and cylindrical sponges

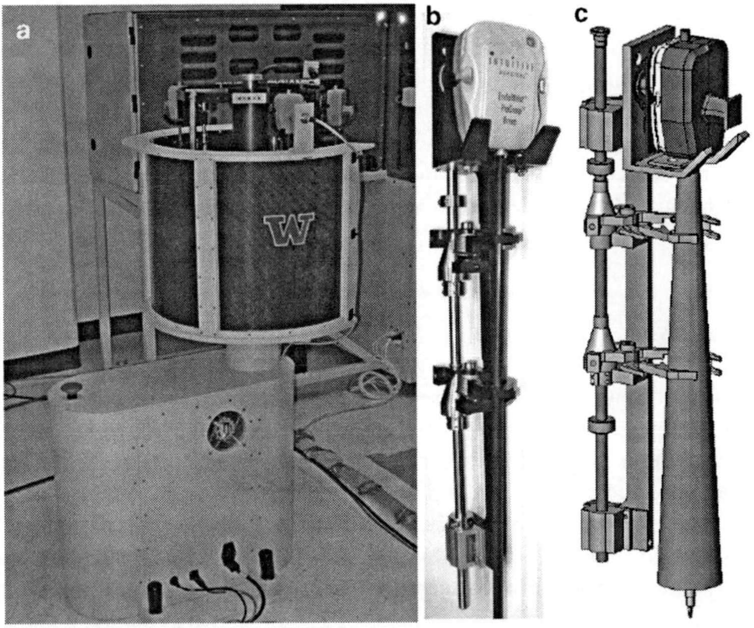

Fig. 3.16 Tool rack subsystem (**a**) overview; (**b**) tool holder grasping a surgical tool; (**c**) CAD display of tool holder grasping a surgical tool under permissible angular misalignment

removed from the TRS along with the surgical tools it contains for sterilization. All sensors and actuators are located below a sterility barrier and do not touch the tools. Each tool is held by a tool-holder mechanism (see Fig. 3.16b or c) that includes a vertical pushrod to which two conical cams are attached. When the pushrod is actuated, each cam opens two normally-closed, spring-loaded jaws that grasp the shaft of a tool with a tensional stiffness of 0.8 Nm. The tool-holder compliance copes with potential angular misalignment (up to 5°) of the tool (Fig. 3.16c) as it is retrieved from or presented to the TRS by the stiff SNS. Two geared, brushless servo motors (made by Animatics, Inc.) actuate the TRS pushrod in less than 100 ms.

Several sensors are incorporated into the TRS to monitor and verify its performance. A photo interpreter sensor is used to determine the positions of the magazine and the pushrod. A 24-byte RFID tag is attached to each tool and is read by an RFID reader attached to the magazine housing with a 10 mm radial gap. A small video camera, mounted on a flexible shaft, is directed at the tool transfer zone between the TRS and the SNS to monitor any problems with the tool transfer.

TRS Software

The TRS is controlled by high-level and low-level control software. The high-level control software is based on a spread protocol [15] and communicates with the TP system using an XML schema. In addition to providing generic messages regarding its status, alarm, and inventory report, the TRS supports other types of message, such as locating and presenting a requested tool, closing or opening a gripper, and reading a tool RFID.

The TRS low-level control software sends commands over a single RS-232 serial port to the two servo motors that actuate the TRS pushrod. The RS-232 line is arranged in a loop so that each actuator echoes packets down the chain and eventually back to the PC.

TRS Function and Performance

The TRS must be able to (1) dispense and accept a tool that is presented to the TRS within a specified maximum pose misalignment by the stiff SNS, and to absorb the energy resulting from such misalignment, and (2) move and present the desired tool in a given pose and within a given time.

We found experimentally that the TRS has the following misalignment tolerances: 4 mm positional, 2.8° orientational, 5.7–7.8 N/mm linear stiffness, and 0.8 Nm tensional stiffness. These tolerances exceed the worst-case misalignment between the TRS and the SNS by a factor of 2; hence, there can be no damage to the SNS or the TRS as a result of such misalignment. We also found experimentally that the time required for presenting a tool to the SNS (which may be done clockwise or counterclockwise) varies between 0 and 648 $\pm$ 8 ms (see Fig. 3.17), and that the time required for tool release or grasping is 98 $\pm$ 8 ms. These results meet our analytical finding that the TRS must present the requested tool to the SNS in less than 700 ms in order for that tool to reach the SRS in less than 10 s.

3.6.2.5 Supervisory Control Subsystem (SCS)

The primary function of the SCS, developed by the University of Texas, is to provide high-level control of all automated subsystems, primarily those performing tool changing and supply dispensing when requested by the surgeon, and to coordinate these subsystems with the teleoperated SRS. High-level control is achieved by using automated task management and by planning a coordinated execution of tasks to be performed by several subsystems. Multiple surgeon requests may also be queued by the SCS and executed in a sequential order. Secondary responsibilities of the SCS include collision checking, subsystem monitoring for inconsistent/unsafe states, and detection of subsystem alarm conditions.

SCS Task Management and Planning

Safety and proper performance are the most important requirements the SCS must meet. The SCS is in charge of delivering a tool or dispensing supply within a given

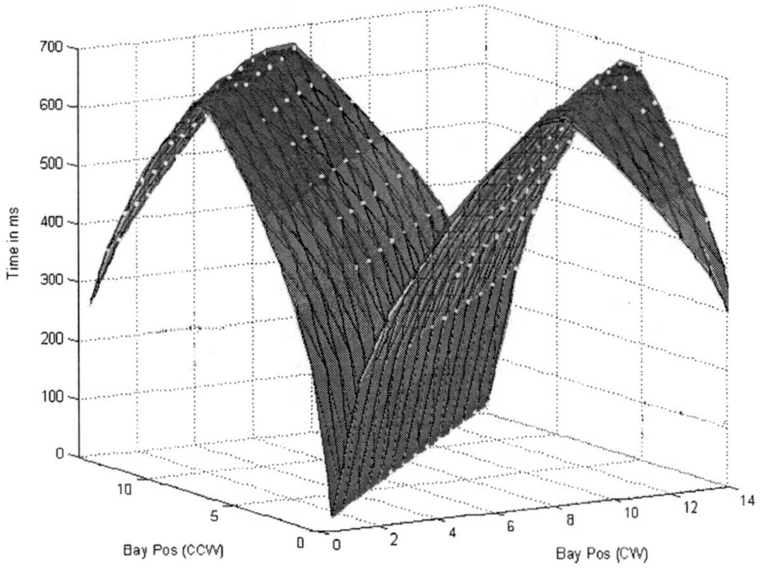

Fig. 3.17 Clockwise and counterclockwise tool-presentation time vs. initial tool position

time duration without endangering the patient or the TP hardware, and handling potential error conditions or unanticipated states that may arise during the task execution.

Automated planning and scheduling algorithms, such as Graphplan or Hierarchical Task Planning [16, 17], were considered early in the project. However, the highly structured nature of the TP tasks made it possible to analyze and write each task plan manually. Conceptually, these task plans may be considered as separate programs, and the SCS may be considered an interpreter that executes these plans. This approach reduces the risk involved in designing a more complex system that would handle recovery from arbitrary error states. Each task plan handles a limited set of the most-likely errors that may arise during execution. Any errors not explicitly handled by the task plan result in a task failure and require administrator intervention.

Similarly, each task plan includes a limited set of most-likely cancellation checkpoints, each allowing the surgeon to terminate execution of a task plan prematurely and return the TP system to a safe operating state. If the surgeon cancels a task anywhere in the plan, the current task execution is continued until it reaches a cancellation checkpoint but the queued tasks are cleared out. For example, Fig. 3.18 shows a plan for tool change that includes one cancellation checkpoint (Node 10). At this point, the SCS terminates the current task and executes a cancellation recovery task to place the TP system in a safe state. Execution of a task plan may also be paused (temporarily) at any point in the plan.

As illustrated in Fig. 3.18, the workflow for a single task plan may be conceptually modeled as a graph of nodes, where each node is shown in a box containing a single command to be executed. Nodes may be executed serially or concurrently. For safety and consistency, a node also entails pre/post conditions that must be met before/after the node is executed. For example, Node 3 (SNS Open Gripper) in Fig. 3.18 entails a pre-condition (the gripper holds no tool) and a post-condition (the gripper is open). If either of these conditions is not met, the task fails.

SCS Function and Performance

The SCS provides plans for the task types shown in Table 3.2. Each surgical task is initiated by a verbal command, voiced by the surgeon through the UIS, which sends an XML message to the SCS. The tool-change and supply-dispense tasks are the only tasks having a requirement for maximum performance time. To meet this requirement, each task plan includes as many subsystem actions in parallel as possible, based on an estimate of the time required to run each action. Robot workspace conflicts also affect the number, concurrency, and sequence of actions in each task plan. The time used by the SCS to execute a task plan is much shorter than that required by other subsystems to execute their commands.

A task may be paused, cancelled, or terminated by either the surgeon or the system administrator. The SCS can add new task types and add or remove new subsystems in a task plan. Task plans can be written and modified independently of changes in the task-management components.

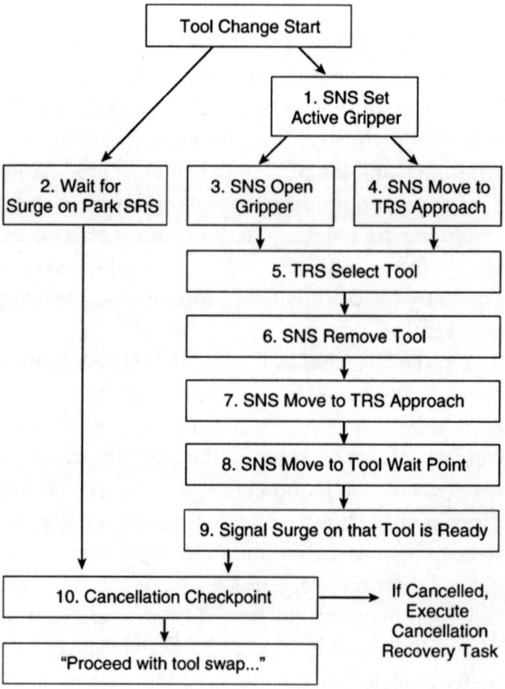

Fig. 3.18 Tool-change plan

Table 3.2 Supervisory control subsystem tasks

Type	Task	Description
Surgical	Change tool	Swap tool on SRS arm with new tool from TRS
	Dispense supply	Provide supply to surgeon from the SDS or fast cache
Surgical pre/post operative	Empty slots	Remove trays from slots and place in waste bin
	Empty SRS arm	Remove tools from SRS and return to TRS
	Populate slots	Populate supply slots with specified supply types
	Populate SRS arm	Populate SRS arms with specified tools
Calibration	Calibrate (automatically or manually)	Calibrate locations of subsystem (PRS, SRS, TRS, SDS, fast cache)

3.6.2.6 Patient Registration Subsystem (PRS)

The PRS holds the patient in a platform, scans the patient surface stereoscopically, and creates a 3-D model that allows the TP robots to move safely around the patient. In Phase I we used an LSTAT (a life-support system made by Integrated Medical Systems, Inc.) as a PRS platform. The LSTAT platform, shown in Fig. 3.19, provides monitoring of vital signs, fluid delivery, and a ventilator.

The PRS serves multiple purposes by providing a platform on which to: (1) achieve patient positioning, comfort, and restraint; (2) accomplish imaging procedures (e.g., CT scan); (3) perform surgical interventions; (4) integrate or interface diagnostic and therapeutic medical devices, a physiological monitor, a ventilator, a defibrillator, and means for suction, infusion, and blood chemistry analysis; (5) apply:

- Machine-vision systems that scan the PRS and patient and generate their 3D poses (which are applicable to collision avoidance and to location of injury sites and anatomical structures for diagnostic and interventional applications)
- Algorithms for closed-loop control of medical devices (such as the ventilator and infusion pumps)
- Collection, storage, and communication (both local and remote) of patient, environment, and system data
- Control of utilities (electrical power, oxygen, etc.) and environmental sensors.

3.6.2.7 Patient Imaging Subsystem (PIS)

The PIS was developed in Phase I of the TP program. The PIS includes a tomographic X-ray facility that is compatible with the TP robots and the LSTAT, and capable of generating CT-like data sets for patient diagnosis and 2-D fluoroscopic data to support interventions. The PIS consists of an X-ray tube mounted on an overhead rail and capable of moving in a plane above the patient, and a large flat-panel X-ray detector embedded in the LSTAT. This configuration allows for acquisition of a sequence of diagnostic 3-D images as the detector is moved in a

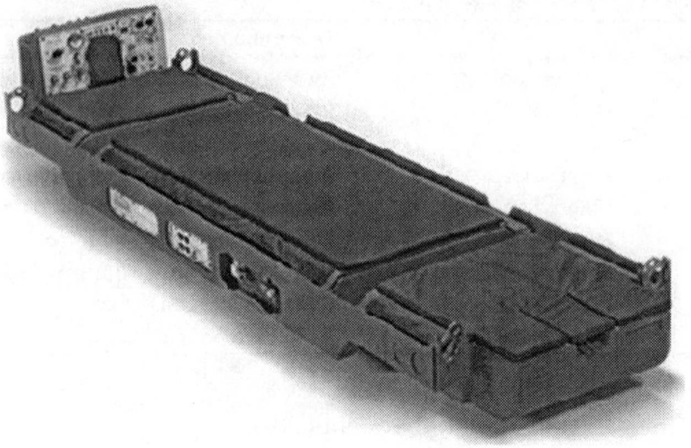

Fig. 3.19 LSTAT platform

2-D grid-like pattern above the patient for 10–30 s. The sequence of images is then reconstructed, using an algorithm similar to that used in a conventional CAT scan, to generate a 3-D image of the patient's interior organs. The topographic imagery obtained is characterized by a high spatial resolution (0.2 mm) in planes parallel to the detector plane and a lower spatial resolution (5 mm) in the depth direction. Figure 3.20 illustrates data acquired in an X-ray test cell using this technique. A key requirement of the PIS is to provide sufficient resolution of soft tissue in order to diagnose various traumatic injuries. Experimental studies have shown that, with the appropriate X-ray technique and use of an anti-scatter grid, we can achieve a soft-tissue contrast resolution that differentiates blood filled organs from the surrounding fatty tissues.

The PIS may also be operated in an interventional mode, allowing the PIS to support minimally invasive procedures, such as stent or shunt placement. Because the patient is not moved after the diagnostic scan, intervention can begin immediately following the scan. This implies that all subsequent 2-D interventional data are automatically co-registered with the initial 3-D diagnostic dataset, allowing for precise SRS navigation and the capability of updating 3-D data as the procedure progresses.

3.6.2.8 Machine Vision Subsystem (MVS)

The MVS will ultimately capture images from cameras positioned around the TP cell and analyze these images in order to track the movement of supplies and tools. During Phase I of the project, the primary function of the MVS was to determine the number of supplies of a given type on each tray moved by the SNS in a normal orientation (horizontally) into and out of the surgical site. The updated count of the tray contents is used by the Medical Encounter Record (MER) component of the RMS (described below) to record the type and number of the supplies consumed

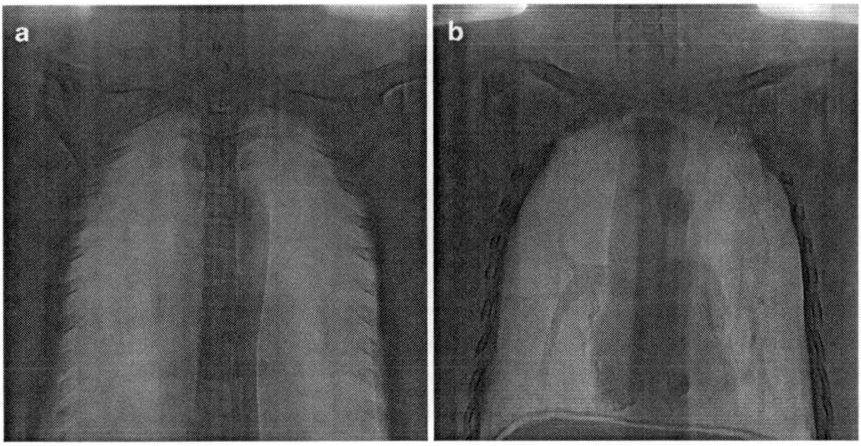

Fig. 3.20 X-ray tomographic reconstruction of torso phantom: (**a**) slice centered on spine; (**b**) slice centered on lungs

during the surgical procedure. The MVS also determines when a supply tray is empty, so it can be disposed.

As noted above (see Fig. 3.15), supply items are placed in individual slots, separated by tray inserts that uniquely accommodate each supply type (shunt kit, sutures, ties, spherical sponges, and cylindrical sponges). When a supply count is requested, the MVS uses XML messaging to determine the type of the supply tray and whether it is being held by the SNS (Fig. 3.21) or by the fast cache (Fig. 3.22). Using this available information increases the speed and reliability of the supply count by the MVS.

Because trays appear in different poses relative to the camera, the counting processing is divided into two stages: tray localization (determining the pose of the tray), and supply count (counting the number of supplies in that tray).

The first stage, tray localization, is facilitated by means of three purple fiducial marks on the end-effector that holds the tray (the gripper of the SNS or each tray-capture mechanism of the fast cache). Distinguishing these fiducial marks from all other marks in the image is performed by means of a "pixel screener" – a software program that examines all the pixels in the image and selects those whose color is sufficiently close to the color (purple) of the fiducial marks. These pixels are then merged into three separate "blobs" that represent the fiducial marks.

The second processing stage, supply count, consists of two steps: (1) registration of a unique set of slot masks onto the tray image, and (2) counting the supplies in that tray. The mask used in the first step is the image of a tray slot viewed normally. For each tray type there is a unique set of polygons that contains images of standard tray slots and ignores extraneous images. There is a one-to-one correspondence between the poses of the tray slots and those of the fiducial marks. An affine transformation based on this correspondence aligns the masks with images of the slots of the tray

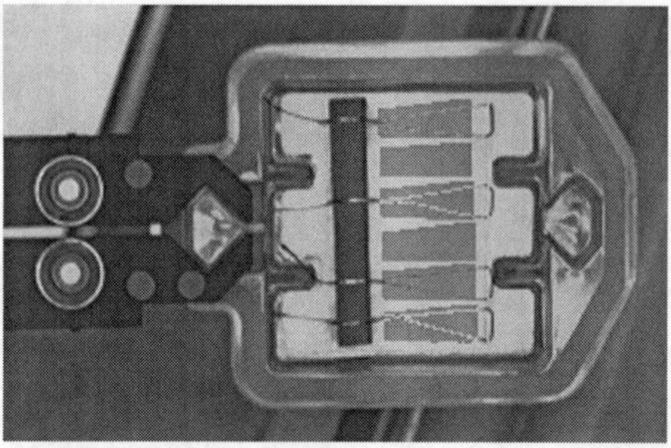

Fig. 3.21 Suture supply tray, held by an SNS gripper with three purple fiducial marks

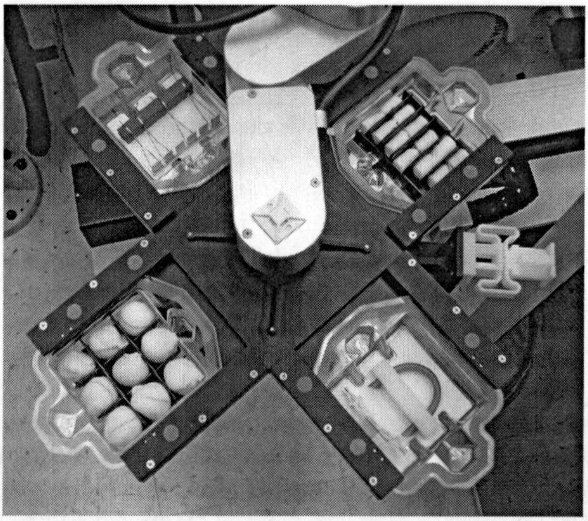

Fig. 3.22 Four supply trays, each held by a capture mechanism with three purple fiducial marks, in the fast cache

type being examined. This alignment enables the software to handle variation in tray pose caused by tray rotation (around its normal), translation, or both.

The second step of this stage uses a pixel screener, unique to a given supply type, to determine if each slot contains that supply. The screener identifies pixels that can belong to the supply's image. Referring to Fig. 3.22, the images of some supplies, such as the sponges, are white on a black tray background; others, such as the sutures and ties, have a white tray background. For these supply types, a screener based on brightness is employed to distinguish between white and black pixels. For other supply types, a hue-based pixel screener is employed to select only pixels whose colors are absent in the image of an empty tray. Such hue-based screener works well for a tray containing a shunt kit, which consists of a blue shunt tube and two yellow vascular clamps. The screened pixels are then merged into blobs, with the further restriction that there can be only one blob per mask and vice versa. A set of algorithms measures the blob dimensions and counts only the blobs that fit into an empirically determined range of lengths and widths. This step rejects small blobs resulting from shadows, specular reflection, and other spurious effects. The number of acceptable blobs is then reported as the supply count.

The entire supply counting process (receiving the request for a count, querying other subsystems for the tray type, localizing the tray in the image, and counting the supplies) is currently completed in 0.52 s or less.

The hardware used for supply counting is a FireWire digital camera (Fire-i™ Board Digital Camera from Unibrain, Inc.). This camera provides industrial quality uncompressed VGA resolution video. A 25 mm telephoto lens is used with this camera to provide the appropriate field of view. The software, written mostly in Java, includes a custom FireWire camera driver.

3.6.2.9 Resource Monitoring Subsystem (RMS)

The RMS, developed by the University of Maryland, automatically records the vast amount of clinical protocols generated in the TP system. The RMS task in Phase I addresses procedural and nursing documentation. Based on interviews with military medics and surgical specialists, we learned that clinical documentation for patients moving from one level of care to another is a critical and often challenging task. With this in mind, we employed a *trauma flow sheet* – a context-aware preoperative information record that automates the creation and processing of care documentation used by civilians and the military. This trauma flow sheet was extracted from the *Advanced Trauma Life Support Manual*, generated by the Committee on Trauma, American College of Surgeons. This digitized document and the underlying information system constitute the Medical Encounter Record (MER) component of the RMS.

The MER monitors and records significant clinical events within the TP, including event times, surgical procedures (e.g., incision, debridement, placement of shunts, and closure), medications, fluids, and other clinical inputs. In addition, the MER also monitors supply and instrument usage, and assists in the final counts of items used. As a trauma flow sheet, the MER is also used to record pre-TP events, such as initial assessment data from medics, and to display personal data, including medical history. The MER uses XML standards and is designed to be compatible with the Armed Forces Health Longitudinal Technology Application.

Figure 3.23 illustrates an intraoperative MER display, one of four primary displays available to medical personnel. This display includes events and their times during resuscitation and stabilization procedures inside the TP.

The MER records significant medical data and video streams and correlates them with a higher-level surgery model. We anticipate that the TP system will have a library of clinical models for the most frequently encountered trauma events on the

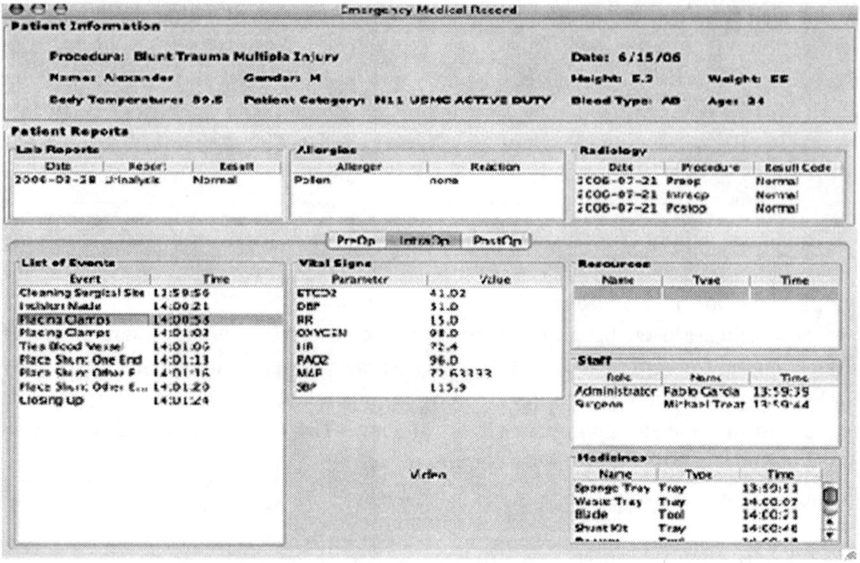

Fig. 3.23 Medical encounter record (MER) display

battlefield, and will correlate each event to the proper model. Low-level data streams will provide basic context information, including medical staff, devices, instruments, supplies, and medications. Patient-monitoring systems and sensors, such as anesthesia machines and pulse oximeters, will provide continuous streams of physiological data. These low-level data streams are processed by an engine that generates higher-level primitive events, such as a patient entering Trauma Pod or the beginning of intubation. A hierarchical knowledge-based event-detection system correlates patient data, primitive events, and workflow data to infer high-level events, such as onset of anesthesia and the beginning of surgery. The processing engine used in the MER is based on the TelegraphCQ software, developed at UC Berkeley, and is particularly suited for continual queries over a large number of data streams. The MER rules engine includes algorithms that analyze physiological data in creating event correlations, such medication effects on heart rate and blood pressure. For Phase I, the latter physiological parameters were extracted from mannequin-based trauma training systems used by the Air Force.

3.6.2.10 Simulator (SIM)

The SIM subsystem, developed by the University of Texas, generates a high fidelity 3D display of the TP, and provides real-time animation of the TP subsystems based on sensory data. The SIM objectives were to provide (1) means for emulating integrated subsystems in the early stages of the TP development, (2) simulated stereoscopic video feedback to the surgeon in the absence of a phantom patient, (3) assistance in the workspace and layout analysis of the TP, and (4) a 3D view of all the subsystems during a TP operation. Specific requirements for the SIM were: (1) high fidelity 3D graphical view based on CAD models of the subsystems, (2) graphics update rate of 30 Hz with stereoscopic output, (3) multiple camera views, (4) high fidelity collision detection, and (5) real-time model update based on sensory data from various TP subsystems.

3.6.2.11 User Interface Subsystem (UIS)

The UIS, developed by SRI International, provides the surgeon with the tools and information required to interact with the TP system in a natural manner. We assume that the user (1) is a qualified surgeon, (2) has been trained to operate the TP system, and (3) speaks English clearly. The UIS is specific to a surgeon; a separate subsystem-management interface is provided for a system administrator. The UIS presents the surgeon with the information needed to conduct surgery and provides visual, verbal, aural, and gesture-based interfaces between the surgeon and the TP system.

The primary function of the *visual interface* (see Fig. 3.24) is to provide a 3D view of the surgical site. The TP system uses standard da Vinci video hardware to generate this view. The view is augmented with icons and text to provide supporting information regarding the status of the TP system. There is also a task list showing current and pending system tasks (supply deliveries and tool changes). An inventory list of tools and supplies can be displayed for specific locations (the corresponding subsystem, the patient, and the waste container). Icons cue the surgeon when the TP system is ready to swap a tool or to move a tray of supplies into the surgical zone. Animated simulation of the system can be displayed as a picture-on-picture, informing the surgeon of the robot status and activity.

The *verbal interface* provides the sole means for the surgeon to issue system commands. It uses Dynaspeak® (a high-accuracy, speaker-independent speech-recognition program developed at SRI International) to translate spoken commands into internal text strings, which are then analyzed and acted upon. For example, when the surgeon says "long tip right," Dynaspeak recognizes the spoken command and translates it into the string "deliver long tip forceps to right." The UIS program analyzes that string and issues an Execute-Task command to the SCS, directing it to add that tool change to the task queue.

The *aural interface* augments the visual cues by providing audible cues and information to the surgeon. For example, when the SNS is ready to move a tray from the wait zone to the surgical zone, the aural interface announces "the tray is ready." This interface also audibly alerts the surgeon to error conditions, and audibly responds to inventory queries by the surgeon (e.g., "how many sutures are in the patient?"). The audible information enhances the UIS, but is not essential; a deaf surgeon could use the UIS successfully.

The *gesture-based interface* performs tool change automatically during surgery to relieve the surgeon from teleoperating this task and to perform it faster and more

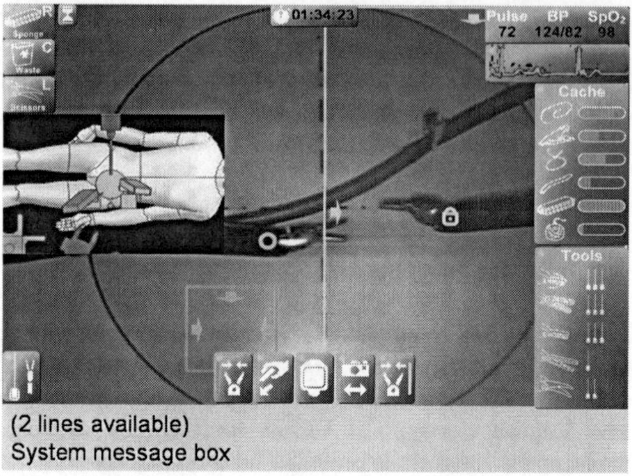

Fig. 3.24 User interface subsystem display (showing all available icons)

safely. The surgeon "gestures" to the system that a tool, currently held by one of the da Vince arms, should be changed by teleoperating the tip of that tool to a prescribed safe location. Consequently, the surgeon loses control of the affected da Vinci arm and the following actions are performed automatically: (1) the affected da Vinci arm moves the tool to a pose reachable by the SNS; (2) the SNS approaches that pose with a new tool, performs a tool change, and withdraws; and (3) the da Vinci arm moves the new tool tip back to the prescribed safe location. At this point, full control is restored to the surgeon, who then continues with the surgery.

3.6.3 Results

Although automatic tool changes and parts dispensing are performed regularly in industrial environments, this is the first time it has been designed to work in a surgical environment. Such an environment presented unique challenges and approaches which are described below.

A complete demonstration of a bowel closure and placement of a shunt on a phantom was successfully conducted on the system, without the need for human assistance during the procedure. The surgeon was observed and interviewed at the end of the procedure to determine the effectiveness of the user interface. Timing measurements were performed throughout the operation and are reported in this section.

3.6.3.1 Challenges and Approaches for the Design of the System

- **Supply Delivery.** *Challenge*: The supply dispenser needs to be able to store sterile surgical supplies, inexpensively packaged to be opened and handled by a robotic manipulator. *Approach*: Designed custom thermoformed supply trays which were sealed sterile in continuous strips and could be opened when pulled by the robot
- **User Interface.** *Challenge*: The surgeon is primarily controlling a teleoperated robot but needs to command the assistance of autonomous robots to complete supply and tool changing tasks. Since both hands are occupied performing surgical tasks, a unique interface is required to control the system without causing a cognitive overload. *Approach*: Designed interface to operate with speech gestures and graphics, without distracting the surgeon from the surgical tasks. Unique protocols were developed to coordinate the interaction between the teleoperated surgical robot and the autonomous robot supporting it (changing tools and dispensing supplies). A supervisory system monitored and prevented impending collisions between manipulators.
- **Tracking of Tools and Supplies.** *Challenge*: Tracking of all supplies, tools and events are an essential part of the surgical process. At the end of the procedure no supplies should be left inside the patient and a patient record should be generated. *Approach*: Developed an information system capable of tracking the traffic of all the supplies and tools in the surgical cell, as well as automatically recording key sequences of events relevant to the medical record
- **Tool Change.** *Challenge*: The design nature of the surgical robot makes it very compliant and tool changes are complicated by the uncertainty on the position of the arms and the deflections induced when contacted by another robot. *Approach*: Designed an autoloader on the surgical arm, capable of loading tools gently placed in the proximity of the arm.
- **Collision Avoidance.** *Challenge*: All the robots physically converge on a very small space at the surgical site where good visualization is essential. *Approach*: Designed distal end effectors to be slim and tightly coordinate the interaction between the robots in this space.
- **Subsystems Interfaces.** *Challenge*: Although all subsystems need to interact for a surgical task, they are likely to be designed as stand-alone units compatible with humans, rather than part of a large tightly coupled system. *Approach*: Designed a flexible architecture, where subsystems are loosely coupled and communicate through high-level interfaces that could be turned into interfaces with human operators. This approach allows the system to function when one or several of the systems are not available, but a human is.
- **Error Handling.** *Challenge*: Errors will occur during the surgical procedure and need to be addressed. *Approach*: Developed methods for detecting errors in the system operation and for recovering from these errors effectively.

3.6.3.2 Observations and Measurements

A complete demonstration of a bowel closure and placement of a shunt on a phantom was successfully conducted on the system, without the need for human assistance during the procedure. The procedure lasted 30 min and was conducted by a surgeon with prior training in the system.

The average time involved in delivering a supply, including depackaging, was 12.7 s with standard deviation of 1 s. The average time involved in changing a tool was 14 s with a standard deviation of 0.4 s. As a point of reference, a typical time required by a human operator to change a tool is about 30 s.

The user interface was very effective in controlling the system, requiring no interventions from the user administrator to complete the procedure. The surgeon was able to conduct the operation smoothly and without delays or interruptions in the flow of the procedure. The interaction with the system felt natural to the surgeon allowing him to concentrate on the surgical tasks.

The ability of the system to queue up several commands was important to avoid interruptions during multiple tool changes or supply deliveries. In the cases where a speech command was not understood by the system the surgeon would get notified to repeat the command. In a few cases the surgeon had to repeat a gesture when interacting with the scrub nurse robot. The small workspace of the surgical robot was a limiting factor when exchanging supplies with the scrub nurse because the view of the surgical site was temporarily blocked.

The count at the end of the procedure was accurate. An automatic video record was generated, time stamped and indexed with the commands issued by the surgeon during the operation.

From the point of view of an observer, the surgical procedure developed a rhythm, similar to what a well trained surgical team is able to accomplish in a surgical suite. Requests were typically issued once without interrupting the procedure and the tasks being conducted by the surgeon had continuity and flow.

3.6.4 Lessons Learnt from Trauma Pod

The goal of this research was to verify the feasibility of remotely conducting a robotic surgical operation, with minimum medical personnel in the surgical site. In Phase I of this research, we have demonstrated the feasibility of conducting simple robotic surgical operations by a remote (human) surgeon and the feasibility of using a robot to automatically change tools and dispense supplies (tasks that are currently performed by a scrub nurse and a circulating nurse).

In the process of developing the TP system, we have learnt the following: (1) *Diagnostics and Error Recovery* – The system diagnostics and error recovery will require the intervention of a system administrator who can operate the system remotely in a manual mode. It is important that the system administrator have the appropriate system granularity and the necessary means for teleoperating subsystems to recover from errors. (2) *Interfaces* – Defining flexible software and hardware interfaces proved critical for achieving overall system integration and allows using subsystems as stand-alone units (3) *Surgical Robot Workspace* – Extending the workspace of the surgical robot beyond the surgical site can ameliorate collision avoidance and may even allow the surgical robot to change tools and acquire supplies by itself. (4) *Number of Robots* – Multiple robots are required to execute or assist in a wide variety of surgical tasks. However, the presence of multiple robots near each other exacerbates the problem of collision avoidance and exponentially increases the complexity of the tasks. (5) *Integrated Information Environment* – Digitizing all the information in the operating and integrating it in some powerful opportunities to automatically correlate data from different sources, such as the vital signs and sequence of events conducted by the surgeon or the anesthesiologist.

Beyond these lessons, there are several issues that need to be resolved before we can perform robotic trauma surgery safely and effectively: (1) *Sterilization* – Cleaning and disinfecting provisions must be incorporated into every subsystem. (2) *Surgical Capabilities* – In order to perform a full surgical procedure, the system must be capable of additional surgical capabilities, such as dissection, as well as integration with the imaging system to perform image guided procedures (3) *TP footprint* – Although the current TP workspace fits an ISO container (2.4 × 2.8 × 12.1 m), we anticipate that this size will have to be decreased for a system deployable in a remote location (4) *Integrate Anesthesia* – Anesthesia needs to be an integral part of the TP system, including the ability to apply intubations and IV lines. (5) *Robustness* – A complex system has many potential failure modes; decreasing the failure rate and handling error recovery fast and safely are critical issues that need to be addressed to achieve system robustness.

In conclusion, we have developed and demonstrated a robotic system that can perform some surgical procedures controlled by a remote surgeon and assisted by an autonomous robot that can change tools and dispense supplies faster and more precisely than human nurses can. We have also demonstrated the feasibility of designing an imaging system compatible with the robotic platform. Future integration of the robotic manipulators with the imaging systems and other sensory technologies will augment what a human surgical team can accomplish.

An integrated system in which all the information, manipulation, and imaging technologies are combined effectively may render significant benefits to the current surgical practice in the field.

3.6.5 Future Vision for Trauma Pod

The ultimate goal of the Trauma Pod system is twofold: (1) produce a rapidly deployable system, capable of performing critical diagnostic and life-saving procedures on the seriously wounded; (2) provide a flexible platform into which new life-saving technologies can be integrated as they emerge from research and development. The TP damage-control interventions will be more invasive and effective than currently available first-aid treatments, and will provide medically necessary measures to allow the wounded to survive evacuation to a site where conventional medical care is provided. The ultimate vision is to provide a system that can perform critical life-saving procedures with the assistance of a corpsman who lacks a high level of medical training. Tasks that require a high level of medical skill will be made as autonomous as permitted by the specific condition of the patient and the complexity of the procedure involved. A remotely located surgeon or intensivist will be able to take control of the system at any point to perform tasks that cannot be automated or executed without their involvement. The interventions performed by the system will involve procedures that preserve life and limb, such as establishing an airway, ventilating the patient, establishing intravenous or intraosseous access for delivery of fluids and medications, performing procedures to achieve hemostasis in non-compressible wounds, manipulating damaged tissues, and providing remote vital signs monitoring. The TP will be used when the timely deployment of skilled medical personnel is not possible or too risky and the patient cannot be evacuated quickly enough to an appropriate medical facility. The TP will be capable of deployment by air, land, or sea, and will function both at the scene on injury and during transport of casualties to a medical facility.

We envision the TP-2 system to consist of three units, which function together or separately: the Imaging Unit, the Procedure Unit, and the Life-Support Unit. A Control Station, located remotely and operated by a medical team, is used to interact with all three units.

Figure 3.25 illustrates the high level architecture for the TP-2 system with its main components and human players. The three units interface with the patient to retrieve diagnostics information and provide life-saving procedures, and are assisted by a corpsman located with the patient. The TP-2 system interacts with a remote physician through a remote interface that presents information and accepts commands. A high level local interface interprets information and presents the result of this automatic interpretation to the corpsman who can then act without needing the specialized medical skills of the remote physician. For example, to verify the airway control has been correctly established, an easily interpreted

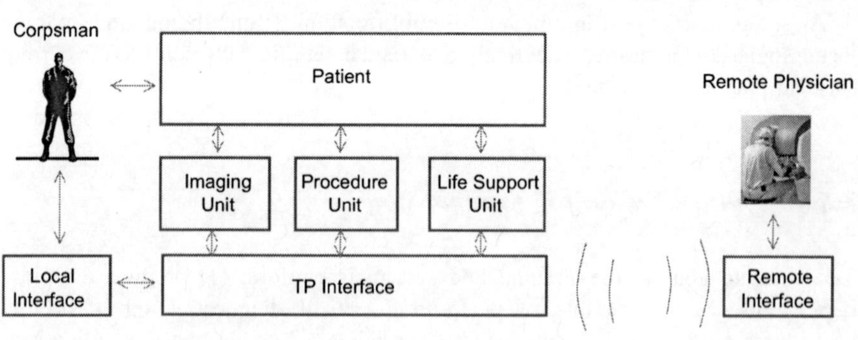

Fig. 3.25 TP-2 high level architecture

indication of airway patency will be provided, rather than the CO_2 curve typically viewed and interpreted by an anesthesiologist in the OR. The goal is to conduct many procedures autonomously with the TP-2 system, requiring only the occasional assistance of a corpsman. For certain diagnostics, complex procedures, or difficult cases, a remote physician will be able to take control and interact with the patient through teleoperation. The autonomous capabilities of the system will make it less susceptible to interruptions in communication and will allow handling of mass casualties more effectively. In the case of mass casualties, a small remote medical team will be able to attend to many patients since each TP system will require only occasional interaction.

3.7 Conclusions

Improvements in the medical care provided to wounded soldiers in recent conflicts have saved many lives. Rapid evacuation to treatment in forward facilities coupled with the use of Critical Air Transport Teams for care en route to level four medical facilities have contributed to the reduction in casualties. However, the availability and appropriateness of such medical care is dependent on the specifics of the conflict, and varies over time.

Widespread use of improved body armor has changed the nature of battlefield injury for coalition forces. Most preventable combat deaths now occur in the first minutes after injury, from rapid blood loss and shock due to severing of major vessels by projectiles, or from uncompressible extremity injuries caused by improvised explosive device detonation. Some of these deaths can be avoided by conducting a limited set of medical procedures at the scene of the injury and during initial evacuation – procedures currently beyond the capabilities of the battlefield medic, focused on improving the survival of the wounded during evacuation. Head trauma remains a significant problem, and early diagnosis of head injury for planning evacuation to appropriate treatment is needed.

Life-saving treatment in the combat zone requires the skill and judgment of the trauma surgeon. Fast and accurate triage, initial diagnosis of injuries, and stabilization procedures can all be accomplished remotely through advanced communication, telerobotic and robotic technologies. These technologies electronically project the specialized skills of the clinician to wherever there are wounded.

New technologies to perform diagnosis and deliver therapies can force multiply the effectiveness of medical teams in the field. Each technology opens up new opportunities to improve the quality of care delivered to injured patients and simplify the logistics of setting up a medical system in the field. The summary provided in this chapter provides a glimpse on what is being done by the research community and what is being used currently in the field. As some of these technologies mature and the military digital telecommunications infrastructure develop, more capabilities will be available to provide a medical system in which the quality of care does not depend on where a patient is treated.

References

1. Satava, R.M.: Disruptive visions – the operating room of the future. Surg. Endosc. **17**, 104–107 (2002)
2. Green, P.S., Hill, J.W., Jensen, J.F., Shah, A.: Telepresence surgery. IEEE Eng. Med. Biol. **14**, 324–29 (1995)
3. Taylor, R.H.: Medical robotics. In: Nof, S.Y. (ed.) Handbook of Industrial Robotics, 2nd edn, pp. 1213–1230. Wiley, New York 1999)
4. Howe, R.D., Matsuoka, Y.: Robotics for surgery. Annu. Rev. Biomed. Eng. **1**, 211–240 (1999)
5. Buess, G.F., Schurr, M.O., Fischer, S.C.: Robotics and allied technologies in endoscopic surgery. Arch. Surg. **135**, 229–235 (2000)
6. Cleary, K., Nguyen, C.: State of the art in surgical robotics: clinical applications and technology challenges. Comput. Aided Surg. **6**, 312–328 (2001)
7. Ballantyne, G.H.: Special issue on surgical robotics. Surg. Laparosc. Endosc. Percutan. Tech. **12**(1), (2002)
8. Speich, J.E., Rosen, J.: Medical robotics, In: Wnek, G., Bowlin, G. (eds.) Encyclopedia of Biomaterials and Biomedical Engineering, pp. 983–993. Marcel Dekker, New York 2004
9. Rosen, J., Hannaford, B.: Doc at a distance. IEEE Spectrum, 34–38 (2006)
10. Kapoor, C., Tesar, D.: Integrated teleoperation and automation for nuclear facility cleanup. Ind. Robot. Int. J. **33**(6) (2006)
11. Pholsiri, C., Kapoor, C., Tesar, D.: Real-time robot capability analysis. In: Proceedings of ASME 2005 Design Engineering Technical Conferences and Computers and Information in Engineering Conference, Long Beach, CA (2005)
12. Harden, T., Kapoor, C., Tesar, D.: Obstacle avoidance influence coefficients for arm motion planning, DETC2005–84223. In: Proceedings of ASME 2005 Design Engineering Technical Conferences and Computers and Information in Engineering Conference, Long Beach, CA (2005)
13. Kapoor, C., Tesar, D.: A reusable operational software architecture for advanced robotics. In: Proceedings of 12th CSIM-IFToMM Symposium on theory and Practice of Robots and Arms, Paris, France, July 1998
14. Robotics Research Group.: OSCAR online reference manual. http://www.robotics.utexas.edu/rrg/research/oscarv.2/
15. Amir, Y., Kim, Y., Nita-Rotaru, C., Schultz, J.L., Stanton, J., Tsudik, G.: Secure group communication using robust contributory key agreement. IEEE Trans. Parallel. Distributed. Syst. **15**(5) (2004)
16. Nau, D., et al.: Applications of SHOP and SHOP2. Intell. Syst. IEEE **20**, 34–41 (2005)
17. Blum, A., Furst, M.: Fast planning through planning graph analysis. Artif. Intell. **90**, 281–300 (1997)

Chapter 4
Overcoming Barriers to Wider Adoption of Mobile Telerobotic Surgery: Engineering, Clinical and Business Challenges

Charles R. Doarn and Gerald R. Moses

Abstract Advances in technology yield many benefits to our daily lives. Our ability to integrate robotics, telecommunications, information systems and surgical tools into a common platform has created new approaches in utilizing less invasive means to treat both common and more complex disease states. A significant amount of investment has been made both from government funding and private sector or commercial funding in the research and development of systems in the area of robotic surgery and the application of telesurgery; and this has led to the development of clinically-relevant distribution of surgical expertise using a surgical robot and telecommunication link. This has predominately been in support of government-funded activities. While early work by Jacques Marescaux in Operation Lindberg and the extensive research performed using Intuitive Surgical's da Vinci, SRI's M7 and the University of Washington's Raven has shown tremendous promise in surgical care, there remains a variety of barriers to wider adoption of telerobotic surgery. These barriers are multidisciplinary and often interdisciplinary. Widespread application of telesurgery as a medical force multiplier depends upon resolution of these barriers, which include bandwidth, latency, quality of service (QoS), research, and reimbursement. The following summarizes how telesurgery has developed, what the challenges are and how they are being ameliorated for wider adoption.

4.1 Introduction

Over the last 20 years tremendous strides have been made in utilizing less invasive means to treat both common and more complex disease states. Armed with continued advances in technology, surgery has seen the widespread application of laparoscopic surgery and the introduction of robotic surgery. As recent as a

C.R. Doarn (✉)
Departments of Surgery and Biomedical Engineering, University of Cincinnati, 2901 Campus Drive, Cincinnati, OH 45221, USA
e-mail: charles.doarn@uc.edu

J. Rosen et al. (eds.), *Surgical Robotics: Systems Applications and Visions*,
DOI 10.1007/978-1-4419-1126-1_4,

few decades ago, leading figures in surgery disregarded the possibility that now exists – clinically relevant distribution of surgical expertise using a surgical robot and telecommunication link.

Integrating robotic systems, or more correctly, telemanipulation systems, in the practice of surgery, permits the surgeon to perform surgery from the patient's bedside or a distant location. Linking robotics used in surgery with communications – robotic telesurgery – has been shown to be possible and effective [1, 2]. This was first demonstrated in 2001 by Dr. Jacques Marescaux during Operation Lindbergh. He operated on a patient in France while he was in New York City [3]. Telesurgery has been subsequently developed through a robust international research effort. Telesurgical systems have also been used to mentor and proctor complex surgeries.

Much of the research that has been accomplished in the area of telesurgery has been funded by the United States Army's Telemedicine and Advanced Technology Research Center (TATRC). This research spans a variety of extreme environments and a variety of systems, including the Defense Advanced Research Project Agency [DARPA] – funded Trauma Pod (managed by TATRC), Intuitive Surgical's da Vinci Surgical System, SRI's M-7, and the University of Washington's RAVEN. There is continued interest in telesurgical capabilities that can deliver damage control and surgical subspecialty care throughout roles of care.

Widespread application of telesurgery as a medical force multiplier depends upon resolution of various challenges that include bandwidth, latency, quality of service (QoS), research, and reimbursement.

4.2 Definitions

Telesurgery can be defined in many ways and by many different people. Simply put its root words come from Greek words 'tele' – meaning far off and 'cheiourgia', which means working by hand. Today, telesurgery incorporates a wide variety of technologies, which often trump one another regarding definition. For example, in the mid 1990s, telemanipulators and telesurgery were used to describe the translation of hand movements from a master controller to end effectors on a slave a short distance away, perhaps 5 m via a hard-wired connection.

We define telesurgery as the ability to use information technology, telecommunications, and robotic systems to enable surgical intervention, where surgeon and patient are separated by a distance. Furthermore, it can also be used to describe educational and professional assessment techniques, surgical discussion among remote participants, and surgery using telemanipulation and telepresence. Telesurgery includes information technology, telecommunications and surgery. Other related terms include:

- *Teleconsultation* – clinical consultation where clinician and patient are separated by some distance
- *Telementoring* – a clinician or care giver is mentored from a distant site by a subject matter expert (SME)

- *Telepresence* – the manipulation of a device at a distant site in a manner that mimics a human is present
- *Telemedicine* – application of telecommunications in the delivery of healthcare
- *Surgical robot* – a powered, computer-controlled manipulator with artificial sensing that can be programmed to move and position tools to carry out a wide range of surgical tasks (telemanipulator)
- *Telesurgery* – remotely performed surgery through combined use of telecommunications and a surgical robot
- *Distributed* – Any system or component data set that is parsed out to various other systems or devices
- *Telesurgical* – A surgical system, usually robotic, that can be manipulated from a distant site
- *Shared Control* – A system that an individual operates while providing feedback on performance and stability.

4.2.1 Telepresence

Telepresence is a human/machine interface where by the human uses displays and a computer interface to interact with objects and people at a remote site as if physically located at the remote site. It is a sophisticated robotic remote control in which the user operates within a relative virtual reality environment. This has been used extensively in planetary exploration, undersea exploration/operations, military operations, and a host of other applications.

Telepresence surgery offers the expertise for those infrequently performed and technically-demanding procedures. It also offers a solution to surgical manpower shortages in remote and underserved areas. Perhaps, this would lead to improvements in outcomes and over time a decrease in overall costs.

4.2.2 Telementoring

Telementoring is the real-time interactive teaching of techniques by an expert surgeon or other subject matter expert (SME) to a student separated by some distance. Current proctoring and precepting have always been difficult to implement. They have been thought of as an inefficient use of expert surgeons' time.

Initial studies investigated telementoring from an adjacent room and from the separate building with the same institution. These studies found that there was not a statistically significant difference in outcomes. Initially, groups used two way and live video feeds to provide for adequate experience.

A number of groups have demonstrated the safety and feasibility of telementoring across long distances. Telementoring of advanced laparoscopic procedures has been performed from the U.S. to a number of international locations, including Ecuador, Austria, Italy, Singapore, Thailand, and Brazil. These experiences show feasibility of international telementoring from large academic centers to large urban centers within other countries. Telementoring has also demonstrated to climbers at base camp of Mt. Everest for ophthalmological examination. Recently, several groups have demonstrated the utility of telementoring between large academic centers and community-based centers.

Dr. Mehran Anvari and his group at the Centre for Minimal Access Surgery (CMAS) in Hamilton, Ontario established a Surgical Support Network on the Bell Canada Multiprotocol Label Switching (MPLS) VPN in 2002 [1]. The network provided telementoring support over a distance of more than 400 km for community surgeons with minimal laparoscopic experience. The network has allowed for the successful performance of advanced laparoscopic procedures without the high rate of complications frequently reported during the learning curve for laparoscopic procedures.

To be highly effective and engaging, telementoring usually requires high-bandwidth telecommunications, which provide lower latency and quality of service (QoS) as well as enhanced image quality and reduced background noise. Although high-bandwidth connections are essential, Broderick et al., demonstrated that low-bandwidth connections could successfully be used to transmit video. This was true only if the camera movements were slow in order to maintain image quality. Because many remote locations and developing countries do not have access to the latest technologies, flexible scalable low bandwidth solutions are crucial to widespread adoption of telementoring and telesurgery.

In October 2001, the FDA approved CMI's Socrates, the first robotic telemedicine device. Socrates was designed to facilitate telementoring. It allowed the telementor to connect with an operating room and share audio and video signals. Socrates was equipped with a telestrator that annotated anatomy for surgical instruction. Initially, the systems used Integrated Services Digital Network (ISDN). Early interactions between CMI and researchers at NASA's Commercial Space Center – Medical Informatics and Technology Applications Consortium (MITAC) in Richmond, VA led to a transition to Transfer Control Protocol/Internet Protocol (TCP/IP). This permitted the Zeus to be controlled from a remote site. It also permitted the development of the wirelessly-controlled RP-7 robot by InTouch.

The use of the Internet increases the flexibility and scalability of telementoring and telesurgery when compared to a dedicated VPN or other proprietary communications link.

4.3 The Rise of Telesurgical Robotic Systems

During the twentieth century, advances such as informatics, telecommunications, computers, and robotics have ushered in profound changes in medical practice and laid the foundation for great advancements in surgical therapies such as laparoscopy. This was largely due to advances in video technology. With the development of improved video technology, laparoscopy slowly gained acceptance. By the mid 1990s, a minimally invasive surgery (MIS) revolution was underway. MIS was proven to be efficacious, cost-effective, and of great benefit to the patient.

However, limitations in MIS, including 2D imaging, decreased range of motion four degrees of freedom [DOF], tremor, and poor ergonomics, posed challenges to surgeons. Although some of these limitations can be overcome through training and experience, unmet needs led to innovation and introduction of robotics into the practice of surgery.

Surgical telemanipulation systems such as Intuitive Surgical's da Vinci Surgical System® (Sunnyvale, CA) have had significant impact in patient care [4]. Such systems are often referred to as surgical robots. Currently, telemanipulator systems have no autonomous component, and therefore, 'robotics' is a misnomer. However, increasing automation suggest that the concept of 'robotics' is becoming increasingly more relevant. The da Vinci currently provides 3D imaging, as well as improved dexterity via 6 DOF, tremor reduction, and an ergonomic work station.

Robotic surgery systems have been commercially available since the early 1990s. Much of the technologies used in these systems were developed through government initiatives over the past three decades. The Defense Advanced Research Project Agency (DARPA) and the National Aeronautics and Space Administration (NASA) played a key role in the development of surgical robotics. DARPA's work led to Intuitive Surgical's da Vinci Surgical System® and NASA-funded research led to Computer Motion's Zeus platform. These systems permit a surgeon to manipulate tissue where patient and surgeon are separated by some distance. Both of these platforms have played key roles in telesurgery.

Discovery and innovation have been enhanced through mutually beneficial collaboration with government, academia and industry. These relationships have proven critical in the development of telesurgery.

"Surgical robots" are powered, computer-controlled manipulators with artificial sensing that can be programmed to move and position tools to carry out a wide range of surgical tasks. The current clinically-used surgical robot is not a smart medical system. The term "surgical robot" is a misnomer as the da Vinci is without significant task automation, and therefore, is properly described as a telemanipulator.

Surgical robots were proposed in the late Twentieth Century as a means to enable performance of complex surgical procedures, improve quality of surgical care, and permit telesurgery. Research and development began with NASA funding of Computer Motion, Inc (CMI) and DARPA funding of SRI International. SRI International developed the M7 telesurgical system and licensed the minimally invasive component of this research to Intuitive Surgical. CMI and Intuitive Surgical, Inc. waged legal battles regarding intellectual property contained within the Zeus (CMI, Goleta, CA) and the da Vinci systems until they merged in 2004. Since the merger, clinical use of the da Vinci has grown remarkably. There are over 1,000 da Vinci systems worldwide and these have been used to perform approximately 80,000 operations in 2008. The Zeus is no longer commercially available or supported technically.

Robotic systems were designed to address the current limitations evident in laparoscopic surgery, namely decreased range of freedom, tremor, 2D imaging, and fatigue. The earliest systems developed were to hold a camera or mill bone and not the multi-arm units that are currently being used. Engineers that first addressed and designed early robotic systems borrowed heavily from the robotic arm used by NASA on board the U.S. Space Shuttle. The first systems were mere modifications in fact and were used to introduce a stable camera platform.

The first robot approved for clinical use by the U.S. Food and Drug Administration (FDA) was in 1994. It was CMI's Automated Endoscopic System for Optimal Positioning (AESOP). When it was first introduced, the robotic arm was controlled either manually or remotely by a foot peddle or hand control. It was later modified to allow for voice control. The robot attached to the side of the operating table and had a multitude of adapters to allow for any rigid scope to be placed. Various groups conducted studies comparing the AESOP versus the surgical assistant, and found that the AESOP could adequately replace the need for a surgical assistant. It was postulated that the system could offer a cost saving advantage to hospitals.

Although AESOP was found to have some benefit, there were a number of limitations. The robot required certain modifications to accommodate the surgeon's operating style, and most surgeons were reluctant to change their way of operating. In medical centers where surgical residents were present, it was felt that there would be no clear cut cost benefit and there could be a detriment to surgical training.

The next generation systems were master – slave telemanipulators between the surgeon and the patient-side robot. Two systems that were approved by the FDA for clinical applications were the da Vinci and Zeus.

In the early 1990's, DARPA funded SRI to develop a surgical system for deployment on the battlefield to support trauma. This led SRI to the development of technologies that were eventually licensed to Intuitive Surgical, resulting in the development of the da Vinci. The da Vinci is an MIS robot comprised of three components – a master, a slave, and control tower. The surgeon sits the master, an ergonomic workstation, equipped with stereo video both providing observation of the surgical site. The master also has hand controllers, which translate hand and wrist movement to the end effectors. The master controls the slave system. The tower houses the video equipment and the insufflator. The slave has three or four arms. The central arm holds the camera and a variety of laparoscopic instruments (end effectors) are held by the other arms.

These instruments are equipped with an *EndoWrist*®, articulated tips of the instruments that provide 7 DOF. This permits more dexterous movement in surgical tasks such as dissecting and suturing. The da Vinci also offers excellent 3D imaging. The standard scope passes through a 12 mm trocar and the surgical instruments pass through 8 mm trocars. The instruments on the da Vinci model available in the U.S. are partially reusable and allow for a predetermined number of uses (ten for human use, 30 for research), requiring replacement. Development is underway to decrease the diameter of the instrument shaft and make them flexible (snake like) to achieve 5 mm in size. The slave is a large, heavy unit which sits next to the patient bed side. It is not attached to the patient bed.

The Zeus robotic surgery platform joined a family of systems from CMI including the AESOP robotic scope manipulator. In the Zeus platform, the AESOP is used as the camera, and two additional units similar to the AESOP are used to grasp the surgical instruments. The three units are attached to the operating table independently. The surgeon is seated at an ergonomic work station. Similar to the da Vinci, the end effectors articulate. The egg-shaped controllers are not as intuitive or easy to use as the da Vinci controllers. The camera is voice-controlled and in order to see the 3D image projected on a single screen, special polarized glasses must be worn. The control and quality of the images are not as good as those within the da Vinci.

Although each platform had telecommunication capabilities, the Zeus was initially adapted for long distance transmission of video and robotic control data. Through TATRC-funded research with Intuitive Surgical, the da Vinci Classic (first generation) system was modified in this way. The da Vinci platform required an extensive reconfiguration and temporary replacement of computer boards in the control system. This in part resulted in a new design for the da Vinci Si, which is more robust and does not require additional telecommunication modification for telesurgery. The latest version of this system, the da Vinci Si, has a dual console for telementoring, collaboration, and education. This capability is a result of collaborative research with academia, industry, and the U.S. Army's Telemedicine and Advanced Technology Research Center (TATRC).

4.3.1 Development of Telesurgery

Initially, NASA and Department of Defense (DoD) research focused on telesurgery. Telesurgery is defined as remote surgical care provided through combined use of a telecommunication link and a surgical telemanipulator. In September 2001, the first successful telesurgery occurred when Dr. Marescaux controlled a Zeus robot from New York City to remove a gallbladder of a patient in Strasbourg, France over an 8 Megabit per second (Mbps), 155 ms latency, fiber optic network. In 2003, Dr. Mehran Anvari controlled a Zeus robot in Hamilton, Ontario to perform 25 complex laparoscopic procedures in patients in North Bay, Ontario over a 45 Mbps, 144 ms Bell Canada virtual private network (VPN).

Through a series of TATRC-funded research grants, telesurgery research and development shifted to other telesurgical systems: the da Vinci Classic, SRI International's M7, the University of Washington's BioRobotics Laboratory (BRL) RAVEN, and the University of Nebraska's 'In Vivo' Surgical Robots. Initially, researchers performed the first da Vinci telesurgery over the public Internet in March 2005. This was collaborative telesurgery in which surgeons at two physically separated sites simultaneously operated on a pig. Subsequently, University of Cincinnati and University of Washington researchers deployed the smaller, lighter weight RAVEN in the desert to perform mobile robotic telesurgery in May 2006.

To overcome satellite communication latency, they successfully used a small Unmanned Airborne Vehicle (UAV)-based communication platform.

Development and evaluation of surgical robotic technology for use in extreme environments began in 2004. Through a TATRC-funded grant, during the NASA Extreme Environment Mission Operations (NEEMO) 7 mission, researchers discovered that the Zeus robot was too large for deployment and use within the confined undersea Aquarius habitat. In the 2006 NEEMO 9 mission, the M7 was deployed in the operationally relevant analog Aquarius and simulated lunar telesurgery was performed. In the 2007 NEEMO 12 mission, the M7 was deployed in Aquarius and autonomously inserted a needle into a simulated blood vessel. In the 2007 NASA C-9 parabolic flight experiments, the M7 was proven flight worthy and demonstrated that acceleration compensation facilitates robotic suturing of simulated tissue during flight.

The initial DARPA investment in telesurgical systems led to the M7 and the da Vinci systems. DARPA recently invested in autonomous, deployable surgical robotics in the Trauma Pod program. The Phase I Trauma Pod program successfully developed semi-autonomous mobile platforms through the integration of telerobotic and automated robotic medical systems. The initial phase included automated functions typically performed by the scrub nurse and circulating nurse; these functions are now performed by semi-autonomous robots working in coordination with a telerobotic surgeon. DARPA continues to consider the Phase II Trauma Pod program. The Phase II program will develop automated robotic airway control, intravenous access, and damage control therapy. Finally, these systems will be miniaturized and incorporated into a tactical platform capable of operating in a battlefield environment.

4.4 Developmental Events

A series of important events transpired during the period between 2002 and 2006. These included some important meetings; such as a research strategic planning meeting, sponsored by TATRC in 2002 related to the Operating Room (OR) of the Future, a workshop sponsored by the Georgetown University Medical Center in March of 2004, entitled OR 2020, The Operating Room of the Future, and an Integrated Research Team (IRT) meeting sponsored by TATRC in September of 2004, entitled Surgical Robotics – The Next Steps. These meetings assisted in identifying the realm of telesurgery as a needed and important topic of advanced technology research.

A panel of 60 national and international experts was convened to draw a roadmap of research and funding needs related to robotic surgery. The panel began by considering the "ideal surgical robot" and articulated a range of requirements needed to achieve advancements in robotic surgery. Targeted research to validate use of the robot to improve outcomes in a specific procedure was suggested as a means to improve acceptance and adoption of robotic surgery. Fear might also have a role in slowing the adoption of surgical robotic technology: patients' fear of robots, surgeons' fear of injuring patients, and hospitals' fear of liability should an error occur. To improve safety and quality, research should examine use of robotic surgical data with a "no fault" policy such as used with data provided by the "black box" used in commercial aircraft accident investigation. It was suggested that business and legal considerations in robotic telesurgery are best pursued in collaboration with experienced organizations such as the American Telemedicine Association (ATA).

Lack of multi/interdisciplinary collaboration was another barrier that the group felt could be overcome by funding specifically targeted to improve interdisciplinary research, design and commercialization. Group members did agree that failure to resolve intellectual property issues would impair and could potentially stop development of robotic surgery.

The working groups concluded that funding in excess of $380M would be required to advance robotic surgical assistants to the point of "crossing the chasm" into early acceptance, from the perspective that we are now at the stage of "early adopters" This places the effort in the realm of "Grand Challenges" on par with the National Nanotechnology Initiative, where many believe it rightly belongs. Early reports do demonstrate that these surgical robots can allow the performance of safer, faster surgery, but the technology is tightly bound to economies of scale as long as the current designs and poor business practices are utilized.

The primary hurdles that need to be overcome in order to even begin addressing the roadmap include funding; the resistance of funding agencies to fund, and academia to support, large-scale, distributed, multidisciplinary teams; the culture and communication barriers between the disparate groups that would need to collaborate; industry's resistance to open architectures and large-scale collaboration. The "Grand Challenge" of developing surgical robotics should begin with a "grand" meeting where the relevant technologies, their current state, and the roadmap are described. While the IRT developed the roadmap, this meeting would develop the policies that will allow the roadmap to move forward. Relevant federal agencies such as NIBIB, NIH, NSF, FDA, NIST, stakeholders from industry, academia and professional and standards organizations would come together to address the hurdles.

4.4.1 Demonstrations of Telesurgery

In addition to convening of meetings, the DOD undertook two aggressive actions to advance technology related to telesurgery. DARPA planned and initiated the Trauma Pod research program to develop and demonstrate technologies that will enable a future generation of battlefield-based unmanned medical treatment systems. When fully developed, the Trauma Pod system will allow a human surgeon to conduct all the required surgical procedures from a remote location using a teleoperated system of surgical manipulators. Automated robotic systems will provide necessary support to the surgeon to conduct all phases of the operation.

Concurrently and separately, TATRC undertook to sponsor several research projects on a lesser scale that collectively advanced technology for telesurgical robotics and demonstrated the feasibility of remote telesurgery. These demonstrations included the following: the conduct of the first transcontinental telesurgery in the US, the conduct of collaborative experiments with NASA within the NASA Extreme Environments Mission Operations (NEEMO) program, the refinement of prototype microsurgery equipment as a model for portable surgery systems, robotic laser tissue welding, robotic replacement for surgical scrub technicians, control of time-delayed telesurgery, and the use of high altitude platforms for transmission of telesurgery signals.

The importance of these demonstrations and the earlier exploratory funding efforts lies in the attempt by DOD to motivate civilian funding institutes and other government funding agencies to assume responsibility for supporting telesurgery as a vital instrument of the national healthcare system. Although technical challenges to telesurgery must be overcome, the real barriers consist of perceptions that cultural, regulatory, reimbursement and safety issues are "too hard" to overcome.

4.5 The Current State of Telesurgery

There has been significant activity and milestones in telesurgery over the past two decades. TATRC or other government entities have funded the majority. The experience gained in each of these activities has yielded discovery and highlighted specific areas where concentration of effort must take place. There have been several earlier reported events that can be highlighted as precursors to telesurgery. Table 4.1 highlights the activities in a timeline. The early events have been characterized as telesurgery; they however, are considered more telemedicine. In addition, there have been numerous meetings and workshops concerning telesurgery that have been concomitant. This research has steadily increased our understanding and has led to new developments and initiatives. Research

Table 4.1 Telesurgery timeline

Year	Event
1964	Early bird – live surgical case observation
2001–2009	Various meetings and integrated research team initiatives
2001	Operation Lindbergh
2001–2004	Canadian telesurgery initiatives
2005–2007	DARPA trauma pod phase I
2005	NASA extreme environment mission operations (NEEMO) 7
2005	First U.S. transcontinental telesurgery – da Vinci
2006	NEEMO 9
2006	High altitude platforms for mobile robotic telesurgery (HAPsMRT)
2007	NEEMO 12

outcomes have led to new capabilities such as Intuitive's da Vinci Si, progressive steps toward semi-autonomous functions, and new robust robotic platforms.

4.5.1 Early Bird

Telesurgery is not a completely novel concept. Since the development of the telecommunication industry, medicine has willingly embraced the potential applications that it provides. In 1964, Dr. Michael DeBakey, a cardiothoracic surgeon in Houston, Texas performed the first televised carotid endarterectomy. The surgery was broadcast on a private network to a room of medical professionals located in Geneva, Switzerland, while Dr. DeBakey was in the operating room at Methodist Hospital in Houston, TX. This event marked the first medical use of America's first telecommunication satellite, Early Bird, launched by NASA in 1964.

4.5.2 Operation Lindbergh

September 7, 2001, marked a major milestone in telesurgery. On this date, Professor Jacques Marescaux, Director of IRCAD/EITS in Strasbourg, France, conducted Operation Lindbergh. This seminal event created a dedicated network between New York City and Strasbourg, France. It was supported by French Telecom and CMI. The Zeus TS workstation was located in New York City and the robotic arms and patient were located in France. Dr. Marescaux and a team of surgeons successfully performed a cholesysectomy on a 68 year old female patient.

Before proceeding, Dr. Marescaux and his group had certain challenges that needed to be addressed. At that time no one had made an attempt to perform telesurgery over a significant distance. The concerns focused on the reliability of telecommunication lines and issues of latency. The feasible distance was thought to be only a couple hundred miles.

The first series of experiments estimated the maximum time delay compatible with the safe performance of telemanipulations at about 300 ms. They were able to achieve a mean time delay of 155 ms over transoceanic distances using a dedicated ATM link with bandwidth of 6–8 Mbps. This allowed the group to perform six laparoscopic cholecystectomies on porcine without complication.

4.5.3 Canadian Efforts

In 2003, Dr. Mehran Anvari, director of McMaster University's CMAS, performed one of the first hospital-to-hospital procedures, a laparoscopic Nissen Fundoplication. Using the Zeus TS platform, Dr. Anvari has performed more than 20 operations to date on patients in Ontario's North Bay located some 400 km away. CMAS has made the practice of telesurgery a reality. Dr. Anvari has sponsored training from his clinical site through telementoring programs and telepresence. This work was made possible by the favorable environment in Canada and with significant support from McMaster University, Bell Canada, and CSA.

Dr. Anvari's clinic in Hamilton, Ontario is connected to the facility in North Bay through a Bell Canada-provided network with significant bandwidth (45 Mbps MPLS VPN). The network had a measured latency of 135–104 ms. Adequate bandwidth was dedicated to Dr. Anvari's research and operational initiatives. This work demonstrated that surgery in a rural center was comparable to the quality of surgery in a large teaching facility.

Recently, the Anvari group has formed a consortium of academic and industry partners to evaluate telementoring and telerobotic surgery to extend the reach of surgeons at major teaching hospitals to successfully perform emergency surgical procedures on patients in remote settings and extreme environments. Engaging the expertise from the Canadian space technologies industry they are developing a new class of Image-Guided Automated Robot. They plan to adapt this technology to develop a mobile system for telementoring and telerobotic surgery through integration of wireless and satellite telecommunication, digital imaging, advanced physiologic sensors and robotic technologies. This creative approach may serve as a model for others to engage industry directly in the development of telesurgery systems.

4.5.4 NASA Extreme Environment Mission Operations

The military has a keen interest in research in extreme environments. They place men and women in such environments. NASA has similar mission characteristics. Through a partnership with the National Oceanic and Atmospheric Administration (NOAA), and the University of North Carolina at Wilmington, the NEEMO Project was established. NEEMO uses NOAA's Aquarius habitat off the coast of Key Largo, FL to conduct research in an extreme environment. The habitat also serves the U.S. Navy. It serves as an ideal laboratory for evaluating technologies and procedures for remote medical care in extreme environments (Fig. 4.1).

TATRC funded three different missions that focused on telementoring and telesurgery. During NEEMO 7, NEEMO 9 and NEEMO 12, research on the concepts of telesurgery and the tools – robotics and telecommunications was conducted. A key research question for these NEEMO missions was how well an operator could perform telerobotic surgical tasks when there is a time delay. Some latencies of 200–500 ms can be adapted too by the user(s).

Each mission successfully evaluated evolutionary steps from telementoring to remote control of the robotic system to semiautonomous functions. Each of these missions was also a collaboration of the U.S. Navy and the U.S. Air Force, as well as industry.

Fig. 4.1 Aquarius Habitat on the Atlantic Ocean floor off the coast of Key Largo, Florida. Courtesy of the U.S. Navy

They were relevant to the military in that the research identifies telecommunication and robotic needs and limitations in this environment that can help define evolving requirements. The TATRC-funded NEEMO missions served as 'Technology Accelerators.'

4.5.4.1 Neemo 7

The objective of this mission was to serve as a proof-of-concept for telementored surgical care in an extreme environment.

Components of CMI's AESOP were deployed in airtight bags/canisters and transported through 70 feet of water to the Aquarius Habitat. Once in the habitat, the system was set up and tested. Communications was accomplished between the habitat and CMAS via an MPLS VPN. This connection permitted interaction between the isolated crew (in the habitat) and a surgeon (M Anvari) located in Canada. The telecommunications network between the NURC and the habitat was via microwave with a total bandwidth of 10–15 Mbps. The performance of the network was characterized by a latency of 50–750 ms. There was noticeable jitter.

During this mission, Dr. Anvari mentored the crew through laparoscopic removal of gallbladder using the Zeus platform. A single arm of the Zeus was transported to the habitat but did not work well. Dr. Anvari felt that the mission was an overall success, but certain challenges still needed to be addressed such as robot size (bulkiness), packet loss, and overcoming jitter. Subsecond latency was shown to be overcome with technique and technology for telementoring.

4.5.4.2 Neemo 9

The NEEMO 9 mission objectives were to see if the crew could assemble a small scale, functional robotic platform that could be remotely controlled using the Internet as the communication system. The SRI M-7 robot was deployed to the habitat in the same way as in the NEEMO 7 mission. Once deployed, Dr. Anvari controlled the robot from over 1,300 miles away (Canada) and performed telemanipulated wound closure. A 2 s delay was introduced. This allowed for Dr. Anvari to see the feasibility of robotic surgery under these conditions.

In order to accomplish this task, the M7 was modified and deployed in airtight bags/canisters to the Aquarius habitat. Once in the habitat, the system was set up and tested. Connectivity from the habitat to shore was done wirelessly with approximately 10 Mbps and latency of 250 ms to seconds. Telesurgical tasks involved telementoring of several activities and the utilizing the M7. This resulted in better understanding of how what activities and tasks could be accomplished through automation, how to continue refining the size (foot print) of the robotic system, and the demonstration of the multi-functionality of the surgical robot.

4.5.4.3 Neemo 12

The NEEMO 12 involved two different surgical robotic systems; the M7 and the RAVEN. Each system was deployed to the habitat, setup, tested, and evaluated. Surgeons located in different parts of the U.S. were able to manipulate both systems easily across the Internet.

A key objective of the M7 deployment was to answer a fundamental question related to the ability to remotely control the robot, which was outfitted with an ultrasound probe on one arm and needle on the other. The surgeon, located in Nashville (live demonstration during the 2007 American Telemedicine Association [ATA] annual meeting), successfully drove the robotic arms in the habitat. A phantom blood vessel in the habitat was scanned using a Sonosite ultrasound probe, the remote surgeon then instructed the robot to insert the needle in to the blood vessel. This demonstration was the world's first semi-autonomous ultrasound-guided needle insertion. This event was conducted in TATRC's booth in the convention center. It was accomplished with COL Jonathan Jaffin, COL Karl Friedl, COL Ronald Poropatich and a number of other TATRC officials.

The RAVEN was also manipulated from Nashville and Seattle to conduct a series of SAGES Fundamentals of Laparoscopic (FLS) tests. Communications was accomplished using the Internet. A HaiVision high end CODEC was evaluated as well. The RAVEN used iChat for the Macintosh platform. The telecommunications that supported the overall link was a wireless Spectra 5.4 GHz bridge with 30 Mbps with a noticeable latency of 500–1,000 ms. This mission was deemed high successful.

4.5.5 First Transcontinental Telesurgery in the U.S

A partnership was created between the University of Cincinnati, HaiVision, Intuitive Surgical, Walter Reed Army Medical Center, and Johns Hopkins to evaluate the da Vinci system as a telesurgery platform. The da Vinci Classic control station was located at UC and the end effectors (patient side) was located at Intuitive Surgical's labs in Sunnyvale, CA. The two sites were connected via the Internet. In March 2005, a nephrectomy was performed on an anesthetized pig. This was repeated again in April 2005 between Denver, CA and Sunnyvale, CA, during the ATA meeting using the public Internet with no QoS guarantees. These experiments represented the first true telesurgery in the U.S., the first telesurgery using the da Vinci Surgical System, the first stereoscopic telesurgery, the first robotic collaborative telerobotic surgery (i.e., both local and remote robot surgeon consoles were used), and the first telesurgery over the Internet using non-dedicated bandwidth. This work led to development of the da Vinci Si, which permits telesurgery.

4.5.6 High Altitude Platforms for Mobile Robotic Telesurgery (HAPsMRT)

Telesurgery cannot be accomplished without a robust telecommunications system. In the summer of 2006, researchers from UC partnered with the University of Washington's BRL, HaiVision and Aerovironment to conduct a series of experiments designed to evaluate wireless communications and mobile robotic surgery. Through a TATRC-funded grant, UC conducted the HAPsMRT project.

A corner stone to HAPsMRT was the utilization of an asset normally used on the battlefield that supports telecommunications. This device, a UAV manufactured by Aerovironment, provided wireless communications so that the UW's U.S. Army-funded, RAVEN could be manipulated from a distant site, where surgeon and robot are separated. The experiment took place in the high desert (an extreme environment) in southern California.

Fig. 4.2 UW's RAVEN robot and Aerovironments UAV in the high desert of Simi Valley, California

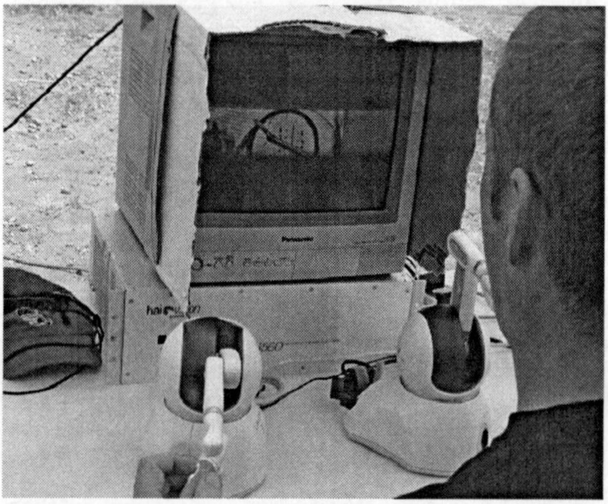

Fig. 4.3 Surgeon manipulating the RAVEN remotely in the high desert of Simi Valley, California

The controllers (master) and robot (slave) were deployed in this extreme environment. The two units were separated by a distance of approximately 200 feet. The UAV was launched in the experimental field. Flying approximately 500 feet above the test field, the UAV provided significant bandwidth (approximately 1.2 Mbps) for communication. This permitted a surgeon for the first time ever to manipulate a robotic system remotely and wirelessly using a UAV. The maximum range of these experiments was approximately 1 mile (Figs. 4.2 and 4.3).

4.5.7 NASA C9A Flight

Researchers from UC collaborated with SRI and NASA personnel to evaluate the deployment and utilization of a modified M7 on NASA's C9A aircraft. The goal was to get the system flight ready and evaluate the acceleration compensation capabilities of the system. The M7 was repackaged into a more suitable system for use on an aircraft. The C9A flies a parabolic profile, which provides approximately 25 s on zero gravity. During this phase of flight, research can be conducted in a variable g flight environment. Military trauma surgeons from United States Air Force (USAF) Center for Sustainment of Trauma and Readiness Skills (C-STARS) Critical Care Air Transport (CCAT) participated in the flight.

During this experiment, the M7 was affixed to the floor of the aircraft. The M7 controllers were remotely located adjacent to the experimental setup. The M7 was equipped with acceleration compensation. Over a series of days, researchers evaluated simple surgical tasks on the system to evaluate the feasibility of using robotic technology to improve access to and quality of surgical care during flight. The acceleration compensation was shown to work effectively. In addition, surgeon and non surgeon performance were evaluated (Fig. 4.4).

Robotic surgery and telesurgery can interject expert surgical care into remote extreme environments and thereby serve as a key component of future military medical care from the battlefield to critical care transport to geographically disperse medical facilities. In addition, such a capability can serve as an effective tool in addressing medical care needs in long duration spaceflight missions. As a critical element of a smart medical system, supervisory-controlled autonomous therapeutics represents a foundation of evolving medical care in these extreme environments.

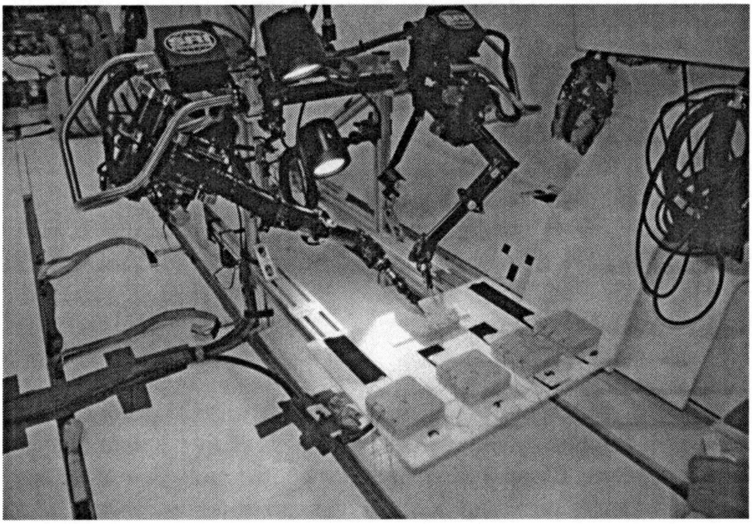

Fig. 4.4 Experimental layout of the SRI's M7 on NASA's C-9 aircraft. Courtesy of NASA

4.5.8 Battlefield Operation

The Army Medical Department (AMEDD) has characterized its mission in the following areas (A) provide, sustain, and enhance soldier health, (B) train, develop, and equip, and (C) delivery leading edge health services. The research presented above has focused on each of these elements. Over the past 10 years, there has been an increase in the use of robotics being deployed on the battlefield. While these devices take many shapes and missions, not to date have been surgical robotics. Research however continues in the development of deployable systems. The joint vision of robotics as a requirement is to develop systems that adapt, integrate, new robotic technologies to treat patients in fixed and mobile medical facilities and to locate, identify, assess, treat, and rescue battlefield casualties under hostile conditions. The technology and operational capabilities developed will be used as medical force multipliers.

4.5.9 Spaceflight Operations

Robotics and autonomous systems have been part of NASA's exploration portfolio for nearly five decades. While each system increases in complexity and fidelity, robotic surgical activities have only been conducted in a research environment. Robotic systems are deployed on the International Space Station and the surface of Mars. The European Space Agency has also worked with IRCAD/EITS to explore new technologies for surgical care.

4.6 Challenges and Barriers

The authors ascribe to the potential advantage to healthcare from a mobile robotic telesurgery system and specify barriers to the employability and acceptance of such a system. We believe that a collaborative effort could design a portable robotic system for telesurgery and develop that system through successful animal trials. Recent advances in engineering, computer science and clinical technologies have enabled prototypes of portable robotic surgical platforms. Specific challenges remain before a working platform is suitable for animal trials, such as the inclusion of image-guidance and automated tasks.

Barriers exist to the development of a mobile robotic surgical platform. These include technical challenges of refinement of robotic surgical platforms, reduction of weight, cube, complexity and cost, and expansion of applications of technology to several procedures. Clinical challenges involve the protection of patient rights and safety, selection of surgical procedures appropriate for the system, the application of surgical skill to evaluate hardware and the application of surgical lore to software programs. Finally, business challenges include resolution of intellectual property considerations, legal liability aspects of telesurgery, patient safety and Health Insurance Portability and Accountability Act (HIPAA), reimbursement and insurance issues, FDA approval of the final product and development of a commercialization plan.

The conduct of telesurgery is not without challenges and barriers. The following are key challenges and barriers to broader application and adoption of telesurgery. For telesurgery to reach its potential there are certain challenges that must first be addressed. Although the practice of telemedicine has been well established, telesurgery has some unique challenges that must be taken into consideration.

4.6.1 Engineering Challenges

There are a number of engineering challenges, including design, material, maintenance, etc. The focus of our attention here is on the following.

4.6.1.1 Technology

Telesurgery is highly dependent on a number of successfully implemented technologies in several disciplines. These include robotics, telecommunications and information systems. Each component has a high technology readiness level (TRL) in the range of 7–9. This level signifies a proven system. As an integrated system, the TRL is in the 3–5 range signifying a concept or experimental system. While some integrated systems are ready for clinical trials, further development and validation is required prior to the TRL increasing to an acceptable level. The TRLs are based on the DoD's Defense Acquisition Guidebook (July 2006) and DOD Directive 5000.1. NASA and the FAA also follow similar guidelines.

When considering a system, several factors must be included in the analysis of what technologies are integrated. These include need, access to technology (price and availability), access to maintenance capabilities, redundancy, reliability, interoperability, and ease-of-use. As in telemedicine, this important step is required to adequately define the requirements, which in turn drive the choices of technology.

4.6.1.2 Access

There are varying levels of access in the integration of telesurgery, including access to systems, access to spare parts, and access to maintenance.

Telesurgical systems may be widely distributed in a future scenario whether on the same area of the battlefield theater or in a rural hospital. They may be linked to surgical expertise in a distributed model or at a central location. The system(s) may have some portability to them, but nevertheless the patient must be brought to the system.

Regarding maintenance and spare parts, this must be readily available and easily accessible to accommodate needs as they arise.

4.6.1.3 Redundancy

Redundancy in telesurgery includes both personnel and systems. Two systems linked together, where surgeon is in one location and patient is another requires personnel redundancy at the patient side, where surgeon will have to intervene in case of communication failure. While this will be a challenge initially, it will be resolved as the systems become more autonomous and reliable.

A second area of redundancy requires a reliable and robust communication system, which insures QoS and avoids inadvertent loss of signal during telesurgery. This is a significant cost issue that must be overcome to permit wider adoption.

4.6.1.4 Reliability

A successful telesurgery system will rely on a variety of systems and people. The robotic system and the communication system (and link) must be reliable. The systems cannot fail during surgical procedures. With proper design and operations, this has not been an issue. The surgical teams must also be able to provide effective surgical care should systems fail or communication is lost.

4.6.1.5 Interoperability

All systems and components should be interoperable. Both the expert site (location of surgeon) and the far site (patient location) must have system components that can cooperatively interact with other medical systems. Telesurgery systems should be capable of robust interaction with supporting technologies such as imaging. Furthermore, all operating room systems should use hardware and software standards that facilitate interoperability. This requires manufacturers to work closely together (similar to Continua [www.continua.com]) so that components seamlessly interact with one another.

4.6.1.6 Maintenance

Telesurgical systems are large and complex and must be maintained. Like any system, maintenance of system components must be accomplished by highly skilled individuals. Scheduled maintenance, training of those personnel, standardized procedures and spare parts must be institutionalized. It cannot be ad hoc. Development and integration into other systems must include this key component.

4.6.1.7 Ease-of-Use

Any system built and deployed must not be difficult to utilize. While all systems, used in the operating room theater, have varying levels of complexity, most require advanced training; they must be capable of being utilized by staff – not additional vendor-specific personnel, which adds more overhead to the operating room. In telemedicine, a key disadvantage has been that simple systems such as cameras and video-teleconferencing are often challenging to operate and therefore, cause user frustration and fatigue. Telesurgery systems must be plug-and-play. This often results in highly-specialized healthcare workers shying away from the use of helpful technology and in some cases eliminating interest in further development.

4.6.1.8 Haptics: Man Machine Interfaces

The ability to project a sense of touch is an important attribute in telesurgery. As the surgeon has moved farther away physically from the surgical site with laparoscopy and robotic surgery, the sense of touch or tactile feedback is no longer present. Current telesurgery systems do not have haptics due to instability above relatively low latency. Haptics is of potential value in telesurgery and has been an area of active research. Force feedback, through servo motors, accomplishes some levels of tactile or haptics. As telesurgery evolves, technology will keep pace and the sense of touch should continue to be evaluated as a potential tool in performing robotic surgical intervention.

4.6.1.9 Telecommunications

Telesurgery is heavily dependent on availability and reliability of a telecommunication network with significant bandwidth. The telecommunications and information technology network must be able to ensure that there will be minimal degradation of picture, minimal or zero latency and high data quality. A high QoS is necessary for telesurgery as opposed to telemedicine. Low packet loss, limited jitter, etc. is important so that robotic system can safely and consistently operate at a distance.

For example, when the distant surgeon moves a controller and the surgical tool moves in the body cavity milliseconds later, this is low latency. Surgeons can compensate for delays of 200–300 ms without any substantial decrease in performance. Delays that are longer than 500 ms are associated with significant decrement in surgeon performance and this not acceptable. Each communication modality has minimal delays. To illustrate further, consider a news service on duty somewhere on the opposite sides of the world. When two individuals are talking to one another, there is a short delay. This is due to the speed of communication. Often communications are accomplished via multiple satellite hops, network segments, and routers, each of which contributes to latency.

Satava addressed this early in the 1990s as robotic surgery was becoming more of a reality. Previous research reported by Marescaux, Anvari, and Broderick et al. has addressed latency in increasingly complex experimental testbeds.

Several current network topologies have been available and evaluated for telesurgery. These include Integrated Services Digital Network (ISDN), ATM, MPLS VPN, Transmission Control Protocol (TCP)/Internet Protocol (IP) User Datagram Protocol (UDP), wireless microwave, digital radio waves, and unmanned airborne vehicles (UAV).

ISDN is a circuit-switched telephone network which enables transmission of digital voice and data over ordinary telephone wires. It provides better quality and higher speeds. ISDN lines are readily available throughout most metropolitan areas and allows for short delays. ISDN lines are considered suitable, but not ideal and its use in the U.S. is waning. It is secondary to low bandwidth per line and lack of scalability.

ATM is a cell relay data link layer protocol which encodes data traffic into small fixed-sized cells. ATM is a connection-oriented technology, in which a connection is established between the two endpoints before the actual data exchange begins. It provides dedicated high bandwidth connections that are reliable and safe. It is readily available throughout the world and is the only technology that guarantees a predefined QoS. Dr. Marescaux used ATM in Operation Lindberg in September 2001.

VPNs via MPLS networks are well suited to telesurgery as it provides traffic isolation and differentiation without substantial overhead. These networks are widely available and used for medical (and banking) applications. It was successfully used in Anvari's Canadian series of experiments.

TCP/IP is the basic communication language or protocol of the Internet. IP is responsible for moving packet of data from node to node. TCP is responsible for verifying the correct delivery of data from one point to another using hand shaking dialogues. Robot commands are typically sent via TC/IP.

UDP is a simple transmission model (as compared top TCP) without handshaking dialogues for guaranteeing reliability, order of packet arrival, or data integrity. Audio and video are often sent via UDP.

Wireless Microwave radio systems are used to transmit and receive information between two points using line-of-sight configurations. A typical microwave radio consists of a digital modem for interfacing with digital terminal equipment, a radio frequency (RF) converter (carrier signal to a microwave signal) and an antenna to transmit and receive. This has been used during the NEEMO telesurgery research.

UAV is a remotely piloted aircraft, often called a drone. They range in size from tiny insect-sized to full-size aircraft. Such aircraft provide varying levels of bandwidth and low latency secondary to relatively low flight altitudes. UAVs were successfully used for the first time in surgery during the HAPsMRT research.

Bandwidth is the rate of data transfer, throughput or bit rate measured in bits per second. Bandwidth is determined by the type and capacity of the medium used such as fiber optic, copper or wireless.

QoS represents a guaranteed level of performance regarding data flow across a network. Examples include bit rate, delay, jitter, packet dropping probability and bit error rate.

Latency is the time it takes for a packet to traverse a network. Latency is a major factor in telesurgery performance. Past telesurgery research has focused on reduction and mitigation from the effects of latency.

Jitter is the inconsistency or variation in the time packets arrive at a destination. Jitter is often caused by network congestion or dynamic route changes.

4.6.2 Non-Technical Challenges

There are numerous non technical challenges, including personnel, cost, licensing, credentialing, FDA approval, etc. Each of these plays a significant role in the broader adoption of telesurgery.

4.6.2.1 Personnel

The integration of technological innovation can disrupt standard operations, by changing the process and structure of tasks performed. Changing operations and culture involves personnel from each step of the process. If telesurgery is to take hold in the clinical environment, personnel from all areas of the healthcare must be involved in the design and implementation.

The organizational culture of the institution or the operating department can impede or halt adoption of disruptive technology. Personnel must go through training of some level depending on the individual's role. This is an important issue. The real challenge or barrier is the paradigm shift that results. The literature often reports on the struggles of implementation due to the unacceptability of new technologies that are purported to make things easier.

While these challenges are present, they can be managed with appropriate curriculum, training, and involvement of key stakeholders in development and targeted application of technology.

4.6.2.2 Cost

Today, healthcare costs are 18% of the gross domestic product in the U.S. There is significant interest in healthcare reform at all levels. In the short term, telesurgery will not have a positive impact on cost as there are many costs associated with telesurgery. These include the cost of the robotics systems, the cost of telecommunication, the cost of personnel, the cost of infrastructure, the cost of training personnel and the cost of research and development.

In addition to cost, a growing concern in American surgery and worldwide is the increasing shortage of general surgeons. This shortage is borne from a number of issues, including a growing population and fewer individuals pursuing general surgery careers. Both the American Board of Surgery and the American College of Surgeons recognize this as a major crisis. This will cause patients to wait longer for surgical treatment, which will have a deleterious effect on healthcare costs and quality. In the long term, distributed and automated surgical care should improve quality, access and cost of care.

Cost of Technology

Telesurgery utilizes sophisticated and complex technology. The cost is often very expensive with the robotic components being the most expensive. For example, the current cost of the da Vinci Si (Intuitive Surgical) is approximately $1.3M, which is accompanied by an annual service contract of $135,000 for a minimum of 5 years or an additional $675,000. In order to optimize the da Vinci, an updated operating suite, capable of supporting networking and telecommunications, must be used. This is also an additional cost. While reduced costs for individual patients may be of some benefit, telesurgery must be proven to be of clinical benefit as well.

There are also costs associated with other ancillary technologies that are used to support telesurgery. These systems include other robotic systems, technologies that drive communications (routers, coder/decoder [CODECs], switches, etc.), computer peripherals (displays, storage devices, etc.), and other interface devices that are used in the operating environment.

Although academic centers may well be able to afford such expenditures, smaller community hospital may not. One of the potentials of telesurgery is the ability to offer telementoring and teleproctoring to rural community settings, and thereby allow expert consultation to people without the cost and burden of moving them far from their homes and families to larger centers. As more systems are deployed and there are more competitors, capital investment and operating costs will be modified. In order to realize the potential of telesurgery, the issues of scalable, cost effective robotics and telecommunications must be addressed.

Cost of Communication

The concept of telesurgery cannot be realized without a robust and reliable telecommunications system. While the availability of bandwidth has increased worldwide and the cost has decreased due to strong economies of scale, it is still a significant cost component of doing telemedicine and telesurgery. However, the cost for copious amounts of bandwidth to support telesurgery is still prohibitive. The Asynchronous Transfer Mode fiber optic communications system used to support Operation Lindbergh (transoceanic telesurgery event in 2001) was in excess of a million dollar, primarily because it was dedicated link for that specific event. The cost of communications must be reasonable for telesurgery to be a significant adjunct. Telesurgery utilizes significantly more bandwidth than other forms of telemedicine and requires low latency and high QoS. The cost of telecommunications for telemedicine is insignificant because the requirements are lower than those for telesurgery. These requirements to support telesurgery add substantial overhead, which drives the cost of communications higher.

The cost of telecommunications includes not only the bandwidth but the various routers, switches, and interface devices at both the transmission site and the receiving site, patient site and surgeon site, respectively.

Cost of Personnel

Personnel to support telesurgical procedures are comprised of a wide variety of highly trained and highly skilled individuals. This includes surgeons at both sites (remote surgeon and patient side). Technical personnel are required at both sites to provide technical support for telecommunication systems and the robotic systems, including preparation, testing nominal operations, and troubleshooting. The patient side requires a surgical team to support the surgeon. At this stage in the evolution of telesurgery, the number of individuals necessary for success is high. It is envisioned that this number will decline once the systems are more robust and the level of comfort and reliability are acceptable. As indicated above, the growing shortages of doctors will create a higher cost structure due to high demand and limited supply. Simply put, the cost of care will increase due to perceived shortages.

Another cost associated with personnel is that of training. All individuals involved in the application of telesurgery will require some form of training based on their assigned duties. The cost of this training must be considered both in creation of curriculum and delivery.

Cost of Inaction (Opportunity Cost)

Today, whether in the military or civilian communities, there are more surgeons retiring than there are being trained. Coupled with projected increase in surgical need in a growing and aging population, there is an ever worsening shortage of surgeons. This shortage is especially severe in rural and extreme environments. As the aforementioned technologies become more integrated in the practice of medicine (telemedicine) and surgery (telesurgery), these challenges will be met.

4.6.2.3 Liability

Telesurgery permits expert consultation and participation. State and international borders may be crossed and jurisdictional conflicts may occur. Complications are an inevitable part of surgery. If these occur, the burden-of-proof to delineate physician error or technical fault must be established. In order for telesurgery to be embraced, physicians must be confident that liability will be accurately accessed. The Society of American Gastrointestinal Endoscopic Surgeons (SAGES) has suggested deferment of clinical implementation until the technology has been validated. In today's litigious society, this will be an important hurdle to overcome.

Not only are state laws and national laws an issue, key international laws complicate the situation. Some effort has been put forth to address these issues, including those involved in teleradiology. These early efforts have laid a foundation from which to move forward.

Organizations such as the Centers for Medicare & Medicaid (CMS) and the Joint Commission on the Accreditation of Healthcare Organizations (JCAHO) are looking at these issues. To date the focus has been on telemedicine and its role in healthcare. Specifically, they have focused on the originating site retaining responsibility for overseeing the care and safety of the patient. JCAHO has released statements regarding this in statement MS13.01.01 with regard to credentialing and privileging process at the originating sites. While there are no specific guidelines for telesurgery from either CMS or JCAHO, the development of them will be based on what currently is available for telemedicine and what is under development. While these issues are different with respect to military medicine, it is nevertheless a significant issue for wider adoption of telesurgery in support of multinational forces.

While telesurgery is still a novel approach, liability will evolve based on need and empirical outcomes.

4.6.2.4 Licensing and Credentialing

Telesurgery is an emerging field, and guidelines for establishing minimum system configuration, proficiency, and competency have not been established. A curriculum and practice guidelines need to be established, whereby the participants must perform a standardized set of procedures that are recorded and measured for performance. After completing these tasks the participant is granted competency. In order for this process to be respected, it is recommended that one of the reputed surgical societies develop and sanction these guidelines. Licensing would allow for a benchmark that can be looked to for excellence. Licensing for medical practice is a state's rights issue and therefore, is controlled by states individually. In addition, organizations like CMS and JCAHO weigh in as well. The recent debate on national healthcare in the U.S. will impact how this is reformed.

All medical establishments (hospitals and ambulatory care organization) in the U.S. are accredited by the JCAHO. The FDA approves all medical devices used in surgery. Each practicing surgeon must also have a credential file in order to practice at a facility. The JCAHO has not addressed telesurgery to any great detail but they have addressed telemedicine. They work closely with CMS as well in addressing new and challenging issues on delivery of healthcare. These guidelines include originating sites responsibilities, performance metrics, adverse outcomes of a sentinel event, and or complaints. The eventual incorporation of telesurgery in clinical practice will require further review of this process and CMS will be a part of this process due to reimbursement issues.

4.6.2.5 Ethics

Even thought the surgeon and patient are separated in telesurgery, the same ethics apply. In fact, there must be more vigilance as patient data, images and outcomes may be controlled from a distance site. Much work must be done in this area to ensure that patient privacy is maintained and that all participants are aware and operate under commonly agreed to ethical standards, that of the patient-physician relationship and fulfilling the needs of the patient. Van Wynsberghe and Castmans have provided a very cogent review of ethics in telesurgery. The moral aim of medicine and surgery continue just using advanced technologies.

4.6.2.6 User Acceptance

Telesurgery in a military setting is challenged by extreme environments, trauma applications and culture. However, secondary to this, the military has funded much of the research and development in this area. The military believes the investment in and incorporation of these new tools will add significant value. While the Trauma Pod's next phase has not been funded to date, the military is nevertheless interested in moving nascent technologies forward to enable better healthcare for the war-fighter.

Application of telesurgery has been successfully demonstrated in a number of settings. While the initial results are very promising, user acceptance both from surgeon and patient must be realized. Of course key military command personnel also must accept this technology for future operational use. Military use requires doctrine, requirements and development of robust technology via additional research and development.

Wider adoption in non military settings will be achieved through education, experience and involvement through the entire life cycle – inception to integration to utilization. Integration of this capability is revolutionizing the status quo; changing legacy systems and protocols requires all levels of personnel involvement and education, including the patient.

4.6.2.7 Financial

In civilian medicine, capital equipment, operations, and maintenance must be offset by revenues. Today, medical systems are very expensive. The cost of these will come down with increase in availability and use.

Many telemedicine and most telesurgery initiatives have been funded by large government grants. While this funding is key to initiating and building a capability, they often do not provide sustainability. With the impending projected shortages of surgeons, physicians, nurses and allied health professionals and the growing need of an aging population, telemedicine and telesurgery services will be significant adjuncts in meeting these challenges. Cost effective, sustainable business models must be developed.

Further development of telesurgery requires additional research and development, including animal experiments, human clinical trials, and finally, FDA approval. From a technical perspective and in clinical evaluations both in animals first and then in humans, this will require a significant investment. A robust research agenda, matched to specific unmet needs, must be promulgated. As telesurgery is clinically proven to improve quality, access and cost of care, sustainable use within government and civilian health systems will be become widespread.

Reimbursement of telemedicine, and certainly telesurgery are challenging particularly in the U.S. primarily due to current policies, regulations, and reimbursement schemas. While this is being resolved for telemedicine, telesurgical reimbursement awaits clinical use to prod the system.

Investment in telesurgical systems, especially in research and development is very expensive. This investment must be made though if telesurgery is to move forward. Any investment of this size must be predicated on unmet needs and perceive value added. If the investment is shown to save lives, then it may be deemed worthwhile and be of benefit in both military and civilian medicine.

4.6.2.8 Research Data – Evaluation

To date, there has been significant research conducted in telesurgery. However, more research is required to further develop systems, protocols, procedures and techniques. Further research and development will evolve from animal trials to human clinical trials. The data garnered from this research will help define and streamline future research initiatives. The great challenge and barrier to further research and development is sustainable funding through government grants and/or investment by industry.

4.6.2.9 Animal Trials

A robust research effort is required. To date, limited animal work has been performed. Further acute and large animal studies must be conducted to confirm efficacy and safety of specific systems. These animal studies provide solid quantitative data on the system and surgeon performance that will further our understanding and implementation of robust telesurgery capabilities. Animal work will help with the development of appropriate procedural improvements.

4.6.2.10 Human Clinical Trials

Once the systems have been successfully evaluated and validated in animal trials, human clinical trials will be designed and undertaken. Rigorous clinical trials to validate safe and efficacious use of systems for surgeon/health system adoption and FDA approval must be conducted. These must be conducted as multi-center trials, which provide a strong, scientific platform for evaluating all components of a telesurgery system in a true clinical setting. A large number of varied clinical cases would have to be conducted to be statistically significant. To date only a handful of human tests have been performed, including those by Marescaux and Anvari.

4.7 A Strategic Solution

Establish a set of research recommendations, criteria, and milestones. Enable research and strategic investment that matches unmet need.

4.8 Conclusions

Over the past three decades, basic, fundamental and applied research in a variety of disciplines has resulted in the development of systems capable of supporting surgical intervention where patient and surgeon are separated. From its earliest beginnings in the 1990s, telesurgery has rapidly grown from a wired system, where the system is in close proximity to the patient, to intercontinental demonstrations. This research has identified needs and limitations in systems and devices, which in turn has help define evolving requirements.

A significant amount of research and discovery in telesurgery has involved a key group of individuals and organizations. While this group is relatively small, it has predominately been driven by funding from TATRC, which has used these efforts as a technology accelerator. Many of these initiatives have resulted in highly effective collaborations between government, academia, and industry. Each activity has led to progressively more autonomous functions. From the wired system, called telesurgery in the mid 1990s by Intuitive Surgical to the UC-lead research in the high desert of California using a UAV, long distance, wireless surgery is possible. Scientific and technological advances will continue to shape medical decision making

The major challenges to wider adoption are being addressed in other areas of telemedicine, robotics, and telecommunications. Further research is necessary to address those challenges, including latency, animal trials and human trials. While some challenges can be ameliorated by technique, some must be overcome by new technology.

While doing an extensive literature search and interacting with subject matter experts, it is clear that the notion of providing surgical care to a remotely-located patient is at hand. Several rather expensive reports have been produced that highlight industrial capabilities in the field of robotic surgery. The peer-reviewed literature is extensive; and the concomitant scholarly text and technical reports provide a unique overview and in-depth review of where telesurgery is and where it is going.

We believe that this chapter will add value as reference on the subject of telesurgery. Clearly the technologies and capabilities that telesurgery can bring to the delivery of surgical care for the warfighter are significant.

This chapter will serve as a summary of telesurgery at a point in time. Continued debate on healthcare reform and national needs will drive the next generation of research initiatives and eventual integration across the spectrum of medicine and healthcare.

Acknowledgements Much of this chapter was gleaned from a detailed report that was prepared for TATRC. This report was authored by Mr. Charles R. Doarn and Dr. Timothy J. Broderick. In addition, we acknowledge the work done by those individuals and institutions mentioned here in for their outstanding contributions to this emerging field.

References

1. Intuitive Surgical. www.intuitivesurgical.com (2009) Last accessed on 1 July 2009
2. WebSurg – History of Robotic Surgery. www.websurg.com/robotics/history.php (2008) Last accessed on 9 Oct 2008
3. MD/Consult – Classification of Surgical Robots – Townsend: Sabiston Textbook of Surgery, 18th Edn. www.mdconsult.com/das/book/body/106987206-2/0/1565/191.html (2008) Last accessed October 9, 2008
4. Technology Readiness Assessment Deskbook (DOD – May 2005)
5. Defense Acquisition Guidebook (DOD – July 2006)
6. DOD Directive 5000.1
7. Autonomous Military Robotics: Risk, Ethics, and Design. CalPoly report to the Office of Naval Research (ONR). Award #N00014-07-1-1152. 2009
8. Curley, K., Broderick, T., Marchessault, R., Moses, G., Taylor, R., Grundest, W., Hanly, E., Miller, B., Gallagher, A., Marohn, M.: Integrated research team final report: surgical robotics – the next steps. Telemedicine and Advanced Technology Research Center (TATRC) Report No. 04–03. United States Army Medical Research and Materiel Command, 2004
9. Doarn, C.R., Broderick, T.J.: Robotic Surgery in Flight. C-9 and Other Microgravity Simulations NASA/TM-2008-214765, pp. 12–18 (2008)
10. Doarn, C.R., Broderick, T.J.: Interim report to USAMRMC – TATRC – Grant W81XWH-07-2-0035. Advanced Center for Telemedicine and Surgical Innovation (ACTSI), 2008
11. Doarn, C.R., Broderick, T.J.: Supplemental (Final) report to USAMRMC – TATRC – Grant W81XWH-05-2-0080, High Altitude Platforms for Mobile Robotic Telesurgery, 2008
12. Doarn, C.R., Broderick, T.J.: Task report USAMRMC – TATRC – Grant W81XWH-07-2-0035. Robotic Surgery in Flight (NASA DC-9 Parabolic Flight Report), 2007
13. Doarn, C.R., Broderick, T.J.: Final report to USAMRMC –TATRC – Grant W81XWH-07-2-0039, NASA Extreme Environment Mission Operation (NEEMO) 12 Robotic Telesurgery: Technology Evaluation and Performance Assessment, 2008

14. Broderick, T.J., Doarn, C.R.: Final report to USAMRMC – TATRC – Grant W81XWH-06-1-0084, Science Support – NASA Extreme Environment Mission Operation (NEEMO) 9: Evaluation of Robotic and Sensor Technology for Surgery in Extreme Environments, Oct 2006
15. Broderick, T.J., Doarn, C.R., Harnett, B.: Interim report to USAMRMC – TATRC – Grant W81XWH-05-2-0080, High Altitude Platforms for Mobile Robotic Telesurgery, 2006
16. Townsend: Sabiston Textbook of Surgery, 18th Edn. Elsevier, Saunders (2007)
17. Kumar, S., Marescaux, J.: Telesurgery. Springer, New York, NY (2007)
18. Satava, R.: CyberSurgery: Advanced Technologies for Surgical Practice Ed. Wiley-Liss, New York, NY (1998)
19. Speich, J.E., Rosen, J.: Medical Robotics. Encyclopedia of Biomaterials and Biomedical Engineering. pp. 983–993. Marcel Dekker, New York, NY (2004)
20. Moses, G.R., Doarn, C.R., Hannaford, B., Rosen, J.: Overcoming Barriers to Wider Adoption of Mobile Robotic Surgery: Engineering, Clinical and Business Challenges. National Institute for Standards and Technology. In: Performance Metrics for Intelligent Systems (PerMIS) Workshop at Gaithersburg, MD. Conference Proceedings. NIST SP 1090. pp. 293–296 (2008)
21. Moses, G.R., Doarn, C.R.: Barriers to wider adoption of mobile telerobotic surgery: engineering, clinical and business challenges. In: Westood, J.D. et al. (eds.) Medicine Meets Virtual Reality 16, Vol. 132, pp. 309–312. IOS Press, Amsterdam (2008)
22. Lum, M.J.H., Rosen, J., King, H., Friedman, D.C.W., Donlin, G., Sankaranarayanan, G., Harnett, B., Huffman, L., Doarn, C., Broderick, T.J., Hannaford, B.: Telesurgery via unmanned aerial vehicle (UAV) with a field deployable surgical robot. In: Westood, J.D. et al. (eds.) Medicine Meets Virtual Reality 15, Vol. 125, pp. 313–315. IOS Press, Amsterdam (2007)
23. Anvari, M.: Reaching the rural world through robotic surgical programs. Eur. Surg. ACA Acta Chir. Austriaca **37**(5), 284–292 (2005)
24. Ballantyne, G.H.: Robotic surgery, telerobotic surgery, telepresence, and telementoring – a review of early clinical results. Surg. Endosc. **16**, 1389–1402 (2002)
25. Bar-Cohen, Y., Mavroidis, C., Bouzit, M., Dolgin, B., Harm, D.L., Kopchok, G.E., White, R.: Virtual reality robotic telesurgery simulations using MEMICA Haptic System. Proceedings of SPIE's 8th Annual International Symposium on Smart Structures and Materials, Newport, CA. pp. 4329–4347, 3–8 Mar 2001
26. Bennett, R. Kirstein, P.T.: Demonstrating minimally invasive therapy over the internet. Int. J. Med. Inform. **47**, 101–105 (1997)
27. Bolt, D., Zaidi, A. Abramson, J., Somogyi, R.: Telesurgery: advances and trends. Univ. Toronto. Med. J. **82**(1), 52–54 (2004)
28. Broderick, T.J., Russell, K.M., Doarn, C.R., Merrell, R.C.: A novel method for visualizing the open surgical field. J. Laparoendosc. Adv. Surg. Tech. **12**(4), 293–298 (2002)
29. Broderick, T.J., Harnett, B.M., Doarn, C.R., Rodas, E.B., Merrell, R.C.: Real-time Internet connections: implications for surgical decision making in laparoscopy. Ann. Surg. **234**(2), 165–171 (2001)
30. Challacombe, B., Kavoussi, L., Patriciu, A., Stoianovici, D., Dasgupta, P.: Technology insight: telementoring and telesurgery in urology. Nat. Clin. Pract. Urol. **3**, 611–617 (2006)
31. Cone, S.W., Gehr, L., Hummel, R., Rafiq, A., Doarn, C.R., Merrell, R.C.: Case report of remote anesthetic monitoring using telemedicine. Anesth. Analg. **98**(2), 386–388 (2004)
32. Costi, R., Himpens, J., Bruyns, J., Cadiere, G.: Fundoplication: from theoretic advantages to real problems. JACS **197**(3), 500–507 (2003)
33. Cregan, P.: Surgery in the information age. eMJA **171**, 514–516 (1999)
34. DeBakey, M.E.: Telemedicine has now come of age. Telemed. J. **1**(1), 4–5 (1995)
35. Demartines, N., Otto, U., Mutter, D., Labler, L., von Weymarn, A., Vix, M., Harder, F.: An evaluation of telemedicine in surgery. Arch. Surg. **135**(7), 849–853 (2000)
36. Dobrosavljevic, S., Welter, R.: TeleMEDian – telesurgery and telemedicine via satellite networks. Int. Congr. Ser. **1230**(6), 1038–1043 (2001)
37. Doarn, C.R., Kevin Hufford, K., Low, T., Rosen, J., Hannaford, B.: Telesurgery and robotics. Telemed. E-Health **13**(4), 369–38 (2007)

38. Doarn, C.R., Anvari, M., Low, T., Broderick, T.J.: Evaluation of teleoperated surgical robots in an enclosed undersea environment. Telemed. E-Health **15**(4), 325–335 (2009)
39. Doarn, C.R., Hufford, K., Low, T. Rosen, J., Hannaford, B.: Telesurgery and robotics: a roundtable discussion. Telemed. J. E-Health **13**(4), 369–380 (2007)
40. Doarn, C.R.: Telemedicine in tomorrow's operating room: a natural fit. Semin. Laparosc. Surg. **10**(3), 121–126 (2003)
41. Doarn, C.R., Fitzgerald, S., Rodas, E., Harnett, B., Praba-Egge, A., Merrell, R.C.: Telemedicine to integrate intermittent surgical services in to primary care. Telemed. J. E-Health **8**(1), 131–137 (2002)
42. Gandas, A., Draper, K., Chekan, E., Garcia-Oria, M., McMahon, R.L., Clary, E.M., Monnig, R. Eubanks, S.: Laparoscopy and the internet – a surgeon survey. Surg. Endosc. **15**, 1044–1048 (2001)
43. Graschew, G., Rakowsky, S., Balanour, P., Schlag, P.M.: Interactive telemedicine in the operating theatre of the future. J. Telemed. Telecare **6**(Suppl 2), 20–24 (2000)
44. Hall, J.C.: The internet: from basics to telesurgery. ANZ. J. Surg. **72**(1), 35–39 (2002)
45. Hanly, E.J., Marohn, M.R., Schenkman, N.S., Miller, B.E., Moses, G.R., Marchessault, R., Broderick, T.J.: Dynamics and organizations of telesurgery. Eur. Surg. ACA Acta Chir Austriaca **37**(5), 274–278 (2005)
46. Harnett, B.M., Doarn, C.R., Rosen, J., Hannaford, B., Broderick, T.J.: Evaluation of unmanned airborne vehicles and mobile robotic telesurgery in an extreme environment. Telemed. J. E-Health. **14**(6), 539–544 (2008)
47. Holt, D., Zaidi, A., Abramson, J., Somogyi, R.: Telesurgery: advances and trends. Univ. Toronto Med. J. **82**(1), 52–54 (2004)
48. Karamanookian, H.L., Pande, R.U., Patel, Y., Freeman, A.M., Aoukar, P.S., D'Ancona, A.G.: Telerobotics, telesurgery and telementoring. Pediatr. Endosurg. Innovative. Tech. **7**(4), 421–425 (2003)
49. Klein, M.I., Warm, J., Riley, M., Matthews, G., Gaitonde, K., Donovan, J.F., Doarn, C.R.: Performance, stress, workload, and coping profiles in 1st year medical students' interaction with the endoscopic/laparoscopic and robot-assisted surgical techniques. Conference Proceedings Annual Meeting of the Human Factors and Ergonomics Society (2009)
50. Latifi, R.: Editorial – telesurgery. Eur. Surg. ACA Acta Chir. Austriaca **37**(5), 267–269 (2005)
51. Latifi, R., Ong, C.A., Peck, K.A., Porter, J.M., Williams, M.D.: Telepresence and telemedicine in trauma and emergency care management. Eur. Surg. ACA Acta Chir. Austriaca **37**(5), 293–297 (2005)
52. Lee, S., Broderick, T., Haynes, J., Bagwell, C., Doarn, C., Merrell, R.: The role of low-bandwidth telemedicine in surgical prescreening. J. Pediatr. Surg. **38**(9), 1181–1183 (2003)
53. Lum, M.J., Rosen, J., Lendvay, T.S., Wright, A.S., Sinanan, M.N., Hannaford, B.: Telerobotic Fundamentals of laparoscopic surgery (FLS): Effects of time delay – a pilot study. Conf. Proc. IEEE Eng. Med. Bio. Sci. **2008**, 5597–5600 (2008)
54. Malassagne, B., Mutter, D., Leroy, J., Smith, M., Soler, L., Marescuax, J.: Teleeducation in surgery: European institute for telesurgery experience. World J. Surg. **25**(11), 1490–1494 (2001)
55. Marescaux, J., Leroy, J. Gagner, M. Rubino, F., Mutter, D., Vix, M., Butner, S.E., Smith, M.K.: Transatlantic robotic assisted remote telesurgery. Nature **413**, 379–380 (2001)
56. Marescaux, J., Rubino, F.: The Zeus robotic system: experimental and clinical applications. Surg. Clin. N. Am. **83**, 1305–1315 (2003)
57. Marescaux, J., Rubino, F.: Robotic surgery: potentials, barriers, and limitations. Eur. Surg. ACA Acta Chir. Austriaca **37**(5), 279–283 (2005)
58. Newlin, M.E., Mikami, D.J., Melvin, S.W.: Initial experience with the four-arm computer-enhanced telesurgery device in foregut surgery. JLAST **14**(3), 121–124 (2004)
59. Melvin, W.S., Needleman, B.J., Krause, K.R., Schneider, C., Wolf, R.K., Michler, R.E., Ellison, E.L.: Computer-enhanced robotic telesurgery initial experience in foregut surgery. Surg. Endosc. **16**, 1790–1792 (2002)

60. Merrell, R.C., Doarn, C.R.: Meeting summary: a department of defense agenda for development of the surgical suite of tomorrow – implications for telemedicine. Telemed. J. E-Health **9** (3), 297–301 (2003)
61. Merrell, R.C.: Telemedicine in surgery. Eur. Surg. ACA Acta Chir. Austriaca **37**(5), 270–273 (2005)
62. Mohamed, A., Rafiq, A., Panait, L., Lavrentyev, V., Doarn, C.R., Merrell, R.C.: Skill performance in open videoscopic surgery. J. Surg. Endosc. **20**(8), 1281–1285 (2006)
63. Moeller, M.G., Santiago, L.A., Karamichalis, J., Mammen, J.M.V.: New ways of practicing surgery: alternatives and challenges. Bull. Am. Coll. Surg. **92**(7), 51–58 (2007)
64. Organ, C.H.: The impact of technology on surgery. Arch. Surg. **134**, 1175–1177 (1999)
65. Panait, L., Rafiq, A., Mohamed, A., Doarn, C.R., Merrell, R.C.: Surgical skill facilitation in videoscopic open surgery. J. Laparoendosc. Adv. Surg. Tech. **13**(6), 387–395 (2003)
66. Panait, L., Doarn, C.R., Merrell, R.C.: Applications of robotics in surgery. Chirurgia (Bucur) **97**(6), 549–555 (2002)
67. Pande, R.U., Patel, Y., Powers, C.J., D'ancona, G., Karamanoukian, H.L.: The telecommunication revolution in the medical field: present applications and future perspective. Curr. Surg. **60**(6), 636–640 (2003)
68. Paul, D.L., Pearlson, K.E., McDaniel, R.R.: Assessing technological barriers to telemedicine: technology-management implications. IEEE Trans. Eng. Manage. **46**(3), 279–288 (1999)
69. Paulos, E., Canny, J.: Delivering real reality to the world wide web via telerobotics. Proceedings of the 1996 IEEE International Conference on Robotics and Automation, 1694–1699 (1996)
70. Rafiq, A., Moore, J.A., Zhao, X., Doarn, C.R., Merrell, R.C.: Digital video capture and synchronous consultation in open surgery. Ann. Surg. **239**(4), 567–573 (2004)
71. Rafiq, A.R., Moore, J.A., Doarn, C.R., Merrell, R.C.: Asynchronous confirmation of anatomical landmarks by optical capture in open surgery. Arch. Surg. **138**(7), 792–795 (2003)
72. Rainer, W.G.: Invited critique: Rural Surgery. Arch. Surg. **135**(2), 124 (2000)
73. Riva, G., Gamberini, L.: Virtual reality in telemedicine. Telemed. E Health **6**(3), 327–340 (2000)
74. Rodas, E.B., Latifi, R., Cone, S., Broderick, T.J., Doarn, C.R., Merrell, R.C.: Telesurgical presence and consultation for open surgery. Arch. Surg. **137**(12), 1360–1363 (2002)
75. Rosser, J.C. Jr, Young, S.M., Klonsky, J.: Telementoring: an application whose time has come. Surg. Endosc. **21**(8), 1458–1463 (2007)
76. Rosser, J.C. Jr, Herman, B., Giammaria, L.E.: Telementoring. Semin. Laparosc. Surg. **10**(4), 209–217.(2003)
77. Rosser, J.C.J.R., Gabriel, N., Herman, B., Murayama, M.: Telementoring and teleproctoring. World. J. Surg. **25**(11), 1438–1448 (2001)
78. Rovetta, A., Sala, R., Wen, X., Togno, A.: Remote control in telerobotic surgery. IEEE Trans. Syst. Man. Cybern. A Syst. Hum. **26**(4), 438–444 (1996)
79. Satava, R.M.: Emerging technologies for surgery in the 21st Century. Arch. Surg. **134**, 1197–1202 (1999)
80. Satava, R.: Surgical robotics: the early chronicles: a personal historical perspective. Surg. Laparosc. Endosc. Percutan. Tech. **12**(1), 6–16 (2002)
81. Satava, R.M.: Disruptive visions: a robot is not a machine. Surg. Endosc. **18**, 617–620 (2004)
82. Satava, R.M.: Telesurgery, robotics, and the future of telemedicine. Eur. Surg. ACA Acta Chir. Austriaca **37**(5), 304–307 (2005)
83. Schlag, P.M., Moesta, T., Rakovsky, S., Graschew, G.: Telemedicine – the new must for surgery. Arch. Surg. **134**, 1216–1221 (1999)
84. Seymour, N.E.: Virtual reality in general surgical training. Eur. Surg. ACA Acta Chir. Austriaca **37**(5), 298–303 (2005)
85. Smithwick, M.: Network options for wide-area telesurgery. J. Telemed. Telecare **1**(3): 131–138 (1995)

86. Sterbis, J.R., Hanly, R.J., Herman, B.C., Marohn, M.R., Broderick, T.J., Shih, S.P., Harnett, B., Doarn, C.R., Schenkman, N.S.: Transcontinental telesurgical nephrectomy using the da Vinci robot in a porcine model. Urology **71**(5), 971–973 (2008)
87. Taylor, R.H., Stoianovici, D.: Medical robotics in computer-assisted surgery. IEEE Trans. Robot. Autom. **19**(5), 765–781 (2003)
88. Gourley, S.: Long distance operators: Robotic and telecommunications take the surgeons skills anywhere in the world. Popular Mechanics Sep. 64–67 (1995)
89. van Wynsberghe, A., Gastmans, C.: Telesurgery; an ethical appraisal. J. Med. Ethics **34**, e22 (2008)
90. Varkarakis, I., Rais-Bahrami, S., Kavoussi, L., Stoianovici, D.: Robotic surgery and telesurgery in urology. Urology **65**(5), 840–46 (2005)
91. Whitten, P., Mair, F.: Telesurgery vs telemedicine in surgery – an overview. Surg. Technol. Int. **12**, 68–72 (2004)

Part II
Systems

Chapter 5
Accurate Positioning for Intervention on the Beating Heart Using a Crawling Robot

Nicholas A. Patronik, Takeyoshi Ota, Marco A. Zenati, and Cameron N. Riviere

Abstract Heart failure resulting from myocardial infarct, oxygen-deprived tissue death, is a serious disease that affects over 20 million patients in the world. The precise injection of tissue-engineered materials into the infarct site is emerging as a treatment strategy to improve cardiac function for patients with heart failure. We have developed a novel miniature robotic device (HeartLander) that can act as a manipulator for precise and stable interaction with the epicardial surface of the beating heart by mounting directly to the organ. The robot can be delivered to and operate within the intrapericardial space with the chest closed, through a single small incision below the sternum. The tethered crawling device uses vacuum pressure to maintain prehension of the epicardium, and a drive wire transmission motors for actuation. An onboard electromagnetic tracking sensor enables the display of the robot location on the heart surface to the surgeon, and closed-loop control of the robot positioning to targets. In a closed-chest animal study with the pericardium intact, HeartLander demonstrated the ability to acquire a pattern of targets located on the posterior surface of the beating heart within an average of 1.7 ± 1.0 mm. Dye injections were performed following the target acquisitions to simulate injection therapy for heart failure. HeartLander may prove useful in the delivery of intrapericardial treatments, like myocardial injection therapy, in a precise and stable manner, which could be performed on an outpatient basis.

Keywords Medical robotics · Minimally invasive surgery · Cardiac surgery · Robotic crawler · Mobile robotics · Navigation · Positioning · Accuracy

---

C.N. Riviere (✉)
Robotics Institute, Carnegie Mellon University, 5000 Forbes Ave, Pittsburgh, PA 15213, USA
e-mail: camr@ri.cmu.edu

DOI 10.1007/978-1-4419-1126-1_5, © Springer Science+Business Media, LLC 2011

5.1 Introduction

Heart failure, or the inability of the heart's pumping function to keep up with the demands of the body, affects more than 20 million patients in the world, with approximately 400,000 newly diagnosed patients and 50,000 deaths annually [1]. This disease is typically caused by myocardial infarction, the death of heart muscle tissue from oxygen starvation, resulting from the narrowing or blockage of one of the coronary arteries that supply the myocardium with blood. Although there is currently no known cure for heart failure, the injection of tissue-engineered materials (e.g., stem cells or biopolymers) into the infarct area to improve heart function is emerging as a therapeutic strategy for post-myocardial infarct heart failure [2–5]. Myocardial injection therapy is currently dominated by catheter-based approaches that access the inside of the heart through small incisions into the larger vasculature of the arm or leg. Although this transcatheter technique has low associated access morbidity, there are several advantages to performing the injections directly into the outside epicardial surface of the heart: clear detection of target infarct lesions, decreased likelihood of cerebrovascular complications [6], and superior site specific efficacy [7]. The major drawback to direct epicardial injection therapy is the lack of dedicated minimally invasive access technology, causing it to be performed only in conjunction with other procedures requiring a full sternotomy or thoracotomy. These invasive approaches incur high associated morbidity, despite the intrinsically simple and noninvasive nature of the injection procedure. The DaVinci™ robotic surgical system (Intuitive Surgical, Mountain View, CA) could be used for minimally invasive epicardial injection [8], but it requires multi-port placement and lung deflation, and does not readily facilitate the precise control of injection placement or depth with its rigid endoscopic instrumentation. A dedicated technology for precise and stable interaction with the heart that balances treatment efficacy and minimal invasiveness would benefit direct epicardial injection therapy.

To address this need, we have developed a novel miniature robotic device (HeartLander) that can act both as a mobile robot for navigation and as a manipulator for positioning tasks on the epicardial surface of the beating heart. This device could facilitate the delivery of minimally invasive therapy through a single incision below the sternum. HeartLander operates within the intrapericardial space – the thin fluid layer above the epicardial heart surface and beneath the pericardial sac that encloses the heart (Fig. 5.1). Previous prototypes have demonstrated safe remote-controlled navigation over the beating heart [9, 10]. This paper describes the addition of real-time 3-D localization, a semi-autonomous control system for the acquisition of targets, and an onboard tool for direct epicardial injection into the myocardium. We present an evaluation of these components through the delivery of an injection pattern onto the posterior left ventricle of a beating porcine heart through a single port approach with the chest closed.

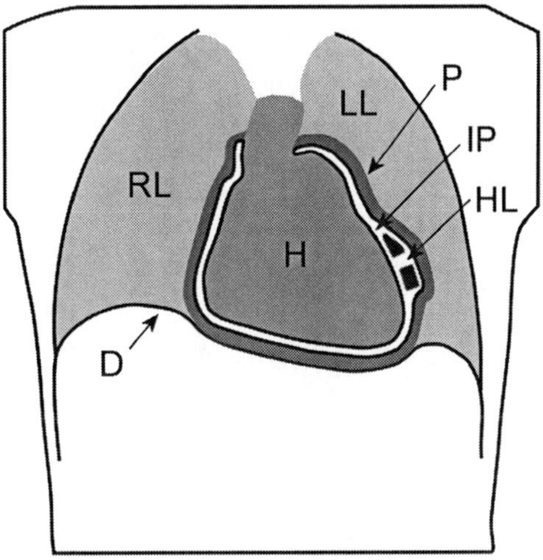

Fig. 5.1 Coronal illustration of HeartLander (*HL*) in the intrapericardial space (*IP*). The surrounding organs include the heart (*H*), pericardium (*P*), right lung (*RL*), left lung (*LL*), and diaphragm (*D*). The sizes of the intrapericardial space and pericardium have been greatly exaggerated for clarity. The robot tether is not shown

5.2 System Design

5.2.1 Overview of HeartLander

The HeartLander system consists of a surgical user interface, supporting robotic instrumentation, and a tethered crawling device or crawler (Fig. 5.2). The surgeon interacts with the system through a joystick and standard computer interface, which features a graphic display of the current location of the robot on the heart surface. The support system uses external motors, vacuum pumps, and a computer control system to translate the commands of the surgeon into the appropriate actions of the crawler. The miniature tethered crawling device is the therapeutic end effector of the robotic system, and is located within the intrapericardial space of the patient. This tethered design allows the crawler to be miniature, lightweight, electrically passive, and disposable.

The envisioned intrapericardial therapies that would be facilitated by Heart-Lander, such as myocardial injection therapy, require multiple treatments to be made at precise locations on the epicardial surface of the heart. In clinical use, the surgeon would first place the crawler on the epicardium through a percutaneous incision below the sternum and subsequent pericardial incision. The robot would then semi-autonomously navigate to the specified target pattern area, and acquire each individual target using a series of positioning motions. A single treatment would be performed by the surgeon after each target acquisition.

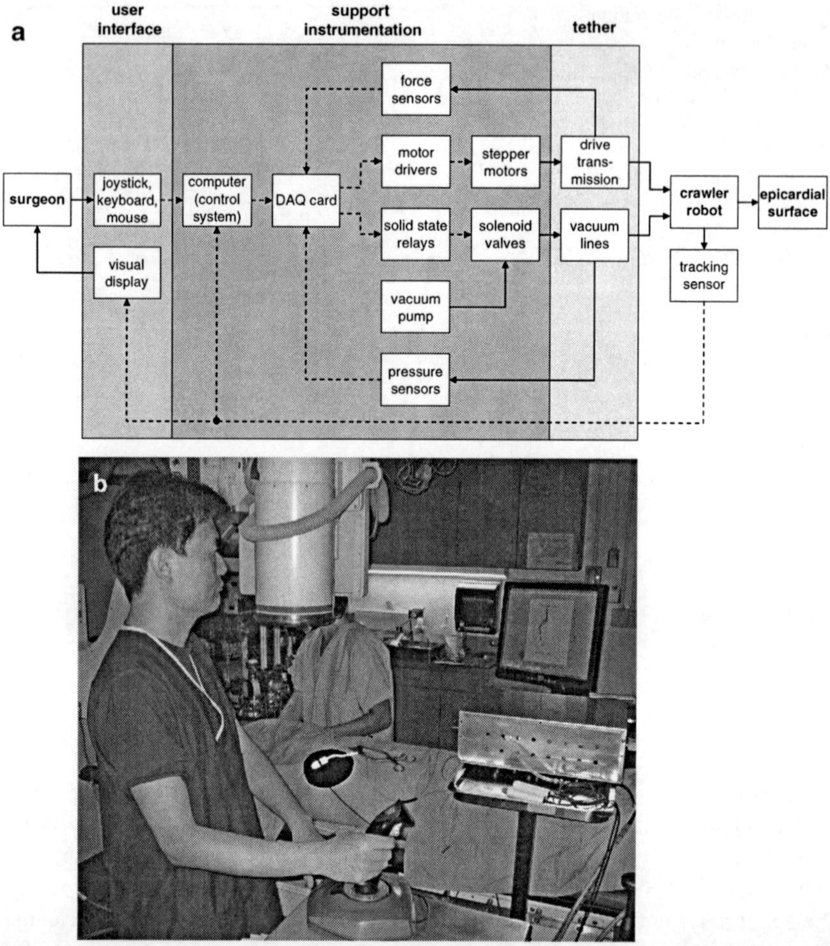

Fig. 5.2 (**a**) Schematic diagram of the HeartLander system, which comprises the user interface, the support instrumentation, and the tethered crawling robot. Mechanical flow is indicated by *solid lines*, while data flow is shown by *broken lines*. (**b**) Photograph of the operating room during animal testing, showing the surgeon interacting with the HeartLander system. The approximate heart shape and robot location inside the body have been illustrated for clarity

5.2.2 *Crawler Design*

The tethered crawler consists of two bodies (front and rear) that each contain an independent suction pad for prehension of the epicardium using vacuum pressure (Fig. 5.3b). Each suction pad has a cylindrical shape, with a diameter of 6.0 mm and a depth of 3.5 mm, which is integrated into the body. Thin latex skirts surround the periphery of the suction pads to help create a vacuum seal with the epicardium. The crawler bodies are each 5.5$\times$ 8 $\times$ 8 mm (H $\times$ W $\times$ L), and are made of

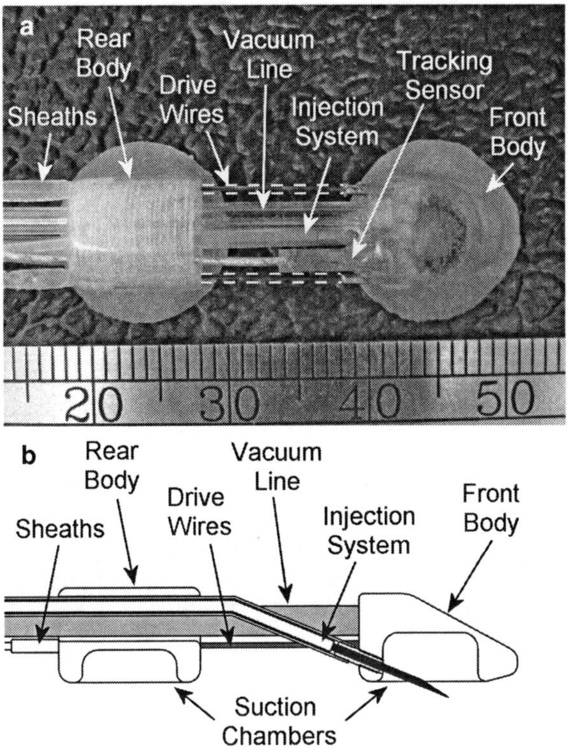

Fig. 5.3 (**a**) Photograph of the top of the HeartLander crawling robot. (**b**) Illustration of the profile of the crawling robot, with a cutaway along the median coronal plane to show the suction pads and injection system

a durable plastic using stereolithography fabrication. These dimensions include a 2-mm diameter working channel, and allow the robot to fit through an 8-mm diameter port. The drive transmission that actuates the crawler is composed of two super-elastic nitinol wires (0.3 mm in diameter) that are attached to the front body and sheathed within lengths of low-friction plastic tubing that are attached to the rear body (Fig. 5.3). The wires slide freely within the plastic sheaths when driven by the support system motors. The lengths of exposed wire between the body sections (L_1 and L_2) determine the distance (L) and angle (β) between the crawler bodies (Fig. 5.4). Vacuum pressure is regulated up to 450–600 mmHg by computer-controlled valves in the support system, and is supplied to the suction pads via vacuum lines.

The inchworm-like locomotion of the crawler is generated by the computer control system by regulating the wire lengths between the crawler bodies and the vacuum pressure in the corresponding suction pads (for more details, see [9]). The vacuum line pressures and drive wire forces are recorded and used offline in the analysis of locomotion performance. During both navigation and positioning, the inverse kinematics of the crawler are used by the control system to calculate the

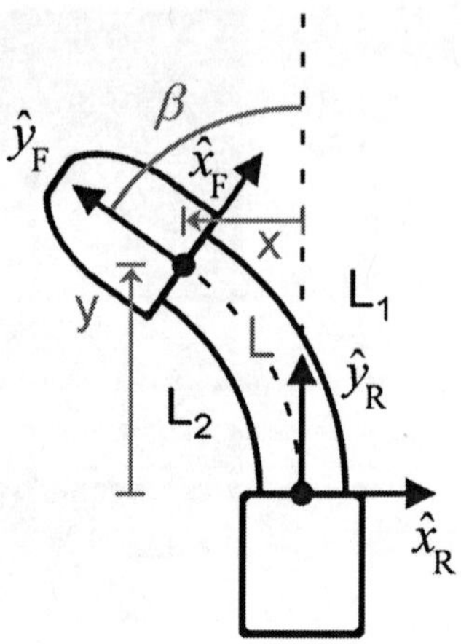

Fig. 5.4 Illustration of the front crawler body turning to the left, with kinematic axes and variables shown. The Cartesian location (x, y) of the front body with respect to the rear body is controlled by the exposed wire lengths L_1 and L_2. The resulting bearing angle β is also shown

wire lengths (L_1 and L_2) required to move the front body to the desired Cartesian location (x, y) with respect to the rear body (Fig. 5.4). A full development of the inverse kinematics can be found in [11].

5.2.3 Remote Injection System

A remote injection system for performing myocardial injections from the intrapericardial space has been developed to fit within the working channel of the HeartLander crawler. When the crawler is in motion, the 27 gauge needle is safely housed inside the working channel of the front body. When the crawler front body reaches the desired target location, the needle is extended into the tissue that has been drawn into the active front body suction pad (Fig. 5.3b). The proximal end of the needle injection system is connected to a syringe. The depth of the needle penetration into the tissue is set by an adjustable mechanical constraint within the range of 1–5 mm.

5.2.4 Electromagnetic Tracking and Localization

The graphic display and robot localization for control use the data from an electromagnetic tracking system (microBIRD, Ascension Technologies, Burlington, VT) [12]. The position and orientation of a miniature tracking sensor, located on the front body of the crawler (Fig. 5.3a), are measured in real time with respect to a magnetic transmitter attached to the operating table. This method of tracking does not require a line of sight between the sensor and the transmitter, and is thus well suited for tracking tools located inside the body. The tracking system has a reported static translational and angular accuracy of 1.4 mm and 0.5°, respectively, within the transmitter workspace. The reported translational and angular resolution are 0.5 mm and 0.1°, respectively.

The graphic display generates a 2-D projection of the 3-D location of the robot on a geometric representation of the heart surface, from the current view angle specified by the surgeon. The creation of the geometric heart model is described in Section 5.3.2. During the target acquisition task, the 3-D location of the target pattern is also displayed. The surgeon uses the display to select the current target, and monitors the progress of the crawler front body during the acquisition. This display is the only form of visualization during the porcine testing, because the robot is inserted through a small incision with the chest closed.

We wish to localize the robot with respect to the moving reference frame of the beating heart, rather than a traditional stationary reference frame. The crawler naturally lies within this heart-based reference frame because it passively moves with the portion of the epicardium to which it is attached at any given time. The tracking system, however, measures the sensor position and orientation with respect to the stationary magnetic transmitter, thus requiring a transformation to the heart-based reference frame. In order to perform this transformation, we filter the physiological motion out of the tracking data in real time. A third-order Chebyshev Type II low-pass filter with a 20-dB cutoff frequency at 1.0 Hz is used to remove the heartbeat motion from the tracking data. To remove the respiration component of the physiological motion, we then use a series of two second-order IIR notch digital filters, with notch frequencies at the primary (0.23 Hz) and secondary harmonics (0.46 Hz) of the ventilation rate. Notch filters are appropriate for this task, because the respiration rate is precisely set by the ventilator to 0.23 Hz. In offline testing, this filtering method attenuates the tangential physiological motion component by 81%, from 7.3 ± 1.2 to 1.4 ± 0.5 mm. The time delay caused by the filtering was 1 s. More details can be found in [11].

5.2.5 Control

The positioning control system enables the surgeon to acquire each target in the specified pattern, in a semi-autonomous manner. The surgeon first defines the number of targets for the pattern, along with their positions relative to on another. The surgeon then specifies the position and orientation of the target pattern on the heart surface model, using the graphic display of the user interface as a reference. The targets are thus fixed in space, and defined with respect to the electromagnetic tracking system. The crawler is then placed on the epicardial surface by the surgeon, and the target acquisition task proceeds. The surgeon selects the current target from the pattern with the user interface, and presses the joystick to begin the acquisition. The control system then uses the crawler inverse kinematics to align the front body and the current target based on the 3-D location of the robot measured from the tracking sensor and the fixed location of the current target.

Throughout the target pattern acquisition, the rear body remains fixed to the epicardium unless the control system determines that taking a step toward the current target is necessary. Target acquisition proceeds as long as the surgeon presses the joystick, until the target is acquired within the distance specified by the surgeon though the user interface. The surgeon monitors the progress of the robot toward the each of the targets with the graphic display. This semi-autonomous control paradigm allows the surgeon to control the entire system at a high level, while allowing the computer control system to perform the low-level calculations for the motions for the robot.

5.3 Experimental Protocol

An animal study was performed under Institutional Review Board approval to evaluate the ability of HeartLander to accurately acquire a target pattern on the epicardium of a porcine beating heart in a semi-autonomous manner. Direct epicardial injection therapy was simulated by injecting dye into the myocardium using the remote injection system, which also required HeartLander to maintain a stable platform for safe interaction with the beating heart.

5.3.1 Animal Preparation

A healthy Yorkshire swine (body weight of 40 kg) was anesthetized and placed in a supine position. A small subxiphoid incision (40 mm) and pericardial incision (15 mm) were created to access the apex of the left ventricle. Blood pressure and electrocardiogram were continuously monitored through the trials. All testing was performed with the chest closed and pericardium intact through a subxiphoid approach. The vacuum pressure range was maintained within the range of 500–550 mmHg, which is considered to be safe by the FDA for suction-based mechanical stabilization of the epicardial surface [13].

5.3.2 Heart Surface Model Construction

Prior to testing, a stationary wire frame model of the heart surface was generated for the graphic display of the user interface. This animal-specific geometric model provided a reference to localize the robot on the heart surface during the closed-chest animal testing. Although the surface model did not contain any fine anatomical detail, the surgeon was able to use the shape and size of the model to estimate the general anatomical regions of the heart (e.g., the left ventricle). To capture the gross structure of the heart surface, the surgeon traced the beating heart with a probe equipped with an electromagnetic tracking sensor at the distal tip. The probe was inserted into the intrapericardial space through the subxiphoid incision. The position data recorded from the tracking probe were then used by the custom HeartLander graphic software to create a wire frame computer model of the heart surface. This tracing method did not account for the displacement or deformation of the heart surface caused by heartbeat and respiration, but provided only an approximation of the geometric shape of the heart surface. Because the heart surface model was defined with respect to the tracking system reference frame, it was automatically registered properly to the robot tracking sensor and the targets.

5.3.3 Positioning Protocol

The positioning study tested the ability of HeartLander to quickly and accurately acquire multiple targets located in a specified local pattern. A circular pattern, with a 15-mm diameter, of seven evenly spaced targets and one central target was chosen for its clinical significance in defining a boundary (Fig. 5.5). The number of targets and target spacing were set by the surgeon based on the size and shape of the posterior left ventricle as determined from the heart surface model. The posterior left ventricle was selected for the target pattern due to its clinical significance in heart failure, and the difficulty with which it is accessed using conventional endoscopic approaches. The surgeon set the acceptable error range for target acquisition to 1.0 mm, which was measured by the control system as the distance between the 3-D planned target location and the location of the robot tracking sensor. This level of accuracy was considered to be sufficient for direct epicardial injection therapy for heart failure. After each successful target acquisition, the front body was locked onto the epicardium with active suction and the surgeon performed an injection of dye into the myocardium with the remote injection system. The surgeon proceeded in this manner until all targets in the pattern were acquired

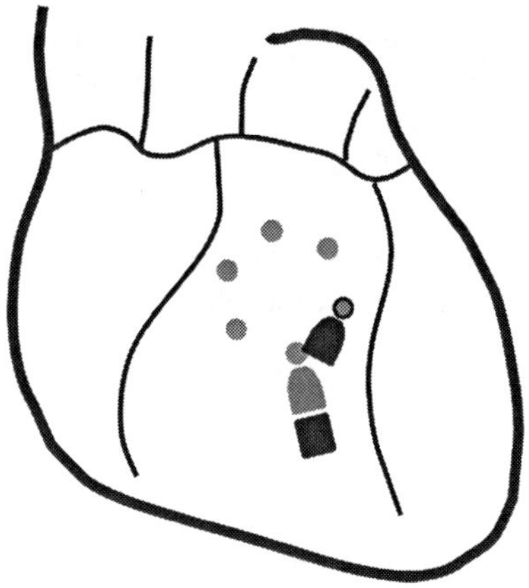

Fig. 5.5 Illustration of the positioning task. The robot makes a series of precise motions of the front body to each of the targets in the specified pattern, shown by the *red circles*. The initial location of the front body is shown in *grey*. The rear body remains stationary throughout the task

and marked with dye. In addition to demonstrating injection capability, the dye marks served as visible landmarks showing the location of the robot following target acquisition. These marks were used to evaluate the accuracy of the positioning system on the excised porcine heart following the trial.

5.3.4 Postmortem Study

The animal was euthanized at the end of the trial, in accordance with the acute protocol approved for the study. The heart was first examined in situ to assess damage to non-cardiac structures, and was then removed. Gross visual inspection was performed from the epicardial vantage to surrounding structures in the mediastinum. The dye mark locations were photographed, and the regions of heart tissue at the injection sites were analyzed histologically for injury.

5.4 Results

The animal tolerated the testing well until planned euthanasia. No adverse hemodynamic or electro-physiological events (e.g., hypotension, fatal arrhythmia, bleeding) were noted during the trial. The surrounding structures were intact upon postmortem examinations. There were no gross or histological injuries to the heart due to the robot motion, suction prehension, or injections.

The surgeon completed the heart surface tracing in approximately 5 min, while viewing the resulting point cloud generation on the graphic display of the user interface to ensure sufficient coverage. The resulting wire frame heart model created by the graphic display software contained sufficient geometric information of the heart surface for the surgeon to identify the general anatomical regions of the heart.

The surgeon was able to use the HeartLander system to administer dye injections to all target locations within the circular pattern on the posterior left ventricle. The target acquisitions were performed solely with motions of the front body, and did not require steps to reposition the rear body. The average acquisition time for each individual target was 23 ± 15 s, excluding the time required for dye injection. Following each target acquisition, an injection of 0.1 ml of oil-based dye was injected into the myocardium using the remote injection system through the working channel of the HeartLander crawler. With the needle penetration depth set to 3 mm, the average dye penetration depth was 3 ± 0.5 mm. The duration of the entire target acquisition and injection testing protocol was 25 min.

In evaluating the accuracy of the HeartLander system in the positioning task, we had two independent measures of the robot locations following target acquisition: the tracker-based sensor positions measured by the control system, and the dye marks visible on the surface of the excised heart. In comparing these two measures to determine positioning accuracy, we considered both the absolute and relative components. We defined the "absolute" accuracy as the degree to which the target pattern was positioned and oriented in a similar manner between the tracker-based positions on the heart surface model, and the dye marks on the excised porcine heart. We defined the "relative" accuracy as the degree to which the shape of the target pattern was similar between the tracker-based measurements and dye marks on the porcine heart. Separating the accuracy into these independent components allowed us to make a more detailed assessment of the HeartLander positioning system.

5.4.1 Absolute Accuracy

The location and orientation of the tracker measurements and dye marks on the posterior heart surface were qualitatively compared in the assessment of absolute accuracy. The pattern of tracker measurements following the target acquisitions, along with the heart surface model, was oriented to display the posterior view of the heart (Fig. 5.6a). Similarly, the dye mark pattern and excised heart were photographed with posterior surface of the heart normal to the camera (Fig. 5.6b). The tracker image and dye photograph were scaled based on known distances in each image, and were oriented so the contours of the hearts aligned. Figure 5.6c shows the tracker measurement image overlaid on the photograph of the excised heart. This figure demonstrates good qualitative agreement between the geometric shapes of the heart surface model and the excised heart. The figure also shows good qualitative agreement between the locations of the tracker measurements of the robot following the target acquisitions and the dye marks on the excised porcine heart.

5.4.2 Relative Accuracy

The relative errors between the locations of the individual tracker measurements and dye marks were calculated in the assessment of the relative positioning accuracy. Four separate measurements of the target patterns were used in the calculation of the relative accuracy:

- (A) the planned target locations defined with respect to the electromagnetic tracking system,
- (B) the tracker measurements of the robot immediately following target acquisition,
- (C) the tracker measurements of the robot during dye injection, and
- (D) the dye marks visible in photographs of the excised porcine heart.

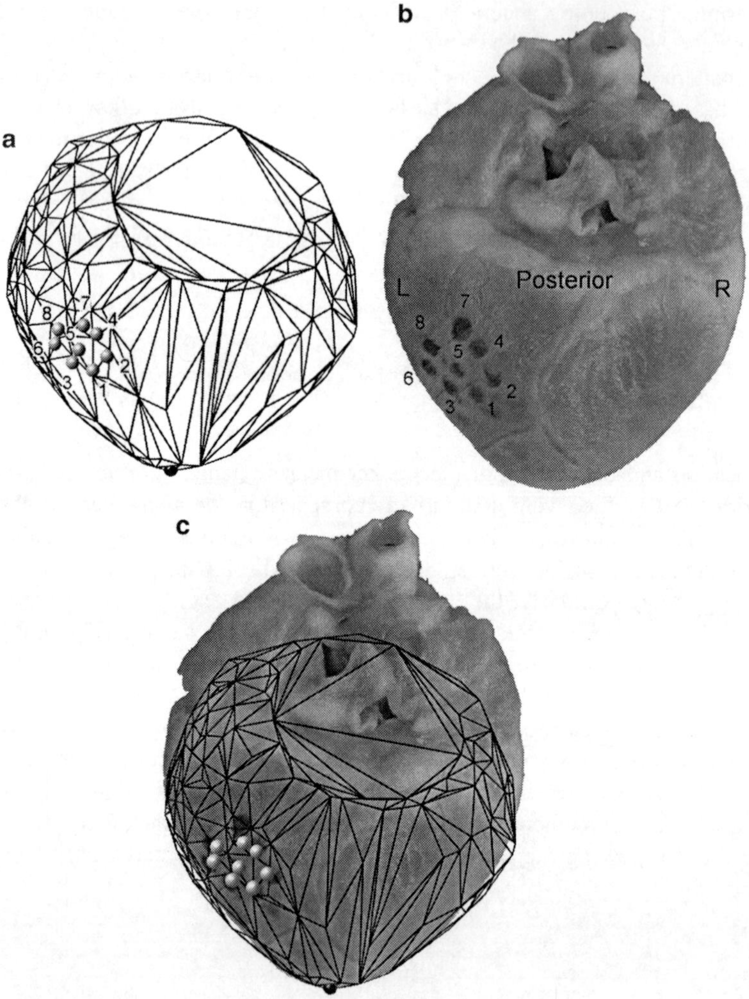

Fig. 5.6 (**a**) The heart surface model and tracker measurements of the robot sensor at the dye injections. (**b**) Photograph of the excised porcine heart and dye mark pattern. (**c**) The heart surface model and injection pattern overlaid on the injection photograph

The difference in the robot state between methods B and C was that the front body did not grip the epicardium during the positioning motions (B), but did grip the epicardium during injection (C). The planned target locations (A) and tracker measurements (B, C) were generated from a 2D projection of the tracker data with a view angle normal to the least-squares best-fit plane through the 3-D tracker data at the time of the dye injections (Fig. 5.7a). The dye mark locations (D) were generated using a photograph of the excised porcine heart with the camera oriented normal to the surface of the heart on which the dye pattern was visible (Fig. 5.7b).

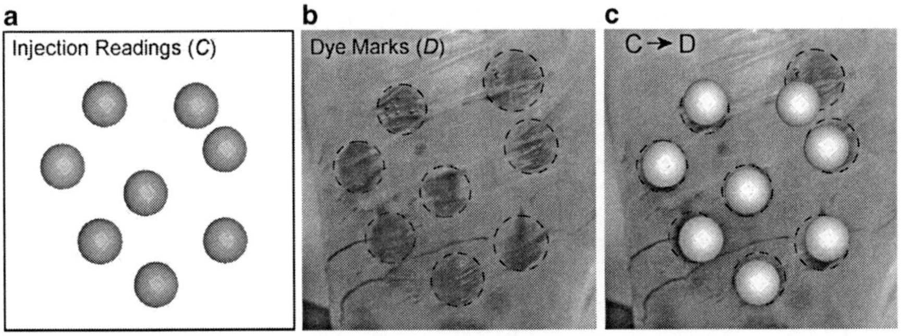

Fig. 5.7 (**a**) The tracker measurements of the robot sensor at the dye injections. (**b**) Photograph of the dye mark pattern (with each mark outlined) on the excised porcine heart. (**c**) The tracker measurements of the robot locations during the dye injections, overlaid on the dye mark pattern

The location of the robot at each injection site was estimated as the center of the circular dye region on the tissue, which was validated in bench testing.

All tracker-based patterns and the dye mark photograph were scaled according to known distances in both images. The locations of the tracker-based patterns were properly registered to one another because they were all measured with respect to the tracking system reference frame. The dye mark pattern from the photograph was aligned with the tracker-based patterns by calculating the translation and rotation that minimized the sum of root-squared distances between the individual targets (Fig. 5.7c). This alignment of the tracker-based and dye mark patterns was reasonable as we were only considering the relative accuracy of the system in this analysis; i.e., the shape of the target patterns.

The total relative error between each planned target location and dye mark (A→D) was the cumulative error between the planned target and the robot tracker at the target acquisition (A → B), the robot tracker at the target acquisition and at the dye injection (B → C), and the robot tracker at the dye injection and the physical dye mark on the heart tissue (C → D). All errors were calculated as the 2D resultant distances between the individual targets in the aligned images, and were reported individually and averaged over all targets. The 2D locations of the targets calculated using all four methods can be found in Fig. 5.8, while the individual and averaged error values are in Table 5.1. The average total relative error from the planned targets to the dye marks on the heart tissue was 1.7 ± 1.0 mm.

5.5 Discussion and Conclusion

In the present study, HeartLander demonstrated semi-autonomous acquisition of a specified target pattern, followed by dye injections, on the posterior surface of a beating porcine heart through a subxiphoid approach with no adverse physiological events. It is noteworthy that surgeons have some difficulty accessing this surface

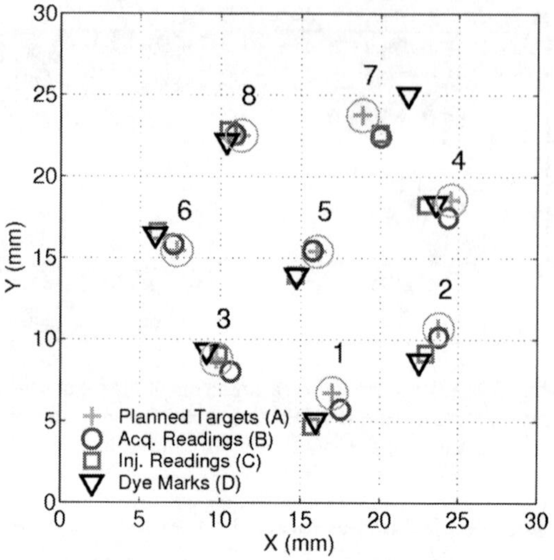

Fig. 5.8 Comparison of the locations of all tracker-based patterns (planned targets, acquisition readings, and injection readings) and the dye marks

Table 5.1 Values of each of the three sources of relative error, and the total relative error for the positioning task

Target no.	Target distance (mm)	A → B error (mm)	B → C error (mm)	C → D error (mm)	A → D error (mm)	Duration (s)
1	5.0	1	1.9	0.4	1.7	13
2	9.8	0.4	0.8	0.6	2.3	28
3	9.8	1	1	0.8	0.9	23
4	15.9	0.9	1.6	0.7	0.3	9
5	12.5	0.3	1.5	0.1	1.6	21
6	15.9	0.4	1.1	0.3	1.9	20
7	19.5	1.5	1.0	3.0	3.6	56
8	5	0.3	0.3	0.7	1.1	11
Mean ± Std		0.7 ± 0.4	1.1 ± 0.5	0.8 ± 0.9	1.7 ± 1.0	23 ± 15

even under full sternotomy. The absolute accuracy of the tracker measurements of the robot location and the photograph of the dye marks on the excised porcine heart was qualitatively good. This validates that the heart surface model serves as a geometric representation of the heart surface that agrees with the surface of the true porcine heart. It also validates the use of the surface model and electromagnetic tracking system to identify the general anatomical location of the robot on the true heart.

The relative accuracy between the planned target locations and the tracker measurements of the robot following target acquisition (A → B) had an average value of 0.7 mm, which was below the range set by the surgeon in the control system (1 mm). The total relative accuracy between the planned target locations and the dye marks (A → D), however, increased to an average of 1.7 mm. Although this value exceeds the error limit set within the control system, these results demonstrate the most accurate positioning on the posterior beating heart surface through a minimally invasive approach known to these authors. The additional relative error from B → C was due to the gripping of the epicardium following target acquisition, which is measured by the tracking system and can therefore be reduced with modifications to the control system. The additional error from C → D may have resulted from the reported static inaccuracy of the electromagnetic tracking system (1.4 mm), or from uneven distribution of the dye into the myocardium. This issue must be explored in future work. It may be that the large error associated with target seven was due to this uneven distribution of the dye into the myocardium, causing us to miscalculate the location of the front body during dye injection.

By adhering directly to the epicardium, and thus passively synchronizing with the heart, the 3-D locations of the tip of the injection needle and the epicardial surface remained aligned. Stabilization of the distance between the needle and the heart surface resulted in reproducible needle penetration depths, while stability tangential to the heart surface ensured that the needle did not tear the tissue. Injection methods not synchronizing with the heart motion (e.g., manual or endoscopic tool injection) suffer with regard to both of these issues. In this manner, HeartLander does not require cardiac stabilization for injection, which significantly reduces the risks of hemodynamic impairment and fatal arrhythmia that can result from the use of commercial mechanical stabilizers to immobilize the surface of the heart. Filtering the physiological motion from the tracking data does not, however, address the local deformation between points on the heart surface that occur during the cardiac cycle. Despite this fact, we were able to demonstrate high relative positioning accuracy (1.7 mm) because these deformations decrease as the distance between the robot and the current target decreases. In the future, this physiological deformation error can be further reduced by gating the tracking data – using real-time ECG and air flow measurements – so that only data collected during specified phases of the cardiac and respiratory cycles are used in the control system.

The most crucial limitation of our current system is the lack of anatomical information in the graphic display of the user interface. This deficit presents a serious safety concern when navigating on and interacting with the epicardial surface, and reduces our ability to define the absolute positioning accuracy of the HeartLander system. Accordingly, one of our future research aims is to register preoperative anatomical imaging (e.g. CT or MRI) with real-time tracking techniques. The surface model presented in this study was only used by the surgeon to estimate the general anatomical location of the robot based on the known geometry of the heart surface. Similarly, the positioning targets were defined relative to one another, rather than to specific anatomical landmarks on the epicardium. In the future, it will be necessary to quantitatively evaluate the absolute accuracy of the positioning system. This evaluation will become possible with proper registration of preoperative imaging and real-time tracking.

There are myriad opportunities for therapeutic applications within the intrapericardial space; including, but not limited to, cell transplant therapy for heart failure, left atrial appendage ligation, epicardial ablation, device-based mitral valve repair, and epicardial pacemaker lead placement for cardiac resynchronization therapy [2, 14, 15]. Recently, several minimally invasive approaches, such as traditional and robotic-assisted thoracoscopy, have been reported for these therapies [8, 16, 17]. These approaches, however, require multiple port placements under general anesthesia with double lumen ventilation. In addition, commercially available rigid endoscopic instruments intrinsically limit the operative field; specifically, it is difficult to access the posterior of the heart, and changing operative sites may require additional incisions and reinsertion of instrumentation. Alternatively, the insertion of HeartLander employs a subxiphoid approach. This approach is a useful method to access the intrapericardial space because there are no significant anatomic barriers, and it requires only a single port. Deploying a robot with navigation and positioning capabilities through this approach has the potential to obviate general endotracheal anesthesia, lung deflation, and full sternotomy without sacrificing treatment efficacy. In this manner, HeartLander has the potential to enable cardiac procedures on an outpatient basis.

Acknowledgments This work was supported in part by the National Institutes of Health (grant no. R56 HL0078839) and a NASA GSRP Fellowship (no. NNG05GL63H).

References

1. Hunt, S.A., Abraham, W.T., Chin, M.H., Feldman, A.M., Francis, G.S., Ganiats, T.G., Jessup, M., Konstam, M.A., Mancini, D.M., Michl, K., Oates, J.A., Rahko, P.S., Silver, M.A., Stevenson, L.W., Yancy, C.W., Antman, E.M., Smith, Jr., S.C., Adams, C.D., Anderson, J.L., Faxon, D.P., Fuster, V., Halperin, J.L., Hiratzka, L.F., Jacobs, A.K., Nishimura, R., Ornato, J.P., Page, R.L., Riegel, B.: ACC/AHA 2005 guideline update for the diagnosis and management of chronic heart failure in the adult: a report of the American College of Cardiology/American Heart Association task force on practice guidelines (writing committee to update the 2001 guidelines for the evaluation and management of heart failure): developed in collaboration with the American College of Chest Physicians and the International Society for Heart and Lung Transplantation: endorsed by the Heart Rhythm Society. Circulation **112**, e154–e235 (2005)
2. Christman, K.L., Lee, R.J.: Biomaterials for the treatment of myocardial infarction. J. Am. Coll. Cardiol. **48**, 907–913 (2006)
3. Huang, N.F., Yu, J., Sievers, R., Li, S., Lee, R.J.: Injectable biopolymers enhance angiogenesis after myocardial infarction. Tissue. Eng. **11**, 1860–1866 (2005)
4. Kohler, B.-U., Hennig, C., Orglmeister, R.: The principles of software QRS detection. IEEE Eng. Med. Biol. Mag. **21**, 42–57 (2002)
5. Orlic, D., Kajstura, J., Chimenti, S., Jakoniuk, I., Anderson, S.M., Li, B., Pickel, J., McKay, R., Nadal-Ginard, B., Bodine, D.M., Leri, A., Anversa, P.: Bone marrow cells regenerate infarcted myocardium. Nature **410**, 701–705 (2001)
6. Segal, A.Z., Abernethy, W.B., Palacios, I.F., BeLue, R., Rordorf, G.: Stroke as a complication of cardiac catheterization: risk factors and clinical features. Neurology **56**, 975–977 (2001)

7. Freyman, T., Polin, G., Osman, H., Crary, J., Lu, M., Cheng, L., Palasis, M., Wilensky, R.L.: A quantitative, randomized study evaluating three methods of mesenchymal stem cell delivery following myocardial infarction. Eur. Heart. J. **27**, 1114–1122 (2006)
8. Ott, H.C., Brechtken, J., Swingen, C., Feldberg, T.M., Matthiesen, T.S., Barnes, S.A., Nelson, W., Taylor D.A.: Robotic minimally invasive cell transplantation for heart failure. J. Thorac. Cardiovasc. Surg. **132**, 170–173 (2006)
9. Patronik, N.A., Zenati, M.A., Riviere, C.N.: Preliminary evaluation of a mobile robotic device for navigation and intervention on the beating heart. Comput. Aided Surg. **10**, 225–232 (2005)
10. Riviere, C.N., Patronik, N.A., Zenati, M.A.: Prototype epicardial crawling device for intrapericardial intervention on the beating heart. Heart. Surg. Forum. **7**, E639–E643 (2004)
11. Patronik, N.A., Ota, T., Zenati, M.A., Riviere, C.N.: A miniature mobile robot for navigation and positioning on the beating heart. IEEE Trans. Robot. **25**, 1109–1124 (2009)
12. Schneider, M., Stevens, C.: Development and testing of a new magnetic-tracking device for image guidance. In: Proceedings of SPIE Medical Imaging 2007: Visualization and Image-Guided Procedures, vol. **6505**, 65090I (2007)
13. Borst, C., Jansen, E.W., Tulleken, C.A., Grundeman, P.F., Mansvelt Beck, H.J., van Dongen, J.W., Hodde, K.C., Bredee, J.J.: Coronary artery bypass grafting without cardiopulmonary bypass and without interruption of native coronary flow using a novel anastomosis site restraining device ("Octopus"). J. Am. Coll. Cardiol. **27**, 1356–1364 (1996)
14. Cappato, R., Calkins, H., Chen, S.A., Davies, W., Iesaka, Y., Kalman, J., Kim, Y.H., Klein, G., Packer, D., Skanes, A.: Worldwide survey on the methods, efficacy, and safety of catheter ablation for human atrial fibrillation. Circulation **111**, 1100–1105 (2005)
15. Rivero-Ayerza, M., Theuns, D.A., Garcia-Garcia, H.M., Boersma, E., Simoons, M., Jordaens, L.J.: Effects of cardiac resynchronization therapy on overall mortality and mode of death: a meta-analysis of randomized controlled trials. Eur. Heart. J. **27**, 2682–2688 (2006)
16. Pruitt, J.C., Lazzara, R.R., Dworkin, G.H., Badhwar, V., Kuma, C., Ebra, G.: Totally endoscopic ablation of lone atrial fibrillation: initial clinical experience. Ann. Thorac. Surg. **81**, 1325–1330 (2006); discussion 1330–1331
17. Zenati, M.A.: Robotic heart surgery. Cardiol. Rev. **9**, 287–294 (2001)

Chapter 6
Miniature In Vivo Robots for NOTES

Shane M. Farritor, Amy C. Lehman, and Dmitry Oleynikov

Abstract Eliminating all external incisions would be a significant step in reducing the invasiveness of surgical procedures. Accessing the peritoneal cavity through a natural orifice, as in Natural Orifice Translumenal Endoscopic Surgery (NOTES), promises distinct patient advantages, but is surgically challenging. Performing laparoscopic surgeries through a single transumbilical incision is also gaining renewed interest as a potential bridge to enabling NOTES. Both of these types of surgical procedures are inherently limited by working with multiple instruments through a constrained insertion point. New technologies are necessary to overcome these limitations and provide the surgeon with adequate visual feedback and triangulation. Miniature in vivo robots provide a unique approach by providing a platform that is completely inserted into the peritoneal cavity to enable minimally invasive surgery. This chapter describes the design and feasibility testing of miniature in vivo robots that can provide stable visualization and manipulation platforms for NOTES and single incision surgery.

Keywords Cholecystectomy · In vivo · Laparoscopy · LESS · Miniature · Minimally invasive surgery · Natural orifice · NOTES · Robot · Single incision

6.1 Introduction

Performing surgical procedures using minimally invasive approaches is well established, with laparoscopy now being the standard of care for many routinely performed surgical procedures. While replacing a large open incision with three to five small incisions offers significant advantages, focus remains on further reducing

S.M. Farritor (✉)
Department of Mechanical Engineering, University of Nebraska-Lincoln, N104 SEC, Lincoln, NE 68588-0656, USA
e-mail: sfarritor@unl.edu

J. Rosen et al. (eds.), *Surgical Robotics: Systems Applications and Visions*,
DOI 10.1007/978-1-4419-1126-1_6,

the invasiveness of these procedures. Natural Orifice Translumenal Endoscopic Surgery (NOTES) is a new approach to abdominal surgery that promises to reduce the invasiveness of surgical procedures by accessing the surgical target through a natural orifice. Theoretically, the elimination of external incisions avoids wound infections, further reduces pain, and improves cosmetics and recovery times [1, 2]. While NOTES may be the ultimate goal of minimally invasive surgery, significant surgical challenges have led to an increased interest in Laparoendoscopic Single-Site (LESS) surgery as an important step, and possibly a bridge to NOTES [3].

Accessing the peritoneal cavity through a natural orifice or a single abdominal incision are appealing methods from the perspective of the patient. However, both methods are surgical challenging. It is difficult to have multiple instruments passing simultaneously through a natural orifice or an incision while maintaining adequate manipulation and visualization capabilities. Current endoscopic and laparoscopic instrumentation are inadequate for NOTES and LESS procedures. New technologies are necessary that can overcome these challenges and provide the surgeon with adequate visual feedback and triangulation. This chapter describes the design and in vivo feasibility testing of miniature in vivo robots that are a novel approach to overcoming the current instrumentation limitations for NOTES and LESS.

6.2 Background

Accessing the surgical environment through small incisions, such as in laparoscopy, is inherently limited in scope as compared to open procedures where the surgeon can directly view and manipulate within the surgical environment. Working with long, rigid tools through access ports in the abdominal wall limits the motion of the tools and provides only a two dimensional image of the surgical environment [4, 5]. While laparoscopy is the preferred intervention for many routine interventions, such as cholecystectomy, these constraints have contributed to the limited application of laparoscopic techniques to more complex procedures.

6.2.1 Robotic Assistances for Minimally Invasive Surgery

Visualization and dexterity limitations for minimally invasive surgery are being addressed through the application of robotics. The Automated Endoscopic System for Optimal Positioning (AESOP) was the first robotic device to receive Food and Drug Administration approval for direct surgical manipulation in laparoscopy and was introduced in the mid-1990s for controlling a laparoscopic camera for surgical procedures [6]. The daVinci® (Intuitive Surgical, Sunnyvale, CA) system is a more advanced tele-robotic device that enables a surgeon located at a remote workstation to control robotic arms that hold the laparoscopic instruments. The surgical dexterity is enhanced through capabilities including wristed action, motion scaling, and tremor reduction. Further, a stereoscopic image of the surgical environment is displayed at the hands of the surgeon, creating the illusion that the surgical tools are extensions of the surgeon's hands [6]. However, the universal use of the daVinci® system has remained limited primarily due to its large size, high cost, and the diminished impact of the dexterous improvements for performing less complex surgical procedures.

6.2.2 Natural Orifice Translumenal Endoscopic Surgery and Laparoendoscopic Single-Site Surgery

Accessing the abdominal viscera through a natural orifice may be the ultimate goal in reducing the invasiveness of surgical procedures. Natural Orifice Translumenal Endoscopic Surgery (NOTES) can be performed as a pure procedure using a single opening or as a combined procedure using multiple orifices. Hybrid procedures can also be performed using a natural orifice in conjunction with conventional transabdominal ports [7]. The feasibility of NOTES was initially demonstrated in multiple animal model studies including peritoneal exploration with liver biopsy, gastrojejunal anastomosis, organ resection, and transvesical thoracoscopy [8–11]. NOTES procedures have also been performed with success in humans including appendectomy and cholecystectomy [12–15].

Laparoendoscopic Single-Site (LESS) surgery, is another type of procedure closely related to NOTES that is gaining renewed interest. LESS procedures are performed using multiple instruments introduced though a single small transabdominal incision. LESS has been reported for cholecystectomy and appendectomy since 1998 [16, 17], but with limited momentum due to technical limitations of conventional instrumentation [18]. With recent advances in instrumentation and access methods, there is a renewed interest in single small incision surgery, with multiple procedures being performed in humans including simple nephrectomy [19, 20], pyeloplasty [19], and cholecystectomy [21, 22]. For LESS, the surgical target is often accessed using a specialized port, such as the Uni-X single Port Access System (Pnavel Systems, Cleveland, OH) or the QuadPort (Advanced Surgical Concepts, Bray, Ireland) that are introduced through a transumbilical incision. These systems incorporate individual ports for a laparoscope, insufflation, and specialized instruments.

6.2.3 Robotic Assistants and Instrumentation for NOTES and LESS

There is great potential for new NOTES and LESS approaches, but the technology remains in evolution [7]. Much of the work for addressing the visualization and manipulation limitations for NOTES is based on the flexible endoscopy platform. Some work focuses on developing locomotion systems for navigation of hollow cavities, such as the colon and esophagus, using methods including inchworm devices [23], rolling stents [24], and adhesion [25]. Other work is focused on improving distal tip dexterity. For example, master-slave robotic systems are being developed that use long-shafted flexible instruments with multiple degrees of freedom with standard gastroscopes or endoscopes to improve distal tip maneuverability [26, 27]. Also, the TransPort EndoSurgical Operating Platform (USGI Medical, San Clemente, CA), is a commercially available four-channel platform scope that uses Shapelock^{TM} technology to lock the base of the endoscope for stability while also allowing for distal tip maneuverability [28].

The need for adequate triangulation for LESS is currently being addressed primarily through the use of bent or articulating instruments. However, bent tools necessitate that tissue dissection, retraction, and cautery be performed with the contralateral hand as compared to laparoscopic surgery [18] and often results in tool collisions. Alternatively, articulating tools such as RealHand (Novare Surgical Systems, Inc., Cupertino, CA) and Autonomy Laparo-Angle (Cambridge Endoscopic Devices, Inc., Framingham, MA) provide a seven degree of freedom maneuverability allowing for easier tissue manipulation. The daVinci S (Intuitive Surgical, Sunnyvale, CA) system has also been used to perform transumbilcial single port radical prostatectomy, dismembered pyeloplasty, and right side radical nephrectomy [29]. These procedures were performed using a multi-channel single access port. These studies described limited intracorporeal tool collisions and improved surgical dexterity.

While these specialized instruments and robotic systems mitigate the constraints of working through a single incision or natural orifice, limitations are inherent in approaches that require working with multiple instruments in a confined space. Even with specialized tools for LESS, instrument collisions internally and externally remain problematic. Also, flexible endoscopy approaches to NOTES instrumentation remain limited by the size and geometry of the natural orifice. An advanced laparoscopic skill set is requisite to the continued advancement of LESS and NOTES using existing instrumentation and robotic systems [18]

Novel approaches to instrumentation are necessary for the universal adoption of NOTES and LESS for performing minimally invasive surgery. One method is the use of miniature robots that are completely inserted into the peritoneal cavity through a natural orifice or a single incision. Unlike the externally actuated devices discussed previously, these robots are not constrained by the entrance incision once inserted into the peritoneal cavity. A transabdominal Magnetic Anchoring and Guidance System (MAGS) including intra-abdominal cameras and retraction instruments are currently being developed to assist in NOTES and LESS procedures [30, 31]. Once inserted into the peritoneal cavity, these devices are attached and positioned within the peritoneum using magnetic interactions with external hand-held magnets. Similarly, insertable monoscopic and stereoscopic imaging devices with multiple degrees of freedom are also being developed for minimally invasive surgery [32, 33].

6.3 In Vivo Robots for Visualization

Miniature in vivo robots provide a platform for visualization that is completely inserted into the peritoneal cavity through a single incision or a natural orifice approach. In contrast to existing flexible endoscopy tools used as the primary visualization for many NOTES procedures, in vivo robots provide visual feedback for the surgeon that is not constrained by the axis of the endoscope or the insertion point. Further the surgeon can reposition the robot throughout a procedure to provide visualization from arbitrary orientations within each quadrant of the peritoneal cavity. Two basic designs of in vivo robots for minimally invasive surgery have been developed, as shown in Fig. 6.1. The first is a mobile in vivo platform that provides a remotely controlled movable platform for visualization and task assistance. The second is a peritoneum-mounted robot that provides visualization from a birds-eye perspective.

6.3.1 Mobile Camera Robot

The basic design of a mobile camera robot, shown in Fig. 6.1 left, consists of two wheels that are driven independently using permanent magnet direct current motors that provide forward, reverse, and turning motions. A tail prevents counter-rotation and allows the robot to reverse directions. A helical design for the wheel profile was developed based on viscoelastic modeling together with benchtop and in vivo testing results [34, 35]. A helical wheel profile has demonstrated maneuverability on all of the pelvic organs, including the liver, spleen, small and large bowels, as well as capabilities of climbing deformable surfaces that are two to three times its height. No visible tissue damage has resulted from the in vivo testing with this robot. Various designs of the mobile robot incorporate on-board cameras for visualization and can also include end effectors, such as a biopsy grasper, to provide task assistance capabilities.

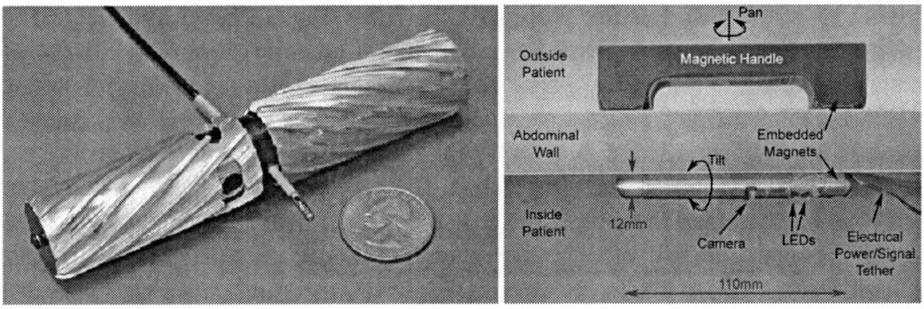

Fig. 6.1 Mobile (*left*) (© 2006 IEEE) and peritoneum-mounted (*right*) imaging robots [36, 40]

6.3.2 Peritoneum-Mounted Imaging Robot

Peritoneum-mounted robots provide primary visualization for minimally invasive procedures [36]. The robot contains an external housing with magnets used for fixation to the interior abdominal wall. These magnets interact with an external magnetic handle to hold the robot to the internal abdominal wall. A schematic of the external magnetic handle, the internal imaging robot, and the abdominal wall is shown in Fig. 6.1 right. Once attached, the robot can be grossly repositioned by maneuvering the magnetic handle along the exterior surface of the abdomen, with the robot being repositioned correspondingly below. The inner housing of the robot contains a camera and two ultra-bright LEDs for lighting. The camera can be rotated to provide a tilting action using a permanent magnet direct current motor that is also contained in the inner housing of the robot. This robot is tethered for power and communications.

6.3.3 In Vivo Results

The in vivo robots for visualization have been demonstrated in multiple survival and non-survival procedures in a porcine model. These procedures were all performed at the University of Nebraska Medical Center with experimental protocols approved by the institutional review committee.

Mobile Camera Robot. The mobile robot platform has been used to provide the sole visualization for performing a laparoscopic cholecystectomy in a porcine model [34]. For this procedure, the robot was introduced into the peritoneal cavity through a specialized trocar. After the robot was completely inserted, this port could then be used with a standard laparoscopic tool to provide additional task assistance capabilities, such as retraction. The mobile robot with a biopsy end effector has also demonstrated the feasibility of performing a single port biopsy using a mobile robot platform, as shown in Fig. 6.2.

A mobile robot has also been inserted into the peritoneal cavity using a transgastric approach [37]. For this procedure, the robot was advanced into the gastric cavity through an overtube with the assistance of an endoscope. The robot then maneuvered within the gastric cavity and was advanced into the peritoneal cavity. The robot demonstrated abilities to explore the peritoneal cavity and was removed through the esophagus using its tether.

Peritoneum-Mounted Imaging Robot. The peritoneum-mounted imaging robot has been demonstrated in multiple survival and non-survival procedures in a porcine model in cooperation with other in vivo robots and endoscopic and laparoscopic tools [36]. In the first procedure, the peritoneum-mounted imaging robot and a mobile robot were inserted into the peritoneal cavity through a standard trocar. The robots provided remotely repositionable platforms to provide visual feedback to the surgeon for the performance of a laparoscopic cholecystectomy, as shown in Fig. 6.3. A subsequent

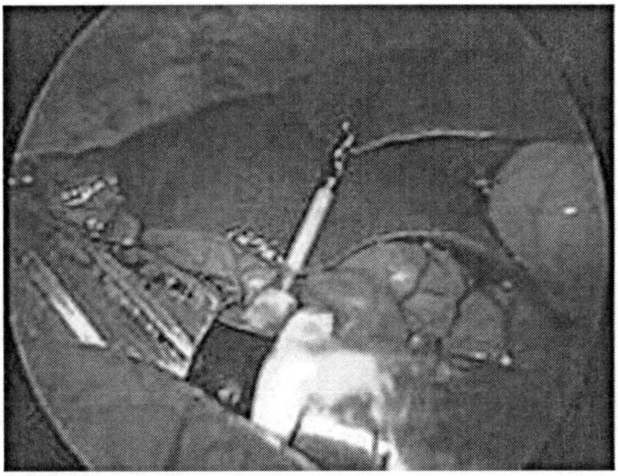

Fig. 6.2 Mobile biopsy robot performing liver biopsy as viewed from laparoscope. (With kind permission from Springer Science+Business Media: [42])

series of three survival cholecystectomies were successfully performed in a porcine model using a similar mobile camera robot and a peritoneum-mounted imaging robot.

In the third cooperative robot procedure, a peritoneum-mounted imaging robot, a lighting robot, and a retractor robot demonstrated the feasibility of inserting multiple robots with differing capabilities into the peritoneal cavity using a transgastric approach. For this procedure, the imaging robot was advanced into the peritoneal cavity through an overtube with the assistance of an endoscope. The robot was then magnetically coupled to the interior upper abdominal wall to provide visualization from an upper perspective. The retractor robot and the lighting robot were then inserted and demonstrated imaging and task assistance capabilities.

6.4 In Vivo Robots for Manipulation

6.4.1 Design Constraints

Definition of the forces, velocities, and workspace required for performing laparoscopic surgical procedures are necessary for the successful design of a manipulator robot for minimally invasive surgery. Available data for laparoscopic procedures are given almost exclusively for the forces applied by the surgeon at the tool handle instead of the actual forces applied to the tissues. Work by the BioRobotics Lab at the University of Washing uses a device called the BlueDRAGON to measure forces and motions applied by surgeons while performing various laparoscopic procedures [38, 39]. The raw data from these procedures provides useful information for determining the design requirements for a dexterous robot for

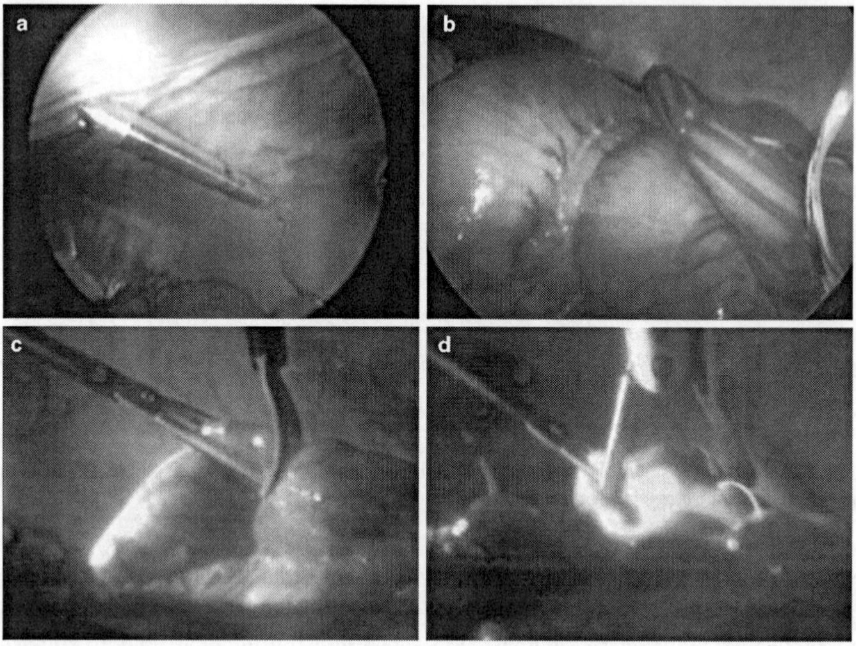

Fig. 6.3 Peritoneum-mounted (**a**) and mobile camera (**b**) imaging robots provide video feedback (**c**, **d**) during cooperative laparoscopic procedure [36, 40]

manipulation. Based on this work, it was determined that the robot should be able to apply forces along and perpendicular to the tool axis of 10 and 5 N, respectively. Also, the angular velocities about the tool axis and the perpendicular axes should be on the order of 1 and 0.4 rad/s, respectively [40]

6.4.2 Conceptual Designs

A dexterous in vivo robot for manipulation is designed to be analogous to the use of standard laparoscopic tools for performing minimally invasive the surgery. The basis of the robot design is to replace two laparoscopic tools with dexterous arms that have similar linear and angular velocity capabilities, and the ability to apply sufficient forces and torques as described above. The robot system, shown in Fig. 6.4, consists of the in vivo robot that is completely inserted into the peritoneal cavity and an external surgical control console. The robot must be flexible for insertion into the peritoneal cavity through the complex geometry of the natural lumen, and once inserted provide a stable platform for tissue manipulation and visualization.

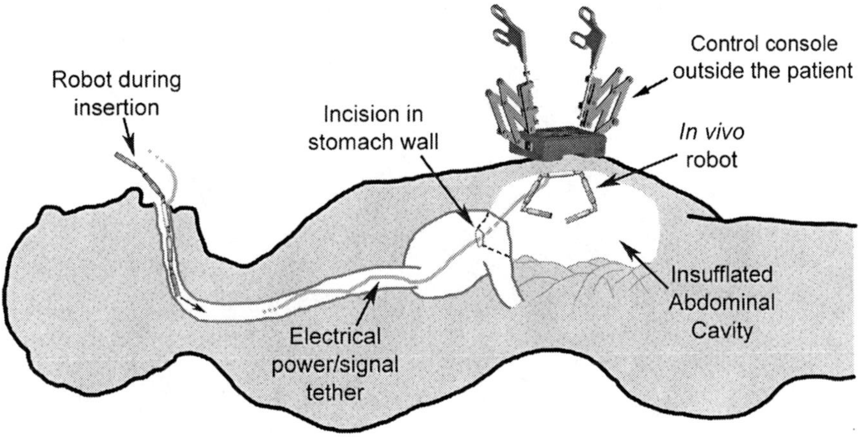

Fig. 6.4 Conceptual design of in vivo manipulator robot system for NOTES and LESS

6.4.3 Prototype Robot Designs

Prototypes of the dexterous manipulator robot have two prismatic arms that are each connected to a central body at a rotational shoulder joint. Each forearm has either a grasper or a cautery end effector for tissue manipulation. The body contains a stereo camera pair to provide visual feedback and an ultrabright LED for lighting. Magnets contained in the central body of the robot interact with magnets in the surgical control console to attach the robot to the interior of the abdomen and allow for gross positioning of the robot internally. This method of attachment to the abdominal wall enables the surgeon to arbitrarily position the robot throughout a surgery to provide improved visualization and manipulation capabilities within each quadrant of the peritoneal cavity [41].

The robot has two configurations to allow flexibility for insertion and also to provide stability for tissue manipulation. The robot can be changed from an articulation configuration to an insertion configuration, as shown in Fig. 6.5, by disconnecting the shoulder joint linkages. In the insertion configuration, the shoulder joints freely rotate allowing for natural orifice insertion. Once the robot has fully entered the peritoneal cavity, the shoulder linkages are reconnected allowing articulation.

6.4.4 Kinematic Models and Analysis

The dexterous manipulator robot is represented with the kinematic model of a two degree-of-freedom planar robot with a rotational shoulder joint and a prismatic arm joint [40]. The prototype NOTES robot is shown overlaid on the robot schematic with details of the shoulder joint expanded in Fig. 6.6. The joint variables are pitch,

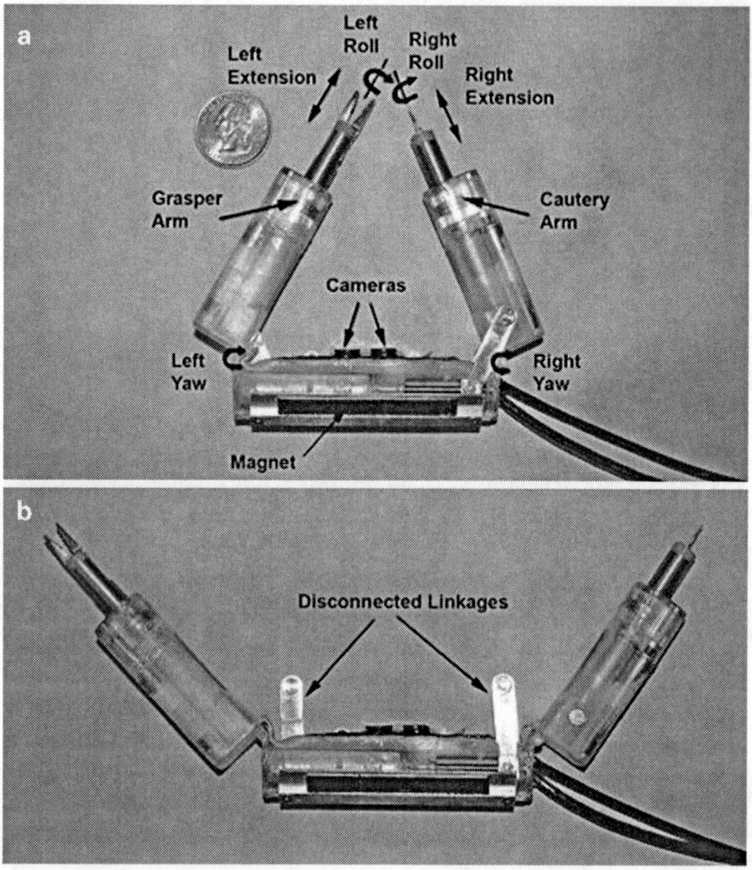

Fig. 6.5 Design of prototype dexterous manipulator robot in articulation (**a**) and insertion (**b**) configurations (©2008 IEEE)

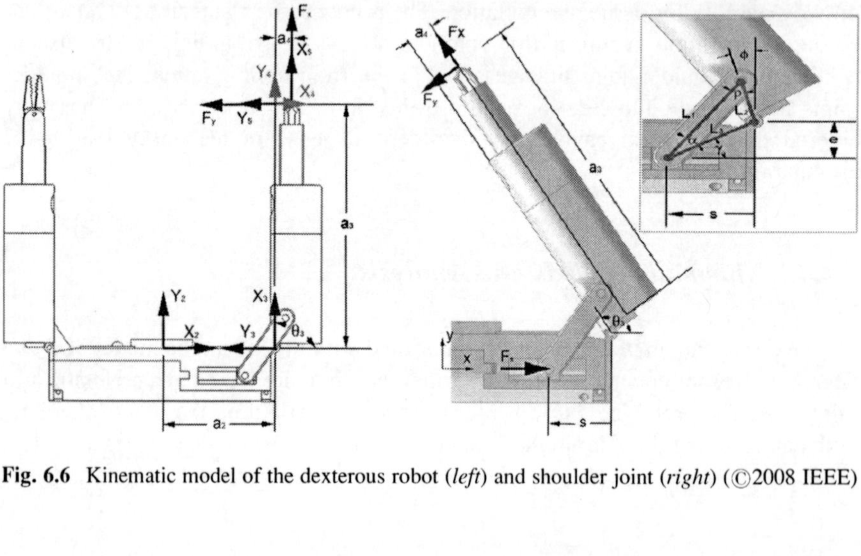

Fig. 6.6 Kinematic model of the dexterous robot (*left*) and shoulder joint (*right*) (©2008 IEEE)

yaw, and lower arm extension, denoted by α_1, θ_3, and a_3, respectively. A universal frame {0} is aligned with the gravity gradient, and the parameter, α_0, defines the rotation of the robot with respect to the universal frame. The parameter, α_1, defining the angle of rotation of the robot cameras with respect to frame {1}, is zero for this robot. The constants, a_2 and a_4, define the half body width and the end effector offset with respect to the shoulder joint, respectively.

The necessary joint torques were determined assuming a half body width, a_2, of 40 mm, a prismatic link mass of 30 g, a workplane, α_0, of 40° below level, and lower extension, a_3, and shoulder yaw, θ_3, ranges of 64–93 mm and 75–154°, respectively. The maximum joint torque necessary at full extension of the lower arm was determined to be 522 mNm. The necessary input force, F_s, to apply the necessary joint torque within the dexterous workspace defined as the range of joint angle, α_3, from 110–154°, was then determined. The shoulder joint articulation method for this robot uses a slider mechanism constrained to move in the x-direction that is coupled to the robotic arm, a_3, by a link L_1, with rotational degrees of freedom at each end. The maximum slider input force required was determined to be 67 N, assuming an overall efficiency of 50%.

6.4.5 User Interfaces

One version of the user interface for the dexterous miniature robot consists of two joystick controls that provide control for rotation of the shoulder joint and forearm extension, as shown in Fig. 6.7 [41]. Two pushbuttons on the left joystick are used to open and close the grasper jaws. The cautery is activated using a foot pedal. The video from one of the robot cameras is displayed on the LCD screen located between the two joysticks. The back of the control console contains magnets that interact with magnets in the body of the robot to hold the robot to the interior

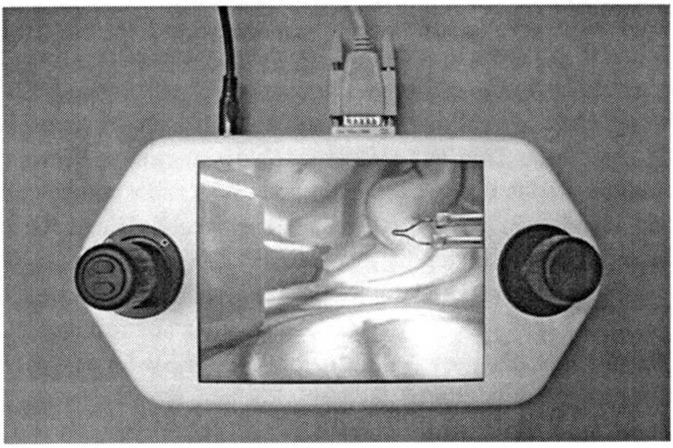

Fig. 6.7 User interface for dexterous manipulator robot. (With kind permission from Springer Science+Business Media: [41])

abdominal wall directly below the screen on the user interface. This provides for an intuitive understanding of the robot position within the peritoneal cavity. Moving the control console along the exterior surface of the abdomen effectively repositions the robot internally.

An alternative design for the user interface uses two modified laparoscopic tools in place of the joysticks. In this design, the robot arms located internally act as extensions of the external laparoscopic handles. This interface also includes an LCD display for video feedback and a foot pedal for cautery activation. This user interface is currently used to provide on-off control of the degrees of freedom for the robot.

6.4.6 In Vivo Results

Multiple non-survival procedures in a porcine model have been performed using iterations of this prototype dexterous manipulator robot design [41]. The basic surgical procedure initiated with accessing the peritoneal cavity through the upper gastrointestinal tract. An overtube was inserted through a transesophageal incision and advanced through the esophagus and into through a transgastric incision made using an endoscope. The robot was then configured for insertion and advanced through the overtube and into the peritoneal cavity using a standard therapeutic endoscope. The endoscope provided supplementary visualization and observation of the robot throughout the procedure. The focus of two of the procedures was to evaluate robot functionality. For these surgeries, the robots were inserted into the peritoneal cavity through a single transabdominal incision with supplementary visualization and retraction being provided by laparoscopic tools.

For each procedure, the robot was lifted from the floor of the abdomen following insertion using the interaction of magnets embedded in the robot and in the external surgical control console. The robot was then grossly maneuvered in the proper orientation for manipulation of the gallbladder. Once in position, the grasper end effector was used to grasp the cystic duct. The cautery end effector was next moved into position to begin the dissection and the cautery was activated. The procedure continued through iterations of this stretch and dissect task with repositioning of the robot as necessary throughout to perform a cholecystectomy. Views of the robot from a laparoscope used to observe the operation of the robot and from the on-board robot cameras are shown in Figs. 6.8 and 6.9, respectively. At the end of each procedure the robot was retrieved by its tether.

These procedures demonstrated the feasibility of using a dexterous in vivo robot platform to perform LESS or NOTES from essentially a laparoscopic platform. The configurations of the robot provided sufficient flexibility for insertion through a natural orifice or a transabdominal incision. Further, the placement of the robot cameras improved tool triangulation and the surgeon's understanding of the surgical environment. The robot design also enabled the application of sufficient off-axis forces for tissue retraction and dissection.

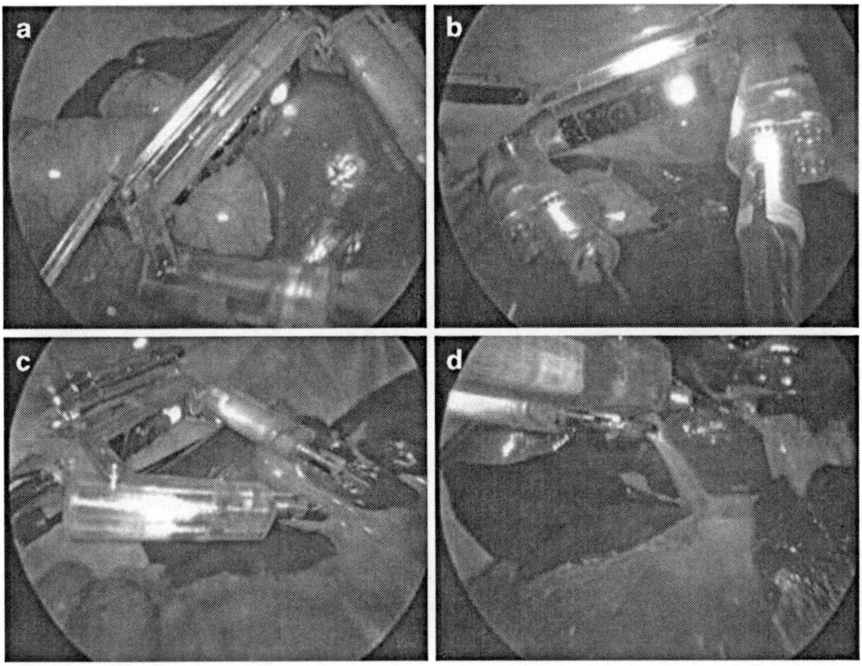

Fig. 6.8 Views of the robot in the peritoneal cavity from the laparoscope. (With kind permission from Springer Science+Business Media: [41])

6.5 Conclusions and Future Work

Central to the widespread conversion of many surgeries in the peritoneal cavity to a less invasive NOTES or LESS approach is the further development of devices that provide the surgeon with a stable multi-tasking platform for visualization and dexterous tissue manipulation. Many of the instruments that are currently being developed are based on the design of existing laparoscopic or flexible endoscopy tools. However, these instruments remain limited by working simultaneously through a confined entrance point.

Miniature in vivo robots that are completely inserted into the peritoneal cavity through a natural orifice or a single incision provide a novel approach for addressing the constraints of LESS and NOTES. These devices can be used inside the peritoneum without the typical constraints of the access point. Miniature robots can provide the surgeon with a repositionable visualization platform using peritoneum-mounted or mobile camera robots. Also, miniature robots with dexterous end effectors, including cautery and grasping, can provide a stable platform for off-axis tissue manipulation and visualization. Further, multiple devices can be inserted through the same incision without being limited by simultaneously working through a natural orifice or single incision.

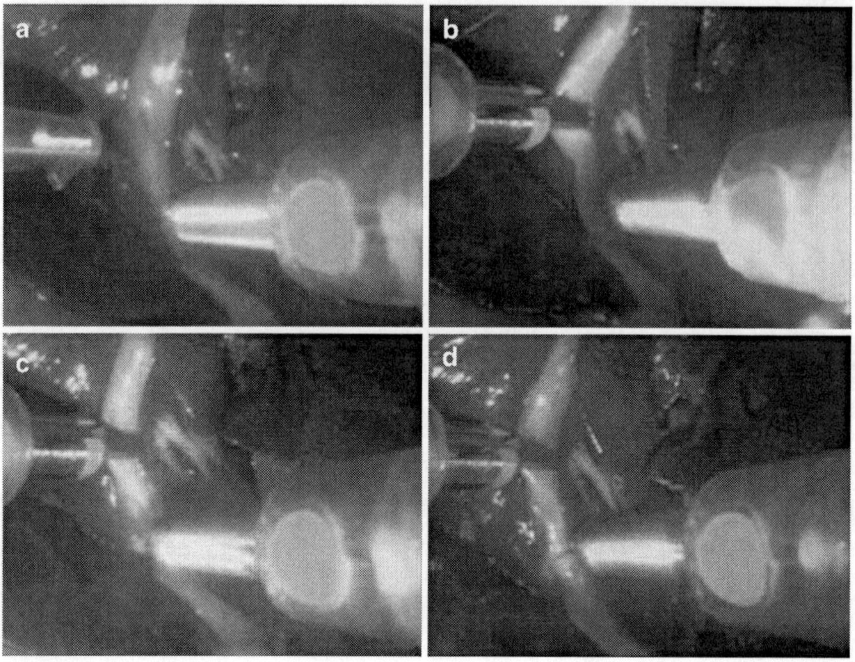

Fig. 6.9 Dissection of cystic duct as viewed from robot camera. (With kind permission from Springer Science+Business Media: [41])

Although barriers remain to the universal application of the NOTES and LESS approaches for performing abdominal procedures, continued developments in robotic technologies promise to provide an improved platform for intuitive visualization and dexterous tissue manipulation to better enable complex procedures using these less invasive approaches. Continuing work to the miniature in vivo robotic platform includes improving robot dexterity and speed while also reducing the size of the robot.

References

1. Kalloo, A.N., Rattner, D., Brugge, W. et al.: ASGE/SAGES working group on natural orifice translumenal endoscopic surgery white paper. Gastrointest. Endosc. **62**, 199–203 (2005)
2. Ko, C.W., Kalloo, A.N.: Per-oral transgastric abdominal surgery. Chin. J. Dig. Dis. **7**, 67–70 (2006)
3. Tracy, C.R., Raman, J.D., Cadeddu, J.A., Rane, A.: Laparoendoscopic single-site surgery in urology: where have we been and where are we heading? Nat. Clin. Pract. Urol. **5**, 561–568 (2008)
4. Tendick, F., Jennings, R., Tharp, G., Stark, L.: Perception and manipulation problems in endoscope surgery. In: Taylor, R.H., Lavallee, S., Burdea, G.C., Mosges, R. (eds.) Computer Integrated Surgery: Technology and Clinical Applications. MIT Press, Cambridge (1996)

5. Treat, M.: A surgeon's perspective on the difficulties of laparoscopic surgery. In: Taylor, R.H., Lavallee, S., Burdea, G.C., Mosges, R. (eds.) Computer Integrated Surgery: Technology and Clinical Applications. MIT Press, Cambridge (1996)
6. Satava, R.M.: Surgical robotics: the early chronicles. Surg. Laparosc. Endosc. Percutan. Tech. **12**, 6–16 (2002)
7. Gettman, M.T., Box, G., Averch, T. et al.: Consensus statement on natural orifice transluminal endoscopic surgery and single-incision laparoscopic surgery: heralding a new era in urology? Eur. Urol. **53**, 1117–1120 (2008)
8. Bergstrom, M., Ikeda, K., Swain, P., Park, P.O.: Transgastric anastomosis by using flexible endoscopy in a porcine model (with video). Gastrointest. Endosc. **63**, 307–312 (2006)
9. Kalloo, A.N., Singh, V.K., Jagannath, S.B., Niiyama, H., Hill, S.L., Vaughn, C.A., Magee, C.A., Kantsevoy, S.V.: Flexible transgastric peritoneoscopy: a novel approach to diagnostic and therapeutic interventions in the peritoneal cavity. Gastrointest. Endosc. **60**, 114–117 (2004)
10. Lima, E., Henriques-Coelho, T., Rolanda, C., Pego, J.M., Silva, D., Carvalho, J.L.: Transvesical thoracoscopy: a natural orifice translumenal endoscopic approach for thoracic surgery. Surg. Endosc. **21**, 854–858 (2007)
11. Merrifield, B.F., Wagh, M.S., Thompson, C.C.: Peroral transgastric organ resection: a feasibility study in pigs. Gastrointest. Endosc. **63**, 693–697 (2006)
12. Bessler, M., Stevens, P.D., Milone, L., Parikh, M., Fowler, D.: Transvaginal laparoscopically assisted endoscopic cholecystectomy: a hybrid approach to natural orifice surgery. Gastrointest. Endosc. **6**, 1243–1245 (2007)
13. Rao, G.V., Reddy, D.N., Banerjee, R.: NOTES: human experience. Gastrointest. Endosc. Clin. N. Am. **18**, 361–370 (2008)
14. USGI Medical (2007) USGI announces first NOTES transgastric cholecystectomy procedures using the USGI endosurgical operating system performed by Dr. Lee Swanstrom at Legacy Hospital in Portland, OR. Updated 2007. http://www.usgimedical.com/news/releases/062507.htm. Retrieved 17 Sep 2007
15. Zorron, R., Filgueiras, M., Maggioni, L.C., Pombo, L., Carvalho, G.L., Oliveira, A.L.: NOTES transvaginal cholecystectomy: report of the first case. Surg. Innov. **14**, 279–283 (2007)
16. Esposito, C.: One-trocar appendectomy in pediatric surgery. Surg. Endosc. **12**, 177–178 (1998)
17. Piskun, G., Rajpal, S.: Transumbilical laparoscopic cholecystectomy utilizes no incisions outside the umbilicus. J. Laparoendosc. Adv. Surg. Tech. A **9**, 361–364 (1999)
18. Kommu, S.S., Rane, A.: Devices for laparoendoscopic single-site surgery in urology. Expert Rev. Med. Devices **6**, 95–103 (2009)
19. Desai, M.M., Rao, P.P., Aron, M. et al.: Scarless single port transumbilical nephrectomy and pyeloplasty: first clinical report. BJU Int. **101**, 83–88 (2008)
20. Rane, A., Rao, P., Rao, P.: Single-port-access nephrectomy and other laparoscopic urologic procedures using a novel laparoscopic port (R-port). Urology **72**, 260–263 (2008)
21. Gumbs, A.A., Milone, L., Sinha, P., Bessler, M.: Totally transumbilical laparoscopic cholecystectomy. J Gastrointest. Surg. **13**, 533–534 (2009)
22. Hodgett, S.E., Hernandez, J.M., Morton, C.A., et al.: Laparoendoscopic single site (LESS) cholecystectomy. J Gastrointest. Surg. **13**, 188–192 (2009)
23. Phee, L., Accoto, D., Menciass, A. et al.: Analysis and development of locomotion devices for the gastrointestinal tract. IEEE Trans. Biomed. Eng. **49**, 613–616 (2002)
24. Breedveld, P., van der Kouwe, D.E., van Gorp, M.A.J.: Locomotion through the intestine by means of rolling stents. In: ASME Design Engineering Technical Conference, Salt Lake City, UT, pp. 963–969 (2004)
25. Menciassi, A., Dario, P.: Bio-inspired solutions for locomotion in the gastrointestinal tract: background and perspectives. Philos. Trans. R. Soc. Lond. **361**, 2287–2298 (2003)
26. Abbott, D.J., Becke, C., Rothstein, R.I., Peine, W.J.: Design of an endoluminal NOTES robotic system. In: IEEE/RSJ International Conference on Intelligent Roots and Systems, San Diego, CA (2007)

27. Phee, S.J., Low, S.C., Sun, Z.L. et al.: Robotic system for no-scar gastrointestinal surgery. Int. J. Med. Robot. Comput. Assist. Surg. **4**, 15–22 (2008)
28. Swanstrom, L.L., Whiteford, M., Khajanchee, Y.: Developing essential tools to enable transgastric surgery. Surg. Endosc. **22**, 600–604 (2008)
29. Kaouk, J.H., Goel, R.K., Haber, G.P. et al.: Robotic single-port transumbilical surgery in humans: initial report. BJU Int. **103**, 366–369 (2009)
30. Park, S., Bergs, R.A., Eberhart, R., et al.: Trocar-less instrumentation for laparoscopy: magnetic positioning of intra-abdominal camera and retractor. Ann. Surg. **245**, 379–384 (2007)
31. Scott, D.J., Tang, S.J., Fernandez, R. et al.: Completely transvaginal NOTES cholecystectomy using magnetically anchored instruments. Surg. Endosc. **21**, 2308–2316 (2007)
32. Hogle, N.J., Hu, T., Allen, P.K., Fowler, D.L.: Comparison of monoscopic insertable, remotely controlled imaging device with a standard laparoscope in a porcine model. Surg. Innov. **15**, 271–276 (2008)
33. Hu, T., Allen, P.K., Nadkami, T., et al.: Insertable stereoscopic 3D surgical imaging device with pan and tilt. In: 2nd Biennial IEEE?RAS-EMBS International Conference on Biomedical Robotics and Biomechantronics, Scottsdale, AZ, pp. 311–317 (2008)
34. Rentschler, M.: In vivo abdominal surgical robotics: tissue mechanics modeling, robotic design, experimentation and analysis. University of Nebraska-Lincoln thesis, Biomedical Engineering (2006)
35. Rentschler, M., Dumpert, J., Platt, S. et al.: Modeling, analysis, and experimental study of in vivo wheeled robotic mobility. IEEE Trans. Robot. **22**, 308–321 (2006)
36. Lehman, A.C., Berg, K.A., Dumpert, J. et al.: Surgery with cooperative robots. Comput. Aided Surg. **13**, 95–105 (2008)
37. Rentschler, M.E., Dumpert, J., Platt, S.R. et al.: Natural orifice surgery with an endoluminal mobile robot. Surg. Endosc. **21**, 1212–1215 (2006)
38. Lum, M.J.H., Trimble, D., Rosen, J., et al.: Multidisciplinary approach for developing a new minimally invasive surgical robotic system. In: The first IEEE/RAS-EMBS International Conference on Biomedical Robotics and Biomechatronics, pp. 841–846 (2006)
39. Rosen, J., Brown, J.D., Chang, L., et al.: The BlueDRAGON – a system for measuring the kinematics and dynamics of minimally invasive surgical tools in vivo. In: IEEE International Conference on Robotics and Automation, Washington DC, pp. 1876–1881 (2002)
40. Lehman, A.C., Wood, N.A., Dumpert, J., et al.: Dexterous miniature in vivo robot for NOTES. In: 2nd Biennial IEEE/RAS-EMBS Conference on Biomedical Robotics and Biomechatronics, Scottsdale, AZ, pp. 244–249 (2008)
41. Lehman, A.C., Dumpert, J., Wood, N.A. et al.: Natural orifice cholecystectomy using a miniature robot. Surg. Endosc. **23**, 260–266 (2009)
42. Lehman, A.C., Rentschler, M.E., Farritor, S.M., Oleynikov, D.: The current state of miniature in vivo laparoscopic robotics. J. Robot. Surg. **1**, 45–49 (2007)

Chapter 7
A Compact, Simple, and Robust Teleoperated Robotic Surgery System

Ji Ma and Peter Berkelman

Abstract The utility of current commercial teleoperated robotic surgery systems is limited by their high cost, large size, and time-consuming setup procedures. We have developed a prototype system which aims to overcome these obstacles by being much smaller, simpler, and easier to set up and operate, while providing equivalent functionality and performance for executing surgical procedures. The prototype system is modular and each component manipulator is approximately 2.5 kg or less, so that they system is easily portable and each manipulator can be individually positioned and fixed in place by hand to a rigid frame above the operating table. All system components and materials are autoclaveable and immersible in fluids, so that each manipulator can be sterilized and stored by the standard operating procedures used for any other surgical instrument, and no sterile draping is required. The system is described and results of untrained user trials performing standard laparoscopic surgery skill tasks are given.

7.1 Introduction

The development of minimally invasive surgical techniques has been a great benefit to patients due to reduced trauma and risk of infection when compared to open surgical procedures. Minimally invasive surgery is much more difficult for the surgeon, however, as dexterity is reduced when handling long, thin instruments through a keyhole incision, and visibility is reduced by the necessity of using a rigid endoscope and video monitor to display the internal tissues and instruments to the surgeon during the procedure.

Teleoperated robotic surgical systems aim to regain and enhance the dextrous capabilities of surgeons to perform minimally invasive procedures by appropriate scaling between motions of the teleoperation masters and slaves, reducing the

P. Berkelman (✉)
Department of Mechanical Engineering, University of Hawaii-Manoa,
2540 Dole St, Honolulu, HI 96822, USA
e-mail: peterb@hawaii.edu

J. Rosen et al. (eds.), *Surgical Robotics: Systems Applications and Visions*,
DOI 10.1007/978-1-4419-1126-1_7,

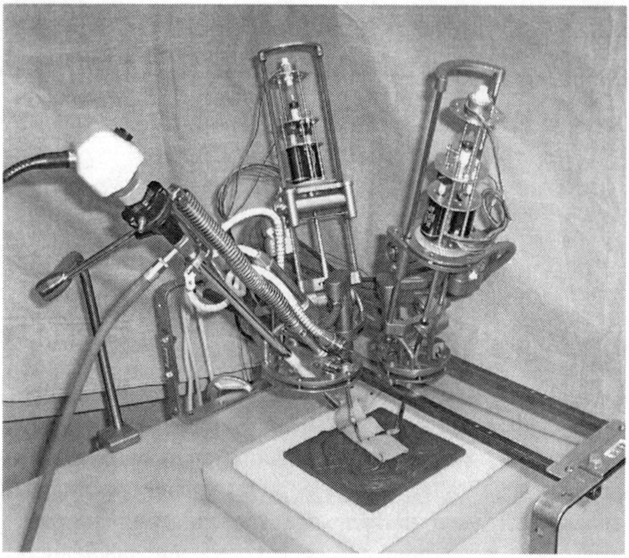

Fig. 7.1 Endoscope and instrument manipulators

manual tremor of the surgeon, and enabling direct control of instrument tips by eliminating the reversed motions between manual minimally invasive surgery instrument handles and tips caused by the instruments pivoting about a fulcrum at the incision point. Commercial systems are being used in a steadily increasing number of hospitals, however in practice their use is often limited to specific cases and procedures, due to complex, costly, and time-consuming setup and maintenance procedures.

At the University of Hawaii we have developed a compact prototype system which aims to address the shortcomings of the first generation of robotic surgery systems by being much smaller, simpler, and easier to set up and use than current commercial systems. The manipulators are shown attached to a rigid frame clamped above a table in Fig. 7.1. The most important features of our system are that it is autoclaveable and immersible in fluids for sterility and cleaning, its small size for easy setup and use, and its modular design so that any component of the system may be easily added, removed, or replaced at any time during procedures.

7.1.1 Disadvantages of Current Systems

The most widely used teleoperated robotic system for minimally invasive surgery is currently the *da Vinci*, from Intuitive Surgical Systems [5]. Although it is commercially successful and approved for a number of surgical procedures, it is large and costly, the layout of the operating room must be reconfigured to accommodate its size, and specialized procedures are necessary for the setup and use of the system. The complete system may be difficult to fit into standard operating rooms, and typically an entire room must be permanently devoted to robotic system regardless of its frequency of use, as it is too massive to be regularly moved between rooms. All parts of the robotic system within the operating field must be carefully enclosed in sterile drapes before each procedure, which adds to the setup time. Furthermore, it may be necessary to replan standard procedures so that the external arms carrying the surgical instruments do not collide. The large size of the surgical manipulators partially obstruct the access and view of the patient by the surgical staff. An immersive teleoperation master console with a stereo display provides a high degree of telepresence for operating instruments inside the patient, but peripheral awareness of the exterior condition of the patient is lost due to the physical separation and the immersive display.

The *ZEUS* from Computer Motion [13], which is currently unavailable due to merging with Intuitive Surgical, was smaller than the da Vinci yet its large manipulator bases attached to the operating table and its use of serial jointed robotic arms to manipulate instruments also somewhat reduces free access to the patient from all directions by the surgical staff and may require careful preplanning and positioning to avoid collisions between the arms during a given procedure. The *Laprotek* from [3] system from Endovia and the *RAVEN* [8] developed at the University of Washington are also both somewhat smaller and less obtrusive than the da Vinci and ZEUS systems. These two systems mainly rely on actuation by cable tendons, which is a possible source of difficulty due to wear, stretching, and variable tension and friction.

7.1.2 Advantages of Smaller, Simpler Modular Systems

Reducing the size and complexity of a robotic surgery system provides many complementary advantages. First, small, lightweight instrument and endoscope manipulators permit them to be fixed directly above the patient, allowing full access to the patient from all sides and not occupying any floor space in the operating room. Each manipulator can be easily positioned with one hand and clamped in place on a rigid frame, and is small enough to fit inside a typical hospital autoclave and stored in a cabinet. Individual manipulators can easily be added, removed, or replaced in the system at any time during surgical procedures, to switch between different types of instruments, to manually operated instruments, or to convert to an open procedure in only minutes.

The small size and light weight of the mechanisms also result in reductions in the motor torques required to counteract gravity loads. With lower motor torque requirements, miniature brushless motors may be used, integrated directly into the manipulator mechanism with a minimal amount of gear reduction and drivetrain components. The use of miniature motors further reduces the total mass of the manipulators, and the simplicity of the drivetrains improves the reliability of the manipulators as the number of potential failure points is reduced, no lubrication or other maintenance procedures are necessary, and stretching and wear of actuation cables are not a cause for concern.

Finally, the reduced mass and motor torque requirements in the manipulators improve the safety of the system. Any unintended collisions between manipulators or impacts with patient anatomy are less likely to cause damage, as the kinetic energy from the moving mass of the manipulators would be more easily dissipated.

7.1.3 University of Hawaii System Components

A schematic diagram of the compact laparoscopic surgery robot system is shown in Fig. 7.2, showing all the components of the system and the communications links between them. The components of the compact surgery robot system include the following:

Teleoperation Masters:	The masters receive the surgeons hand motions then change these motions into position signals. The masters use PHANToM Omni haptic device which can detect the motion of the surgeons hand. Miniature brushless DC motors are used.[1]
Instrument Manipulators:	The instrument manipulator follows the masters motion signal to realize translation and rotation motion of surgical instrument which is attached on the manipulator.
Surgical Instruments:	The surgical instruments provide serials of tools for surgery. The surgical instrument can be attached to or detached from the instrument manipulator rapidly during instrument exchange in surgical procedure.
Endoscope Manipulator:	The structure of endoscope manipulator is same as the instrument manipulator except its lower force/torque requirement for moving endoscope and one less DOF. The endoscope manipulator moves according to voice command or foot pedal input of the surgeon or moves semi-autonomously to trace the object in the surgical site.
Motor Controllers:	Motor controllers are used to drive the DC brushless motors in the manipulators and the instruments.[2]
Video Feedback:	A video signal is acquired from the endoscope and the video feedback is provided by a monitor for the surgeon and nurses to observe the surgical site. In the current surgery robot protope, an endoscope

[1] MicroMo Brushless DC.

[2] MicroMo MCBL 3003.

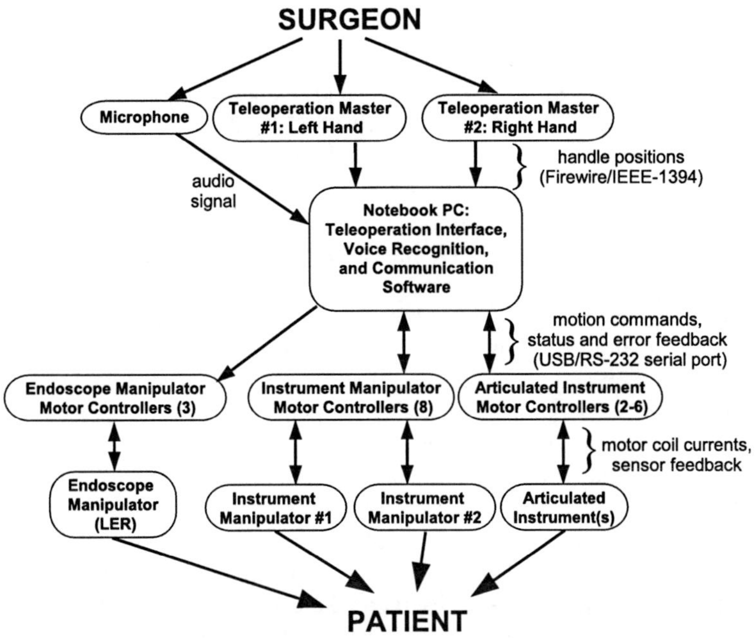

Fig. 7.2 Schematic of communication between modular components of surgical robot system prototype

	and a TV monitor are used to provide the surgeon with a 2-D video feedback.
Endoscope Command:	Since the surgeons hands are occupied by two masters, a voice command recognition system and a set of foot switches are used to control the endoscope manipulator.
Central Controller:	The control software includes hardware device drivers, control software and the user interface, and executes on a PC in a compact enclosure with a touchscreen and running Windows XP. The system control software is described in further detail in [9].

A second enclosure contains additional motor controllers, power supplies, and a multichannel serial port output card to send commands from the high-level PC control to individual motor controllers. The entire system is modular and portable and any of the components may be added, removed, replaced at any time, converting between teleoperated minimally invasive surgery, manual minimally invasive surgery, or to open procedures. Total setup time from packing cases to operation is less than 15 min. The endoscope and instrument manipulators, and the teleoperated instruments described in detail in Sects. 2, 3, and 4.

7.2 Endoscope Manipulator

Early development and testing of the endoscope manipulator used in our teleoperated system was carried out at the TIMC-IMAG laboratory of Grenoble, France, and is described in detail in [2]. The current version has been commercialized by EndoControl S.A. of Grenoble as the *ViKY* (Vision Kontrol EndoscopY) [6] and is shown in Fig. 7.3. The ViKY system has undergone extensive human trials with over 300 procedures performed as a robotic assistant to position and hold an endoscope during minimally invasive surgical procedures performed manually. The device has received CE marking for use in Europe, and was approved by the FDA in December 2008 for use in the United States.

The endoscope manipulator consists of an annular base placed on the abdomen, a clamp to hold the endoscope trocar, and two joints which enable azimuth rotation and inclination of the endoscope about a pivot point centered on the incision. A compression spring around a telescoping shaft with an internal cable which is wound around a motorized spool control the insertion depth of the endoscope. Control of the robot is simple and straightforward, as the motion of each motor directly corresponds to the horizontal and vertical motion and zoom of the endoscope camera image. As a result, no kinematic calculations, initialization procedures, or homing sequences need to be performed to operate the robot. No calibration procedure is necessary and the manipulator is ready to be used immediately after being powered on. The endoscope can be quickly removed at any time for cleaning or replacement, and the robot can be removed while leaving the trocar in place in the abdomen. A set of foot switches and a voice command recognition system are available as user control interfaces to enable the surgeon to control the endoscope manipulator motions while holding surgical instruments in both hands.

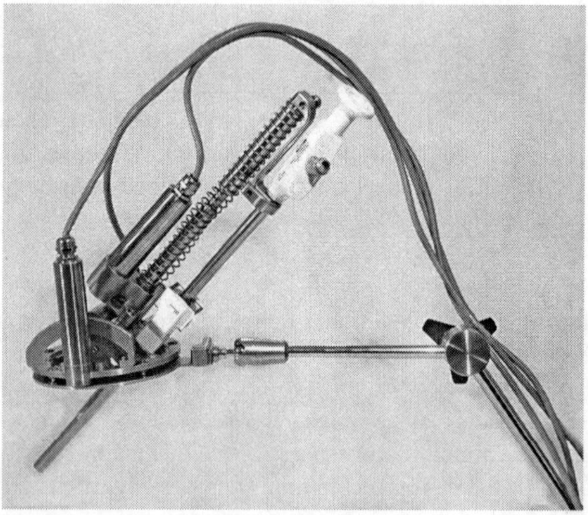

Fig. 7.3 ViKY endoscope manipulator

The manipulator is approximately 800 g. Rigid laparoscopic endoscopes with video cameras are typically 300–500 g. The base diameter is 110 mm and the minimum height is approximately 100 mm.

To meet the sterility and hygiene requirements of the operating room without using sterile drapes, all components are autoclaveable and waterproof. In addition, the surfaces of all moving parts must be accessible to cleaning brushes for rapid cleaning after procedures without using tools for disassembly. To satisfy this requirement, the rotating ring in the base of the manipulator is supported on a set of smooth, rotating pinions rather than a ball bearing.

Earlier commercial endoscope manipulators include the *Aesop* [4] from Computer Motion Inc., and the *EndoAssist* [1] from Prosurgics, formerly Armstrong Healthcare. Both of these manipulators are serial jointed robotic arms, with large massive bases and passive gimbal joints where the endoscope is attached, to allow the endoscope to pivot about the incision point.

7.3 Instrument Manipulation

The basic structure and design of the instrument manipulators in the system are similar to the endoscope manipulator described in the previous section. However the performance requirements of instrument manipulators are more demanding with respect to accuracy, sensitivity, response time, stiffness, and forces generated, and an additional actuated degree of freedom is required to rotate the instrument shaft.

The instrument manipulators move a platform to which various different surgical instruments may be attached. Teleoperation masters control the motions of the instrument manipulators and instruments together. The instrument manipulators and teleoperation masters are described in this section and the articulated robotic surgical instruments developed for the instrument manipulators are described in the next section.

7.3.1 Instrument Manipulators

An instrument manipulator is shown in Fig. 7.4 holding a manual minimally invasive surgical instrument. Figure 7.5 shows a manipulator with no attached instrument.

A rack-and-pinion drive is used in the instrument manipulators to control the instrument insertion depth instead of the cable-and-spring mechanism of the endoscope manipulator due to the greater precision and forces required in the instrument manipulators. The rack-and-pinion motor and a fourth motor to rotate the instrument shaft are both built into the upper platform of the instrument manipulators. The base of the instrument manipulators is smaller in diameter than the endoscope

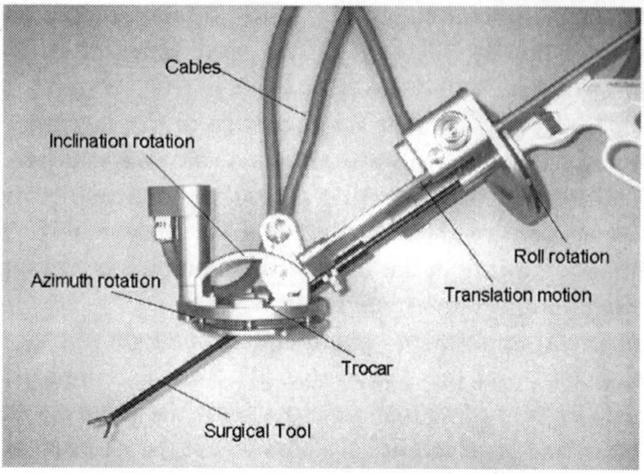

Fig. 7.4 Instrument manipulator holding manual instrument

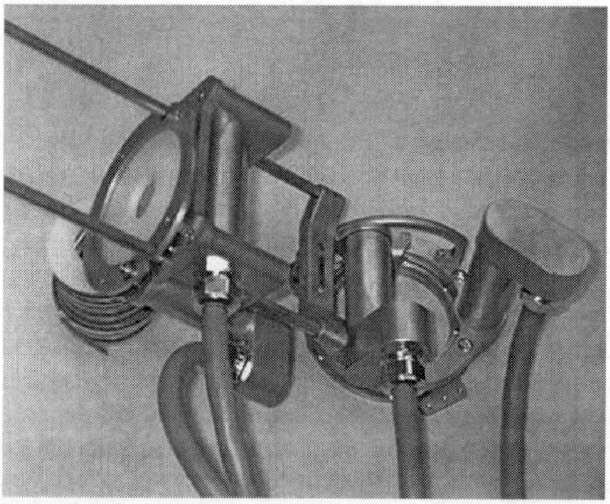

Fig. 7.5 Instrument manipulator without instrument

manipulator in order to more easily place multiple robots next to each other on the abdomen of the patient.

If the motors are disabled or powered off, their motions are backdriveable, which is useful during initial device setup or in case of a motor failure. Each manipulator is particularly compact and lightweight at 1.8 kg, 100 mm in diameter and 360 mm in height.

A pair of commercial haptic interfaces as teleoperation masters [11] A grasper handle with gripping force feedback, shown in Fig. 7.6, is used in place of the original stylus to provide the user a more ergonomic operation of the wrist and more precise and delicate gripper force control which can avoid tissue trauma caused by large forces from the gripper. In the initial development, the two buttons on the handle of the haptic device were used to control the gripper to open and close. But

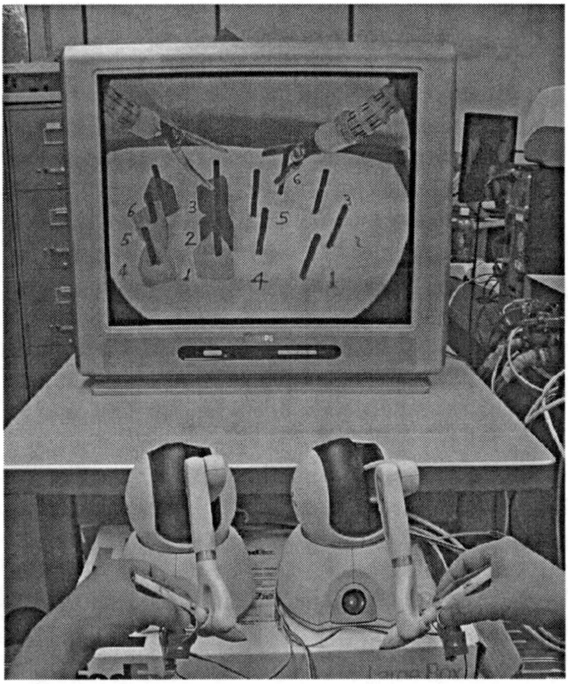

Fig. 7.6 Teleoperation master console

we found induced motions in the system are caused by the forces required to press the buttons. Precise opening and closing of a gripper or scissors is difficult to control precisely using these on/off buttons only.

7.4 Articulated Wrist Instruments

We have produced instruments with articulated, motorized wrists at the active ends to improve surgical dexterity compared to standard manual instruments in minimally invasive surgery. The ability to bend an instrument tip up to 90° in any direction, teleoperated from the master console, enables the tip to approach and contact tissues from a wide range of angles, can avoid occlusions and obstructions, and provides better access for gripping and cutting, and especially suturing, in which the tip of a needle must follow a helical path. The use of articulated wrists in our system provides the full teleoperated robotic surgical assistance functionality equivalent to current commercial systems. Although the current prototype instrument wrists in development have not yet been tested to withstand repeated autoclave cycles, all the motors and materials used in the instrument wrists are commercially available in autoclaveable versions.

7.4.1 Wire-Driven Flexible Spines

Instead of using cable-driven rotational joints in the instrument wrists, our instruments use a wire-driven flexible spine for bending, in an approach similar to Van Meer [12]. This actuation method is easier to fabricate, does not require small bending radii in actuation cables, and is free from kinematic singularities.

The articulated instruments provide an additional two degrees of freedom in wrist motion to increase dexterity and a single degree of freedom for tool actuation. In Fig. 7.7, four wrist plates (1.8 mm height and 6 mm diameter) and three spheres (3mm diameter) are stacked together with each sphere between two plates. The wrist plates have drilled holes for eight Nitinol wires (0.33 mm diameter) to pass through. Four Nitinol wires, actuated in antagonistic pairs, are used for driving the wrist rotations and the other four wires are passive to provide the wrist more axial rigidity.

Figure 7.8 shows the fabricated wrist with a gripper tip. The total length of the flexible part of the wrist is 12 mm. A conventional manual surgical tool tip (5 mm diameter) is attached to the end of the wrist. The opening and closing of the gripper is driven by a ninth superelastic Nitinol alloy wire (0.43 mm diameter) which passes through the pierced spheres and wrist plates. The wrist bending motion range is 90 degrees in all directions. The diameter of the instrument is 6 mm. The materials of the wrist are biocompatible, low cost, and autoclaveable. During instrument exchange in a surgery procedure, different instruments can be detached from and attached to the instrument manipulator quickly. In the current prototype as shown in Fig. 7.9, the driving motors and the instrument are assemble together. The total length of the modular surgical instrument prototype is 450 mm and the weight is 380 g.

Miniature brushless DC are used for actuation of the two perpendicular wrist bending directions and for the instrument gripper or scissor. To convert the motor shaft rotations to the linear motions required for wire actuation and increase their force capabilities, miniature worm gear assemblies are used for the antagonistic

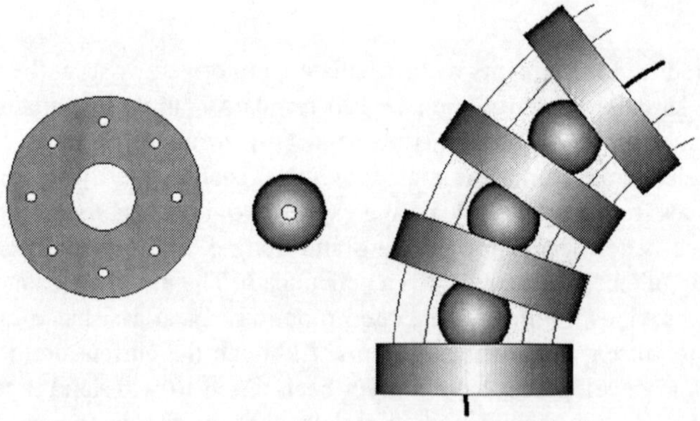

Fig. 7.7 Flexible spine structure for articulated instrument wrist

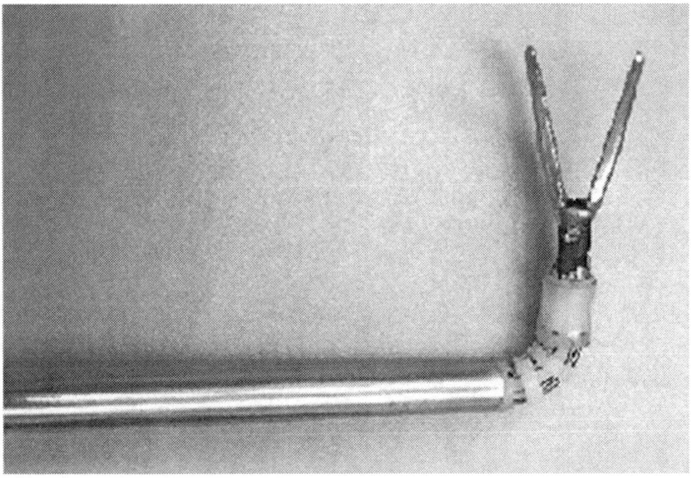

Fig. 7.8 Surgical instrument wrist and gripper

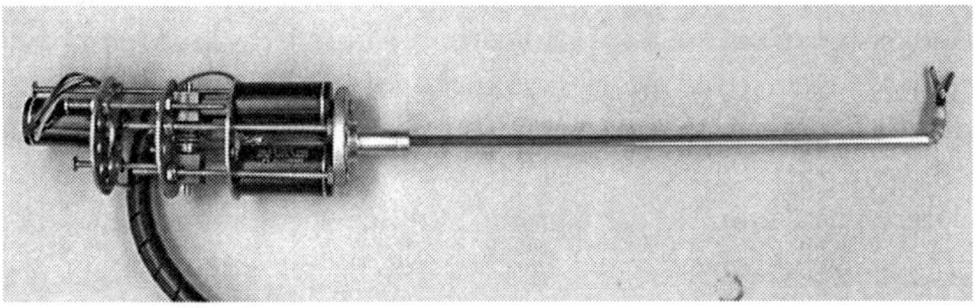

Fig. 7.9 Complete surgical instrument

actuation of wrist bending and a miniature lead screw is used for the gripper/scissor actuation. The motors and gear reductions at the top of a robotic instrument are shown in Fig. 7.10.

7.4.2 Grip Force Feedback

In typical manual laparoscopic surgery, surgeons can feel approximate contact, gripping, and cutting forces from instrument and tissue interactions, but the force sensations are masked by friction and backlash in the manual instrument handle and the trocar. In current robotic surgery without force feedback, the surgeons have no contact sensation from their hand and must judge the forces applied to tissue based on the tissue deformation seen in the video feedback and their surgical experience.

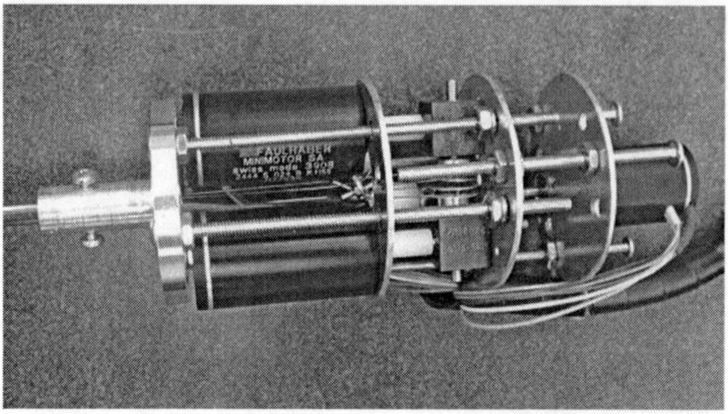

Fig. 7.10 Actuation motors and reduction mechanisms at head of instrument

Aside from the force information from instrument and tissue interaction, the grasping force of the gripper is also important because too much grasping force will damage tissue and it is difficult to judge the grasping force from tissue deformation. It is difficult to sense forces at the tip of the surgical instrument due to the miniature size, the harsh environment inside the body, and biocompatibility and sterility requirements. In the design of our surgery instrument, we installed a load sensor between the gripper motor lead screw and the actuation wire to indirectly measure grasping force. This measure is approximate, however, as the measured friction between the driving wire and the wrist varies from 0.5 N to 2 N as the wrist bends from 0° to 90°.

A miniature voice coil,[3] linear bearing, and linear potentiometer were added to the gripper control mechanism on the handles of the teleoperation masters, as shown in Fig. 7.11. The voice coil can provide a continuous force of 2.3 N and peak forces up to 7.4 N. Due to the friction and hysteresis in the motion gripper actuation wire, it is difficult to provide user force feedback which is both stable and useful to the user for gripping tasks. Further investigation and testing is necessary to filter the sensed force signal and compensate for friction to stably provide realistic grip force feedback to the teleoperator.

7.5 User Trials

The system has undergone sets of preliminary trials in the laboratory to compare its performance to manual laparoscopic surgery and published results from other systems when operated by untrained users. Student volunteers were recruited to perform standardized surgical skills tasks including placing rings on pegs with a single instrument, tracing motion trajectories, peg transfers with two instruments,

[3] MotiCont Inc.

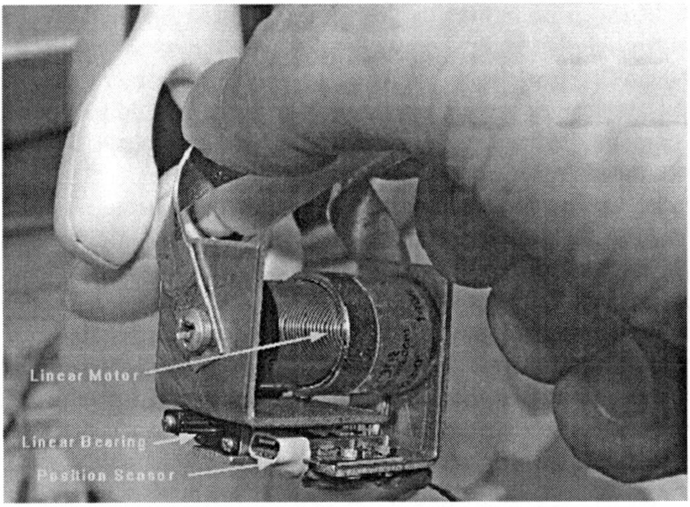

Fig. 7.11 Grip force feedback actuator assembly in teleoperation handle

suturing with knot tying, and precision cutting. The two tasks Sect. 5.1 were performed with instruments with fixed straight wrists, to simplify the teleoperation interface and reduce the training and familiarization period for the operators, and the tasks in Sect. 5.2 were performed with the articulated wrists as described in Sect. 4.

7.5.1 Comparison to Manual Instruments

In order to compare the performance of our surgical robotic system with typical manual MIS instrument operation, user testing experiments were designed and performed. Two tasks were performed by participants using both the manual MIS instrument and the teleoperated robotic system. The positions of the instrument incision point, the task object, the monitor and the user were arranged to be the same for both the manual and robot instrument operations. An optical motion capture system[4] was used to capture the motions of the surgical instrument tips for both the manual and the robot task operations. Further detail regarding these comparative user experiments is provided in [10].

Four infrared LEDs were affixed to the handle of each surgical instrument to enable tracking of the instrument tip position at a 30 Hz sample rate. This motion tracking technique accounts for all motion from hysteresis, vibration, and deformation in the manipulator support clamps, support frame, and mechanism, except for flexing of the instrument shaft between the handle and tip.

[4] Optotrak Certus, Northern Digital Inc.

Before the experiments, each participant was allowed 2–15 min to practice until the participant felt that he or she was familiar with the operations. In the experiments, the novice participants preferred to spend much more time practicing the manual surgical instrument operation than in the robotic teleoperation. The practice time for participants in manual operation was about 5–10 min while the practice time in robotic operation was only about 1–2 min.

Two tasks were used to compare manual laparoscopic instrument operation and teleoperated robotic operation. Each task required one hand only.

Task 1: Pick and Place Use the gripper instrument to pick up nine rings and place them on pegs. The pegs are arranged in a grid with 20 mm separation and the rings were lined up against the edge of the tray containing the pegs before each task. The execution time results are shown in Fig. 7.12, indicating that manual instrument operation takes more time than robot teleoperation except for subjects 5 and 6. It was observed that in teleoperation, the motion of the instrument tips paused while grasping the rings, but in manual operation rings could be easily grasped while moving the instrument. The average grasping time was approximately 21% of total robot teleoperation time.

Task 2: Trajectory Following Use the gripper instrument to follow the given trajectory indicated on a horizontal plane. A 20 mm square was plotted on graph paper as the given trajectory. The participants were asked to control the tip of the surgical gripper instrument to move along the square trajectory. Each participant was asked to follow the trajectory in the clockwise direction for five circuits. Figures 7.13 and 7.14 are sample instrument tip trajectories for manual and

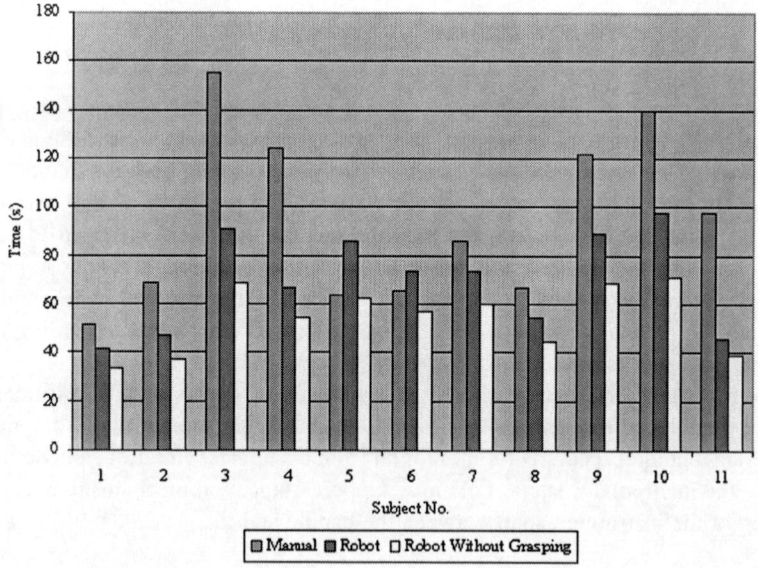

Fig. 7.12 Manual and teleoperated ring placement task completion times

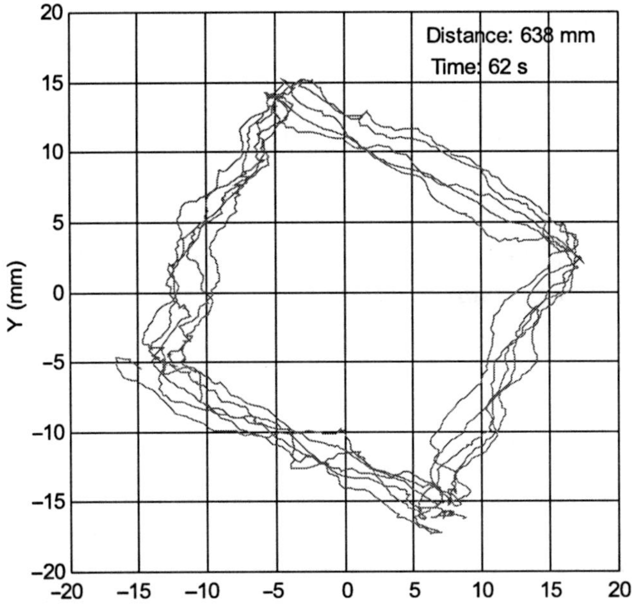

Fig. 7.13 Manual instrument tip trajectory

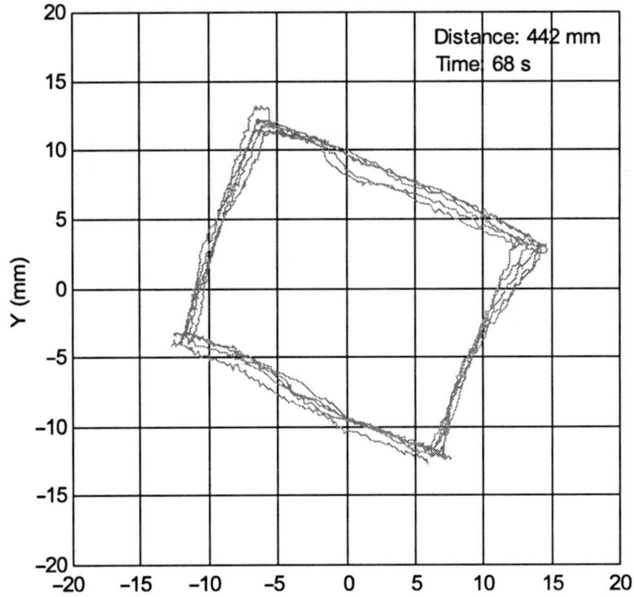

Fig. 7.14 Teleoperated instrument tip trajectory

teleoperated instruments. The trajectory of the manual operation is much coarser due to hand tremors and vibrations. In the teleoperation data, although there are some tremors from the teleoperation master device, the trajectory is much more accurate and smoother than manual instrument operation.

7.5.2 Comparison to Results of Other Robotic Surgery Systems

The Fundamentals of Laparoscopic Surgery (FLS) curriculum is a standard surgery teaching and training curriculum which was created in the late 1990s by the Society of American Gastointestinal Endoscopic (SAGES). The FLS curriculum includes both cognitive as well as psychomotor skills. The FLS tasks have been used to teach, practice and evaluate the skills of thousands of surgeons. The FLS curriculum provides a structured and repeatable means to evaluate surgical skills, and the use of the FLS tasks in the field of surgical robotics can allow researchers from different groups to have a common basis for objective evaluation and comparison of their systems. Sample FLS tasks were performed by volunteers to evaluate our surgery robot system.

There are five tasks in the FLS curriculum: peg transfer, precision cutting, placement and securing of ligating loop, simple suture with extracorporeal knot, and simple suture with intracorporeal knot. Two of the tasks, placement/securing of a ligating loop and extracorporeal knot tying, require special tools and therefore these two tasks are not included in the current evaluation. The other three FLS tasks are the peg transfer, precision cutting, and simple suture with intracorporeal knot tasks. Standard commercial surgical training task kits were used.[5] In the main study, only the peg transfer task is used because the task is relatively easy for untrained novice users. The precision cutting and suture task were performed by the first author who has some experience in teleoperating the surgical robot system.

In the peg transfer task as shown in Fig. 7.15, a peg board with 12 pegs and 6 rings are manipulated by two gripper instruments. The user is asked to use the left gripper to grasp a ring from the left peg board, transfer the ring to the right gripper in midair, then place the object on the right peg board. After all six rings are transferred to the right peg board, the process is reversed, and the user transfers all the rings to the left peg board. In this task, the total number of object transfers is twelve, including six transfers from left to right and six transfers from right to left.

In the suturing and knot tying task, a Penrose drain is attached on a foam block as shown in Fig. 7.16. The instruments required in this task are two needle drivers. The user is required to place a suture precisely through two marks on the Penrose drain. At least three throws including one double throw and two single throws must be

[5] Simulab Inc.

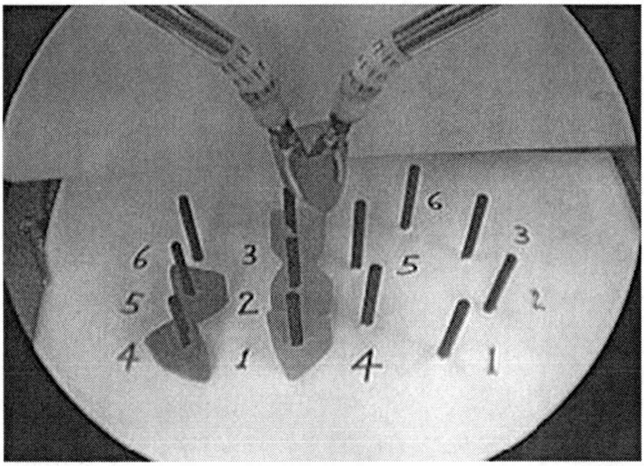

Fig. 7.15 FLS peg transfer task

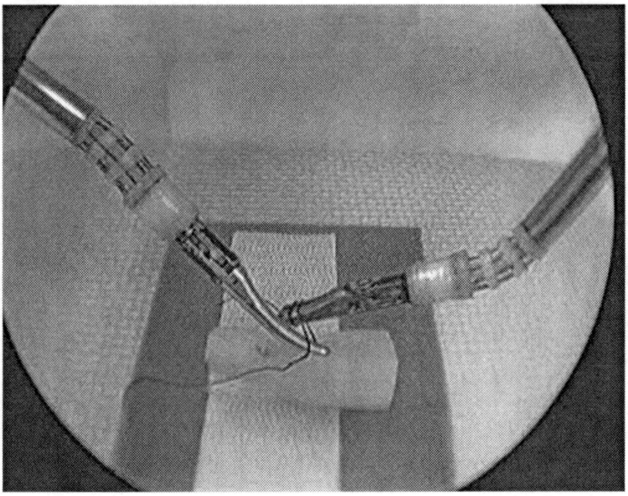

Fig. 7.16 FLS suturing task

placed on the suture. The task requires needle placement, needle transferring, suturing and knot tying skills.

Figure 7.17 shows the cutting task. The tools needed for this task are one gripper and one scissors. A 100 $\times$100 mm gauze piece with a circle pattern is secured in view of the endoscope. The user is asked to cut out the circle along the line precisely. The gripper is used to hold the gauze and to place the gauze the best angle for cutting. This task is designed to use hands in a complimentary manner.

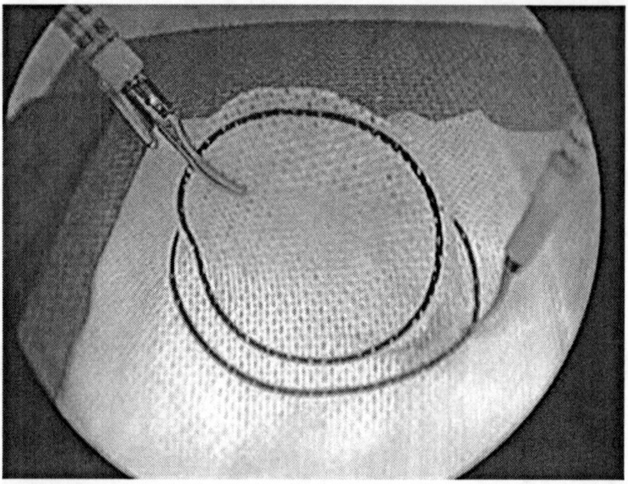

Fig. 7.17 FLS precision cutting task

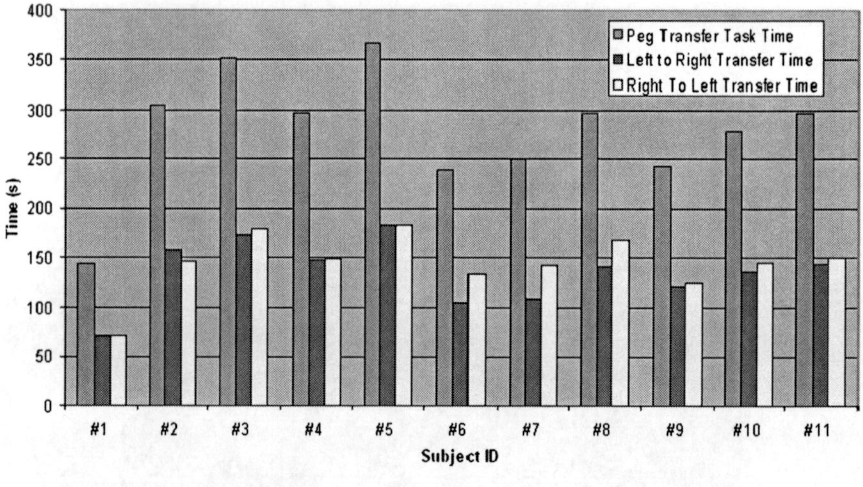

Fig. 7.18 Manual and teleoperated peg transfer task completion times

The results of the peg transfer task executed by volunteer subjects are shown in Fig. 7.18. Each user performed three repetitions of the peg transfer task, and the task times for right-to-left and left-to-right transfers were recorded separately. The cutting and suture with knot tasks have been performed by the first author. The average task completion tasks for each of the three tasks are given in Table 7.1 and compared with published results from the *da Vinci* system from Intuitive Surgical Systems and the *RAVEN* system from the University of Washington [7]. As the number of users is small in each case and variations in experience levels are

Table 7.1 Comparison with other robotic FLS task completion times

Robot system	Peg transfer task	Suture task	Cutting task
da Vinci, Intuitive Surgical Systems	126 s	169 s	208 s
RAVEN, University of Washington	443 s	578 s	–
University of Hawaii	279 s	214 s	280 s

not accounted for, these comparisons are not necessarily indicative of the relative performance or ease of use of each system, however.

7.6 Conclusion

The prototype system is currently complete and fully functional. The endoscope manipulator is the most mature component of our system, in that it has been separately commercialized, has undergone human trials for use with manual instruments, and has obtained CE marking for use in Europe and FDA approval for use in the United States. The instrument manipulators were fabricated with a similar design, components, and materials, so that they are also autoclaveable and immersible in fluids. The teleoperated instruments currently carried by the instrument manipulators can be made to be autoclaveable and waterproof by encapsulating the actuation mechanism and substituting autoclaveable versions of the motors, wiring insulation, and connectors.

The system is modular and portable, it is regularly transported for demonstrations and testing and can be set up completely in less than 15 min. Preliminary user testing using the complete system has been carried out to validate the utility of the system in a laboratory setting. Further testing and validation remains to be done, by users with surgical training performing more realistic and detailed surgical procedures.

Acknowledgements We are grateful for the assistance of Alpes Instruments SA of Meylan, France for the fabrication of the instrument manipulators, and to EndoControl SA of Grenoble, France, for supplying the endoscope manipulator used in the system. Research support has been provided by the NIH under grant #5R21EB006073, "Development of Compact Teleoperated Robotic Minimally Invasive Surgery" and by the University of Hawaii-Manoa College of Engineering and Department of Mechanical Engineering.

References

1. Aiono, S., Gilbert, J.M., Soin, B., Finlay, P.A., Gordon, A.: Controlled trial of the introduction of a robotic camera assistant (EndoAssist) for laparoscopic cholecystectomy. Surg. Endosc. **16**(9), 1267–1270 (2002)

2. Berkelman, P., Cinquin, P., Boidard, E., Troccaz, J., Letoublon, C., Long, J.A.: Development and testing of a compact endoscope manipulator for minimally invasive surgery. Comput. Aided Surg. **10**(1), 1–13 (2005)
3. Franzino, R.J.: The Laprotek surgical system and the next generation of robotics. Surg. clin. North Am. **83**(6), 1317–1320 (2003)
4. Geis, W.P., Kim, H.C., Brennan, E.J., Jr., McAfee, P.C., Wang, Y.: Robotic arm enhancement to accommodate improved efficiency and decreased resource utilization in complex minimally invasive surgical procedures. In: Medicine Meets Virtual Reality: Health Care in the Information Age, pp. 471–481. San Diego (1996)
5. Guthart, G.S., Salisbury, J.K.: The Intuitive (TM) telesurgery system: Overview and application. In: International Conference on Robotics and Automation, pp. 618–621. IEEE, San Francisco (2000)
6. Long, J.A., Cinquin, P., Troccaz, J., Voros, S., Berkelman, P., Descotes, J.L., Letoublon, C., Rambeaud, J.J.: Development of miniaturized light endoscope-holder robot for laparoscopic surgery. J. Endourol. **21**(8), 911–914 (2007)
7. Lum, M.: Quantitative performance assessment of surgical robot systems: Telerobotic FLS. Ph.D. thesis, University of Washington (2008)
8. Lum, M., Trimble, D., Rosen, J., Fodero, K., II, King, H., Sankaranarayanan, G., Dosher, J., Leuschke, R., Martin-Anderson, B., Sinanan, M.N., Hannaford, B.: Multidisciplinary approach for developing a new minimally invasive surgical robotic system. In: International Conference on Biomedical Robotics and Biomechatronics. IEEE-RAS/EMBS, Pisa (2006)
9. Ma, J., Berkelman, P.: Control software design of a compact laparoscopic surgical robot system. In: International Conference on Intelligent Robots and Systems, pp. 2345–2350. IEEE/RSJ, Beijing (2006)
10. Ma, J., Berkelman, P.: Task evaluations of a compact laparoscopic surgical robot system. In: International Conference on Intelligent Robots and Systems, pp. 398–403. IEEE/RSJ, San Diego (2007)
11. Massie, T.H., Salisbury, J.K.: The phantom haptic interface: A device for probing virtual objects. In: Dynamic Systems and Control, pp. 295–299. ASME, Chicago (1994)
12. Meer, F.V., Giraud, A., Esteve, D., Dollat, X.: A disposable plastic compact wrist for smart minimally invasive surgical tools. In: International Conference on Intelligent Robots and Systems, pp. 919–924. IEEE/RSJ, Alberta (2005)
13. Reichenspurner, H., Demaino, R., Mack, M., Boehm, D., Gulbins, H., Detter, C., Meiser, B., Ellgas, R., Reichart, B.: Use of the voice controlled and computer-assisted surgical system zeus for endoscopic coronary artery surgery bypass grafting. J. Thorac. Cardiovasc. Surg. **118**, 11–16 (1999)

Chapter 8
Raven: Developing a Surgical Robot from a Concept to a Transatlantic Teleoperation Experiment

Jacob Rosen, Mitchell Lum, Mika Sinanan, and Blake Hannaford

8.1 Introduction

For decades surgery and robotics were progressing along two parallel paths. In surgery, minimally invasivesurgery (MIS) revolutionized the way a significant number of surgical interventions were performed. Minimally invasive surgery allows the surgeon to make a few small incisions in the patient, rather than making one large incision for access. This technique allows for significantly faster recovery times, less trauma, and decreased pain medication requirements for the patient.

In robotics, teleoperation integrated the human into robotic systems. Only in the last decade have surgery and robotics reached a level of maturity that allowed safe assimilation between the two in a teleoperation mode for creating a new kind of operating room with the potential for surgical innovation long into the future [1].

A detailed historical overview of surgical robotics is beyond the scope of this chapter. The reader may refer to several published papers, which collectively may provide a comprehensive survey of the field of surgical robotics and its applications in various sub-disciplines of surgery and medicine [2–17]. The remaining of this section will provide a brief overview of key systems and millstones of the research activities in the field of surgical robotics and telesurgery.

One of the earliest applications of robotics in surgery in mid 1980s included a modified Puma 560 which was used as a positioning device to orient a needle for biopsy of the brain on a 52 year-old male [18]. In parallel research efforts at IBM were focused on a bone cutting robot with clinical application in total hip-replacement – a system later know as the ROBODOC [19]. The late 1980s also brought on a revolution in surgical intervention. Jacques Perrisat, MD, from Bordeaux, France presented a

J. Rosen (✉)
Department of Computer Engineering, Jack Baskin School of Engineering,
University of California Santa Cruz, 1156 High Street, Santa Cruz, CA 95064, USA
e-mail: rosen@ucsc.edu

J. Rosen et al. (eds.), *Surgical Robotics: Systems Applications and Visions*,
DOI 10.1007/978-1-4419-1126-1_8, © Springer Science+Business Media, LLC 2011

video clip at SAGES (Society of American Gastrointestinal Endoscopic Surgeons) of the first laparoscopic cholecystectomy (gall bladder removal). Minimally invasive surgery techniques greatly influenced the approaches that roboticists have taken toward robot assisted interventions. These key events set the stage for an introduction of surgical robotics into a clinical setup.

Two ground braking systems were developed in academia in the mid to late 1990s both using a four bar mechanism: the *Silver and the Black Falcon* [20] and the *Laparoscopic Telesurgery Workstation* [21]. Intuitive Surgical Inc. and Computer Motion Inc. both produced commercially available FDA-approved surgical robot systems for MIS. Computer Motion's Zeus surgical robot held a surgical tool on a SCARA-type manipulator. The Intuitive Surgical Inc. da Vinci R uses a an extended parallel 4-bar mechanism. The deVinci synthesized two sub systems: (1) The surgeon console (master) including a 3D vision was originally developed by SRI as part of a system known as the M7 – A surgical robot for open surgery; (2) the surgical robot itself along with the wristed tools were based on the Black Falcon developed by MIT. In 2003, after years of litigation and counter-litigation over intellectual property rights, the two companies merged under the name Intuitive Surgical Inc. (ISI). There are currently several hundreds da Vinci systems in use throughout the world.

Telesurgery on a human patient was accomplished on September 9, 2001 by Marescaux and Gagner. In collaboration with Computer Motion, they used a modified Zeus system to teleoperate between New York City and Strasbourg, France under a 155 ms time delay using a dedicated Asynchronous Transfer Mode (ATM) communication link [22, 23]. Several key non-clinical teleoperation experiments were conducted with both the RAVEN, and the M7 in extreme environment (desert, underwater, and zero gravity) demonstrated the capabilities of surgical robotic systems to deliver surgical expertise along large distance with a combination of wired and wireless communication [24–27].

In Asia, a group from the University of Tokyo has recently been working on a new telesurgery system [19, 28], and has completed laparoscopic cholcystectomy on a porcine model between sites in Japan, and more recently between Japan and Thailand [1, 2]. In Europe, the Laboratoire de Robotique de Paris at University of Paris, (LRP) uses a spherical mechanism similar to the RAVEN. This robot moves the port in addition to the tool. This allows to embed force sensors in the device that give a direct reading of the forces at the tool tip, instead of the combined interaction forces of the tool/tissue and trocar/abdomen.

The Light Endoscopic Robot (LER) was developed at the University of Hawaii, Manoa. This device was designed to guide an endoscopic camera, but is now capable of holding disposable endoscopic graspers as well as a tool with wrist articulation [29, 30].

The following chapter describes the research efforts lasted for more the a decade in developing a fully operational surgical robot – the RAVEN, based on a profound collaboration between surgeons and engineers. It will cover: (1) the surgical specification of the system that was based on quantitative measurements in an animal lab as well as the operating room (2) the mechanism kinematic optimization based on these specs; (3) the system integration efforts; and (4) the experimental results of system performance in multiple field and lab experiments of teleoperation under time delay.

8.2 Design of the Surgical Robot

8.2.1 Clinical Requirements

The Blue Dragon is a passive device instrumented with sensors that are capable of measuring the surgical tool position and orientation in space, along with the forces, and torques applied on the minimally invasive tools by the surgeon hands and the corresponding tissues (Fig. 8.1). Using the Blue Dragon, an extensive database was collected of in-vivo tissue handling/examination, dissection, and suturing tasks performed by 30 surgeons operating on a swine models [31–36]. The data is summarized in Table 8.1.

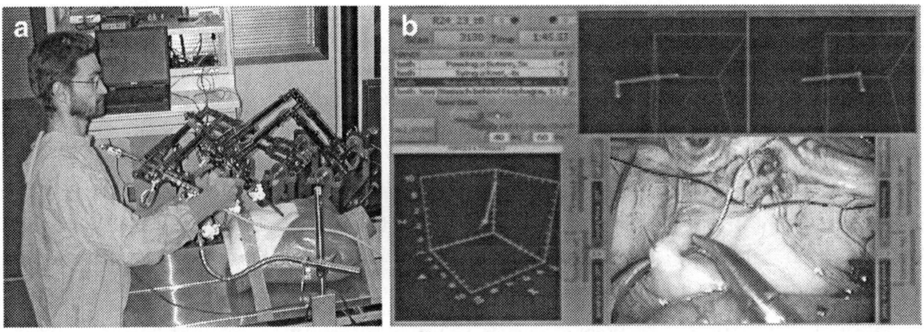

Fig. 8.1 The Blue Dragon system: (**a**) The system integrated into a minimally invasive surgery operating room. (**b**) Graphical user interface

Table 8.1 Design requirements based on in-vivo surgical measurement. Reference frames oriented such that x-axis points superior/inferior, y-axis is lateral/medial, z-axis straight up

Quantity	Symbol	Units	Value
Position/orientation	$\Delta\theta_x$	[Deg]	53.80
	$\Delta\theta_y$	[Deg]	36.38
	$\Delta\theta_z$	[Deg]	148.09
	ΔR	[m]	0.1027
	$\Delta\theta_g$	[Deg]	24.08
	ΔX	[m]	0.1026
	ΔY	[m]	0.0815
	ΔZ	[m]	0.0877
Velocity	ω_x	[rad/s]	0.432
	ω_y	[rad/s]	0.486
	ω_z	[rad/s]	1.053
	V_r	[M/s]	0.072
	ω_g	[rad/s]	0.0468
Peak force	F_x	[N]	14.729
	F_y	[N]	13.198
	F_z	[N]	67.391
	F_g	[N]	41.608
Torque	T_x	[Nm]	2.394
	T_y	[Nm]	1.601
	T_z	[Nm]	0.0464

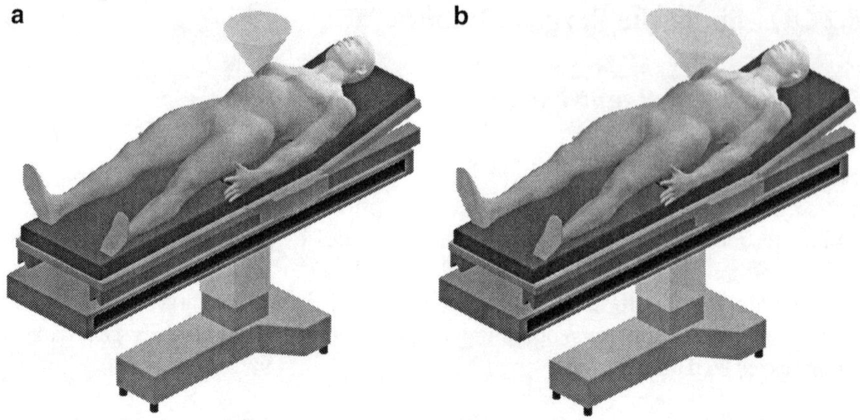

Fig. 8.2 Workspace definitions of the surgical robot in MIS (**a**) dexterous workspace – High dexterity region defined by a right circular cone with a vertex angle of 60and contains 95% of the tool motions based on in-vivo measurements. (**b**) Extended Dexterous Work Space – An elliptical cone with a vertex angle of 60–90 represents the reachable workspace such that any organ in the abdomen can be reached by the endoscopic tool

Analysis of this data indicated that, 95% of the time, the surgical tools were located within a conical range of motion with a vertex angle of 60 (termed the dexterous workspace, DWS). A measurement taken on a human patient showed that, in order to reach the full extent of the abdomen, the tool needed to move 90° in the mediolateral (left to right) and 60° in the superior/inferior direction (head to foot) – Fig. 8.2a. The extended dexterous workspace (EDWS) was defined as a conical range of motion with a vertex angle of 90and is the workspace required to reach the full extent of the human abdomen without reorientation of the base of the robot – Fig. 8.2b. These parameters, obtained through surgical measurement, served as a basis for the kinematic optimization of the RAVEN spherical mechanism [15, 18].

8.2.2 Kinematics of a Spherical Mechanism

8.2.2.1 Description of the Mechanism and Frame Assignments

In the class of spherical mechanisms, all links' rotation and translation axes intersect at a signal point or at infinity, referred to as the mechanism's remote center of rotation. Locating the remote center at the tool's point of entry to the human body through the surgical port, as typically done in minimally invasive surgery (MIS), constitutes a point in space where the surgical tool may only rotate around it but not translated with respect to it. The selected spherical mechanism has two configurations in the form of parallel (5R – Fig. 8.3a) and serial (2R – Fig. 8.3b) configurations. The parallel mechanism is composed of two serial arms: Arm 1 – Link13 and Link35 and Arm 2 – Link24 and Link46 joined by a stationary base defined by Link12 through Joints

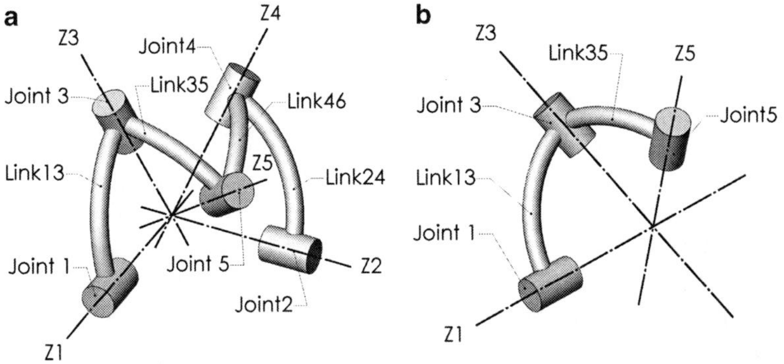

Fig. 8.3 Spherical mechanism – link and joint coordinate frame assignment (**a**) Parallel Manipulator (**b**) Serial Manipulator

1 and 2. The parallel chain is closed at the two collinear joints, 5 and 6. Arm 1 and 2 will be further referred to as the even and odd side of the parallel manipulator respectively. The spherical serial mechanism includes only one arm (Arm 1) with the same notation as the odd side of the parallel mechanism. By definition, the links of both mechanisms are moving along the surface of a sphere with an arbitrary radius R. All the rotation axes of the links intersect at the center of the sphere. The kinematic analysis is independent of the sphere's radius, however, from a practical perspective; the radius of the sphere should be kept as small as possible to minimize the overall size of the mechanism and the dynamic effects. As configured for MIS, the end-effector of the mechanism is inserted through Joint 5. Both the parallel and serial configurations are considered a 2-DOF mechanism. Any two joint angles determine the orientation of the end-effector in space; we chose θ_1 and θ_2 in the parallel configuration and by θ_1 and θ_3 in the serial configuration. For the purpose of kinematic analysis, the center of the sphere serves as the origin for all reference frames of the mechanism's links. Thus, the transformations between the mechanism links' coordinate frames can be expressed as pure rotations.

Following the Denavit–Hartenberg (D–H) frame assignment convention (Table 8.2), frames are assigned to the mechanism joints such that the Z-axis of the nth frame points outward along the nth joint [28]. Following the numbering scheme set for by Ouerfelli and Kumar [25] the frames have odd numbers along Arm1 (Frames 0′, 1, 3 and 5) and even numbers along Arm2 (Frame 0″, 2, 4 and 6) – see Fig. 8.2a. The end-effector frame is Frame 5. Frame 0′ is oriented such that the z-axis points along joint 1 and the y-axis points to the apex of the sphere. The link angle, α_{i+1} expresses the angle between the ith and (i + 1)th axis defined by the mechanism geometry. The rotation angle θ_i defines the variable joint angle as a between the rotation axis i − 1 and axis i.

For the parallel mechanism, when $\theta_1 = 0$ and $\theta_2 = 0$, Link13 and Link24 lie in planes perpendicular to the plane created by intersecting axes z1 and z2. There are two possible solutions to the orientations of Link35 and Link46 such that the mechanism is closed. The default orientation will be the orientation for which

$\theta_6 < 180°$. This will be referred to as the 'end-effector in' orientation. For the serial mechanism, when $\theta_1 = 0$ Link13 lies in a plane perpendicular to the plane created by intersecting axes z1 and x1. When $\theta_3 = 0$, Link35 is folded back onto L_{13}.

The transformation matrices between frames (8.1) are based on the DH-parameter notation that is summarized in Table 8.1. Because all the joints' translation parameters (a_i, d_i) are zero, the kinematic problem is reduced to describing the orientation of the end-effector in space. Thus, the generalized frame transformation matrix is reduced from the typical 4 × 4 translation and rotation matrix, to a 3 × 3 rotation matrix (8.1). The short notation for trigonometric functions *sin* and *cos* as *c* and *s* will be used throughout the manuscript.

$$ {}^{i-1}_{\ i}R = \begin{bmatrix} c\theta_i & -s\theta_i & 0 \\ s\theta_i c\alpha_{i-1} & c\theta_i c\alpha_{i-1} & -s\alpha_{i-1} \\ s\theta_i s\alpha_{i-1} & c\theta_i s\alpha_{i-1} & c\alpha_{i-1} \end{bmatrix} \tag{8.1} $$

8.2.2.2 Forward Kinematics

Given the mechanism parameters (α_{i-1}, θ_i) the forward kinematics defines the orientation of the end-effector ${}^{0'}u$ expressed in Frame 0′. The following sections describe the forward kinematics of the parallel and serial mechanisms.

Parallel Manipulator

Given the two joint angles θ_1 and θ_2, to find the end-effector orientation the full pose of the mechanism must first be solved. A geometric or analytical approach may be utilized in order to solve for parallel mechanism joint angles θ_3, θ_4 and θ_5.

The first step in the geometric solution is posing Link13 and Link24 based on joint rotation angles θ_1 and θ_2. The second step is to sweep links Link35 and Link46 through their range of motion by varying the joint angles θ_3 and θ_4. A closed chain parallel manipulator is formed at the intersection of the two paths swept by joints 5 and 6.

The analytical approach was used, which involves taking a closed-loop coordinate transformation around the 5R parallel manipulator, starting and ending at the same frame (Frame 1). The rotation matrix around a closed chain is equal to an identity matrix I formed as a series of transformation matrices. Using the rotation matrices ${}^{i-1}_{\ i}R$ around the closed chain and the specified joint angles θ_1 and θ_2 (8.2) sets up a system of three equations and three unknown joint angles, θ_3, θ_4 and θ_6.

$$ I = {}^{1}_{1}R = {}^{1}_{3}R\,{}^{3}_{5}R\,{}^{5}_{6}R\,{}^{6}_{4}R\,{}^{4}_{2}R\,{}^{2}_{0}R\,{}^{0}_{0''}R\,{}^{0''}_{1}R \tag{8.2} $$

The resultant matrix equation is given in terms of $sin\theta_i$ and $cos\theta_i$ (i = 3, 4, 6) In order to solve for the two solutions of this system the following trigonometric relation is needed

$$s\theta_i = \pm\sqrt{1 - c^2\theta_i} \quad (8.3)$$

Once all the joints angles ($\theta_1 \ldots \theta_6$) are known, the end-effector orientation can be determined by utilizing the forward kinematics expression of one serial manipulator subset. See the following section for details.

Serial Manipulator

Given the two joint angles θ_1 and θ_3, the forward kinematics of the 2R serial mechanism define the orientation of the end-effector. Using the rotation matrices (${}^{i-1}_{i}R$) – (1) and the joint parameters (Table 8.1), the coordinate transformation from the base to the end-effector is expressed as

$$ {}^{0'}_{5}R = {}^{0'}_{1}R\,{}^{1}_{3}R\,{}^{3}_{5}R \quad (8.4)$$

One may note that tool roll, θ_5, is not represented in (4). As a result, rather than expressing the entire orientation of the end-effector frame, it is sensible to express a vector that is collinear with the end-effector axis, Z_5. This vector is expressed in Frame 5 as ${}^{5}u_z = [0\ 0\ 1]^T$. The end-effector axis, Z_5 vector in Frame $0'$ (${}^{0'}u$) is given by (5) using (4). The vector ${}^{0'}u$ has its origin at the center of the sphere, and it points along the mechanism's end-effector, representing the orientation of the tool attached to the end-effector.

$$ {}^{0'}u = \begin{bmatrix} {}^{0'}u_x \\ {}^{0'}u_y \\ {}^{0'}u_z \end{bmatrix} = {}^{0'}_{5}R \begin{bmatrix} 0 \\ 0 \\ 1 \end{bmatrix} = \begin{bmatrix} c\theta_1 s\theta_3 s\alpha_{35} + s\theta_1 c\theta_3 c\alpha_{13} - s\theta_1 s\alpha_{13} c\alpha_{35} \\ s\theta_1 s\theta_3 s\alpha_{35} - c\theta_1 c\theta_3 c\alpha_{13} - c\theta_1 s\alpha_{13} c\alpha_{35} \\ c\theta_3 s\alpha_{13} s\alpha_{35} + c\alpha_{13} c\alpha_{35} \end{bmatrix} \quad (8.5)$$

8.2.2.3 Inverse Kinematics

Given the mechanism parameters (α_{i-1}) and the end-effector orientation ${}^{0'}u$ expressed in Frame $0'$ the inverse kinematics defines the mechanism joint angles (θ_i). Multiple solutions are available for any ${}^{0'}u$ in the workspace for both parallel and serial mechanism configurations. However, due to practical considerations of the physical joint limits, all the solutions are eliminated except one.

Parallel Manipulator

For any ${}^{0'}u$, the inverse kinematics result in four different solutions. The "elbow up/down" combinations of the two serial arms correspond to the four solutions that define the unique poses of the mechanism.

Using the Z component ${}^{0'}u_z$ (the third line of ${}^{0'}u_z$ – (5)) results in

$$c\theta_3 = \frac{{}^{0'}u_z - c\alpha_{13}\, c\alpha_{35}}{s\alpha_{13}\, s\alpha_{35}} \tag{8.6}$$

The two solutions for θ_3 are as follows:

$$\theta_{3a}\ ,\ \theta_{3b} = a\tan 2\left(\pm\sqrt{1 - c^2\theta_3}, c\theta_3\right) \tag{8.7}$$

Using the expression for ${}^{0'}u_x$, ${}^{0'}u_y$ (the first and second lines of ${}^{0'}u$ – (5)) and the previous result, then solving for $\sin\theta_1$ and $\cos\theta_1$, results in (8).

$$c\theta_1 = \frac{{}^{0'}u_x\, s\theta_3\, s\alpha_{35} - {}^{0'}u_y\,(c\theta_3\, c\alpha_{13}\, s\alpha_{35} - s\alpha_{13}\, c\alpha_{35})}{(s\theta_3\, s\alpha_{35})^2 + (c\theta_3\, c\alpha_{13}\, s\alpha_{35} - s\alpha_{13}\, c\alpha_{35})^2}$$

$$s\theta_1 = \frac{{}^{0'}u_y\, s\theta_3\, s\alpha_{35} + {}^{0'}u_x\,(c\theta_3\, c\alpha_{13}\, s\alpha_{35} - s\alpha_{13}\, c\alpha_{35})}{(s\theta_3\, s\alpha_{35})^2 + (c\theta_3\, c\alpha_{13}\, s\alpha_{35} - s\alpha_{13}\, c\alpha_{35})^2} \tag{8.8}$$

$$\theta_1 = atan2\ (s\theta_1, c\theta_1)$$

Thus, the inverse kinematic equations provide two solutions to the pose of the odd side of the manipulator, θ_{1a} and θ_{3a}, and θ_{1b} and θ_{3b}.

The even side of the manipulator is solved in a similar fashion. The end-effector z-axis in Frame 5 vector expressed in Frame $0'$ (${}^{0'}u$) is first translated into Frame $0''$.

$$0''u = {}^{0''}_{0'}R\,{}^{0'}u \tag{8.9}$$

$$c\theta_4 = \frac{{}^{0''}u_z - c\alpha_{24}\, c\alpha_{46}}{s\alpha_{24}\, s\alpha_{46}} \tag{8.10}$$

$$\theta_{4a}\ ,\ \theta_{4b} = atan2\left(\pm\sqrt{1 - c^2\theta_4}, c\theta_4\right) \tag{8.11}$$

$$c\theta_2 = \frac{{}^{0''}u_x\, s\theta_4\, s\alpha_{46} - {}^{0''}u_y(c\theta_4\, c\alpha_{24}\, s\alpha_{46} - s\alpha_{24}\, c\alpha_{46})}{(s\theta_4\, s\alpha_{46})^2 + (c\theta_4\, c\alpha_{24}\, s\alpha_{46} - s\alpha_{24}\, c\alpha_{46})^2}$$

$$s\theta_2 = \frac{{}^{0''}u_y\, s\theta_4\, s\alpha_{46} + {}^{0''}u_x(c\theta_4\, c\alpha_{24}\, s\alpha_{46} - s\alpha_{24}\, c\alpha_{46})}{(s\theta_4\, s\alpha_{46})^2 + (c\theta_4\, c\alpha_{24}\, s\alpha_{46} - s\alpha_{24}\, c\alpha_{46})^2} \tag{8.12}$$

$$\theta_2 = atan2(s\theta_2, c\theta_2)$$

Table 8.2 Parallel and serial manipulator D–H parameters. The serial manipulator represents a subset of the parallel manipulator denoted by the even link number. Its D–H parameters are marked by a gray background (see Fig. 8.3 for details)

i − *1*	i	I + 1	α_{i-1}	θ_i
0′	1	3	*0*	θ_1
1	3	5	$-\alpha_{13}$	θ_3
3	5	–	α_{35}	$\theta_5 = 0$
0′	0″	2	α_{12}	*0*
0″	2	4	*0*	θ_2
2	4	6	$-\alpha_{24}$	θ_4
4	6	5	α_{46}	θ_6

Table 8.3 Four solutions to the inverse kinematic problem for parallel manipulator (Two solutions to the inverse kinematic problem for serial manipulator are shaded in *gray*)

Pose 1	θ_{1a}	θ_{3a}	θ_{2a}	θ_{4a}
Pose 2	θ_{1a}	θ_{3a}	θ_{2b}	θ_{4b}
Pose 3	θ_{1b}	θ_{3b}	θ_{2a}	θ_{4a}
Pose 4	θ_{1b}	θ_{3b}	θ_{2b}	θ_{4b}

Providing two solutions to the pose of the even side of the manipulator, θ_{2a} and θ_{4a}, and θ_{2b} and θ_{4b}. The complete solution of the inverse kinematic problem for all four configurations is summarized in Table 8.3.

Serial Manipulator

The serial mechanism may be considered as a subset of the parallel mechanism, and therefore the inverse kinematics of the odd side of the parallel mechanism is the solution to the inverse kinematics of the serial mechanism. The two solutions to the inverse kinematics problem for the serial manipulator are summarized in Table 8.2 highlighted in gray.

8.2.2.4 Jacobian Matrix

The Jacobian matrix ${}^{A}J$, the mapping between joint velocities ($\dot{\theta}_i$) and end-effector angular velocities (${}^{j}\omega_5$), can be expressed with respect to any frame (A) associated with the mechanism. The advantage of expressing the Jacobian matrix in the end-effector frame (Frame 5) – (${}^{5}J$) is that the last row of the matrix is [0 0 1] for all poses and joint velocities. Therefore, for computational purposes, the Jacobian matrix (${}^{5}J$) may be reduced by one dimension using only the upper 2 × 2 submatrix of the full 3 × 3 Jacobian. This truncated version of the Jacobian maps the two controlled joint velocities ($\dot{\theta}_1$, $\dot{\theta}_2$ of the parallel mechanism and $\dot{\theta}_1$, $\dot{\theta}_3$ of the serial mechanism) to end-effector angular velocity. This version of the Jacobian will later be used for calculating the manipulator isotropy.

Parallel Manipulator

The Jacobian matrix of the parallel mechanism, expressed in Frame 5 (${}^5J_{par}$) is a 3×3 matrix mapping the input joint velocities ($\dot{\theta}_1$, $\dot{\theta}_2$, $\dot{\theta}_5$) to the end-effector angular velocities (${}^5\omega_5$, (12)). The method we used for obtaining the Jacobian matrix is to develop its inverse form using ${}^5J_{serial}$ for the even and odd sides. From these we can construct ${}^5J_{par}^{-1}$ and then invert this expression.

$$\begin{bmatrix} {}^5\omega_{5x} \\ {}^5\omega_{5y} \\ {}^5\omega_{5z} \end{bmatrix} = \left[{}^5J_{Par}\right]_{3x3} \begin{bmatrix} \dot{\theta}_1 \\ \dot{\theta}_2 \\ \dot{\theta}_5 \end{bmatrix} \tag{8.13}$$

As previously explained the upper 2×2 submatrix of the Jacobian matrix ${}^5J_{par}$ maps the two input joint velocities ($\dot{\theta}_1$, $\dot{\theta}_2$) to the end-effector angular velocities (${}^5\omega_5$) of pitch and yaw. Eliminating tool roll,

$$\begin{bmatrix} {}^5\omega_{5x} \\ {}^5\omega_{5y} \end{bmatrix} = \left[{}^5J_{par}\right]_{2x2} \begin{bmatrix} \dot{\theta}_1 \\ \dot{\theta}_2 \end{bmatrix} \tag{8.14}$$

We used a standard recursive formulation of link $i + 1$ angular velocity expressed in frame $i + 1$ (${}^{i+1}\omega_{i+1}$) to derive the Jacobian matrices of the odd and even numbered arms, treated as independent serial manipulators ${}^5J_{SerialOdd}$ and ${}^5J_{SerialEven}$. Note that by definition, joint $i + 1$ is always rotated along its Z axis denoted by a unit vector $\hat{z}_{i+1}$ Using recursive (15) and tracing the odd arm and even arms of the mechanism from Frame 5 to the base frame results in (16) and (17).

$${}^{i+1}\omega_{i+1} = {}^{i+1}_{i}R\,{}^i\omega_i + \dot{\theta}^{i+1}_{i+1}\,\hat{z}_{i+1} \tag{8.15}$$

$${}^5\omega_5 = {}^5_1R \begin{bmatrix} 0 \\ 0 \\ 1 \end{bmatrix} \dot{\theta}_1 + {}^5_3R \begin{bmatrix} 0 \\ 0 \\ 1 \end{bmatrix} \dot{\theta}_3 + \begin{bmatrix} 0 \\ 0 \\ 1 \end{bmatrix} \dot{\theta}_5 \tag{8.16}$$

$${}^5\omega_5 = {}^5_2R \begin{bmatrix} 0 \\ 0 \\ 1 \end{bmatrix} \dot{\theta}_2 + {}^5_4R \begin{bmatrix} 0 \\ 0 \\ 1 \end{bmatrix} \dot{\theta}_4 + \begin{bmatrix} 0 \\ 0 \\ 1 \end{bmatrix} (\dot{\theta}_5 + \dot{\theta}_6) \tag{8.17}$$

which is simplified into (18) and (19) to get the Jacobian matrix of the odd and even numbered sides

$$\begin{bmatrix} {}^5\omega_{5x} \\ {}^5\omega_{5y} \\ {}^5\omega_{5z} \end{bmatrix} = \left[{}^5J_{Serial\,Odd}\right] \begin{bmatrix} \dot{\theta}_1 \\ \dot{\theta}_3 \\ \dot{\theta}_5 \end{bmatrix}$$

$$= \left[\begin{bmatrix} {}^5_1R \end{bmatrix} \begin{bmatrix} 0 \\ 0 \\ 1 \end{bmatrix} \quad \begin{bmatrix} {}^5_3R \end{bmatrix} \begin{bmatrix} 0 \\ 0 \\ 1 \end{bmatrix} \quad \begin{bmatrix} 0 \\ 0 \\ 1 \end{bmatrix} \right] \begin{bmatrix} \dot{\theta}_1 \\ \dot{\theta}_3 \\ \dot{\theta}_5 \end{bmatrix} \tag{8.18}$$

$$\begin{bmatrix} {}^5\omega_{5x} \\ {}^5\omega_{5y} \\ {}^5\omega_{5z} \end{bmatrix} = \left[{}^5J_{SerialEven}\right] \begin{bmatrix} \dot{\theta}_2 \\ \dot{\theta}_4 \\ \dot{\theta}_5 + \dot{\theta}_6 \end{bmatrix}$$

$$= \left[\begin{bmatrix} {}^5_2R \end{bmatrix} \begin{bmatrix} 0 \\ 0 \\ 1 \end{bmatrix} \quad \begin{bmatrix} {}^5_4R \end{bmatrix} \begin{bmatrix} 0 \\ 0 \\ 1 \end{bmatrix} \quad \begin{bmatrix} 0 \\ 0 \\ 1 \end{bmatrix} \right] \begin{bmatrix} \dot{\theta}_2 \\ \dot{\theta}_4 \\ \dot{\theta}_5 + \dot{\theta}_6 \end{bmatrix} \tag{8.19}$$

By inverting these equations we obtain

$$\begin{bmatrix} \dot{\theta}_1 \\ \dot{\theta}_3 \\ \dot{\theta}_5 \end{bmatrix} = \left[\begin{bmatrix} {}^5_1R \end{bmatrix} \begin{bmatrix} 0 \\ 0 \\ 1 \end{bmatrix} \quad \begin{bmatrix} {}^5_3R \end{bmatrix} \begin{bmatrix} 0 \\ 0 \\ 1 \end{bmatrix} \quad \begin{bmatrix} 0 \\ 0 \\ 1 \end{bmatrix} \right]^{-1} \begin{bmatrix} {}^5\omega_{5x} \\ {}^5\omega_{5y} \\ {}^5\omega_{5z} \end{bmatrix}$$

$$= \left[{}^5J_{Serial\,Odd}\right]^{-1} \begin{bmatrix} {}^5\omega_{5x} \\ {}^5\omega_{5y} \\ {}^5\omega_{5z} \end{bmatrix} \tag{8.20}$$

$$\begin{bmatrix} \dot{\theta}_2 \\ \dot{\theta}_4 \\ \dot{\theta}_5 + \dot{\theta}_6 \end{bmatrix} = \left[\begin{bmatrix} {}^5_2R \end{bmatrix} \begin{bmatrix} 0 \\ 0 \\ 1 \end{bmatrix} \quad \begin{bmatrix} {}^5_4R \end{bmatrix} \begin{bmatrix} 0 \\ 0 \\ 1 \end{bmatrix} \quad \begin{bmatrix} 0 \\ 0 \\ 1 \end{bmatrix} \right]^{-1} \begin{bmatrix} {}^5\omega_{5x} \\ {}^5\omega_{5y} \\ {}^5\omega_{5z} \end{bmatrix}$$

$$= \left[{}^5J_{Serial\,Even}\right]^{-1} \begin{bmatrix} {}^5\omega_{5x} \\ {}^5\omega_{5y} \\ {}^5\omega_{5z} \end{bmatrix} \tag{8.21}$$

The first rows of (20) and (21) define $\dot{\theta}_1$ and $\dot{\theta}_2$ respectively as a function of end-effector angular velocity ${}^5\omega_5$ and the inverted Jacobian matrices of the even and odd arms respectively. For k = 1, 2, 3, the kth column of the Jacobian

$$\dot{\theta}_1 = \left[{}^5J_{SerialOdd}{}^{-1}(1,k)\right] \left[{}^5\omega_{5x} \quad {}^5\omega_{5y} \quad {}^5\omega_{5z}\right]^T \tag{8.22}$$

$$\dot{\theta}_2 = \left[{}^5J_{Serial\mathrm{Even}}{}^{-1}(1,k)\right] \left[{}^5\omega_{5x} \quad {}^5\omega_{5y} \quad {}^5\omega_{5z}\right]^T \tag{8.23}$$

As previously explained, the benefit of expressing the Jacobian matrix in the end-effector Frame 5 is the following simple representation of the joint angular velocity.

$$\dot{\theta}_5 = [0 \quad 0 \quad 1]\left[{}^5\omega_{5x} \quad {}^5\omega_{5y} \quad {}^5\omega_{5z}\right]^T \tag{8.24}$$

Assembling individual joint velocities into matrix form allows us to formulate the inverse Jacobian for the parallel manipulator

$$\begin{bmatrix} \dot{\theta}_1 \\ \dot{\theta}_2 \\ \dot{\theta}_5 \end{bmatrix} = \begin{bmatrix} {}^5J_{SerialOdd}{}^{-1}(1,k) \\ {}^5J_{SerialEven}{}^{-1}(1,k) \\ 0 \quad 0 \quad 1 \end{bmatrix} \begin{bmatrix} {}^5\omega_{5x} \\ {}^5\omega_{5y} \\ {}^5\omega_{5z} \end{bmatrix} \tag{8.25}$$

The relationship between the two input joint angular velocities ($\dot{\theta}_1$and $\dot{\theta}_2$) and the angular velocity of the end-effector, considering only the pitch and yaw of the tool (${}^5\omega_{5x}$, ${}^5\omega_{5y}$) allows us to analyze only the upper 2×2 sub-matrix of the Jacobian Inverse. For k = 1, 2

$$\begin{bmatrix} \dot{\theta}_1 \\ \dot{\theta}_2 \end{bmatrix} = \begin{bmatrix} {}^5J^{-1}_{SerialOdd}(1,k) \\ {}^5J^{-1}_{SerialEven}(1,k) \end{bmatrix} \begin{bmatrix} {}^5\omega_{5x} \\ {}^5\omega_{5y} \end{bmatrix} \tag{8.26}$$

Inverting this inverse Jacobian gives the Jacobian matrix for the parallel mechanism (14).

Serial Manipulator

Using the odd numbered side of the parallel mechanism, the Jacobian matrix of the serial mechanism can be derived from (18). Based on the previous justification, the upper 2×2 submatrix of the full 3×3 Jacobian relates the controlled axes of motion to the end-effector velocity:

$$\begin{bmatrix} {}^5\omega_{5x} \\ {}^5\omega_{5xy} \end{bmatrix} = [J_{5truncated}]_{2x2} \begin{bmatrix} \dot{\theta}_1 \\ \dot{\theta}_3 \end{bmatrix} \tag{8.27}$$

8.2.3 Kinematic Optimization

Experimental results led to the definition of engineering requirements as derivatives from the clinical requirements (see Sect. 8.2.1). The optimization process was developed to define the mechanism parameters (link angles α_{13}, α_{35}, α_{24}, α_{46}, α_{12}) so that the workspace of the mechanism covers the entire EDWS while providing high dexterity manipulation within the DWS. The goal of a high performance yet compact mechanism was achieved by careful selection of the scoring criteria.

Up to this point, the analysis has been purely mathematical. The manipulator could move through singularities, fold on itself and solve for arbitrary poses without regard to how a physically implemented device might accomplish this. Based on mechanical design of the serial manipulator the range of motion of the first joint angle is 180°(0° $<\theta_1<$180°) and the range of motion of the second joint is 160° (20° $<\theta_3<$180°). By adding these joint limits into the optimization, the true reachable workspace of a physically implemented device could be analyzed.

8.2.3.1 Mechanism Isotropy

The Jacobian matrix allows one to analyze the kinematic performance of a mechanism. One performance metric using the Jacobian matrix is Yoshiakawa's manipulability measure, specified in [20]. The manipulability measure most commonly used ranges from 1 to infinity.

$$\omega = \sqrt{\det(J(\theta)J^T(\theta))} \tag{8.28}$$

In contrast, mechanism isotropy is a scoring criterion that ranges from 0 to 1. Isotropy is defined in (29) as the ratio between the lowest eigenvalue ($\lambda_{\min}$) and the highest eigenvalue ($\lambda_{\max}$) of the Jacobian [37]. Note that the isotropy is a function of the mechanism's input joint angles (θ_i). For the serial mechanism, these angles are θ_1, θ_3 and for the parallel mechanism, θ_1, θ_2

$$ISO(\theta_i) = \frac{\lambda_{\min}}{\lambda_{\max}} \quad ISO \in \langle 0, 1 \rangle \tag{8.29}$$

Our analysis uses isotropy as the measure of performance because numerically the range of scores was easier to deal with. Given a design candidate (serial mechanism link angles α_{13} and α_{35} or parallel mechanism link angles α_{13},α_{35},α_{24},α_{46},α_{12}), for every mechanism pose there is an associated isotropy value in the range of 0–1. An isotropy measure of 0 means the mechanism is in a singular configuration and has lost a degree of freedom so there is a direction in which it can no longer move. An isotropy measure of 1 means the mechanism can move equally well in all directions.

Once the kinematic equations and a performance measure are defined, one can evaluate the performance of a particular design candidate at each point its workspace. The integration of the isotropy score over the DWS or EDWS is used as one component of a scoring function for the specific design candidate.

8.2.3.2 Scoring Criteria and Cost Function

Each design candidate must be assigned a single score so that the best overall manipulator design can be selected. Three individual criteria including (1) an integrated average score (2) a minimal single score (3) the cube of the angular link length are incorporated into the composite score, expressed in (31).

In order to analyze the mechanism, the hemisphere was discretized into points distributed equally in azimuth and elevation. Each point is associated with a different area of the hemisphere based on elevation angle. One measure of how well a manipulator performs is to calculate the isotropy at each point, multiply by its corresponding area, then sum all of the weighed point-scores over the the DWS or EDWS. This provides an average performance of the design candidate over the workspace.

The set (K) of all possible cones (k) on the hemisphere with an azimuth angle (σ) and elevation angle (ξ) is

$$K = \{k(\sigma, \zeta) : 0 < \sigma 2\pi, 0 \zeta \pi/4\} \tag{8.30}$$

The set of discrete points ($k^p_{\sigma,\zeta}$)contained in the cone ($k_{\sigma,\zeta}$) is

$$k^p_{\sigma,\zeta} \subset k_{\sigma,\zeta} \tag{8.31}$$

Each point has an associated isotropy and area. Thus the scoring functions are

$$S_{sum} = \underset{K}{MAX} \left\{ \sum_{k^p_{\sigma,\zeta}} ISO(\theta_1, \theta_3) * Area(\sigma, \zeta) \right\} \tag{8.32a}$$

$$S_{\min} = \underset{K}{MAX} \left\{ \underset{k^p_{\sigma,\zeta}}{MIN} \left(ISO(\theta_1, \theta_3) \right) \right\} \tag{8.32b}$$

There are many orientations of the DWS or EDWS with respect to the hemisphere, noted as the set K. Each design candidate has a S_{sum} value that corresponds to the highest summed isotropy score for that design.

The limitation of a summed isotropy score is that singularities or workspace boundaries could exist within a region that has a good summed score. The minimum isotropy value within the cone intersection area is an indicator of the worst performance that can be expected over that area. For each design candidate the highest minimum isotropy score on the set of all cones, K, will be referred to as S_{min}.

While a dynamic analysis is not explicitly addressed in this study, there is a relationship between mass and size of a robot with dynamic performance. A design with greater link angles will have a larger reachable workspace and generally better

S_{sum} and S_{min} values. The drawback to larger link angles is reduced link stiffness and higher mass and inertia (worse dynamic performance) compared to shorter links. Based on long beam theory the mechanism stiffness is inversely proportional to the cube of the sum of the link length (expressed as angles in the spherical case) [28]. As suggested by the experimental findings, in surgery the mechanism needs to be operated in the smallest possible workspace that will satisfy the motion requirements of the surgical tool. Thus, the optimization goal is to maximize the kinematic performance over the surgical workspace while minimizing the link length.

Our composite design candidate performance function takes into account all three individual criteria and is defined as follows

$$\phi = \frac{S_{sum} \cdot S_{\min}}{(\alpha_{13} + \alpha_{35})^3} \tag{8.33}$$

A requirement of the optimization is that over the target workspace, the mechanism does not encounter any singularities or workspace boundaries. By multiplying the summed isotropy by the minimum isotropy, candidates that fail to meet the requirement have a score of zero. By dividing by the cube of the sum of the link angles, the score penalizes mechanism compliance and inertia. Thus, in a scan of the potential design space, the peak composite score should represent a design with maximum average performance, a guaranteed minimum performance and good mechanical properties.

8.2.3.3 Optimization Algorithm

Considering the target workspace to be the DWS, the 60°cone's orientation in azimuth and elevation was varied in order to obtain the best orientation of the DWS for that design candidate. Optimizing for the EDWS, which is an elliptical cone, would add another design parameter, namely cone roll angle. Introducing an additional parameter will increase execution time of the optimization by an order of magnitude. By considering the target workspace to be a 90° cone that encompasses all roll angle orientations of the EDWS, an additional design parameter is avoided. This 90° cone will further be referred to as a superset of the EDWS. Using a superset of the EDWS could result in larger links than necessary; a design that can reach 60° in one direction and 90° in an orthogonal direction may satisfy the EDWS cone but not the superset 90°cone.

The algorithm for evaluating a design candidate is as follows: (1) Define a hemisphere with points distributed evenly in azimuth and elevation (2) Calculate inverse kinematics and isotropy for each point (3) Define a target cone (DWS or superset of the EDWS) (a) Move cone around the hemisphere (b) At each orientation of target cone, calculate integrated isotropy and minimum isotropy within the cone (4) Save the best minimum and integrated scores for that design candidate

Parallel Manipulator

The parallel manipulator is composed of five links. The optimization considered a symmetric device therefore taking into account three mechanism parameters; Base Angle (α_{12}), Link1 ($\alpha_{13} = \alpha_{24}$) and Link2 ($\alpha_{35} = \alpha_{46}$). Link1 and Link2 were varied from 32° to 90° in 2° increments and α_{12} from 0° to 90° in 2° increments for a total of 40,500 design candidates. The hemisphere was discretized into 900 points, distributed evenly in azimuth and elevation angles. The optimization can be summarized:

$$max\ \phi\ (\alpha_{12}, Link1, Link2)\ \begin{cases} 0^o < \alpha_{12} < 90^o \\ = 32^o < Link1 < 90^o \\ 32^o < Link2 < 90^o \end{cases} \tag{8.34}$$

The composite score, ϕ, of the parallel mechanism as a function of the link length angles Link1 ($\alpha_{13} = \alpha_{24}$), Link2 ($\alpha_{35} = \alpha_{46}$) and a base link length of $\alpha_{12} = 0$ is plotted in Fig. 8.4 with a target workspace, the DWS (Fig. 8.5a) and the EDWS superset (Fig. 8.5b). These plots of Link1 and Link2 versus score for a fixed base angle will be referred to as 'optimization design subspace plots'. Searching through the entire design space, optimizing on the DWS indicates a peak composite score of 0.1319 was achieved by a parallel manipulator with link angles $\alpha_{13} = \alpha_{24} = 38°$ (Link 1) and $\alpha_{35} = \alpha_{46} = 50°$ (Link 2) and base angle 0°. In contrast, running the same optimization but using the superset of the EDWS as the target workspace gave link angles $\alpha_{13} = \alpha_{24} = 52$ (Link 1) and $\alpha_{35} = \alpha_{46} = 72$ (Link 2) and base angle 0with a score of 0.0960.

An important parameter for the parallel mechanism is the base angle, α_{12}. A sequence of plots (Fig. 8.6) of the parallel mechanism composite score for different bases angles $\alpha_{12} = 90°–0°$ shows the overall dependency between the mechanism's best design in terms of link length and the base angle. This sequence of plots indicated that the performance of the parallel mechanism increased as the base angle decreased.

The maximal performance score as a function of the base angle is depicted in Fig. 8.5. This result indicates that the performance score decreases as the base angle increases in the range of 0–70° with a minimal value at a base angle equal to 70°. In the range of 70–90° a reversed behavior is observed in which the performance score increases as the base angle decreases.

In order to demonstrate how base angle, α_{12}, effects the mechanism workspace, a mechanism with fixed link lengths was chosen ($\alpha_{13} = \alpha_{24} = \alpha_{35} = \alpha_{46} = 60$) and the base angle was varied. The workspace of each design is plotted as a sequence of figures (Fig. 8.8). The gray region the reachable workspace and black is the unreachable or singular region with the area of highest isotropy in dark gray. With a 0base angle, the region of highest isotropy is a large strip whereas for 90and 45base angles it is a point.

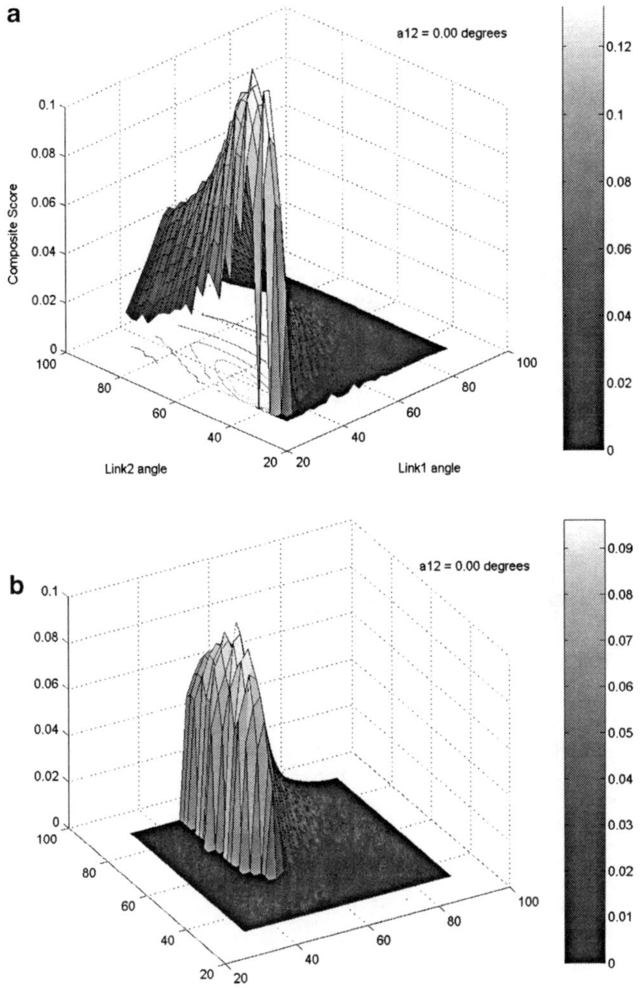

Fig. 8.4 The Composite score ϕ of the parallel mechanism as a function of the link lengths angles Link1 ($\alpha_{13} = \alpha_{24}$), Link2 ($\alpha_{35} = \alpha_{46}$) and a base link length of $\alpha_{12} = 0$: (**a**) Optimization design subspace plot with base angle 0° for DWS (**b**) Optimization design subspace plot with base angle 0°for EDWS superset

Serial Manipulator

The Serial mechanism is composed of two links. The optimization considered all combinations of α_{13} and α_{35} from 16°to 90°in 2increments for a total of 1,444 design candidates. The hemisphere was discretized into 3,600 points, distributed evenly in azimuth and elevation.

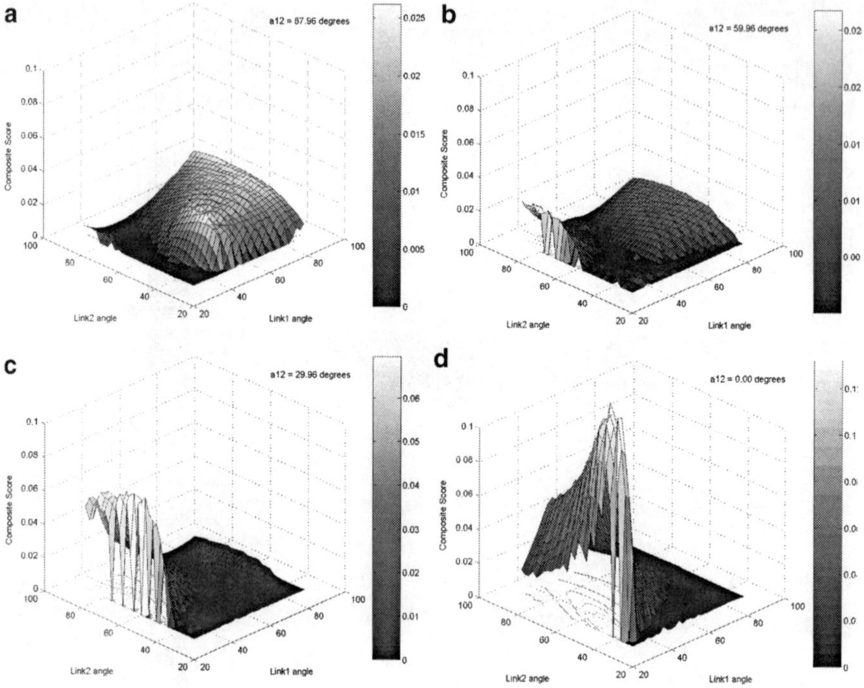

Fig. 8.5 The composite score ϕ of a parrallel mechanism as a function of the link lengths angles Link 1($\alpha_{13} = \alpha_{24}$) Link 2 ($\alpha_{35} = \alpha_{45}$) with varying base angle $\alpha_{12} = 90°$(**a**); 60°(**b**); 30°(**c**); 0°(**d**)

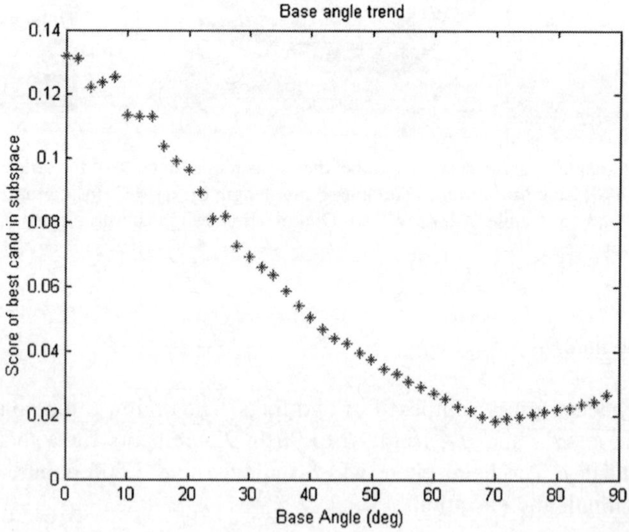

Fig. 8.6 The composite score φ of a parallel mechanism as a function of the link lengths angles Link1 ($\alpha_{13} = \alpha_{24}$), Link2 ($\alpha_{35} = \alpha_{46}$) with varying base angle $\alpha_{12} = 90$ (**a**); 60 (**b**); 30 (**c**) 0 (**d**)

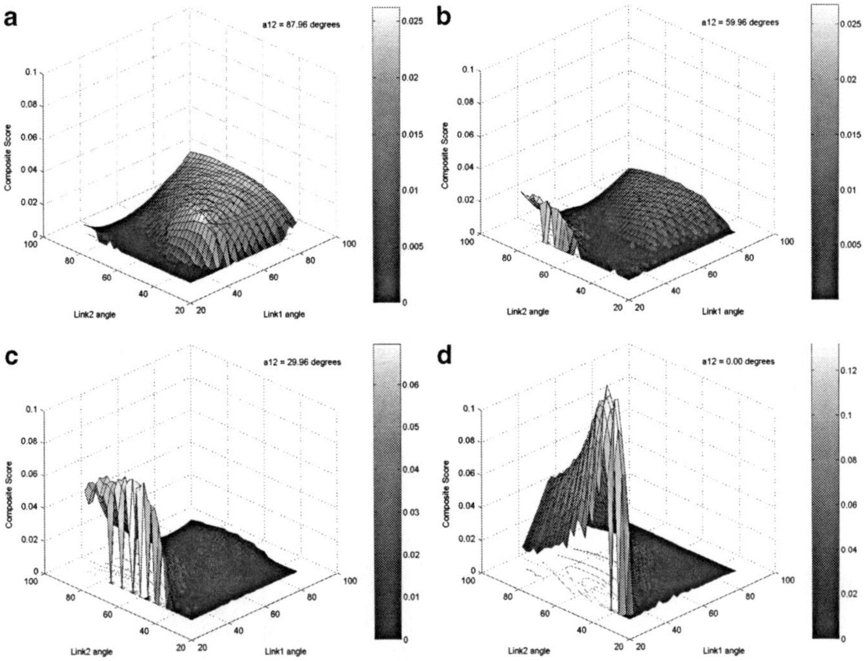

Fig. 8.7 Parallel mechanism peak performance for a fixed link length as a function of the mechanism base angle

$$max\ \phi\ (\alpha_{13}, \alpha_{35}) \quad \begin{cases} 16^o < \alpha_{13} < 90^o \\ 16^o < \alpha_{35} < 90^o \end{cases} \tag{8.35}$$

Using the scoring criteria and hemisphere point resolution of 2°, a numerical scan of the serial mechanism design space was performed using all the combinations of link angles α_{13} and α_{35} in the range of 16–90°. For the serial manipulator optimized for the DWS, the best design was achieved with link angles of $\alpha_{13} = 52$(Link 1) and $\alpha_{35} = 40$(Link 2) and with a composite score of 0.0520 (Fig. 8.9a). In contrast, running the same optimization for the EDWS superset indicated that the optimal mechanism design has link angles $\alpha_{13} = 90$(Link 1) and $\alpha_{35} = 72$(Link 2) with a score of 0.0471 (Fig. 8.9b).

The difference in the results is not unexpected but it does pose an interesting dilemma. If one chooses the design that optimizes on a 90° cone, the resulting design might satisfy the requirements for a larger number of applications. However, this design has lower overall performance and larger links. A compromise between the requirements to maximize performance in the DWS and minimizing the size of the mechanism while being able to reach the EDWS was achieved by selecting the highest performing design over the DWS subject to the requirement that it also can reach the EDWS superset. Considering all the potential mechanism design candidates for the DWS (Fig. 8.9a) and eliminating those that cannot reach the EDWS

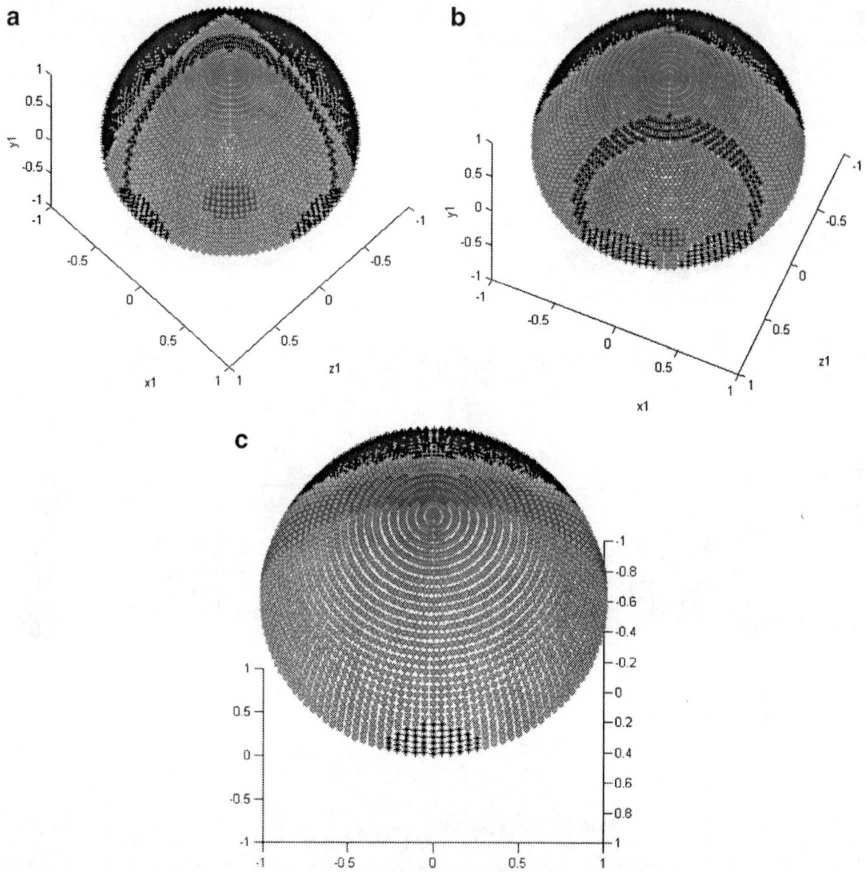

Fig. 8.8 The parallel mechanism workspace link lengths: $\alpha_{13} = \alpha_{24} = \alpha_{35} = \alpha_{46} = 60°$as a function of base angle $\alpha_{12} = 90°$(**a**); 45°(**b**); 0°(**c**)

give the subset plotted in Fig. 8.9e. The highest performing design from this subset had link angles of $\alpha_{13} = 74°$ and $\alpha_{35} = 60°$ and a score of 0.0367.

8.3 System Architecture and Integration

The system includes two main sub systems: (1) the surgical console (master) and (2) the surgical robot (slave). To enable teleportation the master and the slave may be connected through various network layers for example wired, wireless, and hybrid communication links. Figure 8.10 provide an overview of the system architecture. The surgeon initiates the movement of the robot by moving the taylus of a haptic master input device. The position of the stylus is sensed by position sensors embedded in the master joints and acquired by the A/D converter that is

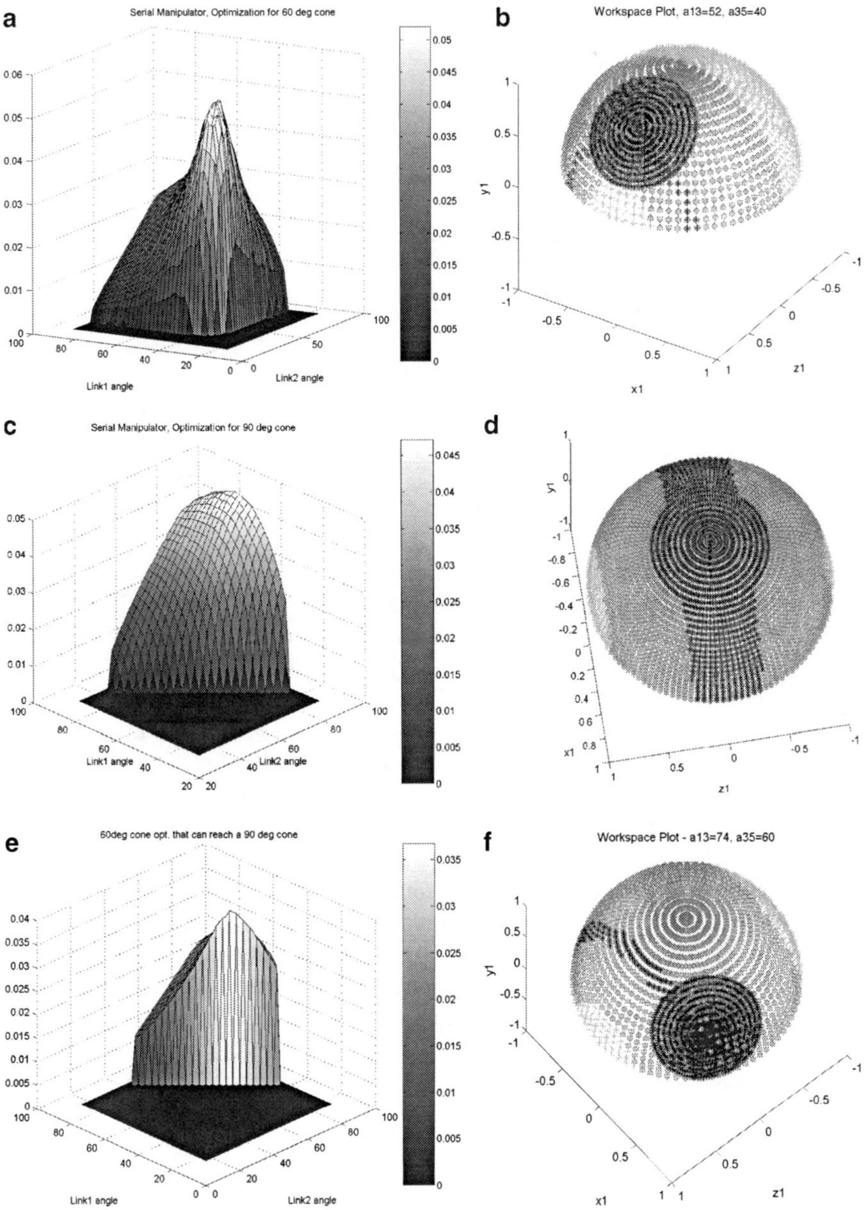

Fig. 8.9 The composite score φ of a serial mechanism as a function of the link length angles Link1 ($\alpha_{13} = \alpha_{24}$), Link2 ($\alpha_{35} = \alpha_{46}$) along with the workspace of the mechanism with the best performance for each design criterion (peak of a) DWS, a cone with a vertex angle of 60° (**b**) Workspace plot for optimal mechanism (**a**) with link angles $\alpha_{13} = 52°$ and $\alpha_{35} = 40°$ (**c**) EDWS, a cone with a vertex angle of 90° (**d**) Workspace plot for optimal mechanism (peak of c) with link angles $\alpha_{13} = 90°$ and $\alpha_{35} = 72°$, (**e**) subset of (a) which can reach a cone with a vertex angle of 90°. (**f**) Workspace plot for optimal mechanism (peak of e) with link angles $\alpha_{13} = 74°$ and $\alpha_{35} = 60°$. For subplots c,d,f, the workspace plots show the hemisphere in green, the reachable workspace in purple, and the orientation of the best cone in black, with the strip of maximum isotropy also in black

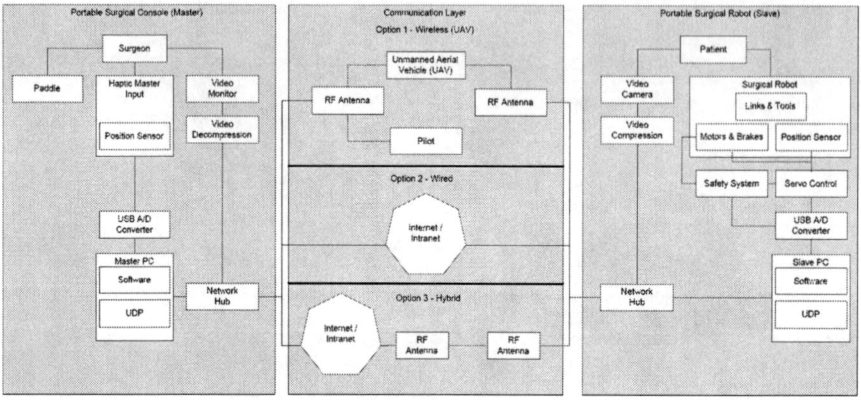

Fig. 8.10 Overview of the system architecture including the surgical console (master) and surgical robot (slave) connected through a communication layer with three different options that were studying as part of the experimental evolution of the system (see Sect. 8.4)

connected to the Master PC via USB. Using a UDP protocol the position command is transmitted through the network layer to the remote site and received by the slave PC. Using the inverse kinematics, the position command is translated into joint command and sent via the D/A to the servo controllers. The servo controllers generate voltage commands to the DC actuators of the surgical robot which in turn move the robot to the commanded position. A video stream of the surgical site is first compressed in the remote site by either software or hardware and then stream through the network layer to the surgical console. In the surgical console, the video following its decompression is presented to the surgeon on a monitor. A foot paddle controlled by the surgeon allows to engage and disengage the master and the salve by activating the brakes. Indexing is the term describing the process in which the surgeon disengage from the robot, reposition the input devices, and reengaged with to robot to continue the operation. Indexing is enabled by brakes mounted on the motors of the robot, which fix the position and the orientation of the robot in space during the time when the robot is disengaged from the surgical console.

8.3.1 Surgical Manipulators

The seven-degree-of-freedom (7-DOF) cable-actuated surgical manipulator, shown in Fig. 8.5, is broken into three main pieces: the static base that holds all of the motors the spherical mechanism that positions the tool and the tool interface. The motion axes of the surgical robot are: (1) the shoulder joint (rotational); (2) the elbow joint (rotational) (3) tool insertion/retraction (translational) (4) tool rotation (rotational) (5) tool grasping (rotational) (6) tool wrist-1 actuation (rotational) (7) tool wrist-2 actuation (rotational).

The first four joint axes intersect at the surgical port location, creating a spherical mechanism that allows for tool manipulation similar to manual laparoscopy. The mechanism links are machined from aluminum, and are generally I-section shapes with structural covers. These removable covers allow access to the cable system, while improving the torsional stiffness of the links. The links are also offset from the joint axis planes, allowing for a tighter minimum closing angle of the elbow joint.

The RAVEN utilizes DC brushless motors located on the stationary base, which actuate all motion axes. Maxon EC-40 motors with 12:1 planetary gearboxes are used for the first three axes, which see the highest forces. The first two axes, those under the greatest gravity load, have power-off brakes to prevent tool motion in the event of a power failure. The fourth axis uses an EC-40 without a gearbox, and Maxon EC-32 motors are used for the remaining axes. Maxon DES70/10 series amplifiers drive these brushless motors. The motors are mounted onto the base via quick-change plates that allow motors to be replaced without the need to disassemble the cable system.

The cable transmission system comprises a capstan on each motor, a pretension adjustment pulley, various pulleys to redirect the cables through the links, and a termination point to each motion axis. The shoulder axis is terminated on a single partial pulley. The elbow axis has a dual-capstan reduction stage terminating on a partial pulley. The tool insertion/ retraction axis has direct terminations of the cables on the tool holder. The tool rotation, grasping, and wrist cables are terminated on capstans on the tool interface. The cable system transmission ratios for positioning the tool tip are as follows. (1) Shoulder: 7.7:1 (motor rotations:joint rotations); (2) Elbow: 7.3:1 (motor rotations:joint rotations); (3) Insertion: 133:1 (radians:meters).

Each axis is controlled by two cables, one for motion in each direction, and these two cables are pre-tensioned against each other. The cables are terminated at each end to prevent any possibility of slipping. The cable system maintains constant pretension on the cables through the entire range of motion. Force and motion coupling between the axes is accommodated for in the control system.

Laser pointers attached to the shoulder and elbow joints allow for visual alignment of the manipulator relative to the surgical port. When the two dots converge at the port location, the manipulator is positioned such that its center of rotation is aligned with the pivot point on the abdominal wall. The power-off brakes can be released by flipping a switch located on the base. The brakes are normally powered by the control electronics, but also have a battery plug-in for easy set-up and breakdown when the system is not powered. ABS plastic covers were created on a three-dimensional printer to encapsulate the motor pack thereby protecting actuators, encoders and electrical wiring. Figure 8.11b shows the complete patient site.

8.4 Experimental Evaluation

The methodology is divided into two sections: field experiments and lab experiments that were performed with Raven [24–27, 38–43]. The field experiments were used in part to define latencies associated with different configurations of

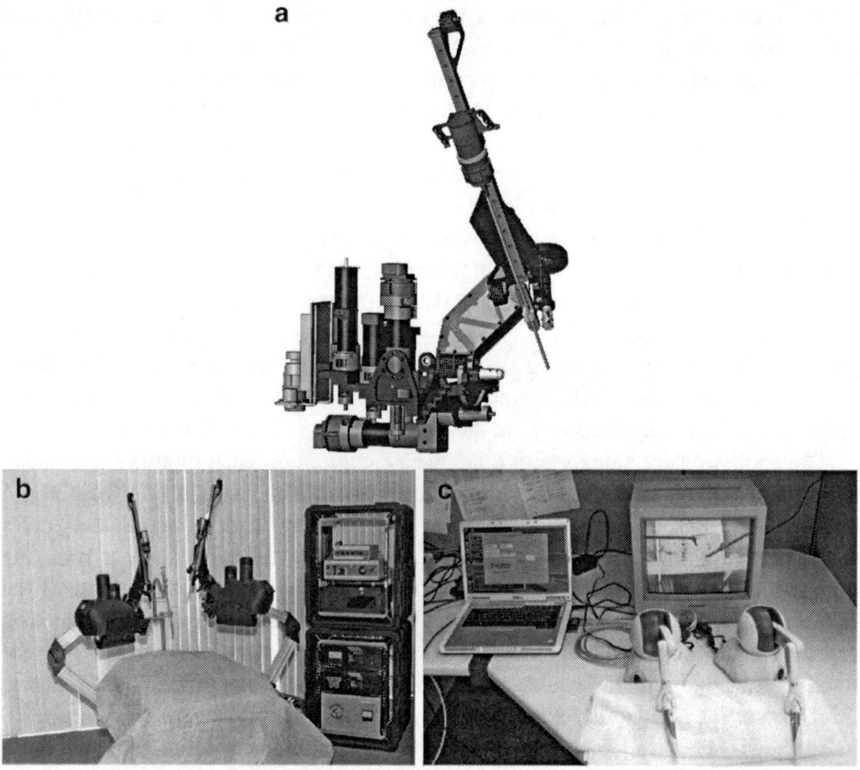

Fig. 8.11 (**a**) RAVEN-Surgical Robotic system. CAD rendering of surgical manipulator shown with plastic covers removed. (Mass: 12.3 kg folded dimensions 61 × 53 × 38 cm extended dimensions: 120 × 30 × 38 cm. (**b**) The RAVEN patient site (Slave) and (**c**) the surgeon site (Master)

network architectures. Based on this information discrete and fixed time delays were selected and emulated in controlled lab experiments.

8.4.1 Preliminary Experimental Evaluation

An adjustable passive aluminum mock-up was fabricated to model the kinematics of the spherical manipulator in parallel and serial configurations (Fig. 8.12 c, d). The mock-up was designed such that a real MIS tool with a 5 mm shaft could pass through the distal joint and allowed for varying the link angles (angles between Joints 1 and 2, and 2 and 3). In a dry-lab set-up, a number of kinematic configurations were compared utilizing a plastic human torso used for training (Simulab, Seattle, WA) to assess range of motion and potential collision (Fig. 8.12a, b). The results indicated that the parallel configuration had a limited workspace with kinematic singularities contained within that workspace (Fig. 8.12a). It was found that mechanisms with large link angles were subject to robot–patient collisions.

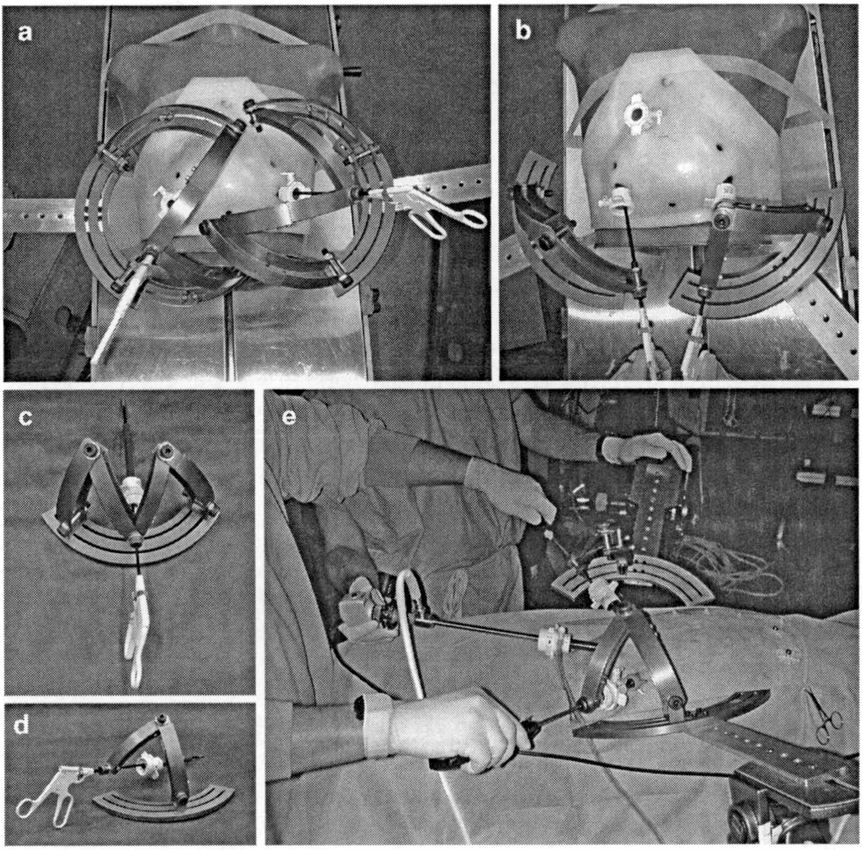

Fig. 8.12 Aluminum Mock-up with adjustable link length and base length. The surgical endo-scopic tool is inserted into a guide located at mechanism apex in a configuration that allow to test different design candidates in a real minimally invasive surgical setup. (**a**) Parallel Configuration (**b**) Serial Configuration (**c, d**) Two configurations tested in a MIS setup. (**e**) Testing the make up in an animal lab. Although the robot manipulator will be teleoperated, in this evaluation the surgeon guides these passive mock-ups in order to evaluate the range of motion

Using two parallel mechanisms created potential self-collisions and constrained the distance between the two mechanisms, even with the link angles chosen as small as possible. Similar problems and constraints were observed for a hybrid configuration with a combination of one serial and one parallel mechanism. The best performance in terms of avoiding self-collision and robot–patient collision was achieved with both mechanisms in serial configuration (Fig. 8.12b).

Using two serial configurations in an experimental surgery with an animal model, surgeons confirmed that the serial 2-link with the dimensions given by the analytical optimization provided sufficient range of motion and did not suffer from self-collision, robot–robot collision, or robot–patient collision during both gross and dexterous manipulations (Fig. 8.12e).

8.4.2 *Field Experiments*

8.4.2.1 Flied Experiments

Seven field experiments were conducted with Raven with various network architectures (wired and wireless) and a wide spectrum of physical distances (Table 8.4). In experiment No. 1 (HAPs/MRT) the system was deployed in two remote sites in desert-like conditions in Simi Valley, CA, while utilizing an Unmanned Aerial Vehicle (UAV) as a wireless node between the sites. In Experiment No. 2, (Imperial College London Collaboration), 6 (Surgical Robot Summer School) and 7 (Tokyo Tech Collaboration), the surgical robot located in Seattle was teleoperated from different sites around the world using commercial internet. In experiments 4 and 5 the surgical robot was deployed in the Aquarius – an undersea habitat, located 3.5 km off-shore in the Florida Keys, and teleoperated from Seattle, WA as well as the National Undersea Research Center, at Key Largo, FL using a combination of wired and wireless communication as part of NASA's NEEMO (NASA Extreme Environment Mission Operations) 12 mission (Fig. 8.13).

8.4.2.2 Lab Experiments

System Setup

In a real teleoperation, physical distance and a real network separate the patient and surgeons sites with time varying delays (Fig. 8.14). When a surgeon makes a gesture using the master device, motion information is sent through the network to the Patient Site with a network time delay (T_n). The manipulator moves and the

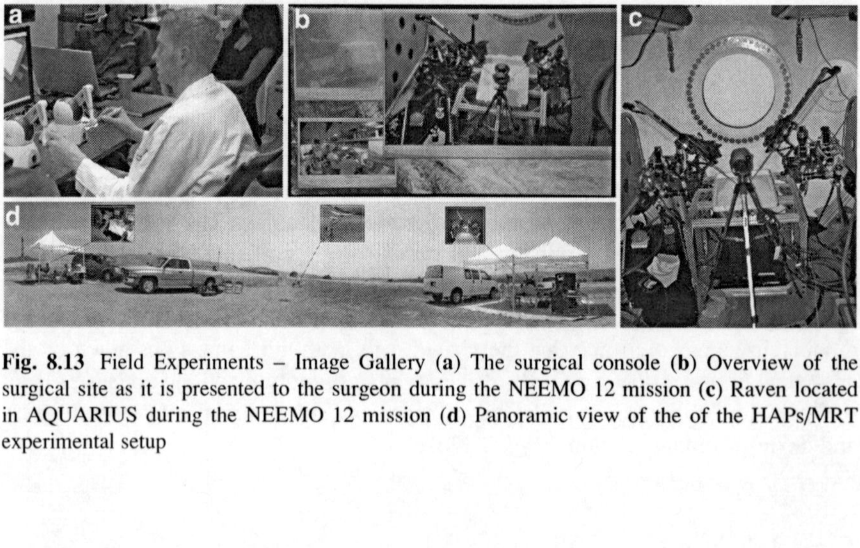

Fig. 8.13 Field Experiments – Image Gallery (**a**) The surgical console (**b**) Overview of the surgical site as it is presented to the surgeon during the NEEMO 12 mission (**c**) Raven located in AQUARIUS during the NEEMO 12 mission (**d**) Panoramic view of the of the HAPs/MRT experimental setup

Table 8.4 Summary of field experiments with Raven

No.	Acronym	Patient site (Slave) surgical robot	Surgeon site (Master) surgical console	Communication layer video	Communication layer network architecture	Time delay (ms) ([a])	Distance (km]
1	HAPs/MRT	Field – simi valley, CA	Field – simi valley, CA	HaiVision, Hai560	Wireless via UAV	16	0.5
2	*ICL*	Seattle, WA	London, UK	iChat/Skype	Commercial internet	172	7,700
3	*Animal Lab*	Seattle, WA	Seattle, WA	Direct S-Video	LAN	1	0
4	*NEEMO 12 – Aquarius*	Aquarius (underwater), Key Largo, FL	Seattle, WA	HaiVision, Hai1000	Commercial internet – seattle WA to key Largo, FL microwave Comm. – Key Largo, FL to Aquarius	76	4,500
5	*NEEMO 12 – NURC*	NURC (land), Key Largo, FL	Seattle, WA	HaiVision, Hai200	Commercial internet	75	4,500
6	*Surgical Robot – Course*	Seattle, WA	Montpellier, France	*iChat/Skype*	*Commercial Internet*	170	8,500
7	*Tokyo Tech Collaboration*	Seattle, WA	Tokyo, Japan	iChat/Skype	Commercial Internet	150	7,600

[a]The time delay refers to the latency in sending position commands between the muster and the salve. The latency regarding the video transmission, compression and decompression was not recorded

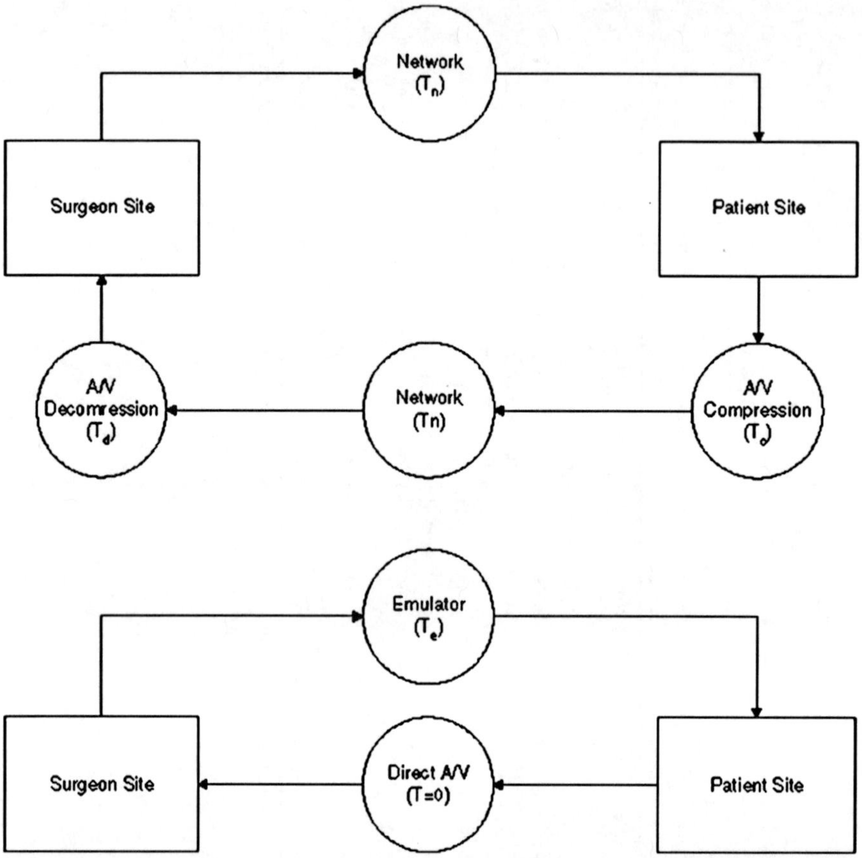

Fig. 8.14 Simplified teleoperation communication flow. Real teleopetaion in the field experiments (*top*) – Detailed diagram is depicted in Fig. 8.1. A setup for the lab experiments with emulated time delay (*bottom*)

audio/video (a/v) device observes the motion. Digital a/v is compressed (T_c), sent from the Patient Site to the Surgeon Site through the network (T_n), then decompressed (T_d) and observed by the surgeon. The surgeon has experienced a total delay $T = 2T_n + T_c + T_d$, from the time (s)he made the gesture to the time that action was observed. In the simulated teleoperation the Surgeon and Patient sites are not separated by physical distance but are connected through a Linux PC with two network cards running NISTNET that emulates a real network. This emulator allows the experimenter to adjust the average packet delay between the Surgeon and Patient sites. The a/v feed is connected directly from the camera at the Patient Site to the monitor at the Surgeon Site through S-video eliminating any delay due to compression/decompression. The surgeon experiences a total delay, T_e due to the emulator, from the time (s)he made the gesture to the time that action was observed.

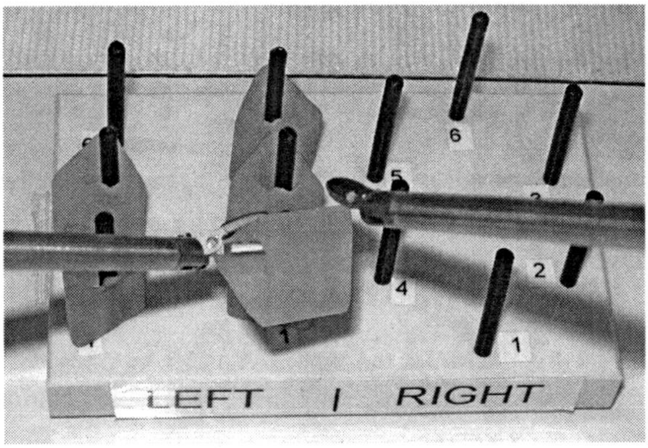

Fig. 8.15 The Society of American Gastrointestinal and Endoscopic Surgeons (SAGES) Fundamental Laaproscopic Skill (FLS) Block Transfer task board set up with the RAVEN

Experimental Design

The Society of American Gastrointestinal and Endoscopic Surgeons (SAGES) developed a curriculum for teaching the Fundamentals of Laparoscopic Surgery (FLS) which includes both cognitive and psychomotor skills. The skills assessment consists of five tasks. The FLS skills tasks have been validated to show significant correlation between score and postgraduate year and are considered by many the "gold standard" in surgical skill assessment. The Block Transfer is one out of these five tasks that emulate tissue handling and manipulation.

The experimental task consists of moving six blocks, one at a time, from the left side of the FLS peg board (Fig. 8.15) to the right side and back to the original position in a sequential predefined order (total of 12 transfers). In our experiment, the completion time as well as the tool tip trajectory were recorded. The three treatments of the experiment included emulated delays of 0, 250 and 500 ms presented to the subjects in randomized order. Each experimental treatment was conducted three times by each subject (nine times total) in a randomized order.

The subjects performed the training tasks first with no delay then with 250 ms delay in order to learn how to teleoperate the RAVEN and minimize the learning effects during the execution of the experimental protocol. Within 1 week from the start of their training, they returned to perform the time delayed block transfer experiment.

Subjects: Definition of the Population

Fourteen subjects, five surgeon and nine non-surgeons, ages ranging from 18 to 43, participated in this study under University of Washington Human Subjects Approval Number 01-825-E/B07.

8.4.3 Results

8.4.3.1 Field Experiments

Given the stochastic nature of the network there is a specific distribution of packet delay. Figure 8.16 depicts the distribution of the delay during the NEEMO experiments given a transmission rate of 1 K. The distribution of the latency is in the range of 63–95 ms with a pick at 78 ms. Table 8.1 defines the average latencies during all the field experiments. One should note that these latencies represent only the delay in sending position command from the master to the slave (T_n). The latency due to the digital a/v compressed (T_c) and decompressed (T_d) which is usually larger and hardware/software dependent and is estimated to be in the range of 200 ms (hardware compression/decompression) in the HAPs/MRT and about 1 s for the commercial internet (software compression/decompression).

Figure 8.17 summarizes the mean completion time for a single expert surgeon (E1) who participated in multiple field and lab experiments performing the block transfer task. In each of the first 3 weeks of training, E1 performed three repetitions of the Block Transfer in the lab environment with effectively no delay. There is a learning effect as E1's mean time improved from week to week. During the NEEMO mission E1 completed a single repetition of the task with the RAVEN in Aquarius and another single repetition with it on-shore in NURC Key Largo, FL. For comparison, E1, who uses a da Vinci clinically was able to complete the block transfer task in about 1 min using the da Vinci, taking only slightly longer with the stereo capability disabled.

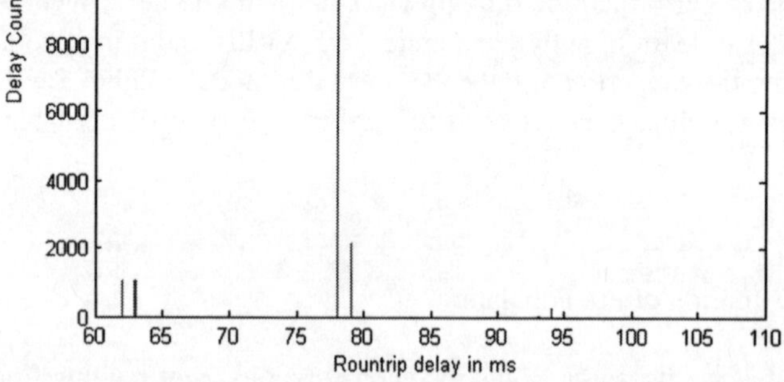

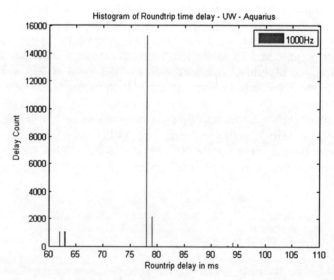

Fig. 8.16 Histogram of number of packets with respect to delay between Seattle WA and Aquarius, Key Largo, FL at a transmission/receiving frequency of 1 K

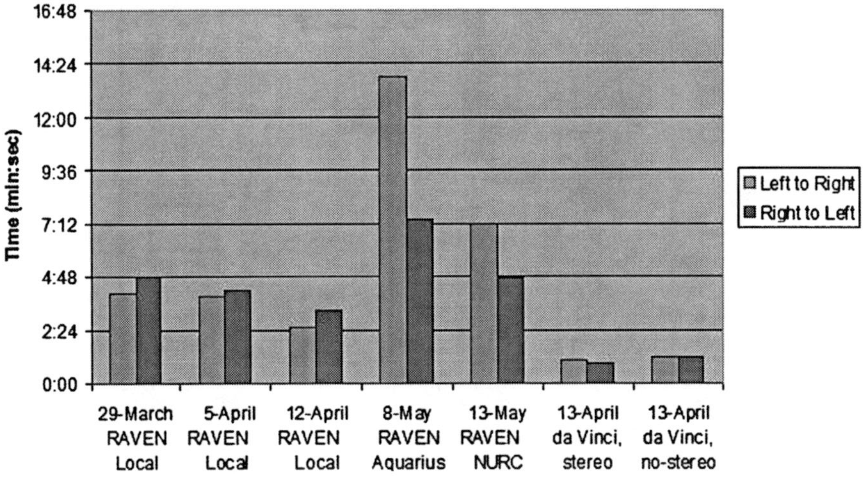

Fig. 8.17 Average block transfer completion times of a single expert surgeon during local training on the RAVEN as well as during the NEEMO mission. Completion times using an ISI da Vinci are included for comparison

8.4.3.2 Lab Experiments

A two-way analysis of variance (ANOVA) was used to determine the effect of the time delays (0, 250, 500 ms) and the subject group (surgeon, non-surgeons) on the three performance parameters: (1) block transfer completion time, (2) the number of errors (dropping the block – recovered and unrecovered); (3) tool tip path length as the response variables (time delays and group type). In general both the completion time and the tool tip trajectory length monotonically increased as the time delay increased (Fig. 8.18). The ANOVA analysis indicated that the difference in mean block transfer time as well as the tool tip path length between each of the three treatments (0, 250, and 500 ms delay) are statistically significant ($\alpha < 0.02$) – Fig. 8.17. While the stated objective of the task was to minimize errors some errors occurred but the number of errors in response to delay effect and surgeon effect were not significant. It is possible that if the task was more technically challenging, the frequency or severity of errors would start to differentiate between subjects with more and less skill. The difference in mean block transfer completion time between surgeons and non-surgeons is not statistically significant.

8.5 Conclusion and Discussion

This chapter provides a comprehensive overview of the development, integration and the experimental evaluation of a surgical robotics system and it application in telesuergry.

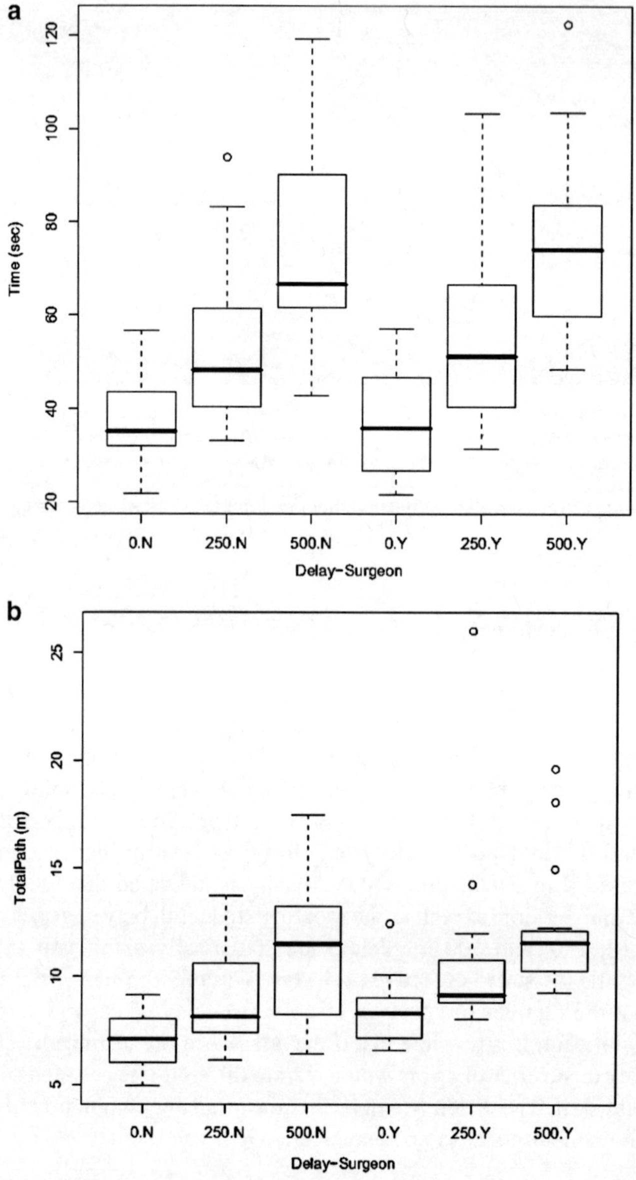

Fig. 8.18 The effect of the block transfer completion time (**a**) and the tool tip path (**b**) are significantly different ($\alpha < 0.02$) for the three time delays (0, 250, 500 ms), for both the surgeon (Y) and the non-surgeon (N)

The design of surgical robotic is based in part on transformation of clinical requirements that are translated to a set of quantitative specifications. The quantitative specifications for the surgical robot were obtained in part during in-vivo experiments with animal models and instrumented surgical tools. These specifications define the workspace of the surgical tools along with and the loads applied on the tissues through the surgical tools in a MIS setup. The quantitative specifications enable a formal mechanism optimization that led to the smallest mechanism with the highest manipulability which meets the clinical requirements. As apart of the optimization process the kinematic equations of parallel, and serial spherical manipulators with link length angles less than 90°. The optimization cost function balanced a guaranteed minimum isotropy and summed isotropy over the target workspace with minimal total link length in order to yield a very compact, high-dexterity mechanism. Given the definitions of the DWS as the highly dexterous workspace and the EDWS as a reachable workspace associated with MIS, the parallel manipulator with the highest composite score had link angles of $\alpha_{13} = \alpha_{24} = 52°$ (Link 1) and $\alpha_{35} = \alpha_{46} = 72°$ (Link 2) and base angle $\alpha_{12} = 0°$. Given the same constrains the serial manipulator with the highest composite score is the one with link angles of $\alpha_{13} = 74°$ (Link 1) and $\alpha_{35} = 60°$ (Link 2).

The base angle α_{12} of the parallel manipulator has a profound influence on the workspace of the mechanism as well as its composite score. The reachable workspace of the parallel manipulator can be defined as the intersection of the workspaces of the left serial link and the right serial link arms acting as independent serial manipulators. When the base angle is zero, these two workspaces have a maximum overlap and therefore more space in which to place the required workspace cone that led to the highest performance score.

As the base angle, α_{12} was varied from 90° to 0° the performance score first decreased, then increased, with a maximal value at a base angle of 0° and a minimal value at a base angle of 70°. In order to understand the relationships between the performance score as a function of the base angle, Link1 and Link2 were fixed at angle of 60° ($\alpha_{13} = \alpha_{24} = \alpha_{35} = \alpha_{46} = 60°$) and the base angle varied. A sequence of plots shows how the singular region that bisects the workspace moves with varying base angles(Fig. 8.8). When the base angle is 90°, the overlap between the odd and even sides is smaller, and the singular region is near the edge of the parallel workspace. Decreasing the base angle to 45°increases the overall workspace but moves the singular region more toward the middle of the workspace, thereby decreasing the usable workspace because the cone must not contain any singularities. Decreasing the base angle to 0°results in maximum workspace and moves the singularity down below the joint limits guaranteeing the reachable workspace that contains no singularities.

Optimization of the parallel mechanism shows that for a fixed base angle, better performance would be achieved by a mechanism design where the angle of Link2 is greater than that of Link1 ($\alpha_{13} < \alpha_{35}$ and $\alpha_{24} < \alpha_{46}$). Based on a spherical geometry, if Link2 is greater than Link1 and the base angle is 0, then the parallel mechanism cannot be put into a singular configuration.

The results for the serial mechanism showed good performance in a small form factor. For any serial design there exists a strip of maximum isotropy across the reachable workspace. This corresponds to a pose where θ_3 is some specific value and θ_1 can be varied. As the link lengths get larger the strip of best performance pushes away from the base (located at Joint 1), yet remains roughly centered in the overall workspace (Fig. 8.9b, d, f).

Fabricating passive aluminum mock-ups of the mechanism under study was critical to the evaluation of this class of manipulator for MIS applications. Through evaluation of various combinations of parallel and serial configurations, it was determined that two serial manipulators provide the smallest footprint and the fewest collisions in the common cases where two or more mechanisms occupy the limited space above the patient body. Based on the kinematic optimization, the link angles of these serial manipulators should be 74°for the first link and 60°for the second link. This configuration and design of manipulators results in the most compact mechanism that will perform MIS procedures with high dexterity in the entire workspace required.

The system integration of a surgical robot is an immense engineering effort that was resulted in an open architecture surgical robotic system that enables a wide spectrum of studies in surgical robotics and teleoperation. The chapter provides an overview of lab and field experiments studying the effect of time delay on surgical performance. A subset of a standard set of tasks (Fundamentals of Laparoscopic Surgery – FLS), which adopted by the MIS surgical community many surgical residency training programs was used in the majority of the field experiments and all the lab experiments. Block transfer emulating tissue handling along with suturing and knot tying defined the subset of the FLS that was adopted for the telerobotics experiments. Time delay is an embedded characteristic of any network. As indicated in the field experiments the latency related to the compression/and decompression of the audio/video (a/v) is significantly larger then the latency related to the transmission of position commands between the master and the slave. As such the a/v transmission latency is the limiting factor that determines the overall performance of the system. The results acquired in the lab experiments indicated a degradation of teleoperation performance characterized by increased completion time and tool path length, which were used as performance measures.

8.6 Visions

The premise of surgical robotics is to deliver high level of dexterity and enhance view of anatomical structures that is either too small or difficult to access (expose) for conventional surgical tools, the surgeons' fingers and a line of site. An approach to this anatomical structure may use either an open surgery technique, MIS or minimally access techniques that were otherwise. Aviation is often used a sources of inspiration for surgery. In both of these frameworks the pilot and the surgeon have to deal with complex system with high level situation awareness to communicate with additional individuals such as Co-Pilot and air traffic control personal in the aviation domain, and nurses (circulation nurse or sterile nurse) along with co-surgeons, and anesthesiologist in the surgery domain in addition to the immediate interaction with the system which is the airplane in aviation and the surgical site during surgery. Regardless of the many similarities between the two domains one should note that there is one fundamental difference between them. In aviation the goal is to fly an airplane from point A to Point B with minimal interaction with the environment. However in surgery the goal is to change the environment (anatomical structure) in order to alter its function.

Although in small aircraft the pilot control the airplane through the stick which is analogues to a surgeon holding a surgical tool, automation is a concept that is widely used in a form of an automatic pilot in commercial and military airplanes. Automation in this context removes the human form the low-level operation (e.g. correcting a drift in the airplane flight path or manipulating the soft tissue due to deformation in surgery) and reposition the pilot or the surgeon as a high-level decision maker using supervisory control mode with interrupted intervention in critical steps of the operation.

With the introduction of a surgical robot automation in surgery can be implemented at two different levels: (1) system services and (2) surgical operation. The Trauma Pod project as envisioned by Dr. Richard Satava and its translation into practice as part of the a phase 1 DARPA project (see for details the chapter by Pablo Garcia) demonstrated that services to the surgical robot along with overall operating room (OR) management can be fully automated. The end-result was a fully automated operating room in which the only human in the OR may be only the patient. A centralized robotic surgical nurse provided services to the surgical robot such as tool dispensing from an automatic tool changer and changing, as well as equipment dispensing and disposing from an automatic equipment dispenser, vital signs monitoring, and overall supply chain management (inventory monitoring) – actions that took place upon voice command by a surgeon interacting with the surgical console.

A far as services to the surgical robot are concerned (tool change and equipment dispensing) centralized solution was used in a form of a single surgical nurse. As a result the performance of the entire system in providing these two services is dictated by the performance of the surgical nurse alone. An alternative approach in which each surgical robot is attached to another robotic arm in a micro–macro configuration, allows each surgical robot to be an independent unit that may serve itself (replacing it own tools and picking its own supplies) – Fig. 8.19. In this configuration the nurse function is distributed among all the robotics arms allowing each one to function autonomously.

The micro–macro approach to surgical robotics in which a gross manipulator carry a high dexterous manipulator is inspired by the human arm and hand in which the human arm serves as the gross manipulator positions the wrist in space for high dexterity manipulation by the hand. This approach may allow to design small and fast responding manipulator with high manipulability in a limited workspace which occupy only a subset of the surgical site (micro) and mount it on a land slow manipulator with a large workspace (macro). The macro manipulator may move the base of the micro manipulator in case it will sense the micro manipulator operates close to the edge of its workspace. The movement is obviously transparent to the surgeon who should not be involved nor aware of these adjustments of workspaces. This distributed architecture from the functional perspective may also allow to distribute the surgical robotics arm around the patients in a way that will avoid robot to robot collision.

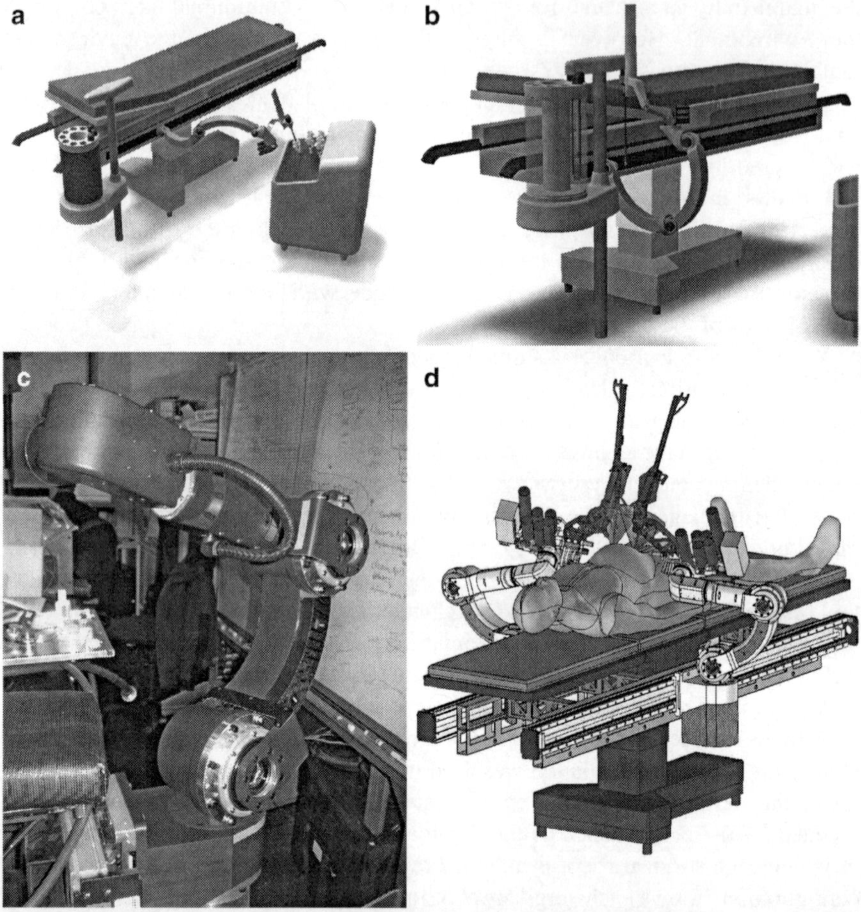

Fig. 8.19 Distributed approach of automating services in the operating room. CAD rendering of a gross positioning arm (macro arm) carrying a surgical robotic arm (micro) replacing it own tool (**a**) or picking its own supply (**b**). A design of a macro arm named the "C-arm" (**c**). CAD rendering showing the Raven system (micro) carried by the C-arm (macro)

Automating the sub-tasks of the surgical procedure itself is another aspect that may contribute to the process in which the surgeons can be removed from his or her position of manually controlling the surgical robot to a position where he or she supervise the operation. In the context of tissue manipulation the surgeon may point to the target position and the robot will execute the command autonomously given constraints regarding obstacle avoidance and stress thresholds. One may envision a scenario in which the surgeon define the beginning and the end of the suture and the surgical robot operating in an autonomous mode stitch the tissue and tie a knot. One of the limiting factors to execute such a task autonomously is our limited capabilities in simulating and predicting the response of soft tissue to external loads. Once the robot interacts with the tissue its geometry changes significantly. Our limited capabilities in predicting the soft tissue response are due to the non-linear, and un-isotropic nature of soft tissues. Another futuristic scenario may involve automatic edge detection and dissection of a tumor based on preoperative scans or biomarker that may be injected into the tissue.

Introducing a robotic system into the OR and utilizing such a system clinically was a major milestone for the field of surgical robotics. Emerging technologies based on scientific discoveries will continue to flow into the surgical robotic systems and provide better tool in the hands of the surgeons for delivering high quality healthcare.

Acknowledgment Development of the RAVEN was supported by the US Army, MRMC, grant number DAMD17-1-0202. The HAPs/MRT project was supported by the US Army TATRC, grant number W81XWH-05-2-0080. The surgical robot development was a collaborative team effort of students and faculty members in engineering and medicine who developed the system over the years including: Mitchell J.H. Lum (EE), Hawkeye King (EE), Diana C.W. Friedman (ME), Denny Trimble (ME), Thomas S. Lendvay (Children's Hospital – Seattle), Andrew S. Wright (Surgery), Mika N. Sinanan (Surgery). The authors would like to thank the HAPs/MRT collaborators at the University of Cincinnati, AeroVironment, and HaiVision as well as our collaborators in London at Imperial College and at Tokyo Tech. The NEEMO XII participation has been supported by the US Army TATRC grant number W81XWH-07-2-0039. The authors would like to thank the NEEMO collaborators from University of North Carolina at Wilmington, US Navy, National Undersea Research Center, National Oceanographic and Atmospheric Administration, NASA, and in particular, Dr. Timothy Broderick, Charles Doarn, and Brett Harnett from the University of Cincinnati.

References

1. Satava, R.: Disruptive visions: the operating room of the future, Surg. Endosc. **17**(1), 104–107 (2003)
2. Satava, R.M.: Surgical robotics: the early chronicles: a personal historical perspective, Surg. Laparosc. Endosc. Percutan. Tech. **12**, 6–16 (2002)
3. Satava, R.M.: Emerging technologies for surgery in the 21st century, Arch. Surg. **134**, 1197–1202 (1999)
4. Howe, R.D., Matsuoka, Y.: Robotics for surgery. Ann. Rev. Biomed. Eng. **1**, 211–240 (1999)
5. Buess, G.F., Schurr, M.O., Fischer, S.C.: Robotics and allied technologies in endoscopic surgery, Arch. Surg. **135**, 229–235 (2000)
6. Davies, B.: A review of robotics in surgery. Proc. Instn. Mech. Eng. **214**(H), (2000)
7. Clearly, K., Nguyen, C.: State of the art in surgical robotics: clinical applications and technology challenges. Comput. Aided Surg. **6**, 312–328 (2001)
8. Bloom, M.B., Salzberg, A.B., Krummel, T.M.: Advanced technology in surgery. Curr. Probl. Surg. **39**(8), 733–830 (2002)
9. Taylor, R.H., Stoianovici, D.: Medical robotics in computer-integrated surgery, Robotics and Automation. IEEE Trans. Robot. Autom. **19**(5) 765–781 (2003)
10. Lanfranco, A.R., Castellanos, A.E., Desai, J.P., Meyers, W.C.: Robotic surgery a current perspective. Ann. Surg. 239(1), 14–21 (2004)
11. Louw, D.F., Fielding, T., McBeth, P.B., Gregoris, D., Newhook, P., Sutherland, G.R.: Surgical robotics: a review and neurosurgical prototype development. Nuerosurgery **54**(3), 525–537 (2004)

12. Speich, J.E., Rosen, J.: Medical robotics, In: Wnek, G., Bowlin, G. (eds.) Encyclopedia of Biomaterials and Biomedical Engineering, pp. 983–993. Marcel Dekker, New York (2004)
13. Binder, J., Bräutigam, R., Jonas, D., Bentas, W.: Robotic surgery in urology: fact or fantasy? BJU Int. **94**(8), 1183–1187 (2004)
14. Korb, W., Marmulla, R., Raczkowsky, J., Mühling, J., Hassfeld, S.: Robots in the operating theatre – chances and challenges, Int. J. Oral. Maxillofac. Surg. **33**(8), 721–732 (2004)
15. Diodato, Jr, M.D. Prosad, S.M., Klingensmith, M.E., Damiano, Jr, R.J.: Robotics in surgery. Curr. Probl. Surg. **41**(9), 752–810(2004)
16. Camarillo, D.B., Krummel, T.M., Salisbury, J.K.: Robotic technology in surgery: past, present, and future. Am. J. Surg. **188**(Suppl to October 2004), 2S–15S (2004)
17. Curley, K.C.: An overview of the current state and uses of surgical robots. Oper. Tech. Gen. Surg. **7**(4), 155–164 (2005)
18. Kwoh, Y.S. Hou, J.L. Jonckheere, E.A., Hayall S.: A robot with improved absolute positioning accuracy for ct guided stereotactic brain surgery. IEEE Trans. Biomed. Eng. **35**(2), 153–161 (1988)
19. Takahashi, H., Warisawa, S., Mitsuishi, M., Arata, J., Hashizume, M.: Development of high dexterity minimally invasive surgical system with augmented force feedback capability. Biomedical Robotics and Biomechatronics, 2006. BioRob 2006. In: The First IEEE/RAS-EMBS International Conference on, pp. 284–289 (2006)
20. Madhani, A.J. Niemeyer, G., Salisbury, Jr, J.K.: The black falcon: a teleoperated surgical instrument for minimally invasive surgery. In: Proceedings of the Intelligent Robots and Systems, vol. 2, pp. 936–944 (1998)
21. Cavusoglu, M.C. Tendick, F., Cohn, M., Sastry, S.S.: A laparoscopic telesurgical workstation. IEEE Trans. Robot. Autom. **15**(4), 728–739 (1999)
22. Marescaux, J.: Transatlantic robot-assisted telesurgery. Nature **413**, 379–380 (2001)
23. Marescaux, J. Leroy, J., Rubino, F., Smith, M., Vix, M.,Simone, M., Mutter, D.: Transcontinental robot-assisted remote telesurgery: feasibility and potential applications advances in surgical technique. Ann. Surg. **235**(4), 487–492 (2002)
24. Rosen, J., Hannaford, B.: Doc at a distance. IEEE Spectrum, 34–38 (2006)
25. Sankaranarayanan, G., Hannaford, B., King, H., Ko, S.Y., Lum, M.J.H., Friedman, D., Rosen, J., Hannaford, B.: Portable surgery master station for mobile robotic surgery. In: Proceedings of the ROBOCOMM, the first International, conference on Robot Communication and Coordination, Athens, Greece, Oct 2007
26. Brett, H., Doarn, C., Rosen, J., Hannaford, B., Broderick, T.J.: Evaluation of unmanned airborne vehicles and mobile robotic telesurgery in an extreme environment, Telemed. e-Health **14**(6) 534–544 (2008)
27. Lum, M.J.H., Friedman, D.C.W., Sankaranarayanan, G., King, H., Fodero, II, K., Leuschke, R., Hannaford, B., Rosen, J., Sinanan, M.N.: The RAVEN – a multidisciplinary approach to developing a telesurgery system. Int. J. Robot. Res. Special Issue: Medical Robotics Part I 28 (9), 1183–1197 (2009)
28. Mitsuishi, M., Arata, J., Tanaka, K., Miyamotoamd, M., Yoshidome, T., Iwata, S., Hashizume, M., Warisawa, S.: Development of a remote minimally-invasive surgical system with operational environment transmission capability. In: Robotics and Automation, 2003. Proceedings. ICRA'03. IEEE International Conference, vol. 2, pp. 2663–2670 (2003)
29. Berkelman, P., Ji, M.: The University of Hawaii teleoperated robotic surgery system. In: Intelligent Robots and Systems, 2007. IROS 2007. IEEE/RSJ International Conference on, pp. 2565–2566, 29, 2 Nov 2007
30. Berkelman, P., Ji, M.: A compact, modular, teleoperated robotic minimally invasive surgery system. In: Biomedical Robotics and Biomechatronics, 2006. BioRob 2006. The First IEEE/RAS-EMBS International Conference (2006)
31. Rosen, J., Brown, J.D., Chang, L., Sinanan, M.N., Hannaford, B.: Generalized approach for modeling minimally invasive surgery as a stochastic process using a discrete Markov model, IEEE Trans. Biomed. Eng. **53**(3), 399–413 (2006)

32. Rosen, J., MacFarlane, M.,Richards, C., Hannaford, B.,Pellegrini, C., Sinanan, M.N.: Surgeon/ Endoscopic Tool Force-Torque Signatures in the Evaluation of Surgical Skills During Minimally Invasive Surgery, Studies in Health Technology and Informatics – Medicine Meets Virtual Reality, vol. 62, pp. 290–296. IOS Press, Amsterdam (1999)
33. Rosen, J., Solazzo, M.,Hannaford, B., Sinanan, M.N.: Objective Laparoscopic Skills Assessments of Surgical Residents Using Hidden Markov Models Based on Haptic Information and Tool/Tissue Interactions, Studies in Health Technology and Informatics – Medicine Meets Virtual Reality, vol. 81, pp.417–423. IOS Press, Amsterdam (2001)
34. Rosen, J., Brown, J.D., Barreca, M., Chang, L., Hannaford, B., Sinanan, M.N.: The Blue DRAGON – a System for Monitoring the Kinematics and the Dynamics of Endoscopic Tools in Minimally Invasive Surgery for Objective Laparoscopic Skill Assessment, Studies in Health Technology and Informatics – Medicine Meets Virtual Reality, vol. 85, pp.412–418. IOS Press, Amsterdam (2002)
35. Rosen, J, Brown, J.D., Chang, L., Barreca, M., Sinanan, M.N., Hannaford, B.: The blue DRAGON – a system for measuring the kinematics and the dynamics of minimally invasive surgical tools in-vivo. In: Proceedings of the 2002 IEEE International Conference on Robotics & Automation, Washington DC, USA, 11–15 May 2002
36. Rosen, J., Chang, L., Brown, J.D., Hannaford, B., Sinanan, M.N., Satava, R.: Minimally invasive surgery task decomposition – etymology of endoscopic suturing, studies in health technology and informatics – medicine meets virtual reality, vol. 94, pp. 295–301. IOS Press, Amsterdam (2003)
37. Taylor, R.H., Paul H.A., Brent, D., Mittelstadt et al.: Robotic Total Hip Replacement in days. Proc 11 th IEEE Medicine & Biobgy Conft. pp. 887–892, Seattle, 1989
38. Lum, M.J.H., Rosen, J., King, H., Friedman, D.C.W., Donlin, G., Sankaranarayanan, G., Harnett, B., Huffman, L., Doarn, C., Broderick, T., Hannaford, B.: Telesurgery via unmanned aerial vehicle (UAV) with a field deployable surgical robot. In: Proceedings of Medicine Meets Virtual Reality (MMVR 15), pp. 313–315, Long Beach CA, 6–9 Feb 2007
39. Lum, M., Friedman, D., King, H., Donlin, R., Sankaranarayanan, G., Broderick, T., Sinanan, M., Rosen, J., Hannaford, B.: Teleoperation of a surgical robot via airborne wireless radio and transatlantic internet links. In: The 6th International Conference on Field and Service Robotics, Chamonix, France, 9–12 Jul 2007
40. Lum, M.J.H., Friedman, D.C.W., Sankaranarayanan, G., King, H., Wright, A., Sinanan, M.N., Lendvay, T., Rosen, J., Hannaford, D.: Objective assessment of telesurgical robot systems: telerobotic fls, medicine meets virtual reality (MMVR 16) pp. 263–265, Long Beach CA, Jan 29–Feb 1 2008
41. Lum, M.J.H., Rosen, J., Lendvay, T.S., Wright, A.S., Sinanan, M.N., Hannaford, B.: Tele-Robotic fundamentals of laparoscopic surgery (FLS): effects of time delay – pilot study. In: 30th Annual International Conference of the IEEE Engineering in Medicine and Biology Society EMBS, pp. 5597–5600, Vancouver, British Columbia, Canada, 20–25 Aug 2008
42. Lum, M.J.H., Rosen, J., Lendvay, T.S., Sinanan, M.N., Hannaford, B.: Effect of Time Delay on TeleSurgical Performance. In: IEEE International Conference on Robotics and Automation, Kobe, Japan, 12–17 May 2009
43. Lum, M.J.H., Rosen, J., King, H., Friedman, D.C.W., Lendvay, T., Wright, A.S., Sinanan, M.N., Hannaford, B.: Teleopeartion in Surgical Robotics – Network Latency Effects on Surgical Performance. In: 31th Annual International Conference of the IEEE Engineering in Medicine and Biology Society EMBS, Minneapolis MN, Sept 2009

Chapter 9
The *da Vinci* Surgical System

Simon DiMaio, Mike Hanuschik, and Usha Kreaden

Keywords *AESOP*® · clinical indication clearance history · Computer Motion, Inc. · *da Vinci*® · device clearance history · endoscope control manipulator · Food and Drug Administration (FDA) · Intuitive Surgical, Inc., · laparoscopic surgery · master manipulator · patient-side manipulator · robotic surgery history · safe medical device act · surgical console · telemanipulation · tele-robotic surgery · *Zeus*®

9.1 Introduction

In this chapter we describe the evolution of the *da Vinci* surgical system from the very early days of Intuitive Surgical, through to 2009. In order to provide context, we begin with a short summary of the origins of telerobotic surgery itself. This involves a unique convergence of technologies and clinical needs, as well as several groups of individuals who independently recognized the role that robotics and telepresence technologies could play in medicine. The regulatory landscape has played – and continues to play – an important role in the development and use of surgical devices such as *da Vinci*. In this chapter, we describe some of the important aspects of device regulation and how they affect the deployment of medical devices such as ours. It should be noted that this story is told from the Intuitive Surgical perspective and is not intended to be exhaustive. Nevertheless, we hope that it provides some insight into the unique process of invention and development that has resulted in a marriage of technology and medicine that benefits hundreds of patients each day.

S. DiMaio (✉)
Intuitive Surgical Inc., 1266 Kifer Road, Sunnyvale, CA, USA
e-mail: simon.dimaio@intusurg.com

J. Rosen et al. (eds.), *Surgical Robotics: Systems Applications and Visions*,
DOI 10.1007/978-1-4419-1126-1_9,

9.2 The Origins of Telerobotic Surgery

The story of telerobotic surgery involves the convergence and intersection of two very different technologies and the emergence of a completely new approach to minimally-invasive surgery. The first of these technologies emerged in the 1940s and was called "telemanipulation." Robert A. Heinlein's 1942 science fiction short story, titled "Waldo," described a glove and harness device that allowed the lead character, Waldo Farthingwaite-Jones – born frail and weak, and unable to lift his own body weight – to control a powerful mechanical arm by merely moving his hand and fingers. It was not long before these kinds of remote manipulators – popularly known as "waldoes" – were developed in the real world for moving and manipulating hazardous radioactive materials. These devices made use of cables and linkages to allow an operator to safely stand on one side of a thick, leaded glass window while maneuvering a mechanical manipulator located in a radioactive environment on the other side of the glass partition. In the 1950s, Raymond Goertz and his contemporaries explored applications that required remote control over greater distances. They replaced the mechanical linkages with electrical sensors, signals and actuators, thus allowing for greater separation between the operators and their robotic "slaves," as well as for more complex applications [1] (Fig. 9.1).

In the medical arena, it is thought that the earliest ideas on endoscopic techniques date back as far as an Arabian physician named Albukasim (936–1013 A.D.). However, it was not until the early 1800s that Phillip Bozzini developed some of the first practical methods for observing the inside of the living human body – he used a light-guiding instrument that he named the *Lichtleiter* to examine the urinary tract, the

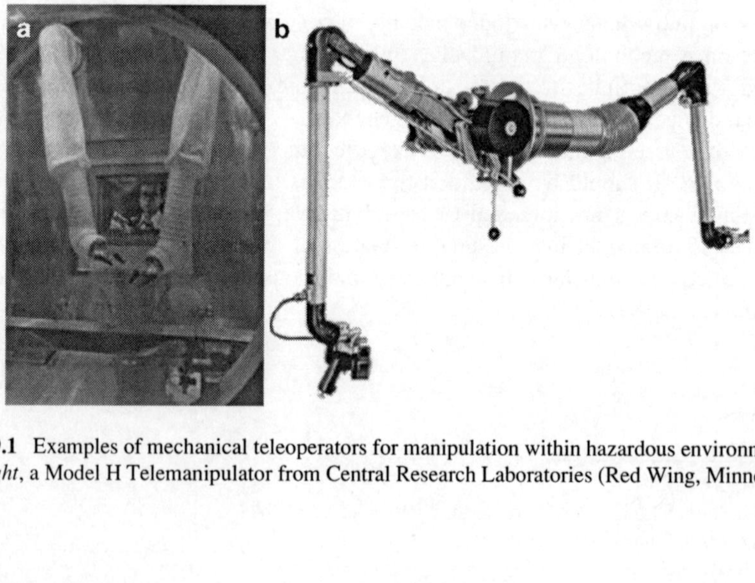

Fig. 9.1 Examples of mechanical teleoperators for manipulation within hazardous environments. At *right*, a Model H Telemanipulator from Central Research Laboratories (Red Wing, Minnesota)

rectum, and pharynx. The development of endoscope-like devices and their applications continued into the 1980s, at which time the emergence of Charge Coupled Devices (CCDs), video electronics and display technologies revolutionized the field and led to laparoscopic techniques for minimally invasive surgery, culminating in the first laparoscopic cholecystectomy in 1987, by French physician Mouret [2].

It would be another decade before the emergence of telerobotic minimally-invasive surgery—a decade in which some key pieces of technology began to emerge. The principles of telemanipulation had been further advanced for application to hazardous material handling, deep sea and space exploration. In the 1980s – with the rapid advancement of microelectronics and computing – virtual reality techniques started to develop the concept of immersive environments, with technologies such as head-mounted displays, graphically rendered synthetic 3D worlds, and haptic interfaces. By the late 1980s and early 1990s, laparoscopy had blossomed; however, limitations in the technique were being reached. The tools that were being employed for manual laparoscopy worked well for relatively simple surgical procedures that involved the excision of tissue and basic tissue closure – procedures such as laparoscopic cholecystectomy (removal of the gall bladder), oophorectomy (removal of an ovary or ovaries), and simple hysterectomy (removal of the uterus) were broadly adopted. At the same time, applications requiring complex reconstruction failed to adopt these laparoscopic techniques in a significant way. Sophisticated mechanisms, such as staplers and other tissue closure devices were developed; however, these still did not allow laparoscopic techniques to gain traction in the most complex of surgical procedures.

During the late 1980s and early 1990s, several groups began to see the potential cross-over between the telemanipulation and virtual reality principles that had been developed for material handling and remote exploration, and the challenges that were being experienced in minimally-invasive laparoscopic surgery. These groups observed the limitations of minimally-invasive surgical techniques at the time and recognized the impact that telerobotics could have in enabling new capabilities and applications. Below are just a few of the groups who first began to marry telerobotic technologies with minimally-invasive surgical techniques, and that would ultimately influence the development of the *da Vinci* Surgical System:

- Dr. Russell Taylor and his group, then at the IBM Watson Research Center (Yorktown Heights, NY), in collaboration with Dr. Mark Talamini, a general surgeon from Johns Hopkins University, developed the Laparoscopic Assistant Robotic System (LARS) – a "third hand" that allowed surgeons to manipulate a laparoscopic endoscope under joystick control [3].
- At the University of California at Santa Barbara, Dr. Yulun Wang developed a robotic system for NASA and later used it to manipulate an endoscope for laparoscopic surgery in 1992. This become the seed for the Automated Endoscopic System for Optimal Positioning (*AESOP*) system and a company called Computer Motion, Inc. Dr. Wang worked with Dr. Carlos Gracia, a leading endoscopic surgeon at San Ramon Regional Medical Center (San Ramon, CA) [4].

- At MIT, Professor Kenneth Salisbury and his students developed innovative human–machine interfaces and haptics. One of his graduate students, Dr. Akhil Madhani, developed the "Black Falcon" as part of his doctoral work: a teleoperated surgical instrument with force feedback. Another alumnus of the Salisbury lab, David Brock collaborated with Dr. Gary Rogers, a general surgeon from the Boston University Medical Center (Boston, MA), and together they created a company called Brock–Rogers Surgical in 1996 (renamed endoVia Medical, Inc., in 2002) [5].
- Dr. Hari Das at the Jet Propulsion Laboratory (Pasadena, California) worked with ophthalmic surgeon Steve Charles. Funded by NASA, their Robot Assisted Microsurgery (RAMS) workstation also made the connection between telerobotics and minimally-invasive surgery [6].
- Professor Brian Davies and his team at Imperial College (London, UK) developed robotic mechanisms for prostate and neurosurgical applications – their I.C. PROBOT prostate surgery device was trialed in the early 1990s [7].
- Professor Blake Hannaford and his group at the University of Washington (Seattle, Washington) began experimenting with teleoperation and haptics for minimally-invasive surgical applications in the mid-1990s [8].
- Phil Green, then at SRI International (Menlo Park, CA), in collaboration with surgeons at Stanford University, as well as army surgeon Dr. John Bowersox developed the "telepresence surgery system," a device that would later contribute key components to early *da Vinci* prototypes.

9.2.1 Early Funding Sources

Several of the early pioneers in telerobotic surgery shared common funding from military sources. In the United States, Dr. Richard Satava, a program manager for the Defense Advanced Research Projects Agency (DARPA), became interested in the idea of robot-assisted battlefield surgery, and began funding telerobotics research programs in the early 1990s. The team at SRI – after starting with their own internal funding and funds from an NIH grant in the late 1980s – was one of the first to begin receiving support from DARPA. Dr. Yulun Wang received initial funding from DARPA to develop his early voice-activated robotic camera manipulator, while Dr. Salisbury and his group at the MIT Artificial Intelligence Lab were also funded by DARPA initiatives at this time.

The military vision for this technology was the idea that mobile telerobotic systems could be used to perform surgery on wounded soldiers immediately at the front lines of the battlefield, under the control of surgeons located out of harm's way, at remote locations. While this model of front-line surgery shifted toward front-line stabilization and rapid evacuation during the Iraq conflict, these DARPA funding initiatives can be credited with helping to provide a significant portion of the early support for telesurgery research.

9.2.2 The Intuitive Surgical Timeline

In 1994, Dr. Frederick Moll became interested in the telerobotic system that had been developed by SRI. He left, his employer at the time, Guidant to attempt to raise venture capital on his own and in 1995 was introduced to Rob Younge, co-founder of Acuson Corporation (a manufacturer of diagnostic ultrasound equipment). Fred Moll, Rob Younge and John Freund, a former venture capitalist from Acuson, collaborated to write a business plan and succeeded in raising initial venture capital for the newly incorporated company – Intuitive Surgical Devices (Fig. 9.2).

This was the beginning of a fast-paced technology development effort. Intuitive Surgical licensed the telepresence surgery technology from SRI and began hiring engineers. An engineering team was in place by April of 1996 and over the next 3 years this team developed three generations of technology prototypes that would support the first set of animal and human trials, eventually culminating in Intuitive's flagship product, the *da Vinci* Surgical System. Along the way, additional technology and staff came from many of the early pioneers in the field, including from Dr. Russell Taylor's laboratory at IBM Research, as well as from Dr. Ken Salisbury's group at MIT.

While awaiting FDA clearance in the United States, Intuitive Surgical began marketing the *da Vinci* System in Europe in 1999. The company raised $46 million in an Initial Public Offering in June, 2000 and a month later was granted FDA approval for applications in general surgery. The following year, in 2001, the FDA cleared the use of the system for thoracoscopic (chest) and radical prostatectomy surgeries.

Shortly before its Initial Public Offering in the year 2000, Intuitive Surgical was sued for patent infringement by its then competitor, Computer Motion – makers of the *Zeus* Surgical System that had launched in 1997. The *Zeus* System was based on their earlier product called *AESOP*, a voice-controlled endoscope manipulator that was the first surgical robotic device to receive FDA approval. The *Zeus* concept was to provide the laparoscopic surgeon with improved precision and tremor filtration. The *da Vinci* approach was somewhat different in the sense that it sought to recreate the feeling of open surgery, but with the potential benefits of minimally-invasive

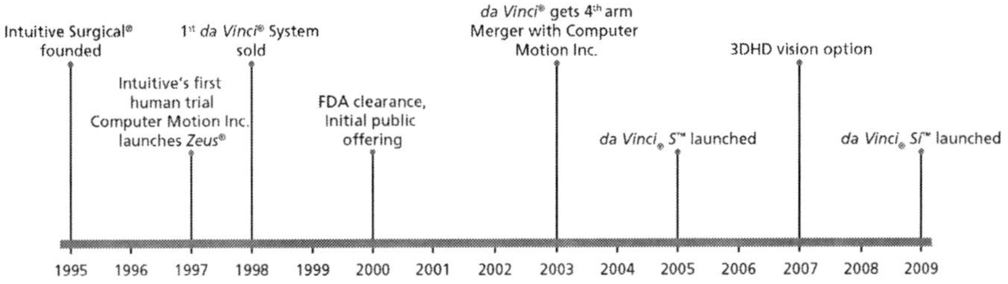

Fig. 9.2 The intuitive surgical timeline

access. To achieve this, the *da Vinci* System had to provide the surgeon with 3D stereo vision, intuitive and dexterous manipulation of surgical instruments inside of the patient. While the *Zeus* system was marketed primarily to laparoscopists, the *da Vinci* System was aimed at the open surgeon. Competition between the companies centered around the differences between these two philosophies. Initially, the *Zeus* was preferred by the general laparoscopic surgeon, while the *da Vinci* was adopted by the open surgeon who did not perform laparoscopic surgery. *Zeus* was smaller, had a lower price point, but was less capable. *da Vinci* was bulky and often accused of being over-engineered. By 1999, Computer Motion had begun to adjust their approach and started moving toward the course that Intuitive Surgical was on.

Shortly after Computer Motion brought forward its lawsuit, Intuitive Surgical counter-sued and the two fledgling companies became embroiled in a legal battle that severely limited growth on both sides. In 2003, as both companies were running out of funds, Intuitive Surgical and Computer Motion agreed to merge, thus ending the litigation between them. At that time, the *Zeus* System was phased out in favor of the *da Vinci* System, because of *da Vinci*'s additional capability.

9.3 The Evolution of the *da Vinci* Surgical System

When Intuitive Surgical made its start toward the end of 1995 the product vision included four key specifications, or product pillars. First and foremost, the system needed to be reliable and failsafe in order to be feasible as a surgical device; second, the system was to provide the user with intuitive control of the instrument; third, the instrument tips were to have six-degree-of-freedom dexterity as well as a functional gripper. The fourth pillar was that of compelling 3D vision. These product pillars supported the ultimate goal to regain several key benefits of open surgery that had been lost in the laparoscopic approach – while maintaining minimal invasiveness – by virtually transposing the surgeon's eyes and hands into the patient in a reliable and effective way. The technology that was licensed from SRI, IBM and MIT provided a starting point for realizing this vision.

In 1995, the SRI prototype system had four-degree-of-freedom instrument manipulators, plus a gripper. It had a simple master interface that allowed the user to command instrument motion, and a control system to intuitively match the motions of the instrument tip with commanded motions of the master interface. Kinematic similarities between the master and slave mechanisms simplified the mathematics behind this intuitive mapping. From this prototype, three generations of prototypes were derived over a period of 3 years, culminating with the *da Vinci* Surgical System – Intuitive Surgical's first commercial product to be shipped to customers.

The *da Vinci* System was named during the very first month of Intuitive Surgical's existence. The renowned renaissance icon, Leonardo da Vinci had combined art, science, anatomy and engineering throughout his career of invention and innovation: a combination that seemed to befit Intuitive's vision. Early prototypes were not named *da Vinci*, as this title was reserved for later. The first prototype was named after Leonardo, but contracted to "Lenny" for fear that this first attempt would fail to meet expectations.

9.3.1 Lenny

Beginning from the SRI prototype, this first design iteration added a wrist at the end of the patient-side manipulator, thus increasing the total number of degrees of freedom at the tip of the instrument from five to seven. At that time, the instrument was not interchangeable, as shown in Fig. 9.3a. The kinematic similarity and intuitive mapping between the master and slave manipulators – as exhibited by the SRI prototype – were maintained. These patient-side manipulators were mounted to the operating table by means of a simple positioning mechanism that could be manually adjusted using a screwdriver and wrench (this is shown in Fig. 9.7a).

In order to visualize of the surgical field, a commercially available Welch Allyn 3D endoscope – shown in Fig. 9.4 – was mounted to an endoscope manipulator that was almost identical to that of the SRI prototype. This endoscope featured two CCD video chips mounted at the tip of the endoscope, each with an image resolution below that of Standard Definition NTSC video, but nevertheless sufficient to provide the user – seated at a control console – with stereo video of the surgical site. A commercial CrystalEyes display system was used to display the stereo video coming out of the Welch Allyn endoscope. This system alternated left- and right-eye video frames on a single video monitor, with the user having to wear a pair of active shutter glasses that alternated left and right eye views in synchrony with the display. The CrystalEyes system is shown in Fig. 9.8a.

The Lenny prototype was completed and taken to animal trials during the summer of 1996, with an initial focus on complex general surgery applications. A great deal of important learning occurred during this first set of in vivo experiments. For example, it was clear from the experience that the six-degree-of-freedom motion

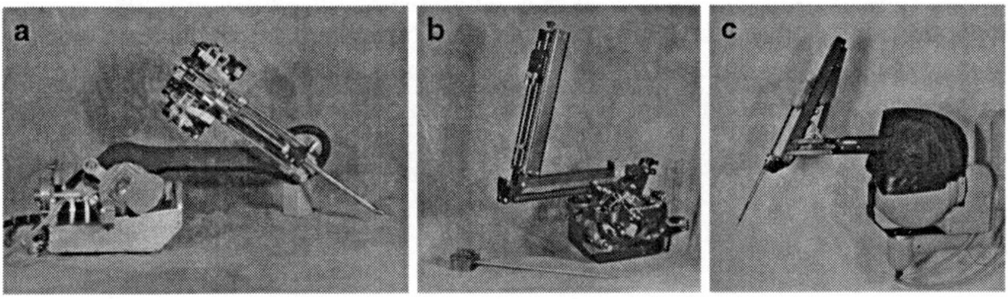

Fig. 9.3 Evolution of the patient-side manipulator (PSM). (**a**) the *Lenny* manipulator that was derived largely from the SRI prototype; (**b**) the *Mona* manipulator with its exchangeable instrument architecture; and (**c**) the *da Vinci* manipulator that ultimately became part of the commercial product

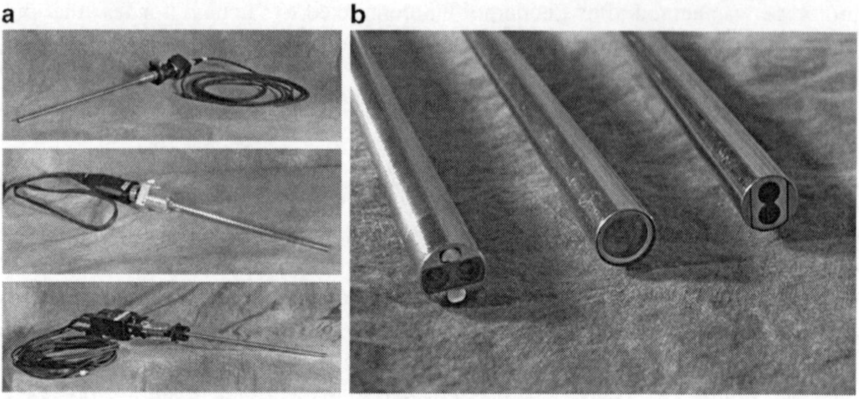

Fig. 9.4 Development of the stereo endoscope. From top to bottom (**a**) or right to left (**b**): the Welch Allyn stereo "chip on stick" endoscope; an Olympus endoscope with single optical train and rear-mounted CCDs; and the *da Vinci* endoscope with dual optical trains and dual three-chip camera heads

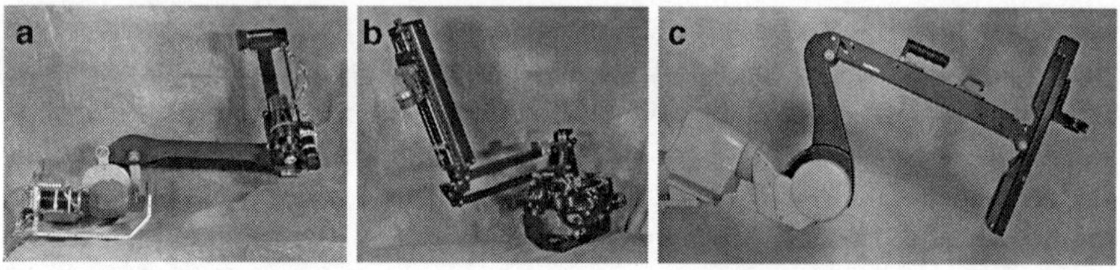

Fig. 9.5 Three generations of the Endoscope Control Manipulator (ECM): (**a**) the manipulator from the Lenny prototype, which was very similar to that of the original SRI system; (**b**) the Mona ECM; and (**c**) the *da Vinci* ECM

afforded by the new wrist, in addition to the intuitiveness of the motion mapping, were providing value that was worth their additional cost and complexity. On the other hand, the system was not reliable – it was fragile – and the visualization provided by the combination of the Welch Allyn endoscope and the CrystalEyes display system was poor. Moreover, the importance of the patient-side manipulator support mechanism and the ability to easily position and reposition the manipulators became clear. Lenny had a relatively short 6–9 month lifespan, but provided the key insights that would set the stage for a second generation prototype (Fig. 9.5).

9.3.2 Mona

Following the lessons learned from Lenny, several subsystems were dramatically redesigned leading up to the Mona prototype, named after Leonardo's renowned painting, the Mona Lisa. These would be the first to be tested during human surgery.

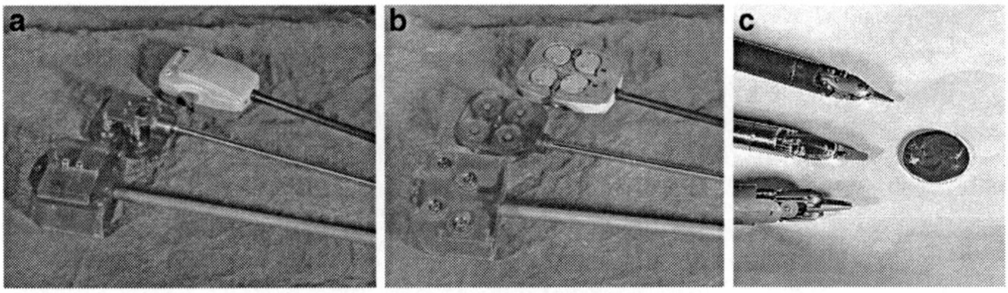

Fig. 9.6 Three iterations of *da Vinci EndoWrist®* instruments with wrists at their tips and a coupling mechanism at their rear that allowed for rapid exchange and the maintenance of a sterile barrier

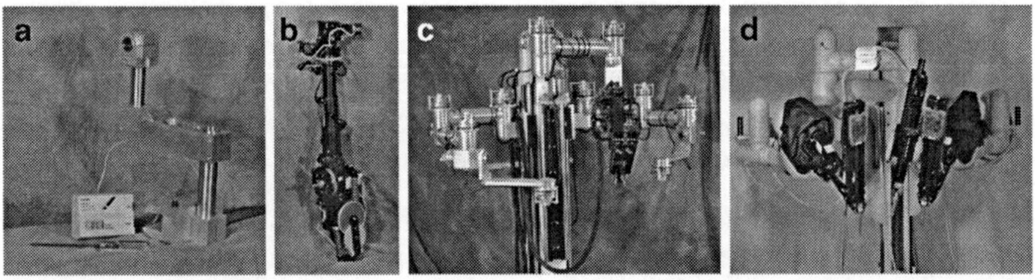

Fig. 9.7 Three generations of setup mechanisms that were used to support and position the instrument manipulators at patient-side. (**a**) The extremely simple mounting that was used to support the Lenny PSMs; (**b**) at early attempt at a repositionable setup arm that was used with the Mona prototype; (**c**) an early prototype of a cart-mounted setup system that provided stable, yet flexible positioning of the PSMs; and (**d**) the *da Vinci* patient-side cart that ultimately became the commercial product

Perhaps the most significant aspect of this re-design was that the patient-side manipulators were completely revised to accommodate an exchangeable sterile instrument architecture. This was a key feature missing in the Lenny prototype and essential for first human use. It meant not only that the system could accommodate many different styles of instruments for different surgical tasks, but also that these instruments could be separated from the non-sterile robotic manipulators and sterilized independently. The instruments themselves (shown in Fig. 9.6) were based on a low-friction cable and pulley design that was influenced by early work by Professor Kenneth Salisbury and his group at MIT.

A second significant focus was on the patient-side manipulator positioning problem that had been identified during experimental work with Lenny. A "setup" mechanism was developed to allow for flexible positioning of three manipulators positioned at the patient side. This mechanism was clamped to rails on the side of the operating table and used a system of gears, springs and linkages to support and dynamically counterbalance the robotic manipulators as they moved above the operating table. Two of these devices were mounted on one side of the table, while a third was mounted on the opposite side. A portion of this mechanism is shown in Fig. 9.7b.

In early 1997, the Mona prototype was used to perform Intuitive Surgical's very first human surgeries at the Saint-Blasius Hospital in Dendermonde, Belgium. This provided valuable clinical experience; however, a number of critical lessons were learned:

- While the new interchangeable instrument architecture was a significant step, the coupling between the instruments and the patient-side manipulators was sensitive to mechanical tolerances and lead to unreliable instrument engagement.
- Improvements in the master and slave interfaces focused new attention on the visualization system, with the realization that both image acquisition and display quality were not sufficient for clear and comfortable stereo viewing of the surgical field.
- Perhaps the most challenging aspect of the Mona prototype was the operation of the setup mechanisms. The table-mounted, counterbalancing mechanism proved to be unstable and inflexible, as well as being too heavy for the table rails to which they were mounted. It was clear that a radical change in the setup approach would be required.

During this period of time, there was also a shift in the clinical landscape. The Nissen Fundoplication procedure, a complex general surgery application, was starting to be performed successfully by manual laparoscopic techniques. At the same time Heartport Inc. – industry pioneers in developing less invasive cardiac surgery products – was struggling with minimally-invasive cardiac surgery. These factors resulted in a shift of Intuitive's focus from general surgery toward cardiac surgery. This new focus, in conjunction with the experience of first human use with the Mona prototype, drove a third generation of prototype, one that would lead to a first commercial product offering.

9.3.3 da Vinci

Early experiments with the Mona prototype had highlighted severe shortcomings with the ability to position the instrument manipulators at the patient side in a flexible yet stable way. Other engineering activities at Intuitive came to a halt in order to focus on this problem of "setup." Several new concepts were developed, first with models constructed using toothpicks, then one-eighth-scale cardboard models and finally full-scale wooden models; several designs were constructed before settling on a design that was to be fabricated in metal. This design was mounted on a free-standing cart. It was bulky and heavy (weighing in at approximately 1,400 pounds), yet it delivered the patient-side manipulators where they needed to be in a flexible way and provided a stable working platform (Fig. 9.7c, d).

Shortcomings with the quality of the vision system were a second major concern at this stage. A new stereo viewer concept was developed, which shifted from the single-display shuttered approach to a dual-display approach, with one dedicated video display for each eye. An early prototype of this concept is shown in Fig. 9.8b;

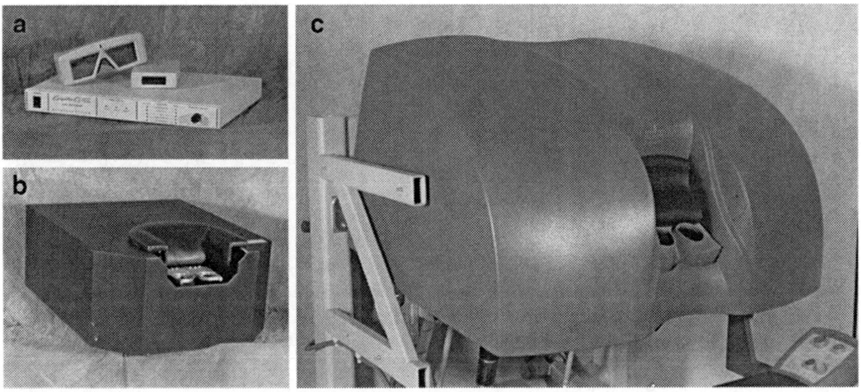

Fig. 9.8 Development of a stereo viewer. (**a**) The CrystalEyes shuttered eyeglass system used in the Lenny and Mona prototypes. A stereo viewer box containing two Sony CRT displays – a prototype that eventually lead to the *da Vinci* stereo viewer design. (**c**) The *da Vinci* high resolution stereo viewer

a pair of Sony CRTs, some mirrors and baffling not only simplified the stereo viewer, but also provided improved display image quality and a more compelling sense of depth and stereo. In order to provide improved video quality on the imaging side, the Welch Allyn endoscope – with its low-resolution distal video chips – was replaced by an Olympus design that used a single optical channel and a dual camera head located at the rear of the endoscope (Fig. 9.4). While image quality was improved, the stereo separation achieved by this design was inadequate. Therefore, it was decided to develop a custom endoscope design with portions of the optics work contracted to a company called Precision Optics Corporation (POC). This was a 12 mm endoscope with two independent sets of 5 mm rod lens optics packaged side-by-side, leading to two three-chip camera heads mounted at the rear of the endoscope.

The problem of temperamental instrument couplings was resolved with a new design based on the principle of the Oldham coupling, which dramatically reduced sensitivity to mechanical tolerances in the instrument and PSM interfaces. On the surgeon console side of the system, there were significant changes in the master interfaces, which evolved from a telescoping design (Fig. 9.9a) to a backhoe design that was kinematically dissimilar to the patient-side manipulators, but with larger range of motion and improved reliability (Fig. 9.9b).

At this stage, the technology had matured to the point that the latest in the succession of prototypes finally earned the name of *da Vinci*, and it was clear that the team saw this as their likely first product offering. The *da Vinci* prototype was taken to human trials in Mexico, France and Germany in 1998 and 1999, with a clinical focus on cholecystectomy and Nissen fundoplication procedures, thoracoscopic internal mammary artery harvesting and mitral valve repair. The setup problem had been resolved, the system now had good 3D vision quality, and the intuitiveness and dexterity of the surgical instrument control were proving to be invaluable. While

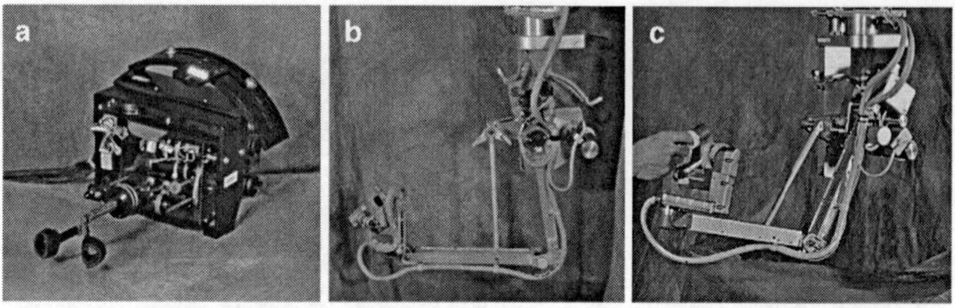

Fig. 9.9 Evolution of the Master Tool Manipulator (MTM) that is manipulated by the surgeon in order to control the patient-side manipulators and surgical instruments. (**a**) A telescoping MTM design that was used in both the Lenny and Mona prototypes; (**b**) a backhoe design that was developed for *da Vinci* in order to provide greater range of motion and improved robustness – this prototype was used for clinical studies with the *da Vinci* prototype; and (**c**) the MTM design that eventually shipped with the *da Vinci* product

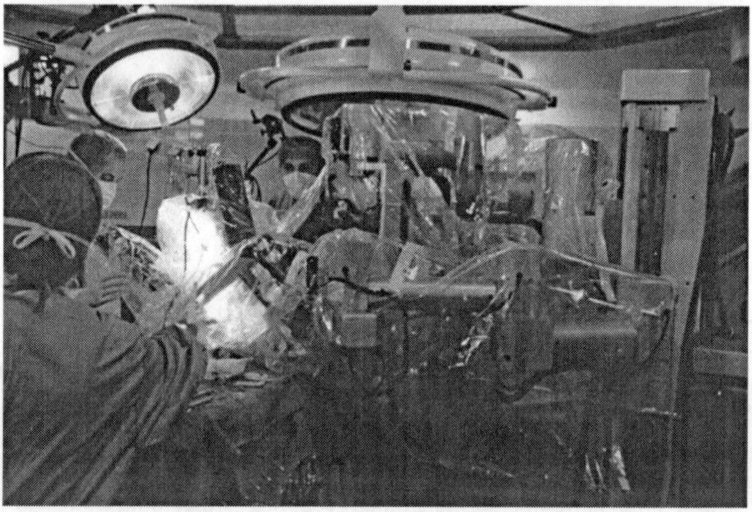

Fig. 9.10 Early cardiac trials in France and Germany during May of 1998

system reliability was still poor (there were frequent system errors and interruptions during early cases), the fault detection mechanisms that had been put into place meant that procedures could be completed safely (Fig. 9.10).

The four key pillars of the product were finally standing. It had taken more than two years from April of 1996 to develop a system with the complexity of a car (approximately 10,000 components) and to ship the very first product to a customer – the Leipzig Heart Center, Germany, in December 1998. Ten systems were shipped during the following year. This number was intentionally limited in order to continue to prove out the product and its market, while not overextending the still-fledgling company that was just beginning to learn how to manufacture and support products in the field (Fig. 9.11).

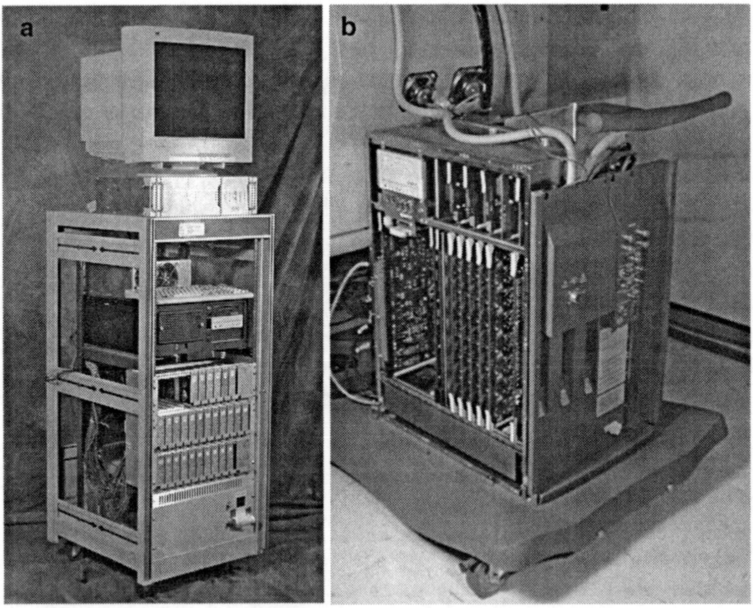

Fig. 9.11 Evolution of the electronics chassis from a PC-based servo and control system to the custom printed circuit assemblies housed within the *da Vinci* surgical console

9.3.4 Continuous Improvement and Development

The next 4 years were focused on stabilizing the reliability of the system and scaling up production and support teams. The evolution that brought an operative surgical robot to the marketplace in 1999 continued after the release of the first product.

On the *da Vinci* platform, slimmer 5 mm diameter instruments (as opposed to 8 mm instruments) were developed based on a novel vertebrae design. In 2003, a fourth arm was added to the patient-side cart in order to give the surgeon greater control over tissue retraction. This included a user interface to swap control among all arms. The suite of surgical instruments was expanded from six to over 50.

The *da Vinci S*™ product release in 2006 focused on improving the patient-side experience by providing a more streamlined and ergonomic design that reduced set up time by half, with just half the number of set-up steps. The slave arms were smaller, lighter, more easily manufactured and serviced, with greatly expanded range of motion. The fourth arm was integrated into the design (as opposed to the retrofit arm that had been added to *da Vinci* in 2003). A smaller, simpler patient cart with fewer degrees of freedom helped to better balance the need for flexibility with simplicity in set-up. Distributed power and control were incorporated in order to minimize cabling. Visualization was improved with WXGA high definition video (1280×768 pixel resolution; roughly equivalent to 720p) and matching image capture equipment. A patient-side touch-panel display and *TilePro*™ multi-input display were added for improved interaction and control. Moreover, significant architectural improvements were put into place to allow for greater reliability and fault tolerance, and faster development.

The latest in the product line, the *da Vinci Si* – launched in April, 2009 – focused the product development on refining the platform to meet the demands of a maturing market. Where the prior model (the *da Vinci Si*) focused on the patient cart, the *da Vinci Si* focused on the surgeon console and vision cart. The surgeon console received improvements in ergonomic adjustability to accommodate a greater range of users comfortably; as well as higher resolution 3D monitors (SXGA), internally cabled master controllers and a simplified user interface. Additionally, the footprint was reduced to make it easier to move around the operating room for small nurses (Fig. 9.12).

The vision cart and associated camera assembly were revised: The touch-screen monitor became widescreen and supported a higher resolution (WXGA+, 1440×900 pixel resolution). Prior to the *da Vinci Si*, the camera controllers were separately controlled, requiring that users perform two steps each time they wanted to adjust one setting (e.g., white balance). Taking all of these vision controls into the system internally made it possible for synchronized 3D vision control. In addition, the size of the camera head was reduced to make it easier to handle and advanced controls were added to simplify vision system setup.

Lastly, *da Vinci Si* was built to support two consoles operating in concert with a single patient-side system. An instrument "give-and-take" paradigm allows surgeons to share control of the *da Vinci* instruments, for the purpose of enhanced surgeon training, as well as for enabling collaborative surgery (Fig. 9.13).

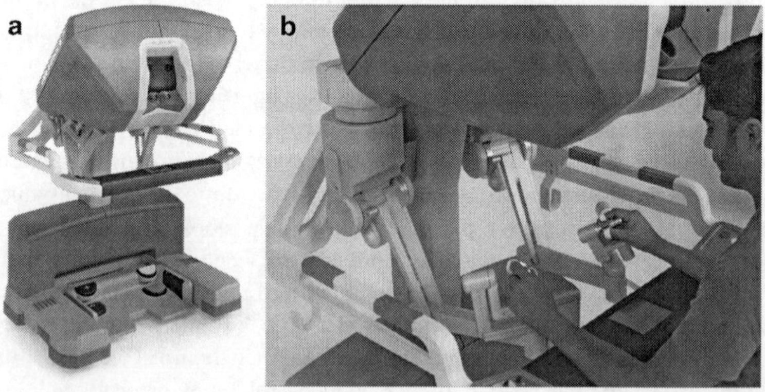

Fig. 9.12 The *da Vinci Si* surgical console (**a**) and redesigned master interfaces (**b**)

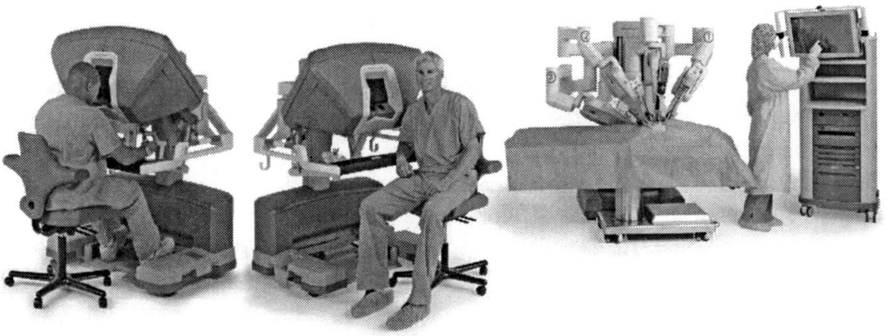

Fig. 9.13 The complete *da Vinci Si* System with dual surgical consoles offered as a optional capability

9.4 The Regulatory Landscape

In the United States, the governing body that regulates medical devices is the U.S. Food and Drug Administration (FDA). This agency was given the authority to regulate medical devices in 1976 under the Medical Device Amendments [9]. In 1990, the Safe Medical Device Act further built upon the Medical Device Amendments to provide a regulatory structure by which medical devices could ensure safe and effective use [10]. In general, devices are classified into three main categories; Class I devices representing the lowest risk, followed by Class II devices and lastly Class III devices which involve those of highest risk. The determination of classification of a medical device depends on the risk it poses to the patient and the user as well as its intended use [11]. Class III devices are typically those that sustain life, such as cardio-pulmonary bypass systems. Class II devices are typically tools whose misuse or failure may cause serious injury. Class I devices are typically those whose misuse or failure are unlikely to result in serious injury.

As pioneers in computer-assisted surgical devices, Intuitive Surgical and Computer Motion were responsible for many first-time decisions made by the FDA for regulating robotic medical devices. One such example was the determination of the classification of the *da Vinci* Surgical System, which was initially evaluated as a Class III device but was later moved to class II by FDA just prior to its approval in 2000. Today, Class II devices include those that fall under a "fly by wire" definition – the type of control system for an airplane or spacecraft, in which the controls are actuated by electrical impulses, as from a computer [12].

Over the past decade regulators have worked hard to understand the numerous computer-assisted systems under consideration for clearance and have only recently created an internal nomenclature by which they are able to categorize a particular device. FDA has created three main categories in which current computer-assisted medical devices fall. The first category is "stereotactic devices or navigational systems" which includes devices for computer-controlled breast biopsies, orthopedic navigation systems and stereotactic systems for navigation in neurology and radiation therapy delivery systems. The second category includes "fly-by-wire systems," where the surgeon's movements are mimicked much like the controls for a pilot. Tele-operators like the *da Vinci* System fall within this category. Systems designed to perform autonomous clinical tasks under indirect supervision fall into the third category of computer-assisted medical devices. Because most computer-aided devices going through the FDA clearance process were developed in the last decade, the nomenclature above is still not well-known outside of industry and FDA [12].

Pulling together the concepts above, the regulatory pathway that a new computer-assisted device will travel through FDA depends upon its architectural type, its risk to the health of the patient and provider, as well as its intended use. Submissions rely upon providing regulatory agencies a combination of clinical data, bench-top data, verification and validation information. In a recent paper by Janda and Buch, a summary of computer-assisted devices by regulatory class and common requirements for testing, validation and clinical data was shared. While there may be exceptions, Table 9.1 provides some guidelines as to the burden of testing and clinical proof required to take a computer-assisted medical device to market.

The *da Vinci* Surgical System was first seen by the U.S. Food and Drug Administration in 1997 and the first clearance received by Intuitive Surgical was for the visualization of tissue using the robotically controlled endoscope. At its inception, the system was classified as a Class III device requiring clinical data demonstrating safety and effectiveness in order to be cleared for active use (i.e., cutting, cauterizing suturing, ligating etc.). A description of the regulatory submissions over the last 10 years pertaining to the *da Vinci* platform evolution discussed earlier in this chapter is shown in Table 9.2. This table does not include the numerous instruments and

Table 9.1 Summary of the classification of computer-assisted devices and their testing requirements

Device type	Regulatory class	Bench testing	Animal testing	Software validation	Clinical data
Preoperative planning	II				
Stereotactic frames	II	X			
Computer-assisted or navigation/intraoperative planning	II	X			
Robotic operating assistants	II or III	X			
Fly-by-wire Systems	II or III	X			
Robots	Unclassified	X	X		

Adapted from Janda and Buch [12]

Table 9.2 Device clearance history

Platform	Year of clearance
da Vinci Surgical System (standard)	2000
4th arm	2003
da Vinci Surgical System (S)	2006
da Vinci Surgical System (Si)	2009

Table 9.3 Clinical indication clearance history

Indication	Year of clearance
Visualization and tissue retraction (da *Vinci*)	1997
Endoscopic visualization (*da Vinci*)	1999
Laparoscopic/general surgery	2000
Thoracoscopic surgery	2001
Radical prostatectomy	2001
Mitral valve repair	2002
ASD closure	2003
Coronary anastomosis For CABG	2004
Cardiac tissue ablation/microwave energy	2005
Broad urology indication	2005
Broad gynecology indication	2005
Pediatric indication	2005
Coronary anastomosis for beating heart CABG	2008
Transoral otolaryngology (ENT)	2009

accessories that have gone through the premarket notification process. In total, there have been 35 separate regulatory submissions to the U.S. FDA regarding the *da Vinci* platform as of December 2009.

As mentioned above, clearance for a Class II device by the FDA is made for a specific set of indications, approved uses of the device for which FDA reviews evidence of the device's safety and efficacy for its intended use. Table 9.3 describes the history of regulatory submissions specific to new clinical indications for the *da Vinci* Surgical System. The clearance of new indications for the *da Vinci* System has required submission of clinical data to the FDA demonstrating its safety and efficacy through the performance of both prospective clinical trials and analysis of retrospective clinical trial data.

As public agencies seek to understand the impact of new technologies on healthcare outcomes and costs, peer-reviewed clinical publications and evidence-based medicine have become increasingly important. Intuitive tracks the creation and publication of peer-reviewed publications on clinical uses of *da Vinci* Surgical Systems. The current library includes over 2,000 PubMed-indexed publications representing multiple surgical specialties, the vast majority of which were researched and written independent of Intuitive. Figure 9.14 shows publications by surgical specialty from 1998 to June 2009. The *da Vinci* clinical library is increasing at a current rate of approximately 90–110 publications per month. While some critics cite the lack of clinical evidence for the efficacy of robotic surgery, the peer-reviewed literature is, in fact, both deep and compelling across many clinical applications.

For companies seeking to commercialize computer-aided interventions, the regulatory process can be daunting. Submissions and communications with regulatory bodies worldwide range from discussions of electronics design, imaging systems and embedded software to clinical trial design and patient outcomes analysis. In the past decade, regulatory agencies worldwide have made investments in learning both the base technologies in our field and in understanding their clinical use. In general,

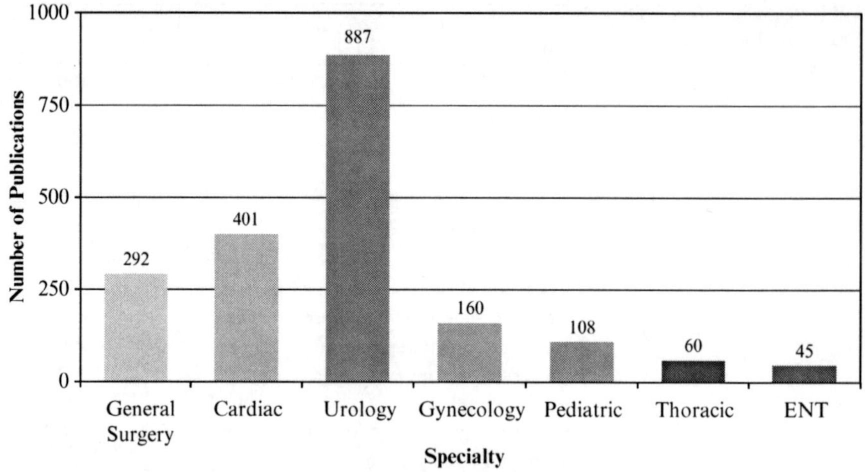

Fig. 9.14 *da Vinci* Publications from 1998 – June 2009 by Surgical Specialty

the level of scientific sophistication in the clinical and technical reasoning and documentation has increased over time. Success in bringing products to the market has required a cross-functional approach (engineers, clinicians and regulatory experts) on the part of both industry and regulatory agencies to examine the opportunities and risks presented by computer-aided intervention.

9.5 Conclusion

The development of telerobotic surgery from science fiction to reality over the last two decades has brought together many different disciplines in medicine, engineering, science and industry. In particular, the clinical and engineering research communities put into place many of the key foundations of this technology and jointly recognized the value of a marriage between their fields – an insight that provided a key catalyst for creating a completely new way of performing surgery. Regulatory bodies, too, had to innovate and adapt their processes in order to strike a balance between the benefits and potential risks of this new technology, with the health and safety of patients being their ultimate priority and a tremendous responsibility.

It is on the shoulders of giants that an industry has been built. As a community we look forward to the innovations that the readers of this book are yet to imagine and to build, in order to power the next revolution in surgery. Multi-disciplinary collaboration has clearly been a key component of the development of our field to-date and will become increasingly important as new capabilities are realized and new applications are explored. This teamwork between clinical scientists, surgeons and OR staff, academic researchers, engineers, regulatory groups and many others will help to transition novel ideas into technologies that will ultimately benefit patients and their families in remarkable new ways.

References

1. Goertz, R., Bevilacqua, F.: A force reflecting positional servo-mechanism. Nucleonics **10**, 43–45 (1952)
2. Spaner, S.J., Warnock, G.L.: A brief history of endoscopy, laparoscopy, and laparoscopic surgery. J. Laparoendosc. Adv. Surg. Tech. A **7**, 369–373 (1997)
3. Taylor, R.H., Funda, J., Eldridge, B., Gruben, K., LaRose, D., Gomory, S., Talamini, M., Kavoussi, L., Anderson, J.: A telerobotic assistant for laparoscopic surgery, IEEE EMBS Mag. **14**, 279–291 (1995)
4. Sackier, J.M., Wang, Y.: Robotically assisted laparoscopic surgery. From concept to development. Surg. Endosc. **8**, 63–66 (1994)
5. Madhani, A., Niemeyer, G., Salisbury, J.: The black falcon: a teleoperated surgical instrument for minimally invasive surgery. In: Proceedings IEEE/Robotics Society of Japan International Conference on Intelligent Robotic System, vol. 2, pp. 936–944 (1998)
6. Schenker, P., Das, H., Ohm, T.: A new robot for high dexterity microsurgery. In: Proceedings of the Conference on Computer Vision, Virtual Reality, and Robotics in Medicine (1995)
7. Davies, B.L., Hibberd, R.D., Ng, W.S., Timoney, A.G., Wickham, J.E.: The development of a surgeon robot for prostatectomies. Proc. Inst. Mech. Eng. H **205**, 35–38 (1991)
8. Rosen, J., Hannaford, B., MacFarlane, M., Sinanan, M.: Force controlled and teleoperated endoscopic grasper for minimally invasive surgery – experimental performance evaluation. IEEE Trans. Biomed. Eng. 46(10), 1212–1221 (1999)
9. Medical Device Amendments of 1976. Public Law 94–295. May 28, 1976, U. S. Congress, Ed
10. The Safe Medical Devices Act of 1990 – FDA. Notice, Fed Regist. **56**, 14111–14113 (1991)
11. Device Classification. In: U.S. Food and Drug Administration
12. Janda, M., Buch, B.: The challenges of clinical validation of emerging technologies: computer-assisted devices for surgery. J. Bone. Joint. Surg. Am. **91**(Suppl 1), 17–21 (2009)

Chapter 10
RIO: Robotic-Arm Interactive Orthopedic System MAKOplasty: User Interactive Haptic Orthopedic Robotics

Benny Hagag, Rony Abovitz, Hyosig Kang, Brian Schmitz, and Michael Conditt

10.1 Introduction

The dream of robots, automatons that can think, move, and act like humans, has been with us for centuries, dating back to Leonardo daVinci, the ancient Greeks, and possibly even earlier. The word robot was introduced by Czech writer Karel Capek in his play R.U.R. (Rossum's Universal Robots), published in 1920. Capek's play describes robots that are very close to the androids of films like *Star Wars* – mechanical, intelligent servants with a purpose to serve their human masters.

The dream of robotics has had many false starts and prophets – it has been intrinsically linked to the development of artificial intelligence, the computing software that can process multiple sensory inputs and develop intelligent assessments of the environment, and even abstract thought. To date, artificial intelligence has been a dismal failure, with the expectation of the field far exceeding the promised delivery. However, we have recently crossed into the beginnings of computational capability powerful enough to contemplate artificial intelligence again. This may be better thought of as *Limited, or Focused Synthetic Intelligence* – where the goal is to not emulate human thought, but to exploit the superior aspects of machine computation where possible. An analogy is the comparison of a bird to a jet: we have failed at developing a birdlike machine, however we have developed machines that can fly farther and faster than any bird. So begins the age of *Synthetic Intelligence*, where machines both augment and complement human intelligence in ways yet to be discovered.

The use of robotics in surgery appears to be, at first glance, one of the most complex uses and ambitions for robotics. One would first contemplate the use of robotics in supposed menial tasks, such as cleaning, or janitorial work. The issue with such supposed menial tasks is that they present significant problems of machines dealing with an unstructured environment, and one where very simple

---

B. Hagag (✉)
MAKO Surgical Corp., 2555 Davie Road, Ft. Lauderdale, FL 33317, USA
e-mail: bhagag@makosurgical.com

J. Rosen et al. (eds.), *Surgical Robotics: Systems Applications and Visions*,
DOI 10.1007/978-1-4419-1126-1_10,

commands must lead to a series of complex and interrelated tasks. There has been some limited success with robotics in this area, but it is very preliminary and subject to a variety of special required conditions (e.g., the Roomba robotic vacuum cleaner). A field of robotics which shows great promise, and can bridge the gap between current computational limitations and full-fledged, general Synthetic Intelligence is human-interactive robotics. This field, dominated by what is called *haptics*, or the study of simulating tactile interactions, offers a unique blend: couple the dexterity of the machine with its limited intelligence with the superior context and general intelligence of the human. This combination is proving out to be a formula for practical success in a number of fields, but it holds specific and uniquely large potential in the field of surgery.

Many types of surgery benefit from less-invasive approaches, ones where the surgeon will have limited direct visualization and even less limited portals into the body for procedural access. The combination of the surgeon's problem-solving and context skill set, with a haptic robotic system's ability to provide tactile interactions, create a whole new toolset for surgery. The coupling of this capability with advanced visualization technologies such as 3D scopes and/or patient specific graphics models taken from imaging systems (CT, MR, X-ray) gives the surgeon a wide array of dexterity and function enhancing capabilities, enabling the execution of techniques which can be quite favorable to a patient. The key existence proofs of such systems exist in the widespread use of Intuitive Surgical's daVinci® teleoperative robotic system, and MAKO Surgical's haptic centered RIO® robotic arm. Both robotic systems have crossed the threshold of proof-of-concept, and are now in practical, every day use in hospitals around the country. The success and surgeon adoption and enthusiasm for such systems centers around the principle concept of keeping the surgeon in control of the robot – of essentially integrating human and machine in complementary and performance enhancing ways. Autonomous robotics may show promise for future applications, but in a very practical sense, human-interactive robotics is no longer a futuristic concept – it is here today.

10.2 State of the Art in Orthopedic Surgical Robots

Recent advances in imaging technology such as computed tomography (CT) and the rigid nature of bony structure have made orthopedics one of the first and a natural application to adapt the robotic technology. Furthermore, the required accuracy of bone preparation to fit orthopedic implants properly in joint replacement surgery for maximizing the implant longevity provides a fertile environment for introducing the robot assisted orthopedic surgery.

This section reviews the surgical robotic systems which are commercially available now or soon to be commercially available in orthopedic application even though there has been extensive research on different robotic systems.

The first commercial surgical robot, ROBODOC, in orthopedic surgery was introduced by Integrated Surgical Systems in the early 1990s. The system was used to prepare the femoral cavity for total hip arthroplasty. An industrial SCARA (Selective Compliance Assembly Robot Arm)-type robot augmented with an additional joint and safety features was used to prepare the femoral cavity for total hip arthroplasty with a high-speed milling device at the end of the robot arm. ROBODOC can be viewed as CAD/CAM robot which is operated in autonomous mode. The robot basically executes the pre-planned cutting path. Due to its autonomous operation, however, the bone must be rigidly clamped in place by a 'fixator' during the milling procedure and cannot be moved after registering the bone with respect to the base of the robot. This constraint significantly changes the nature of a typical orthopaedic procedure, where the joint is regularly manipulated for better visualization and access, and for evaluating soft tissue, range of motion, and the fit of trial implants. The automated cutting proceeds slowly for safety reasons and will stop if any unexpected forces are encountered. Therefore, the soft tissues must also be retracted or resected so that the robot does not encounter them during the bone milling step. The specific patient soft tissue anatomy is not generally imaged or used for planning the cutting paths, and some users have reported increased damage to soft tissues relative to conventional instrumentation. While this system can create accurate bone cuts, it is not interactive; the surgeon simply watches the automatic cutting process. Given that the surgeon has less direct control over the cutting procedures and early adverse events in Germany, regulatory approval and acceptance from orthopaedic surgeons is challenging.

ACROBOT, or Active Constraint Robot, is a robotic device designed for assisting in unicompartmental and total knee arthroplasty and other procedures and is being commercialized by The Acrobot Company, Ltd. It uses an interactive device mounted on a large position controlled robot arm. The surgeon interacts with the device using a force sensing handle mounted near the end of the robot arm. From its name, the robot actively constrains the surgeon to cut within a predefined safe region by providing haptic feedback to the surgeon during the execution phase of the procedure. The latest version of the system, Acrobot Sculptor®, has an improved bone fixation technique by adapting an instrumented mechanical linkage over its early rigid fixation to the robot base and also has add-on passive remote center mechanism joints which enables the tool to be freely orientated while its tool tip is maintained at the same spatial position at the expense of limiting their supporting cutting tool to a spherical cutter only. However, the need to interact with the force sensing handle and the limitation to three haptic degrees-of-freedom limits its application for MIS procedures that require significant dexterity.

BRIGIT, Bone Resection Instrument Guidance by Intelligent Telemanipulator, was originally developed by MedTech S.A. and was acquired by Zimmer in 2006. It is a positioning arm which places the cutting guide for guiding an oscillating saw and surgical drill in total knee replacement surgery, targeting to reduce the instruments and procedure time. BRIGIT provides an imageless system, requiring collecting anatomical landmarks with a robotic arm through a force sensor. Acceptance rate from the surgeon is still challenging, even though it obtained the FDA 510(k) clearance early, due to the fact that the leg needs to be fixed to the robot base during whole procedure.

iBLOCK developed by PRAXIM is a miniature robot with two motorized revolute joints for preparing the femoral implant in total knee arthroplasty, orienting the cutting block for a standard surgical saw while the early prototype, PRAXITELES, provided a side milling tool guide. iBLOCK is laterally mounted on the distal femur with two or three bicortical pins with a two degree-of-freedom passive angle adjustment device which adjusts varus/valgus and internal/external rotational angles with a navigation system. However, it supports only the femoral preparation while the tibia is prepared with a navigated instrument.

Pi GALILEO ACCULIGN, Smith & Nephew, is a software controlled electromechanical device, two motorized linear joints with a navigation system, to position the 5-in-one cutting block to prepare the femoral cut in TKA.

Precision Freehand Sculptor (PFS) is a handheld tool with an optical tracking system which retracts the rotating burr inside a protection guard when a user goes to the forbidden region.

10.3 User Interactive Haptic Robotics

There are a number of reasons supporting the introduction of a user interactive surgical robot. One compelling rationale is the degree of unpredictability of the surrounding soft tissues even though the primary objective structure, bone, in orthopaedic surgery is considered as a rigid body which can provide the convenient reference frame to general autonomous robotic applications. This interaction with soft tissue becomes more critical in minimally invasive surgery (MIS). MIS is the performance of surgery through incisions that are considerably smaller than incisions used in traditional surgical approaches. For example, in an orthopedic application such as total knee replacement surgery, an MIS incision length may be in a range of 4 in. whereas an incision length in traditional total knee surgery is typically in the order of 12 in. As a result of the smaller incision length, MIS procedures are generally less invasive than traditional surgical approaches, which minimizes trauma to soft tissue, reduces post-operative pain, promotes earlier mobilization, shortens hospital stays, and speeds rehabilitation.

One drawback of MIS is that the small incision size reduces a surgeon's ability to view and access the anatomy. As a result, soft tissue management and re-positioning the leg may become necessary during surgery, requiring the surgeon to interact with the environment for optimal joint exposure and safety. Autonomous robots are less suited to clinical application where MIS procedures are preferred. The combination of the human's inherent ability to perceive the complex environment and the robot's nature of reproducible accuracy provide the foundation of human–robot interaction.

For this reason, a new user interactive haptic robotic arm, the RIO® shown in Fig. 10.1, was developed. The design concept of RIO® is not intended to replace a surgeon with robots, but to enhance the surgeon's skills by providing him with intuitive and interactive tools which increase the safety of the patients, relieve the

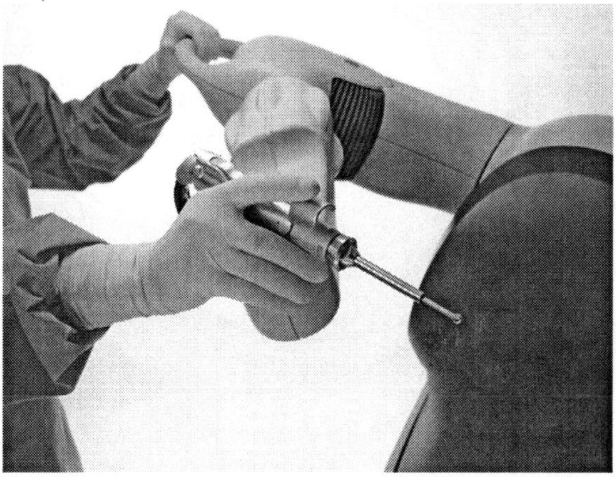

Fig. 10.1 User interactive robot

surgeon's burden to handle the complex surgical environment, and improve surgical outcome. The surgeon can grasp the cutting tool mounted at the end of the arm and can maneuver the arm to interact with the environment. In this way, he or she can feel more comfort level working closely with this robotic arm and as a result, it helps to increase the surgeon's acceptance rate.

One of the appealing features in RIO® is to embrace multimodal sensory feedback.

Nowadays, surgical navigation systems visualize the spatial 3D positions and orientations of tracked objects with respect to the operated anatomy in real time and have become a mainstream in computer-assisted orthopaedic surgery (CAOS). Most interfaces in navigation systems utilize a visual display and an auditory feedback. Unlike visual and auditory feedback, haptic feedback allows a user to both feel and interact with an interface. Haptic, from the Greek word Haphe, means the sense of touch by interacting the mechanical stimulation such as forces and vibrations. The haptic interaction in RIO® is implemented through haptic rendering.

The haptic rendering consists of two main elements, a virtual haptic environment and a haptic device as shown in Fig. 10.2. The virtual haptic environments are computer-generated synthetic environments to emulate a physical world and they can compute the interacting forces in response to the user's interaction with virtual objects. The haptic device is a mechanical device to convey the haptic information to the user.

The primary functions of the RIO® are,

- Measure the human operator motion in the physical world to explore the virtual haptic environment
- Display the haptic information to constrain the human motion according to the interacting forces generated in the virtual haptic environments

The haptic device in RIO® is a highly back-drivable six degree-of-freedom manipulator which simulates the mechanical impedance – measures the pose (position

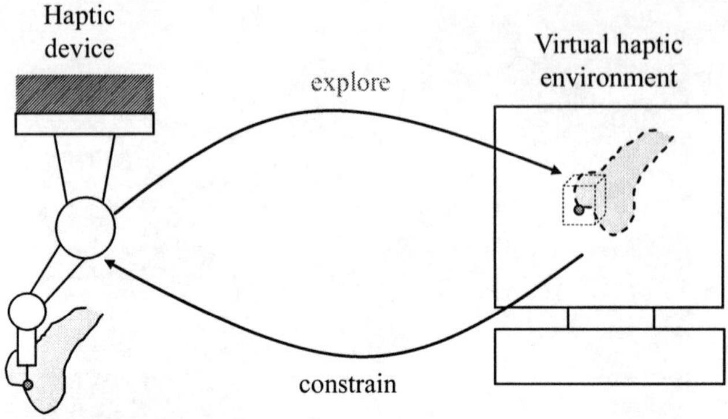

Fig. 10.2 Haptic rendering

and orientation) of the tool tip and applies a wrench (forces and torques) to the tool tip. The back-drivable manipulator permits the surgeon to freely maneuver the cutting tool within the haptic cavity that is designed to match the features of the operated implant. During this freehand cutting procedure, the surgeon can feel the natural cutting dynamics of the cutting tool which can provide another intraoperative assessment on the bone quality. Only when moving beyond the desired cutting boundaries will the system push back against the user, creating haptic guidance boundaries that help to keep the surgeon from cutting outside the planned volumes. This supplementary haptic integration improves the quality of surgical work, provides intuitive feedback to the surgeon, and reduces the surgical time.

The RIO® does not require the bone to be clamped in place. A standard orthopaedic leg holder is used to restrain the leg, combined with bone-mounted trackers to track the bones during the haptic-guided cutting operation. The camera-observed positions of the trackers are used to automatically adjust the position of the haptic environment to compensate for motion of the bony anatomy.

One other novel feature is that the RIO® permits an intraoperative revision of the preoperative plan. In general, orthopaedic surgery is a highly interactive process and a hands-on discipline. Many intraoperative revisions of the preoperative plan are made in the operating room to optimize results and improve the longevity of the implants. For example, variations in bone quality and ligament balancing can make the surgeon decide to revise the plan intraoperatively.

10.4 The MAKO Robotic System Design

The MAKO RIO® Robotic Arm Interactive Orthopedic System (RIO) is comprised of three major hardware components as shown in Fig. 10.3. The Robotic Arm supports the cutting system and provides sufficient dexterity to allow the surgeon to create the desired resections of bone. The Camera Stand supports both the computer

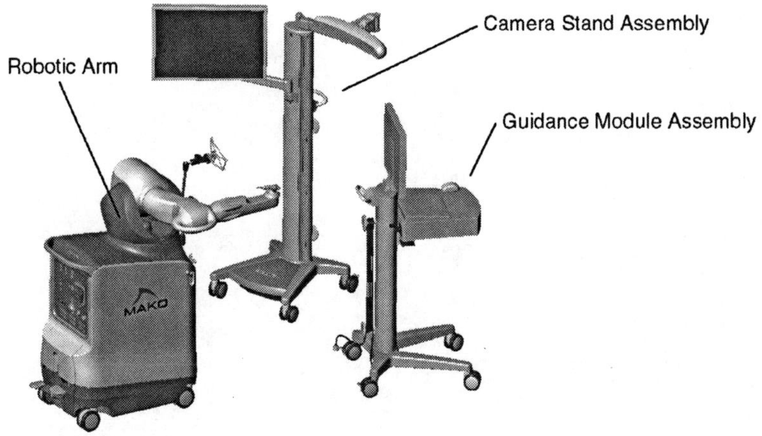

Fig. 10.3 MAKO RIO® robotic arm interactive orthopedic system

monitor used by the surgeon to view the bone resections as well as the localizing camera system for tracking the patient anatomy though the use of tracking arrays mounted to the bone. The Guidance Module is used by the physician's assistant or a surgical technician to assist the surgeon navigating through the implant planning and surgical application. The three components are connected electrically to provide communications and distribute power through the system.

In the MAKOplasty® application, the robotic arm is used in a passive way by the surgeon, meaning the surgeon moves the robotic arm manually to guide the cuts within the planned anatomy defined in the planning portion of the procedure prior to surgery. The robotic arm applies haptic force feedback only if and when the surgeon attempts to move the cutting burr outside the pre-defined surgical plan. If the surgeon attempts to cut outside the plan, the motors, drive systems and control systems inside the robotic arm apply forces against the surgeon's movements to maintain the cutting burr inside the surgical plan.

Figure 10.4 is a screenshot of the RIO® system software showing the model of the patient's femur bone with the planned resection volume in green as well as a portion of the bone already removed. This is the interface the surgeon uses to guide the bone resections. The surgeon essentially cuts all of the green colored bone away until he hits the Planned boundaries shown as white. If the surgeon attempts to move the cutting burr outside of the planned green resection areas, the RIO robotic arm applies a force against the surgeon's hand preventing him from cutting outside the planned boundaries.

As a result of the passive nature in the MAKOplasty® application, one of the most important aspects of the surgeon interactive robotic arm is that it moves freely with low friction and low inertia such that the surgeon does not become fatigued during a surgery – a term called "back-drive-ability". This need for back-drive-ability in the RIO® robotic arm and the need to provide haptic force feedback to the surgeon differentiates the RIO® robotic arm from most industrial robots, which are difficult or impossible to move manually.

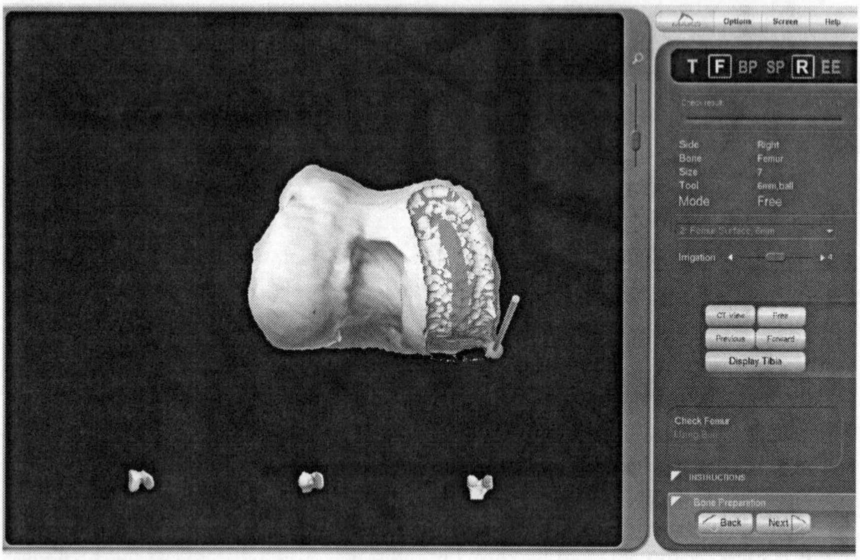

Fig. 10.4 RIO® system software screen showing femur resection

The RIO® robotic arm is also defined as a "serial" robot, meaning the arm joints are configured with the first stage being affixed to ground (a non-movable stationery reference coordinate system), as opposed to a parallel robot arm that may have several joints affixed to ground. In the RIO® robotic arm, the first degree of freedom (arm joint) is affixed to the ground and the remaining five joints are attached one to the other with the input of the next sequential joint being attached to the output of the previous joint, or degree of freedom. The advantage of a serial robot over a parallel robot is an increase in the overall range of motion (reach) of the final output end of the robot arm while providing adequate dexterity to accommodate a wide range of arm positions.

In the RIO® robotic arm, the total range of motion of each joint is designed to accommodate both right-handed and left-handed surgeons, as well as providing sufficient range of motion to perform the worst case surgery envisioned. Figure 10.5 identifies each of the defined degrees of freedom of the RIO® robotic arm.

As described earlier, the RIO® robotic arm is a passive robot that requires the arm be easily moved by the surgeon with very little friction. The arm also must be able to provide haptic force feedback to the surgeon based on the cutting burr location relative to the surgical plan. The robotic arm must also be extremely accurate, with total errors in positioning of the cutting burr less than 1 mm from the desired locations. These requirements are accomplished in part through the use of cable driven systems in the RIO® robotic arm to provide motion control with low friction and negligible amounts of backlash in the mechanisms. Wire cables attach the drive motor for each joint to the rotational output of each joint through a series of pulleys – to reduce the amount of unsupported cable in the system – and tensioning mechanisms, to remove all back-lash.

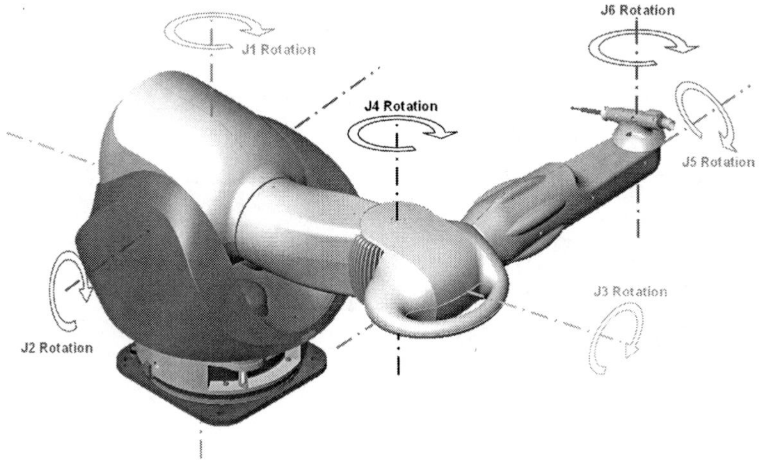

Fig. 10.5 Joint motion definitions of the RIO® robotic arm

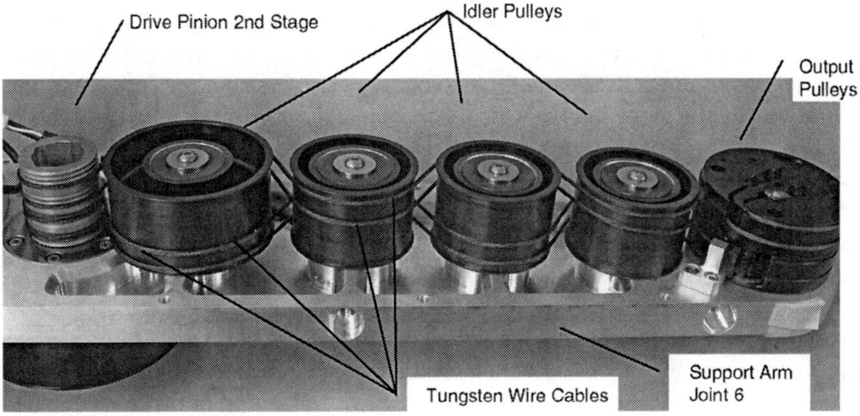

Fig. 10.6 Joint 6 assembly second stage cable drive

Figure 10.6 is an image of one of the cable drive systems used in the second stage portion of the J6 arm joint for RIO®. The drive cables are rigidly affixed to the drive pinion on the left, traversing around a number of idler pulleys to reduce unsupported cable runs, to the output pulley that also incorporates the mechanism to tension the cables to eliminate any slack that could contribute to backlash.

Figure 10.6 also illustrates one of the many safety features designed into the RIO® robotic arm that make the system safe for use in surgery. Since the MAKO RIO® robotic arm is used in medical applications, special attention was made to ensure safe operation of the system. The RIO® system is designed such that a failure of a single component cannot create an unsafe condition or allow unsafe operation of the robotic arm. Shown in Fig. 10.6 is a completely redundant set of

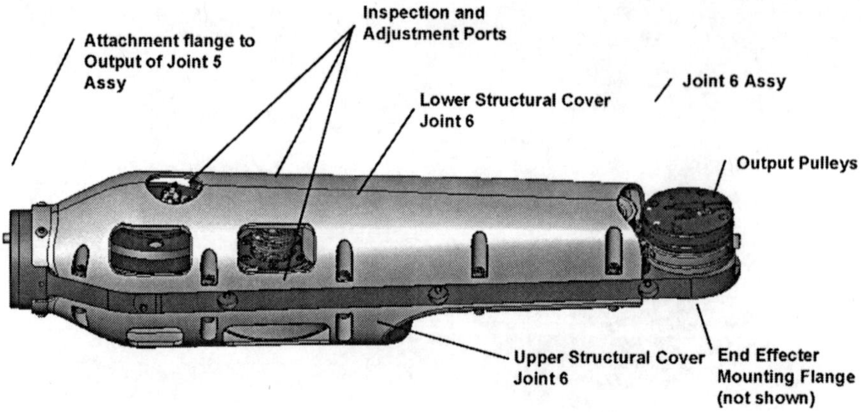

Fig. 10.7 Joint 6 arm assembly

drive cables that allow up to two cables to fail without affecting the accuracy or safety of the robotic arm. An additional safety feature designed into each arm joint of the RIO® robotic arm is the use of a high resolution encoder on the output of the joint as well as a high resolution encoder for the drive motor (input) for the joint. The electronics of the RIO® robotic arm continuously compares the encoder signals of the joint output to the joint input for each joint in the arm. If the encoder counts of the two encoders ever disagree as to the position of the arm joint, the entire robotic arm is disabled from operation. Supporting this safety mechanism is the use of fail-safe mechanical brakes for every joint of the robotic arm. In the event of any failure that could jeopardize the safety of the patient or the surgeon, the arm brakes are engaged, preventing any motion of the RIO® arm.

Accuracy and precision are of paramount importance for the MAKOplasty® procedure, and are provided in the design of the RIO® robotic arm. Accuracy and precision are enabled by using high resolution angular encoders in the joint output stages as described previously, as well as the structural design of the robotic arm to minimize compliance of the arm joints in the planes orthogonal to joint motion which cannot be measured using the high resolution encoders.

Figure 10.7 is a CAD model of the Joint 6 arm assembly illustrating the upper and lower structural arm covers that provide a high degree of bending stiffness to the joint while providing access for inspections and adjustment of the tensioning mechanisms for the drive cables. Figure 10.7 also depicts the modular design of the RIO robotic arm facilitating manufacturing, service repair, and expansion for future applications; each arm joint having all of the functional mechanical and electrical drive elements required to operate the joint assembly residing in the joint module. This enables high volume production methods and makes field service much easier, as entire joints of the robotic arm can be assembled, tested and serviced as separate components.

The RIO® robotic arm ultimately supports, tracks and controls the location of a high speed cutting burr located inside the End Effector Assembly shown in

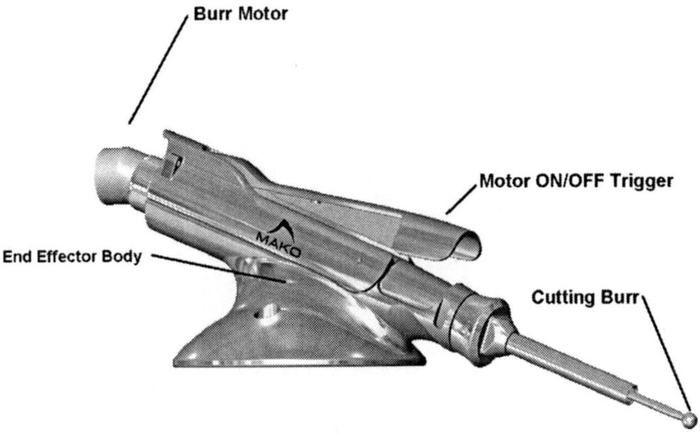

Fig. 10.8 End effector assembly

Fig. 10.8. The surgeon grasps the End Effector Body in his dominant hand to move the cutting burr (as well as the robotic arm) in the planned resection areas of the bone. The cutting burr spins at up to 80,000 RPM, quickly and easily removing the bone volume to be replaced by the implant. A variety of cutting burrs can be used in the system and easily and quickly changed during a surgery. Irrigation is provided to the area of the resection using irrigation tubing attached to the End Effector Assembly (not shown) to cool the bone during cutting, preventing thermal necrosis which can lead to loosening of the prosthetic implant over time.

The RIO® system is designed to be easily and quickly moved in and out of the operating room, allowing quick turn-around between procedures, but also must be very stable during the surgical procedure, as any movement of the robot during the procedure can create error in the location of the bone resection. A lifting mechanism in the base of the enables the RIO® robotic arm to move up onto the casters to allow it to be rolled for positioning, moving, and storage, as well as lowered down on to rigid legs to be stable during a surgical procedure. As shown in Fig. 10.9, lifting the Robotic Arm up onto the casters is accomplished by depressing the foot pump pedal several times, while lowering the Robotic Arm down on the rigid legs is accomplished by depressing the lowering release pedal.

The RIO® robotic arm is powered and controlled by electronics and a real-time operating system running proprietary control software, all of which reside beneath the arm as shown in Fig. 10.10. A dedicated computer is housed in the Card Cage Assembly that runs the real-time software needed for highly accurate and high bandwidth control of the robotic arm. Also included in the Card Cage Assembly are the controller boards required for each of the arm joints.

To support worldwide operation, the RIO® system includes a voltage selectable isolation transformer which converts the power for the system into 120 V AC regardless of the input voltage of the various countries using the system. The RIO® also includes an un-interruptible power supply to keep the system powered and running for up to 10 min in the event of a power interruption during surgery.

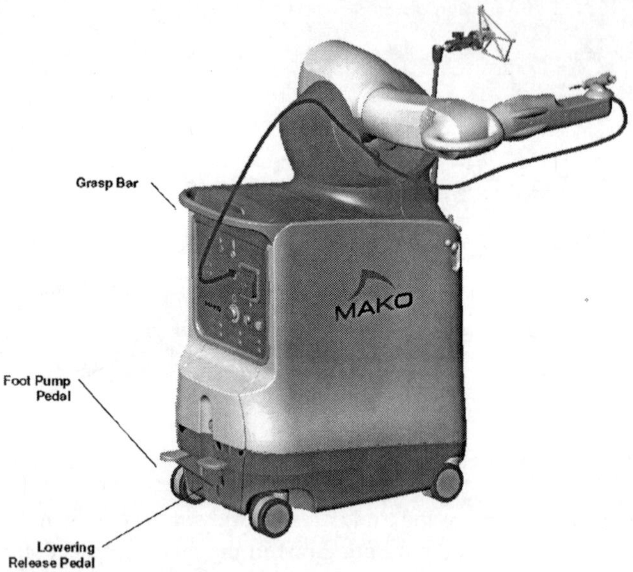

Fig. 10.9 Robotic arm assembly

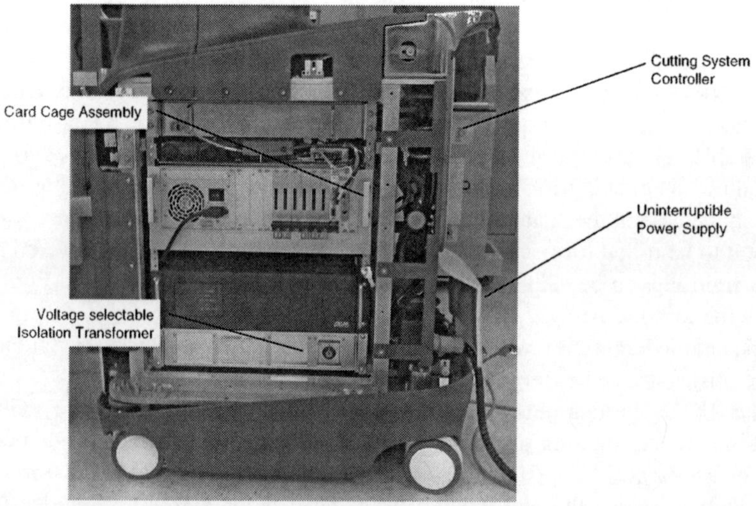

Fig. 10.10 Robotic arm electronics

In summary, the RIO® system is a robotic arm platform designed specifically to enable accurate and precise bone resection for minimally invasive orthopedic implants. It provides flexibility to support multiple orthopedic procedures with built-in mechanisms to ensure safe and effective operation in worldwide markets.

10.5 Surgical Technique

The RIO® system is currently used for implantation of medial and lateral UKA components as well as patellofemoral arthroplasty. The platform allows pre-op planning with the ability to adjust the plan intra-operatively. The robotic arm is under direct surgeon control, and gives real time tactile feedback to the surgeon as the procedure is performed. This surgical platform blends surgical interactive robotics, computer assisted planning and guidance with an intelligent bone cutting/shaping tool through minimally invasive techniques.

10.5.1 Preoperative Imaging

Preoperative CT scans are obtained for all patients. Scans are taken with the patient in a supine position with a motion rod attached to the affected leg. One-millimeter slices are taken at the knee joint with 5-mm slices taken through the hip and ankle. Images are saved in DICOM format and transferred to the software of the RIO® System so that sagittal slices of the distal femur and proximal tibia may be segmented, defined, and recombined to produce 3-dimensional (3-D) models of each bone. Implant models are then positioned on the reconstructed bone models, resulting in patient-specific CT-based planning (Fig. 10.11). CT-based planning is

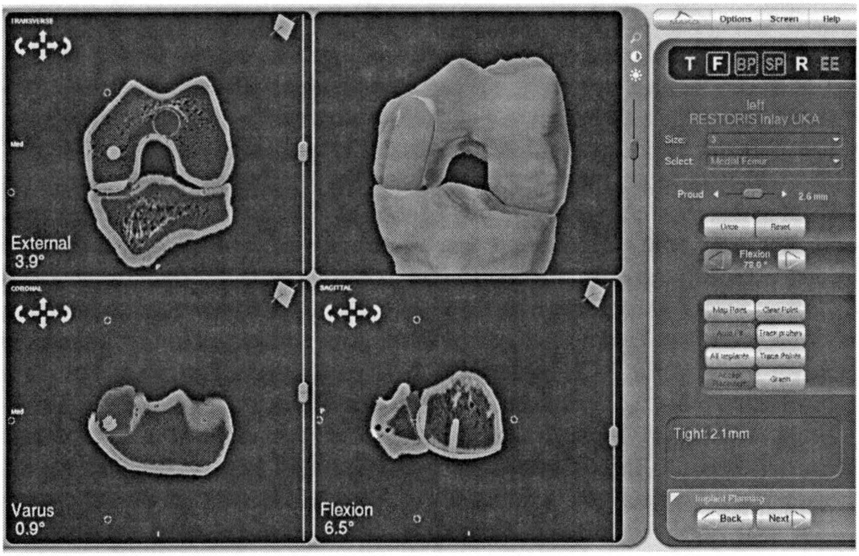

Fig. 10.11 CT based patient specific preoperative planning

limited in that soft tissues cannot be visualized with CT. Consequently, only bony alignment can be achieved pre-operatively, and the plan must be intraoperatively modified to achieve precise gap balancing and long-leg alignment. CT planning allows for assessment of the subchondral bone bed, osteophyte formations, and volume definition of cysts and avascular necrosis.

10.5.2 Preoperative Planning

The preoperative plan is based on four main parameters: metrics of component alignment, 3-D virtual visualization of implant position, intraoperative gap kinematics, and dynamic lower limb alignment assessment. Accurate implant positioning requires integrating into the system the precise dimensions of the femoral and tibial prostheses, with their target positions programmed into the preoperative planning software. The implant computer assisted design (CAD) models are positioned on the 3-D models of the patient's distal femur and proximal tibia, and alignment parameters reported on the computer display unit (currently planning and resection of the patellar implant is done manually). During this step, the surgeon can visualize the predicted implant congruence and attempt to minimize areas of edge loading through adjustments to the plan. Feedback regarding alignment metrics and bony anatomy (e.g., subchondral bony bed, cortical rim) is continuously displayed. Although the implants are not customized to the patient, implant orientation is patient-specific and includes quantitative feedback from both bony and soft-tissue anatomy. Thus, bone resection volumes are defined automatically by the system, and boundaries for the cutting instrument are set to prevent inadvertent surgery to areas outside these predefined zones. The preliminary plan is based on alignment parameters and 3-D visualization of implant position. During surgery, the plan is modified according to gap measurements throughout flexion and dynamic lower limb alignment values. Before surgery, the alignment parameters reported by the robotic system (and recommended by MAKO) are used in combination with parameters supported by the literature. Specifically, tibial slope in the coronal and sagittal planes is carefully controlled. The medial tibial plateau typically has varus with respect to the mechanical axis of the tibia in patients with medial compartment osteoarthritis. Collier and colleagues [1] demonstrated that correction of this varus slope with the tibial implant can improve survivorship. In addition, more than 7° of posterior slope of the tibial component has been shown to increase the risk for ACL rupture [2] It is recommended to place the tibial components in 2–4° of varus and avoiding more than 7° of posterior slope. In patients with ACL deficiency, the posterior sagittal slope of the tibia is maintained between 2 and 5°. Three-dimensional visualization of implant position ensures proper sizing. For example, a 2-mm rim of bone surrounding the pocket created for the inlay tibial component is advocated. This rim can be planned and measured directly on the 3-D model. On the femur, the prosthesis is sized such that coverage is maintained while symmetric flexion and extension gaps are created. In addition, depth of resection can be planned precisely; 3 mm of tibial bone resection is typically planned. This resection depth can be modified according to intraoperative gap kinematics.

10.5.3 Setup

The RIO® is positioned before the patient arrives in the OR. Positioning is based on the knee to be operated on and on surgeon preference (right- or left-hand dominant). Once the system is positioned, the robot base unit is secured with brakes to prevent motion. After conventional positioning and sterile draping of the affected limb, robot registration is performed (Fig. 10.12). The surgeon moves the robotic arm through a defined 3-D path to calibrate its movements while optimizing the RIO® accuracy in the surgical volume and setting the center point for the cutting instrument. The femoral and tibial reference tracking arrays are then attached to the patient. Bone pins are placed in the femur and tibia, and optical arrays are securely attached. The camera is now positioned to track the robot and leg arrays through all ranges of motion (ROMs). Anatomical surface landmarks are registered before the skin is incised, and the leg is put through full ROM while the appropriate valgus load is applied on the joint. After skin incision, small articular accuracy checkpoint pins are inserted on tibia and femur, and the two bone surfaces are registered at these points to match them to the CT models. Incisions can be made as short as 2.25 in. in some patients with minimal strain on soft tissue.

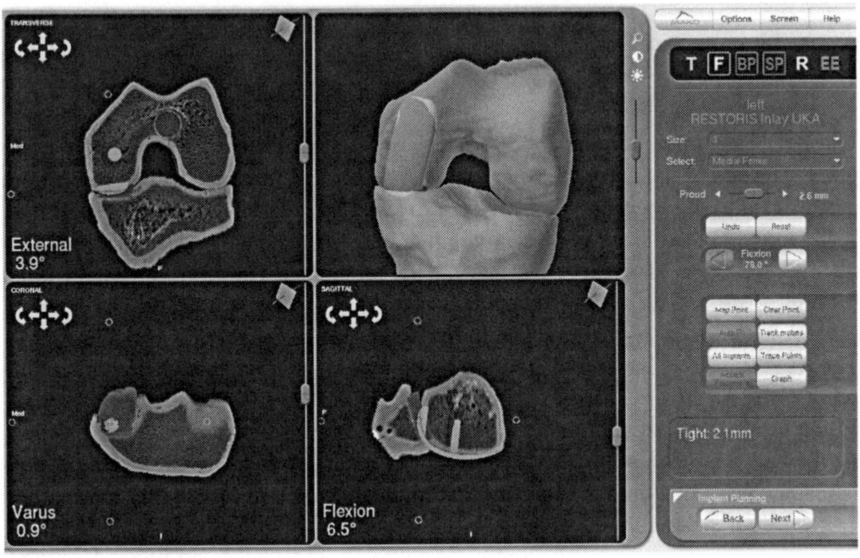

Fig. 10.12 Robot registration

10.5.4 Intraoperative Soft-Tissue Balancing

After intraoperative registration of bony anatomy and implant position target setting, a dynamic soft-tissue gap balancing algorithm is initiated. Virtual kinematic modeling of the knee and intraoperative tracking allow real-time adjustments to be made to obtain correct knee kinematics and soft-tissue balancing. First, osteophytes interfering with medial collateral ligament function are removed, and capsular adhesions interfering with knee function are relieved. As one indication for UKA is a correctable deformity, removal of these impediments makes it possible to achieve correct leg kinematics and tissue tension during passive manipulation throughout the full ROM with an applied valgus stress. Three-dimensional positions of femur and tibia are captured throughout the ROM with the medial collateral ligament properly tensioned. This provides correct bone spacing (extension and flexion gaps) during implant planning such that, after resection and component implantation, knee mechanics will be properly restored throughout the ROM. The articular surfaces of the components are then adjusted to fill that space throughout the ROM. Once optimized, the plan incorporates alignment metrics, implant congruence, and gap kinematics in a highly individualized fashion. Finally, any varus deformity is manually corrected with application of a valgus force to the knee, while lower limb alignment is simultaneously monitored and recorded by the navigation system. As the virtual components are optimized to fill the space necessary to correct this deformity, final lower limb alignment is reliably predicted. We target final lower limb alignment of approximately 2° of varus. Care is taken to avoid undercorrection (final alignment, <8° varus) and overcorrection (final alignment in valgus) of long-leg alignment [3].

10.5.5 Robotic Arm

The RIO® has three components: robotic arm, optical camera, and operator computer. The end of the robotic arm has a full 6 degrees of freedom, and its movement is restricted to the incision site by the 3-D virtual boundaries preset in the software at the time of customized preoperative planning; intraoperative adjustments of that plan are made to ensure correct soft tissue balancing. The optical camera is an infrared system. The system computer runs the software that drives the surgical plan. A high-speed burr is attached to the distal end of the robotic arm (Fig. 10.1). The surgeon moves the arm by guiding its tip within the predefined boundaries. The robot gives the surgeon active tactile, visual, and auditory feedback during burring. While inside the volume of bone to be resected, the arm operates without resisting. As the burr approaches the boundary, auditory feedback (a series of warning beeps) is given, and, when the burr reaches the boundary, the arm resists that motion and keeps the burr within the accepted volume. Thus, the arm effectively is a 3-D virtual instrument set that precisely executes the preoperative plan. In addition, excessive force at the limits of the 3-D cutting volume or rapid movement of patient anatomy immediately stops the cutting instrument to prevent unintentional resection outside the defined implant area. Unlike other active and semi-active robot systems, the RIO® does not require rigid fixation of the robot to the patient. Rather, osseous reference markers track the position of the tibia and the femur. As the bones move during surgery, the haptic 3-D resection volume moves coincidently. During resection, a leg holder is used to keep the limb stable while allowing optimal positioning of the knee to ensure access to the targeted surfaces.

10.5.6 Bone Resection Burr

A hand-powered or foot-pedal-operated high-speed (80,000 rpm) electric burr is used for resection. Burrs of different sizes are used: a 6-mm-diameter spherical burr for rapid removal of major bone material and to allow insertion of the femoral prosthesis post; a 2-mm-diameter spherical burr for fine finishing, including fine-finishing of the edges and corners of the resection area, and a 2 or 1.4 mm router may be used for keel canal preparation. All burring is visualized on a computer screen display, which shows the 3-D models of distal femur and proximal tibia. The models are color-coded and updated in real time based on resection progress, such that the resection area color is different from the color of the surrounding bone. If the robotic arm goes 0.5 mm outside the planned resection area (green), red appears on the display, and the arm stiffens progressively; if the user is pushing the arm any farther outside the green area, the robot will resist such motion (haptic feedback). The user will be warned with a visual and audio feedback and the burr will immediately stop spinning. These safety features are intended to prevent the user from cutting outside of the plan.

10.5.7 Prosthesis Selection

Implant choice depends on surgeon preference and the specific characteristics of the patient's osteoarthritis pattern. Current treatment options include isolated medial UKA, lateral UKA and patellofemoral arthroplasty (PFA) (Fig. 10.13). In addition, bicompartmental arthroplasty may be performed consisting of simultaneous medial UKA and PFA. Two implant types are available for selection of the tibial component, inlay and onlay designs. An inlay tibial prosthesis is an all-polyethylene design and utilizes the patient's formed tibial subchondral sclerotic bone bed to support the tibial component (Fig. 10.14) [4, 5]. The cavity resected utilizes the intact tibial rim for rotational control. The angle of the tibial component is usually within several degrees of varus to reproduce the patient's normal varus inclination. The patient's posterior slope is recreated with a deeper pocket posteriorly. Onlay tibial components utilize the patient's cortical rim for support (Fig. 10.15). The prosthesis is modular with a metal backing. The tibial angle is usually 90° to the mechanical tibial axis in the coronal plane.

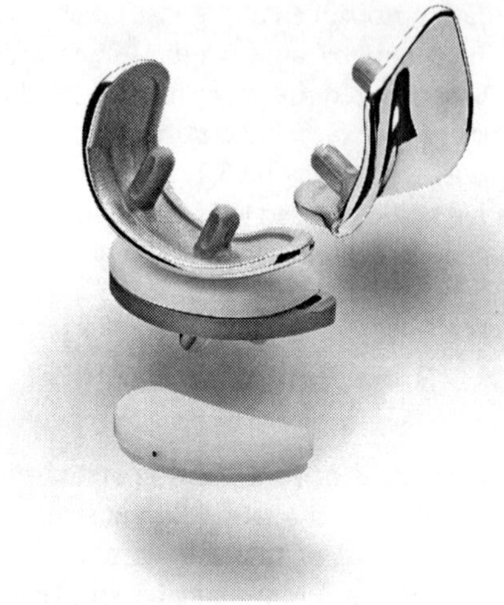

Fig. 10.13 UKA and PFA implants

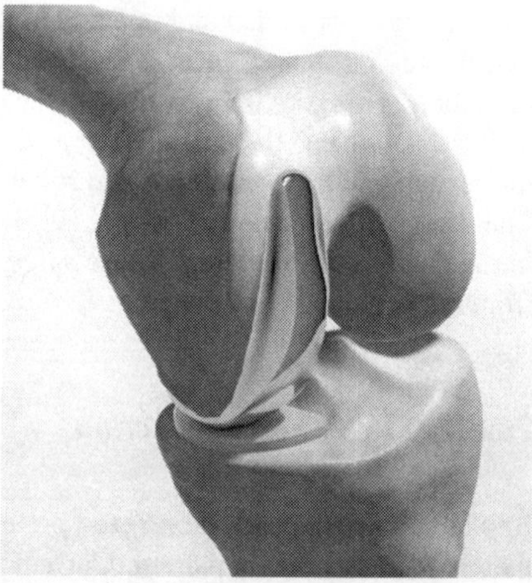

Fig. 10.14 Inlay UKA

10.5.8 Operative Technique

The robotic arm assists the surgeon during defined burring of the tibial and femoral surfaces. The arm helps control depth, width, and length of burring with graphical feedback on the navigation monitor. It is recommended that the arm be used to prepare the tibial cavity before addressing the femoral surface so as to allow easier access to that surface, particularly its posterior side. The arm also allows

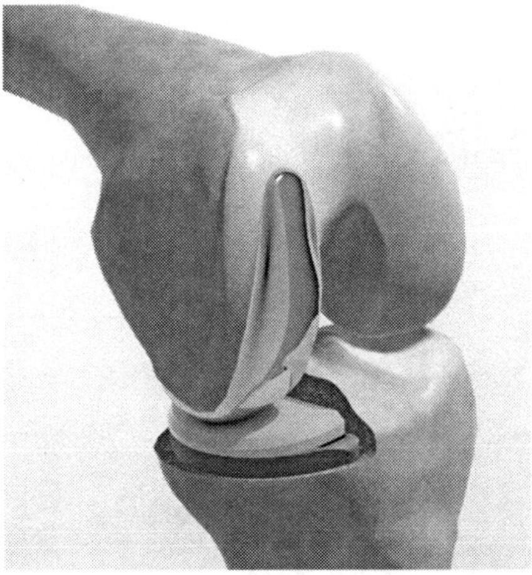

Fig. 10.15 Onlay UKA

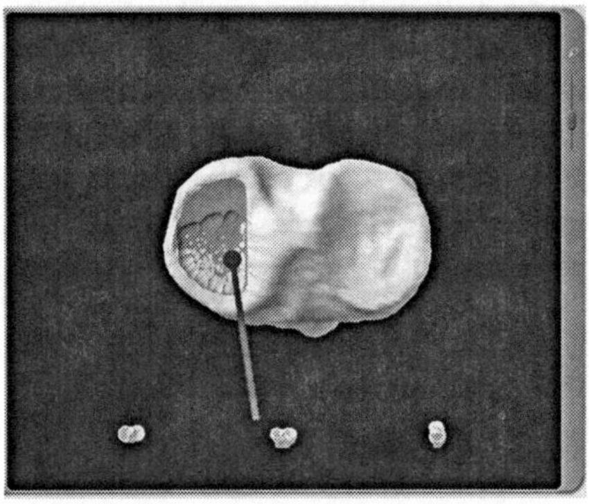

Fig. 10.16 Cavity milling process

intraoperative conversion to a metal-backed onlay implant. With use of only soft-tissue retractors, initial burring of tibial and femoral surfaces (including the femoral post hole) is performed with a 6-mm spherical burr; fine-milling is performed with the 2-mm spherical burr. The femoral keel slot is burred with the 2/1.4-mm fluted router. The navigation screen continuously shows the planned cavity vs. the actual cavity (Fig. 10.16). Once both have been completely milled (Fig. 10.17), femoral and tibial component trials are inserted, and a complete flexion–extension arc is performed. Dynamic long-leg alignment is displayed on the computer monitor so that final alignment can be tracked. Finally, once the implant is satisfactorily positioned, both implant components are cemented, and a final ROM of the knee joint is executed so that original, trial, and final implant kinematics and knee alignment can be compared. Before site closure, the mini-checkpoints and bone reference arrays are removed.

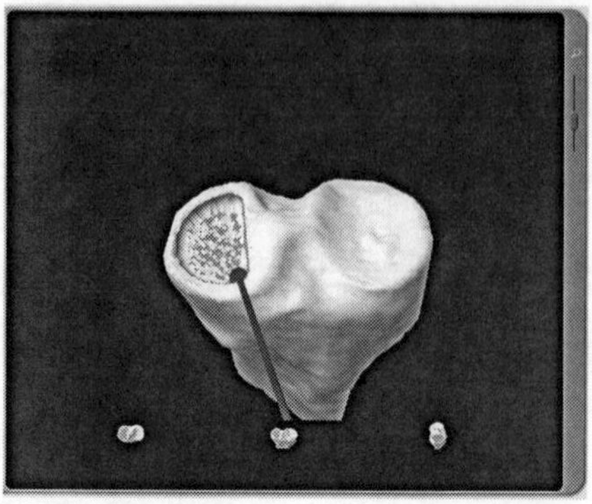

Fig. 10.17 Finished implant cavity

10.5.9 Post-operative Regimen

For UKA and PFA, patients can be discharged at home the same day if cleared by Physical therapy and medically stable. A 23 h overnight stay for pain control, antibiotics and anticoagulation is often used. The patient is mobilized the same day with PT and a CPM is used overnight to begin motion and determine comfort level. The patients usually require minimal therapy over the next several weeks. Surgical teams should focus on quadriceps stability and gait; the range of motion returns quite rapidly. Squats or lunges are initially avoided to minimize stress to the healing pin sites. Follow-up can occur semi annually.

10.6 Clinical Experience

With the recent resurgence of unicompartmental knee arthroplasty (UKA), investigators have overcome previous deficiencies with regard to implant design and patient selection and have more recently sought to improve upon existing surgical techniques. Attempts at minimally invasive surgery (MIS) and accuracy of component alignment are two areas that have received significant attention. UKA performed through a minimally invasive approach has resulted in significant clinical and financial benefits, including shorter length of hospital stay, faster recovery time, and reduced post-operative morbidity, when compared to conventional open approaches. However, the technical challenges of conventional jig-based UKA techniques appear to be magnified with the MIS approach. Impaired visualization has resulted in inferior component alignment and a high incidence of early implant failure in some series. Thus, these two goals have been difficult to achieve simultaneously using manual instrumentation. In addition, there appears to be a significant learning curve with respect to the manual MIS UKA technique, such that a surgeon's initial case series may result in inferior clinical outcomes when compared to subsequent cases.

10.6.1 Accuracy

In a study by Roche et al. [6], postoperative radiographs of 43 MAKOplasty® patients were examined for outliers, as defined by a panel of orthopedic surgeons. The radiographs shown represent a typical series of pre- and post-operative radiographs from one patient. Of the 344 individual radiographic measurements, only four (1%) were identified as outliers (Fig. 10.18).

In a comparative study, Coon et al. [7] examined a cohort of 33 MAKOplasty® patients and 44 standard UKA patients. The coronal and sagittal alignment of the tibial components were measured on post-operative AP and lateral radiographs and compared to the pre-operative plan. The RMS error of the tibial slope was 3.5° manually compared to 1.4° robotically. In addition, the variance using manual instruments was 2.8 times greater than the robotically guided implantations ($p < 0.0001$). In the coronal plane, the average error was 3.3 ± 1.8° more varus using manual instruments compared to 0.1 ± 2.4° when implanted robotically ($p < 0.0001$) (Fig. 10.19).

In a study of the first 20 MAKOplasty® patients at one institution, Sinha et al [8] similarly found that using the robotic arm resulted in extremely accurate and precise reconstruction of the individual patient anatomy. Postoperatively, all femoral components matched their preoperative alignment in terms of varus/valgus and flexion. They also reported a change in the femoral joint line of only 0.4 ± 0.5 mm. On the tibial side, the bone preparation matched the preoperative alignment with respect to posterior slope and varus, but there was slightly higher error in the final tibial component position, indicating that care must be taken such that pressurization and polymerization of polymethylmethacrylate does not change the tibial component position within the prepared cavity.

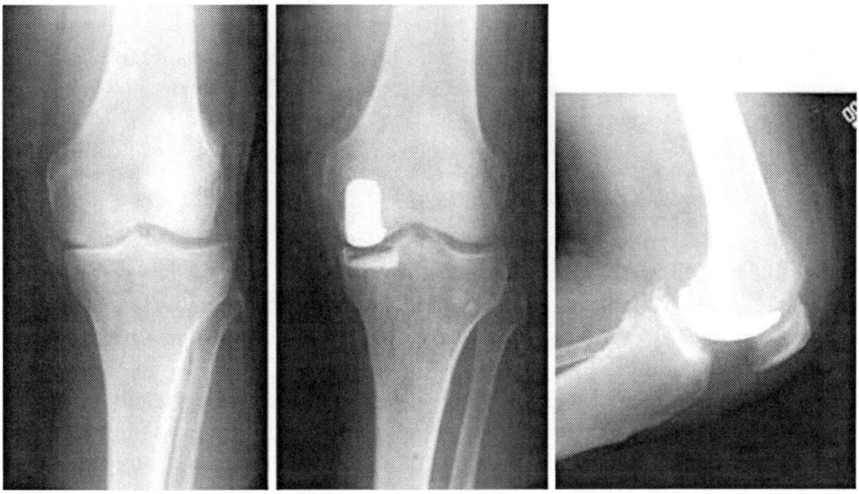

Fig. 10.18 Pre-operative AP X-ray and post-operative AP and lateral views

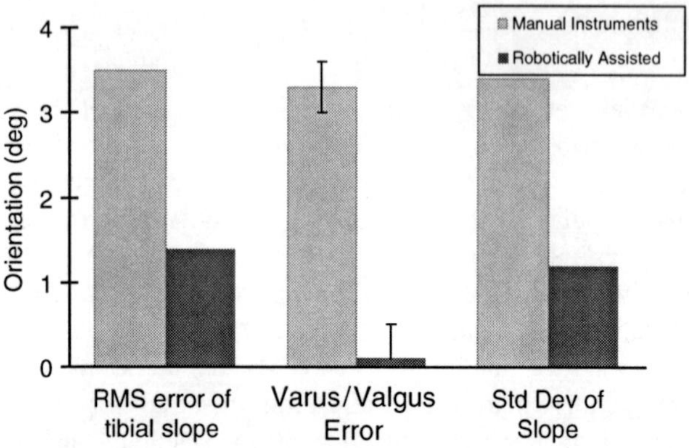

Fig. 10.19 Tibial slope error and variance in UKA using manual instrumentation vs. robotically assisted UKA

10.6.2 Bone Preservation

In a study by Coon et al. [9] comparing MAKOplasty® inlay vs. manual onlay implantations, the average depth of medial tibial plateau resection was significantly less with inlay tibial components (3.7 ± 0.8 mm) relative to onlay tibial components (6.5 ± 0.8 mm, $p < 0.0001$). In a separate study by Kreuzer et al. [10], the depth of resection was compared between a group of 26 MAKOplasty® patients and 16 patients who received an all-poly manual "resurfacing" UKA implant. Average depth of bony medial plateau resection was significantly greater in the standard technique onlay design group (8.5 ± 2.26 mm) compared to the robotically assisted inlay group (4.4 ± 0.93 mm) ($p < 0.0001$). At conversion to a standard TKA, the proposed tibial osteotomy would require medial augmentation/revision components (insert thickness >15 mm) in 75% of the onlay group as compared to 4% of the robotically assisted inlay group ($p < 0.0001$) (Fig. 10.20).

10.6.3 Clinical Outcomes

In the study of 43 MAKOplasty® patients from Roche et al. [6], it was found that the average flexion significantly increased at 3 months post-operatively to 126 ± 6° compared with 121 ± 8° pre-operatively ($p < 0.001$). Post-operative KSS and WOMAC total scores significantly improved from 95 ± 16 to 150 ± 27 ($p < 0.001$) and 41 ± 15 to 21 ± 17 ($p < 0.001$), respectively. Quality of life, as measured by the SF-12 Physical Summary also significantly improved from 30 ± 9 to 39 ± 12 ($p < 0.001$). Robotically guided UKA significantly improved every measured clinical outcome. In a comparative study by Coon et al. [11], while the average length of hospital stay was the same for both onlay (LOS = 1.0 ± 0.2 days) and

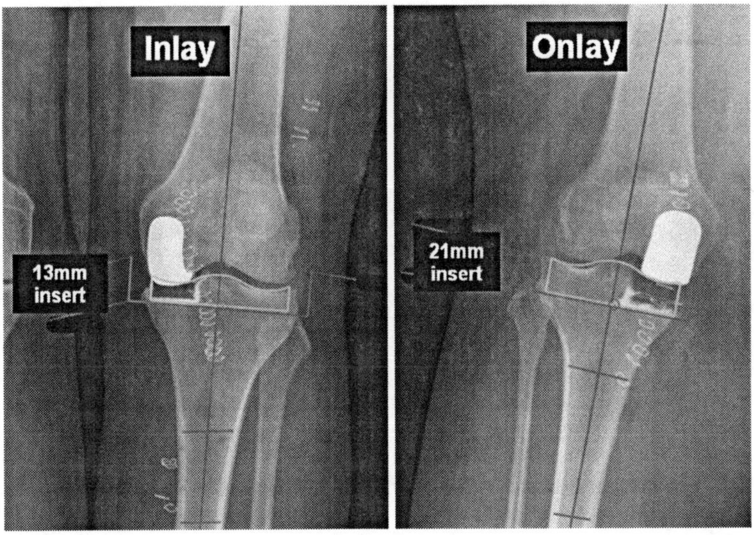

Fig. 10.20 MAKOplasty inlay and manual onlay implantations templated for a TKA with the predicted insert thicknesses as indicated

inlay (LOS = 0.9 ± 0.5 days) UKA procedures, a significantly higher percentage of inlay patients went home the day of surgery (18 vs. 2%, $p < 0.0001$). There was no significant difference in terms of average KSS, change in KSS, or Marmor rating between the two groups at any of the three follow-ups. At 12 weeks, for example, the average increase in the combined KSS was 83.6 in the conventional group and 79.7 in the haptic-guided group ($p = 0.66$). Furthermore, there were no significant differences in the measures that comprise these scores, such as range of motion, pain, and use of assist devices ($p > 0.05$) (Fig. 10.21).

We examined a cohort of 159 MAKOplasty® patients, consisting of 86 females and 73 males with an average follow-up of 1 year (range: 6 weeks to 25 months). The average age at surgery was 69 years with an average BMI of 28.3 kg/m^2. The clinical outcomes of these patients were measured pre-operatively and at follow-ups of 6 weeks, 3 month and 1 year. MAKOplasty® significantly improved all measured clinical outcomes (Fig. 10.22).

10.6.4 Learning Curve

Integrating new technology into the operating room can be associated with a significantly long learning curve, which introduces inefficiency in to the surgeon's practice and the hospital's OR work flow. In a study by Jinnah et al. [12], the surgical times of 781 MAKOplasty® patients, performed by 11 different surgeons, were

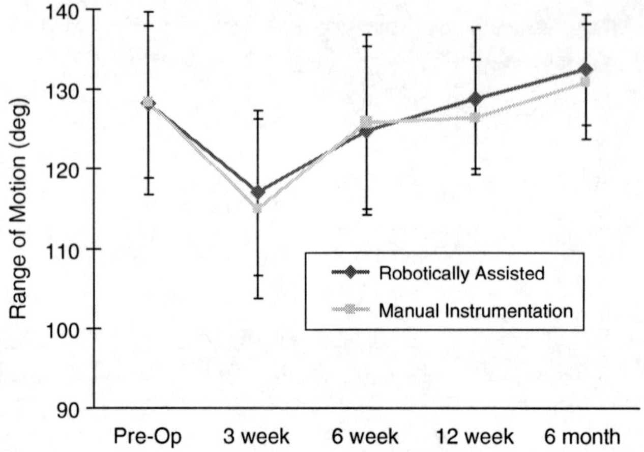

Fig. 10.21 Patient range of motion outcomes in UKA with manual instrumentation vs. robotically assisted UKA. There was no significant difference between the two groups at any of the four follow ups

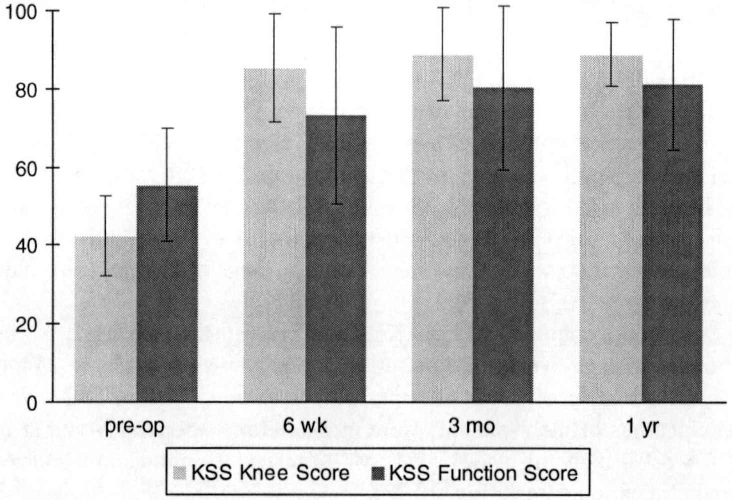

Fig. 10.22 KSS Knee and Functional Scores for 159 MAKOplasty patients. Both scores were significantly improved at all three follow ups

examined. Each surgeon had performed at least 40 surgeries with the new technology. The average surgical time for all surgeries across all surgeons was 55 ± 19 min (range: 22–165 min). The surgeon with the shortest steady state surgical time averaged 38 ± 9 min, while the surgeon with the longest steady state surgical time averaged 64 ± 16 min. The average number of surgeries required to have three surgeries completed within the 95% confidence interval of the steady state surgical time was 14 ± 8 (range: 5–29) (Fig. 10.23).

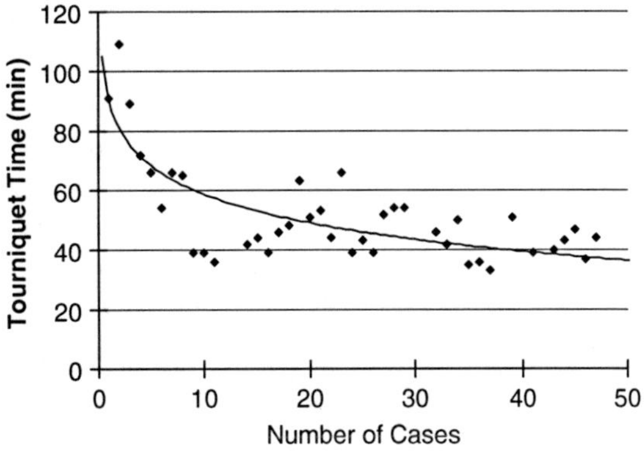

Fig. 10.23 Average MAKOplasty learning curve for new users

10.6.5 Complications

10.6.5.1 Intra-Operative Conversions

To date, 770 procedures have been performed at 18 hospitals by 39 different surgeons. Thirty cases have been converted intra-operatively to either a manual UKA or a TKA, for a 96% success rate. Six were converted to a TKA because of abnormal anatomy or progression of osteoarthritis into the patellofemoral or lateral compartment unrecognized pre-operatively. Three procedures were converted (two to a manual UKA and one to a TKA) due to user error such as ignoring accuracy check warnings or malpositioning of the infrared array on the bone. Twenty-one other procedures were safely and successfully converted due to camera, software or array issues detected by either the surgeon or the multiple internal accuracy/registration checks performed throughout the procedure. Three were finished manually with the intended implant, seven were converted to a manual UKA with a different implant and 11 were converted to a TKA.

10.6.5.2 Revisions

Ten MAKOplasty® patients have been revised, for a clinical failure rate of 1.3% at an average follow-up of 7.7 months (range: 1 day to 30 months). There have been five failures due to loosening of the tibial inlay component, with one of those due to a fall. Four were revised to an onlay UKA and one converted to a standard TKA. Two patients have been revised to a TKA due to infection. Two patients have been revised for persistent pain, one to an onlay UKA and one to a TKA. One patient who initially received an onlay UKA was revised to another onlay UKA one day post-operatively due to the initial pre-operative plan positioning the tibial component too posterior.

10.6.6 *Conclusions*

This new robotically guided UKA procedure provides comprehensive, three-dimensional planning of UKA components, including soft tissue balancing, followed by accurate resection of the femur and the tibia. We have shown that this preparation allows for placement and alignment of the components significantly more accurately and less variably compared to manual, jig-based instrumentation. We have also shown that, when appropriate, resurfacing inlay components resect less bone and provide quicker recovery than standard onlay UKA components. These bone preserving, resurfacing implants provide the potential to convert to an onlay UKA or a standard TKA with no augmentation should failure occur. All patients showed significant improvement in the post-operative function in every functional measurement. Clinical results of this initial series of UKAs using a new haptic-guided surgical technique are comparable to those using established techniques, thus alleviating concerns regarding the acquisition of a new skill set and inferior outcomes during the learning curve, which has been shown to be relatively short. Finally this new procedure has been shown to have a low complication and revision rate. The early use of robotic guidance for unicompartmental knee arthroplasty has been proven to be safe and effective and provides a more precise and accuracy "instrument set" compared to manual jig-based instrumentation. In addition, robotic guidance allows the realization of the clinical benefits of resurfacing implants.

10.7 Summary

We have discussed many of the motivations for surgical robotics, surveyed the current landscape, discussed key design principles in developing a surgical robot, illustrated a human-interactive surgical technique, and described clinical results to date. We have shown that surgical robotics is not only viable and feasible, but it can produce wonderful clinical results that benefit patients in significant ways.

Surgical robotics is no longer in its infancy, yet it is still far from achieving its goal of becoming a fully mature and integrated part of the surgical workflow in every hospital around the world. It is very likely that surgical robotics will be omnipresent in multiple fields in just a few decades – it seems to be a natural progression from the introduction of imaging and computing technology to medicine.

Robotics is enormously complex because the very nature of robotics implies a highly integrated system with emergent properties. These systems, unlike almost any other machine, can take on very human like properties and sometimes even behaviors. The need and demand for machines that can complement, and sometimes emulate human capabilities puts our ability to develop software that has contextual state awareness in the forefront. We have the ability to design and manufacture highly dexterous machines. There are still major limitations in terms of power/size motor ratios and the need for highly efficient, lightweight, portable, and very long lasting battery sources. We will always crave faster processors and more graphics power. However, the need to get our machines to be more aware of us, their surroundings, and how to behave in unexpected situations will continue to grow. We are well on the road to developing a complementary partner for the surgeon – the capabilities of such machines are bound to grow without any limit we can foresee today.

What we should always keep in mind as both a design principle and philosophy, is that these devices should *complement* and serve our purposes, not replace us. We believe that products and technologies that emerge based on this philosophy will be warmly welcomed by users, and become highly successful. Safety systems and inherent safety of the architecture should be a principle for all future systems. The "human in control" principle is clearly one that enables the current success of surgical robotics, and we should tread cautiously in designs that oppose this principle.

Surgical robotics offer the possibility of *democratized medicine* – the possibility of excellent outcomes everywhere, at every hospital. The field also may be one of the best pathways to significant cost reductions and efficiency gains in healthcare – automation often leads to both significantly increased speed without sacrificing quality, and sometimes even improving quality. Best practice techniques may be delivered to each patient in the same manner – allowing patients to experience a consistency of outcomes never before seen.

Today the field is still dominated by technology savvy and pioneering surgeons partnering with innovative companies and research teams. The next level of acceptance will be based on simplification and ease of use, through an understanding the new psychology of human–robot interactions, and by continued validation and publication of positive clinical outcomes.

The reader should note that the dream of surgical robotics is no longer a dream – it is a practical reality in the early twenty-first century. In the same way that flight, television, radio, and computing gripped and changed the twentieth century, the next 100 years may be the era of the robot. We can and should actively shape these technologies to be things that make life better, to promote and prolong our time and enjoyment. We believe that we have established the basic principles and philosophy, if followed, will be the right path for future robotics in medicine.

The authors also wish to thank all of their teams, families, research collaborators, surgeon partners, and patients for being part of a whole new chapter in medicine.

References

1. Collier, M.B., Eickmann, T.H., Sukezaki, F., McAuley, J.P., Engh, G.A.: Patient, implant, and alignment factors associated with revision of medial compartment unicondylar arthroplasty. J. Arthroplast. **21**(6 Suppl 2), 108–115 (2006)

2. Hernigou, P., Deschamps, G.: Posterior slope of the tibial implant and the outcome of unicompartmental knee arthroplasty. J. Bone. Joint. Surg. Am. **86-A**(3), 506–511 (2004)
3. Hernigou, P., Deschamps, G.: Alignment influences wear in the knee after medial unicompartmental arthroplasty. Clin. Orthop. Relat. Res. **423**, 161–165 (2004)
4. DeHaven, K.E.: Repicci II unicompartmental knee arthroplasty. Arthroscopy **19**(Suppl 1), 117–119 (2003)
5. Romanowski, M.R., Repicci, J.A.: Minimally invasive unicondylar arthroplasty: eight-year follow-up. J. Knee. Surg. **15**(1), 17–22 (2002)
6. Roche, M., Augustin, D., Conditt, M.A.: Accuracy of robotically assisted UKA. In: Proceedings of the 21st Annual Congress of the International Society of Technology in Arthroplasty. International Society for Technology in Arthroplasty, Sacramento, CA, p. 175 (2008)
7. Coon, T., Driscoll, M., Conditt, M.A.: Robotically assisted UKA is more accurate than manually instrumented UKA. In: Proceedings of the 21st Annual Congress of the International Society of Technology in Arthroplasty. International Society for Technology in Arthroplasty, Sacramento, CA, p. 274 (2008)
8. Sinha, R.K., Plush, R., Weems, V.J.: Unicompartmental arthroplasty using a tactile guided system. In: Proceedings of the 21st Annual Congress of the International Society of Technology in Arthroplasty. International Society for Technology in Arthroplasty, Sacramento, CA, p. 276 (2008)
9. Coon, T., Driscoll, M., Conditt, M.A.: Does less medial tibial plateau resection make a difference in UKA? In: Proceedings of the 21st Annual Congress of the International Society of Technology in Arthroplasty. International Society for Technology in Arthroplasty, Sacramento, CA, 274 (2008)
10. Kreuzer, S., Driscoll, M., Conditt, M.A.: Does conversion of a UKA to a TKA require medial augmentation? In: Proceedings of the 21st Annual Congress of the International Society of Technology in Arthroplasty. International Society for Technology in Arthroplasty, Sacramento, CA, p. 274 (2008)
11. Coon, T., Driscoll, M., Conditt, M.A.: Early clinical success of novel tactile guided UKA technique. In: Proceedings of the 21st Annual Congress of the International Society of Technology in Arthroplasty. International Society for Technology in Arthroplasty, Sacramento, CA, p. 141 (2008)
12. Jinnah, R., Horowitz, S., Lippincott, C.J., Conditt, M.A.: The Learning Curve of Robotic-Assisted UKA. Submitted to the Institute of Mechanical Engineers, Knee Arthroplasty: From Early Intervention to Revision, (2009)

Chapter 11
Robotic Surgery: Enabling Technology?

Moshe Shoham

Abstract Since its emergence, modern robotics has empowered mankind to reach goals ranging from the hazardous to unfeasible. In the medical field, robots have ushered in an era of minimized invasiveness, improved accuracy, lessened patient trauma and shortened recovery periods.

This chapter offers an overview of currently available medical robots and especially evaluates their technology-enabling capacities. Combination of significantly higher accuracy than conventional free-hand techniques with minimally invasive capability renders robotics an enabling technology. Obviously, dramatic dimensional changes in robots, to levels allowing for their introduction to the body for diagnostic and therapeutic purposes, also designates them to an enabling technology. The few currently available surgical robots are categorized in this chapter according to their enabling potential, along with a presentation of a future micro-robot for in-body treatment.

Keywords Active constraint · Enabling technology · Micro-robots · Minimally invasive · Pedicle screw · Snake robot · Spine surgery · Surgical robots

11.1 Introduction

Evolution of industrial robotics from its early inception in the early 1960s, has advanced this technology to autonomously perform common tasks in a more cost-effective and accurate manner. Its first applications were often toward activities which involved difficult or hazardous working environments. Later, as in the case of

M. Shoham (✉)
Robotics Laboratory, Department of Mechanical Engineering, Technion – Israel Institute of Technology, Haifa, Israel;
Mazor Surgical Technologies, Cesarea, Israel
e-mail: shoham@technion.ac.il

J. Rosen et al. (eds.), *Surgical Robotics: Systems Applications and Visions*,
DOI 10.1007/978-1-4419-1126-1_11,

Unimate, robots were implemented in pick and place manufacturing applications to ease the tedious and boring labor characteristic of production lines. Introduction of remotely operated systems allowed for robot exploitation in execution of tasks otherwise impossible for humans to withstand, such as radioactive, subterraneal or undersea operations. Modern-day robotics integrates sensory and advanced navigational hardware into automative systems, rendering them more adaptable to dynamic environments. In industry, the advanced degrees of repeatability, accessibility and speed offered by robots, yield increased throughput and facilitate quality control.

The potential contribution of robotic precision and delicacy toward revolutionizing medical applications soon became appreciated. In practice, major surgical operations are performed under robotic guidance via less invasive methods, leading to lessened patient trauma and enhanced recovery rates. In addition, development of micro-robotics is currently underway to locate, diagnose and specifically treat diseases from within the body. Yet, although the seeds of introducing robotic talent to surgical practices were planted some 20 years ago, only a handful are in routine use.

Looking back at the 50-year robotic era, one may wonder whether modern robots accomplish missions unconquerable by unassisted humans. When considering extreme examples, such as the Mars Exploration Rover or NASA's Pathfinder and Sojourner operating in outer space, the muser will be convinced that robotics indeed possess enabling features. However, in most cases, robotics simply offers improved consistency and accuracy and performs tasks within shorter periods of time. Economically, accelerated operative capacities and notable operational accuracy are often sufficient enough justification for product marketing. Yet, the intent of the current article is to evaluate the enabling features of the growing medical applications of robotics. While numerous research and development projects aiming at incorporating robots in the medical arena are underway, only those which have matured and are in ongoing use are presented below.

Do surgical robots merely provide highly regulated clinical assistance, as in the case of Computer Motion's (now Intuitive Surgical) laparoscopic camera holder, or can they execute tasks otherwise impossible by free-hand performing surgeons? Are there insurmountable applications limited by human physical capacities that can be achieved by the robot?

11.2 OR-Implemented Surgical Robots

Integrated Surgical Systems Inc. (ISS) made the pioneering breakthrough in robotic-aided surgeries by developing the ROBODOC® designed to provide optimal fit and alignment of hip prosthesis placement procedures. The application's precision was manifested in its ability to accurately direct the robotically held mill along predefined trajectories and was applied in over 20,000 joint replacement cases. After several years of suspended operations, use of the ROBODOC® system was reinitiated by Curexo Technology Corporation with broadened applications for revision surgeries, and received FDA clearance in August 2008. The cumulative results of its clinical application demonstrate enhanced implant fit to the robotically-shaped cavity, with broader implant-to-bone contact areas and less susceptibility to bone severage. Yet, these reports and the long term results did not provide sufficient evidence demonstrating a substantial edge on free-hand techniques. While to date, the advantages of this pioneering robot did not provide ample justification for its standardized integration into operating rooms (OR), reevaluation of its accumulating clinical impact may modify this ruling.

The growing need and demand for robotics supporting less invasive surgeries, a critical feature currently lacking in the ROBODOC® platform, initiated the development of new approaches to surgery-assisting robot design. The emergence of image-guided robotic assistance enabled surgeons to visualize and navigate complex anatomical structures during planning and executing stages. These remarkable advances ushered in an era of heightened accuracy and greater prospects for minimally invasive surgical approaches. More specifically, the Mako Surgical orthopedic device company engineered the RIO® robotic arm in response to the need for tools offering both accuracy and minimally invasive (MIS) surgical approaches for uni-compartmental knee replacement. As in the case of Acrobot's Sculptor® active-constraint robot, this robot is programmed to prevent the surgeon from moving a bone cutting tool beyond a predefined milling area. In this manner, the device keeps the active milling to within specific limits of the pre-planned field and at the same time offers minimally invasive access to the region of interest.

Prosurgics' Pathfinder and Neuromate® robots offer image-guided stereotactic neurosurgery robotics directing surgeons along a predefined path to the specific point of interest within the brain. While these systems claim sub-millimetric accuracy, beyond that of a free-hand surgeon, can they be considered enabling devices?

Robotically-guided radiotherapy provides accurate non-invasive treatment as reported for the CyberKnife and Gamma Knife streotactic radiotherapy systems which direct focused beams to their target tissue while correcting for natural patient breathing motions. Similarly, BrainLAB's Novalis Tx radiosurgery shaped-beam technology directs treatment beams to tumors in an accurate and non-invasive manner. Insightech's Exablate® provides a highly controlled and targeted method for uterine fibroid removal through noninvasive MRI-guided focused ultrasound. These systems offer accurate, non-invasive treatment and can therefore be designated enabling technologies.

Intuitive Surgical's *da Vinci*, currently the most widespread surgical robot, replicates and scales down surgeon hand motion and eliminates tremor leading to heightened accuracy. It allows dexterous manipulation of surgical tools within the body cavity, through small ports, thereby allowing for combined accuracy and minimal invasiveness. In these respects, the *da Vinci* robot system enables new operational technology otherwise impossible by the free-hand surgeon. Moreover, as the design of the teleoperating *da Vinci* also allows the surgeon to operate from a remote console physically separated from the robotic arms performing the surgery, future uses can be made in such remote locations as space, undersea or battlefields and will also globally extend expert abilities.

When looking back at the short history of surgical robotics, it seems that accuracy enhancement alone might not suffice marketing justification. Rather, the accelerating commercial competition and technological expertise demand advanced qualifications of such robotic systems, including significant minimizing invasiveness features, or decreased radiation exposure. Yet, marketing success will still not always reflect the degree to which a robotic device is enabling.

11.3 The SpineAssist Robot

Consider Mazor's SpineAssist robot (Figs. 11.1 and 11.2) developed by a team including the author of this chapter, designed to direct surgeons to accurate locations along the vertebra in efforts to enhance procedural accuracy of pedicle screw insertion, tumor resection or biopsy. This robot presents marketing advantages through its high accuracy, minimal invasiveness, and reduced radiation exposure [1–7]. It has performed in over 1,000 clinical cases, with more than 4,000 implants and reported no permanent neurological deficits, as compared to 2–5% reported in the literature for pedicle screw insertions [8–10]. Moreover, in 49% of the cases, surgeons acting with SpineAssist positioning preferred percutaneous over open approaches for screw insertions, in contrast to the 5% of non-robotically guided surgeries rate reported throughout Europe [11]. While these contributive elements do not mandate designation of SpineAssist an enabling technological advance, specific clinically complex cases where anatomical landmarks are missing and when free-hand surgeries are viewed as high-risk, as in severe cases of scoliosis and repeated revision surgeries, transform this positional platform into an enabling device.

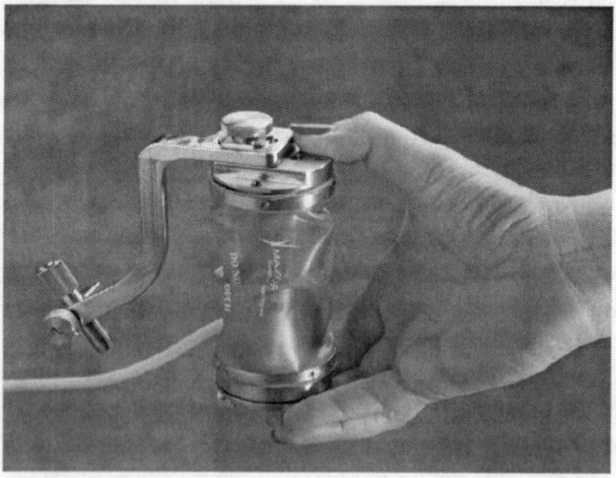

Fig. 11.1 SpineAssist surgical robot

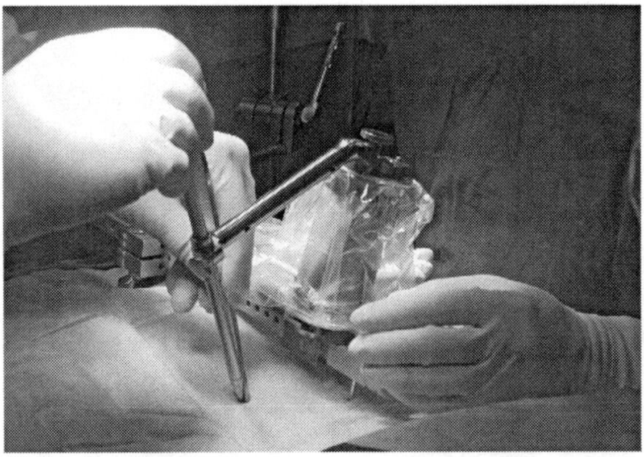

Fig. 11.2 SpineAssist assembly in a less invasive pedicle screw insertion. A guiding tube attached to the robot arm guides the surgical tool along a predefined trajectory

Fig. 11.3 Spinal fixation by pedicle screws and stabilizing rods vs. GO-LIF screw

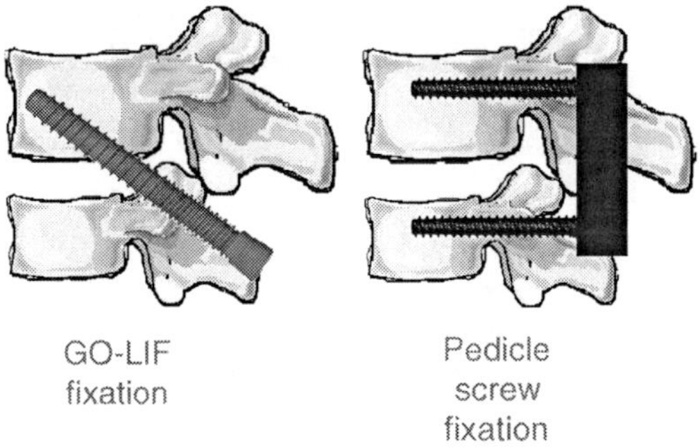

A novel, developing application of SpineAssist may merit its characterization as an enabling technology. The Guided Oblique Lumbar Interbody Fusion (GO-LIF) procedure is a unique approach to vertebra fixation, such that two diagonally inserted screws connect and stabilize neighboring vertebrae. The required trajectory comprises delicate extension of implants both anteriorly and superiorly, crossing the inferior vertebra as well as the interbody disc space into the superior vertebra. This strategy requires instrumentation of two screws alone, in place of the four screws and two stabilizing rods otherwise necessary in spinal fusion procedures performed according to typical pedicle screws stabilization protocols (Figs. 11.3 and 11.4). However, the close proximity to the spinal cord and nerve roots render such a procedure neurologically risky. Thus, the GO-LIF approach would be inapplicable under unguided free-hand percutaneous surgical technique. SpineAssist's innate ability to accurately move along predefined trajectories along with its bone-attached structure, can be exploited in such cases to high degrees. In applications of this nature, the SpineAssist robot can be unequivocally classified as an

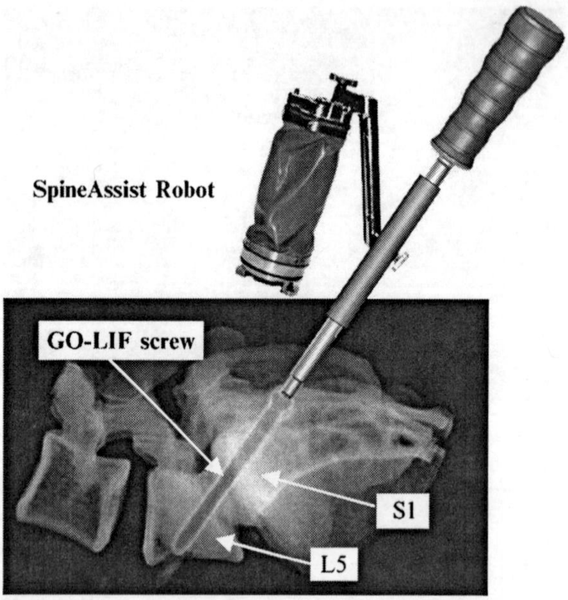

Fig. 11.4 GO-LIF screw trajectory guided by SpineAssist robot. Fixation of L5-S1 vertebrae

enabling technology by enabling percutaneous GO-LIF procedures with the added advantage of precision while considerably minimizing degree of invasiveness.

In collaborative efforts with Clevelend Clinic, fine-tuning of the SpineAssist platform to the GO-LIF implant technique has been effectuated. A preclinical cadaver fitting study has shown that GO-LIF fixation of 23/24 screws yielded a mean deviation of 1.3 ± 0.2 mm when compared to the preoperative plan, with no encroachment on nerve roots [12]. This early study, along with the results of 50 recent live cases, demonstrated that SpineAssist-guided GO-LIF is percutaneously attainable under robotic guidance.

11.4 Micro-Robots

With the advance of robotically-guided surgeries, along with the advent of sensors and micro-manufacturing abilities, much research is being devoted to development of micro-robots, designed to enter, traverse and operate in areas too small or too dangerous for humans or large robots. Micro-robots are being engineered to provide enhanced visualization of the surgical environment and dexterous task assistance unconstrained by entry port incision limits.

The objectives of robot miniaturization encompass design of those which are inserted through small ports to operate from within the body under external supervisory control. Other micro-robots have been prepared to autonomously propel from within, without the need for actuation from the small entrance port.

Future supervisory-controlled robotic enabling technology will provide accessibility via minimally invasive procedures, to locations far beyond human hand reach. Development of such devices have been described in the planning of miniature snake-like robots such as Carnegie Mellon's CardioArm designed to dynamically negotiate complex three-dimensional configurations with possible implications in cardiac treatment [13]. Accurate catheter placement is advanced by Hansen's Sensei® Robotic Catheter System by translating hand motions at the workstation to full catheter control within the heart [14], while the Stereotaxis Magnetic Navigation System applies external magnetic fields to maneuver the working tips of catheters, guidewires and other magnetic devices in interventional procedures. Similar principles are being employed by Johns Hopkins/Columbia robotic engineers in preparation of miniature snake robots designed to perform in narrow throat regions or in the delicate areas of the eye [15, 16]. In addition, preliminary experience describing *da Vinci*-based support of 30 NOTES (Natural Orifice Translumenal Endoscopy) procedures in porcine has been recently published [17].

Much research in this fascinating field of autonomous micro-robots focuses on applications within the human body, e.g. [18–20]. However, to the best of the author's knowledge, no routine clinical use of micro-robots has been made to date. The ViRob, for example, engineered by Shvalb, Salomon and the author of this chapter (Fig. 11.5), is 1 mm in diameter and actuated by an external magnetic field, stimulating it to crawl through cavities. The vast, still theoretical, applications of the ViRob can encompass drug or catheter delivery, tissue sampling, blood vessel maintenance or imaging of internal organs. Other works have described a modular crawler which successfully explored the gastric cavity, collected liver biopsies and enhanced imaging resolution for surgical procedures performed in porcine models [21]. Similarly, autonomously propelled robots have been developed to explore hollow cavities such as the colon or esophagus with "inch-worm"-based locomotion technique [22].

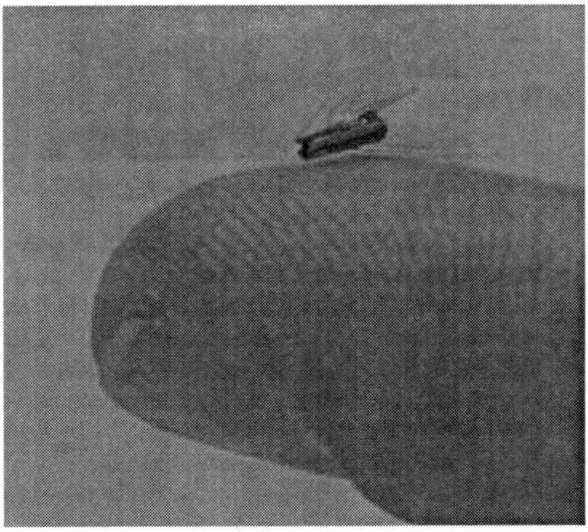

Fig. 11.5 ViRob – a crawling micro-robot

Due to their size and dexterity, miniature robots bear great potential when applied toward delicate surgical applications. Moreover, cooperative robots can be simultaneously inserted and manipulated to allow for maximal visualization, precision and overall procedural efficacy. Many future surgical applications are expected to demand precision extending beyond human hand capacities, rendering accuracy an adequate promoting factor for novel surgical technologies. In this manner, upon materialization, micro-robotics will transform to an enabling technological discipline.

11.5 Summary

Surgical robotics offers enhanced dexterity, increased accuracy and minimal invasiveness directed toward reducing patient trauma and improving clinical results. Significant robotic advances have simplified complex surgical procedures and have broadened the range of conditions treatable in the modern operating room. However, labeling of surgical robots as technology enablers requires the combination of significantly enhanced accuracy in conjunction with the option of minimally invasive surgical approaches. In parallel, in cases of complex microsurgery, accuracy extending beyond human limits also encapsulates the enabling features of surgical robots. In the future, robots will be designed to work in vivo at the level of single cells, in arenas beyond the surgeon's reach. In this manner, technologically enabling surgical robots will be engineered to distinguish between tumor and healthy cells, clear blocked blood vessels, deliver drugs or perform biopsies.

Equipping the surgeon's armament with such enabling devices will allow for development of new surgical procedures solely based on the high precision, miniaturization and enhanced accessibility features provided by robots. While once categorized as science fiction, robotics is slowly advancing the medical field by enabling the otherwise unconquerable.

References

1. Shoham, M., Burman, M., Zehavi, E., Joskowicz, L., Batkilin, E., Kunicher, Y.: Bone-mounted miniature robot for surgical procedures: concept and clinical applications. IEEE Trans. Robot. Autom. **19**(5), 893–901 (2003)
2. Shoham, M., Lieberman, I.H., Benzel, E.C., Togawa, D., Zehavi, E., Zilberstein, B., Fridlander, A., Joskowicz, L., Brink-Danan, S., Knoller, N.: Robotic assisted spinal surgery – from concept to clinical practice. J. Comput. Aided Surg. **12**(2), 105–115 (2007)
3. Togawa, D, Kayanja, M., Reinhardt, M.K., Shoham, M., Balter, A., Friedlander, A., Knoller, N., Benzel, E.C., Lieberman, I.H.: Bone-mounted miniature robotic guidance for pedicle screw and translaminar facet screw placement: part 2 – evaluation of system accuracy. Neurosurgery. 60, 129–139 (2007)(Operative Neurosurgery 1)
4. Dietl, R., Barzılay, Y., Kaplan, L., Roffman, M.: Miniature robotic guidance for vertebral body augmentation. Minim. Invasive. Spinal. Tech. (Serial Online) **1**, 2 (2008)

5. Lieberman, I.H., Hardenbrook, M, Wang, J, Guyer, R.D, Khanna, A.J.: Radiation exposure using miniature robotic guidance for spinal surgery. Spine J. **7**, 5 (2007)
6. Pechlivanis, I., Kiriyanthan, G., Engelhardt, M., Scholz, M., Lucke, S., Harders, A. Schmieder, K.: Percutaneous placement of pedicle screws in the lumbar spine using a bone mounted miniature robotic system, first experiences and accuracy of screw placement. **Spine J.** 34, 4 (2009)
7. Zaulan, Y., Alexandrovsky, V., Khazin, F., Silberstein, B., Roffman, M., Bruskin, A.: Robotic assisted vertebroplasty: our experience with a novel approach to the treatment of vertebral compression fractures. In: **World Society for Endoscopic Navigated and Minimal Invasive Spine Surgery (WENMISS) Annual Congress, London, UK** (2008)
8. Castro, W.H., Halm, H., Jerosch, J., Malms, J., Steinbeck, J., Blasius, S.: Accuracy of pedicle screw placement in lumbar vertebrae. Spine **21**, 11 (1996)
9. Schulze, C.J., Munzinger, E., Weber, U.: Clinical relevance of accuracy of pedicle screw placement, a computed tomographic-supported analysis. Spine **23**, 20 (1998)
10. Schwender, J.D. Holly, L.T., Rouben, D.P., Foley, K.T.: Minimally invasive transforaminal lumbar interbody fusion (TLIF): technical feasibility and initial results. J. Spinal Disord. Tech. **18**, S1–S6 (2005)
11. iData Research Inc. Europe Statistics (2009)
12. Selvon, S.C., Lieberman, I.H.: Guided oblique lumbar interbody fusion (go-lif): a surgical anatomic study of fixation. Computer Assisted Orthopedic Surgery (CAOS), Boston (2009)
13. Ota, T., Degani, A., Zubiate, B., Wolf, A., Choset, H., Schwartzman, D., Zenati, M.A.: Epicardial atrial ablation using a novel articulated robotic medical probe via a percutaneous subxiphoid approach. Innovations **1**(6), 335–340 (2007)
14. Kanagaratnam, P., Koa-Wing, M, Wallace, D.T, Goldenberg, A.S, Peters, N.S, Davies, D.W: Experience of robotic catheter ablation in humans using a novel remotely steerable catheter sheath. J. Interv. Card. Electrophysiol. **21**, 1 (2008)
15. Simaan, N., Xu, K, Wei, W, Kapoor, A, Kazanzides, P, Taylor, R, Flint, P: Design and integration of a telerobotic system for minimally invasive surgery of the throat. Int. J. Robot. Res. **28**, 1134–1153 (2009)
16. Wei, W, Goldman, R.E., Fine, H.F., Chang, S., Simaan, N : Performance evaluation for multi-arm manipulation of hollow suspended organs. IEEE Trans. Robot. **25**, 1 (2009)
17. Haber, G.P., Crouzet, S., Kamoi, K, Berger, A, Monish, A, Goel, R, Canes, D, Desai, M, Gill, I, Kaouk, J.H : Robotic NOTES (Natural Orifice Translumenal Endoscopic Surgery) in reconstructive urology: initial laboratory experience. Urology **71**, 996–1000 (2008)
18. Kosa, G, Shoham, M, Zaaroor, M : Propulsion method for swimming micro-robots. IEEE Trans. Robot. **2**, 1 (2007)
19. Yesin, K.B, Vollmers, KNelson,B.J: Modeling and control of untethered biomicrorobots in a fluidic environment using electromagnetic fields. **Int. J. Robot. Res. 25**, 5–6 (2006)
20. Stefanini, C, Menciassi, A, Dario, P : Modeling and experiments on a legged microrobot locomoting in a tubular, compliant and slippery environment. **Int. J. Robot. Res. 25**, 5–6 (2006)
21. Rentschler, M.E, Dumpert, J, Platt, S.R, Farritor, S.M, Oleynikov, D: Natural orifice surgery with an endoluminal mobile robot. Surg. Endosc. 21, 1212–1215 (2007)
22. Phee, L, Accoto, D, Menciassi, A, Stefanini, C, Carrozza, M, Dario, P : Analysis and development of locomotion devices for the gastrointestinal tract. IEEE Trans. Biomed. Eng. **49**, 613–616 (2002)

Chapter 12
Enabling Medical Robotics for the Next Generation of Minimally Invasive Procedures: Minimally Invasive Cardiac Surgery with Single Port Access

Howie Choset, Marco Zenati, Takeyoshi Ota, Amir Degani, David Schwartzman, Brett Zubiate, and Cornell Wright

Abstract Minimally invasive cardiac surgery (MICS) is an evolving strategy aimed at delivering the desired form of cardiovascular therapy with the least change in homeostasis, ideally matching the same degree of invasiveness of percutaneous cardiac interventions. Cardiac surgery is different from other surgical procedures because the large sternotomy incision required to access the heart requires general endotracheal anesthesia (GETA) and the heart–lung machine that is required for open-heart surgery (e.g. valve repair) adds further morbidity. We have developed a novel, highly articulated robotic surgical system (CardioARM) to enable minimally invasive intrapericardial therapeutic delivery through a subxiphoid approach. The CardioARM is a robotic surgical system consisting of serially connected rigid cylindrical links housing flexible working ports through which catheter-based tools for therapy and imaging can be advanced. The CardioARM is controlled via a computer-driven user interface which is operated outside of the operative field. We believe single port access to be key to the success of the CardioARM. We have performed preliminary proof of concept studies in a porcine preparation by performing epicardial ablation.

Keywords Minimally invasive cardiac surgery · Single port surgery · Snake robotics · Surgical robotics

12.1 Introduction

Minimally invasive interventions have the potential to revolutionize surgical practice by offering reduced pain, faster recovery, and fewer complications. We believe the key to achieving such potential is to eliminate the need for multiple (up to 8) ports by using a single port entry. Single port access approaches may facilitate

H. Choset (✉)
The Robotics Institute, Carnegie Mellon University, Pittsburgh, PA 15213, USA
e-mail: choset@cs.cmu.edu

J. Rosen et al. (eds.), *Surgical Robotics: Systems Applications and Vision*,
DOI 10.1007/978-1-4419-1126-1_12,

existing procedures but perhaps more importantly, they will enable new ones, and at a lower cost. This reduced cost has the added benefit of making therapies available to a larger portion of the general public.

A key feature requirement for single port entry is the need for dedicated robotic technology that can operate without access limitation and full feedback from a single entry point. Although they have great visual feedback, conventional surgical robots, such as the Intuitive Surgical's DaVinci SystemTM, are not adequate for single port entry because the three or four large robot arms manipulate linear chopstick laparoscopic-like devices which have limited access to line-of-sight regions from the ports. A small articulated device or a miniature mobile/crawling unit, not a conventional robot, is key to accessing many anatomical targets from a single port.

We have developed a novel, highly articulated robotic surgical system (CardioARM) to enable minimally invasive intrapericardial therapeutic delivery through a subxiphoid approach. The principle benefit of the CardioARM is that it has many internal articulated degrees of freedom to virtually provide unlimited but controllable flexibility which allows access to anatomic targets deep in the body without disturbing surround tissue. The CardioARM also has working channels through which catheter-based tools for therapy and imaging can be advanced. In six experimental subjects, CardioARM was introduced percutaneously via subxiphoid access. A commercial 5 Fr radiofrequency ablation catheter was introduced via the working port, which was then used to guide deployment. In all subjects, regional ("linear") left atrial ablation was successfully achieved, without complications. Based on these preliminary studies, we believe that CardioARM promises to enable deployment of a number of epicardium-based therapies. Minimally invasive cardiac surgery is one application for this technology; we are currently investigating the use of this robot for natural orifice transluminal endoscopic surgery (NOTES).

12.2 Relationship to Prior Work

12.2.1 Minimally Invasive Cardiac Surgery

Two approaches in the 1990s attempted to make cardiac surgery less invasive. First, the MIDCAB (Minimally Invasive Direct Coronary Artery Bypass) procedure involved a single vessel coronary bypass on the anterior surface of the heart on a beating heart through a small anterior thoracotomy [1]. Although the MIDCAB procedure catalyzed the minimally invasive movement in cardiac surgery, it now constitutes a minority of procedures; however, it did evolve into the current OPCAB (off pump coronary artery bypass grafting) procedure in which multi-vessel bypass is performed on a beating heart through a median sternotomy incision [2]. Second, the Port Access approach attempted totally endoscopic coronary artery bypass surgery on an arrested heart through ports but still using cardiopulmonary bypass [3]. Because of the complexities involved with the challenge of endoscopic microvascular reconstruction of coronary vessels, the totally endoscopic approach was prohibitive.

Computer-assisted telemanipulation systems like Intuitive Surgical's DaVinci™ Surgical System are being evaluated in prospective randomized trials as potential enabling technology for endoscopic cardiac surgery for such procedures as closure of atrial septal defects, mitral valve repair and coronary artery bypass; however, the prohibitive cost and the complexity of the technology limit the potential for widespread adoption [4]. Furthermore, totally endoscopic cardiac procedures require multiple (3–4) ports in the left, right or both chest and require special anesthetic techniques with double-lumen endotracheal tubes in order to collapse alternatively the left or the right lung [5].

An innovative MICS approach for epicardial interventions is subxiphoid videopericardioscopy (SVP) pioneered by Zenati [6] (Fig. 12.1). The SVP is designed to minimize the changes in patient's homeostasis by: (1) requiring only one port (<20 mm diameter), (2) not requiring general endotracheal anesthesia (as no invasion of pleural spaces is necessary), (3) not requiring the use of the heart–lung machine. This new original approach has the potential for redefining MICS, allowing a truly minimally invasive access to the beating heart and bridging to interventional cardiologists and electrophysiologists as potential users as well as minimally invasive surgeons. Dr. Zenati has used this approach for epicardial left heart pacing lead implantation for resynchronization [6, 7].

The subxiphoid videopericardioscopy (SVP) device (Guidant Corporation, Santa Clara, California) is the only dedicated technology available for endoscopic video exploration of the pericardial cavity; Guidant received approval for clinical use of this device by the Food and Drug Administration (510k number K023629, November 12, 2002). The SVP device is an elongated instrument made of stainless steel that contains two contiguous circular lumens. Its aggregate maximal diameter is 16 mm. The superior lumen, 4.2 mm in diameter, is used to house a 4 mm diameter endoscope (Scholly Fiberoptic, Denzlingen, Germany) and is enclosed distally by a conical, transparent plastic tip. The inferior lumen, 7.3 mm in

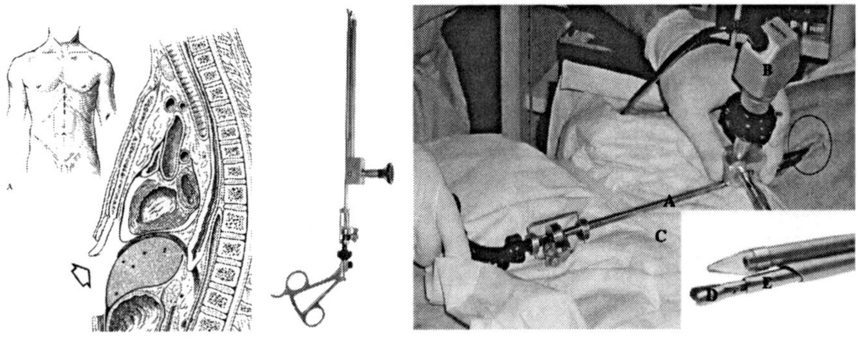

Fig. 12.1 (*left*) Subxyphoid access, (*middle*) Guidant's videopericardioscope (SVP), Images from SVP in porcine model

diameter, is open-ended and used as a "working port" with ability to accommodate a 7 mm diameter instrument.

One major problem associated with the present configuration of the SVP device is its rigidity and hence the significant potential for compression of the beating heart with trigger of life-threatening arrhythmia. Most of the anatomical targets for videopericardioscopy are located in remote areas of the pericardium, away from the entry point in the pericardium below the xiphoid process. A number of cardiovascular therapies based on the epicardium could be effectively delivered by the SVP approach if these physical constraints (rigidity, lack of maneuverability) could be solved, including but not limited to epicardial ventricular and atrial ablation for life-threatening arrhythmias [8, 9], left atrial appendectomy [10, 11], pulmonary vein electrical isolation [12], myocardial revascularization, cell transplantation [13, 14], injection of myocardial growth factors [15], gene therapy [16].

12.2.2 Snake Robots

The highly articulated mechanism development in our work is born out of Hirose's seminal 1972 work in which he developed mechanisms to crawl like snakes [17]. Inspired by this work, others developed the hyper-redundant manipulator and some basic techniques to position the end-effector (i.e., perform inverse kinematics) [18]. These works, and others that followed, were generally limited to one plane and could not travel in three-dimensional volumes.

Many spatial snake robots have a fixed based and typically are a serial chain of linearly actuated universal joints stacked on top of each other. Takanashi developed at NEC a new two-DOF joint for snake robots that allowed a more compact design. This joint used a passive universal joint to prevent adjacent bays from twisting while at the same time allowing two degrees of freedom: bending and orienting. This universal joint enveloped an angular swivel joint, which provided the two degrees of freedom. The universal joint being installed on the outside rendered the joint too bulky. Researchers at Jet Propulsion Laboratory (JPL) "inverted" Takanashi's design by placing a small universal joint in the interior of the robot. This allowed for a more compact design, but came at the cost of strength and stiffness (backlash). A small universal joint cannot transmit rotational motion at big deflection angles nor can withstand heavy loads.

Over the past 5 years, our group has developed a family of "snake robots" (Fig. 12.2), both fixed-base elephant trunk-like probes [19] and free-crawling robots [20]. The development of these robots was geared toward urban search-and-rescue in collapsed buildings [19, 21]. One robot is a series of innovative two-degree-of-freedom universal joints with a camera at its distal end to report imagery to remote users. The efficacy of this snake robot has been demonstrated in urban search and rescue training scenarios with the Chemical Biological Response Force (Marines) and the Center for Robotics Assisted Search and Rescue (see http://snakerobot.com).

Fig. 12.2 (*left*) Fixed base "elephant trunkbot" on a mobile base (*right*) Free crawling snake robot

12.2.3 Novel Forms of Actuation

For MICS, a snake robot would require an outer diameter of less than 10 mm. The many degrees of articulation that furnish a small snake robot with its enhanced capabilities also offer its main research challenges: constructing a maneuverable device with many degrees of freedom in a small space. Much prior work, therefore, has been geared toward developing novel actuators that are small and hopefully strong and robust to operate an articulated device.

Numerous works have been presented on active catheters and endoscopes, most actuated by non-conventional actuation technology such as shape memory alloys (SMA) actuators (Tohuko University, Olympus Optical Co). SMA spring and wire actuation has been implemented by Hirose to detect contact forces; the overall accuracy of the device is 2.3 mm [17]. Another endoscopic active device developed was an 8-mm diameter worm-like mechanism formed by a sequence of segments articulated to each other by SMA driven pin joints [22]. This device was specifically designed to explore the intestine with a camera. A 2.8-mm diameter active catheter was developed based on silicon micromachining [23]. This multilane manipulator is connected by joints made of SMA, fixed at equilateral triangular locations to allow bending in several directions. Unfortunately, SMA-based designs generate indirect heating, which can cause collateral tissue damage. Several other papers have reported additional endoscopic, SMA-based, tools [24–29]. However, disadvantages of these tools include its relatively low stiffness and its requirement of high activation voltage causing heat to be generated.

An alternative to SMAs are electrostrictive polymer artificial muscles (EPAM), which have been used to create snake-like endoscopic robot composed of several blocks joined by a concentric spine [30]. In addition, another snake-like manipulator using EPAM has been designed with a special actuator design, allowing control of curvature [31]. A different activation concept involves a 5 mm diameter two-degrees-of-freedom tool driven by super-elastic nitinol (NiTi) wires [32]. However, wire actuation, SMA, and EPAM actuation all become a challenge with robots having multiple degrees of freedom due to limited space inside the robot's mechanical envelope. For this reason, the previous work was not able to achieve a complex curve, such as an "S" shape, in a three-dimensional space and therefore was limited to a confined luminal tube-like environment.

12.3 CardioArm Mechanism

The CardioARM is composed of 50 rigid cylindrical links serially connected by four cables. Two adjacent links can rotate approximately ±10° relative to each other. The current distal apparatus is 10 mm in diameter, 300 mm in length with 105 degrees of freedom. A novel feature of this mechanism is that all of the links do not have to be individually controlled and hence this device is sometimes called a "follow-the-leader" mechanism. When the user specifies inputs for the distal tip of the robot, all the other links follow its location. Since the distal apparatus is capable of preserving its previous three-dimensional (3-D) configuration, in a sense the CardioARM traces a curve in three dimensions. The radius of curvature of the distal apparatus is 35 mm at minimum. The maximum speed of forward and reverse movement is up to 20 mm/s. See Fig. 12.3.

A feeding mechanism, rigidly attached to the operating room table via an adjustable support mechanism, inserts the CardioARM through a small incision

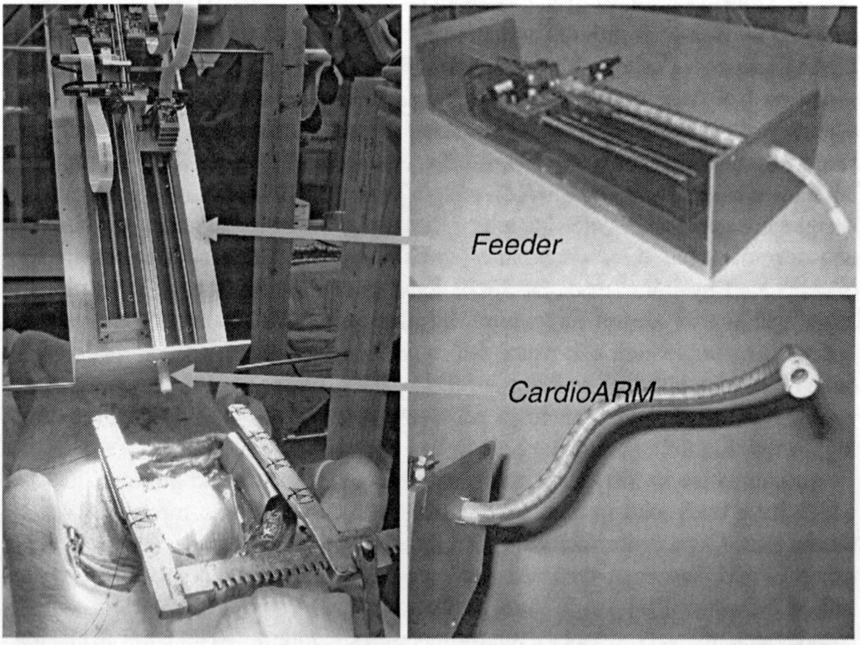

Fig. 12.3 Feeder mechanism (*upper left and right*) inserting CardioARM (*lower right*) into a pig (*lower left*)

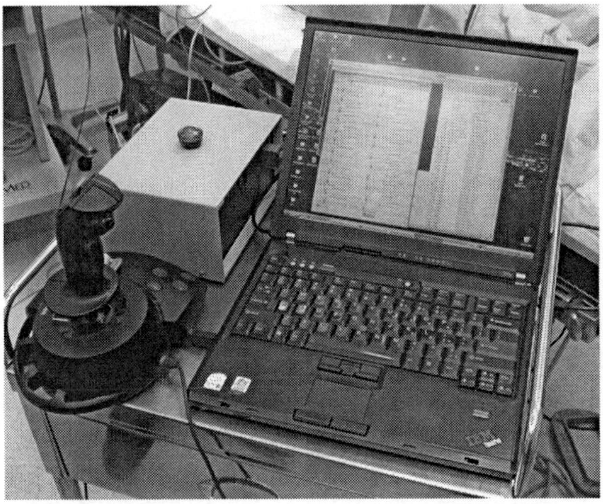

Fig. 12.4 CardioARM user interface

or port. The feeder contains four motors that pull on the cables to marionette the CardioARM. The feeder also has two motors to push and pull the CardioARM. Finally, the feeder houses all of the electronics that drive the probe and interface it with a user input device, such as a joystick, and computers to take high level commands to direct the CardioARM. See Fig. 12.4.

Control of the CardioARM requires a toggle and a straight-forward inverse kinematics procedure: The toggle selects direction with a button on the joystick to choose between forward or reverse motion. In forward mode, the joystick heading angle is converted into cable tensions via simple inverse kinematics to steer the CardioARM. The cable lengths are set via their associated motors as the motor pushing the mechanism engages. The process repeats as the joystick is in use. In reverse mode, the controller simply invokes the tensioning/advancing commands in reverse order causing the CardioARM to retract along the path it originally followed. The operator uses a 2-DOF joystick to control the distal link together with a button to control forward/backward motions.

We are able to pass catheter-based tools for therapy and imaging through the CardioARM. Visualization is provided by an on-board optical 15K bundle fiber scope with an integrated light guide, 65° FOV, 640 × 480 CCD camera. (fiber: FIGH-30-850N, Myriad Fiber Imaging Technology, Dudley, MA, camera: EO-2AN, Edmund Optics, Barrington, NJ).

12.4 Experiments

The CardioARM has been tested (Fig. 12.5) in 15 large (35–45 kg) healthy Yorkshire pigs of either sex at the Surgical Research Laboratory of the University of Pittsburgh [33, 34]. Initially, Zenati performed preclinical feasibility testing of

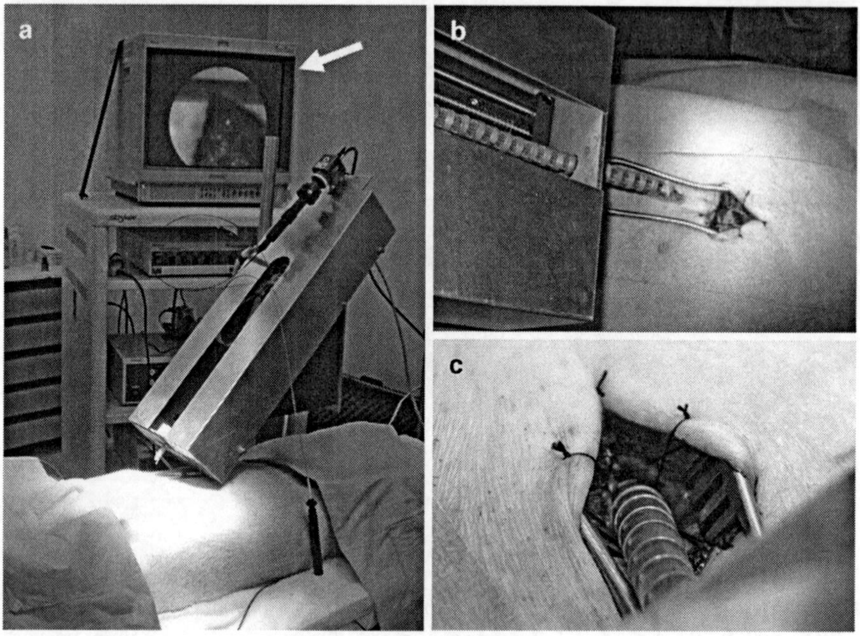

Fig. 12.5 CardioARM mount, display, and insertion into porcine model

the CardioARM prototype in the open-chest intrapericardial environment. We then performed preclinical testing accessing the intrapericardial environment through the subxyphoid.

The thoracic experiments were performed in the anesthetized animals following a median sternotomy. A 15 mm opening was created on the pericardium at the junction with the diaphragm; through this opening, the CardioARM was remotely guided to slide between the pericardium and the anterior wall of the right ventricle on the beating heart. After opening of the pericardium, the CardioARM was guided through the transverse sinus of the pericardium. This path was successfully completed both with the CardioARM entering through a subxiphoid access (Fig. 12.6), and from a right thoracic port. An endoscopic biopsy forceps was advanced through the CardioARM's working channel to simulate a pericardial biopsy; several specimens of pericardium were successfully harvested through the CardioARM (Fig. 12.6-left).

Furthermore, we were able to insert through the CardioARM's working channel a customized EndoloopTM (Ethicon Endosurgery), and simulate an LAA ligation (Fig. 12.6-right). The endoscopic biopsy forceps were found to be a useful adjunct to presentation of the LAA for optimal placement of the EndoLoop at its base. After the completion of the experiment, absence of any gross epicardial damage to the heart was confirmed by visual inspection. No adverse hemodynamic or electrocardiographic interference induced by the motion of the CardioARM was detected.

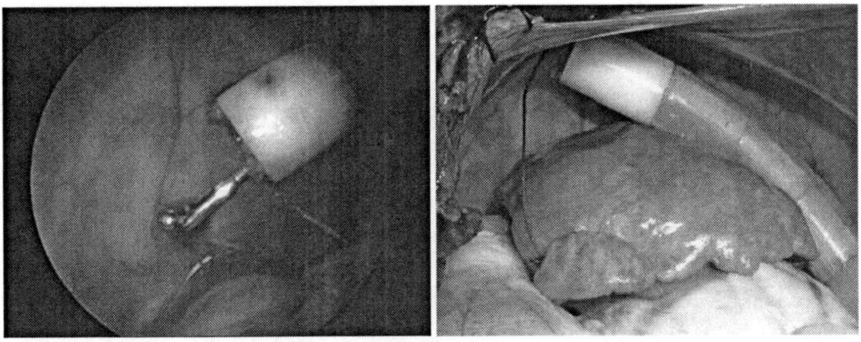

Fig. 12.6 Inserting tools through CardioARM's working channel: (*left*) Using biopsy forceps to perform pericardium biopsies; (*right*) Using Endoloop to ligate the LAA

Our initial experiments verified that we can indeed perform an ablation in hard to reach portions on a live beating heart (in a porcine model). The CardioARM system was introduced through a 15-mm subxiphoid incision into the pericardial space. Upon reaching the base of the left atrial appendage, a 7 Fr commercially available ablation catheter was introduced through one of the ports (Fig. 12.7), and under direct vision, epicardial ablation was performed. Epicardial transmural lesions were confirmed by histopathology. A detailed account of these experiments was published on the official journal of the International Society for Minimally Invasive Cardiac Surgery [35].

Further experiments entailed navigating to multiple portions of a live beating heart (in a porcine model) and performing ablations [36]. Six healthy large swine were anesthetized and placed in a supine position. A small subxiphoid skin incision (20 mm in length) and pericardiotomy (15 mm in diameter) were created under direct visualization. CardioARM was mounted on a surgical table (Fig. 12.5a) in a position for easy insertion through the subxiphoid incision. The distal apparatus of CardioARM was introduced into the pericardial space under the surgeon's control while watching a monitor display of the on-board optic fiber view (Fig. 12.5a). First, navigation trials to acquire several anatomical targets (i.e., right atrial appendage, superior vena cava, ascending aorta, left atrial appendage, transverse sinus from the left side, and atrioventricular groove in the posterior wall of the heart) were performed. When one target was acquired, the distal apparatus retracted to the initial position (i.e., the subxiphoid incision), then moved to another target. Following the navigation trials, left atrial ablation trials were performed. Once the tip of CardioARM was positioned at the vicinity of a target on the left atrium, a 5 Fr radiofrequency ablation catheter (Biosense Webster, Diamond Bar, CA) was passed through a working port of CardioARM, and a linear ablation lesion was created on the left atrial epicardium. A radiofrequency energy generator (Stockert 70, Biosense Webster) was set to deliver a power of 30 W for 30 s per lesion. Blood pressure and electrocardiogram were monitored throughout the trials. The animals were euthanized at the end of the trials, and postmortem examination was performed.

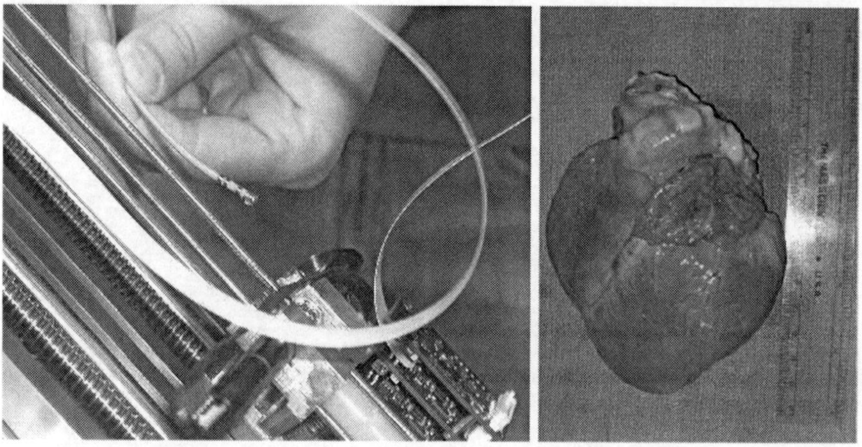

Fig. 12.7 (*left*) 7 Fr irrigated ablation catheter before insertion into CardioARM. (*right*) Two discrete ablation lesions are clearly visible on the LAA

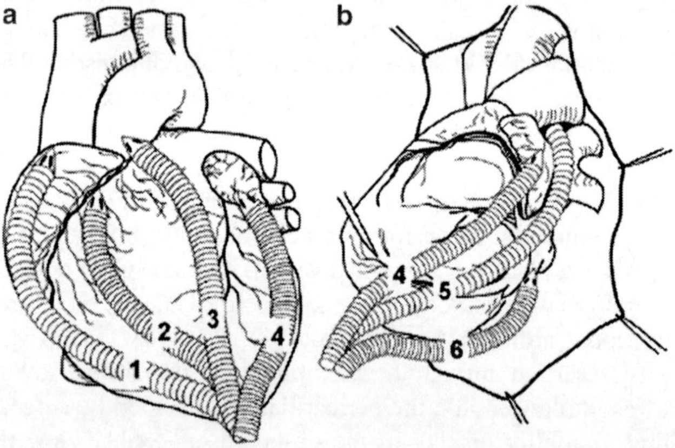

Fig. 12.8 The accomplished courses of the distal apparatus of CardioARM in the navigation trials. (**a**) Front view. (**b**) Left lateral view. #1: superior vena cava, #2: right atrial appendage, #3: ascending aorta, #4: left atrial appendage, #5: transverse sinus, #6: atrioventricular groove

All animals tolerated the procedures until their elective euthanasia. In the navigation trials, the distal apparatus of CardioARM followed a complex 3-D path from the subxiphoid incision along the ventricular wall to each target (Fig. 12.8). All navigation targets were acquired without complications (e.g., fatal arrhythmia, hypotension, bleeding). The on-board camera provided adequate visualization for navigation (Fig. 12.9a).

In the ablation trials, a linear lesion composed of several consecutive "dot-to-dot" lesions at the base of the left atrial appendage was successfully completed (Figs. 12.9b, c). No adverse event was noted during the trials. There was no injury

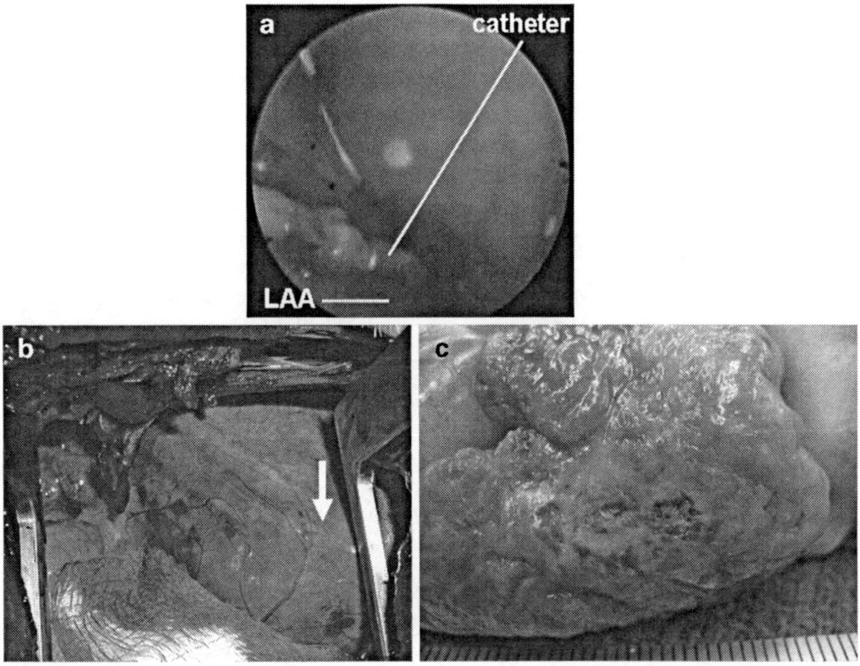

Fig. 12.9 (**a**) On-board view during the epicardial ablation trials (**b**) The distal apparatus of CardoARM is seen through the pericardium (*arrow*). The tip of the robot is navigated to the left atrial appendage. (**c**) A linear "dot-to-dot" lesion at the base of the left atrial appendage of the excised heart

due to the positioning of the robot and the tool's manipulation to the surrounding mediastinal structures (i.e., phrenic nerve, lung, pulmonary artery) upon postmortem examinations.

12.5 Conclusion

Despite the success of commercially available robotic systems aimed at minimally invasive cardiac surgery, current technology has significant limitations. The ability to operate in highly confined and dynamic spaces is of particular concern in cardiac surgery. The CardioARM was developed in an effort to provide dedicated intrapericardial therapeutic delivery with single port access (i.e., subxiphoid approach) and it has the potential to navigate the entire surface of the heart with visualization. Moreover, the CardioARM can accommodate any commercially available catheter-based tools through the working ports of the robot (up to 8 Fr for the current CardioARM model). Therefore, it is technically feasible to perform not only epicardial ablation but also epicardial injection, biopsy, mapping, and left atrial appendage ligation.

Perhaps the greatest feature of the CardioARM is its ability to preserve its shape in 3-D space during navigation. This feature is distinctively different from general endoscopic devices which rely on a static shaft and the ability to only control the tip. The "shape-keeping" ability of CardioARM is especially important in the pericardial space, where there is concern about interference with the beating heart.

Our future work will focus on improving the maneuverability of the device by decreasing the diameter and the radius of curvature of the mechanism. Moreover, haptic feedback will be incorporated to signal interaction between the device and its surrounding tissue to improve operator controllability. We are also seeking to develop new technology that will make the CardioARM a platform for natural orifice transluminal endoscopic surgery.

References

1. Zenati, M., Domit, T.M., Saul, M., Gorcsan, J., Katz, W.E., Hudson, M., Courcoulas, A.P., Griffith, B.P.: Resource utilization for minimally invasive direct and standard coronary artery bypass. Ann. Thorac. Surg. **63**, S84–S87 (1996)
2. Puskas, J.D., Williams, W.H., Mahoney, E.M., Huber, P.R., Block, P.C., Duke, P.G., Staples, J.R., Glas, K.E., Marshall, J.J., Leimbach, M.E., McCall, S.A., Petersen, R.J., Bailey, D.E., Weintraub, W.S., Guyton, R.A.: Off-pump vs conventional coronary artery bypass grafting: early and 1-year graft patency, cost and quality of life outcomes. JAMA **291**, 1841–1849 (2004)
3. Stevens, J.H., Burdon, T.A., Peters, W.S., Siegel, L.C., Pompili, M.F., Vierra, M.A., St Goar, F.G., Ribakove, G.H., Mitchell, S.R., Reitz, B.A.: Port access coronary artery bypass grafting: a proposed surgical method. J. Thorac. Cardiovasc. Surg. **111**, 567–573 (1996)
4. Zenati, M.A.: Robotic heart surgery. Cardiol. Rev. **9**, 287–294 (2001)
5. Falk, V., Diegler, A., Walther, T., Autschbach, R., Mohr, F.W.: Developments in robotic cardiac surgery. Curr. Opin. Cardiol. **15**, 378–387 (2000)
6. Zenati, M.A., Bonanomi, G., Chin, A.K., Schwartzman, D.: Left heart lead implantation using subxiphoid videopericardioscopy. J. Cardiovasc. Electrophysiol. 14, 949–953 (2003)
7. Leclercq, C., Kass, D.A.: Retiming the failing heart: principles and current clinical status of cardiac resynchronization. J. Am. Coll. Cardiol. **39**, 194–201 (2002)
8. Schweikert, R.A., Saliba, W.I., Tomassoni, G., Marrouche, N.F., Cole, C.R., Dresing, T.J., Tchou, P.J., Bash, D., Beheiry, S., Lam, C., Kanagaratna, L., Natale, A.: Percutaneous pericardial instrumentation for endo-epicardial mapping of previously failed ablations. Circulation **108**, 1329–1335 (2003)
9. Sosa, E., Scanavacca, M., D'Avila, A., Oliveira, F., Ramires, J.A.F.: Nonsurgical transthoracic epicardial catheter ablation to treat recurrent ventricular tachycardia occurring late after myocardial infarction. J. Am. Coll. Cardiol. **35**, 1442–1449 (2000)
10. Zenati, M.A., Schwartzman, D., Gartner, M., Mc Keel, D.: Feasibility of a new method for percutaneous occlusion of the left atrial appendage. Circulation **106**, II-619 (2002)
11. Zenati, M.A., Nichols, L., Bonanomi, G., Griffith, B.P.: Experimental off-pump coronary bypass using a robotic telemanipulation system. Comput. Aided Surg. **7**, 248–253 (2002)
12. Lee, R., Nitta, T., Schuessler, R.B., Johnson, D.C., Boineau, J.P., Cox, J.L.: The closed heart MAZE: a nonbypass surgical technique. Ann. Thorac. Surg. **2**, 1696–1702 (1999)
13. Li, R.K., Jia, Z.Q., Weisel, R.D., Merante, F., Mickle, D.A.G.: Smooth muscle cell transplantation into myocardial scar tissue improves heart function. J. Mol. Cell. Cardiol. **31**, 513–522 (1999)

14. Thompson, R.B., Emani, S.M., Davis, B.H., van den Bos, E.J., Morimoto, Y., Craig, D., Glower, D., Taylor, D.A.: Comparison of intracardiac cell transplantation: autologous skeletal myoblasts versus bone marrow cells. Circulation **108**(1), II264–II271 (2003)
15. Ujhelyi, M.R., Hadsall, K.Z., Euler, D.E., Mehra, R.: Intrapericardial therapeutics: a pharmacodynamic and pharmacokinetic comparison between pericardial and intravenous procainamide delivery. J. Cardiovasc. Electrophysiol. **6**, 605–611 (2002)
16. Losordo, D.W., Vale, P.R., Isner, J.M.: Gene therapy for myocardial angiogenesis. Am. Heart. J. **138**(2), S132–S141 (1999)
17. Hirose, S.: Biologically Inspired Robots: Serpentile Locomotors and Manipulators. Oxford University Press, Oxford (1993)
18. Chirikjian, G.S., Burdick, J.W.A.: Modal approach to hyper-redundant manipulator kinematics. IEEE Trans. Robot. Autom. **10**(3), 343–354 (1994)
19. Wolf, A., Brown, H.B., Casciola, R., Costa, A., Schwerin, M., Shamas, E., et al.: A mobile hyper redundant mechanism for search and rescue tasks. In: IEEE/RSJ International Conference on Intelligent Robots and Systems (IROS), vol. **3**, pp. 2889–2895 (2003)
20. Lipkin, K., Brown, I., Peck, A., Choset, H., Rembisz, J., Gianfortoni, P., et al.: Differentiable and piecewise differentiable gaits for snake robots. In: IEEE/RSJ International Conference on Intelligent Robots and Systems (IROS), pp. 1864–1869 (2007)
21. Wolf, A., Choset, H.H., Brown, B.H., Casciola, R.W.: Design and control of a mobile hyper-redundant urban search and rescue robot. Adv. Robot. **19**(3), 221–248 (2005)
22. Kuhl, C., Dumont, G.: Virtual endoscopy: from simulation to optimization of an active endoscope. In: Proceedings of the Modeling & Simulation for Computer Aided Medicine and Surgery, pp. 84–93 (2002)
23. Lim, G., Minami, K., Yamamoto, K., Sugihara, M., Uchiyama, M., Esashi, M.: Multi-link active catheter snake-like motion. Robotica **14**(5), 499 (1996)
24. Nakamura, Y., Matsui, A., Saito, T., Yoshimoto, K.: Shape-memory-alloys active forceps for laparoscopic surgery. In: IEEE International Conference on Robotics and Automation, pp. 2320–2327 (1995)
25. Ikuta, K., Nolata, M., Aritomi, S.: Hyper redundant active endoscope for minimally invasive surgery. In: Proceedings of the First Symposium on Medical Robotics and Computer Assisted Surgery, pp. 230–237 (1994)
26. Ikuta, K., Nokata, M., Aritomi, S.: Biomedical micro robot driven by miniature cybernetic actuator. In: IEEE International Workshop on MEMS, pp. 263–268 (1994)
27. Ikuta, K., Tsukamoto, M., Hirose, S.: Shape memory alloys servo actuator system with electric resistance feedback and application for active endoscope. In: Proceedings of the IEEE International Conference on Robotics and Automation, pp. 427–430 (1988)
28. Dario, P., Paggeti, N., Troisfontaine, E., Papa, E., et al.: A miniature steerable end effector for application in an integrated system for computer assisted arthroscopy. In: Proceedings of the IEEE International Conference on Robotics and Automation, pp. 1573–1579 (1997)
29. Reynaerts, D., Peirs, J., Van Brussel, H.: Shape memory micro-actuation for a gastro intestinal intervention system. Sens. Actuatators **77**, 157–166 (1999)
30. Kornbluh, R.D., Pelrine, R., Eckerle, J., Joseph, J.: Electrostrictive polymer artificial muscle actuators. In: Proceedings of the IEEE International Conference on Robotics and Automation, pp. 2147–2154 (1998)
31. Frecker, M.I., Aguilera, W.M.: Analytical modeling of a segmented unimorph actuator using electrostrictive P (VDF-TrFE) copolymer. Smart. Mater. Struct. **13**(1), 82–91 (2004)
32. Simaan, N., Taylor, R., Flint, P.: A dextrous system for laryngeal surgery. In: IEEE International Conference on Robotics and Automation, 351–357 (2004)
33. Zenati, M.A., Wolf, A., Ota, T., Degani, A., Choset, H.: Percutaneous subxiphoid left atrial appendage obliteration with an innovative highly articulated teleoperated catheter. In: Innovation in Intervention: i2 Summit, Atlanta (2006)
34. Degani, A., Choset, H., Wolf, A., Ota, T., Zenati, M.A.: Percutaneous intrapericardial interventions using a highly articulated robotic probe. In: The First IEEE/RAS-EMBS

International Conference on Biomedical Robotics and Biomechatronics, (BioRob 2006). Best paper award, pp. 7–12 (2006)
35. Ota, T., Degani, A., Zubiate, B., Wolf, A., Choset, H., Schwartzman, D., Zenati, M.A.: Epicardial atrial ablation using a novel articulated robotic medical probe via a percutaneous subxiphoid approach. Innovations **1**(6), 335–340 (2006)
36. Ota, T., Degani, A., Schwartzman, D., Zubiate, B., McGarvey, J., Choset, H., Zenati, M.A.: A highly articulated robotic surgical system for minimally invasive surgery. Ann. Thorac. Surg. **87**(4), 1253–1256 (2009)

Chapter 13
Wireless Intraocular Microrobots: Opportunities and Challenges

Olgaç Ergeneman, Christos Bergeles, Michael P. Kummer, Jake J. Abbott, and Bradley J. Nelson

Abstract Many current and proposed retinal procedures are at the limits of human performance and perception. Microrobots that can navigate the fluid in the interior of the eye have the potential to revolutionize the way the most difficult retinal procedures are conducted. Microrobots are typically envisioned as miniature mechatronic systems that utilize MEMS technology to incorporate sensing and actuation onboard. This chapter presents a simpler alternative approach for the development of intraocular microrobots consisting of magnetic platforms and functional coatings. Luminescence dyes immobilized in coatings can be excited and read wirelessly to detect analytes or physical properties. Drug coatings can be used for diffusion-based delivery, and may provide more efficient therapy than microsystems containing pumps, as diffusion dominates over advection at the microscale. Oxygen sensing for diagnosis and drug therapy for retinal vein occlusions are presented as example applications. Accurate sensing and therapy requires precise control to guide the microrobot in the interior of the human eye. We require an understanding of the possibilities and limitations in wireless magnetic control. We also require the ability to visually track and localize the microrobot inside the eye, while obtaining clinically useful retinal images. Each of these topics is discussed.

Keywords Eye · Microrobot · Minimally invasive surgery · Magnetic control · Wireless · Tracking · Localizing · Ophthalmoscopy · Intraocular · Luminescence · Coatings · Drug delivery · Drug release · Sensing

---

O. Ergeneman (✉)
Institute of Robotics and Intelligent Systems, ETH Zurich, Tannenstr. 3, CLA H 17.1, 8092 Zurich, Switzerland
e-mail: oergeneman@ethz.ch

J. Rosen et al. (eds.), *Surgical Robotics: Systems Applications and Visions*,
DOI 10.1007/978-1-4419-1126-1_13, © Springer Science+Business Media, LLC 2011

13.1 Introduction

During the past decade, the popularity of minimally invasive medical diagnosis and treatment has risen remarkably. Further advances in biomicrorobotics will enable the development of new diagnostic and therapeutic systems that provide major advantages over existing methods. Microrobots that can navigate bodily fluids will enable localized sensing and targeted drug delivery in parts of the body that are currently inaccessible or too invasive to access.

Microelectromechanical systems (MEMS) technology has enabled the integration of sensors, actuators, and electronics at microscales. In recent years, a great deal of progress has been made in the development of microdevices, and many devices have been proposed for different applications. However, placing these systems in a living body is limited by factors like biocompatibility, fouling, electric hazard, energy supply, and heat dissipation. In addition, the development of functional MEMS devices remains a time-consuming and costly process. Moving microsized objects in a fluid environment is also challenging, and a great deal of research has considered the development of microactuators for the locomotion of microrobots. However, to date the most promising methods for microrobot locomotion have utilized magnetic fields for wireless power and control, and this topic is now well understood [2, 28, 32, 49, 66]. A large number of micropumps have been developed for drug delivery, but as size is reduced diffusion begins to dominate over advection, making transport mechanisms behave differently at small scales. Consequently, future biomedical microrobots may differ from what is typically envisioned.

In this chapter we focus on intraocular microrobots. Many current and proposed retinal procedures are at limits of human performance and perception. Microrobots that can navigate the fluid in the interior of the eye have the potential to revolutionize the way the most difficult retinal procedures are conducted. The proposed devices can be inserted in the eye through a small incision in the sclera, and control within the eye can be accomplished via applied magnetic fields. In this chapter we consider three topics in the design and control of intraocular microrobots: First, we discuss functional coatings – both for remote sensing and targeted drug delivery. Next, we discuss magnetic control, and the ability to generate sufficient forces to puncture retinal veins. Finally, we discuss visually tracking and localizing intraocular microrobots. The eye is unique in that it is possible to observe the vasculature and visually track the microrobot through the pupil.

Throughout this chapter, we consider the assembled-MEMS microrobots as shown in Fig. 13.1, but the conclusions extend to other microrobot designs. Fabricating truly 3-D mechanical structures at the microscale is challenging. With current MEMS fabrication methods, mechanical parts are built using 2-D (planar) geometries with desired thickness. Three-dimensional structures can be obtained by bending or assembling these planar parts, and it has been demonstrated that very complex structures can be built with such methods [33, 59, 65, 66]. The philosophy of designing simple structures with no actuation or intelligence onboard begs the

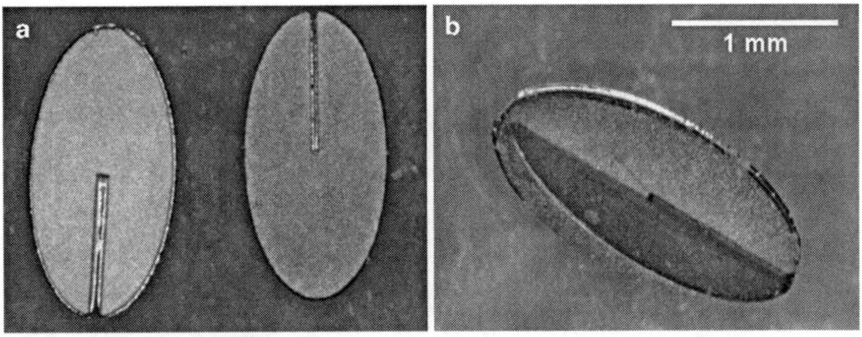

Fig. 13.1 Relatively simple 2-D parts can be assembled into complex 3-D structures (©IEEE 2008), reprinted with permission. The parts shown have dimensions 2.0 mm × 1.0 mm × 42 μm and can be further miniaturized

question: Are these devices microrobots? It may be more accurate to think of these devices as end-effectors of novel manipulators where magnetic fields replace mechanical links, sensing is performed wirelessly, and system intelligence is located outside of the patient. However, this matter of semantics is inconsequential if the goal is to develop functional biomedical microdevices.

13.2 Functionalizing Microrobots with Surface Coatings

We present an alternative approach for the development of biomicrorobots utilizing a magnetic platform and functional coatings for remote sensing and targeted drug delivery (Fig. 13.2). Coatings possessing sensor properties or carrying drugs may be superior to more complicated electromechanical systems. Luminescence dyes immobilized in coatings can be excited and read out wirelessly for detecting analytes or physical properties. Drugs coated on a carrier can be used for diffusion-based delivery and may provide more efficient therapy than microsystems containing pumps. Because of the discrepancy in scaling of volume and surface area, reservoirs built inside microfabricated devices may be insufficient, whereas surface coatings alone may provide sufficient volume. Fabrication of devices utilizing coatings will also be simple compared to systems with many electrical or mechanical components. All of these properties make wireless microrobots consisting of magnetic bodies and functional coatings feasible in the near term.

13.2.1 Biocompatibility Coatings

Magnetic microrobots contain nickel, cobalt, iron, or their alloys. These elements and their alloys are declared to be non-biocompatible. Hence, they are not used in

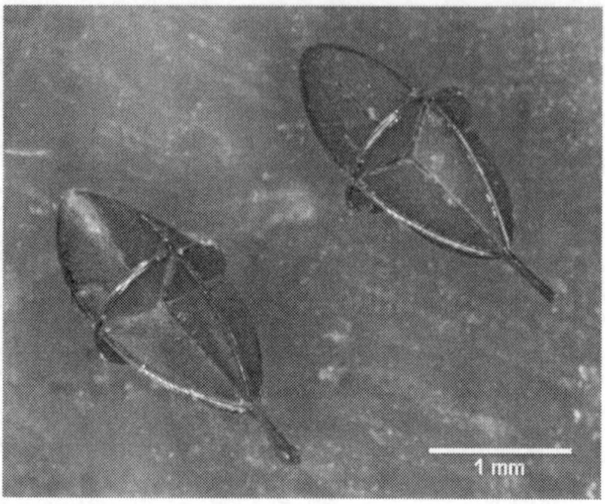

Fig. 13.2 Microrobot utilizing a functional coating (©IEEE 2008), reprinted with permission. *Right*: A bare magnetic microrobot made of thin assembled nickel pieces, based on [66]. *Left*: A microrobot coated with an oxygen-sensitive film

medical devices. Ti and Ti-alloys are used extensively in biomedical applications because of their excellent combination of biocompatibility, corrosion resistance, and structural properties [51]. In order to achieve biocompatibility without sacrificing the magnetic properties, microrobot pieces can be coated with a thin layer of Ti, forming a titanium dioxide layer once exposed to air.

In [17] microrobot pieces made of Ni are coated by Ti with thicknesses of 100 nm, 200 nm, 300 nm, and 500 nm using a DC sputterer. Biocompatibility covers a broad spectrum of non-toxic and non-allergic properties, with various levels of biocompatibility associated with the purpose of a medical device. Biocompatibility tests involve toxicity tests, corrosion tests, and allergy tests. To validate the quality of coatings, against possible faults and crack formations, in vitro direct-contact cell-toxicity tests, in line with ISO 10993-5 8.3 standard, were performed on coated and uncoated microrobots using NIH 3T3 fibroblast cells. Results show that as thin as a 100-nm-thick Ti coating is sufficient to obtain biocompatibility.

13.2.2 Coatings for Remote Sensing

Surface coatings can be used to fabricate minimally invasive wireless sensor devices, such as the intraocular sensor depicted in Fig. 13.3. The proposed device consists of a luminescence sensor film that is integrated with a magnetically controlled platform. This system can be used to obtain concentration maps of clinically relevant species (e.g., oxygen, glucose, urea, drugs) or physiological parameters (e.g., pressure, pH, temperature) inside the eye, specifically in the preretinal area. Effects of specific physiological conditions on ophthalmic disorders can be conveyed.

These devices can also be used in the study of pharmacokinetics as well as in the development of new drug delivery mechanisms, as summarized below [48].

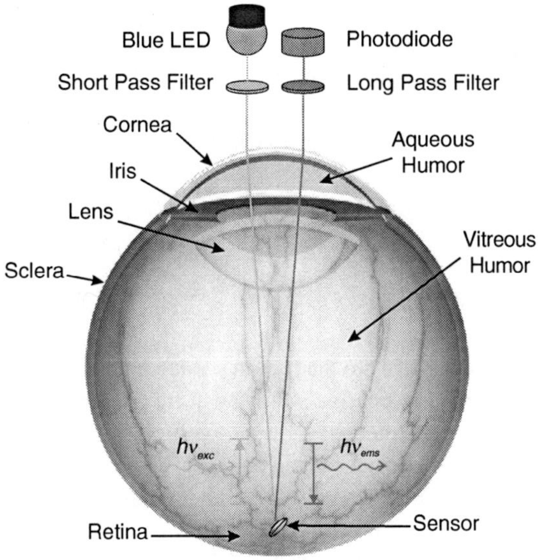

Fig. 13.3 Artist's conception of the magnetically controlled wireless sensor in the eye (©IEEE 2008), reprinted with permission [21]

The study of pharmacokinetics of drugs that diffuse into the eye following intraocular drug injection requires analysis of ocular specimens as they change in time. The risk of iatrogenic complications when penetrating into the ocular cavity with a needle has restricted ocular pharmacokinetic studies on animals and humans. Microdialysis has become an important method for obtaining intraocular pharmacokinetic data and it has reduced the number of animals needed to estimate ocular pharmacokinetic parameter values. However, the insertion of the probe and anesthesia have been shown to alter the pharmacokinetics of drugs. The microrobotic system presented here can replace microdialysis probes for obtaining intraocular pharmacokinetic data as it provides a minimally invasive alternative for in vivo measurements of certain analytes. Concentration as a function of time and position can be obtained by steering the magnetic sensor inside the vitreous cavity. Knowledge of concentration variations within the vitreous will expedite the optimization of drug administration techniques for posterior segment diseases.

13.2.2.1 Luminescence Sensing

Photoluminescence is the emission of electromagnetic radiation (i.e. photons) from a material in response to absorption of photons. The intensity and the lifetime of emission can be decreased by a variety of processes referred to as luminescence quenching.

Optical luminescence sensors work based on quenching of luminescence in the presence of a quencher (i.e., analyte of interest); the decrease in luminescence is related to the quantity of the quencher. A number of devices using this principle have been demonstrated and the basic principles of different methods can be found in [40]. The quenching of luminescence is described by Stern-Volmer equations:

$$\frac{I_0}{I} = 1 + K[Q] \tag{13.1}$$

$$\frac{\tau_0}{\tau} = 1 + K[Q] \tag{13.2}$$

where I_0 and I are the luminescence intensities in the absence and in the presence of quencher, respectively, τ_0 and τ are the luminescence lifetimes in the absence and presence of quencher, respectively, $[Q]$ is the quencher concentration, and K is the Stern-Volmer quenching constant whose units are the reciprocal of the units of $[Q]$. Luminescence dyes with high quantum yield, large dynamic range, and large Stokes shift are preferred for luminescence sensors. To be used as a sensor, these dyes need to be immobilized. They are usually bound to transparent and quencher-permeable supporting matrices such as polymers, silica gels, or sol-gels. Quencher permeability, selectivity, and the luminophore solubility are the important factors for choosing appropriate supporting matrices.

Luminescence sensing can be done either based on luminescence intensity or luminescence lifetime. The main difference between the two methods is that intensity is an extrinsic property whereas lifetime is an intrinsic property. Extrinsic techniques depend on parameters such as the dye concentration, optical surface quality, photo-bleaching, and incidence angle, which change from sample to sample. When the sensor's position changes, the optical path distance (OPD) from the light source to the sensor and back to the photo detector changes. The total amount of light collected by the sensor changes depending on the OPD and orientation. These quantities are hard to control in such a wireless sensor application, limiting the accuracy of this technique. Intrinsic properties do not depend on the parameters described above, making lifetime measurements more promising for wireless microrobotic applications.

There are two methods that are used for measuring luminescence lifetimes: time-domain measurements and frequency-domain measurements. In time-domain measurements the sample is excited with light pulses, and the intensity signal that changes as a function of time is measured and analyzed. In frequency-domain measurements the sample is excited with a periodic signal that consequently causes a modulated luminescence emission at the identical frequency. Because of the lifetime of emission, the emission signal has a phase shift with respect to the excitation signal. The input excitation signal is used as a reference to establish a zero-phase position and the lifetime is obtained by measuring the phase shift between the excitation and emission signals.

13.2.2.2 An Intraocular Oxygen Sensor

The retina needs sufficient supply of oxygen and other nutrients to perform its primary visual function. Inadequate oxygen supply (i.e. retinal hypoxia) is correlated with major eye diseases including diabetic retinopathy, glaucoma, retinopathy of prematurity, age-related macular degeneration, and retinal vein occlusions [25]. Retinal hypoxia is presumed to initiate angiogenesis, which is a major cause of blindness in developed countries [14]. Attempts to test this hypothesis suffer from the current methods of highly invasive oxygen electrodes. Hypoxia is typically present at the end stages of retinal diseases. However, during the early stages, the relation between blood flow sufficiency, vessel patency, and tissue hypoxia are still unknown [55]. The influence of oxygen on these diseases is not well understood and the ability to make long-term, non-invasive, in vivo oxygen measurements in the human eye is essential for better diagnosis and treatment. Measuring the oxygen tensions both in aqueous humor and vitreous humor, and particularly in the pre-retinal area, is of great interest in ophthalmic research.

To address these issues, an intraocular optical oxygen sensor utilizing a luminescence coating has been developed [21]. The sensor works based on quenching of luminescence in the presence of oxygen. A novel iridium phosphorescent complex is designed and synthesized to be used as the oxygen probe. The main advantages of this iridium complex, when compared to other metal complexes, are its higher luminescence quantum yield, higher photo-stability, longer lifetime, stronger absorption band in the visible region, and larger Stokes shift. Polystyrene is chosen as the supporting matrix because of its high oxygen permeability and biocompatible nature. The microrobots are dip-coated with polystyrene film containing luminescence dye, and good uniformity is achieved across the magnetic body. Biocompatibility tests must still be performed on the polystyrene film with embedded dye. If needed, an additional layer of pure polystyrene could be added to isolate the sensing layer.

An experimental setup has been built to characterize the oxygen sensitivity of the sensor. The details of the sensor and characterization setup can be found in [21]. A blue LED is used as the excitation source for the oxygen sensor system and a photodiode is used to detect the luminescence. Optical filters are used to separate the emission signal from the excitation signal. The frequency-domain lifetime measurement approach is used in this work. De-ionized water is used for the dissolved oxygen measurements. The sensor's location in the setup is maintained with a magnet. The distances between the components and the sensor are chosen considering the geometry of the eye. A range of oxygen concentrations is achieved by bubbling air or nitrogen gas. Nitrogen replaces oxygen molecules in the solution, and air provides oxygen. Figure 13.4 shows the Stern-Volmer plot as a function of oxygen concentration. As seen in this figure, a linear model proved to be an excellent predictor ($R^2 = 0.989$) for oxygen concentrations compared to a commercial sensor.

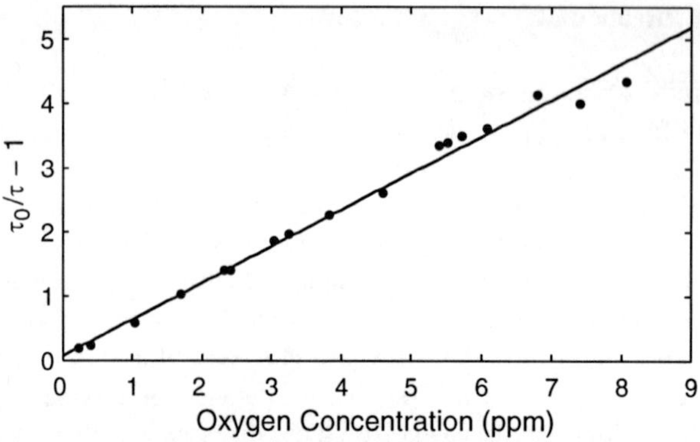

Fig. 13.4 Stern-Volmer plot of the luminescence dyes immobilized in polystyrene film under various oxygen concentrations (©IEEE 2008), reprinted with permission [21]

Fig. 13.5 Measurement of local gradients is possible exciting and reading out from different parts of the microrobot (©IEEE 2008), reprinted with permission

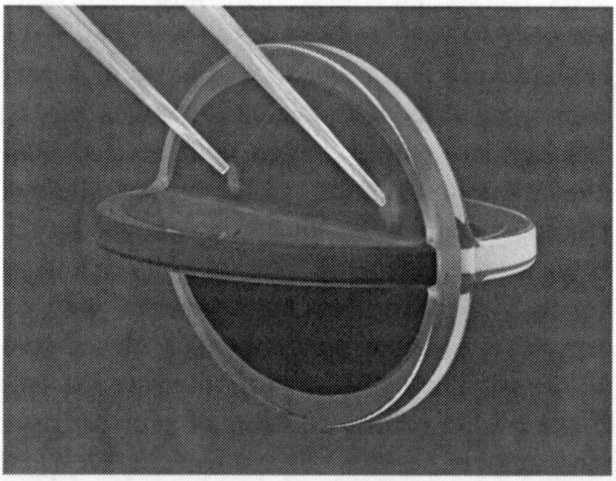

13.2.2.3 Measuring Gradients

It may be desirable to measure spatial gradients in a quantity. This can be accomplished by taking measurements while moving the microrobot. However, these measurements will be separated in time, and the movement of the microrobot could potentially affect the environment, particularly in a low-Reynolds-number regime. It is possible to measure gradients directly with a stationary microrobot. Specific locations on the microrobot can be excited and sensed simultaneously, as depicted in Fig. 13.5. This requires the ability to focus the excitation signal on a specific region of the microrobot. Clearly, this necessitates a greater level of sensing spatial resolution. Alternatively, multiple dyes with different emission spectra can be excited simultaneously, and the emitted signals can be band-pass filtered.

13.2.3 Coatings for Targeted Drug Delivery

The main challenge of ophthalmic drug delivery is to keep desired drug concentrations in the target area for the desired duration, while minimizing the drug levels in the remainder of the body. To date, a variety of drug delivery approaches have been shown to be effective therapeutically. Some salient findings from [48] are summarized below. Ocular delivery can be achieved by topical administration, systemic administration, periocular injections, and intraocular injections. Depending on the target area, drug delivery can be achieved by penetrating through the cornea, conjunctiva, or sclera following topical administration (i.e. eye drops), or across blood-aqueous barrier along with blood-retinal barrier following systemic administration. Only a minute fraction of applied dose reaches the intraocular target area after topical and systemic administration. Drug delivery using gelatin wafers, collagen shields, and soft contact lenses placed on the cornea or in the cul-de-sac have been tested, as well as methods like iontophoresis. Drug delivery for posterior-segment disorders (e.g. diabetic retinopathy, macular degeneration, retinal edema, retinal vein occlusions) has always been a challenge as it requires access to the retina and the choroid. Periocular injections and intraocular injections place the drug closer to the target tissue, overcoming some of the ocular barriers. Many of the drugs used to treat vitreous and retinal disorders have a narrow concentration range in which they are effective, and they may be toxic at higher concentrations. Slight changes in injection conditions (e.g. position, shape) will produce different drug concentrations within the vitreous, and therefore the efficacy of the treatment produced by the drug can be sensitive to injection conditions. Intraocular injections have also been associated with serious side effects, such as endophthalmitis, cataract, hemorrhage, and retinal detachment [26], and long-duration drug delivery is not possible with these methods.

New drug-delivery methods provide many advantages compared to the traditional methods. However, it is not always possible to deliver drugs to the target tissue with existing methods. Superior methods for targeted drug delivery are needed, and robotic assistance in drug delivery will have major benefits. Devices inserted into the aqueous or vitreous cavity bear great potential for drug delivery. Recently, methods to deliver drugs using carriers such as liposomes, gels, and nanoparticles have been evaluated. Methods to achieve desired concentrations over long periods of time using drugs that become active inside the eye (prodrugs) are also being investigated. Controlled-release devices and biodegradable implants can increase the effectiveness of these devices.

At the microscale, diffusion becomes the dominant mechanism for mass transport. Low-Reynolds-number flow is laminar, and the lack of turbulent mixing puts a diffusion limit on drug delivery using micropumps. In [17] an alternative approach to targeted drug delivery is proposed: wireless magnetic microrobots surface coated with drug. The microrobot will be steered to the site of action and it will be kept at this position as the drug is released from the microrobot by diffusion.

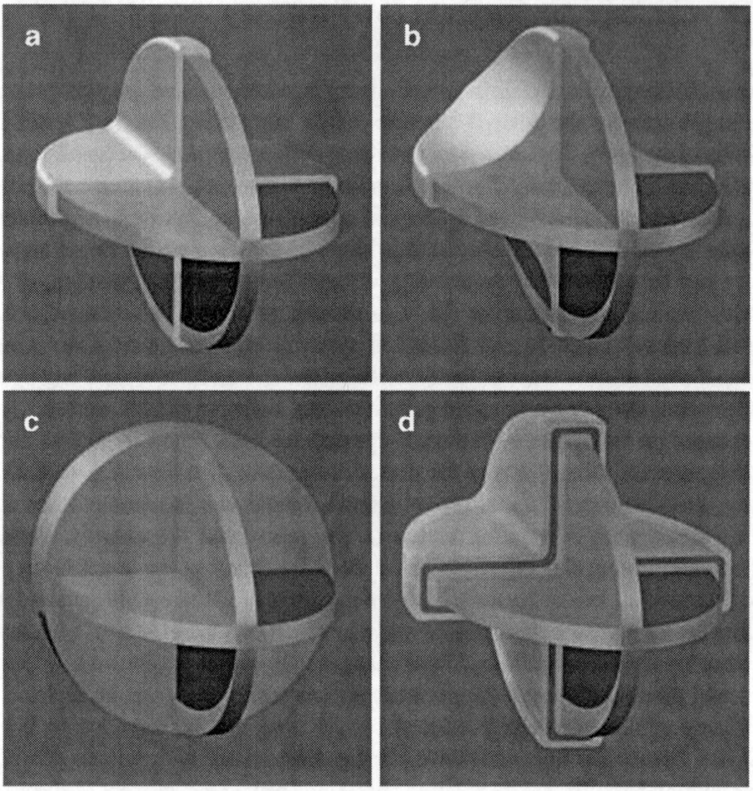

Fig. 13.6 Drug coatings can range from thin surface coatings to coatings that take advantage of the total available volume created by the microrobot structure. Coatings are shown only at the back part of the microrobot (©IEEE 2008), reprinted with permission

13.2.3.1 Quantity of Coated Drug

Carrying drug by surface coating becomes more desirable as size is reduced (Fig. 13.6). Consider an assembled microrobot like those shown in Fig. 13.2. The microrobot can be modeled by two elliptical pieces of magnetic material of length $2a$, width $2b$, and thickness c, and by a circular piece of diameter $2b$ and thickness c. The volume of the magnetic structure is calculated as

$$v_s = \left(2\pi ab + \pi b^2\right)c - (2a + 4b)c^2 + c^3 \tag{13.3}$$

The microrobot has a volumetric footprint of an ellipsoid of volume

$$v_e = \frac{4}{3}\pi ab^2 \tag{13.4}$$

If we consider a coating of drug that fills in the entire ellipsoidal volumetric footprint of the microrobot (similar to Fig. 13.6c), the volume of drug carried is simply the volume of the ellipsoid minus the volume of the magnetic structure

$$v_f = v_e - v_m \tag{13.5}$$

If we consider a single thin surface coating of thickness t, (similar to Fig. 13.6a), the volume of drug carried is given by

$$v_t = \left(4\pi ab + 2\pi b^2\right)t - (8a + 16b - 12c)t^2 + 8t^3 \tag{13.6}$$

Figure 13.7 shows the effect of scaling on the ability to carry drug by surface coating. For even relatively large microrobots, the amount of drug carried on the surface with a single thin coating is comparable to the total volume of the magnetic structure. As the size of the microrobot is reduced, the volume of drug in a single thin coating becomes comparable to the total ellipsoidal volume of the microrobot. In practice, any fabricated reservoir could only amount to a fraction of the total volume of the structure, and the drug would need to be in solution (that is, diluted) in order to be pumped. The ability to surface coat highly concentrated drug increases the benefits of surface coatings even beyond what is observed in Fig. 13.7.

In order to bind more proteins or drugs onto a microrobot of the same surface area, multilayer surface coatings or coatings embedded in different base matrices should be developed (Fig. 13.6d). Among others, hydrogels, agarose, starch micro-capsulations, polymer matrices, liposomes, and biodegradable needles are widely used for making drug delivery matrices that can hold much more drug due to their material properties [15, 50]. These materials can be used to encapsulate drug

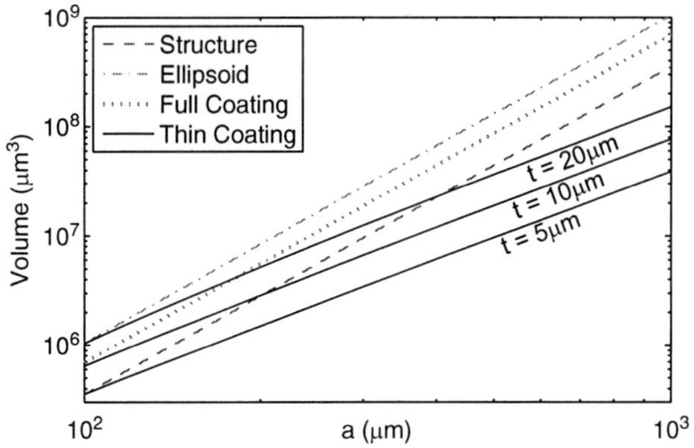

Fig. 13.7 Effect of scaling on volume with microrobots of the type shown in Fig. 13.2 (©IEEE 2008), reprinted with permission. Curves correspond to (3)–(6), respectively, with $b = a/2$ and $c = a/10$

molecules as an outer coating, enabling multilayer coatings. These multilayer coatings can be used to coat multiple drug types on one microrobot, or used to fine tune delivery times or dosage. Alternatively, embedding drug molecules in a porous matrix facilitates slower diffusion and more drug loading capacity. Controlled release of drugs has been demonstrated using intelligent polymers that respond to stimuli such magnetic fields, ultrasound, temperature, and pH. They enable fine tuning of diffusive drug release.

13.2.3.2 Drug Delivery for Retinal Vein Occlusions

Retinal vein occlusion (RVO) is a common retinovascular disease caused by obstruction of blood flow due to clot formation. RVO is among the most common causes of vision loss around the world, with one study reporting a prevalence of 1.6% in adults aged 49 years or older [60]. Various treatment methods for RVO have been proposed and attempted, however to date there is no effective clinical treatment for RVO. Among these methods, prolonged local intravenous thrombolysis (i.e., clot dissolution) with tissue plasminogen activator (t-PA) injection is the most promising treatment [53], based on excessive postoperative complications or inconclusive clinical trials of other methods.

Retinal drug delivery by injections requires precise manipulation that is constrained by the limits of human performance and perception [34]. Retinal veins are small delicate structures surrounded by fragile retinal tissue, and prolonged manual cannulation of retinal veins risks causing permanent damage to the retina. Robotic systems have been proposed to assist with retinal vein cannulation, utilizing robot-assisted surgical instruments that pass through a hole in the sclera as in conventional vitreoretinal surgery [47, 52].

In [17] an alternative approach to RVO treatment is proposed: a wireless magnetic microrobot coated with clot-dissolving t-PA. The microrobot will be steered to the thrombus site as it is tracked visually through the pupil, and will be immobilized in close proximity of the retinal veins. Immobilization can be achieved by puncturing and docking to a retinal vein. Diffusion of t-PA from the surface coating of the microrobot into the clotted region will start clot dissolution. There is strong evidence that t-PA in the preretinal area can diffuse into the retinal vasculature and break clots [27]. Since t-PA is an enzyme, and the clot dissolution reaction rate depends on enzyme reaction rate, long-term release of t-PA is thought to be more effective than bolus injections [43]. The proposed delivery mechanism provides drug release without the need for a micropump, and an efficient therapy using small amounts of t-PA over prolonged periods. Moreover, a microrobot is potentially less invasive than other methods, and has the potential to be left in the eye for extended periods of time, even in an outpatient scenario. However, it is not yet known what quantity of t-Pa is required to effectively dissolve a clot using the proposed method.

13.2.3.3 Preliminary Drug Release Experiments

This section presents the results of preliminary drug release experiments using the untethered microrobot and discusses the feasibility of microrobotic drug delivery. A drug substitute is coated on microrobots in [17], the release kinetics is characterized, and the amount of drug that can be coated in a single layer on a microrobot is quantified.

In order to analyze the release kinetics of a diffusion-based drug delivery microrobot, in vitro experiments are conducted. As the drug molecule substitute, bovine serum albumin (BSA) was chosen. BSA is a plasma protein that can be used as a blocking agent or added to diluents in numerous biochemical applications. BSA is used because of its stability, its inert nature in many biochemical reactions, its representative molecular size, and its low cost.

Four elliptical microrobot pieces of length 900 μm, width 450 μm, and thickness 50 μm are used as the magnetic platform holding the coating. The pieces are made from electroplated nickel and then coated with titanium for biocompatibility. The pieces are first sterilized and then placed in different wells of a 96-well culture plate. A sterilized BSA-solution of 3 mg/mL is prepared and labeled with Alexa-Fluor-546 (Molecular Probes) fluorescent marker. This solution is then mixed with sterilized PBS in order to create solutions with different concentrations of labeled-BSA molecules. Three of the microrobot pieces are dipped in BSA concentrations of 3 mg/mL, 2 mg/mL, 1 mg/mL, respectively, and one is dipped into a pure PBS solution, which contained no BSA, as a control set. The pieces are left in the solutions to allow the BSA to bind to the microrobot. The surface-coating process is done for 12 h at room temperature in a humidity chamber.

Coated microrobot pieces are taken from coating wells and placed in new wells filled with 200 μL PBS each. Following that, the florescence intensity is measured in set time intervals for three days using an automated spectrum analyzer. In this way, the kinetics of diffusion-based drug delivery with surface-coated microrobots are obtained.

Figure 13.8 quantifies the amount of time required to release the drug through diffusion, and it also gives qualitative information about the kinetics of release. It is clear that the concentration of the coating solution does not affect the amount of drug bound to the surface. This provides strong evidence that the amount of drug will be limited by the surface area of the microrobot.

Next, the amount of BSA released from a single layer on a single piece is quantified. The release wells of the culture plates are analyzed in the multiwell plate reader for fluorescence and absorbance values. The BSA standard concentration curve is obtained by preparing a Bradford Assay with ten different known concentrations of BSA in 1:2 dilutions, and analyzing this assay for fluorescence and absorbance. The obtained standard curve is used to calibrate the multiwell plate reader. The fluorescence intensity in the release wells is measured and, using the calibration curve, the amount of BSA released is found to be $2.5 \pm 0.1\,\mu g$.

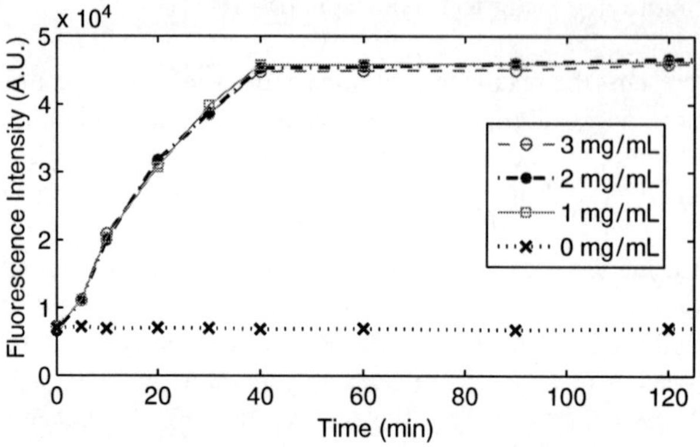

Fig. 13.8 Fluorescence intensity vs. time for the release experiment (©IEEE 2008), reprinted with permission [17]

13.3 Magnetic Control

One approach to the wireless control of microrobots is through externally applied magnetic fields [3]. There is a significant body of work dealing with non-contact magnetic manipulation [28]. Research has considered the 3-D positioning of permanent magnets [29, 36]. Magnetic fields have been used to orient small permanent magnets placed at the distal tips of catheters [1, 62]. Researchers have considered the position control of soft-magnetic beads as well, where a spherical shape simplifies the control problem [6]. The precision control of non-spherical soft-magnetic bodies has also been considered [2]. In addition to magnetic manipulation of simple objects (e.g., beads, cylinders), it is possible to manipulate more complicated shapes. In [66], a soft-magnetic assembled-MEMS microrobot is controlled by applying decoupled magnetic torque and force. Assembled-MEMS microrobots have the potential to provide increased functionality over simpler geometries. Controlled magnetic fields can be generated by stationary current-controlled electromagnets [45, 67], by electromagnets that are position and current controlled [29, 66s], or by position-controlled permanent magnets, such as with the Stereotaxis NIOBE Magnetic Navigation System, or even by a commercial MRI system [44]. In all cases, the rapid decay of magnetic field strength with distance from its source creates a major challenge for magnetic control.

Surgeons have been using magnets to remove metallic debris from eyes for over 100 years [9]. However, there has been no prior work of controlled magnetic manipulation of an object that has been intentionally inserted in the eye. Intraocular procedures are unique among in vivo procedures, as they provide a direct line of sight through the pupil for visual feedback, making closed-loop control possible.

If we want to apply controlled torques and forces to an object with magnetization **M** (A/m) using a controlled magnetic field **B** (T) the resulting equations for torque and force are as follows [35]. The magnetic torque, which tends to align the magnetization of the object with the applied field:

$$\mathbf{T} = v\mathbf{M} \times \mathbf{B} \tag{13.7}$$

in units N ·m where v is the volume of the body in m^3. The force on the object is:

$$\mathbf{F} = v(\mathbf{M} \cdot \nabla)\mathbf{B} \tag{13.8}$$

in units N, where ∇ is the gradient operator:

$$\nabla = \begin{bmatrix} \frac{\partial}{\partial x} & \frac{\partial}{\partial y} & \frac{\partial}{\partial z} \end{bmatrix}^T \tag{13.9}$$

Since there is no electric current flowing through the region occupied by the body, Maxwell's equations provide the constraint $\nabla \times \mathbf{B} = \mathbf{0}$. This allows us to express (8), after some manipulation, in a more intuitive and useful form:

$$\mathbf{F} = v \begin{bmatrix} \frac{\partial}{\partial x}\mathbf{B}^T \\ \frac{\partial}{\partial y}\mathbf{B}^T \\ \frac{\partial}{\partial z}\mathbf{B}^T \end{bmatrix} \mathbf{M} \tag{13.10}$$

The magnetic force in any given direction is the dot product of 1) the derivative of the field in that direction and 2) the magnetization.

We can also express the applied magnetic field's flux density as an applied magnetic field $\mathbf{H}$ with units A/m. $\mathbf{B}$ is related to $\mathbf{H}$ simply as $\mathbf{B} = \mu_0\mathbf{H}$ with $\mu_0 = 4\pi \times 10^{-7}$ T ·m/A, since air and biological materials are effectively nonmagnetic. Both (7) and (8) are based on the assumption that the magnetic body is small compared to spatial changes in the applied magnetic field's flux density, such that the applied flux density is fairly uniform across the body, and $\mathbf{B}$ is the value at the center of mass of the body. It has been verified experimentally that this assumption gives an accurate prediction of magnetic force and torque [2, 49].

If the body of interest is a permanent magnet, the average magnetization $\mathbf{M}$ is effectively independent of the applied magnetic field. The magnetization of a permanent magnet is governed by the remanent magnetization of the material and geometry of the body. This makes the calculation of torque and force acting on a permanent magnetic body straightforward as long as the magnetization and orientation of the body is known. The torque can be increased by increasing the angle between $\mathbf{B}$ and $\mathbf{M}$, up to 90°, or by increasing the strength of $\mathbf{B}$. The force can be increased by increasing the gradients in the applied magnetic field.

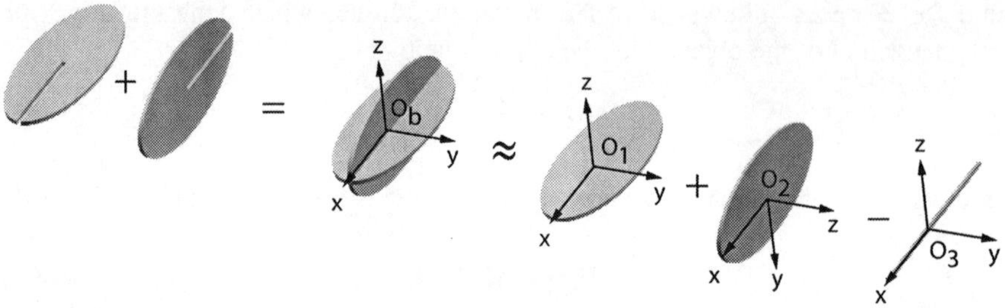

Fig. 13.9 Microfabricated nickel parts (*left*) are assembled to form a microrobot (*middle*). The body frame is assigned to the microrobot arbitrarily. For the computation, rather than using the actual complex shapes of the parts, we consider the microrobot as a superposition of simpler geometries (*right*). The frames of the individual parts are assigned with the x-axis along the longest dimension of the body and with the z-axis along the shortest dimension (©IEEE 2008), reprinted with permission [49]

If the body of interest is made of a soft-magnetic material, the magnetization is a nonlinear function of the applied field, with a magnitude limited by the saturation magnetization for the magnetic material, and can rotate with respect to the body. The governing equations for control are significantly more complex [2]. We sometimes refer to the magnetic moment or magnetic dipole moment, which represent the total strength of a magnet (hard or soft). The magnetic moment is simply the product of the volume v and the average magnetization $\mathbf{M}$.

The magnetization of a body depends on its shape, so bodies made of the same material but having different shapes will have different magnetization characteristics. The magnetization characteristics also differ along different directions within the body. This is known as shape anisotropy. Demagnetizing fields that tend to weaken magnetization create the shape anisotropy. Demagnetizing fields are largest along short directions of the body. A long direction in a body is referred to as an easy axis, since it is a relatively easy direction to magnetize the body. In general, the longest dimension of a soft-magnetic body will tend to align with the direction of the applied field. Other types of anisotropy exist, such as crystalline anisotropy, but these are typically negligible compared to shape effects, even at the scale of microrobots.

In [49] the total force ${}^{b}\mathbf{F}_b$ and torque ${}^{b}\mathbf{T}_b$ on an assembled-MEMS microrobot is computed as the sum of the individual forces and torques on assembled parts. This is achieved by neglecting the magnetic interaction between the individual pieces, and by considering the microrobot as a superposition of simpler geometries as shown in Fig. 13.9, rather than using the actual complex shapes of the parts.

13.3.1 Magnetic Control in Fluids

Microrobots, like microorganisms, operate in a low-Reynolds-number regime. When controlling a magnetic object through Newtonian fluid at low-Reynolds-number regime, the object nearly instantaneously reaches its terminal velocity **V** where the viscous drag force, which is linearly related to velocity through a drag coefficient ψ_{υ}, exactly balances the applied magnetic force **F**:

$$\mathbf{F} = \psi_{\upsilon}\mathbf{V} \tag{13.11}$$

Similarly, the object nearly instantaneously reaches its terminal rotational velocity Ω where the viscous drag torque, which is linearly related to rotational velocity through a drag coefficient ψ_{ω}, exactly balances the applied magnetic torque **T**.

$$\mathbf{T} = \psi_{\omega}\Omega \tag{13.12}$$

If we consider a spherical body of diameter d and a fluid with viscosity η, the translational and rotational drag coefficients are described in Stokes flow (see [64]) as:

$$\psi_{\upsilon} = 3\pi\eta d \tag{13.13}$$

$$\psi_{\omega} = \pi\eta d^3 \tag{13.14}$$

It is clear that velocity is inversely proportional to fluid viscosity with all other parameters held constant.

In [39] hydrodynamic properties of assembled-MEMS microrobots are determined experimentally by placing a microrobot in a known-fluid-filled vial and tracking it by digital cameras as it sinks under its own weight and under different applied magnetic forces. Modeling an assembled-MEMS microrobot as a sphere for the purposes of calculating fluid drag is found to be quite accurate. This also agrees with the typical assumption that fluid drag is insensitive to geometry at low Reynolds number. The coefficient of viscous drag of the microrobot in a Newtonian fluid obtained in [39] is found to be $\psi_{\upsilon} = (1.41 \times 10^{-2}\,\mathrm{N \cdot s/m}) \cdot d$.

The calculations in this section assume a Newtonian fluid, which will be a valid assumption after a vitrectomy has been performed. In the presence of intact vitreous humor, a more complicated fluid model must be considered.

13.3.2 Developing Sufficient Force for Levitation and Puncture

There is interest in determining the amount of force that can be developed wirelessly for the purpose of puncturing retinal veins. A drug delivery method where the microrobot docks to a blood vessel to allow the drug to release over extended periods of time is proposed, as shown in Fig. 13.10. This will require a microneedle to puncture the blood vessel. The magnetic forces on microrobots in applied magnetic fields are well understood. However, the magnitude of forces needed to puncture retinal veins is not available in the literature.

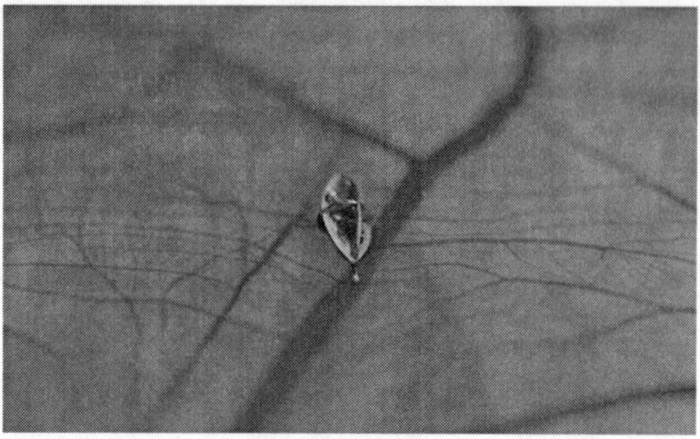

Fig. 13.10 Concept photo of a microrobot docked to a blood vessel for drug delivery (©IEEE 2008), reprinted with permission [17]. The assembled-MEMS microrobot shown is based on [66]

13.3.2.1 Retinal Puncture Forces

In [34], retinal puncture forces together with the scleral interaction forces are measured. However, needle and blood-vessel size, which affect puncture forces, are not specified. In [30], a retinal pick equipped with strain gauges is used to manipulate the retina of porcine cadaver eyes, and the range of forces acquired during a typical procedure is reported. However, the force of an individual retinal vein puncture is not provided. Conducting in vivo experiments on animal eyes is difficult with a high risk of tissue damage, and postmortem experiments may provide inaccurate results due to rapid changes in tissue properties of vessels after death. In [17], forces required for retinal vein punctures are measured and analyzed. Experimental data is collected from the vasculature of chorioallantoic membranes (CAM) of developing chicken embryos. The CAM of the developing chicken embryo has been used by ophthalmologists as a model system for studying photodynamic therapy and ocular angiogenesis. Recently, it was reported that the CAM of a twelve-day-old chicken embryo is a valid test bed for studies on human retinal vessel puncture [41]. The CAM's anatomical features and physiologic and histologic responses to manipulation and injury make it an effective living model of the retina and its vasculature. The vasculature of a twelve-day-old CAM and a human retina have roughly the same diameter and wall thickness (i.e., vessels with 100–300 μm outer diameter). The measurements are done using a capacitive force sensor with an attached microneedle, mounted on a 3-DOF Cartesian micromanipulator. Microneedles were pulled out of 1 mm OD boron-silicate glass pipettes in a repeatable way using a pipette puller and the outer diameters were inspected with a microscope.

There is variance in the force data due to effects that are not accounted for, such as anatomical variance between individual vessels and embryos, the state of the embryo (e.g., blood pressure, temperature), non-Hookean behavior of the vessel walls, errors in measured microneedle diameter, and error in the angle of incidence

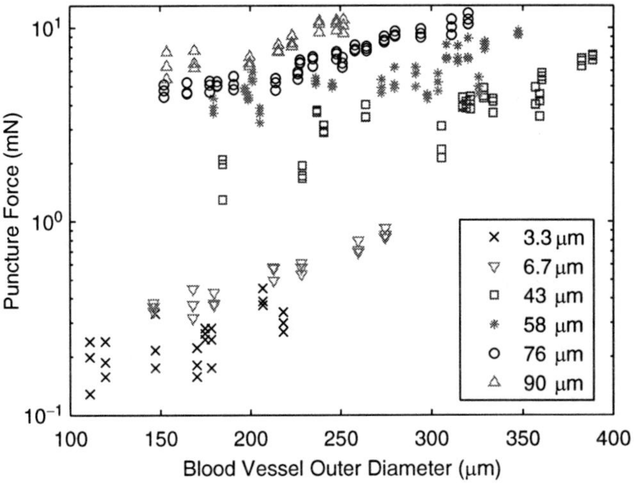

Fig. 13.11 Experimental data of puncture force with blunt tip microneedles vs. vessel diameter for different microneedles ODs (©IEEE 2008), reprinted with permission [17]

of the microneedle with respect to the blood vessel. Despite the variance, the data exhibit clear trends, as shown in Fig. 13.11. It is observed that there is an approximately quadratic trend in blood-vessel diameter and an approximately linear trend in microneedle diameter.

The experiments are performed with blunt-tip needles, so the forces shown in Fig. 13.11 should be taken as upper bounds for required puncture forces. It is known that beveling the needle's tip will significantly reduce the puncture forces [5, 16].

In [30], it is shown that 88% of all tool/tissue interaction forces during vitreoretinal surgery are below 12.5 mN, which corresponds well with the results shown in Fig. 13.11. In [34], higher puncture forces with larger variance than the results in Fig. 13.11 are reported. However, the reported forces include scleral interaction forces, and the force sensor is mounted on a handheld device.

13.3.2.2 Developing Sufficient Force

Let us consider a microrobot assembled from two thin elliptical pieces, as shown in the inset of Fig. 13.12. The volume of this microrobot is given by $v = 2\pi abt - 2at^2$. The force on such assembled microrobots made of Ni and CoNi are measured using the magnetic measurement system described in [39], and the results are shown in Fig. 13.12. The system uses a 40 mm × 40 mm × 20 mm NdFeB magnet with the north and south poles on the largest faces, and a field value of 0.41 T measured in the center of the north pole face. In addition to the measured data, a microrobot made of permanent-magnetic (hard-magnetic) material with a remanence magnetization of 4×10^5 A/m, which is a value that can currently be achieved using microfabrication techniques, is simulated.

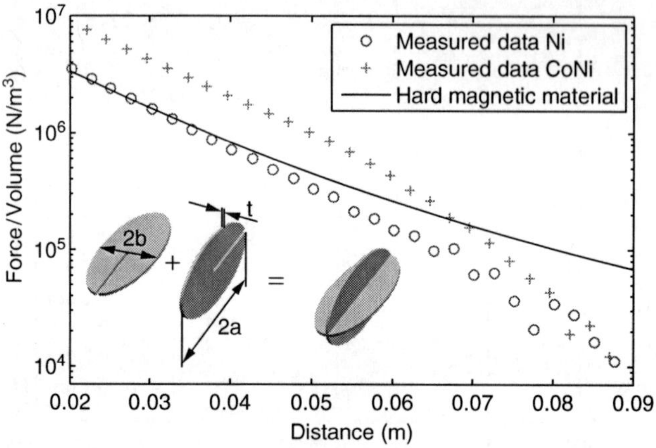

Fig. 13.12 Experimental data of normalized force vs. distance of the microrobot from the magnet's surface (©IEEE 2008), reprinted with permission. The magnet is described in [39]. A simulation of permanent-magnetic material with a remanence of 4×10^5 A/m is also shown. *Inset*: Definition of dimensions of the assembled microrobots used

In Fig. 13.12 we see that magnetic force drops off rapidly with increasing distance between the microrobot and the magnet. Increasing the saturation magnetization of a soft-magnetic material (compare CoNi and Ni) can lead to increases in force in a high-field region. Even relatively poor soft-magnetic materials (Ni) match good permanent magnets in a high-field region since the saturation magnetization values for soft-magnetic materials are typically higher than the remanence magnetization values of permanent magnets. However, as field strength is reduced, and the soft-magnetic microrobots are no longer saturated, they begin to provide similar force, and each provides less than that of the permanent-magnetic material. This is due to the magnetization of the soft-magnetic material, which is a function of the applied field, dropping below the remanence of the permanent-magnetic material.

Let us consider a microrobot with dimensions $a = 1000\,\mu\text{m}$, $b = 500\,\mu\text{m}$, and $t = 100\,\mu\text{m}$, and a volume of $v = 3\times10^{-10}\,\text{m}^3$; this microrobot could be electroplated and assembled, and fit through a 1-mm incision. If we consider the microrobot at a position 70 mm away from the surface of the magnet, soft- and permanent-magnet materials provide approximately the same force of 0.05 mN. For a length comparison, the diameter of the human eye is 25 mm, so this places the surface of the magnet almost three eye diameters away from the microrobot. If we move the microrobot only about 10 mm closer to the magnet, we gain an order of magnitude in our magnetic force, bringing it beyond the level needed for puncturing retinal veins with a blunt-tip needle of a few micrometers in size. As mentioned previously, beveling the needle's tip would reduce the required force even further. The magnet used in the experiment was chosen somewhat arbitrarily; other magnet shapes and sizes can be chosen to project the magnetic field at greater distances.

Under the above considerations, it seems feasible that enough magnetic force can be developed by pulling with magnetic field gradients to puncture retinal veins, provided that the microneedle is made small enough and sharp enough. These demands are attainable with current microfabrication technology. Puncture also requires an intelligent design of the magnetic-field generation system, which will use the superimposed fields of multiple permanent magnets or electromagnets, increasing the ability to generate strong fields at a distance. The choice of soft- or permanent-magnetic material for the microrobot will ultimately depend on the design of the magnetic-field generation system. This issue of force generation is discussed further in the next section.

It has been shown that microrobots that swim using helical propellers that mimic bacterial flagella theoretically have the potential to develop higher forces than obtained with gradient-based force generation at small scales [4]. It has also recently been shown that magnetic helical microrobots can be fabricated and wirelessly controlled [67]. This provides another option for retinal drug delivery.

13.3.3 OctoMag

There are two viable sources for the generation of controlled magnetic fields: permanent magnets and electromagnets. Permanent magnets exhibit a very advantageous volume to field-strength ratio. However, if we are interested in medical applications, electromagnets offer simpler real-time control, and present an inherently safer choice. A system using electromagnets can be implemented such that no moving parts are required to control magnetic field strength. This is important for both patient and medical-personnel safety. In addition, electromagnets are safer in the event of system failure: permanent magnets retain their attractive/repulsive strength in case of sudden power loss, whereas an electromagnetic system becomes inert, and in addition for the case of an intraocular microrobotic agent, the microrobot would slowly drift down under its own weight rather than experiencing uncontrolled forces with the potential of inflicting irreparable damage inside the eye. Using an array of stationary magnetic field sources also simplifies the task of designing a system that will respect the geometry of the human head, neck, and shoulders.

Bearing these considerations in mind, a robotic system called OctoMag for 5-DOF wireless magnetic control of a fully untethered microrobot (3-DOF position, 2-DOF pointing orientation) was developed at ETH Zurich [38]. It is difficult to control torque about the axis of **M** using the simple model in (7), which is why 5-DOF control was achieved as opposed to 6-DOF control. A concept image of how the system would be used for the control of intraocular microrobots can be seen in Fig. 13.13.

13.3.3.1 Control with Stationary Electromagnets

Soft-magnetic-core electromagnets can create a magnetic field that is approximately 20 times stronger than the magnetic field created by air-core electromagnets, for the geometry shown in Fig. 13.13. As opposed to air-core electromagnets, their

Fig. 13.13 Concept image of the OctoMag electromagnetic system: An eyeball is at the center of the system's workspace (©IEEE 2010), reprinted with permission. The electromagnet arrangement accommodates the geometry of the head, neck, and shoulders. The OctoMag is designed for a camera to fit down the central axis to image the microrobot in the eye

individual magnetic fields are coupled, which complicates modeling and control. However, cores made of high-performance soft-magnetic materials impose only a very minor constraint on modeling and control [38].

Within a given static arrangement of electromagnets, each electromagnet creates a magnetic field throughout the workspace that can be precomputed. At any given point in the workspace **P**, the magnetic field due to a given electromagnet can be expressed by the vector $\mathbf{B}_e(\mathbf{P})$, whose magnitude varies linearly with the current through the electromagnet, and as such can be described as a unit-current vector in units T/m multiplied by a scalar current value in units A:

$$\mathbf{B}_e(\mathbf{P}) = \tilde{\mathbf{B}}_e(\mathbf{P})i_e \tag{13.15}$$

The subscript e represents the contribution due to the eth electromagnet. However, although the field $\mathbf{B}_e(\mathbf{P})$ is the field due to the current flowing through only electromagnet e, it is due to the soft-magnetic cores of every electromagnet. With air-core electromagnets, the individual field contributions are decoupled, and the fields can be individually precomputed and then linearly superimposed. This in not the case with soft-magnetic-core electromagnets. However, if an ideal soft-magnetic material with negligible hysteresis is assumed, and the system is operated with the cores in their linear magnetization region, the assumption is still valid that the field contributions of the individual currents (each of which affect the magnetization of every core) superimpose linearly. Thus, if the field contribution of a given electromagnet is precomputed in situ, it can be assumed that the magnetic field at a point in the workspace is the sum of the contributions of the individual currents. This assumption is clearly also valid for air-core electromagnets and the linear summation of fields can be expressed as:

$$\mathbf{B}(\mathbf{P}) = \left[\tilde{\mathbf{B}}_1(\mathbf{P}) \cdots \tilde{\mathbf{B}}_n(\mathbf{P})\right] \begin{bmatrix} i_1 \\ \vdots \\ i_n \end{bmatrix} = \mathcal{B}(\mathbf{P})I \tag{13.16}$$

The 3 $\times n$ $\mathcal{B}(\mathbf{P})$ matrix is defined at each point $\mathbf{P}$ in the workspace, which can either be analytically calculated online, or a grid of precomputed or measured points can be interpolated online. It is also possible to express the derivative of the field in a given direction in a specific frame, for example the x direction, as the contributions from each of the currents:

$$\partial\mathbf{B}(\mathbf{P})\partial x = \left[\frac{\partial\tilde{\mathbf{B}}_1(\mathbf{P})}{\partial x} \cdots \frac{\partial\tilde{\mathbf{B}}_n(\mathbf{P})}{\partial x}\right] \begin{bmatrix} i_1 \\ \vdots \\ i_n \end{bmatrix} = \mathcal{B}_x(\mathbf{P})I \tag{13.17}$$

If we are interested in controlling a microrobot moving through fluid, where the microrobot can align with the applied field unimpeded, rather than controlling torque and force acting on the microrobot, we can simply control the magnetic field to the desired orientation, to which the microrobot will naturally align, and then explicitly control the force on the microrobot:

$$\begin{bmatrix} \mathbf{B} \\ \mathbf{F} \end{bmatrix} = v \begin{bmatrix} \mathcal{B}(\mathbf{P}) \\ \mathbf{M}^T\mathcal{B}_x(\mathbf{P}) \\ \mathbf{M}^T\mathcal{B}_y(\mathbf{P}) \\ \mathbf{M}^T\mathcal{B}_z(\mathbf{P}) \end{bmatrix} \begin{bmatrix} i_1 \\ \vdots \\ i_n \end{bmatrix} = \mathcal{A}(\mathbf{M},\mathbf{P})I \tag{13.18}$$

That is, for each microrobot pose, the n electromagnet currents are mapped to a field and force through a 6 $\times n$ actuation matrix $\mathcal{A}(\mathbf{M},\mathbf{P})$. For a desired field/force vector, the choice of currents that gets us closest to the desired field/force value can be found using the pseudoinverse:

$$I = \mathcal{A}(\mathbf{M},\mathbf{P})^{\dagger} \begin{bmatrix} \mathbf{B}_{des} \\ \mathbf{F}_{des} \end{bmatrix} \tag{13.19}$$

Full 5-DOF control requires a rank-6 actuation matrix $\mathcal{A}$. If there are multiple solutions to achieve the desired field/force, the pseudoinverse finds the solution that minimizes the 2-norm of the current vector, which is desirable for the minimization of both power consumption and heat generation. Note that the use of (19) requires knowledge of the microrobot's pose and magnetization. If the direction of $\mathbf{B}$ does not change too rapidly, it is reasonable to assume that $\mathbf{M}$ is always aligned with $\mathbf{B}$, which means that one need not explicitly measure the microrobot's full pose, but rather, must only estimate the magnitude of **M** and measure the microrobot's position **P**. In addition, if we the magnetic field does not vary greatly across the workspace, it may be reasonable to assume that the microrobot is always located at $\mathbf{P} = \mathbf{0}$ for purposes of control, eliminating the need for any localization of the microrobot.

There are a number of potential methods to generate the unit-current field maps that are required for the proposed control system. One can either explicitly measure the magnetic field of the final system at a grid of points or compute the field values at the grid of points using FEM models. In either case trilinear interpolation is used during real-time control. To generate the unit-current gradient maps using either method, one can either explicitly measure/model the gradient at the grid of points, or numerically differentiate the field data. Alternatively, one can fit an analytical model – for example the point-dipole model [24] – to field data obtained from an FEM model of the final system for each of the unit-current contributions. An analytical field model also has an analytical derivative. These analytical models can then be used to build the unit-current field and gradient maps during run time.

13.3.3.2 System Implementation

Equipped with a general control system using n stationary electromagnets, it is now possible to use this controller in the design of a suitable electromagnet configuration. The singular values of the actuation matrix in (18) provide information on the condition of the workspace and can be used as performance metric in a design optimization [38]. Figure 13.14 shows a physical embodiment of the concept image presented earlier. This prototype setup was designed with a workspace that is large enough that, after experimenting in artificial and ex vivo eyes, could be used for animal trials with live cats and rabbits.

Each electromagnet is completely filled with a core made of VACOFLUX 50 – a CoFe alloy from VACUUMSCHMELZE – with a diameter of 42 mm. This material has a saturation magnetization on the order of 2.3 T, a coercivity of 0.11 mT, and a maximum permeability of 4500 H/m. To prevent temperatures inside the coils to elevate beyond 45°C, every electromagnet is wrapped with a cooling system. The current for the electromagnetic coils is sourced through custom-designed switched amplifiers to reduce the power consumption. Two stationary camera assemblies provide visual feedback from the top and side and allow to extract the 3-D position of a microrobot in the system. For the envisioned intraocular application the visual feedback will have to be produced using a single camera which is detailed in Sect. 4.

An example of the manipulation capabilities of the system can be seen in Fig. 13.15. Automated pose control refers to closed-loop position control with open-loop orientation control. The system exhibits similar performance in a wide array of different trajectories as well as for a variety of robot orientations. The forces this

Fig. 13.14 OctoMag prototype: The system contains eight 210-mm-long by 62-mm-diameter electromagnets (©IEEE 2010), reprinted with permission. The gap between two opposing electromagnets on the lower set is 130 mm. The inset shows a 500-μm-long microrobot of the type described in [66] levitating in a chamber. This is the side-camera view seen by the operator

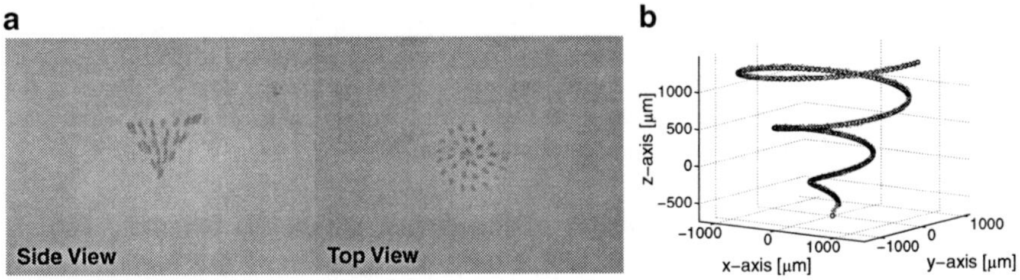

Fig. 13.15 Demonstration of automated pose control (©IEEE 2010), reprinted with permission. Both time-lapse image sequences (**a**) show a 500-μm-long microrobot following a spiral trajectory keeping its orientation constantly pointing at the vertex of the spiral.Way points (*black circle*) and tracker data (*red plus*) are shown in the isometric graph (**b**). Average trajectory-completion time: 33.4 s

system can exert on the tiny Ni microrobot shown in Figs. 13.14 and 13.15 as well as on a larger NdFeB cylinder with a diameter of 500 μm and a hight of 1 mm are tabulated in Table 13.1.

13.4 Issues in Localizing Microrobots

In the previous sections, we have discussed the functionalization of the microdevices and the principles of their control. In order to control these devices near the retina, knowledge of their position is usually required. In the case of untethered magnetic

Table 13.1 Maximum force in OctoMag setup on a small Ni microrobot and a full NdFeB cylinder for various agent orientations (©IEEE 2010), reprinted with permission

	Ni Microrobot					NdFeB Cylinder				
Field Orientation	$\mathbf{F}_{up}$ (μN)	$\mathbf{F}_{down}$ (μN)	$\mathbf{F}_{lat,x}$ (μN)	$\mathbf{F}_{lat,y}$ (μN)	$\mathbf{F}_{lat,xy}$ (μN)	$\mathbf{F}_{up}$ (μN)	$\mathbf{F}_{down}$ (μN)	$\mathbf{F}_{lat,x}$ (μN)	$\mathbf{F}_{lat,y}$ (μN)	$\mathbf{F}_{lat,xy}$ (μN)
z	2.1	1.3	1.1	1.3	1.4	281	221	141	172	200
$-z$	3.3	2.1	1.7	2.0	2.2	281	221	141	172	200
x	2.2	2.2	3.6	3.1	3.9	189	222	213	341	276
xy	2.7	2.7	4.1	3.0	2.5	274	263	263	286	259

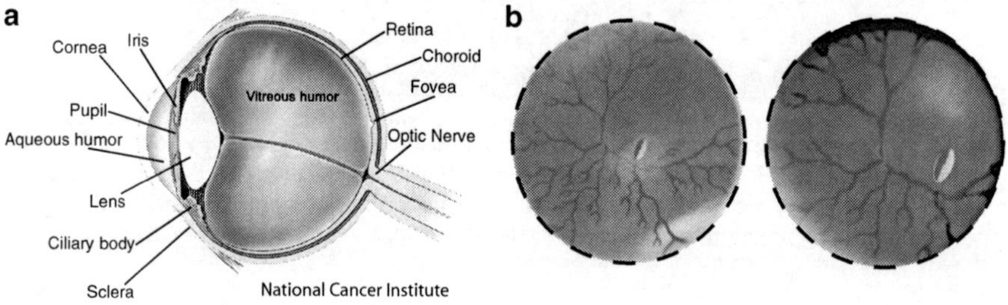

Fig. 13.16 (**a**) Anatomy of the human eye. (**b**) The biomedical microrobot of [66] in the model eye [31]. *The left image* shows the intraocular environment without the eye's optical elements, and *the right image* shows the effect of the model eye optics. Images are taken with an unmodified digital camera (©IEEE 2008), reprinted with permission

devices, knowledge of the position of the device within the magnetic field is necessary for precise control [2, 49]. Since, the interior of the human eye is externally observable, vision can be used to perform 3D localization. In addition, with clear images of the retina and the microrobot, visual feedback can be used to close a visual-servoing loop to correct for any errors in the localization procedure.

Ophthalmic observation has been practices for centuries, with its most critical task being to be able to visualize the human retina with high-definition. Nowadays, clinicians have the ability to acquire magnified images of the human retina using an ever-increasing variety of optical tools that are designed specifically for the unique optical system that is the human eye (Fig. 13.16).

However, keeping intraocular objects that are freely moving in the vitreous humor of the eye (and not just on the retina) constantly in focus is challenging, and the captured images are often blurry and noisy. The unstructured illumination that reaches the interior of the eye, either through endoillumination, transpupilary or transscleral means, can deteriorate the images with uneven brightness and backreflections. Moreover, the microtools that operate in the human eye are generally specular, and have no distinctive color features. For precise localization, robust visual tracking that detect the microrobot in the images is needed. The extracted segmentation information will be used for 3D localization.

The literature lacks algorithms for the localization of untethered intraocular devices. In addition, because the microrobots will be controlled by a magnetic-field-generation system surrounding the patient's head, we are interested in compact solutions that utilize a single stationary camera. In the following sections, we will introduce the first intraocular localization algorithm, using a custom-built stationary camera. Our approach is based on depth-from-focus [20]. Focus-based methods do not require a model of the object of interest, but only knowledge of the optical system. Applied in the eye, they could also localize unknown objects such as floaters. As a result, our analysis need not be considered only in the scope of microrobot localization, but is applicable on any type of unknown foreign bodies.

In the following, we will firstly evaluate different ophthalmoscopy methods with respect to their advantages in imaging and localizing. Then, based on our results, we will introduce a level-set tracking algorithm that successfully segments intraocular microdevices in images. Finally, we will present a method for wide-angle intraocular localization.

13.4.1 Comparison of Ophthalmoscopy Methods

Our results are based on Navarro's schematic eye [22] (i.e. an optical model based on biometric data that explains the optical properties of the human eye). Navarro's schematic eye performs well for angles up to 70° measured from the center of the pupil and around the optical axis. For greater angles, the biometric data of each patient should be considered individually. Simulations are carried out with the OSLO optical lens design software. Throughout this section, the object's depth z is measured along the optical axis. We begin by investigating the feasibility of imaging and localizing intraocular devices using existing ophthalmoscopy methods.

13.4.1.1 Direct Ophthalmoscopy

In a relaxed state, the retina is projected through the eye optics as a virtual image at infinity. An imaging system can capture the parallel beams to create an image of the retina. In direct ophthalmoscopy the rays are brought in focus on the observer's retina [57]. By manipulating the formulas of [56] the field-of-view for direct ophthalmoscopy is found as 10° (Fig. 13.17a).

Every object inside the eye creates a virtual image. These images approach infinity rapidly as the object approaches the retina. Figure 13.18 (solid line) displays the distance where the virtual image is formed versus different positions of an intraocular object. In order to capture the virtual images that are created from objects close to the retina, an imaging system with near to infinite working distance is required. Such an imaging system will also have a large depth-of-field, and depth information from focus would be insensitive to object position (Table 13.2).

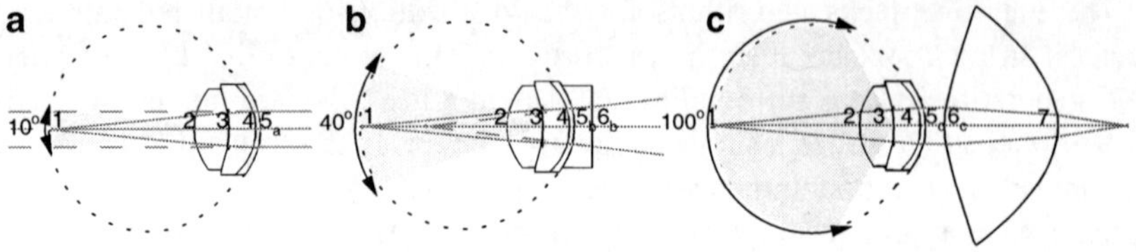

Fig. 13.17 (**a**) Direct ophthalmoscopy with Navarro's schematic eye [22]. (**b**) Ophthalmoscopy with Navarro's schematic eye with a vitrectomy lens [23]. (**c**) Indirect ophthalmoscopy with Navarro's schematic eye with a condensing lens [63] (©IEEE 2008), reprinted with permission

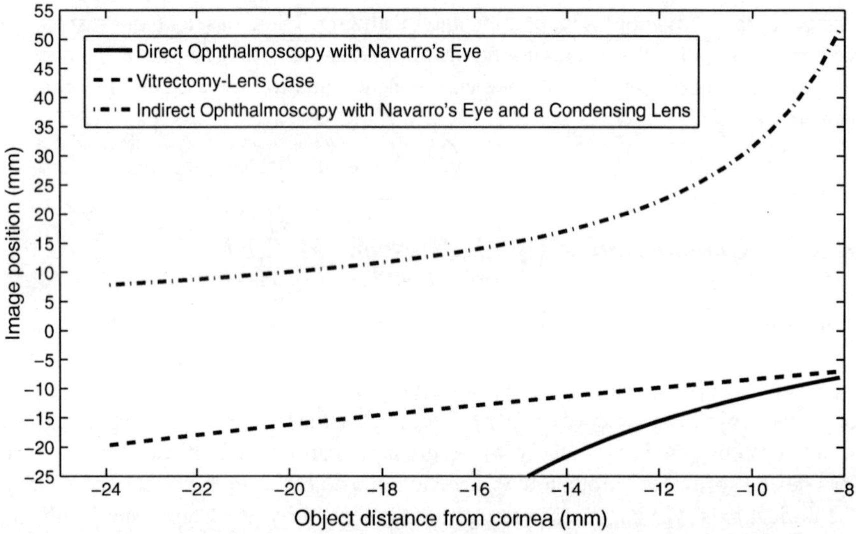

Fig. 13.18 Image position versus intraocular object position for the direct ophthalmoscopy case, the vitrectomy-lens case, and the indirect ophthalmoscopy case. Image distances are measured from the final surface of each optical system (5a, 6b, 7 respectively) (©IEEE 2009), reprinted with permission

13.4.1.2 Vitrectomy Lenses

To visualize devices operating in the vitreous humor of phakic (i.e. intact intraocular lens) eyes, only plano-concave lenses (Fig. 13.17b) need to be considered [57]. Vitrectomy lenses cause the virtual images of intraocular objects to form inside the eye, allowing the imaging systems to have a reduced working distance. Based on data given from HUCO Vision SA for the vitrectomy lens S5. 7010 [23], we simulated the effects of a plano-concave vitrectomy lens on Navarro's eye (Fig. 13.17b). This lens allows for a field-of-view of 40 °, significantly larger than the one obtainable with the method described in Sect. 4.1.1.

As shown in Fig. 13.18 (dashed line), the virtual images are formed inside the eye and span a lesser distance. Thus, contrary to direct observation, imaging with an

Table 13.2 Optical parameters for the systems of Fig. 13.17 (©IEEE 2008), reprinted with permission

Surface	1	2	3	4	5a	5b	5c	6b	6c	7
Radius (mm)	12.00	6.00	−10.20	−6.50	−7.72	−7.72	−7.72	∞	11.65	−9.48
Conic constant	0.00	−1.00	−3.13	0.00	−0.26	−0.26	−0.26	0.00	−9.24	−1.07
Thickness (mm)	16.32	4.00	3.05	0.55	∞	2.00	2.00	∞	13.00	∞
Refraction index	1.336	1.420	1.337	1.376	1.000	1.425	1.000	1.000	1.523	1.000

optical microscope (relatively short working distance and depth-of-field) is possible. The working distance of such a system must be at least 20 mm. As depth-of-field is proportional to working distance, there is a fundamental limit to the depth-from-focus resolution achievable with vitrectomy lenses.

13.4.1.3 Indirect Ophthalmoscopy

Indirect ophthalmoscopy (Fig. 13.17c) allows for a wider field of the retina to be observed. A condensing lens is placed in front of the patient's eye, and catches rays emanating from a large retinal area. These rays are focused after the lens, creating an aerial image of the patient's retina. Condensing lenses compensate for the refractive effects of the eye, and create focused retinal images.

We simulated the effects of a double aspheric condensing lens based on information found in [63]. This lens, when placed 5 mm from the pupil, allows imaging of the peripheral retina and offers a field-of-view of 100°. As a result, it can be part of an imaging system with a superior field-of-view than the ones described in Sect. 4.1.1 and Sect. 4.1.2. The image positions versus the intraocular object positions can be seen in Fig. 13.18 (dashed-dotted line). A sensing system with a short working-distance and shallow depth-of-field can be used in order to extract depth information from focus for all areas inside the human eye. Depth estimation is more sensitive for objects near the intraocular lens, since smaller object displacements result in larger required focusing motions.

Dense CMOS sensors have a shallow depth-of-focus, and as a result, they can be used effectively in depth-from-focus techniques. Based on Fig. 13.18, to localize objects in the posterior of the eye a sensor travel of 10 mm is necessary. A $24 \times 24\,\text{mm}^2$ CMOS sensor can capture the full field-of-view. The simulated condensing lens causes a magnification of 0.78×9 and thus, a structure of 100 μm on or near the retina will create an image of 78 μm. Even with no additional magnification, a CMOS sensor with a common sensing element size of $6 \times 6\,\mu\text{m}^2$ will resolve small retinal structures sufficiently. As a conclusion, direct sensing of the aerial image leads to a high field-of-view, while having advantages in focus-based localization.

In the following, we will use indirect ophthalmoscopy methods for tracking and localizing intraocular microrobots.

13.4.2 Tracking Intraocular Microrobots

To track intraocular microrobots, we use the method presented in [10], which is based on the framework developed in [54]. For successful tracking to occur, one should evaluate the quality of different colorspaces with respect to the biomedical application of interest. For example, in [18, 7, 61] the Hue-Saturation-Value (HSV) colorspace is used, but in [10] it is shown that this is a not suitable colorspace in

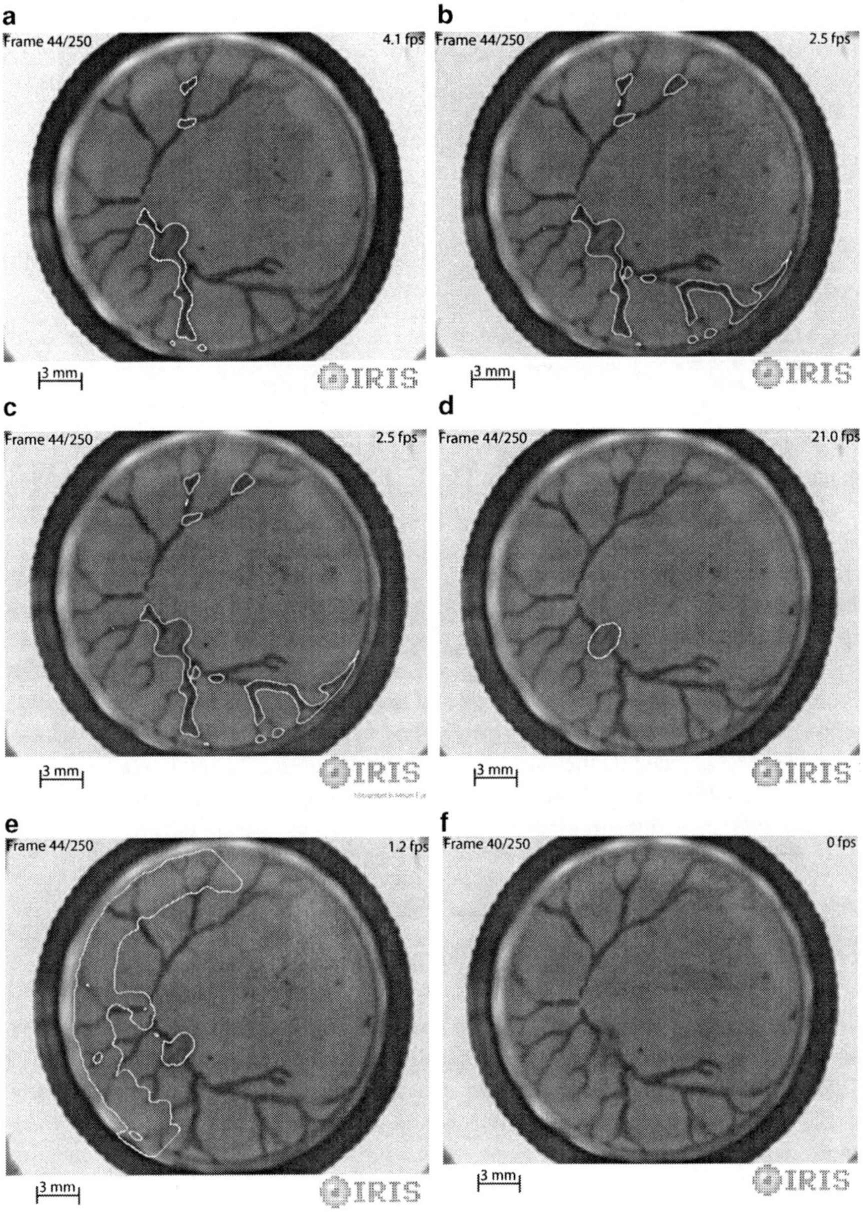

Fig. 13.19 Tracking using (**a**), (**b**) the R-G channels of the RGB colorspace without and with thresholds, respectively, (**c**), (**d**) the Y-V channels of the YUV colorspace without and with thresholds, respectively, (**e**), (**f**) the H-S channels of the HSV colorspace without and with thresholds, respectively (© IEEE 2009), reprinted with permission

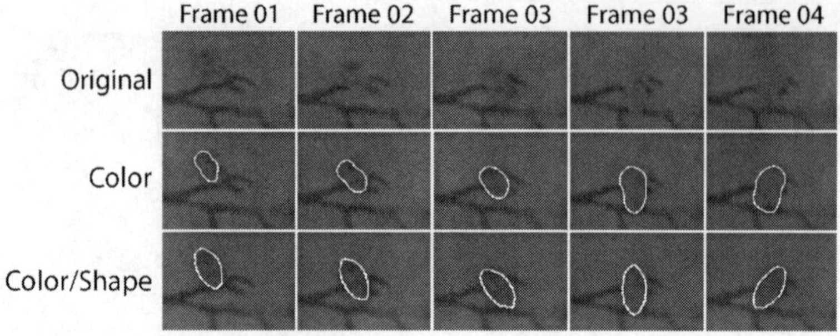

Fig. 13.20 Tracking using color information and color/shape information, for different frame sequences (©IEEE 2009), reprinted with permission

which to track intraocular microdevices. Tracking in the best colorspace ensures reduced vein segmentation; compare Fig. 13.19a with Fig. 13.19c and Fig. 13.19e).

After choosing the appropriate colorspace, thresholds that ensure the maximum object-from-background separation are calculated. These thresholds help vanish the erroneous vein segmentation, and lead to successful detection of the microrobot in the images (Fig. 13.19b, d). If the selected colorspace/channel combination is not of appropriate quality, the thresholds will cause the tracking to fail (Fig. 13.19e).

To further increase the segmentation accuracy, in [10], the statistical shape prior evolution framework of [42] is adapted and used. Using shape information together with the color information ensures diminished vein segmentation, as the misclassifications are discarded by the shape information. Figure 13.20 compares tracking in R-G, and tracking in R-G using shape information; when shape information is incorporated the results are improved.

13.4.3 Wide-Angle Localization

With the tracking method of [10] we can robustly estimate the position of an intraocular microrobot in images, successfully handling cases of occlusion and of defocus. However, in order to perform accurate magnetic control, we need to know its 3D position in the intraocular environment. Here, we present a method for wide-angle intraocular localization using focus information. This method was first introduced by Bergeles *et al.* [12].

13.4.3.1 Theory of Intraocular Localization

As previously stated, the condensing lens projects the spherical surface of the retina onto a flat aerial image. Moving the sensor with respect to the condensing lens focuses the image at different surfaces inside the eye, which we call *isofocus*

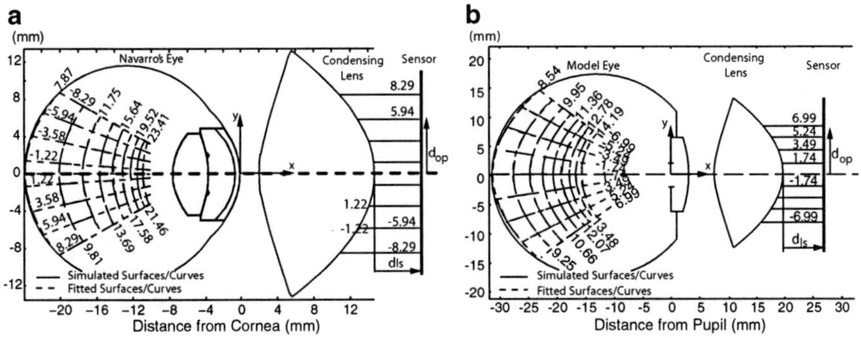

Fig. 13.21 Simulation of the isofocus surfaces and isopixel curves for (**a**) indirect ophthalmoscopy with Navarro's eye, and (**b**) indirect ophthalmoscopy with the model eye [31] (©IEEE 2009), reprinted with permission. The different isofocus surfaces correspond to the distance from the lens to the sensor (d_{ls}), for uniform sensor steps of (**a**) ~ 1.95 mm, and (**b**) ~ 0.7 mm. The isopixel curves correspond to pixel distances from the optical axis (d_{op}), for uniform steps of (**a**) ~ 2.25 mm, and (**b**) ~ 1.75 mm

surfaces. The locus of intraocular points that are imaged on a single pixel is called an *isopixel* curve. Figure 13.21a shows a subset of these surfaces and curves and their fits for the system of Fig. 13.17c. The position of an intraocular point is found as the intersection of its corresponding isopixel curve and isofocus surface.

The location of the isofocus surfaces and isopixel curves are dependent on the condensing lens and the individual eye. The optical elements of the human eye can be biometrically measured. For example, specular reflection techniques or interferometric methods can be used to measure the cornea [46], and autokeratometry or ultrasonometry can be used to measure the intraocular lens [37]. Then, the surfaces and curves can be accurately computed offline using raytracing. The sensitivity of the isopixel curve and isofocus surfaces calculation with respect to uncertainties in the knowledge of the parameters of the different optical elements is examined in [13]. In theory there is an infinite number of isofocus surfaces and isopixel curves, but in practice there will be a finite number due to the resolution of sensor movement and pixel size, respectively.

The density of the isofocus surfaces for uniform sensor steps in Fig. 13.21a demonstrates that the expected depth resolution is higher for regions far from the retina. The isopixel curves show that the formed image is inverted, and from their slope it is deduced that the magnification of an intraocular object increases farther from the retina. As a result, we conclude that both spatial and lateral resolutions increase for positions farther from the retina.

The isofocus surfaces result from the optics of a rotationally symmetric and aligned system composed of conic surfaces. We therefore assume that they are conic surfaces as well, which can be parametrized by their conic constant, curvature, and intersection with the optical axis. Since the isofocus surfaces correspond to a specific sensor position, their three parameters can also be expressed as functions of the sensor position.

The isopixel curves are lines, and it is straightforward to parametrize them using their slope and their distance from the optical axis at the pupil. Each isopixel curve corresponds to one pixel on the image, and its parameters are functions of the pixel's offset (measured from the image center) due to the rotational symmetry of the system. For the 2D case, two parameters are required.

In Fig. 13.22a–d the parametrizing functions of the isofocus surfaces and isopixel curves are displayed. The conic constant need not vary (fixed at -0.5) because it was observed that the surface variation can be successfully captured by the curvature. For each parameter, we fit the least-order polynomial that captures its variability. The parametrizing functions are injections (Fig. 13.22a–d), and thus, 3D intraocular localization with a wide-angle is unambiguous.

13.4.3.2 Localization Experiments

As an experimental testbed, we use the model eye [31] from Gwb International, Ltd. This eye is equipped with a plano-convex lens of 36 mm focal length that mimics the compound optical system of the human eye. Gwb International, Ltd. disclosed the lens' parameters so that we can perform our simulations. We also measured the model's retinal depth and shape.

The optical system under examination is composed of this model eye and the condensing lens of Fig. 13.17c, where the refraction index was chosen as 1.531. The simulated isofocus surfaces and isopixel curves of the composite system, together with their fits, are shown in Fig. 13.21b. Based on these simulations, we parametrize the isofocus surfaces and the isopixel curves (Fig. 13.22e–h). The behavior of the parameters is similar to the one displayed in Fig. 13.22a–d for Navarro's schematic eye. We assume an invariant conic constant of -1.05, because the variability of the surfaces can be captured sufficiently by the curvature.

In order to calibrate the isofocus surfaces for their intersection with the optical axis, we perform an on-optical-axis depth-from-focus experiment on the aligned optical system. We use a Sutter linear micromanipulation stage to move a checkerboard calibration pattern in the model eye with 1 mm steps, and estimate the in-focus sensor position. The best focus position is estimated with techniques described in [58]. The estimated sensor positions with respect to different object depths can be seen in Fig. 13.23. The uncalibrated model fit is displayed with a solid line, and, as can be seen, calibration is needed.

In the model eye, we can calibrate for the relationship between the in-focus sensor position and the depth of the object using the full set of data points. However, such an approach would be clinically invasive as it would require a vitrectomy and a moving device inside the eye. The only minimally invasive biometric data available are the depth and shape of the retina that can be measured from MRI scans [8]. Assuming that there are accumulated errors that can be lumped and included as errors in the estimated image and object positions, it is shown in [11] that by using a first-order model of the optics, calibration using only the depth of the retina is possible. By adapting this method to our framework, we are able to

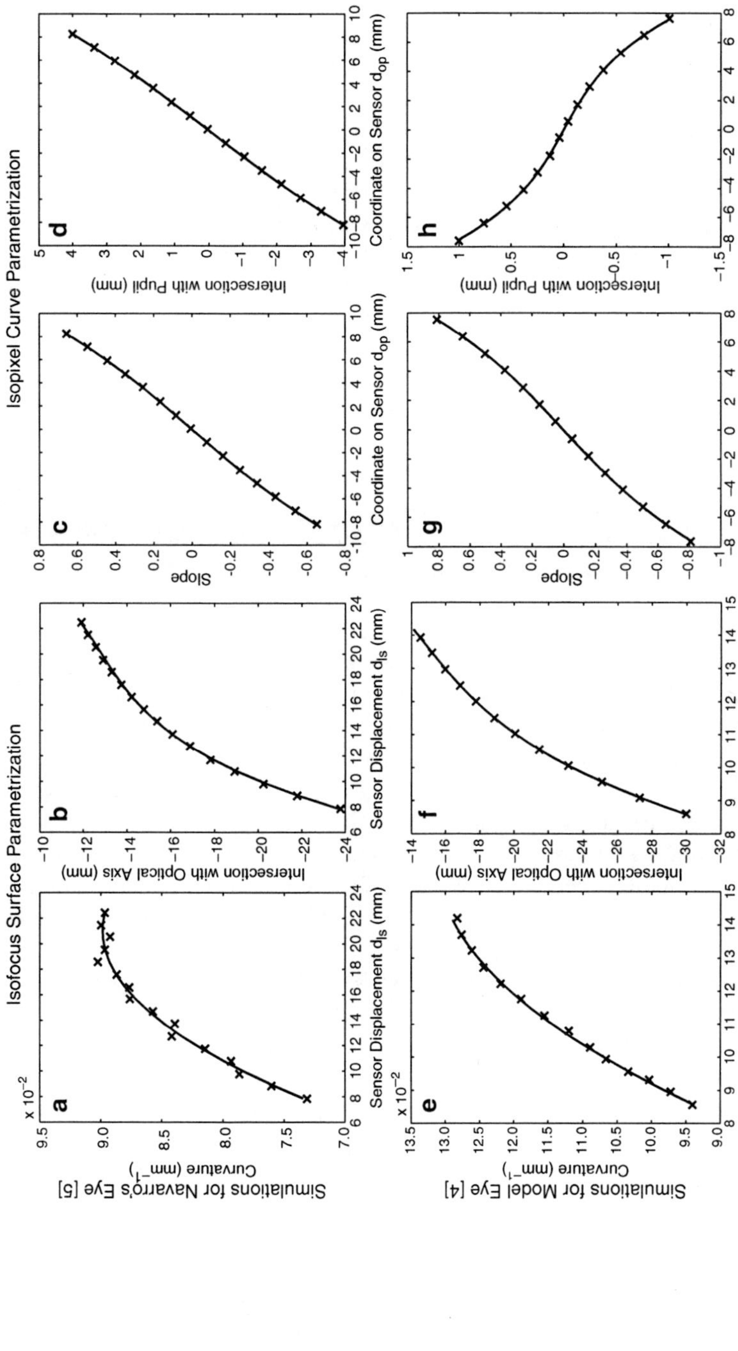

Fig. 13.22 *top row:* Parametrization polynomials for the system of Fig. 13.21a, and *bottom row:* for the system of Fig. 13.21b (©IEEE 2009), reprinted with permission. Isofocus surface parametrization: **(a)**, **(b)**, **(e)**, **(f)** Fitted 2nd- and 3rd-order polynomials for the curvature and for the intersection with the optical axis. Isopixel curve parametrization: **(c)**, **(d)**, **(g)**, **(h)** Fitted 3rd-order polynomials for the line slope and for the intersection with the pupil

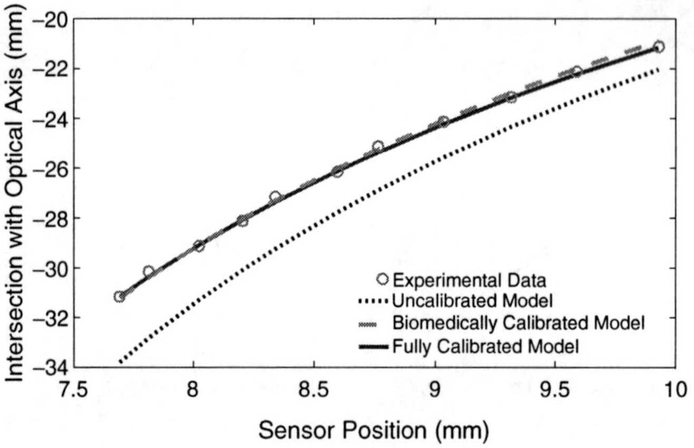

Fig. 13.23 Model fits for the function describing the intersection of the isofocus surfaces with the optical axis. Biometric calibration errors: mean = 159 μm, std = 94 μm (©IEEE 2009), reprinted with permission

biometrically calibrate for the parameters of the polynomial that describes the intersection of the isofocus surfaces with the optical axis. The resulting fit can be seen in Fig. 13.23.

The remaining two parameters of the isofocus surfaces control the shape of the isofocus surfaces but not their position. The condensing lens is designed to create a flat aerial image of the retinal surface, and our experiments have shown that we can use it to capture an overall sharp image of the model eye's retina. Therefore, we conclude that there exists an isofocus surface that corresponds to the retinal surface, and we consider it as the first surface. From Fig. 13.21b we see that the first isofocus surface does indeed roughly correspond to the retinal shape (mean error = 371 μm). As a result, calibration for the conic constant and the curvature is not needed. If our model was not accurately predicting the shape of the retina, then we would calibrate the parameters of the first isofocus surface so that is has exactly the same shape as the retina.

To estimate the validity of the presented wide-angle localization algorithm, we consider points in the model eye for various angles with respect to the optical axis and various distances from the pupil. In Fig. 13.24, the results using the proposed wide-angle localization algorithm are displayed. For comparison, we also show the results based on the paraxial localization algorithm presented in [11] for angles up to 10° from the optical axis. The paraxial localization results deteriorate as the angles increase. However, the localization method proposed here can be used for regions away from the optical axis with high accuracy.

To fully control untethered devices, their position and orientation (i.e. their pose) must be estimated. Until now, we have only addressed position estimation. For the estimation of orientation, methods exist for perspective projection systems [19]. These methods work by projecting an estimate of the 6-DOF pose of a CAD model

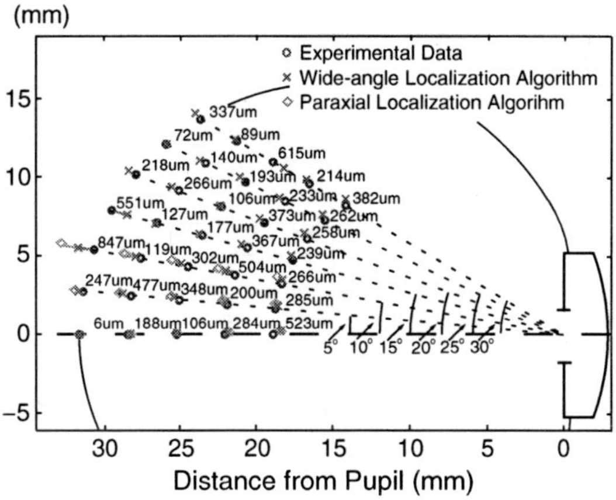

Fig. 13.24 Localization experiment showing the validity of the proposed algorithm. Errors: mean = 282 μm, std = 173 μm (©IEEE 2009), reprinted with permission

of the device on the image, based on a calibrated camera projection matrix, and then adapting the estimate of the pose until projection agrees with the perceived image. It may be possible to apply similar methods for pose estimation of intraocular objects. However, the estimates of the pose would be projected on the image by making use of the isopixel curves. This topic is left as future work.

13.5 Conclusions

In this chapter we considered three topics in the design and control of intraocular microrobots. First, we discussed functional coatings – both for remote sensing and targeted drug delivery. We developed a wireless oxygen sensor, and we presented a method to deliver clot-busting tPA to retinal veins. Next, we discussed magnetic control. We showed that it is technically feasible to generate sufficient forces to puncture retinal veins. Finally, we showed that it is possible to visually track and then localize intraocular microrobots with a single stationary camera. The localization results can be utilized in the magnetic control system. The groundwork has been laid, and it won't be long before intraocular microrobots change the way that the most difficult retinal procedures are done.

Acknowledgments This research is supported by the NCCR Co-Me of the Swiss National Science Foundation. The authors would like to thank the many people that have contributed to this research: Berk Yeşin, Karl Vollmers, Zoltan Nagy, Brad Kratochvil, Ruedi Borer, Görkem Dogangil, Kamran Shamaei, and George Fagogenis, as well as Heike Hall from ETH Zurich, and Mohammad K. Nazeeruddin from Ecole Polytechnique Fédéral de Lausanne. The authors also would like to thank Justus Garweg, M.D. from the Swiss Eye Institute in Bern and Carsten Framme, M.D. of the Inselspital Bern for their guidance on vitreoretinal applications for untethered microrobots.

References

1. Stereotaxis (2007). URL http://www.stereotaxis.com
2. Abbott, J.J., Ergeneman, O., Kummer, M.P., Hirt, A.M., Nelson, B.J.: Modeling magnetic torque and force for controlled manipulation of soft-magnetic bodies. IEEE Trans. Robot. **23** (6), 1247–1252 (2007)
3. Abbott, J.J., Nagy, Z., Beyeler, F., Nelson, B.J.: Robotics in the small, part I: Microrobotics. IEEE Robot. Automat. Mag. **14**(2), 92–103 (2007)
4. Abbott, J.J., Peyer, K.E., Cosentino Lagomarsino, M., Zhang, L., Dong, L.X., Kaliakatsos, I.K., Nelson, B.J.: How should microrobots swim? Int. J. Rob. Res. **28**(11–12), 1434–1447 (2009)
5. Abolhassani, N., Patel, R., Moallem, M.: Needle insertion into soft tissue: A survey. Med. Eng. Phys. **29**, 413–431 (2007)
6. Amblard, F., Yurke, B., Pargellis, A., Leibler, S.: A magnetic manipulator for studying local rheology and micromechanical properties of biological systems. Rev. Sci. Instrum. **67**(3), 818–827 (1996)
7. Ascari, L., Bertocchi, U., Laschi, C., Stefanini, C., Starita, A., Dario, P.: A segmentation algorithm for a robotic micro-endoscope for exploration of the spinal cord. In: IEEE Int. Conf. Robotics and Automation, pp. 491–496 (2004)
8. Atchison, D.A., Pritchard, N., Schmid, K.L., Scott, D.H., Jones, C.E., Pope, J.M.: Shape of the retinal surface in emmetropia and myopia. Invest. Ophthalmol. Vis. Sci. **46**(8), 2698–2707 (2005)
9. Bach, M., Oschwald, M., Röver, J.: On the movement of an iron particle in a magnetic field. Doc. Ophthalmol. **68**, 389–394 (1988)
10. Bergeles, C., Fagogenis, G., Abbott, J.J., Nelson, B.J.: Tracking intraocular microdevices based on colorspace evaluation and statistical color/shape information. In: Proc. IEEE Int. Conf. Robotics and Automation, pp. 3934–3939 (2009)
11. Bergeles, C., Shamaei, K., Abbott, J.J., Nelson, B.J.: On imaging and localizing untethered intraocular devices with a stationary camera. In: Proc. IEEE Int. Conf. on Biomedical Robotics and Biomechatronics (2008)
12. Bergeles, C., Shamaei, K., Abbott, J.J., Nelson, B.J.: Wide-angle intraocular imaging and localization. In: Int. Conf. Medical Image Computing and Computer Assisted Intervention, vol. 1, pp. 540–548 (2009)
13. Bergeles, C., Shamaei, K., Abbott, J.J., Nelson, B.J.: Single-camera focus-based localization of intraocular devices. IEEE Trans. Biomed. Eng. **57**, 2064–2074 (2010)
14. Berkowitz, B., Wilson, C.: Quantitative mapping of ocular oxygenation using magnetic resonance imaging. Magn. Reson. Med. **33**(4), 579–581 (1995)
15. Cao, X., Lai, S., Lee, L.J.: Design of a self-regulated drug delivery device. Biomed. Microdevices **3**(2), 109–118 (2001)
16. Davis, S.P., Landis, B.J., Adams, Z.H., Allen, M.G., Prausnitz, M.R.: Insertion of microneedles into skin: Measurement and prediction of insertion force and needle fracture force. J. Biomech. **37**, 1155–1163 (2004)
17. Dogangil, G., Ergeneman, O., Abbott, J.J., Pane, S., Hall, H., Muntwyler, S., Nelson, B.J.: Toward targeted retinal drug delivery with wireless magnetic microrobots. In: Proc. IEEE/RSJ Int. Conf. on Intelligent Robots and Systems, pp. 1921–1926 (2008)
18. Doignon, C., Graebling, P., de Mathelin, M.: Real-time segmentation of surgical instruments inside the abdominal cavity using a joint hue saturation color feature. Real-Time Imaging **11** (5-6), 429–442 (2005)
19. Drummond, T., Cipolla, R.: Real-time visual tracking of complex structures. IEEE Trans. Pattern Anal. Mach. Intell. **24**, 932–946 (2002)
20. Ens, J., Lawrence, P.: An investigation of methods for determining depth from focus. IEEE Trans. Pattern Anal. Mach. Intell. **15**(2), 97–108 (1993)

21. Ergeneman, O., Dogangil, G., Kummer, M.P., Abbott, J.J., Nazeeruddin, M.K., Nelson, B.J.: A magnetically controlled wireless optical oxygen sensor for intraocular measurements. IEEE Sens. J. **8**(1), 29–37 (2008)
22. Escudero-Sanz, I., Navarro, R.: Off-axis aberrations of a wide-angle schematic eye model. J. Opt. Soc. A. **16**(8), 1881–1891 (1999)
23. FCI Ophthalmics: (2009) S5.7010 planoconcave lens [Online]. Available: http://www.fci-ophthalmics.com/html/retina.html#lenses
24. Furlani, E.P.: Permanent Magnet and Electromechanical Devices. Academic Press, San Diego, California (2001)
25. Galloway, N.R., Amoaku, W.M.K., Galloway, P.H., Browning, A.C.: Common Eye Diseases and their Management, third edn. Springer, London (2005)
26. Geroski, D.H., Edelhauser, H.F.: Drug delivery for posterior segment eye disease. Invest. Ophthalmol. Vis. Sci. **41**(5), 961–964 (2000)
27. Ghazi, N.G., Noureddine, B.N., Haddad, R.S., Jurdi, F.A., Bashshur, Z.F.: Intravitreal tissue plasminogen activator in the management of central retinal vein occlusion. Retina **23**, 780–784 (2003)
28. Gillies, G.T., Ritter, R.C., Broaddus, W.C., Grady, M.S., Howard, M.A. III, McNeil, R.G.: Magnetic manipulation instrumentation for medical physics research. Rev. Sci. Instrum. **65**(3), 533–562 (1994)
29. Grady, M.S., Howard, M.A. III, Molloy, J.A., Ritter, R.C., Quate, E.G., Gillies, G.T.: Nonlinear magnetic stereotaxis: Three-dimensional, in vivo remote magnetic manipulation of a small object in canine brain. Med. Phys. **17**(3), 405–415 (1990)
30. Gupta, P.K., Jensen, P.S., de Juan, E. Jr.: Surgical forces and tactile perception during retinal microsurgery. In: Proc. Int. Conf. Med. Image Comput. and Comput.-Assisted Intervention, pp. 1218–1225 (1999)
31. Gwb International, Ltd.: (2009) Model Eye (2mm pupil) [Online]. Available: http://www.gwbinternational.com/model_eye.htm
32. Ishiyama, K., Arai, K.I., Sendoh, M., Yamazaki, A.: Spiral-type micro-machine for medical applications. J. Micromechatronics **2**(1), 77–86 (2003)
33. Iwase, E., Shimoyama, I.: Multistep sequential batch assembly of three-dimensional ferromagnetic microstructures with elastic hinges. J. Microelectromech. Syst. **14**(6), 1265–1271 (2005)
34. Jagtap, A.D., Riviere, C.N.: Applied force during vitreoretinal microsurgery with handheld instruments. In: Proc. IEEE Int. Conf. Engineering in Medicine and Biology Society, pp. 2771–2773 (2004)
35. Jiles, D.: Introduction to Magnetism and Magnetic Materials. Chapman and Hall, London (1991)
36. Khamesee, M.B., Kato, N., Nomura, Y., Nakamura, T.: Design and control of a microrobotic system using magnetic levitation. IEEE/ASME Trans. Mechatron. **7**(1), 1–14 (2002)
37. Kirschkamp, T., Dunne, M., Barry, J.C.: Phakometric measurement of ocular surface radii of curvature, axial separations and alignment in relaxed and accommodated human eyes. J. Ophth. Phys. Opt. **24**(2), 65–73 (2004)
38. Kummer, M.P., Abbott, J.J., Kratochvil, B.E., Borer, R., Sengul, A., Nelson, B.J.: OctoMag: An electromagnetic system for 5-DOF wireless micromanipulation. In: Proceedings of the International Conference on Robotics and Automation (2010)
39. Kummer, M.P., Abbott, J.J., Vollmers, K., Nelson, B.J.: Measuring the magnetic and hydrodynamic properties of assembled-MEMS microrobots. In: Proc. IEEE Int. Conf. Robotics and Automation, pp. 1122–1127 (2007)
40. Lakowicz, J.R.: Principles of Fluorescence Spectroscopy, 2nd edn. Kluwer Academic/Plenum Publishers (1999)
41. Leng, T., Miller, J.M., Bilbao, K.V., Palanker, D.V., Huie, P., Blumenkranz, M.S.: The chick chorioallantoic membrane as a model tissue for surgical retinal research and simulation. Retina **24**(3), 427–434 (2004)

42. Leventon, M.E., Grimson, W.E.L., Faugeras, O.: Statistical shape influence in geodesic active contours. IEEE Int. Conf. Computer Vision and Pattern Recognition **1** (2000)
43. M. K. Tameesh *et al.*: Retinal vein cannulation with prolonged infusion of tissue plasminogen activator (t-PA) for the treatment of experimental retinal vein occlusion in dogs. Am. J. Ophthalmol. **138**(5), 829–839 (2004)
44. Mathieu, J.B., Beaudoin, G., Martel, S.: Method of propulsion of a ferromagnetic core in the cardiovascular system through magnetic gradients generated by an MRI system. IEEE Trans. Biomed. Eng. **53**(2), 292–299 (2006)
45. Meeker, D.C., Maslen, E.H., Ritter, R.C., Creighton, F.M.: Optimal realization of arbitrary forces in a magnetic stereotaxis system. IEEE Trans. on Magnetics **32**(2), 320–328 (1996)
46. Mejia-Barbosa, Y., Malacara-Hernandez, D.: A review of methods for measuring corneal topography. J. Opt. Vis. Sci. **78**(4), 240–253 (2001)
47. Mitchell, B., Koo, J., Iordachita, I., Kazanzides, P., Kapoor, A., Handa, J., Hager, G., Taylor, R.: Development and application of a new steady-hand manipulator for retinal surgery. In: Proc. IEEE Int. Conf. Robotics and Automation, pp. 623–629 (2007)
48. Mitra, A.K.: Ophthalmic Drug Delivery Systems. Marcel Dekker Inc., New York (2003)
49. Nagy, Z., Ergeneman, O., Abbott, J.J., Hutter, M., Hirt, A.M., Nelson, B.J.: Modeling assembled-MEMS microrobots for wireless magnetic control. In: Proc. IEEE Int. Conf. Robotics and Automation, pp. 874–879 (2008)
50. Park, J.H., Allen, M.G., Prausnitz, M.R.: Polymer microneedles for controlled-release drug delivery. Pharm. Res. **23**(5), 1008–1018 (2006)
51. Ratner, B.D., Hoffman, A.S., Schoen, F.J., Lemons, J.E. (eds.): Biomaterials Science: An Introduction to Materials in Medicine, second edn. Elsevier Academic Press (2004)
52. Riviere, C.N., Ang, W.T., Khosla, P.K.: Toward active tremor canceling in handheld microsurgical instruments. IEEE Trans. on Robotics and Automation **19**(5), 793–800 (2003)
53. Shahid, H., Hossain, P., Amoaku, W.M.: The management of retinal vein occlusion: Is interventional ophthalmology the way forward? Br. J. Ophthalmol. **90**, 627–639 (2006)
54. Shi, Y., Karl, W.C.: Real-time tracking using level sets. IEEE Int. Conf. Computer Vision and Pattern Recognition **2**, 34–41 (2005)
55. Shonat, R., Kight, A.: Oxygen tension imaging in the mouse retina. Ann. Biomed. Eng. **31**, 1084–1096 (2003)
56. Smith, G., Atchison, D.A.: The eye and visual optical instruments. Cambridge University Press, Cambridge (1997)
57. Snead, M.P., Rubinstein, M.P., Jacobs, P.M.: The optics of fundus examination. Sur. Ophthalm. **36**(6), 439–445 (1992)
58. Sun, Y., Duthaler, S., Nelson, B.J.: Autofocusing in computer microscopy: selecting the optimal focus algorithm. J. Microsc. Res. Tech. **65**(3), 139–149 (2004)
59. Syms, R.R.A., Yeatman, E.M., Bright, V.M., Whitesides, G.M.: Surface tension-powered self-assembly of microstructures—the state-of-the-art. J. Microelectromech. Syst. **12**(4), 387–417 (2003)
60. Tang, W.M., Han, D.P.: A study of surgical approaches to retinal vascular occlusions. Arch. Ophthalmol. **118**, 138–143 (2000)
61. Tjoa, M., Krishnan, S., Kugean, C., Wang, P., Doraiswami, R.: Segmentation of clinical endoscopic image based on homogeneity and hue. IEEE Int. Conf. Engineering in Medicine and Biology pp. 2665–2668 (2001)
62. Tunay, I.: Modeling magnetic catheters in external fields. In: Proc. IEEE Int. Conf. Engineering in Medicine and Biology Society, pp. 2006–2009 (2004)
63. Volk, D.A.: Indirect ophthalmoscopy lens for use with split lamp or other biomicroscope (1998). U.S. Patent 5,706,073
64. White, F.M. (ed.): Fluid Mechanics, 3rd edn. McGraw-Hill Inc., New York (1994)
65. Yang, G., Gaines, J.A., Nelson, B.J.: Optomechatronic design of microassembly systems for manufacturing hybrid microsystems. IEEE Trans. Ind. Electron. **52**(4), 1013–1023 (2005)

66. Yesin, K.B., Vollmers, K., Nelson, B.J.: Modeling and control of untethered biomicrorobots in a fluidic environment using electromagnetic fields. Int. J. Rob. Res. **25**(5-6), 527–536 (2006)
67. Zhang, L., Abbott, J.J., Dong, L.X., Kratochvil, B.E., Bell, D., Nelson, B.J.: Artificial bacterial flagella: Fabrication and magnetic control. Appl. Phys. Lett. **94**(064107) (2009)

Chapter 14
Single and Multiple Robotic Capsules for Endoluminal Diagnosis and Surgery

Arianna Menciassi, Pietro Valdastri, Kanako Harada, and Paolo Dario

Abstract The present chapter illustrates robotic approaches to endolomuninal diagnosis and therapy of hollow organs of the human body, with a specific reference to the gastrointestinal (GI) tract. It gives an overview of the main technological and medical problems to be approached when dealing with miniaturized robots having a pill-like size, which are intended to explore the GI tract teleoperated by clinicians with high precision, flexibility, effectiveness and reliability. Considerations on different specifications for diagnostic and surgical swallowable devices are presented, by highlighting problems of power supply, dynamics, kinematics and working space. Two possible solutions are presented with details about design issues, fabrication and testing: the first solution consists of the development of active capsules, 2–3 cm^3 in volume, for teleoperated diagnosis in the GI tract; the second solution illustrates a multiple capsule approach allowing to overcome power supply and working space problems, that are typical in single capsule solutions.

Keywords Active locomotion · Biomechatronics · Biorobotics robotic surgery · Capsule endoscopy · Endoluminal surgery · Gastrointestinal endoscopy · Robotic endoscopy · Legged locomotion · Meso-scale robotics · Minimally invasive gastroscopy · Modular robotics · Reconfigurable robotics

14.1 Introduction

The current generation of operating robots, which are really master–slave manipulators based on existing industrial prototypes, will be replaced by a second generation that meets more closely the requirements of minimally invasive and endoluminal access therapy and surgery [1].

A. Menciassi (✉)
Scuola Superiore Sant'Anna, Viale Rinaldo Piaggio 34-5602, Pisa, Italy
e-mail: arianna.menciassi@sssup.it

J. Rosen et al. (eds.), *Surgical Robotics: Systems Applications and Visions*,
DOI 10.1007/978-1-4419-1126-1_14,

The need for an advanced way to perform surgery is motivated by the progress of diagnostic techniques (e.g. imaging technologies and screening tests), which allow to discover pathologies at a very early stage and to treat them when they are limited to a small area of the human body or even to few cells.

In this rapid evolution scenario, the most disruptive technologies consisted of self-propelled robots for the investigation of hollow organs of the human body [2], robotic mini-capsules to be swallowed naturally by the patients and teleoperated from outside [3], micro-robots able to perform specific operations inside the human body [4], flexible endoscopes enabling bimanual operations [5].

In all the previous examples, biorobotic technologies have been essential for devising practical solutions, for designing systems centered around humans (i.e. the patient and the medical doctor), and for restoring the link between the operator and the patients, link that has become more and more weaker from traditional surgical robots to current biomedical microrobots.

The authors have been developing for the last decade microrobotic devices intended for a wireless inspection and treatment of the human body, giving priority to the diagnosis and therapy of the gastrointestinal (GI) tract, that naturally allows an endoluminal access, both from the mouth and from the anus. Starting from a more diagnostic and screening approach, based on robotic pills with vision and active locomotion abilities but limited therapeutic functions, they moved recently to a more surgical-oriented approach, that has been demonstrated to be feasible by exploiting multiple capsule solutions.

This chapter intends to give an overview on miniaturized endoscopic and therapeutic devices for the GI tract, from the design to the testing phases. Specifically, the chapter illustrates typical problems in modeling, dimensioning and conceiving micro-robots for endoluminal diagnosis and surgery.

The chapter is divided in four sections: this first section presents also a medical overview of the problems related to diagnosis and therapy in the GI tract. The second section focuses on wireless capsules for diagnostic purposes and simple therapeutic applications, by presenting three different case studies: legged capsules for colon inspection, swimming capsules for stomach diagnosis and clip release capsules for simple therapeutics delivery. The third section illustrates an original approach based on multiple swallowing capsules to be combined and reconfigured inside the GI tract in order to build up a more complicated structure for inspection, therapy delivery, and complete surgical tasks. Finally, conclusions are reported in the last section.

14.1.1 Medical Overview and Organs Features

The GI tract is a long tube from mouth to anus that can reach 8 m in adults. The diameter of this hollow organ ranges from 3 cm in the esophagus, up to 10 cm in the stomach, with an average diameter of the small intestine and large intestine respectively of 4 and 7 cm. Normally, the GI tract is collapsed and each object (i.e. food or device) that is introduced in the GI tract is completely wrapped by intestinal tissues and it is moved forward by peristaltic waves. During traditional endoscopy, tissue dilatation by air insufflation is extremely important in order to obtain a correct distension of the intestinal wall and a complete visualization of the internal mucosa. The development of wireless robotic systems for the inspection and the intervention in the GI tract cannot proceed by neglecting the specific anatomy of the organs. Inspection and intervention require mandatorily the distension of the tissue: this distension can be obtained in the large intestine by mechanical means integrated into a robotic capsule (e.g. small legs, protruding hooks, etc.), but it cannot be easily obtained in the stomach because of its large diameter. On the other hand, studies on human volunteers [Marc O. Schurr, personal communication] have demonstrated that the stomach can be fully expanded by ingesting transparent liquid (e.g. Poly-Ethylene Glycol, PEG), thus dividing the dilatation task from the active locomotion task with major advantages in terms of actuating forces and required energy, as better specified in the following section.

Based on the above considerations, the upper GI tract (UGI tract) and the lower GI tract (LGI tract) can be featured as follows.

The UGI tract possesses a large diameter that cannot be fully expanded by wireless devices without insufflating from outside. On the other hand, it can be expanded by liquid ingestion, thus opening interesting possibilities for developing swimming capsules for diagnostic purposes. In addition, pathologies of the UGI tract and in particular stomach cancers located in the upper side of the stomach (Fundus and Cardia) require more complicated and precise surgical tasks and the operation site cannot be approached by a single capsule with a limited positioning accuracy and limited volume. Consequently, endoluminal surgical robots composed by several reconfigurable capsules can represent a viable alternative to laparoscopic surgery of the stomach.

The LGI tract is more difficult to be investigated by traditional endoscopy (this is true especially for the small intestine). Thus, traditional endoscopic interventions in the LGI tract are quite simple (e.g. polyp resections or biopsy), and they do not involve complex bi-manual operations. In addition, the limited average diameter of the LGI tract opens the possibility to distension solutions based on single capsules or on a maximum of two capsules. Finally, while liquid ingestion can contribute dramatically to stomach dilatation, it cannot contribute to intestine dilatation since all remaining fluids get absorbed in this final part of the LGI. Thus colonic tissue distension must be obtained with different means (e.g. locomotion mechanisms with the twofold task of locomotion and tissue distension).

The size, speed, and safety requirements for any capsule robot are primarily determined by medical considerations. These are obtained from physicians in terms of general objectives. Such objectives can then be taken into account when considering capsule design.

Medical considerations for capsule robot design include:

- *Size*. Ideally, a capsule robot should be small enough to swallow. However, "swallowable" is somewhat challenging to define, because the maximum swallowable size varies from person to person. However, since commercial pill cameras (e.g., the Given Imaging PillCam which is 11 mm in diameter and 26 mm long) have been used in extensive clinical testing, any device that matches their dimensions can be considered swallowable.

- *Speed*. A standard colonoscopy is completed in approximately 20 min to 1 h, while a gastroscopy usually takes 30 min. It is desirable for a locomoting robot to travel through the target organ in a similar time period.
- *Safety*. Contact with the walls of the target organ should cause no more damage than the standard endoscopic procedure.
- *Pain reduction*. Air insufflation during standard colonoscopy causes abdominal pain for the patient. Similarly, the presence of a rigid tube in the oesophagus during gastroscopy is ill tolerated by the patient. For this reason, the capsule should be wireless and have a locomotion system able to propel it without insufflation.
- *Functionality*. Physicians must be able to visually observe the interior of the target organ. More advanced goals include obtaining biopsies and delivering treatments directly.

14.2 Robotic Capsules for Diagnosis and Simple Therapy

This section presents the design and development of capsules for diagnosis of the GI tract, both the LGI tract (the legged capsule, Sect. 14.2.1) and the UGI tract (the swimming capsule, Sect. 14.2.2). In addition, this section introduces also a capsule device able to perform therapy in the GI tract, although in very specific cases and with limited dexterity. This capsule is mainly intended for clip release in case of internal wounds and it will be presented in Sect. 14.2.3. All this capsular platforms will be discussed in Sect. 14.2.4.

14.2.1 Diagnostic Capsule for the Colon: The Legged Capsule

14.2.1.1 Design Considerations

The first problem to approach for developing a capsule for inspecting the colon under an accurate supervision and teleoperation by the endoscopist is to provide it with active on-board locomotion means.

A preliminary analysis of locomotion issues in the gut has been performed by P. Dario et al. in [6]: it has been outlined how effective locomotion in a slippery and deformable substrate – such as the human gut – must take into account the biomechanics of the tissue, which is an extremely compliant, non-linear, visco-elastic material, typically covered by a thick (up to 2 mm) layer of lubricant mucus, with a friction coefficient as low as 10^{-3}. Taking into account the guidelines provided by this preliminary study, a legged locomotion system was investigated [7] as possible solution to the problem of locomotion in the gut. In fact, legged locomotion offers several advantages in terms of:

- Good control of the trajectory thus allowing the capsule to pass over critical areas without touching them.
- Better adaptability to the environment: thanks to legs, the capsule is adequate to propel in anatomically and biomechanically different areas (stomach, small and large intestine) featured by different average diameter.
- Simplified adhesion: by localizing the contact points in small areas (tip of each leg) larger contact pressures can be achieved, thus producing a significant local deformation. In this way high friction coefficient can be reached in the contact points, thus improving the leg lever effect.

According to this analysis, a legged locomotion system for propelling an endoscopic capsule in the LGI tract should possess the following features:

- Two sets of legs, one in the front and one in the rear part of the capsule. The rear set of legs must produce a thrust force for propulsion; the frontal set has the function both to fix the capsule in its position (when the rear legs are retracted) and to steer the capsule when approaching an intestine curve.
- At least one active degree of freedom (DoF), for moving the leg in the longitudinal direction along the capsule body, and one passive DoF at the leg knee, adjusting the leg to the compliance of the tissue. The ability of the capsule to propel in the gut is strictly connected to the number of active DoF of its legs: a larger number of DoF improve the locomotion performance, but reduces the feasibility of a working prototype, due to the complexity of the actuation system and to its low efficiency in miniaturized size.
- A propulsion force large enough to distend the tissue normally collapsed over the capsule body.

The typical force necessary for locomotion has been estimated thanks to simulations based on the theoretical model described in [8]. In fact, in order to accomplish the design of an efficient legged locomotion system, simulations of the environment and of the capsule interacting and deforming such an environment can help very much. Simulations of the legged capsule locomotion have been performed by considering the following issues:

- A capsule with four legs (in the rear part of the capsule) and with a frontal balloon (used for helping the intestine dilatation) of about 3 cm in diameter.
- A tubular, compliant and slippery tissue as walking environment, with an average diameter of 4 cm.
- A *rower* locomotion gait: the rear legs make an arc shaped course performing a spanning angle of about 120° (Fig. 14.1).

In this condition, the simulation has revealed that:

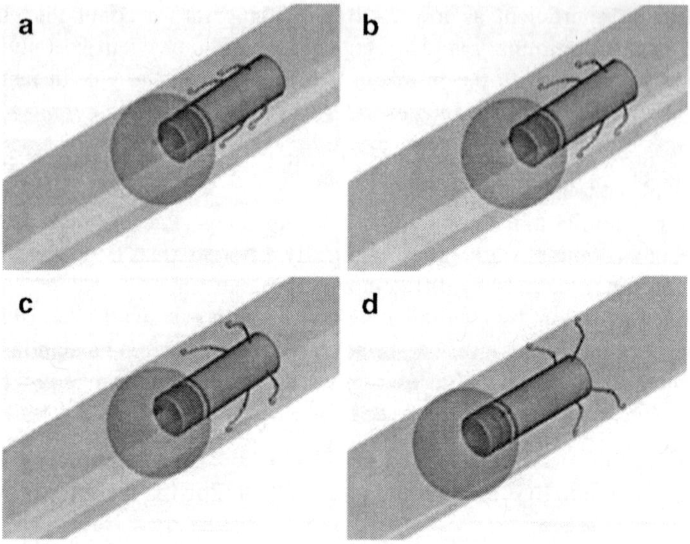

Fig. 14.1 Different positions of the legs in the simulation (© 2008 IEEE), reprinted with permission [11]

- A 0.25 N force per leg is necessary for distending the tissue and propelling the capsule forward.
- In the simulated locomotion an *accordion* effect has been observed due to the lack of a stopping mechanism in the frontal part of the capsule body, which could fix the capsule in a specific point when the rear legs are in the closing phase.

14.2.1.2 Technical Solutions

The development of a legged endoscopic microcapsule for screening the LGI tract is strictly connected to the availability of miniature and reliable actuation systems.

The main features for an actuation system to be integrated in a legged endoscopic capsule are the following:

- The compactness, for not increasing the size of the device whose overall volume must be approximately 4–5 cm^3
- The flexibility in terms of degrees of freedom (each leg should be provided with one independent actuator)
- Be low power, being limited the power stored on-board the capsule and being low efficiency the wireless power transmission
- Be intrinsically safe and biocompatible

By considering these basic requests, the authors initially selected Shape Memory Alloy (SMA) wires, analyzed and modeled as in [9], as first actuation system. Experimental tests revealed that the SMA actuation system is able to propel the capsule only in a non-collapsed gut (i.e. insufflation from outside was necessary in order to reduce the resistance of the tissue to the capsule advancement), because of the low force provided by SMA based actuators at this small scale [10]. Moreover, some problems related to heat transfer during functioning have been observed: heat transfer was too low and, consequently, it was not possible to achieve high frequencies of the gait cycle (no more than ~2.7 cycles/min).

Another problem related to the SMA based version of the capsule was the power consumption: a large amount of energy (6.64 J for cycle) is required to obtain a feasible spanning angle (100°) with an exploitable propulsion force at the leg's tip of only 0.08÷0.09 N. Finally, the SMA based solution was not reliable, making difficult to perform a large number of experimental tests with the same prototype without breaking problems.

For these reasons, a different actuation system for the legged capsule based on one or more traditional electromagnetic motors has been selected [11]. The authors developed a legged locomotion system embedded in a capsule (with a volume of about 4.5 cm^3) and actuated by a brushless minimotor. This device, represented in Fig. 14.2, is provided with four superelastic legs, allowing large stroke advancement in the gastrointestinal tract, and a CMOS frontal camera, for diagnostic purposes.

A set-back of the 4-leg device is the slipping effect at the end of each leg action cycle. This was only partly balanced by the friction enhancers on the surface of the frontal balloon.

In order to avoid this effect, a second frontal set of legs has been considered.

Consequently, the 8-leg endoscopic capsule has been developed that includes two microelectromagnetic motors and a gear system for actuating the legs (four in the frontal part and four in the rear part of the capsule body) to obtain self propulsion inside the LGI tract. The outer dimensions are 12 mm in diameter and 40 mm in length; a camera and an illumination system are included. The capsule is connected to the external world by a small cable that is used for control and for a power supply. All locomotion and visualization functionalities are onboard the

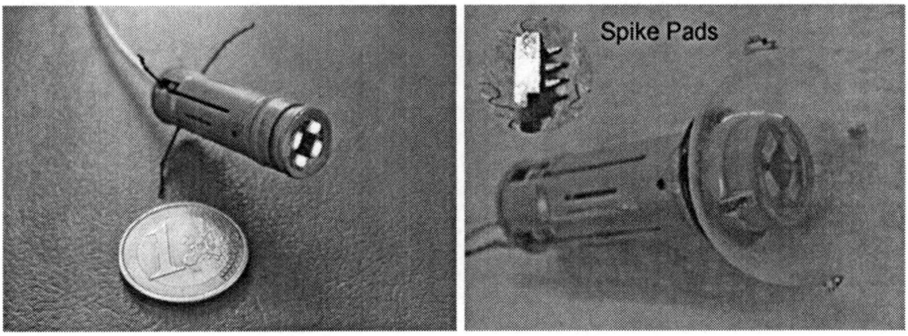

Fig. 14.2 *Left*: overall view of the four legged capsule prototype. *Right*: capsule with frontal balloon and a detail of the spike pads (© 2008 IEEE), reprinted with permission [11]

capsule. Actuation mechanisms are designed to independently open the two leg sets in opposite directions, on the basis of their different functions (Fig. 14.3).

This additional set of legs counteracts the unintended backwards motion of the capsule, as demonstrated in [12] and detailed in the following section.

Despite the successes of the 4-leg and the 8-leg prototypes in demonstrating the promise of legged locomotion, a number of limitations remained. These include dimensions incompatible with swallowing, and some difficulty traversing flexures in the intestine, which we hypothesize was due to both capsule size and leg placement. A new design was developed and described in [13], being it significantly smaller; its mechanical components now match the dimensions of commercial pill-cams. It also has 12 legs which are axially nearer the center of the capsule to enhance turning. Leg tips when open are now also nearly equally radially offset from one another, which aids in tissue distention (improving camera images) and enhances locomotion (providing more evenly distributed points of contact with the intestine). This new 12-leg design, pictured in Fig. 14.4, features two leg sets of six legs each. This design is an important step toward an eventual pill-based

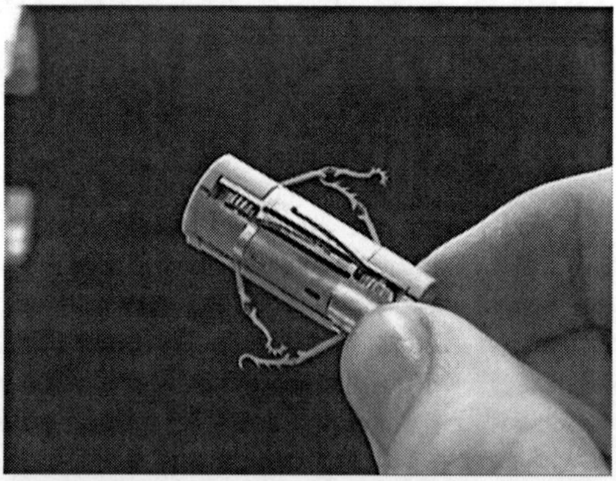

Fig. 14.3 A picture of the complete legged capsule integrating the legs

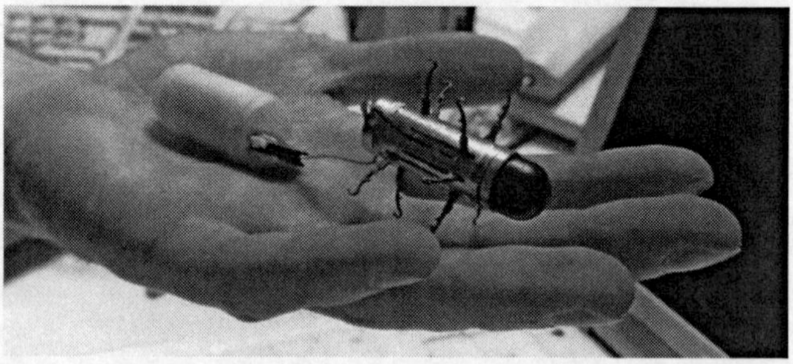

Fig. 14.4 Prototype of 12-leg endoscopic capsular robot. The plastic rear module contains a battery power supply, and will be optimized and miniaturized in future studies (© 2009 IEEE), reprinted with permission [13]

colonoscopy system, mechanically enabling legged locomotion in a pill-sized device for the first time, and including enough legs to uniformly distend collapsed colon tissue, enhancing visualization and locomotion efficiency. Details about fabrication of this complex miniaturized robot are given in [14].

The mechanical module of the capsule alone is 25 mm in length in our design, and the length of the capsule is 29 mm when the end caps and electronics they house are included. If we also consider the volume required by a snap-on vision system for capsular endoscopy, typically 300 mm^3, the total length will be 33 mm. At these dimensions, it is likely that only a certain percentage of the population will be able to swallow the complete device.

Limitations and potentialities of this design are discussed below.

14.2.1.3 Legged Capsule Tests

Four different types of superelastic legs have been designed and tested with the 4-leg prototype, with the objective to identify the best leg configuration for capsule locomotion. Experimental results demonstrate that the device can travel in the digestive tract with a typical speed ranging between 10 and 40 mm/min. Further details are reported in [11].

The feasibility proof of the 8-leg capsule was pursued by extensive testing and by following a rigorous medical protocol. The testing session was organized into repetitive ex vivo trials and in vivo tests. The repetitive tests, represented in Fig. 14.5, were performed for collecting reproducible data in various small series of individual experiments in standardized conditions, thus defining the best locomotion parameters. In vivo tests were performed in a porcine colon: the capsule,

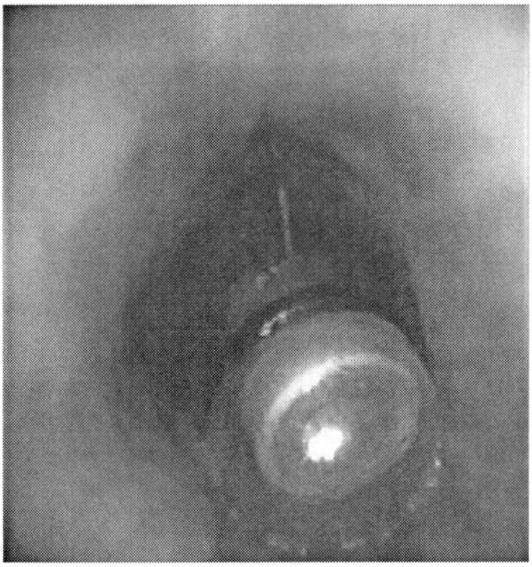

Fig. 14.5 Colonoscopic image of the capsule inside the LGI phantom model

inserted transanally, traveled upward in the oral direction for 15 cm in about 5 min, against peristalsis.

A dedicated set of experiments designed to evaluate the locomotion ability of the 12-leg capsule was performed using a closed, straight ex vivo model. The experimental setup was the same as that used to evaluate previous legged capsule prototypes, namely a fixture capable of holding a tubular structure such as the gut at both ends. The colon specimen was fixed at both ends of the testbed, as shown in Fig. 14.6. This setup also made it possible to adjust the path that the capsule traverses by changing the height and the distance between the fixtures. We used a configuration where the colon was freely suspended with its lumen collapsed. These tests were performed with a wired power supply.

Using this setup, we conducted an experiment to compare tissue distension capabilities of the capsule against those of our previous 8-leg design. Increasing the number of legs and modifying their position so that the leg sets do not align axially, distributes the radial forces more evenly, thus reducing the potential for foot slippage and minimizing possible irritation to the lumen wall. As is clearly visible in Fig. 14.7, the 12-leg capsule (Fig. 14.7b) distended the colon wall in a more uniform manner than the previous 8-legged capsule (Fig. 14.7a).

Another set of experiments was carried out using a lower gastrointestinal tract phantom model, consisting of an anatomical model of the abdominal, chest, and pelvic cavities, with additional accessories for the simulation of organs (e.g. liver, spleen, and sphincter). In addition, the model has fixtures aligned in the shape of human mesentery for the attachment of ex vivo animal intestine. Fresh porcine colon, obtained and preserved as described in the previous subsection, was attached alongside the fixtures. Once fixed, the colon can be set up to simulate typical anatomical characteristics, such as the angles and alignment of the sigmoid curve and the sharpness of the left colonic flexure. We used the human-like large bowel geometry model to verify the capsule capability in realistic human-like conditions, as represented on the left side of Fig. 14.8.

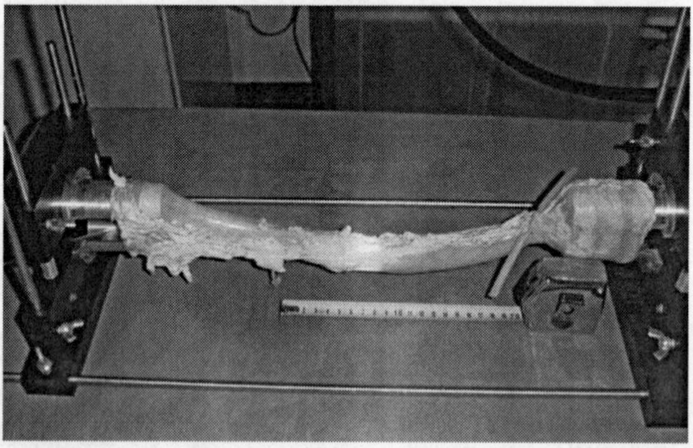

Fig. 14.6 Straight phantom model (© 2009 IEEE), reprinted with permission [13]

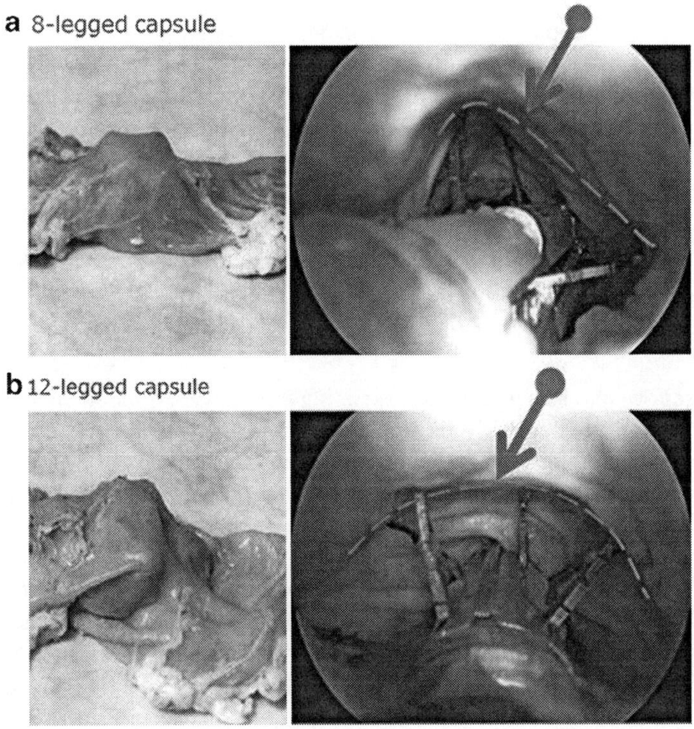

Fig. 14.7 Evaluation of colon wall distension when four (**a**) or six (**b**) legs are included in a single leg set. External views of the capsules in the colon are shown in the *left images*, while endoscopic views of the capsules inside the lumen are shown on the *right* (© 2009 IEEE), reprinted with permission [13]

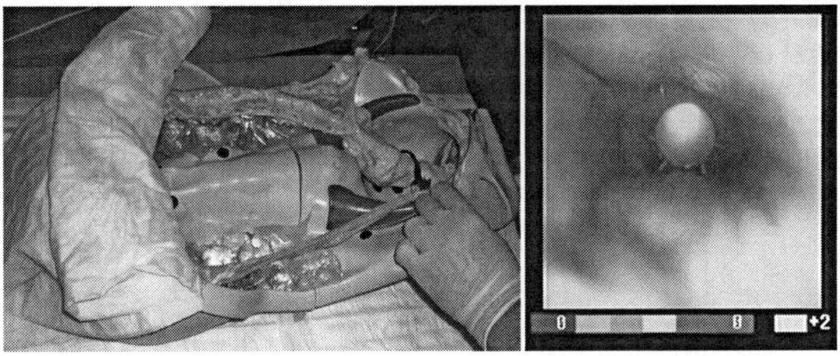

Fig. 14.8 Illustration of the Lower GI phantom model (*left*) and endoscopic view of the capsule during locomotion in the colon (*right*) (© 2009 IEEE), reprinted with permission [13]

The capsule was able to propel itself through all the parts of the lower GI phantom model, including the hepatic and splenic flexures, thus demonstrating improved capability over the previous 8-leg capsule prototype. The capsule crawled with an average speed of 5 cm/min. Considering a mean length of 140 cm for the entire colon, a full passage would take less than 30 min and is thus within the time frame of a standard colonoscopy. An endoscopic picture of the capsule crawling in the colon can be seen on the right hand side of Fig. 14.8.

14.2.2 Diagnostic Capsule: Swimming Capsule for the Stomach

14.2.2.1 Medical Prerequisites

The concept we propose in [15] for distending the stomach is based on an extended bowel cleaning protocol as it is implemented before any regular colonoscopy. A common scheme for preparation comprises the intake of 2–3 l of PEG solution on the evening before the procedure and 1–2 l in the morning [16, 17]. Based on the initial results described in the introduction, a serial trial among healthy volunteers was conducted to evaluate the feasibility of liquid distension with PEG. The study protocol was reviewed by the Institutional Review Board (IRB) and informed consent of the participants was gathered. Eleven subjects underwent a preparation with 2 l of PEG solution on the day before the procedure. Additional 1 l was administered in the morning, 2 h before the examination. Capsule endoscopy was performed using a PillCam SB (Given Imaging, Yoqneam, Israel). After swallowing the capsule, 0.5 l of PEG solution were administered. The volunteers underwent different posture changes to provoke position changes of the capsule. After 30 min, the intake of 0.5 l of PEG solution was repeated. The participants were asked for discomfort on a subjective scale; when discomfort was reported, the procedure was stopped. The feasibility of the procedure was determined by a predefined set of parameters which included scales for wall visualization, debris and recognition of predefined areas of key interest and bubble formation. Of the ten subjects that completed preparation, visualization of the target areas was achieved in nine cases (90%). It was preliminarily concluded that the oral ingestion of 0.5–1 l of liquid may achieve the desired result without significant discomfort for the subject. Based on these findings, PEG solution was considered adequate for this study [18]. Nevertheless, other visible light transparent agents, such as fiber solutions [19], can be investigated in the future in order to improve patient's acceptability.

14.2.2.2 Technical Solution

The use of a liquid to distend the stomach facilitates the development of an active locomotion system for a wireless endoscopic capsule robot. In empty and undistended state, the stomach is collapsed and represents a virtual cavity: thus an endoscopic robot should first distend the tissue and then perform imaging, as described in [20]. Large forces are necessary to achieve effective distension, which requires an amount of power supplied to the device that is impossible to store inside a swallowable volume. Liquid distension of the stomach enables the use of smaller and less energy demanding actuators, used just for locomotion and not for distension.

Given a target video pill size (typically 26 mm in length and 11 mm in diameter, as the Given Imaging SB) and the hydrodynamic properties of PEG solution, viscosity effects are negligible if compared to inertial forces, thus traditional fluidodynamics is applicable to describe the capsule motion. Thus, we designed a locomotion system [21] inspired by submarines, exploiting the high thrust capability offered by propellers at this scale. Furthermore, propellers can be placed at the back of the capsule and embedded within protective structures, thus avoiding protruding parts.

Usually, a propeller is made up of sections of helicoidal surfaces which act together "screwing" through the liquid environment. The number of blades per propeller usually varies from one to five. Since both thrust efficiency on one hand and induced vibrations on the other increase with the number of blades, three-blade propellers are commonly used as a compromise.

A major and undesirable effect of propeller actuation is the induced roll torque. The rotation of a propeller on its own axis induces an opposite rotation of the stator. In an endoscopic capsule which holds the motor, this would induce the capsule to roll, according to the conservation principle of angular momentum. The induced roll torque should be avoided for endoscopic applications, since stable camera images are highly desirable. An effective way to prevent this effect is to activate pairs of counter-rotating propellers. To achieve this goal, particular care must also be devoted to the winding of the single propellers. Having a left winding propeller rotating clockwise and a second one, featuring a right winding, going anticlockwise would balance the roll torque, thus resulting in a net forward propulsion. For this reason the proposed device activates an even number of actuators both during forward motion and steering. Four propellers are active if the capsule has to move forward, while to achieve steering in one direction, the two propellers located on the opposite side must be activated. The schematic in Fig. 14.9 shows that in

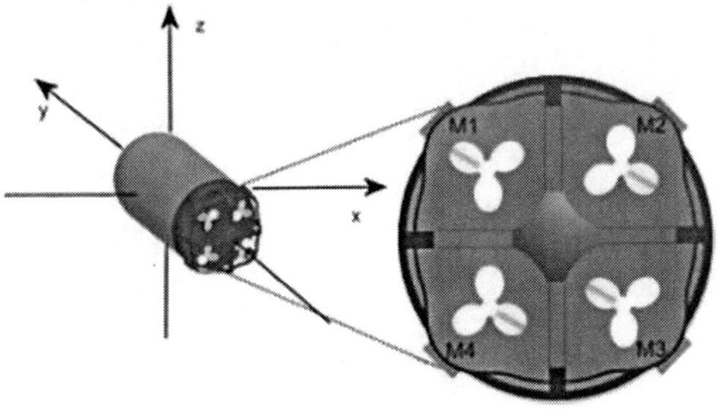

Fig. 14.9 Schematic representation of the capsule and its four propellers (© 2009 Taylor & Francis) [15]

order to propel the capsule upward, actuators M3 and M4 must be operated, while M1 and M2 are off.

The proposed propellers activation strategy allows for a reliable three dimensional locomotion if the capsule has a neutral buoyancy. Furthermore, the capsule is able to hold its position when the propellers are idle, thus saving energy for static target observation. In order to achieve a neutral buoyancy, the Archimedes force and the weight force of the capsule in a fluid must be considered:

$$F_A = \rho_{flu}\, g\, V_{flu} \quad (14.1)$$

$$F_W = \rho_{cap}\, g\, V_{cap} \quad (14.2)$$

where ρ_{flu} is the fluid density and ρ_{cap} is the capsule density, while *g* stands for the gravity acceleration, V_{flu} is the volume of fluid displaced and V_{cap} the volume of the capsule. Three different conditions may occur when the capsule is completely submerged ($V_{flu} = V_{cap}$):

1. $\rho_{flu} < \rho_{cap} \rightarrow$ th capsule would sink.
2. $\rho_{flu} > \rho_{cap} \rightarrow$ the capsule floats on the liquid.
3. $\rho_{flu} = \rho_{cap} \rightarrow$ the capsule is in equilibrium, holding its position.

Therefore the density of the capsule must be properly trimmed in order to equal the fluid density, thereby obtaining a neutral buoyancy for the device. This can be achieved by selecting the proper compromise between capsule weight and volume.

Hardware Overview

The whole system is composed of two main functional units, namely the wireless robotic pill and the human machine interface (HMI), as represented in Fig. 14.10.

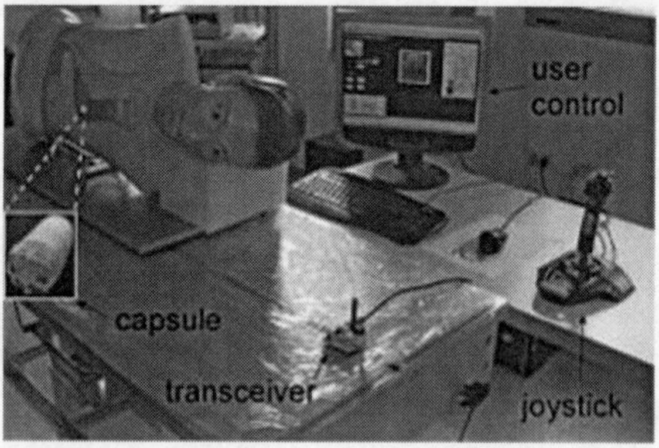

Fig. 14.10 The external unit of the system and the endoscopic capsule located in a upper torso simulator (© 2009 Taylor & Francis) [15]

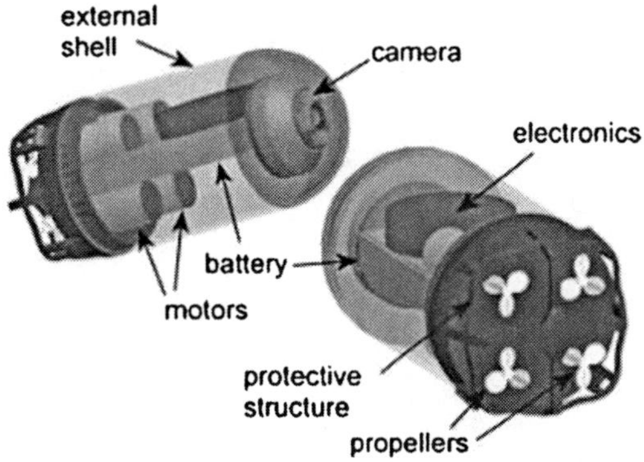

Fig. 14.11 3D sketch of the capsule and its internal components (© 2009 Taylor & Francis) [15]

The endoscopic capsule is composed of several functional sub-modules integrated inside a biocompatible shell. The capsule is wirelessly connected to the user console, where the endoscopist can look at the real-time video stream coming from the capsule image sensor, while controlling its motion by a joystick interface.

The capsular device, represented in Fig. 14.11, is composed by an actuation unit, with four motors each connected to a single propeller, a vision module, an electronic board, including a wireless microcontroller and motor drivers, and a rechargeable battery.

All the aforementioned units were assembled inside a cylindrical shell. The total volume of the current device is 15 mm in diameter and 30 mm in length, with a shell thickness of 0.5 mm. These dimensions can be further reduced down to a fully swallowable size exploiting smaller motors, as detailed below. The shell is composed of three main parts: a middle cylinder, a spherical cup on the front side, to host the vision module, and a rear dome, where the propellers are located. Once those parts were assembled together, all the junctions were sealed with epoxy glue in order to guarantee waterproofing. Rubber rings were used to seal the cavities where the motor shafts are located. The front and the rear domes are functionally and structurally different. The first one is a transparent semi-sphere designed to house a vision system, comprising illumination, optics and camera. The space allocated for the video module is 450 mm^3, currently more than the volume of the PillCam camera (without other components). This space was left empty in the current prototype, since the main purpose here was to design and validate the locomotion unit. A functional video unit will be integrated as a future step. The rear part of the capsule is designed to connect the propellers to the final section of the motor shafts, while preventing potentially harmful collisions between the helixes and the organ walls.

As explained above, pairs of left and right winding three blade propellers were selected for capsule locomotion. The diameter of the single helix is 3.8 mm and each individual blade is 1.50 mm long. This dimensional range is sufficiently large to consider inertial forces dominant over viscous ones, (i.e. high Reynolds numbers).

The outer shell and the propellers were manufactured in acrylic material (VisiJet XT 200) by rapid prototyping (Invision Si2 by Inition). Specifically, the acrylic

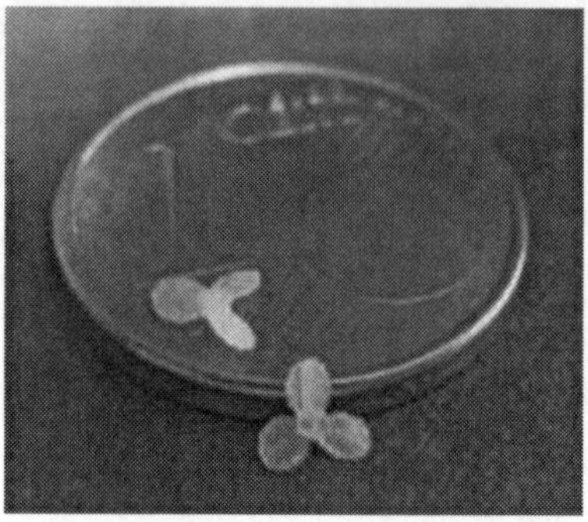

Fig. 14.12 Prototypes of 3-blades propellers (© 2009 Taylor & Francis) [15]

material is composed by urethane acrylate polymer (35–45%) and triethylene glycol dimethacrylate ester (45–55%). To improve biocompatibility of the next prototypes, we will adopt medical grade PEEK (Polyether Ether Ketone), thus also achieving a significative improvement in mechanical properties and structural robustness. The three blade propellers are illustrated in Fig. 14.12.

Due to the low forces required for navigating in a liquid environment, the required torques can also be generated by relatively small electromagnetic motors. Direct current (DC) motors (MK04S-24, Didel) were selected mainly for their small size, low cost and easy operation. Each of the motors is 4 mm in diameter and 8 mm in length with a weight of 0.7 g. Simple on/off control can be implemented with a single digital line per motor.

A wireless microcontroller (μC) (CC2430, Texas Instruments), mounted on a specifically developed circular electronic board (9.6 mm in diameter, 2.3 mm thick, 0.28 g of weight) together with motor drivers, was used to enable bidirectional communication and control of the actuators. The miniaturized board, already tested reliably through in vivo experiments [22], guarantees a very low power consumption and safe levels of electromagnetic energy emission. A 31 mm long whip antenna departs from the board and is embedded in the lateral part of the capsule shell. Motor control is based on pulse width modulation (PWM) technique, thus allowing to vary the speed of each motor by setting the duty-cycle of a digital signal provided by the μC. Proper connections were already developed in the board for vision and illumination control, in order to be used once those subsystems will be integrated in the device. The code implemented on the μC for wireless motor control allows a real-time operation of the device with a minimum refresh time of 0.15 s.

In every wireless active device, powering is a crucial issue both in terms of volume occupied on board and in terms of operative lifetime. Wireless power supply of the swimming capsule was investigated and preliminary results are reported in [23]. This approach is able to provide an infinite operative lifetime, but it requires a bulky external equipment which can hardly be used in a traditional clinical scenario. Thus, we used Lithium Ion Polymer batteries (LiPo batteries), having the highest energy density (200 Wh/kg) available for off-the-shelf

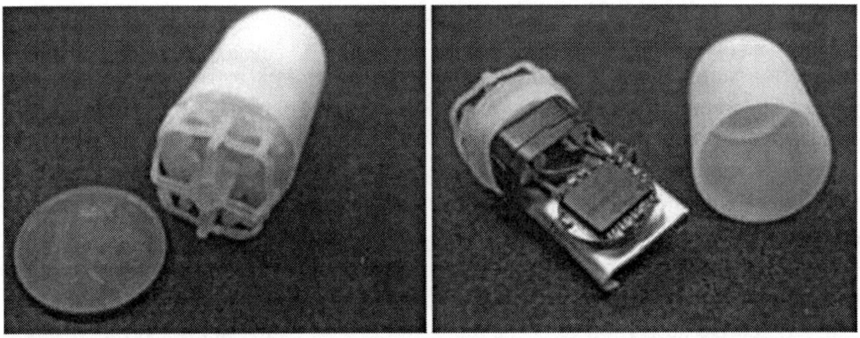

Fig. 14.13 Assembled capsule (*left*); view of the internal components (*right*) (© 2009 Taylor & Francis) [15]

components. In particular we used the LP20 from Plantraco, which is a 3.7 V LiPo cell with a nominal capacity of 20 mAh, a weight of 0.75 g, and very small dimension (17 × 10 × 3 mm). It is able to deliver a peak current of 400 mA, thus allowing to drive all the motors at the same time at full speed. A unipolar magnetic switch (A110× Family, Allegro MicroSystems, Inc) was integrated in the capsule and used together with a permanent magnet to keep the wireless device in an idle state, thus preventing a premature draining of current from the battery. Once the magnet is removed from the capsule, the μC exits from the sleep mode and establishes a wireless communication with the remote console.

The final prototype is reported in Fig. 14.13. As clearly visible from both Figs. 14.11 and 14.13, a considerable amount of free space (1–1.5 cm^3) is available for reducing capsule size towards current swallowable passive devices.

External Console

In order to remotely operate the robotic capsule, a purposely developed external transceiver must be placed nearby the patient, as shown in Fig. 14.10, and connected to a Universal Serial Bus (USB) port of a personal computer (PC). This device is composed by a CC2430EM module from Texas Instruments, allowing bidirectional wireless communication with the capsule, and a USB/serial converter (FT232R from FTDI) to properly interface the telemetric module with the USB port of the PC. Since the external unit has no dimensional constraints, a 2.4 GHz Swivel antenna (Titanis, Antenova) was used to guarantee good communication performances. In terms of total polarization, the radiation pattern of this antenna is almost uniform in all the three Cartesian planes.

In order to prevent interferences among different capsules, each of them is programmed with a unique numerical identifier. The graphical HMI allows to control the desired device by changing the identification number ("Unit ID") inside the panel named "power control". The selected capsule can be placed in idle mode by switching the "power" button. Information coming from the remote device about the battery status ("battery status" panel), the quality of the wireless link ("wireless connection status") and a feedback about the propellers activation are also displayed on screen ("speed and direction indicator "panel). Here, virtual LEDs corresponding to a particular direction (up, down, left, right) turn on when the capsule is moving. A vertical indicator reports the actual propellers speed, in addition to a numerical value given for each actuator. The HMI is also able to manage real-time video streaming, thus allowing the direct view of the gastric walls ("visual feedback" panel). The user can control the capsule motion through a commercial triaxial joystick (Cyborg evo, Saitek), commonly used for flight simulators. Selective activation of propellers is controlled by the main lever, depending on its inclination (e.g. referring to Fig. 14.9, M1 and M2 are activated by pushing the joystick in the forward direction). Furthermore, the speed of the active propellers is proportional to the lever inclination up to a limit that can be adjusted in the "speed settings" panel of the HMI. The main button of the joystick triggers a battery status request, while other functions can be assigned to remaining buttons as well, depending on the peculiar application.

14.2.2.3 Experimental Results

Propellers thrust capability was experimentally evaluated in order to quantify the speed and steerability in a liquid environment consisting of PEG solution [15]. Diagnostic speed (i.e. the normal speed that should be utilized to perform a gastroscopy, thus allowing a good control in a small volume) is between 0 and 5–7 cm/s and can be set from the graphical HMI by setting the PWM control signal to 5–10% of duty cycle.

Traditional gastroscopy performed by the endoscopist using a gastroscope takes several minutes, that are required to advance the endoscopic tube along the oesophagus to stomach and to inspect the tissue (5–15 min, depending on the patients' behaviour and medical doctor's ability, and not considering pre-diagnosis time). An effective inspection of a volume comparable with the distended gastric cavity can be achieved with the proposed device in a similar timeframe, but with a dramatically reduced invasiveness. In terms of battery lifetime, the capsule is able to be actively controlled for more than 30 min at diagnostic speed (1.5 cm/s) with a mean current consumption below 40 mAh. During this time, a complete scan of a volume comparable to a human stomach can be achieved by directing the capsule frontal part towards all the regions of the inner surface of such a volume. Current consumption decreases to less than 1 μAh in power down mode, when the capsule is maintained in idle mode by the magnetic switch. This allows the capsule to be maintained in idle mode until the moment of medical examination, similarly to Given Imaging capsules. Once the gastric inspection is over, the motor can be switched off and the capsule would naturally proceed toward the anus.

Tridimensional locomotion was qualitatively and quantitatively evaluated. Once capsule neutral buoyancy (as visible in Fig. 14.14) was assessed, free swimming

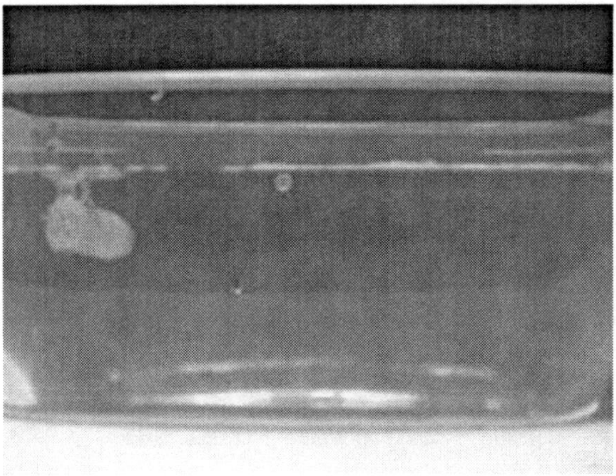

Fig. 14.14 Neutral buoyancy with idle motors (© 2009 Taylor & Francis) [15]

was performed in tanks of different sizes and in flexible low density polyethylene (LDPE) containers, having a volume comparable with a liquid distended stomach (15 × 8 × 8 cm, that is 0.9 l). The liquid medium was PEG solution, according to medical requirements described before.

It was possible to control and orient the capsule head towards all the regions of the inner surface of the tank, running it at 1.5 cm/s speed within 30 min from the beginning of the procedure. For a quantitative evaluation of the steering, the steering radius was assessed. The steering or turning radius of a steerable device is the radius of the smallest circular turn radius that the device is able to achieve. The capsule has a normal tight turning radius varying from 2 to 4 cm, depending on capsule speed, in every direction. Afterwards, several circular targets were placed inside a liquid filled tank (25 × 18 × 11 cm) at different coordinates to quantify tridimensional locomotion accuracy and steerability. The targets (Fig. 14.15) are 25 mm in diameter rings located at three different height stands (43, 65, 80 mm from the basement to the centre of the ring). Few training time is required to have a precise control of the capsule. Ten beginners users were able to move the capsule to any desired area inside the tank after an average of 5 min of use. The mean speed of the capsule during those trials was 1.5 cm/s. Each of the users was able to control the capsule through at least one target, as shown in Fig. 14.16, within 30 min of practice. This kind of test aimed to assess the robustness of control and the steering ability of the device.

Further tests were performed on two 40 kg female pigs. The experiments were carried out in a specialized experimental animal facility, with the assistance and collaboration of a specially trained medical team in compliance with the regulatory issues related to animal experiments.

The capsule was introduced into the stomach through a traditional endoscopic procedure, using a commercially available capsule delivery device (AdvanCE™, US Endoscopy) with a purposely developed distal container, as represented in

Fig. 14.15 Detail of the ring-shape targets and comparison with the capsule (© 2009 Taylor & Francis) [15]

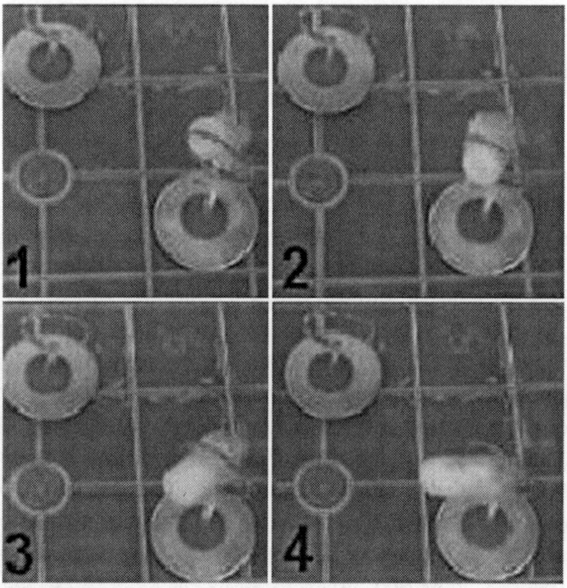

Fig. 14.16 Consecutive frames displaying a complete passage through the medium-height target (© 2009 Taylor & Francis) [15]

Fig. 14.17. After capsule releasing, stomach was distended with liquid PEG, according to the medical protocol.

During the first session, the wireless link was not continuous, thus causing an unreliable capsule control. This occurred mostly when the capsule was moved to the dorsal wall of the stomach, thus maximizing the distance between the capsule and the external receiver. In order to improve capsule performance, the on-board

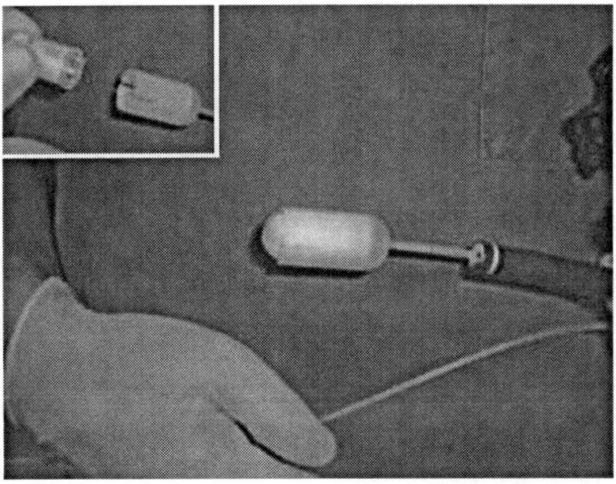

Fig. 14.17 Capsule insertion into the releasing device (© 2009 Taylor & Francis) [15]

antenna was positioned in the centre to prevent contact with the shell and length was doubled, passing from a fourth to half of the wavelength. Then, a reliable wireless link was successfully established. This enabled a real-time propellers activation and control, thus performing different locomotion tasks. The procedure was followed real-time by the endoscope in order to assess the effective controllability of the device towards different areas. The capsule was successfully directed from the pylorus to the cardia at low speed (1.5 cm/s), thus assuring enough stability for images acquisition. The total duration of the procedure was below 30 min. Further experiments will be performed exploiting the direct view from the camera that will be integrated as a future step. Figure 14.18 shows two endoscopic pictures taken during animal experimentation.

14.2.3 Capsule for Simple Therapeutic Tasks: Clipping Capsule

Performing surgical tasks by using a single swallowable capsule is very challenging, consequently we have to define precisely which therapeutic tasks to address, because different therapeutic tasks require different capsule versions. However, only very simple tasks (e.g. biopsy) can be possible given the very limited dexterity of single capsules.

The device proposed in [24], illustrated in Fig. 14.19a, has cylindrical shape, with a diameter of 12.8 mm and a length of 33.5 mm, and embeds an electromagnetic motor, a bidirectional wireless communication platform, and four permanent magnets arranged symmetrically on the external surface. A purposely developed mechanism, actuated by the on-board motor, allows the releasing of a surgical clip once the proper command is issued by the external human machine interface (HMI), running on a personal computer (PC) (Fig. 14.19b). A dedicated transceiver,

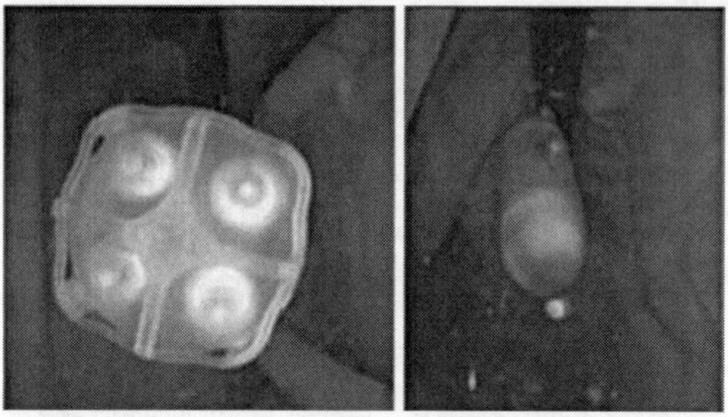

Fig. 14.18 Animal experiments: Four propeller rotating (*left*); frontal-upper view (*right*) (© 2009 Taylor & Francis) [15]

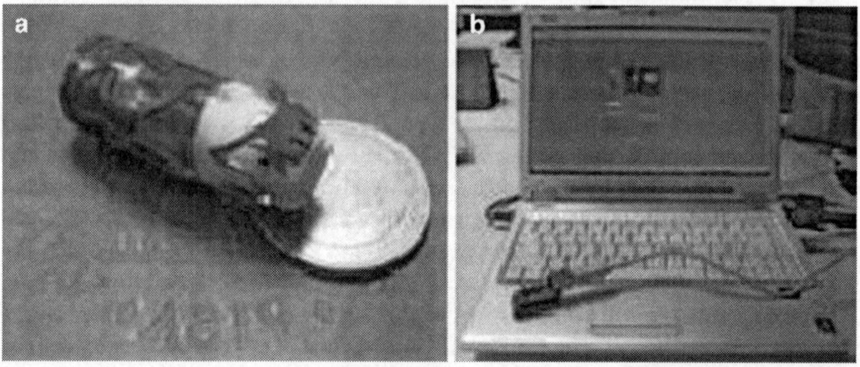

Fig. 14.19 (**a**) A picture of the clipping capsule having the OTSC clip loaded. (**b**) HMI and USB transceiver (*bottom left corner*) allowing wireless communication with the capsule (© 2008 Georg Thieme Verlag KG), reprinted with permission [24]

connected to a standard Universal Serial Bus (USB) port of a PC, allows wireless communication with the robotic capsule.

The surgical clip [25, 26] used in this work ("Over-the-scope" clip (OTSC)), is made of Nitinol, a biocompatible superelastic shape-memory alloy, and it is designed to be placed over the tip of a flexible endoscope without limiting the field of view of the vision system. It is important to underline that a slightly different design of the mechanism actuated by the motor would allow the deployment of different clips [27] or the accomplishment of a different surgical task, such as biopsy or controlled drug delivery, depending on the intended application.

Although the current capsule prototype does not include a vision system, proper space (about 300 mm^3) has been allocated in the capsule front side, in between the clip jaws, for the future integration of a camera module having almost the same volume of the Given Imaging SB1 one [28]. This would not affect the final size of the device.

In the devised procedure, first the clip is loaded onto the capsule, then the device is swallowed by the patient. Once it reaches the target area, the position of the capsule is adjusted toward the precise location to be treated by external magnetic steering. An external permanent magnet (Sintered NdFeBmagnets, B&W Technology & Trade GmbH, China) is currently used for this purpose. It is a cylinder with a diameter of 60 mm and a length of 70 mm, placed on a hydraulic passive arm and controlled manually by the medical operator, in the fashion of an ultrasound probe. The arm allows high precision rotations and tridimensional movements of the magnet in the space surrounding the patient's abdomen. Thanks to this kind of external locomotion, the capsule can be maneuvered through the entire colon with high precision and good reliability.

The proper positioning of the capsule can be assessed thanks to the on board vision system or by external imaging systems, e.g. fluoroscopy. Once the clipping capsule is properly positioned toward the target spot, the releasing command can be issued by the human machine interface (HMI) and wirelessly transmitted to the capsule, which immediately deploys the clip. As soon as this procedure is accomplished, the capsule can be magnetically driven away from the operative scenario and left in the gastrointestinal tract for a natural expulsion.

In a first initial set of ex vivo phantom trials, we assessed the functionality of the wireless therapeutic capsule and the steering ability by means of a magnetic arm. Afterwards, we validated the therapeutic capsule with an in vivo experiment.

14.2.3.1 Experimental Results

Lower GI Phantom Model

The lower gastrointestinal tract (LGI) phantom model is a standard training phantom as illustrated in Fig. 14.20a. It consists of an anatomical model of the abdominal, chest and pelvic cavities with additional accessories for the simulation of organs (e.g. liver, spleen and sphincter). In addition, the model has fixtures aligned in the shape of human mesenteries for the attachment of ex vivo animal intestine. Fresh porcine colon was attached alongside of the fixtures. Once fixed, the colon was set up to simulate typical anatomical characteristics such as the angles and alignment of the sigmoid curve and the sharpness of the left colonic flexure.

As mentioned above, the external magnet was fixed at the end of a massive metal arm-like framework, visible in Fig. 14.20b. The purpose of the arm was to facilitate a simple positioning of the magnet by the medical doctor.

The robotic capsule was prepared for introduction into the colon by using a latex cover in order to secure waterproofing. The capsule was inserted into the sigmoid section of the LGI phantom and steered by means of the external magnet to a specific target, identified by a surgical suture at a distance of 3 cm before the left flexure. After approaching the target, the capsule head was pushed upright into the colonic wall. The pushing manoeuvre was performed by reducing the distance between the external magnet and the capsule. Then the releasing signal was issued

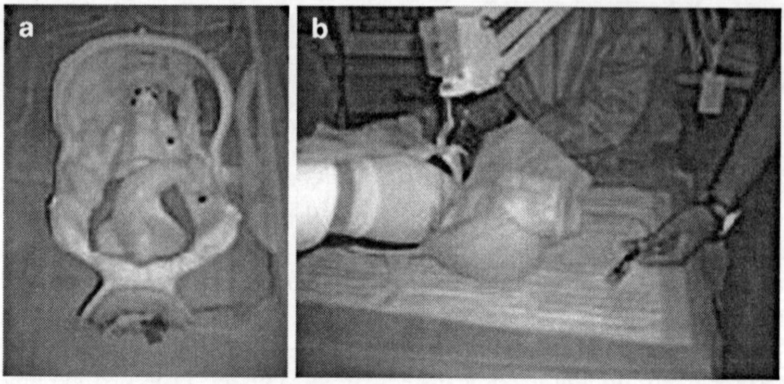

Fig. 14.20 (**a**) LGI phantom model before the experiment. (**b**) LGI phantom model during the experiment, with the external permanent magnet and the transceiver clearly visible (© 2008 Georg Thieme Verlag KG), reprinted with permission [24]

through the HMI and the clip was delivered in place. For guidance and observation, a conventional flexible endoscope (Pentax Medical, 102 Chestnut Ridge Road, Montvale, NJ, 07645-1856 USA) was introduced into the colon and manoeuvred to the suture knot.

The wireless therapeutic capsule was successfully assessed and showed all functions were working. The wireless releasing of clip, locomotion and positioning of capsule was demonstrated in the experiment. The releasing of the clip was instantly performed. The locomotion was effective and fast. The positioning of the capsule was precise and the designated target could be reached. It took from 3 to 4 min to get the desired location, while the positioning of the capsule up straight to the marker was accomplished in a time ranging from 2 to 3 min. Two sessions of five trials each were performed and all clips were successfully released.

In vivo Animal Model

After completion of the phantom trials, the capsule was assessed in an in-vivo experimental session, with the aim of testing the effectiveness of approaching a bleeding in the colon, finely positioning the capsule before releasing the clip, remotely controlling the deployment of the surgical clip, and finally moving the wireless device away from the operating scenario.

The feasibility study was performed with a domestic female 50 kg pig. The experiments were performed in an authorized laboratory, with the assistance and collaboration of a specially trained medical team, in accordance to all ethical considerations and the regulatory issues related to animal experiments. The capsule was observed by using a flexible endoscope, which was maintained quite far from the operative location, in order not to affect positioning and locomotion of the wireless capsule.

After intravenous sedation of the animal and preparation of the bowel by water enemas, the capsular device was inserted transanally up to 15 cm from the anus. A bleeding lesion was induced by using a biopsy grasper (Fig. 14.21a). The capsule was moved towards the target by external magnetic locomotion (Fig. 14.21b). As soon as the capsule was correctly positioned (Fig. 14.21c, d), the releasing command was delivered, the clip was placed (Fig. 14.21e) and the capsule was moved away by magnetic control (Fig. 14.21f).

The capsule locomotion and positioning was as good as in the ex vivo experiment. The clip was released successfully on the desired target. After releasing, the capsule was easily moved away by magnetic locomotion to the rectum. The clip was still in place until the end of the experiment. The amount of grasped tissue is satisfactory, even if lower than in the case the standard endoscopic clip releaser [26] is used. A single in vivo trial was performed: in fact, since our main goal was to assess the feasibility of the proposed approach. Further in vivo tests will be performed once we will have a vision system integrated on the capsule.

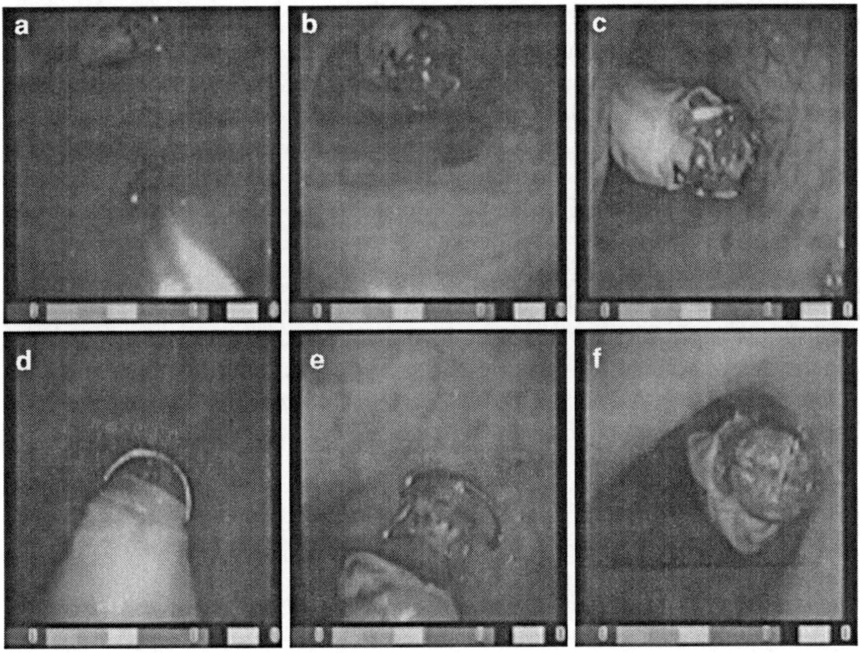

Fig. 14.21 (**a**) Biopsy grasper inducing a bleeding lesion in the colon wall; (**b**) the capsule approaching the target; (**c**) magnetic positioning of the capsule on the target; (**d**) capsule before clip releasing; (**e**) clip was applied; (**f**) capsule without clip moving away from the target (© 2008 Georg Thieme Verlag KG), reprinted with permission [24]

14.2.4 Discussion on Capsules for Diagnosis and Simple Therapy

14.2.4.1 Diagnostic Capsule for the Colon: The Legged Capsule

Power is an important consideration for a capsule robot such as ours. Up to now, the capsule was connected to external power using wires. Eliminating them is an important challenge in future capsule robot design. Average power required for locomotion of our capsule was experimentally measured to be 430 mW, and the vision system we have used on previous capsules requires 180 mW for real-time streaming image transmission at video graphics array (VGA) resolution. Considering a 3.3 V supply, the average current required by the capsule under full load will be 184 mA. This implies that a 100 mAh battery would provide enough energy to complete an entire 30 min colon transit.

Note, however, that this calculation is somewhat conservative, since peristalsis will assist the capsule's motion, and may even be intentionally harnessed during portions of the journey that are not of interest to the desired diagnostic procedure. Commercial batteries having a nominal capacity of 120 mAh would provide a margin of safety and the ability to capture additional images of other parts of the GI tract. The smallest examples of such batteries (e.g., the TLM-1030 from Tadiran, Israel) are generally approximately the same size as the capsule itself (10 mm diameter by 30 mm long). Such a battery is encapsulated in a trailing plastic module shown attached to the capsule in Fig. 14.4. Whether this bimodular power solution will be accepted by both the medical community and patients themselves is an important issue that the authors are currently carefully evaluating.

Another promising potential solution is wireless power delivery, as described in [23]. This approach could be implemented by using an external coil together with three small coils integrated inside the locomotion module. A continuous energy transfer can be used to feed the different subsystems or can be stored onboard in a rechargeable battery or a capacitor.

Another area of potential future research is in gait design. The locomotion gait used in this paper is essentially open-loop motion coordination, where time and angular delays account for dynamic effects in the gait, allowing them to die out before the next step. A dynamic model of foot–colon interaction has the potential to improve capsule locomotion, and will be studied in future work.

14.2.4.2 Diagnostic Capsule: Swimming Capsule for the Stomach

The use of PEG to distend the stomach and the relaxation of constraints for powering and torque, enables the integration of all the required components in a swallowable volume. Inside an external pill-shaped shell, the system comprises four motors and propellers, a wireless μC for control and telemetry, a rechargeable battery and a magnetic switch. The capsule is able to move in all directions inside the stomach under wireless control by the medical doctor. From a dedicated HMI, the endoscopist is able to watch a real-time image stream, coming from the camera that will be integrated on board, and to control the capsule towards interesting areas, by using a triaxial joystick. The HMI also allows to set the essential operative parameters and to get the battery status, the wireless link quality and a feedback about propellers activation. The proposed system was tested in a stepwise approach comprising in vitro, ex vivo and in vivo experiments. In vitro trials consisted of qualitative and quantitative performance evaluation. Freestyle swimming was feasible inside tanks with different sizes and in flexible LDPE containers. High precision of movements can be achieved by the joystick control, enabling a fine locomotion of the capsule through small ring-shaped targets. Preliminary tests on ten subjects were performed to assess HMI intuitiveness. Ex vivo trials confirmed the feasibility of propelled capsule solution, allowing smooth locomotion inside an explanted porcine stomach. Afterwards, animal experiments were performed, and tridimensional locomotion of the capsule inside the stomach was tested. The capsule size was small enough to be safely introduced through the mouth and the esophagus of a 40 kg pig using a readily available endoscopic capsule delivery system. Some problems related to the telemetric link quality were encountered and successfully solved during development and testing phases. Further in vivo tests will be necessary to validate the system once the vision module is integrated on board.

Regarding future work, several steps are still required to have a full functioning device. Further miniaturization of the overall capsule can be achieved by using DC brushless miniaturized motors (SBL02-06, Namiki), 2 mm in diameter and 6.5 mm in length with a weight of 0.12 g, thus reducing the volume of the four motors down to 82 mm^3 rather than 402 mm^3. The control of brushless motors can be achieved with the same control board used for the current prototype. Our next main target is to include a real-time video module on the front part of the capsule.

The imaging system must be light and compact, providing a sufficient depth of focus to observe the entire gastric cavity with a good resolution (VGA) and at high frame rate (at least 20 frames per second to control the active motion, according to medical advices). An adjustable-focus system would enable the acquisition of sharp images regardless of distance between the camera and the target, also allowing a better understanding of depth for control [29]. The imaging module integration will lead to the development of an autonomous and innovative endoscopic system for the stomach. A full and controlled inspection of this peculiar district, at the best of authors' knowledge, is still not covered by any commercial solution for capsular endoscopy. Once the design, including the camera, will be fully defined, different fabrication techniques, such as shape deposition manufacturing (SDM), as in [30], may be investigated in order to improve waterproofing and decrease fabrication costs.

The development of a wireless device for active capsular endoscopy in the stomach holds great promise for improving patient comfort during gastroscopic exams and might thereby increase the number of people undergoing targeted screening programs. The development of such device would pave the way to new perspectives regarding non-invasive diagnostic procedures inside the stomach.

14.2.4.3 Therapeutic Capsule: Clipping Capsule

To the best of the authors knowledge [24], presents for the first time a completely wireless therapeutic capsule for treatment of diseases in the GI tract. In this particular application, wireless clip releasing was demonstrated, however the same technological platform can be easily adapted to other tasks by simply modifying the internal mechanism.

The current version of the capsule (cylindrical shape having diameter of 12.8 mm and length of 33.5 mm) doesn't have a swallowable size yet. However, thanks to a fully scalable design [31], the next prototype will be almost half of the size, thus enabling oral ingestion. Vision system has not been implemented in the current prototype, but the required volume has been allocated in the capsule for future integration without any significant increase in size.

The results reported in this paper look promising for a wireless treatment of bleeding in the GI tract; however we need further investigation to understand if the released clip may have an effective haemostatic capability in real situations.

The proposed device can also be used to mark suspect areas of the GI tract or for NOTES (Natural Orifice Transluminal Endoscopic Surgery) procedures as a port-closure device. For this purpose the capsule must be equipped with tissue manipulation tool in order to approximate the tissue into the clip.

A possible scenario that can be envisaged may take advantage of pre-operatory images, acquired by magnetic resonance imaging or by a tomography scanner, together with robotic steering by the external permanent magnet [32]. In this way the patient would have just to swallow the surgical pill and then a fully automated procedure may be started, thus minimizing invasiveness and discomfort. This approach may have a disruptive potential similar to WCE, since it would enable for the first time capsular surgery by swallowable untethered devices.

Furthermore, the proposed approach would enable to stitch or clip the endoscopic capsule to the wall of the stomach so that prolonged examinations of bleeding ulcers or varices become possible. Long-term endoscopy with wireless endoscopes attached to the wall of the gut seems an obvious way to improve the management of gastrointestinal bleeding and other disorders, as envisaged in [33].

14.3 Multi-Capsule System for UGI Tract Internal Surgery

The single capsule approach is being studied mainly for the pathologies in the LGI tract and the diagnosis in the UGI tract. In order to enhance the dexterity of the conventional endoscopic tools and perform screening and interventions in the UGI tract (especially in the stomach), many advanced endoscopic devices have been developed worldwide. The new surgical procedure called Natural Orifice Translumenal Endoscopic Surgery (NOTES) is also accelerating the development of innovative endoscopic devices [4, 34]. These devices are promising for the future development of minimally invasive and endoluminal surgery. However, advance endoscopes with some miniaturized arms have poor dexterity; the diameter of such arms must be small to be inserted through an endoscopic channel (2–4 mm in general), resulting in a small force generated at the tip.

Based on the above motivations, a reconfigurable modular robotic system is proposed to overcome the intrinsic limitations of a single-capsule or endoscopic approach [35]. In the proposed system, miniaturized robotic capsules are ingested and assembled in the stomach cavity, and the assembled robot can change its configuration according to the target location and target task. The multi-capsule approach therefore facilitates the delivery of more components inside a body cavity that has small entrance and exit, i.e. stomach. The robotic reconfiguration allows the precise positioning of the vision and therapeutic tools which can cooperate to perform complicated surgical tasks.

The multi-capsule modular surgical robot is interesting as a feasible application of self-reconfigurable modular robots as well as an innovative surgical robot. After the first modular robot was introduced in 1988 [36], many self-reconfigurable modular robots have been studied worldwide [37, 38]. The objective of the self-reconfigurable modular robots is to be robust and adaptive to the working environment. Therefore, most of them are designed for exploration or surveillance tasks, thus having no strict constrains of the number of modules, modular size and working space. So far, there have been no reconfigurable modular robots for surgical use reported in the literature to the best of the authors' knowledge. Having a specific application and the GI tract environment, the proposed system raises many issues that have not been discussed well in the modular robotic field. For example, the modular miniaturization down to the ingestible size is one of the most challenging goals for the medical application.

14.3.1 Clinical Setting and Constraints

The clinical target for the multi-capsule surgical robotic system can be the entire GI tract, i.e. the esophagus, the stomach, the small intestine and the colon. Among the many GI tract pathologies that can benefit from the multi-capsule features, biopsy of early cancer on the upper side of the stomach (the Fundus and the Cardia) was defined as the surgical task to be focused on.

Stomach cancer is the second leading cause of cancer death worldwide [39] and stomach cancer located on the upper side of the stomach has a worse outcome in terms of the 5-year survival ratio [40]. Thus, early diagnosis of the cancer utilizing an endoluminal robotic device may lead to a better prognosis. In addition, this target is good for demonstrating the advantages of the multi-capsule modular approach. The large space in the stomach, whose volume is 1,400 ml when distended, provides the working space to assemble the ingested robotic capsules and change the topology of the assembled robot inside (i.e. self-reconfiguration). The assembled robot can change its topology to reach the upper side of the stomach, while a single capsule is not capable of reaching this district due to the absence of this kind of locomotion ability. Having multiple functional capsules such as a camera and forceps, the assembled robot may perform precise surgical tasks that a single capsule endoscope is not capable of conducting.

Concerning the clinical constraints, each capsule should be small enough to be swallowed and pass through the whole GI tract as described in the specifications for a single capsular device. Each module should be minimized down to the size of commercial endoscopic capsules (11 mm in diameter and 26 mm in length) to be used in the clinical cases.

14.3.2 Robotic Scheme

Because the proposed multi-capsule surgical robotic system has new features that have rarely been studied in surgical robotics or modular robotics, two robotic schemes are being proposed and investigated in parallel: the first one is the homogeneous scheme [41] where all modules are identical except for one or two surgical or diagnostic modules (Fig. 14.22a). In this scheme, the identification and control of each module can be allocated after all modules are connected in any sequence. The advantage of this scheme is the simplicity in assembly, while the disadvantage is that it is effective only for simple tasks. The second robotic scheme is the heterogeneous scheme where each module can be different (some of them can be identical as well). As an example, Fig. 14.22b illustrates modular topology consist of a central branching module, structural modules and functional modules for diagnosis and/or intervention. The assembly process for this robotic scheme is more difficult because the assembling sequence should be realized as planned. On the other hand, this scheme may provide more dexterous manipulation to achieve complex surgical tasks.

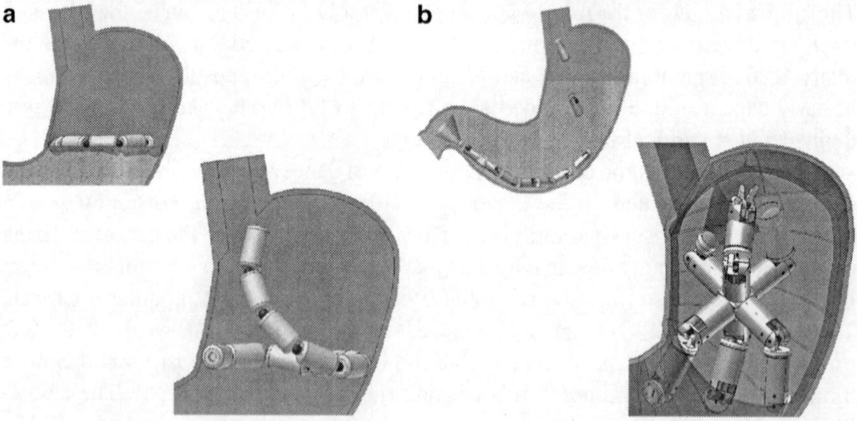

Fig. 14.22 Multi-capsule robotic scheme (**a**) homogeneous scheme (**b**) heterogeneous scheme

The design, prototyping and evaluation of the robotic modules only for the heterogeneous scheme are described below.

14.3.3 Proposed Surgical Procedures

Figure 14.23 shows the surgical procedures for the proposed multi-capsule surgical robot. Prior to the surgical procedure, the patient drinks a liquid to distend the stomach. Next, the patient ingests 10–15 robotic modules, and then the swallowed modules complete the assembling process before the liquid naturally drains away from the stomach, which is in 10–20 min. Magnetic self-assembly using permanent magnets in a liquid has been chosen since its feasibility has been demonstrated in literature [42]. Soon after the assembly, the assembled robot configures its topology as planned based on a preoperative planning. Self-reconfiguration can be achieved if necessary by repeated docking and undocking of the modules (the undocking mechanism and electrical contacts between modules are necessary but they have not been implemented yet in the presented design). The assembly, the robotic configuration and the surgical tasks are controlled via wireless bidirectional communication with an external console operated by the surgeon. Additional modules having different interventional functions and/or modules containing an extra

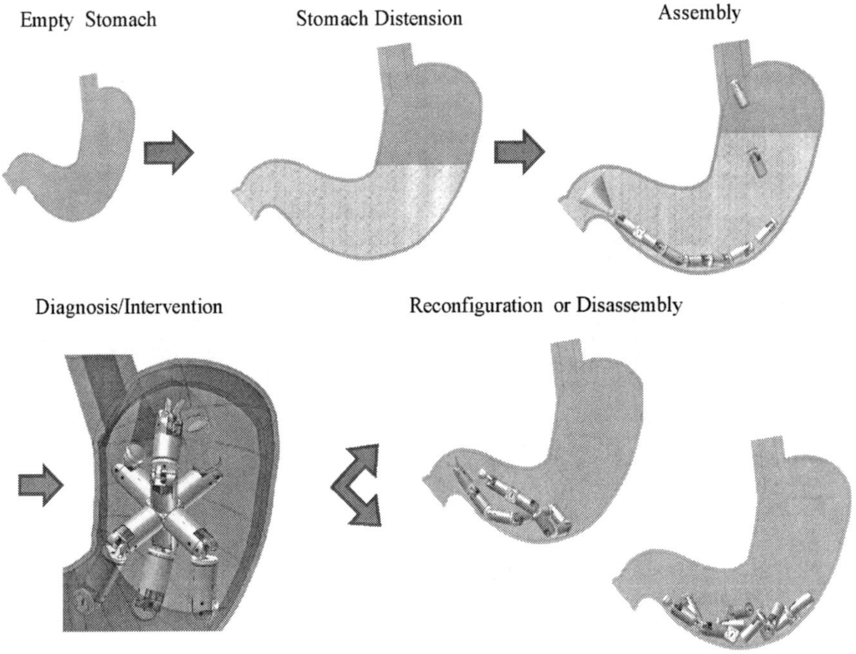

Fig. 14.23 Procedures for the multi-capsule surgical robotic system

battery can be added later to the robotic structure even during the operation. The modules after use can be detached and discarded if it's not necessary in the following procedures. In the same way, the module can be easily replaced with the new one when it is broken. After the surgical tasks are completed, the robot reconfigures itself to a snake-like shape, for example, to pass through the pyloric sphincter and travel in the LGI tract, or completely disassembles itself into individual modules. One of the modules can bring a biopsy tissue sample out of the body for the detailed examinations afterwards.

14.3.4 Design and Prototyping of the Structural Module

Figure 14.24 shows the design of the structural module that the authors have conceived for the heterogeneous scheme. The module has two DOFs (±90° of bending and 360° of rotation). It contains a Li-Po battery (20 mAh, LP2-FR, Plantraco Ltd., Canada), two brushless DC motors of 4 mm in diameter and 17.4 mm in length (SBL04-0829PG337, Namiki Precision Jewel Co. Ltd., Japan) and a custom-made motor control board [43] capable of wireless control.

Since the available motor driver board for the selected motor (SSD04, Namiki Precision Jewel Co., Ltd., 19.6 × 34.4 × 3 mm) has an inappropriate size for the ingestible device, a new control board equipped with wireless control was designed and developed in-house as described above. The implemented algorithm enables the control of the selected brushless DC motor in Back Electro-Motive Force (BEMF) feedback mode or slow speed stepping mode. When the stepping mode is chosen, the selected motor can be driven with a resolution of 0.178°.

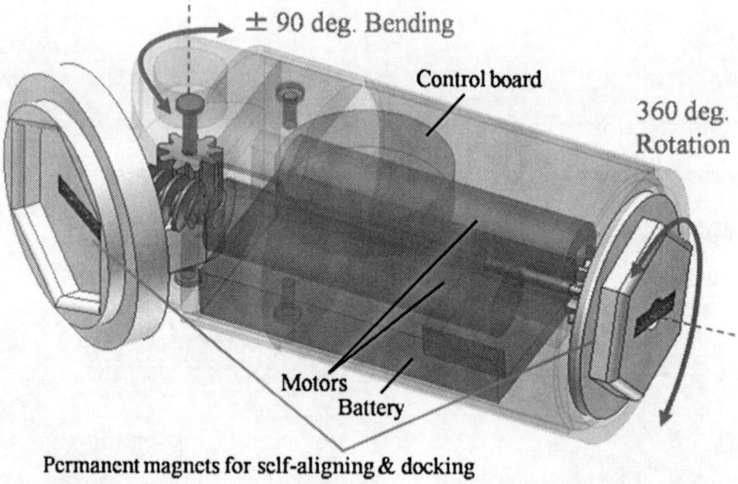

Fig. 14.24 Design of the structural module (© 2009 IEEE), reprinted with permission [48]

The battery capacity carried by each module may differ from one to another depending on the available space inside the module from 10 to 50 mAh; a 20 mAh Li-Po battery was chosen for the presented design. Continuous driving of the selected motor on its maximum speed using a 20 mAh Li-Po battery was shown to last up to 17 min. One module does not withdraw power continuously by turning on its motors all the time because the bending and rotation mechanisms can maintain their position when there is no current to the motor thanks to the high gear reduction of the motor (337:1) and non-backdrivability of the worm gear used for the bending mechanism. A module consumes more power during actuation but very less while it is in stand-by mode.

The stall torque of the selected brushless DC motor is 10.6 mN m and the speed is 112 rpm controlled by the above mentioned controller. The bending mechanism is composed of a worm and a spur gear (9:1 gear reduction), whereas the rotation mechanism is composed of two spur gears (no gear reduction). All gears are made of Nylon and were purchased from DIDEL (DIDEL SA, Switzerland) to shorten the time necessary for prototyping. The width, the length and the diameter of the central holes of the gears were modified by additional machining afterward.

Two permanent magnets (Q-05-1.5-01-N, Webcraft GMbH, Switzerland) are attached at each end of the module to help with self-alignment and modular docking. The hexagonal shape at each end restricts the rotational motion after being docked. The adequate choice of the magnets is being investigated to maximize the possibility of self-assembly and force to maintain docking.

The module dedicated for docking/undocking using permanent magnets is under development [44], intended to be integrated in the modular scheme. Implementing a mechanical docking/undocking mechanism to the current module is also being investigated for rigid docking. The electrical connection between modules is necessary for reconfiguration and it will be implemented in the future.

The size of the fabricated prototype is 15.4 mm in diameter and 36.5 mm in length, that is still to be miniaturized, and the total weight is 5.6 g. Figure 14.25 shows the components and a fabricated prototype. The casing is made of acrylic plastic and it was fabricated by a 3D printing machine (Invison XT 3-D Modeler, 3D systems, Inc., USA).

14.3.5 Design and Prototyping of the Biopsy Module

Figure 14.26 shows the design and the prototype of the biopsy module used as a functional module. The grasping mechanism has a worm and two spur gears, allowing wide opening of the grasping parts. The grasping parts can be hidden in the casing to avoid tissue damage during ingestion (Fig. 14.26 Left). The motor and other components used for this biopsy module are same as the abovementioned structural module.

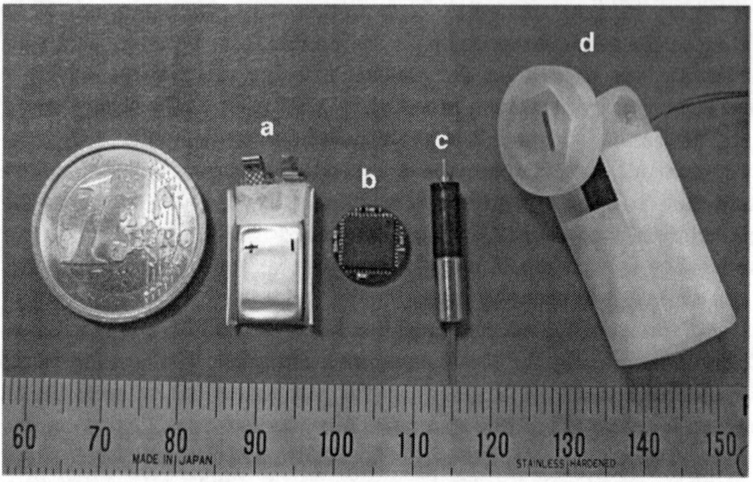

Fig. 14.25 Components of the prototype ((**a**): Li-Po battery, (**b**): control board, (**c**): motor, (**d**): casing) (© 2009 IEEE), reprinted with permission [48]

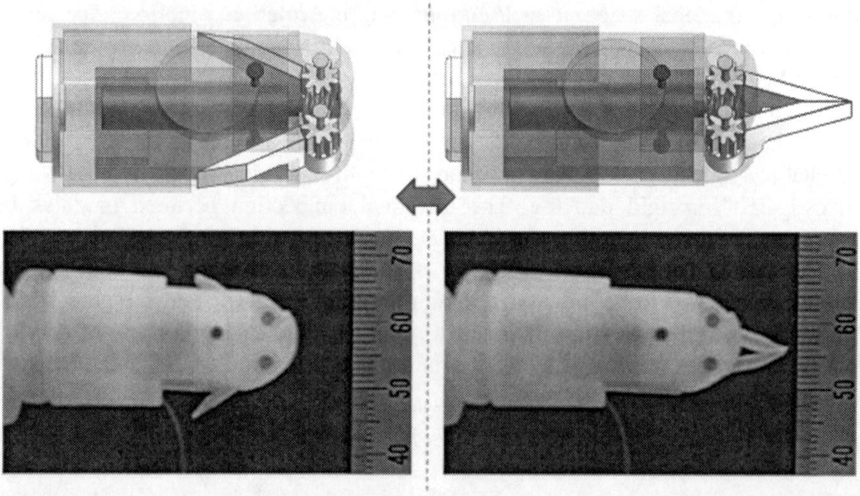

Fig. 14.26 Design and prototype of the biopsy module (*Left*: for ingestion, *Right*: for operation) (© 2009 IEEE), reprinted with permission [48]

This biopsy module can generate more force and grasp bigger tissue samples compared to conventional endoscopic forceps thanks to its bigger diameter. This is one of the big advantages that only this multi-capsule approach can provide.

In the conventional endoscopy, forceps are inserted through endoscopic channels that are parallel to the direction of endoscopic view. On the contrary, the biopsy module can be positioned relatively to the endoscopic view. This feature enhances operability of the functional modules.

14.3.6 Assembled Robot

Possible topologies using a set of 12 modules are shown in Fig. 14.27. This example of the assembled robot consists of one central branching module, one camera module (to be developed), one biopsy module, one storage module for biopsy tissue sample, and eight structural modules. Using the same set of the modules, the assembled robot can configure itself into different topologies.

Central branching modules can facilitate a topology having more than one arm, and a functional module either for diagnosis or intervention can be attached to each arm. These arms with different tools can cooperate to perform a complicated surgical task, resulting in enhanced dexterity of the assembled robot.

14.3.7 Testing and Experiments

The performance of the bending and rotation DOFs of the structural module were measured using a protractor as illustrated in Fig. 14.28. In Fig. 14.29, the bending angle was changed up to ±90° with steps of 10° for three times in succession. The

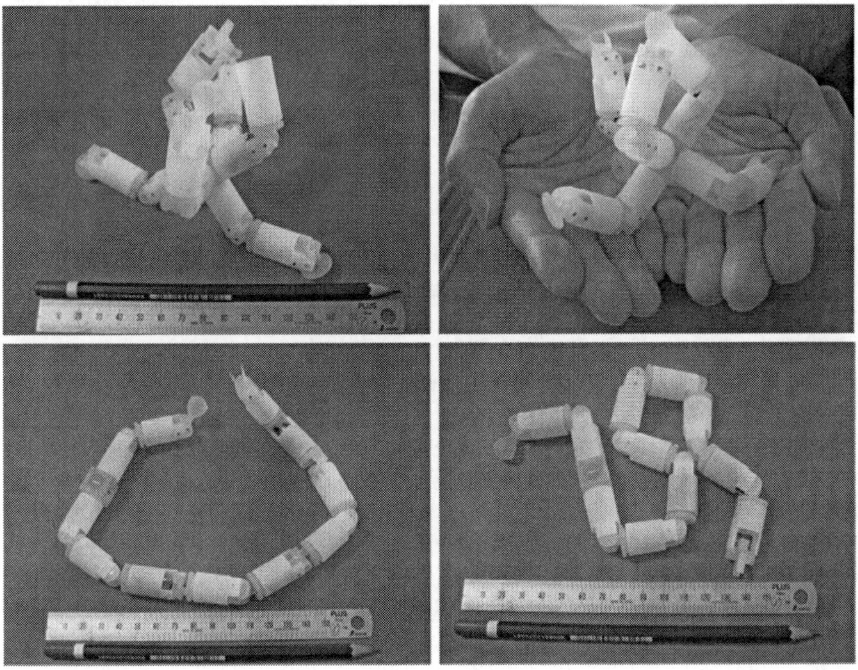

Fig. 14.27 Assembled modules and possible topologies (© 2009 IEEE), reprinted with permission [48]

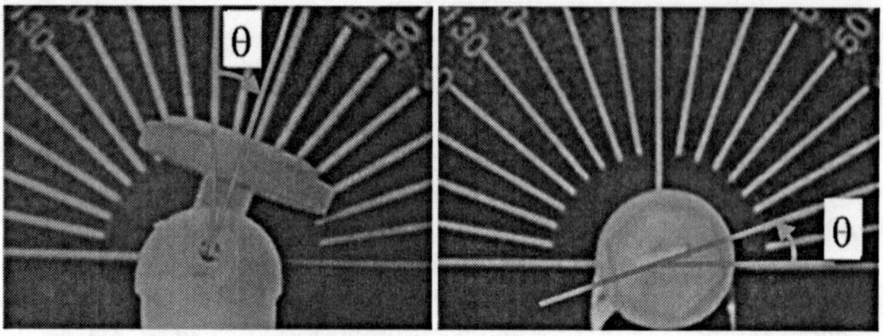

Fig. 14.28 Experimental setup for the angle measurement (*Left*: bending angle, *Right*: rotation angle) (© 2009 IEEE), reprinted with permission [48]

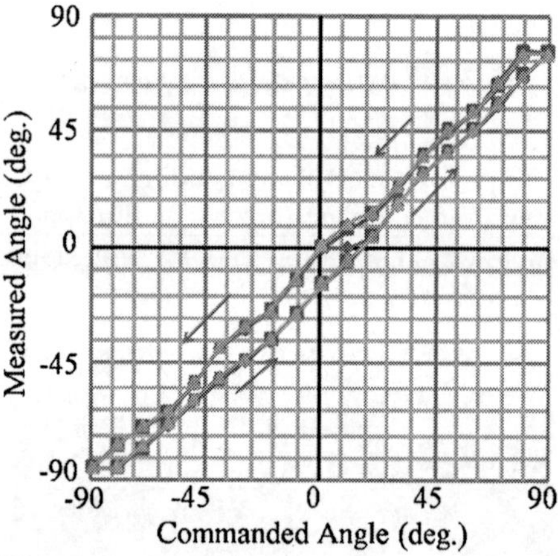

Fig. 14.29 Bending angle (±90° with steps of 10° for three times in succession) (© 2009 IEEE), reprinted with permission [48]

measured range of the bending angle was from −86.0 to +76.3° and the maximum error was 15.8°. The rotation angle was increased from 0 to 180° with steps of 45° for three times in succession as shown in Fig. 14.30, and the measured range of the rotational angle was from 0 to 166.7° with a maximum error of 13.3°.

The difference between the driven angle and the measured angle is due to the backlash of the commercial Nylon gears. The casing is made of acrylic plastic and fabricated by 3D printer, resulting in low precision in assembly. Regardless of the errors and the hysteresis shown in Fig. 14.29, the repeatability is high enough for the intended use for both DOFs. These results indicate that the accuracy of each motion can be improved by changing the materials of the gears and the casing. Since the motor itself can be controlled with a resolution of 0.178°, more precise surgical task can be achieved by adequate choice of the materials and better fabrication.

In addition to the angle measurements, both bending and rotation torque were measured. The maximum torque was measured by attaching cylindrical parts with permanent magnets one by one at each end of the module (Fig. 14.31) until the

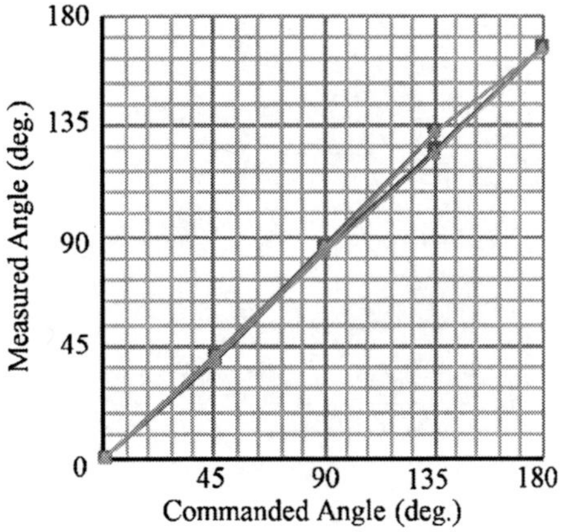

Fig. 14.30 Rotation angle (180° with steps of 45° for three times in succession) (© 2009 IEEE), reprinted with permission [48]

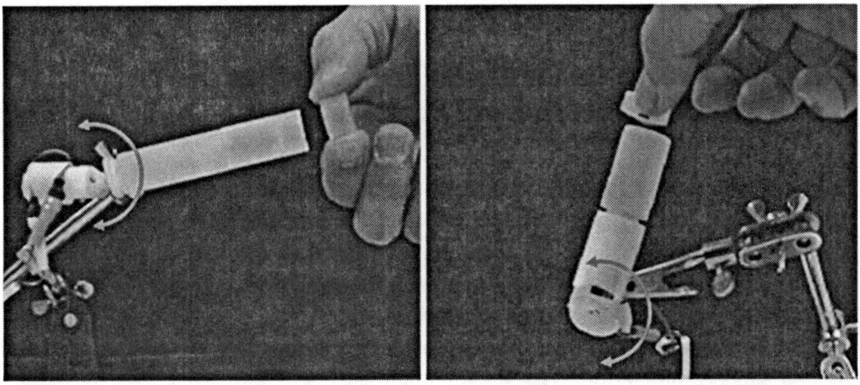

Fig. 14.31 Torque measurement (*Left*: bending, *Right*: rotation) (© 2009 IEEE), reprinted with permission [48]

bending/rotational motion is stopped. The measured maximum bending torque is 6.5 mN m and the maximum rotation torque is 2.2 mN m. The torque measurement includes qualitative tests for assessing the ability of combined modules to move each other and to generate adequate forces for reconfiguration tasks. Figure 14.32 shows one module lifting up two modules attached to its bending mechanism. Torque of 4.5 mN m was required to perform this task.

The generated torque, that is very important for self-reconfiguration and surgical tasks, was big considering the robotic size but it was limited due to the fabrication problems. The thin walls of the casing made of acrylic plastic were easily deformed and caused friction between the parts. The use of the commercial Nylon gears and additional machining on them account for less accuracy of positioning and low gear transmission efficiency. The casing using metal or PEEK and tailor-made metal gears with high precision would improve the mechanical rigidity and performance.

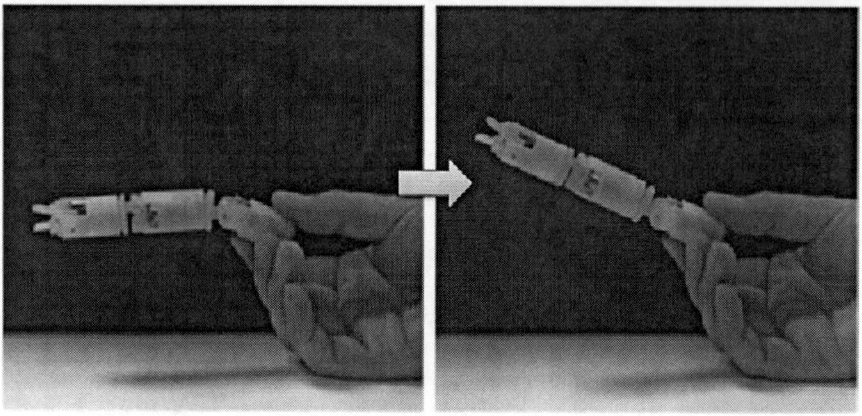

Fig. 14.32 Lift up of the two modules (© 2009 IEEE), reprinted with permission [48]

14.3.8 Discussion on the Multi-Capsule System for UGI Tract Internal Surgery

Multi-capsular surgical robotic system has been proposed for endoluminal surgery and the design, prototyping and evaluation of the modules have been reported in this section. Although there are some issues related to fabrication problems, the initial results show that the modular approach for endoluminal surgery is promising as a surgical device in the next generation.

The advantages of multi-capsule approach are summarized as follows. The one advantage is customization to each patient and each surgical target. The size of the GI tract, location of the target area and required tools for diagnosis/intervention differ from each patient. Based on pre-operative images, required number of modules, optimum topology and work space of the assembled robot can be calculated to be customized to each patient. As a configuration planner, interval analysis has been studied to find the suitable topology that the modules can assemble to, given the task requirements [45]. The topologies can be customized even during the operation by self-reconfiguration in the GI tract. If there are more than one target locations in the GI tract (for example, biopsy from two locations in the stomach), the robot can change its topology to be optimized to the target task and working district, also by changing the functional robotic modules. Another big advantage accounts for the size and number of the functional devices that can be brought into the GI tract. The diameter of the functional robotic module can be same as that of the conventional endoscope and it is far bigger than the diameter of an endoscopic forceps (2–4 mm). The function of endoscopic device is usually limited by the size constraints, but in the multi capsule approach, the volume of one capsule can be dedicated to each device and several devices can be used at one time when central branching modules are used. This allows the assembled robot perform complicated interventional tasks that cannot be done with endoscopic devices. Redundancy is also good features of the multi-capsule approach. If the assembly fails, another robotic module can be added until the planned topology can be assembled. If the battery is run out or a robotic module is broken, new or additional module can be added to the robot even during the operation. This feature enhances the reliability as a surgical system.

Although the feasibility of the multi-capsule approach has been confirmed, there are many things to be improved. Miniaturization of the robotic modules and improved fabrication using different materials are the first priorities. Electrical contacts between modules are necessary and reliable docking/undocking mechanisms need to be developed. Reliable self-assembly, pre-operative planning, power management, localization, intuitive interface device, on-board sensors are key technologies necessary for the system integration.

The surgical system utilizing multi-modular approach is very challenging, but the advantages that the patient can benefit from are obvious. Further development of this system would lead to scarless screening and intervention with high precision and dexterity.

14.4 Conclusions

This chapter presented several microrobotic solutions for diagnosis and intervention in the GI tract, by paying attention to the different possible tasks to be performed and to the different anatomies of the organs to be inspected or treated. Based on these considerations, the adequateness and usability of single or multiple swallowable capsules have been discussed. For both approaches (i.e. single and multiple capsules), the possibility to swallow the single pill or the multiple assembling pills in sequence is a unique prerequisite distinguishing our endoluminal approach to diagnosis and therapy. While single capsule prototypes for diagnosis – based on active locomotion systems allowing teleoperation and controllability by the operator – are already at a good level of engineering and testing, the reconfigurable surgical system based on multiple assembling capsules is at the proof of concept level, although some modules for swallowable capsules have been already dimensioned, designed and tested.

On the other hand, one of the major limitations of endoluminal capsules for diagnosis and endoluminal reconfigurable robots for internal surgery is power supply.

Current research activities are focusing on hybrid approaches for diagnostic capsule locomotion, thus merging the advantages of external magnetic propulsion of endoluminal devices with the advantages of miniaturized low power mechanisms for local management of collapsed tissues.

Also for endoluminal reconfigurable surgical robots, a more practical solution – applicable in a real medical scenario – could be constituted by the combination of small chains of pre-linked internal modules with many degrees of freedom and with a wired connection to the external world; this solution could guarantee more safety for the patient and more flexibility for different operating scenario. The possibility to help the robot stability by external magnetic forces is also an interesting opportunity to be explored in this challenging medical field [46, 47].

Acknowledgments This work was supported in part by the European Commission, in the framework of the ARES (Assembling Reconfigurable Endoluminal Surgical system) and VECTOR (Versatile Endoscopic Capsule for gastrointestinal TumOr Recognition and therapy) European Projects, and in part by the Intelligent Microsystem Center (IMC-KIST, Seoul, South Korea) in the framework of the OPTIMUS project. The authors would like to thank Professor Alfred Cuschieri for his medical consultancy. The authors are grateful to Dr. E. Susilo and Ms. S. Condino for their invaluable technical support and Mr. N. Funaro for the manufacturing of the prototypes. The authors thank Dr. D. Oetomo, University of Melbourne, Australia and Mr. Z. Nagy, ETH Zurich, Switzerland for technical discussion.

References

1. Cuschieri, A., Melzer, A.: The impact of technologies on minimally invasive therapy. Surg. Endosc. **11**, 91–92 (1997)
2. Phee, L., Accoto, D., Menciassi, A., Stefanini, C., Carrozza, M.C., Dario, P.: Analysis and development of locomotion devices for the gastrointestinal tract. IEEE Trans. Biomed. Eng. **49**, 613–616 (2002)
3. Hu, C., Meng, M.Q.H., Mandal, M.: Efficient magnetic localization and orientation technique for capsular endoscopy. Int. J. Inf. Acquisition **2**, 23–26, (2005)
4. Lehman, A.C., Rentschler, M.E., Farritor, S.M., Oleynikov, D.: The current state of miniature in vivo laparoscopic robotics. J. Robot. Surg. **1**, 45–49 (2007)
5. Yonezawa, J., Kaise, M., Sumiyama, K., Goda, K., Arakawa, H., Tajiri, H.: A novel double-channel therapeutic endoscope ("R-scope") facilitates endoscopic submucosal dissection of superficial gastric neoplasms. Endoscopy **38**, 1011–1015 (2006)
6. Dario, P., Ciarletta, P., Menciassi, A., Kim, B.: Modeling and experimental validation of the locomotion of endoscopic robots in the colon. Int. J. Robot. Res. **23**(4–5), 549–556 (2004)
7. Menciassi, A., Stefanini, C., Gorini, S., Pernorio, G., Dario, P., Kim, B., Park, J.O.: Legged locomotion in the gastrointestinal tract problem analysis and preliminary technological activity. IEEE Int. Conf. Intell. Robots.Syst. **1**, 937–942 (2004)
8. Stefanini, C., Menciassi, A., Dario, P.: Modeling and experiments on a legged microrobot locomoting in a tubular, compliant and slippery environment. Int. J. Robot. Res. **25**(5–6), 551–560 (2006)
9. Dutta, S.M., Ghorbel, F.H.: Differential hysteresis modeling of a shape memory alloy wire actuator. IEEE/ASME Trans. Mechatron. **10**(2), 189–197 (2005)
10. Gorini, S., Quirini, M., Menciassi, A., Pernorio, G., Stefanini, C., Dario, P.: A novel sma-based actuator for a legged endoscopic capsule. In: Proceedings of IEEE/RAS-EMBS International Conference on Biomedical Robotics and Biomechatronics – BioRob (2006)
11. Quirini, M., Menciassi, A., Scapellato, S., Stefanini, C., Dario, P.: Design and fabrication of a motor legged capsule for the active exploration of the gastrointestinal tract. IEEE/ASME Trans. Mechatron. **13**(2), 169–179 (2008)
12. Quirini, M., Scapellato, S., Menciassi, A., Dario, P., Rieber, F., Ho, C.N., Schostek, S., Schurr, M.O.: Feasibility proof of a legged locomotion capsule for the GI tract. Gastrointest. Endosc. **67**(7), 1153–1158 (2008)

13. Valdastri, P., Webster, R.J. III, Quaglia, C., Quirini, M., Menciassi, A., Dario, P.: A new mechanism for meso-scale legged locomotion in compliant tubular environments. IEEE Trans. Robot. **25**(5), 1047–1057 (2009)
14. Quaglia, C., Buselli, E., Webster, R.J. III, Valdastri, P., Menciassi, A., Dario, P.: An endoscopic capsule robot: a meso-scale engineering case study. J. Micromech. Microeng. **19**(10), 105007 (2009)
15. Tortora, G., Valdastri, P., Susilo, E., Menciassi, A., Dario, P., Rieber, F., Schurr, M.O.: Propeller-based wireless device for active capsular endoscopy in the gastric district. MITAT **18**(5), 280–290 (2009)
16. Arezzo, A.: Prospective randomized trial comparing bowel cleaning preparations for colonoscopy. Surg. Laparosc. Endosc. Percutan. Tech. **10**, 215–217 (2000)
17. Schanz, S.: Bowel preparation for colonoscopy with sodium phosphate solution versus polyethylene glycol-based lavage: a multicenter trial. Diagn. Ther. Endosc. 713521 (2008)
18. Rieber, F., Tognoni, V., Cenci, L., di Lorenzo, N., Schurr, M.O.: Capsule endoscopy of the entire GI tract. In: SMIT annual meeting (2008)
19. Hebert, J.J., Taylor, A.J., Winter, T.C., Reichelderfer, M., Weichert, J.P.: Low attenuation oral GI contrast agents in abdominal-pelvic computed tomography. Abdom. Imaging **31**, 48–53 (2006)
20. Quirini, M., Menciassi, A., Scapellato, S., Dario, P., Rieber, F., Ho, C., Schostek, S., Schurr, M.O.: Feasibility proof of a legged locomotion capsule for the GI tract. Gastrointest. Endosc. **67**, 1153–1158 (2008)
21. Dario, P., Menciassi, A., Valdastri, P., Tortora, G.: Dispositivo endoscopico wireless a propulsione autonoma per esplorazione gastrica. Italian Patent BI865F/FMB/fpd (2008)
22. Valdastri, P., Menciassi, A., Dario, P.: Transmission power requirements for novel zigbee implants in the gastrointestinal tract. IEEE Trans. Biomed. Eng. **55**, 1705–1710 (2008)
23. Carta, R., Lenaerts, B., Thoné, J., Tortora, G., Valdastri, P., Menciassi, A., Puers, R., Dario, P.: Wireless power supply as enabling technology towards active locomotion in capsular endoscopy. Biosens. Bioelectron. **25**(4), 845–851 (2009)
24. Valdastri, P., Quaglia, C., Susilo, E., Menciassi, A., Dario, P., Ho, C.N., Anhoeck, G., Schurr, M.O.: Wireless therapeutic endoscopic capsule: in-vivo experiment. Endoscopy. **40**, 979–982 (2008)
25. Kirschniak, A., Kratt, T., Stuker, D., Braun, A., Schurr, M.O., Konigsrainer, A.: A new endoscopic over-the-scope clip system for treatment of lesions and bleeding in the GI tract: first clinical experiences. Gastrointest. Endosc. **66**, 162–167 (2007)
26. Schurr, M.O., Hartmann, C., Ho, C.N., Fleisch, C., Kirschniak, A.: An over-the-scope clip (OTSC) system for closure of iatrogenic colon perforations: results of an experimental survival study in pigs. Endoscopy **40**, 584–588 (2008)
27. Raju, G.S., Gajula, L.: Endoclips for GI endoscopy. Gastrointest. Endosc. **59**, 267–79 (2004)
28. Given Imaging Ltd. website. http://www.givenimaging.com
29. Cavallotti, C., Piccigallo, M., Susilo, E., Valdastri, P., Menciassi, A., Dario, P.: An integrated vision system with autofocus for wireless capsular endoscopy, Sens. Actuators. A Phys. **156** (1), 72–78 (2009)
30. Cheng, Y., Lai, J.: Fabrication of meso-scale underwater vehicle components by rapid prototyping process. J. Mater. Process. Technol. **201**, 640–644 (2008)
31. Valdastri, P., Quaglia, C., Menciassi, A., Dario, P., Ho, C.N., Anhoeck, G., Schoesteck, Rieber F., Schurr, M.O.: Surgical clip releasing wireless capsule, European patent application 08425604.9, filed on 16/09/2008
32. Ramcharitar, S., Patterson, M.S., van Geuns, R.J., van Meighem, C., Serruys, P.W.: Technology insight: magnetic navigation in coronary interventions. Nat. Clin. Pract. Cardiovasc. Med. **5**, 148–156 (2008)
33. Swain, P.: The future of wireless capsule endoscopy. World J. Gastroenterol. **14**, 4142–4145 (2008)

34. Bardaro, S.J., Swanström, L.: Development of advanced endoscopes for natural orifice transluminal endoscopic surgery (NOTES). Minim. Invasive Ther. Allied Technol. **15**(6), 378–383 (2006)
35. The ares (assembling reconfigurable endoluminal surgical system), Project Website http://www.ares-nest.org (2006)
36. Fukuda, T., Nakagawa, S., Kawauchi, Y., Buss, M.: Self organizing robots based on cell Structures – CKBOT. In: IEEE International Workshop on Intelligent Robots, pp. 145–150 (1988)
37. Yim, M., Shen, W., Salemi, B., Rus, D., Moll, M., Lipson, H.: Klavins, E., Chirikjian, G.: Modular self-reconfigurable robot systems [Grand Challenges of Robotics]. IEEE Robot. Autom. Mag. **14**(1), 865–872 (2007)
38. Murata, S., Kurokawa, H.: Self-reconfigurable robots. IEEE Robot. Autom. Mag. **14**, 71–78 (2007)
39. World Health Organisation, Fact sheet n.297, Online: http://www.who.int/mediacen-ter/factsheets/fs297 (2006)
40. Pesic, M., Karanikolic, A., Djordjevic, N., Katic, V., Rancic, Z., Radojkovic, M.: Ignjatovic, N., Pesic, I.: The importance of primary gastric cancer location in 5-year survival rate. Arch. Oncol. **12**, 51–53 (2004)
41. Harada, K., Susilo, E., Ng Pak, N., Menciassi, A., Dario, P.: Design of a bending module for assembling reconfigurable endoluminal surgical system. In: Proceedings of the 6th International Conference of International Society of Gerontechnology (ISG'08), Pisa, Italy, pp. ID–186, 4–7 June 2008
42. Nagy, Z., Oung, R., Abbott, J.J., Nelson, B.J.: Experimental investigation of magnetic self-assembly for swallowable modular robots. In: Proceedings of IEEE/RSJ International Conference on Intelligent Robots and Systems, pp. 1915–1920, 22–26 Sept 2008
43. Susilo, E., Valdastri, P., Menciassi, A., Dario, P.: A miniaturized wireless control platform for robotic capsular endoscopy using advanced pseudokernel approach. Sens. Actuators A Phys. **156**(1), 49–58 (2009)
44. Nagy, Z., Abbott, J., Nelson, B.: The magnetic self-aligning hermaphroditic connector: a scalable approach for modular microrobotics. In: Proceeding of IEEE/ASME International Conference Advanced Intelligent Machatronics, pp. 1–6, Zurich, (2007)
45. Oetomo, D., Daney, D., Harada, K., Merlet, J.P., Menciassi, A., Dario, P.: Topology design of surgical reconfigurable robots by interval analysis. In: IEEE International Conference on Robotics and Automation (ICRA2009), pp. 3085–3090 (2009)
46. ARAKNES Project Website www.araknes.org (2008)
47. Lehman, A.C., Dumpert, J., Wood, N.A., Redden, L., Visty, A.Q., Farritor, S., Varnell, B., Oleynikov, D.: Natural orifice cholecystectomy using a miniature robot, Surg. Endosc. **23**(2), 260–266 (2009)
48. Harada, K., Susilo, E., Menciassi, A., Dario, P.: Wireless reconfigurable modules for robotic endoluminal surgery. In: IEEE International Conference on Robotics and Automation. ICRA '09, pp. 2699–2704 (2009)

Chapter 15
Visual Guidance of an Active Handheld Microsurgical Tool

Brian C. Becker, Sandrine Voros, Robert A. MacLachlan, Gregory D. Hager, and Cameron N. Riviere

Abstract In microsurgery, a surgeon often deals with anatomical structures of sizes that are close to the limit of the human hand accuracy. Robotic assistants can help to push beyond the current state of practice by integrating imaging and robot-assisted tools. This paper demonstrates control of a handheld tremor reduction micromanipulator with visual servo techniques, aiding the operator by providing three behaviors: "snap-to", motion scaling, and standoff regulation. A stereo camera setup viewing the workspace under high magnification tracks the tip of the micromanipulator and the object being manipulated. Individual behaviors are activated in task-specific situations when the micromanipulator tip is in the vicinity of the target. We show that the snap-to behavior can reach and maintain a position at a target with Root Mean Squared Error (RMSE) of 17.5 ± 0.4 μm between the tip and target. Scaling the operator's motions and preventing unwanted contact with non-target objects also provides a larger margin of safety.

Keywords Medical robotics · Microsurgery · Visual servoing · Machine vision · Tremor · Eye surgery · Vitreoretinal surgery · Micromanipulation · Accuracy

15.1 Introduction

Micromanipulators aid surgical operations by providing extremely precise movements on a small scale. Features such as remote control, force feedback, motion scaling, and virtual fixtures enable advanced behaviors that an unassisted

C.N. Riviere (✉)
Robotics Institute, Carnegie Mellon University, 5000 Forbes Ave, Pittsburgh, PA 15213, USA
e-mail: camr@ri.cmu.edu

J. Rosen et al. (eds.), *Surgical Robotics: Systems Applications and Visions*,
DOI 10.1007/978-1-4419-1126-1_15, © Springer Science+Business Media, LLC 2011

human would find difficult to replicate. Of particular interest are microsurgery and cell micromanipulation, where very delicate operations must be performed precisely on structures with cross sections varying from millimeters down to microns. Tools like the Steady-Hand [1] help surgeons by suppressing tremor or involuntary hand movement on the order of 50–100 micrometers (μm).

The precision of the surgical gesture and the comfort of the surgeon during the operation can also be improved by exploiting domain-specific knowledge, using preoperative data to design augmented reality systems for biomicroscopy [2, 3] and real-time tracking of surgical instruments and anatomical targets [2] to further improve the precision of the gesture. This information can then incorporate more intelligent behavior into the micromanipulator, such as commanding the tip to reach a target or avoiding anatomical areas that could lead to complications in the surgery. For example, Li et al. [4] derive a controller for gene injection into a 90 μm-diameter oosperm under 100× magnification using a single camera to servo the micro-injector using visual feedback. With a stereo camera setup, the system developed in [5] servos the micromanipulator to inject a sesame seed viewed under a 20× microscope.

Our lab has developed a fully handheld active micromanipulator called Micron whose capabilities include basic tremor suppression and motion scaling [6]. The central problem addressed in this paper is the control of the endpoint of a manipulator to perform tasks relative to an observed point or surface in space. We propose to introduce three behaviors that incorporate domain-specific knowledge and visual feedback to give the surgeon guidance in specific tasks: snap-to, motion scaling, and standoff regulation. One of the main challenges in doing so is that the optical system involved is a microscope: at such high magnifications, the standard calibration techniques commonly used, such as [7], yield unsatisfactory results.

In the remainder of this section, these three behaviors and their workings are explored.

15.2 Background

Micron (Fig. 15.1) is a handheld micromanipulator with piezoelectric actuators built into the handle of the tool. The actuators can position the endpoint, or tip, within a roughly cylindrical workspace 1,500 μm in diameter and 500 μm long. An external measurement system called ASAP (Apparatus to Sense of Position) supplies Micron

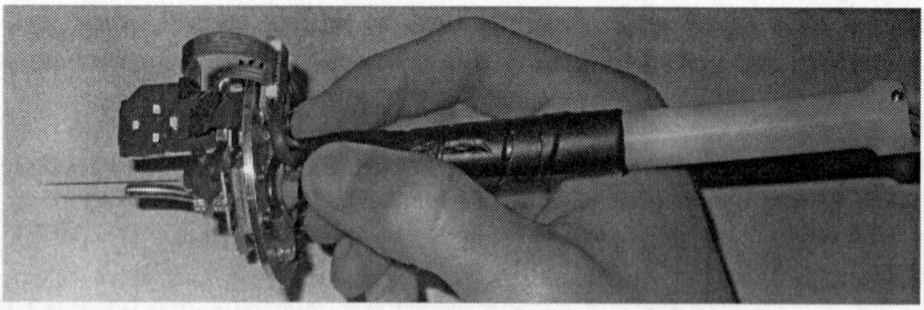

Fig. 15.1 Fully handheld active Micron micromanipulator

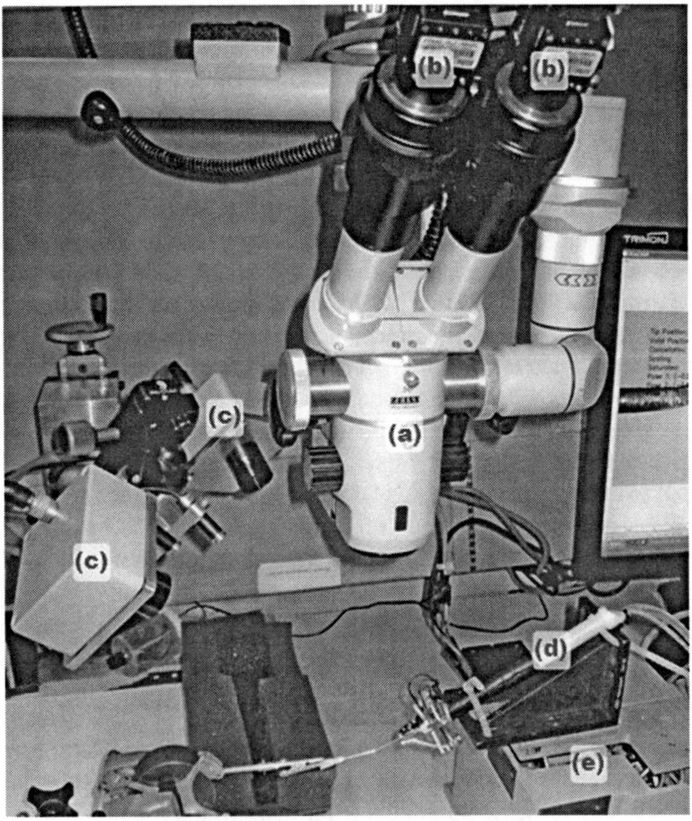

Fig. 15.2 System setup. (**a**) Microscope, *middle center*. (**b**) Cameras, *top right*. (**c**) ASAP measuring sensors, *middle left* (**d**) Micron, *bottom right*, attached to (**e**) Hexapod micropositioner, *lower right*

with real-time position and pose information accurate to ~4 ± 2 μm [8]. In addition to the X, Y, and Z location of the Micron tip, a 3 × 3 rotation matrix defining the pose of the instrument can be obtained. Another useful measurement available is the center position, or where the tip would be nominally pointing if the piezoelectric actuators were not active. ASAP uses position-sensitive detectors (PSD's) to detect four pulsed LEDs mounted inside diffuse spheres on the shaft of the instrument. This allows Micron to perform basic tremor suppression and motion-scaling functionalities. See Figs. 15.1 and 15.2 for the setup.

15.2.1 Problem Definition

The ASAP measurement system provides very fine positioning information of the instrument and allows for general tremor compensation via filtering in the frequency domain. However, it has no knowledge of what an operator sees in the microscope. Thus, the current Micron system cannot attempt to keep the tip aligned with a vein or to maintain a safe distance from anatomical features. The core issue addressed here is controlling the tip of Micron based on observations made with a stereo camera to effect three behaviors: snap-to, motion scaling, and standoff regulation.

The first behavior, snap-to, involves guiding the Micron tip to a 3D point in space and maintaining the tip at that location. Snap-to can be thought of as a more constrained version of tremor suppression and is useful when the tip must be held stable, e.g., for injections. The second behavior, motion scaling, is more practical when very precise movements are needed. Every movement made by the operator is scaled by a user-defined factor, thus reducing positioning errors. The finite range of the actuated tip limits when motion scaling can be applied, so it is only turned on when in the vicinity of the target. The final behavior, standoff regulation, serves as a preventative measure against accidental, unwanted contact. In this mode, Micron can attempt to avoid bringing the tip in contact with points by actively maintaining a preset distance from pre-defined "off-limits" areas. Each behavior has unique properties that can benefit a surgeon in different circumstances.

15.2.2 Visual Servoing

is a popular approach to guiding a robotic appendage or manipulator using visual feedback from cameras [9]. Given a target pose or position that the robot is to reach, the goal of visual servoing is to minimize the following error:

$$e(t) = s(m(t), a) - s^* \tag{15.1}$$

where s^* represents the desired positions, $s(m(t),a)$ the measured positions, $m(t)$ the measured feature points in the image, and a any external information needed (such as camera parameters). In general, $s(m(t),a)$ is usually the endpoint, or tip, of the manipulator. The tip position is determined from $m(t)$, the image coordinates of the tracked tip, and a, the camera mapping between world space and the image coordinates. In practice, s^* may not be known a priori and may instead be calculated from image features as well.

The approach used to minimize the error $e(t)$ can be done in one of two general ways: position-based visual servoing or image-based visual servoing [9, 10]. Image-based servoing uses an *image Jacobian* or *interaction matrix* to convert errors measured in the image directly into a velocity the robot should attempt to maintain. In contrast, position-based servoing treats the camera system as a 3D position sensor and measures the error in the task space rather than in the image. In the case of controlling only the translation of a tool, as is the case with the 3-degree-of-freedom (3DOF) Micron micromanipulator, it has been shown that these two approaches are equivalent [11]. Therefore, this research uses the position-based visual-servoing approach because of the well-defined positioning system provided by ASAP and the ability to run calibration routines online. Furthermore, implementing the standoff regulation requires distance metrics in 3D space.

15.2.3 Novelty Considerations

Micron has unique characteristics that differentiate the system from a typical visual-servoing setup. First, Micron is not a fully autonomous robot, in that it has a limited range of motion whose reach is determined by the user holding the micromanipulator. The operator may move Micron to a position where the tip of the instrument cannot reach the target. Second, an external measurement system is available which allows for online calibration of the system, either before each run or during the run. Third, unlike a purely closed-loop system, Micron has to account for human dynamics involving the human eye-hand coordination feedback loop. Other research, mainly [12], explicitly considers control by a human user in micromanipulation, but only for the purposes of high-level task sharing and direction. For the majority of this paper, interaction with the human controller will not be explicitly considered, so as to give a more complete characterization of the system's performance in the context of visual servoing.

These novel problems must be eventually resolved, and thus the results will be evaluated and discussed with these considerations in mind.

15.3 System Design

ASAP and Micron are integrated on a real-time LabVIEW® target machine with the ASAP measuring system running in parallel with the controller. The Micron interface and the vision system run on a standard Windows® PC that is networked to the real-time machine to retrieve ASAP positioning information, perform stereo visual servoing, and send control signals back to Micron.

15.3.1 Visual Feedback

Designed for microsurgical work, Micron is operated under a high-power Zeiss OPMI® one microscope with a magnification often exceeding 25× and a visual workspace often only several millimeters in diameter. Two PointGrey Flea2 cameras capturing 800 × 600 video at 30 Hz are mounted to the microscope, providing a stereo view of the workspace. Each camera view is approximately 2 × 3 mm with each pixel corresponding to ∼3.4 μm.

The micromanipulator tip and feature landmarks are tracked in both views, giving a stereo reconstruction of the instrument tip relative to the targets. Since advanced tracking is not the topic of this research, the tip and target are marked with different colored paint for easy visual identification. Tracking is performed with a simple but robust color tracker [13]. The highly optimized and popular Intel® OpenCV library is used for implementing the computer vision techniques.

15.3.2 Micromanipulator Control

The chief control problem is how to use visual servoing in the context of Micron; in other words, given the tracked tip and target position, how can a control signal be derived in world, or ASAP, coordinates? To do so, an understanding of how camera coordinates relate to 3D world ASAP coordinates is needed. The fundamental perspective camera equation is $p = MP$, where P is a 4×1 homogenous point in 3D world coordinates that is projected to p, a homogenous 3×1 image coordinate by the 3×4 camera matrix, M. M is derived in the following section and defines the projective mapping between ASAP and the cameras. A second camera observes the same point P, creating a joint observation system defined by the fundamental perspective camera equation for each camera: $p_1 = M_1 P$ and $p_2 = M_2 P$. These equations can be combined and solved using the homogenous linear triangulation method described in [14]. Thus, as seen in Fig. 15.3, each set of the 2D points in the stereo pair will yield an acceptable back-projected 3D point.

Now that the tracked tip and target image locations have been reconstructed as 3D points in the ASAP workspace, the goal is to drive the error $E = P_{tip} - P_{target}$ to zero by controlling the endpoint velocity of the tip. When the error is zero, the tip and target should be coincident. The velocity can be determined as a proportion λ of the error:

$$v - \lambda(P_{tip} - P_{target}) \qquad (15.2)$$

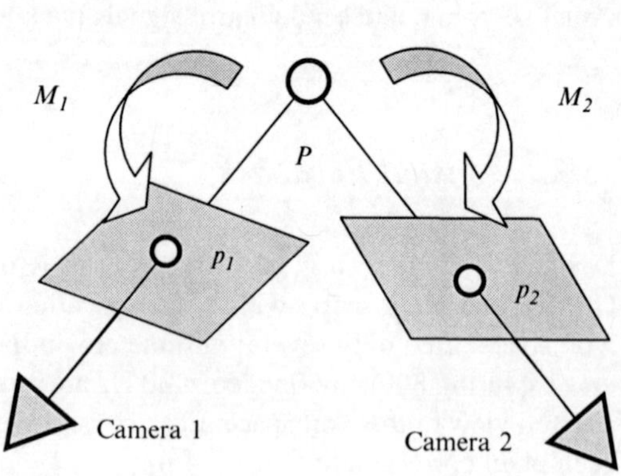

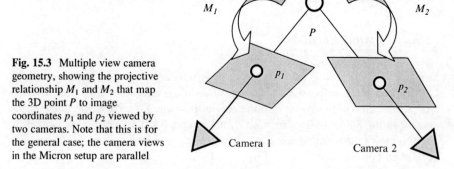

Fig. 15.3 Multiple view camera geometry, showing the projective relationship M_1 and M_2 that map the 3D point P to image coordinates p_1 and p_2 viewed by two cameras. Note that this is for the general case; the camera views in the Micron setup are parallel

Because velocities command the motion of the tip, errors due to errors in the calibration are absorbed with each new velocity calculation and drive E asymptotically to zero even in the presence of calibration errors.

15.3.3 Calibration

Currently, control of Micron operates in the coordinate system defined by the ASAP measurement system. However, the control signals are derived from the tracked tip and target viewed in the stereo camera setup. Furthermore, any preoperative information is usually registered in the image reference system. Thus calibration mappings M_1 and M_2, for cameras 1 and 2 respectively, are needed to transform from pairs of 2D image coordinates p_1 and p_2 to a 3D world coordinate point P.

In visual servoing, only the rotation mapping is important as translation is handled by streaming many sequential velocities to guide the tip in the direction of the target. However, because ASAP provides very accurate positions of the tip in 3D world coordinates, the full perspective mapping can be obtained. Furthermore, the full perspective camera mappings are useful because they can provide the visual trackers a rough estimation of where the tip is in the image. This reduces the amount of processing time required to locate the tip in the image, yielding increased framerates and better control performance.

A corresponding set of 3D world coordinates P_i and 2D image coordinates p_{1i} and p_{2i} form an over-determined system of equations $p_{1i} = M_1P_i$ and $p_{2i} = M_2P_i$, which can be solved by $M_1 = p_1\,P^+$ and $M_2 = p_2\,P^+$ where P^+ denotes pseudo-inversion. A method that is more robust is the Direct Linear Transformation algorithm [14]. This yields the pinhole camera perspective parameters, allowing a two-way mapping between 2D stereo image coordinates and 3D ASAP coordinates.

Calibration can be performed online, using the first 5–60 s of corresponding world and image coordinates in each run. If system positioning, magnification, and focus do not change between runs, calibrations may be reused, since the visual servoing does not require highly accurate absolute calibration [15], which is cumbersome to obtain with a microscope [16]. The calibration routine involves the operator moving the tip randomly through the workspace, including up and down. A typical 60-s calibration yields approximately 2,000 data points, from which outliers are automatically removed via a simple distance metric before the calibration calculations are performed. After calibration, the 3D tip position as measured by ASAP can be projected in the image accurately within 20 pixels and the tracked tip in the stereo images can be reconstructed in 3D space with an absolute mean error of ~200 μm. Over time, the absolute calibration accumulates errors relative to ASAP as the tool pose is changed and the system shifts; however, this is not a problem as the control depends on the relative difference between the tip and target positions.

15.4 Experimental Results

Three different experiments were carried out to evaluate the desired behaviors of snap-to, motion-scaling, and standoff-regulation. First and foremost, the goal is to demonstrate the correctness and accuracy of each behavior. To eliminate any human-in-the-loop influences or disturbances, such as tremor, and achieve very repeatable results, all experiments requiring movement of the handheld micromanipulator were performed with the micromanipulator attached to the very precise (sub-μm) six-axis Polytec PI F-206.S HexAlign™ Hexapod manipulator. Since a machine will be "holding" Micron, it is equally important to evaluate the feasibility of applying these behaviors when a human is operating the micromanipulator. As such, a fourth experiment tested the helpfulness of the snap-to feature in a pointing task involving human tremor and compared it against basic tremor cancellation already implemented in Micron.

15.4.1 Experiment 1: Snap-to

Experiment 1 tested the snap-to functionality for convergence time and accuracy to a stationary point. Human hand motion and tremor are rapid, necessitating a fast response time. The experiment, seen in Fig. 15.4, servos the Micron tip to a 3D point defined by a colored needle tip. Both the micromanipulator and target point were held stationary. The target point on the needle was rigidly held by a clamp and Micron was firmly affixed to the Hexapod; only the Micron tip was actuated. Because the Hexapod was not actively moving the Micron handle, tremor reduction was not used for this experiment.

The experiment was performed from three different starting locations with three identical runs for each location. A summary of several important statistics is listed for each location in Table 15.1. First, the initial distance between the tip and target is listed, as this determines convergence time, i.e., the time required for the tip to converge on the target, where successful convergence is defined as being within

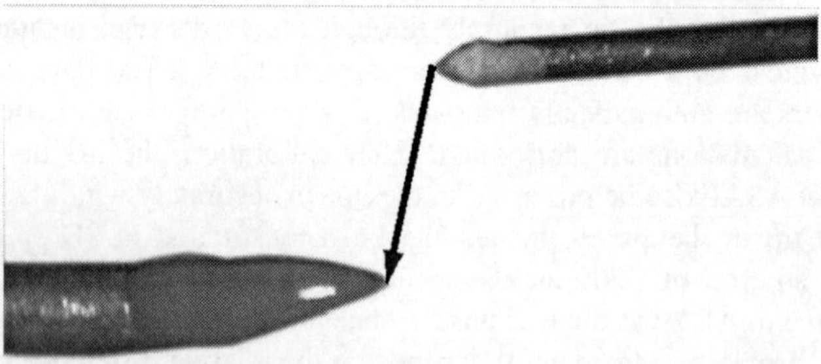

Fig. 15.4 Snap-to target experiment servoing green-color coded Micron instrument tip (200-μm diameter) to a stationary 3D target point defined by a red colored needle tip

Table 15.1 Speed and accuracy of servoing the tip to a 3D target from three different locations. Each location was reached three times to obtain a mean and standard deviation. RMSE measures the distance error between the tip and target for 15 s after convergence

Initial distance (μm)	Convergence Time (s)	Distance RMSE (μm)
375.3 ± 2.5	0.54 ± 0.02	16.6 ± 0.4
449.2 ± 5.6	0.59 ± 0.07	17.5 ± 0.4
677.9 ± 0.8	0.69 ± 0.04	18.3 ± 0.4

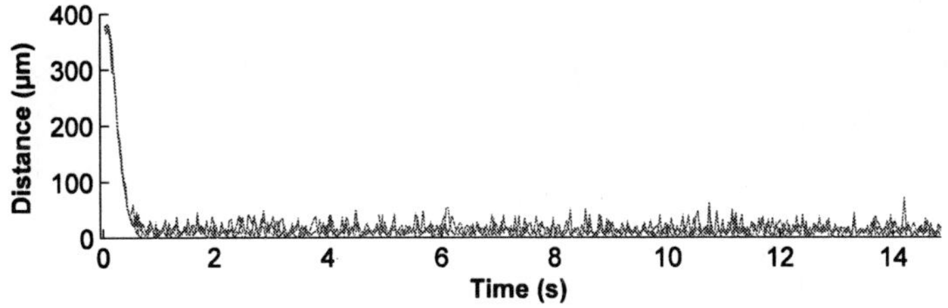

Fig. 15.5 Euclidean distance between Micron tip and target for three runs from the second location

25 μm of the target. Once the tip has converged, the RMSE is calculated from the 3D Euclidean distance between the tip and target during the 15 s after the target has been reached. Fig. 15.5 shows the distance error between the actuated Micron tip and the target for all three runs of the first location. The noise after convergence is caused mostly by small errors in the image trackers magnified by the backprojection into 3D. All runs exhibit similar trajectories and noise patterns.

While the convergence accuracy of the snap-to behavior is high, the convergence time of approximately half a second is rather long. Once converged, the visual servoing can hold a stationary point very well, even in the presence of disturbances. For example, if the whole Micron instrument is moved, it is desirable for the tip position to actuate fast enough that it can still remain steady on the target. If the response is too slow, tremor introduced as the human user operates will result in oscillations as the tip tries and fails to keep moving faster than tremor.

To model tremor-like movement, a disturbance signal was induced by moving the Hexapod base in all three dimensions and in a roughly sinusoidal pattern. This was done only with snap-to, and without any of the tremor-compensation techniques used by Micron. As shown in Fig. 15.6, the Micron tip remained steady even with rather large and rapid changes in position. This indicates that even in the case of rapid tremor motion, the Micron tip can remain "snapped-to" a stationary target very accurately. Experiment 4 expounds more by testing it with real tremor.

15.4.2 Experiment 2: Motion-Scaling

In experiment 2, the micromanipulator was moved at a constant speed in 3D space past the target point using the Hexapod micropositioner. When the tip of the

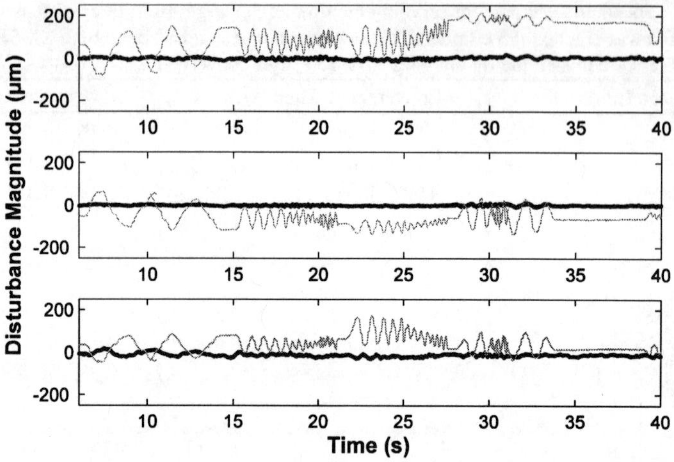

Fig. 15.6 Micron tip position (*thick black line*) snapped-to a target even with large, rapid changes in Micron instrument positioning (*thin green lines*) in X, Y, and Z directions (*top, middle, bottom* respectively)

Table 15.2 Mean measured 3D velocity before coming in range of the target, while the tip is within a 170 μm radius of the target, and after the tip has passed the edge of the circle. Three runs were executed for each scale tested: 1.5, 2, 3

Initial velocity (μm/s)	Scaled velocity (μm/s)	Final velocity (μm/s)	Measured scale
95.5 ± 1.9	62 ± 0.1	96.4 ± 0.7	1.5 ± 0.0
91.6 ± 4.6	43.9 ± 4.0	91.7 ± 8.4	2.1 ± 0.1
95.3 ± 1.0	30.6 ± 0.2	96.9 ± 0.9	3.2 ± 0

instrument was detected to be within an arbitrarily chosen 170-μm radius of the target, the motion-scaling behavior was activated with a scale factor of 1.5, 2, or 3. Thus the constant velocity of the tip should be reduced by a factor of $1/s$ while the tip is in vicinity of the target. The variability from $1/s$ was examined to determine the effectiveness of this technique. As seen in Table 15.2, the motion scaling works very well. There is a slight difference between the desired scale and the actual scale, but the overall variability is low. Velocity variability from time instant to time instant is influenced by the noise level and is difficult to assess accurately. Figure 15.7 shows the slope of the position and the numerically-differentiated velocity (smoothed with a 1 Hz lowpass filter).

15.4.3 Experiment 3: Standoff-Regulation

Experiment 3 validates the ability of the visual-servoing controller to maintain a standoff distance from a 3D target point. In this case, instead of snapping-to, the

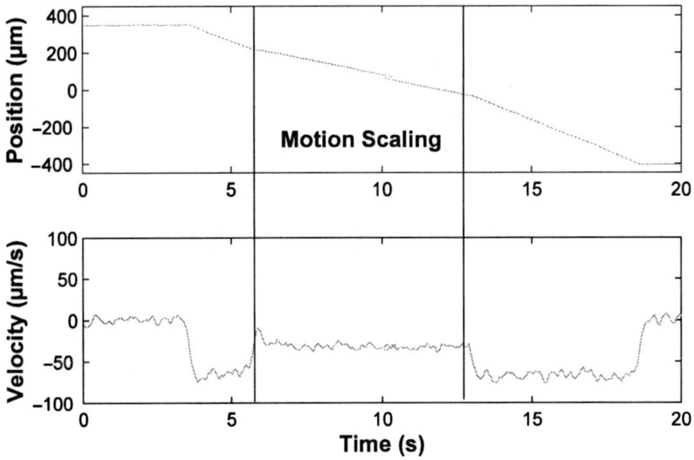

Fig. 15.7 Position and velocity of the Micron tip from one run as the instrument is moved with constant velocity. The middle motion-scaling region represents when the tip is within range of the target

desired action is for the tip to avoid the target. This could be useful in many surgical situations where "keep-away" points or zones might be defined to avoid unwanted and possibly harmful contact with tissue. As with the motion scaling, a spherical volume is defined around the target point in 3D. If the tip comes within 240 µm of the target, a repulsive force is exerted on the tip to guide it away. In this simple example, the repulsive field is modeled after a charged particle; thus the force exerted on the tip is always away from the center of the target. The Hexapod micropositioner is used to move the Micron tip in a straight line approaching the target at three different offset distances. The resulting trajectories can be viewed in Fig. 15.8. The standoff-regulation does push the trajectory of the tip away from the target. Additionally, the trajectory that clearly did not come near the target was unaffected. This simplistic model of a charged particle can be modified to maintain a constant standoff.

15.4.4 Experiment 4: Pointing Task

One common task in micromanipulation is pointing, or keeping the tip steady at some target. Sustained injection of a drug or chemical agent, for example, may be the purpose. To enhance accuracy in this task, it is necessary not only to suppress the "neurogenic" tremor component at 8–12 Hz [17], but also to suppress lower-frequency components as well. One way to accomplish this is to know the target point about which the user is trying to keep the tip steady. Experiment 4 compares a human operator attempting a pointing task unaided, with frequency-based tremor compensation, and with the snap-to image guidance. One subject performed three

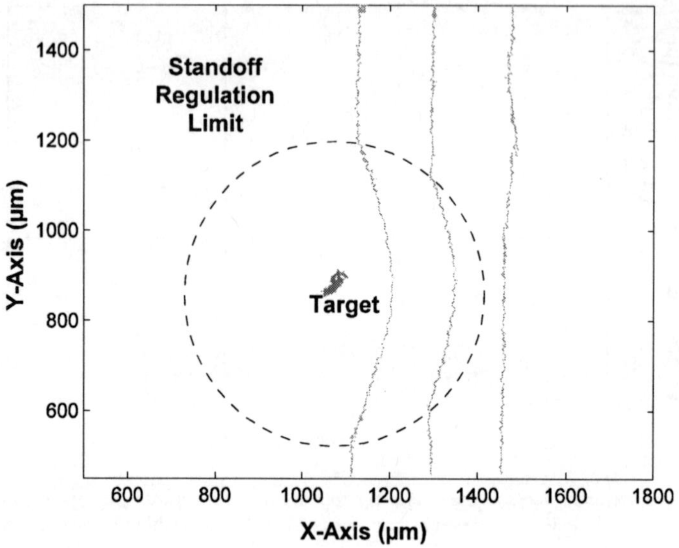

Fig. 15.8 Three tip trajectories, two of which infringe the repulsive field and are therefore pushed away from their paths to maintain separation between the target and the tip

Table 15.3 RMSE distance between the target and Micron tip during a 60-s pointing task

Unaided (μm)	compensation (μm)	Snap-to (μm)
222.9 ± 61.3	175.8 ± 65.9	70.1 ± 25.9

runs for each scenario, attempting to keep the Micron tip steady at the 3D target position for 60 s. The evaluation metric is the RMSE between the target point and the Micron tip. Table 15.3 lists the results. As expected, frequency based tremor compensation performed better than unaided. The snap-to performed even better, although the error was much higher than reported earlier because Micron could not always reach the target point.

15.5 Discussion and Conclusion

We present techniques for vision-based guidance of an active handheld micromanipulator, which yield encouraging initial results in a challenging environment. Using a stereo vision setup with a simple calibration routine, Micron employs visual servoing to increase accuracy in pointing tasks, even over existing tremor-cancellation methods. In a stationary pose, we have shown that finer motor control is possible locally around a target using motion scaling, and a surgeon can define "keep-away" areas to help avoid unwanted tissue contact. A demonstration of Micron with these three behaviors can be viewed in the video accompanying this paper.

One issue not dealt with is the finite manipulator range. If Micron is snapping-to, the user easily can shift the entire manipulator out of range of the target. Because the actuators can only move a finite amount before reaching their limits, additional behavior is needed to compensate. Additionally, to be useful in a completely handheld situation such as a surgical procedure, the controller will need to resolve conflicts between the user's movements and the activated behaviors. For instance, it may be more important to observe the "keep-away" constraints and not damage tissue than to maintain a motion-scaling behavior.

As future work, we plan on developing more realistic ways to track targets and surgical instruments, thereby obviating the current reliance of color tracking. Because magnification and focus can change during an operation, future work will include online routines that detect these changes and update the calibration accordingly through adaptive gains. As for the controller, a more sophisticated control algorithm is needed for standoff regulation. Evaluation work will move toward testing of these behaviors in more realistic settings.

Acknowledgments This work was supported in part by the National Institutes of Health (grant no. R21 EY016359), the American Society for Laser Medicine and Surgery, the ARCS Foundation, and a National Science Foundation Graduate Research Fellowship.

References

1. Mitchell, B., Koo, J., Iordachita, I., Kazanzides, P., Kapoor, A., Handa, J., Hager, G., Taylor, R.: Development and application of a new steady-hand manipulator for retinal surgery. In: IEEE International Conference on Robotics and Automation, pp. 623–629 (2007)
2. Fleming, I., Voros, S., Vagvolgyi, B., Pezzementi, Z., Handa, J., Taylor, R., Hager, G.: Intraoperative visualization of anatomical targets in retinal surgery, In: IEEE Workshop on Applications of Computer Vision (2008)
3. Berger, J.W., Madjarov, B.: Augmented reality fundus biomicroscopy a working clinical prototype. Arch. Ophtalmol. **119**, 1815–1818 (2001)
4. Li, X., Zong, G., Bi, S.: Development of global vision system for biological automatic micro-manipulation system. IEEE Int. Conf. Robot. Autom. **1**, 127–132 (2001)
5. Yamamoto, H., Sano, T.: Study of micromanipulation using stereoscopic microscope. IEEE Trans. Instrum. Meas. **51**, 182–187 (2002)
6. Riviere, C.N., Ang, W.T., Khosla, P.K.: Toward active tremor canceling in handheld micro-surgical instruments. IEEE Trans. Robot. Autom. **19**, 793–800 (2003)
7. Zhang, Z.: A flexible new technique for camera calibration. IEEE Trans. Pattern. Anal. Mach. Intell. 1330–1334 (2000)
8. MacLachlan, R.A., Riviere, C.N.: High-speed microscale optical tracking using digital frequency-domain multiplexing. IEEE Trans. Instrum. Meas. 58(6), 1991–2001 (2008)
9. Chaumette, F., Hutchinson, S.: Visual servo control, part I: basic approaches. IEEE Robot. Autom. Mag. **13**, 82–90 (2006)
10. Hutchinson, S., Hager, G.D., Corke, P.I.: A tutorial on visual servo control. IEEE Trans. Robot. Autom. **12**, 651–670 (1996)
11. Hespanha, J.P., Dodds, Z., Hager, G.D., Morse, A.S.: What tasks can be performed with an uncalibrated stereo vision system? Int. J. Comput. Vis. **35**, 65–85 (1999)

12. Ammi, M., Ferreira, A.: Involving the operator in the control strategy for intelligent tele-micromanipulation. In: IEEE/ASME International Conference on Advanced Intelligent Mechatronics, p. 2 (2003)
13. Yang, M.H., Ahuja, N.: Gaussian mixture model for human skin color and its application in image and video databases. In: Proceedings of the SPIE Storage and Retrieval for Image and Video Databases, vol. **3656**, pp. 458–466 (1999)
14. Hartley, R., Zisserman, A.: Multiple View Geometry in Computer Vision. Cambridge University Press, Cambridge (2003)
15. Hager, G.D.: A modular system for robust positioning using feedback from stereovision. IEEE Trans. Robot. Autom. **13**, 582–595 (1997)
16. Danuser, G.: Photogrammetric calibration of a stereo light microscope. J. Microsc. **193**, 62–83 (1999)
17. Elble, R.J., Koller, W.C.: Tremor. Johns Hopkins, Baltimore (1990)

Chapter 16
Swimming Micro Robots for Medical Applications

Gábor Kósa and Gábor Székely

Abstract We review micro-systems with robotic aspects that are used in medical diagnosis and intervention. We describe the necessary components for a micro-robot and present the state of the art and gaps of knowledge. One of the great challenges in micro robots is the propulsion. Different propulsive strategies and specifically flagellar propulsion is evaluated in this chapter. We analyze the influence of the miniaturization on the micro-robot and try to estimate the future developments in the field.

Keywords Micro-robots · Swimming · Capsule-endoscopy · Positioning-actuator · Manipulation-actuator · Micro-sensor · Flagella · Traveling-wave · Medical-robots · Power source · Scale analysis · Neuro-surgery · MRI · Piezoelectric-actuator · Magnetic-actuator

16.1 Introduction

A micro-system is a system that includes structural components, sensors and actuators whose characteristic size is less than 1 mm. A micro-robot is a micro-system that is able to perform complicated tasks and receive inputs from its environment and react to them. One of the most eminent fields in which micro-systems are utilized is medical diagnosis and intervention. Although micro sensors and micro tools are already used in the bio-medical industry, micro robots are still in research and development stage.

G. Kósa (✉)
Computer Vision Laboratory, Department of Information Technology and Electrical Engineering, ETH Zurich, Switzerland
e-mail: kosa@vision.ee.ethz.ch

J. Rosen et al. (eds.), *Surgical Robotics: Systems Applications and Visions*,
DOI 10.1007/978-1-4419-1126-1_16,

The potential areas for medical application of a micro robot:

1. The gastrointestinal tract (GI). Currently standard endoscopy can reach the upper part of the small intestine through the mouth, throat, esophagus, stomach and duodenum and the lower third of the colon through the rectum. Although there is inconvenience for the patient and the current tools have drawbacks, there is no critical necessity for and untethered swimming micro robot. Capsule endoscopy is the only tool for the inspection the small intestine and the upper colon. Capsule endoscopes are only able to monitor these areas and not intervene such as regular endoscopes. In addition capsule endoscopes cannot control their position and they rely on the natural peristaltic motion of the intestines.
2. The central nervous system (CNS). Both the brain and the spinal cord are immersed in CSF (Cerebro Spinal Fluid), which is produced in the ventricle of the brain and flows out of the ventricular system to the *subarachnoid space* (spinal and cerebral SAS). This space covers all over both the brain and the spinal cord and is connected to the internal ventricular system of the brain with the outer surface of the CNS. Theoretically, this CSF space can be a conduit for introducing imaging and interventional devices that can reach CNS pathology without the need for conventional operative methods.
3. The eye. The eye contains a transparent fluidic gel called the *vitreous humor*. A swimming micro-robot that can use this media to treat the retina. Interventions, such as injection of medicine into the retinal veins, are difficult to perform with by an external approach [3]. Swimming micro robots with sensors and tools for intervention will enable new medical procedures for the treatment of the retina.
4. Thoracoscopy. Thoracoscopy is a minimally invasive surgery for removing tumorous lung nodules. In order to reach the target nodules the *pleural cavity* is inflated with air and the surgeon uses an endoscope to inspect and remove the cists on the lungs. When there are multiple tumorous lung nodules several ports are opened which complicates the procedure. Filling the *pleural cavity* with liquid and using a swimming micro robot will enable treatment of several cists from one port.
5. Fetal surgery [4]. Fetal surgeries are treatment of the fetus still in the womb. One example of such an intervention is treatment of twin-to-twin transfusion syndrome (TTTS). In this intervention an endoscope has to navigate between the twin fetuses and reach the ovule in order to ablate with a laser the mutual blood vessels that cause the transfusion. Using a swimming micro robot will aid in navigation and reduce the risk of injuring the fetuses and the mother.

Although micro-robots currently cannot perform complicated tasks, several micro-systems with robotic features are already commercially available. Such a system is pill camera that diagnoses the gastrointestinal tract, PillCam, manufactured by Given Imaging [5]. The pill-cam has necessary capabilities required from a medical micro-robot such as vision, autonomous power source, communication but it does not have the ability of self positioning and intervention. Other endoscopic capsules have been developed by Olympus [6], Jinshan Science and Technology

Group [7], Intelligent Microsystem Center and RF SYSTEM lab [8]. All capsules are passively driven although GI patented [9] a method to control the position of the capsule by stimulating natural peristaltic contractions of the smooth muscle surrounding the small intestines. Kyungpook National University [10] develops an endoscopic pill driven by electric stimuli similar to the GI patent. SmartPill [11] developed a pill that is able to monitor pressure, temperature and pH in the gastro-interologic track. MiniMitter [12] developed a pill that monitors temperature. The capsules are only diagnostic and do not work in real time. The pills use wireless communication to transmit the images to an external data storage device.

Currently, the research is targeted toward autonomous propulsion for the capsule endoscopy. Several groups are developing swimming capsule endoscopes. Menciassi et al. [13] presented a gastric capsule driven four miniature propellers, powered by batteries and demonstrated a swimming velocity of 21.3 cm/s for an operating period of 7–8 min. RF Systems [8] developed a capsule endoscope for stomach inspection the endoscope is driven and powered externally by a magnetic field.

16.2 Components

The ideal micro robot for medical applications is fully autonomous and it is able to position itself, perform diagnosis with various sensors, locate itself accurately (position accuracy of 1 mm), transfer data and receive commands by wireless communication and perform medical intervention within the human body. Such a system is a goal of research, even though several medical micro devices already been possess some robotics aspects as shown in the previous section.

The subsystems that are necessary for an autonomous swimming micro robot are also under development. In this section we would like to present the essential components necessary to build an autonomous swimming micro robot.

Dario [14] introduced and Ebefors and Stemme [15] developed an essential component list for a micro robot. The authors suggested the following sub-system division: Control Unit (CU), Actuation for Positioning (AP), Power Source (PS), and Actuation for Manipulation (AM). In the case of medical micro robots one should add also the following parts: Sensor Unit (SU) and arguably Communication Transceiver (CT); one can include it in the control unit although the communication is usually developed separately from the control.

The different sub-systems are in different development stages part of them are commercially available, such as CT and CU, part are in advance development stages such as SU and AM and part are still in early development stages. The major challenges in medical micro robots are AP and PS which are the enabling technologies for the full micro system. The efforts and achievements in the different subsystems described above will be summarized below.

16.2.1 Positioning

A micro-device can be placed in the body by a regular open surgery, minimally invasive surgery or by a medical micro robot. A minimally invasive tool such as an active catheters [16] or a steerable endoscopes [17] can be classified as a micro robot especially if it has closed loop control and embedded sensors and actuators. Such micro devices have limited working space compared to an untethered micro-robot.

An untethered micro robot can be driven by natural conduits (such as the GI track nad the blood stream) or positioned by externally induced forces or to be self propelled. Different positioning methods have been proposed for a self propelled medical micro robot: swimming, crawling (advancing by continues contact by a stick and slip mechanism) and legged locomotion.

Several swimming actuators for positioning (AP) of a medical micro robot have been suggested. The actuators use piezoelectric, ICPF (Ionic Conducting Polymer Film), SMA (Shape Memory Alloy) and magnetic driving principles.

Piezoelectric swimming actuator for a micro robots was introduced by Fukuda [18]. The robot was a vibrating tail with a motion enhancing mechanism and it was 55 mm long and not very efficient. Friend et al. [19] developed a piezoelectric micro-motor that rotates a helical tail and currently working on a 0.3 mm micro-robot swimming in the blood stream. Another piezoelectric swimming actuator was developed by Kosa et al. [20] and its working principle and performance will be detailed in the following sections.

ICPF actuators were used by Guo et al. [21] to propel a 30 mm swimming robot with a swimming velocity of 5 mm/s. Guo et al. added also a floating actuator to control the buoyancy of the micro robot and two "walking" actuators to advance along the basin of the vessel. A similar actuator was developed by Vidal et al. [22] based on ECP (Electro Conducting Polymer). Nakabo et al. [23] used IPMC to build a swimming actuator that creates an undulating motion. The actuators overall dimensions were $54 \times 8 \times 0.2$ mm and it achieved a swimming velocity of 1.5 mm/s.

Esashi et al. used SMA wires to steer an active catheter for angioplasty [16]. Cho et al. [24] presented an aquatic propulsion system made of three linked swimming actuator with two SMA actuators at the joints.

Several studies use magnetic forces to propel micro systems. The operational principle of these actuators can be divided into two main types: the one is placing a permanent magnet in an alternating a magnetic field creating rotation or vibration of a propeller and the other is placing a permanent magnet into a magnetic field with constant gradient.

Honda et al. [25] made a 21 mm long helix with a 1 mm^3 SmCo magnet at its head and achieved a swimming velocity of 20 mm/s. Ishiyama et al. [26] used the same principle but the helix was wrapped around a capsule and the permanent magnet was at the center of the capsule. In addition to swimming, the capsule can crawl in the small intestine by a screwing motion.

Dreyfus et al. [27] showed that a linear chain of colloidal magnetic particles linked by DNA and attached to a red cell stabilized and actuated by an external magnetic field can create undulating motion and swim. The filament's length was 24 μm and it achieved a propulsive velocity of 3.9 μm/s. Bell et al. [28] applied the same principle as Honda in a smaller scale and manufactured nano-helices out of a rectangular strip (40 μm length, 3 μm diameter, 150 nm helix strip thickness) and attached it to a magnetic micro bead. The propulsive velocity of the nano-helix was also 3.9 μm/s.

Yesin [29] used external coils to drive with a constant gradient magnetic field and stabilize with a constant magnetic field a 1 mm long ellipsoid micro robot embedded with NdFeB powder. Mathieu et al. [30] used the gradient coil of the MRI to propel 600 μm magnetized steel beads (carbon steel 1010/1020).

Kosa et al. [2, 31] suggested a different magnetic propulsion method using the constant magnetic field of an MRI device. This actuation method will be detailed in the following sections.

An interesting swimming propulsion method is harnessing bacteria to propel a micro robot. Martel et al. [32] used MC-1 bacteria that have natural magnetic polarization in order to manipulate their motion. Behkam and Sitti [33] used *Serratia marcescens* bacteria to move directionally 10 μm spherical beads. Steager et al. [34] propelled a 50 μm triangular structure with *S. marcescens* bacteria.

Another positioning method in the body is crawling or legged propagation. Dario et al. developed several legged capsules for propagation in the colon. The capsules were driven by an SMA wires [35], and later by two miniature DC brushless electromagnetic motors (SBL04-0829 from Namiki Precision Jewel Co., Ltd.), [13]. Kim et al. [36] developed a crawling miniature robot (50 mm long) using SMA wires to create an earthworm like motion.

16.2.2 Power Source

One of the challenges in an autonomous robot is to design a power source that can fulfill the power requirements for a proper operational period. In a medical micro-robot this challenge is further increased by the geometrical limitations and the human body's sensitivity (such as radiation limitations). In order to estimate the performance of a power source two parameters will be used: the one is energy density (the quantity of energy that can be stored in the in a given mass) and battery capacity (the number of hours in which a battery can supply a given current at a given voltage). The two main strategies to power a micro robot are by an internal power source such as batteries and by wireless power transfer such magnetic induction.

There are some excellent reviews on the different power sources for a power autonomous mobile robot [37–41] however other summaries don't take into account the small size of a medical micro robot.

The most eminent and available internal power source is the battery. There are commercially available hearing aid batteries based on zinc-air technology that have small volume (Renata SA ZA-10 Maratone: Diameter of Ø5.8 mm and height of 3.6 mm), capacity of 100 mAh and specific energy of 333 mAh/kg. The drawback of the zinc-air technology is its relatively low power density, about 100 mW/kg and the battery's dependence in an air (a difficulty in a fluidic media). Such batteries fit to applications with long operation time and low consumption current. Quallion's QL0003l is a battery that was developed specifically for a Functional Electrical Stimulation micro-device implanted in the spine. It is based on lithium ion rechargeable technology and has the diameter of Ø2.9 mm and length of 11.8 mm. The energy density of QL0003l is 54 Wh/kg and the capacity is 3 mAh. The power density, 300 mW/kg, of Li-ion technology enables higher currents for shorter operation periods. The specific energy of Li–ion batteries is 200 Wh/kg [42]. Thin film Li-ion batteries are also a potential power source although their energy density is far from the regular battery values [43]. One way to improve the performance of such batteries is to deposit the different layers on a 3D structure thus increasing the surface area and the energy density. Nathan et al. [44], showed that a 3-D battery have a capacity density of 1 mAh/cm^2 (in thin film batteries it is hard to estimate the specific energy thus the capacity is related to the surface area) in comparison with a 2-D battery manufactured in parallel that has 0.04 mAh/cm^2.

An evolving internal power source technology is the fuel cell. The fuel cells have large specific energy of about 1,000 Wh/kg, although they are not able to produce large currents and their power density is about 50 W/kg. Yao et al. [45] developed a 16.4 cm^3 volume fuel cell with all the compliant micro fluidic systems with an energy density of 1,540 Wh/kg and a maximal power supply of 10 mW.

Alternative internal power sources are flywheels and super capacitors. Flywheels theoretically can provide high power but it is difficult to downscale them [46]. Super capacitors [47] are characterized by high power density but low energy density [37, 40]. These make capacitors a complementary power source in a hybrid system which contains an additional power source with high energy density such as a zinc-air battery or a fuel cell.

Wireless transfer of energy by induction is limited to small distances due the fast decay of the magnetic field and its high attenuation in the body. The advantage of wireless transfer methods is that the energy density is infinite. The method has been used to power a micro camera for capsule endoscopy [48] where the system delivered 200 mA. Another option is to use the RF magnetic field of the MRI. Kosa et al. [49] showed that such a device can potentially transfer 1.68 W into a $5 \times 5 \times 5$ mm^3 omni-directional coil.

Another wireless energy transfer method is converting stresses induced by an external ultrasonic source or by natural body vibrations (energy harvesting) into electrical charge by a piezoelectric generator [50]. Such micro devices have been realized in small scales [51, 52] although they were to produce only several μW-s of induced power. In addition to a piezoelectric conversion method the external energy can be converted by a magnetic power generator [53]. Such generators can be downscaled under 1 cm^3 and they achieved a power density of 2.2×10^{-3} W/cm^3 [54].

16.2.3 Control Unit

Control units of micro robots are usually limited to driving circuits for the actuators [55]. Endoscopic capsule that are developed around the world are using IC (Integrated Circuit) design to process the input from the SU (usually only a micro camera) and prepare it for the CT [13, 55]. Casanova et al. [56, 57] an MXS chip for the control of the AP, AM and CT.

The CU have to be specially designed for a medical micro robot because the specific requirements and the necessity of low energy consumption [13]. A good example for the extreme limitations of micro robots is the I-Swarm project [58], in which a control unit was made for a $3 \times 3 \times 3$ mm^3 robot with the power consumption bellow 1 mW.

16.2.4 Communication Transceiver

Implanted medical devices having wireless communications capabilities are used for various applications. Traditionally, magnetic coupling was used for communications between medical devices and external programmers. Magnetic coupling links use large antennas and low frequency and therefore are suitable for relatively large devices needing low communications bandwidth (data rates on the order of a few hundreds bits/second).

Originating from pacemakers and cochlear implants [59], the demand for increased data rates and decrease in power consumption had caused system designers to move up to higher frequencies, where low-power circuits and small antennas can be used and higher bandwidths are available [60].

Endoscopic capsules use standard wireless micro controllers [13] for example Texas Instruments CC2430 or specific ASIC design [8].

One of the most demanding applications in terms of its communication system requirements is the Battery-Power Micro-Stimulator (BPM) for FES (Functional Electrical Stimulation). This device packs all the requirements described above, along with various sensors and actuators and a rechargeable battery into a very small size cylinder [59]. Using a low-power wakeup circuitry and TDMA regime, the system is capable of maintaining bi-directional communications with less than 10 ms latency continuously with an average current consumption as low as 10 μA.

16.2.5 Intervention Micro-Tools

Endoscopes enable diagnosis and intervention from the same port with the same tool. Capsule endoscopes diagnose the GI passively with the SU (usually a camera). In order that a medical swimming micro robot will effectively replace an endoscope it should have tools for intervention, AM. Several groups a re working on AM tools for medical micro robots.

In order to distinguish between malignant and benign tumors the micro robot has to retrieve tissue samples by biopsy. Kong et al. [61] developed a rotational micro biopsy device for the small intestines. Park et al. [62] developed a micro biopsy unit for capsule endoscopes. Rotation of a torsional spring is converted by a mechanism into linear motion of a pair of micro spikes. The tool has a diameter of Ø10 mm and height of 1.8 mm. Byun et al. [63] proposed a barbed needle for micro biopsy made by MEMS technologies. Needle insertion experiments showed that 3 [N] force [64] is needed to penetrate soft tissue which is difficult to achieve with AP in a medical micro robot. An additional difficulty in micro-biopsy is to insure proper contact between the robot and the tissue. One possible solution for these problems is a helical robot that drills itself into the tissue. Ishiyama et al. [65] showed that a magnetic micro robot with a helical drill-head (Ø2 mm diameter and 11.5 mm long) can borrow into bovine tissue.

An additional micro tool that is necessary for AM is a micro holder or gripper. Micro-grippers drawn vast attention as one of the basic building blocks of micro-robotics [66]. A micro-gripper for medical application has to be energy efficient and not fragile. A good example for a micro-gripper for medical application is the piezoelectric micro-gripper developed by Menciassi et al. [67]. The micro-gripper was made of nickel and was actuated by a PZT actuator. The gripper had a strain gauge which was able to monitor the applied force. Micro-grippers can be used in a medical micro robot in various medical tasks such as grasping of retinal membranes [68], placing radioactive agents for cancer therapy (brachytherapy) [69] and controlled release drug delivery micro systems [70].

Other micro tools for manipulation may include micro pumps [71] for drug delivery and various ablation tools although they can be problematic from the power requirements point of view.

16.2.6 Tools for Diagnosis

The research of sensors and sensing system is one of the most developed in MEMS, [72], thus there are many options for integration of SU in a medical micro robot. From the medical point of view the micro robot and the endoscope are extensions of the surgeon's senses during MIS. For the surgeon the most important senses during an intervention are the vision and the tactile (including the force and temperature) senses, for that reason in a medical micro robot a micro camera with LEDs, thermal sensor, and force sensor should be integrated. There is a large variety of CMOS camera chips that are small enough to be integrated into a medical micro robot, in order to use a micro camera for endoscopy special optics has to be added and white LEDS for illumination. An example for a small camera for medical applications is the IntroSpicio™ 115 Camera System manufactured by Medigus [73]. The camera head's size is $1.8 \times 1.8 \times 11\ mm^3$ and its power consumption is 80 mW.

It is very important to know the exact location of the robot in order to navigate in in the body especially in featureless conduits such as the spinal SAS or the small intestine in the GI track. Intra-body tracking method of the robot has be used in order to locate the robot. In most medical applications 5-D localization (cartesian location coordinates and two tilting angles) is suffice. The most promising localization method is magnetic tracking. There are wired magnetic trackers with accuracy under 1 mm [74] that can be used for medical applications but additional research is for integrating them into a wireless micro robot. Nagaoko and Uchiyama developed a 3-D position sensor for an endoscopic capsule [75] the receiving coil of the sensor was Ø6.5 mm with 160 turns of 40 μm copper wire. The planar (X-Y) accuracy of 2.8 + 2.2 mm and 13.4°+ 20.9 (deg) (Average + Standard Deviation) within a 0.4 (m) range from the magnetic field generator.

16.2.7 Example of a Micro Robot for Medical Application

One of the applications for a medical micro robot is neuro-surgery. We designed a swimming micro robot for ventriculostomy. The micro robot effectively demonstrates the different aspects and components detailed above.

The usual approach in ventriculostomy is using a rigid or flexible endoscope. Currently, endoscopes are introduced through a cannula into the lateral ventricles. When using rigid endoscopes, an opening of a diameter of Ø10 mm in the brain is necessary in order to reach the target area. Since the surgeon wants to minimize damage to the brain, he needs to insert the endoscope as close as possible to the surgical target. Such a minimal distance approach, however, can endanger other parts in the brain and may damage them.

Flexible endoscopes are partially solving the problem but they still offer only limited access due to external tethering. An untethered swimming micro robot (about 1–2 cm^2 in size but its components are made by micro-fabrications) can be inserted through a cannula crossing a (potentially remote) functionally blank brain area but still could reach its target area without any strong spatial constraints.

The robot's task is to execute the following intervention:

1. Inserting a Ø6.6 mm cannula into the lateral ventricle through the right-frontal or right-parieto-occipital bone without jeopardizing functionally vital areas (standard procedure for ventriculostomy).
2. Introducing the robot through the cannula into the lateral ventricle as shown in Fig. 16.1a.
3. Navigation in the ventricle to reach the target area as shown in Fig. 16.1b.
4. Inspection of the target area by high resolution imaging and local investigation of the electrical neuronal activity by the robot's built-in electrodes.
5. Targeted intervention: force guided biopsy, local chemotherapy or placement of radioactive agents for brachytherapy. If sufficient power is available, ablation or cauterization can also be carried out.

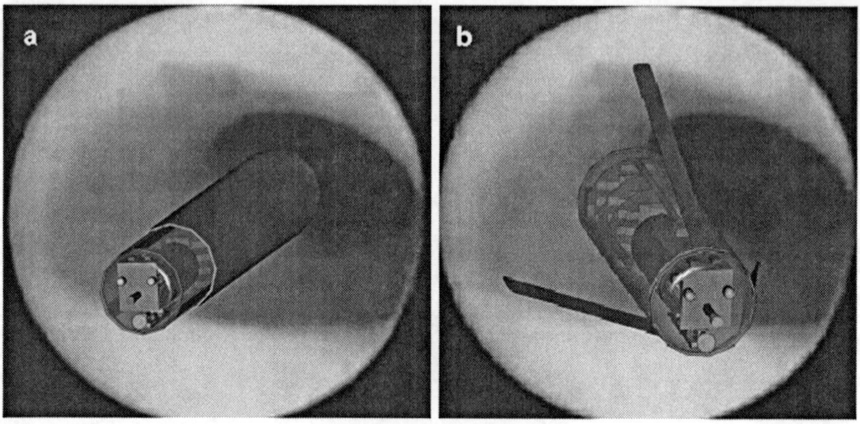

Fig. 16.1 Illustration of the swimming micro robot (**a**) inserted by a cannula into the lateral ventricle and (**b**) swimming to its target area. The background has been generated from an endoscopic image of the lateral ventricles (Copyright 2008 IEEE.)

6. Returning of the robot to the insertion point, and docking to the cannula.
7. Retrieving the robot and the cannula.

The robot has a capsular form and when it is released from the cannula three swimming tail AP actuators pop out to enable maneuvering in five DOF. This swimming robot has a diameter of Ø6.6 mm (determined mostly by the diameter of the batteries) and a length of 31 mm which results in a net volume of 1.1 cm^3. The propulsive units' angle with the axis of the body is 30° and the outer diameter of the robot with open swimming tails is Ø26 mm.

The different components of this robot match the categories of Sect. 16.2.1–16.2.6 in this fashion (See Fig. 16.2):

1. AP – Three flagellar piezoelectric swimming tails, (for detailed description of their operation method see Sect. 16.3).
2. PS – Power source made of batteries or magnetic coils for RF induction.
3. CU – Custom designed integrated circuits (IC) for command and control. The main tasks of the CU here is process the data of the SU, convert it to a proper communication protocol, generate the actuation signals for the AP according to external command.
4. CT – IC for communication and an antenna that is able to transmit the signal out of the body.
5. AM – Micro tools for biopsy, and holding and deploying radioactive agents for brachytherapy.
6. SU – Endoscopic camera and LEDs, a coil for magnetic tracking of the robot and a force sensor array to transmit tactile information and to increase the safety (warning of collision with the ventricle walls).

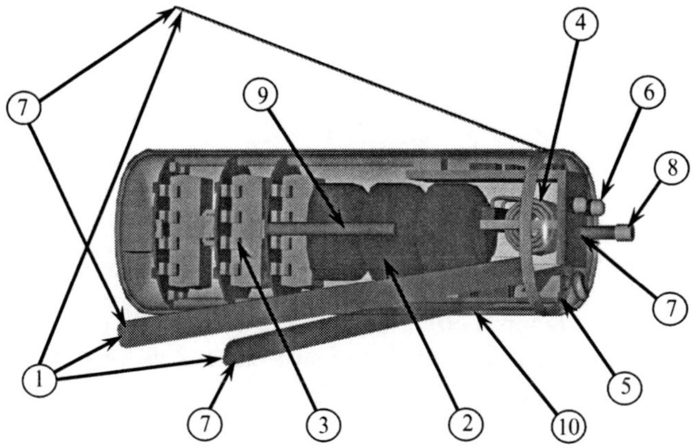

Fig. 16.2 Conceptual assembly drawing of a swimming micro robot for neurosurgery. The components of the robot are: (1) Three swimming tails, (2) Power source (here 3 Renata ZA10 batteries in series), (3) Packaged IC for command control and communication, (4) Antenna, (5) Endoscopic Camera, (6) LEDs, (7) Force sensors, (8) Tool for intervention, (9) Localization sensor (here an Aurora magnetic tracker receiving coil), (10) Casing (Transparent in the drawing to show the internal components) (Copyright 2008 IEEE)

16.3 Flagellar Swimming for Positioning

As it was shown earlier one of the main positioning mechanism for a medical micro robot is a swimming actuator. Micro robots swim in low Reynolds number fluidic regime (Stokes flow), for example a typical 0.1 mm micro robot that swims in water with a velocity of 1 mm/s has a Reynolds number of 0.1. Due the reversibility in low Reynolds number flow (e.g., Stokes flow) the action of swimming micro organisms in nature are different from regular size swimmers [76]. All the micro swimming mechanisms such as spermatozoa [77], cilia [78] and amoeba [79] create in one way or another a traveling wave, advancing in the opposite direction of the micro organism's locomotion. The simplest swimming method for a micro system is flagellar swimming by creating a planar or helical [80] traveling wave in an elastic tail.

In order to achieve propulsion in a swimming tail one has to create undulating motion along it. Such motion is created by rotating an object with non-zero off-diagonal elements in the fluidic resistance tensor, [81] or by deforming the object at different locations [20].

It is easier to create bending vibration in a beamlike structure then rotation in a MEMS actuator. The simplest undulating motion that one can formulate with a vibrating beam is a planar traveling wave. A traveling wave can be decomposed into the natural vibration modes of the beam, $\varphi_k(x)$:

$$
\begin{aligned}
w(x,t) &= w \sin \kappa(x - Ut) = w(\sin \kappa x \cos \kappa Ut - \cos \kappa x \sin \kappa Ut) \\
&= \sum_{k=1}^{\infty} (Cs_k \cos \kappa Ut - Cc_k \sin \kappa Ut)\varphi_k(x) = \sum_{k=1}^{\infty} g_k^{(d)}(t)\varphi_k(x) \\
&= \sum_{k=1}^{\infty} G_k \sin(\Omega t - \Phi_k)\varphi_k(x)
\end{aligned}
\tag{16.1}
$$

w is the amplitude of the traveling wave, κ is the wave number, U is the velocity of the traveling wave, $\varphi_k(x)$ are the vibration modes of a the beam, Cs_k and Cc_k are the coefficients of the decomposition of the $\sin \kappa x$ and $\cos \kappa x$ functions accordingly.

In order to approximate the traveling wave one has to form the desired time functions $g_k^{(d)} = G_k \sin(\Omega t - \Phi_k)$, by setting the phases and amplitudes of the input signal of the actuators.

16.3.1 Theoretical Model

The phases Φ_i for all $i = 1, n$ and amplitudes G_i for all $i = 1, n$ of the actuator's input signal are derived from the solution of the beam model. When the motion and the geometry of the beam is small enough (for example a PZT bimorph with the size of $10 \times 1 \times 0.1\ \text{mm}^3$) one can solve the beam's equations analytically. Otherwise a numerical model of FEA coupled with CFD or an equivalent mass spring model [82] should be used (Fig. 16.3).

The theoretical model of a beam divided into n sections. The field equation of each section is:

$$
\begin{aligned}
&\mathrm{m}_1 \frac{\partial^2 w_1(x,t)}{\partial t^2} + q_1(x,t) + \hat{\mathrm{K}}_1 \frac{\partial^4 w_1(x,t)}{\partial x^4} = 0 \forall x = [0, \alpha_1 L] \\
&\qquad\qquad \vdots \qquad\qquad\qquad\qquad \forall i = 1, 2, \ldots n \\
&\mathrm{m}_n \frac{\partial^2 w_2(x,t)}{\partial t^2} + q_n(x,t) + \hat{\mathrm{K}}_n \frac{\partial^4 w_n(x,t)}{\partial x^4} = 0 \forall\ x = [\alpha_n L, L]
\end{aligned}
\tag{16.2}
$$

$\mathrm{m}_1 = \sum_{\mathrm{j}=1}^{\mathrm{n}} \rho_{\mathrm{ij}} \mathrm{A}_{\mathrm{ij}}$ is the distributed mass of each elastic domain. In each domain, designated by index i, there are n layers. Each layer, designated by index j, has different cross sectional area, A_{ij} (in the case of a rectangular cross section $A_{ij} = b_{ij} t_{ij}$), and density, ρ_{ij}. q_i (x,t) is the distributed force applied by the fluid on each elastic domain. $\hat{\mathrm{K}}_i = \sum_{j=1}^{n} Y_{ij}[(1 + \xi_{ij}) I_{ij} + A_{ij} \mathrm{Z}_{ij}^2]$ is the stiffness of each elastic domain, i. The stiffness depends on the Young modulus Y_{ij}, and in the case of piezoelectric beams on the cross coupling coefficient, ξ_{ij} (see [83] for further details on this coefficient), the cross section inertia I_{ij} (in the case of a rectangular cross section $I_{ij} = b_{ij} t_{ij}^3 / 12$), the cross section area A_{ij} and the distance from neutral axis Z_{ij}. The neutral axis position is fixed if no voltage is applied to the piezoelectric layers and its value is $\mathrm{Z}_{NA} = \sum_{j=1}^{n} \zeta_{ij} A_{ij} Y_{ij} / \sum_{j=1}^{n} A_{ij} Y_{ij}$ when measured from an arbitrary point.

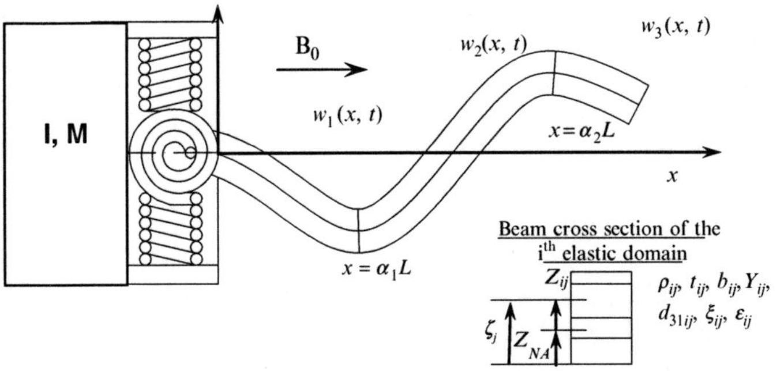

Fig. 16.3 Illustration of the swimming micro robot's tail divided into three sections (Copyright 2007 IEEE.)

The boundary conditions (BC) are linear, K, and angular spring, K_θ, attached to head section with mass, M, and moment of inertia, I, at $x=0$ and free at $x=L$. Spring BC conditions were chosen because of the difficulty of creating clamped boundary conditions in an experimental setup. The boundary conditions are:

$$
\begin{aligned}
&@\ x=0\\
&\hat{\mathsf{K}}_1\frac{\partial^2 w_1(x,t)}{\partial x^2}+\mathsf{M}_{E1}(t)=\mathsf{I}\frac{\partial^3 w_1(x,t)}{\partial x\partial t^2}+K_\theta\frac{\partial w_1(x,t)}{\partial x}\\
&\hat{\mathsf{K}}_1\frac{\partial^3 w_1(x,t)}{\partial x^3}=\mathsf{M}\frac{\partial^2 w_1(x,t)}{\partial t^2}+Kw_1(x,t)-F_{L1}(t)\\
&\quad @\ x=L\\
&\text{and } \hat{\mathsf{K}}_n\frac{\partial^2 w_n(x,t)}{\partial x^2}=-\mathsf{M}_{En}(t).\\
&\quad \hat{\mathsf{K}}_n\frac{\partial^2 w_n(x,t)}{\partial x^3}=F_{Ln}(t)
\end{aligned}
\tag{16.3}
$$

The continuity conditions (CC) between the different elastic sub-domains n and n+1 are:

$$
\begin{aligned}
&@\ x=\alpha_i L\\
&w_i(\alpha_i L,t)=w_{i+1}(\alpha_i L,t)\\
&\frac{\partial w_i(\alpha_i L,t)}{\partial x}=\frac{\partial w_{i+1}(\alpha_i L,t)}{\partial x}\\
&-\hat{\mathsf{K}}\frac{\partial^2 w_i(x,t)}{\partial x^2}-\mathsf{M}_{Ei}(t)=-\hat{\mathsf{K}}_{i+1}\frac{\partial^2 w_{i+1}(x,t)}{\partial x^2}-\mathsf{M}_{Ei+1}(t)\ \forall i=1,2,\ldots,n\\
&\hat{\mathsf{K}}_i\frac{\partial^3 w_i(\alpha_i L,t)}{\partial x^3}+F_{Li}(t)=\hat{\mathsf{K}}_{i+1}\frac{\partial^3 w_{i+1}(\alpha_i L,t)}{\partial x^3}+F_{Li+1}(t)
\end{aligned}
\tag{16.4}
$$

Equations (16.2)–(16.4) is a generic description of actuation by torques, M_{Ei} (t), such as piezoelectric or thermo-elastic bending actuators, or actuation by shear forces, $F_{Li}(t)$, such as a magnetic Lorenz force. The forces and torques are applied at the boundaries, $\alpha_i L$, of the sub-domains.

In the case of a piezoelectric actuator the elastic moment, $\mathsf{M}_{Ei}(t)$, is

$$\mathsf{M}_{Ei}(t) = \sum_{j=1}^{n} Z_{ij} Y_{ij} A_{ij} d_{ij} \bar{E}_{ij}(t) \quad \forall i = 1, 2, \ldots, n. \tag{16.5}$$

d_{ij} is the effective piezoelectric coefficient of the jth layer and $\bar{E}_{ij}(t) = -V_{ij}(t)/t_{ij}$ is the electric field on the j th layer.

In the case of a magnetic actuator the Lorenz force is

$$F_{Li}(t) = N_i b_i I_i(t) B_0 \quad \forall i = 1, 2, \ldots, n \tag{16.6}$$

B_0 is a constant magnetic field in the direction of the beam's longitudinal axis. N_i is the number of turns in the ith coil, b_i is the width of the coil, $I_i(t)$ is the current in the coil.

Figure 16.4 illustrates how the different actuators drive the elastic tail. One should notice that applying a torque or force with the same phase and amplitude nulls out the actuators in the sub-domain boundaries and the actuator behaves as a single beam actuator with the boundary conditions (16.3).

The PDE set (16.2)–(16.4) is solved analytically by the method of separation of variables. A more detailed solution of a piezoelectric and magnetic three sectioned swimming tail is provided in [20] and [31] accordingly.

Setting the proper phases to the input signals of the actuators will result in a traveling wave as shown in Fig. 16.5. Notice the maxima (designated by the arrows) and minima of the traveling wave is advancing along the x axis.

Based on the swimming theory developed by Taylor [84], the propulsion velocity of the swimming tail, U_{MR} is

$$\frac{U_{MR}}{U} = -\frac{1}{2} \frac{w^2 \kappa^2}{\varphi(\beta_1)} \left(K_0(\beta_1) - \frac{\beta_1}{2} \left(K_1(\beta_1) - \frac{K_0^2(\beta_1)}{K_1(\beta_1)} \right) \right). \tag{16.7}$$

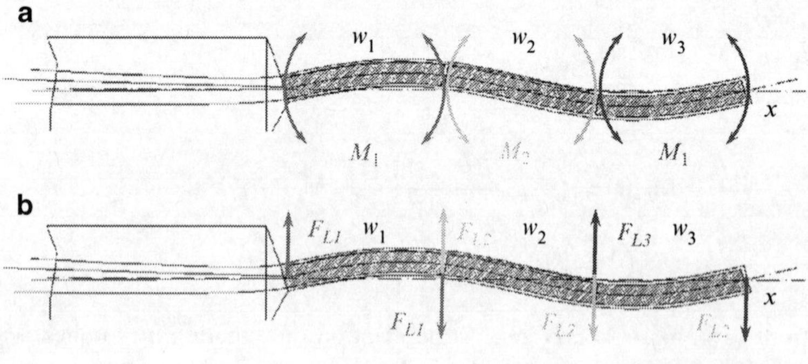

Fig. 16.4 A three section swimming tail with actuators that apply: **(a)** torques M_i or **(b)** forces F_{Li}

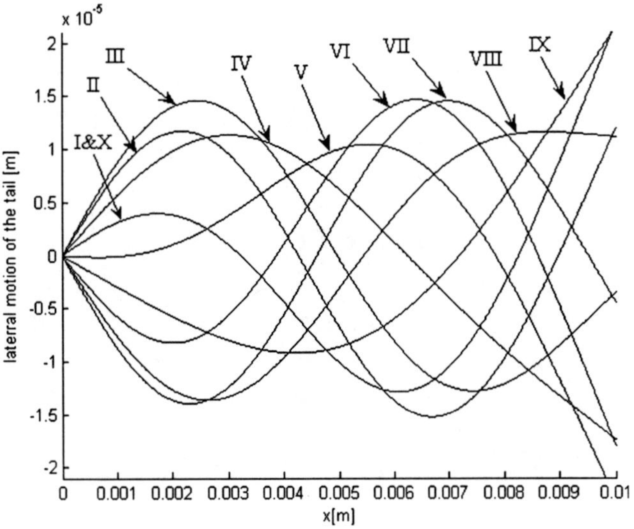

Fig. 16.5 Illustration of the motion of the tail by sequel snapshots of the tail simulation at ten different time-points (I~X) (Copyright 2008 IEEE.)

$\beta_1 = \kappa\hat{c}_0$ is a non-dimensional parameter that characterizes the tails cross section, $\phi(\beta_1) = \beta_1 K_1(\beta_1)\left(\frac{1}{2} + \frac{1}{2}\frac{K_0(\beta_1)}{K_1(\beta_1)} - \left(\frac{K_0(\beta_1)}{K_1(\beta_1)}\right)^2\right) + K_0(\beta_1)$ and $K_i(\bullet)$ are Bessel K functions of the i-th order.

16.3.2 *Swimming Experiments*

We built several swimming tails in order to estimate the effectiveness of the theoretical model. The up – scaled swimming actuators were used in highly viscous fluid in order to simulate properly the swimming conditions of a propulsive micro-system in water. Piezoelectric and magnetic actuation methods were used.

16.3.2.1 Piezoelectric Swimming Actuator

In the initial experiments a commercially available piezoelectric bimorph stripe actuator (Catalog number 40–1055 by APC International Ltd., [23]) was used. The total length of the beam was L = 35 mm (The length of the piezoelectric elements was 31 mm). The thickness of the actuator was T = 0.6 mm. (The thickness of each PZT element was 0.2 mm). The width of the actuator was B = 2.5 mm. The tail was divided into three segments with the lengths: $L_i = [3.9, 10.1, 20.9]$, according to the optimization method described in [85] ($\alpha_1 = 0.125$; $\alpha_2 = 0.45$). The weight of the swimming tail was 0.86 g (Fig. 16.6).

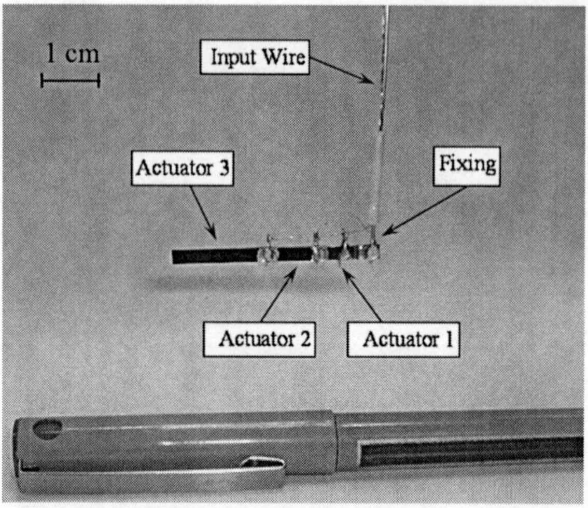

Fig. 16.6 The up-scaled swimming tail (Copyright 2007 IEEE.)

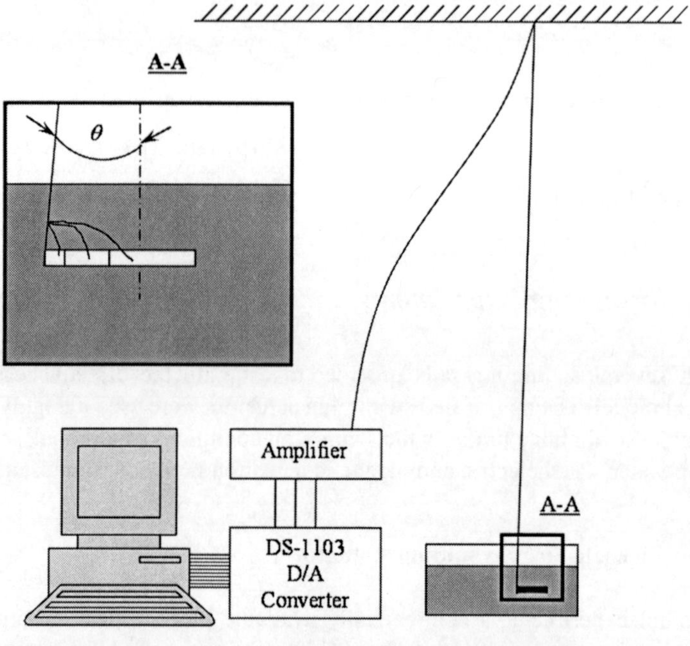

Fig. 16.7 Illustration of the experimental system (Copyright 2007 IEEE.)

The experimental setup included the following instrumentation: PC computer, D/A converter, an amplifier with three channels, input wire, the up-scaled swimming tail and a container with SAE80W90 gear oil (see Fig. 16.7).

The PC supplied three square signals with two delays, which were converted by the D/A converter that had a rise time of 5 μs with 12 bit accuracy. The driving voltage was a square wave with the frequency 2.8 kHz (the third natural frequency), which meant that each square wave period was converted to 71 points. The D/A signal was a trigger for the amplifier and it set the delays between the three actuators. The amplifier supplied a square wave with amplitude of 60 V. The constant amplitude wave is the optimal solution for this beam. The input wire was hanged from 2 m height in order to be able to sense small propulsive forces. The weight of the wire, a parameter needed later for propulsion computation, was measured to be 0.43 g. The tail was submerged into a container with SAE80W90 gear oil, and its depth in the liquid was 5 mm. The viscosity of the oil was measured by a viscometer and the Arrhenius viscosity function was fitted to the results. The absolute viscosity was μ=176.4cP@40$C°$ and μ=607.6cP@20$C°$.

The activation swung out the swimming tail 8 mm from the static equilibrium point where its weight balanced the propulsive force. The phase delays of the three actuators were set according to the theoretical model's prediction:

$$\Phi_1^{(V)} = 0; \Phi_2^{(V)} = 1.06\pi; \Phi_3^{(V)} = 0.26\pi. \tag{16.8}$$

Figure 16.8a shows the initial position of the actuator. The supply was turned on with the phases set to (16.8) (Fig. 16.8a). The actuator reached its maximal distance (Fig. 16.8b) and then the phases were zeroed $\Phi_1^{(V)} = \Phi_2^{(V)} = \Phi_3^{(V)} = 0$, i.e., a harmonic vibration was induced instead of a traveling wave. No propulsion was observed when the beam was actuated by a standing wave (Fig. 16.8c). The experimental results were reproducible and no aging was observed due to heating of the piezo tail.

The propulsive force was calculated from the torque equilibrium around the hanging point of the actuator. The dominant forces in the system are: the propulsive force, $F_p^{(e)}$; the weight of the swimming tail, m_2g; the weight of the input wire, m_1g;

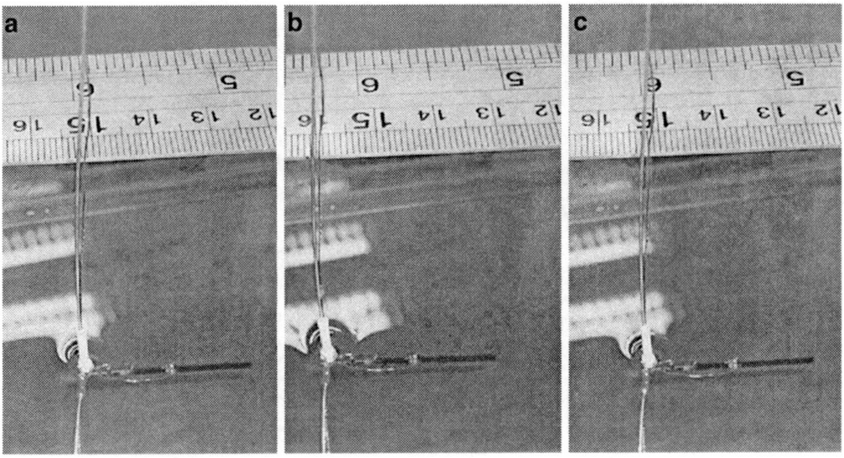

Fig. 16.8 Propulsion experiment: (**a**) power off; (**b**) power on with calculated phase shifts; (**c**) power on with 0 phase shifts (Copyright 2007 IEEE.)

and the floating force, $F_a = \rho_o LBTg$. The propulsive force attained in the experiment was as follows:

$$F_p^{(e)} = \left(m_2 - \rho_o LBT + \frac{m_1}{2}\right) g \sin\left(\frac{\Delta}{l}\right) = 4.03 \cdot 10^{-5}\text{N} \qquad (16.9)$$

Where: $m_2 = 0.86$g is the mass of the tail, $\rho_0 = 910$ kg/m^3 is the density of the fluid, $m_1 = 0.43$g is the mass of the tail, g = 9.81 m/s^2 is gravity, $\Delta = 8$ mm is the maximal distance the tail traveled and $l = 2$ m is the height of the hanging point.

The propulsive force created by the swimming tail equals the drag force of a rigid body with the shape of the tail moved in the fluid with constant velocity, U_{MR}. The propulsive force can be approximated to the drag on an elongated rod as given by [86]:

$$F_p^{(t)} = \frac{2\pi\mu L U_{MR}}{\ln(2L/\hat{c}_0) - 0.80685} = 3.792 \cdot 10^{-5}\text{N} = 0.94 F_p^{(e)} \qquad (16.10)$$

16.3.2.2 Magnetic Coil Swimming Actuator

Another type of swimming tail we utilized was a magnetic coil actuator based on Lorenz force driving. Each coil had an outer length of 7 mm and height of 5 mm (see Fig. 16.9). The dimensions of the inner hole were 5 × 3 mm^2 respectively. Each coil had 100 turns and was wounded using copper magnet wire diameter of 20 μm.

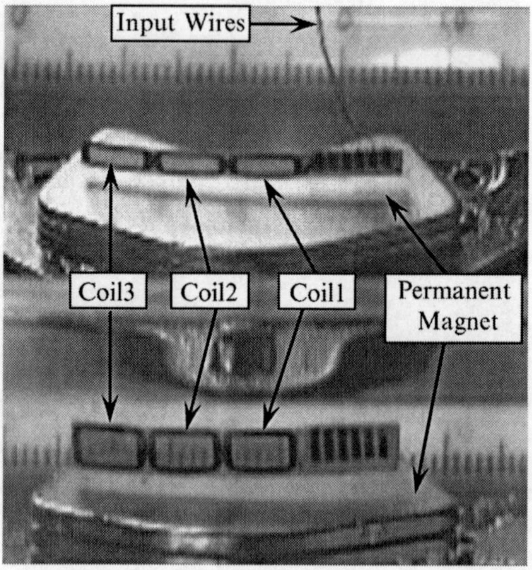

Fig. 16.9 Image of the magnetic swimming tail from two views: profile (*lower image*; reflection from the vessel wall) and half profile (*upper image*; direct trough the water) (Copyright 2008 IEEE.)

Three of such coils were mounted on a 50 μm thick foil. The total length of the tail was 30 mm.

The swimming tail was hung from the height of 1.72 m over a hard disc's permanent magnet that produces a static magnetic field in parallel to the length of the swimming tail of about 0.2 [T]. The input signals that created the largest swimming velocity were:

$$\begin{aligned} I_1 &= i_0 \sin 2\pi\Omega t \\ I_2 &= 2i_0 \cos 2\pi\Omega t \\ I_3 &= -i_0 \sin 2\pi\Omega t \end{aligned} \tag{16.11}$$

$i_0 = 1.12$ mA is the signal amplitude and $\Omega = 10.125$ is the signal frequency. In order to achieve backward swimming we reversed the phases of the signals in (16.11).

The experimental setup is similar to the piezoelectric swimming tails shown in the previous section. The weight of the tail was $m_2 = 0.073$ g and the total weight of the wire was $m_l = 0.29$ g.

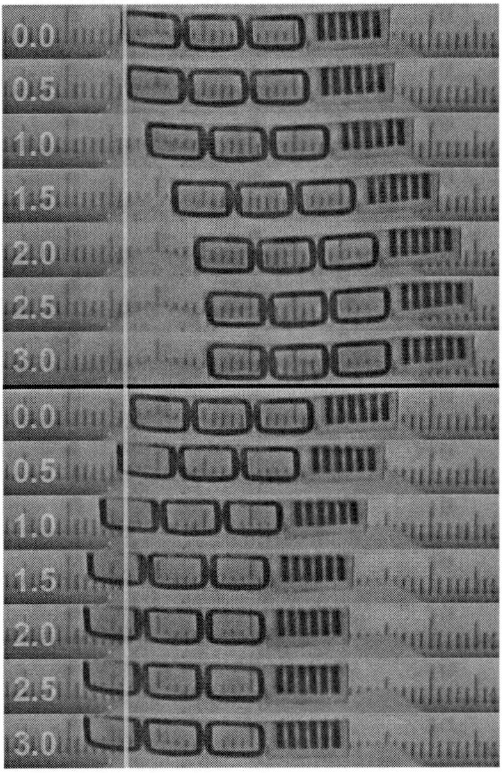

Fig. 16.10 Swimming of the magnetic tail forward (the upper seven images) and backward (the lower five images). The *vertical white line* denotes the initial location of the tail (Copyright 2008 IEEE.)

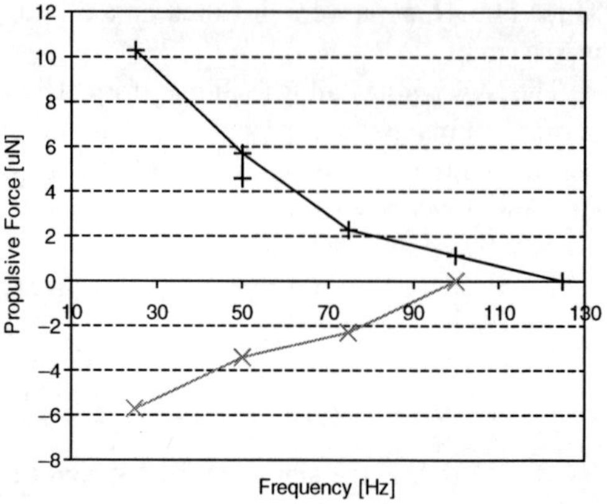

Fig. 16.11 Propulsive force versus frequency in forward (*blue line* with plus markers) and backward (*red line* with × markers) swimming of the magnetic swimming tail (Copyright 2008 IEEE.)

Figure 16.10 illustrates the swimming of the magnetic tail. In this experiment the tail was driven with a frequency of $\Omega = 25$ Hz and amplitude of $i_i = 1.15$ mA. The image shows forward and backward motion of in sequences of 0.5 s.

In this experiment the swimming tail moved 9 mm forward and 5 mm back which are equivalent to a propulsive force of 10.3 μN and 5.7 μN accordingly (calculated from [16.9]).

The magnetic swimming tail was tested in several frequencies from 1 to 150 Hz. Figure 16.10 summarizes the results of those experiments. The lower the frequency the larger the propulsive force. The first natural frequency of the tail was found at $\omega_n = 110$ Hz, we did not found any change in the decay of the swimming velocity at this frequency. As in the piezoelectric swimming tail, whilst the phase difference between the coils is zeroed, the propulsion is zeroed too (Fig. 16.11).

16.4 Scaling of Micro Robots

Pioneered by Trimmer [87] scaling laws are synthesis directives for micro system designer. It is hard to derive general scaling laws for a full system therefore different studies focused their analysis on components of a micro system such as: actuators [14, 87], sensors [88], and power sources [38].

In order to scale down a micro robot one has to consider the system as a whole. It is difficult to derive conclusions for a full robot because the multitude of the degrees of freedom therefore a specific design has to be made and analyzed as shown in Sect. 16.4.1.

A specific medical task such as a swimming micro robot with AP and PS can be analyzed by scaling. Two different swimming methods using magnetic actuator have been compared by Abbott et al. [89]. The one drives a permanent magnet by a magnetic field with constant gradient and the other is based on the propulsion created by the rotation of a helix. The study showed that that using the same external coil as PS, the swimming velocity of helical swimming scales linearly with the radius driving sphere's, $V \propto b$ and the swimming velocity of the gradient driving scales with the square of the radius $V \propto b^2$. This result clearly shows the smaller the magnetic sphere the more advantageous is the helical swimming. The propulsive force F_p is proportional to the $F_p \propto b^2$ in helical and $F_p \propto b^3$ in gradient swimming accordingly.

Analyzing the scaling of a piezoelectric bending actuator (see Sect. 16.3) we assume that the actuator is a bimorph and the length L, width 0.085 L and thickness of 0.005 L. There is a maximal electrical field that the piezoelectric material can use without depolarization thus the driving voltage has to be scaled down according to L too: $\hat{V} \propto L$. The driving torque $M_{Ei}(t)$ scales down $M_{Ei}(t) \propto L^3$ assuming that the material properties (Young modulus, piezoelectric coefficient, dielectric coefficient) are preserved in thin layers. The scaling of the beam's distributed mass, $m \propto L^2$ and elastic stiffness, $\hat{K} \propto L^4$ leads to the linear increase of the natural frequency with the length $f_N \propto 1/L$. The damping coefficient, which is proportional to the distributed damping force exerted by the fluid on the beam $q(x,t)$, remains constant with L. Assuming we are able to drive the swimming tail at its third natural frequency the amplitude of the traveling wave created in the tail scales to $w \propto L$. The velocity of the traveling wave U, remains constant with scaling because the increase of the natural frequency and the decrease of L cancel each other.

That lead us to the conclusion that the swimming velocity in a piezoelectric **swimming tail** V **does not change by the scaling of** L. The propulsive force is scaled linearly with the length, $F_P \propto L$. One has to remember that manufacturing technologies, electronics and power source consideration may limit such a promising result.

The power consumption of the piezoelectric swimming tail is scaled by $S_{AP} \propto L^2$. Assuming power supply of a battery is scaled by the volume $S_{PS} \propto L^3$ there will be minimal volume in which the system of PS + AP will not be able to function. One can overcome this disadvantage by reducing the supply voltage to the actuators: $\hat{V}_{new} \propto L^{3/2}$. The supply voltage is scaled linearly with the swimming velocity, $V \propto L$ but scales by the square with the power requirement $S_{AP} \propto L^3$ consequently the PS and AP scale down equally and the system is balanced.

The magnetic swimming tail does not scale as favorably as the piezoelectric one: $V \propto L^2$ because the current density in the coils, J, has to remain constant. The current in the coils is scaled down by the cross section are of the wire, $I \propto L^2$. The power consumption of the actuator though is scaled down by $S_{AP} \propto L^3$.

If the PS is magnetic induction and the frequency of the supply coil is increased as the tail is scaled down: $\Omega_{PS} \propto 1/L$ the EMF will scale linearly with L, $V_{EMF} \propto L$. In a small coil the resistive component of the coil is dominant and the power transfer by induction scales approximate with the volume of the inducted coil, $S_{PS} \propto L^3$. Thus the scaling of an inductive power source isn't better then the scaling of a battery, nevertheless the amount of energy is unlimited in power transfer.

One has to remember that scaling considerations are only part of the design and generally the manufacturing limitations set the minimal size of a micro system.

16.4.1 Numerical Example

The medical micro robot presented in 16.2.6 is driven by AP made of three piezoelectric swimming tails of the dimensions 20 × 1.7 × 0.2 mm^3. The power consumption of the actuators is:

$$S_{AP} = 3S_{AP}^{(1)} = 3.11.1[mW] = 33.3[mW] \tag{16.12}$$

The power consumption of a commercial micro controller that can perform the tasks needed for this robot is about 20 mW (for example Intel 8051). There is IC specially designed for low power consumption with limited performance [58]. Several designs were able to reduce the power consumption of the CU to 1-2 mW [58, 90], thus

$$S_{PC} = 5[mW] \tag{16.13}$$

CT can be added in a separate chip with low power consumption such as the chip developed by AMF:

$$S_{CT} = 0.014[mW] \tag{16.14}$$

Another solution is to use a single chip that combines CT and CU such as Texas Instruments CC2430:

$$S_{PC} + S_{CT} = 54[mW]. \tag{16.15}$$

The most power consuming components in the SU are the micro camera and the LEDs. The power consumption of a Medigus micro camera for endoscopy is 80 mW and its maximal frame rate is 50 Hz. The power consumption of a Toshiba TLWH1100(T11) led is 126 mW for continous operation. The camera is used at the rate of 10 Hz therefore using a switching circuit for the LED will reduce the SU consumption to

$$S_{SU} = S_{camera} + S_{LED} = 80[mW] + 126[mW] \cdot \frac{10[Hz]}{50[Hz]} = 110[mW]. \tag{16.16}$$

The AM of the robot is a biopsy or holder micro tool. In order to penetrate tissue a force of 3 [N] is needed. The AP cannot create such a force thus the robot cannot "ram" the tissue in order to penetrate and retrieve a sample. One option is using a pre-loaded spring. Another possibility is a chemical transducer. In both cases the power consumed by the triggering mechanism alone and it can be neglected from the general calculation.

The total power demand of the robot is:

$$S_{AP} + S_{CU+CT} + S_{SU} = 120[mW]. \qquad (16.17)$$

A PS made of two ZA-13 zinc-air batteries with a nominal power density of 72 W/kg is able to provide a maximal current of 62 mA at a voltage of 1.4 V. In order to supply the power demand the batteries should supply 85.7 mA. With a power condenser and power tasking the PS can supply such power. With nominal current the a ZA-13 battery's energy can provide a capacity of 310 mAh. Such a capacity can provide an operation period of:

$$TP_N = \frac{Ca}{I_S} = \frac{310[mAh]}{53.5[mA]} = 3.6[h]. \qquad (16.18)$$

This example shows that with proper miniaturization and integration it is possible to realize an untethered micro robot for neurosurgery. The robot combines all the aspects of a medical micro robot (AP, CU, CT, SU and AM).

16.5 Conclusions and Future Directions

This chapter presents potential clinical application for medical micro robots, especially swimming micro robots. The necessary components to assemble such a robot are identified. From the review of the different components several conclusions can be derived:

1. There are available components to assemble a swimming micro robot and there are no specific gaps of knowledge. The challenge is to integrate the components into an operating system and fit it to the medical requirements.
2. The most critical component is the power source of the robot. There is no adequate internal power sources for a micro robot with a volume under 10 mm^3. The only solution is external induction although it is not favorable from the scaling point of view. A promising PS is batteries with new innovative geometries but currently such batteries are still weaker (having lower specific energy) then standard batteries.
3. The highest power consumers in the payload of the robots are the micro-cameras and an additional way to overcome the power requirements is the development of efficient micro camera and lighting.

An interesting area in medical micro robots is the AP. We summarized all the effort done in this area and presented more profoundly positioning by swimming and especially flagellar swimming.

Scaling analysis is a useful design tool to derive generic directives on promising directions for further research. We showed by scaling that a piezoelectric flagellar swimming is promising especially if micro-fabrication of piezoelectric thin layers will be standardized and available.

Another important conclusion is that one has to exploit external resources for the micro robots. For example the endoscopic capsules use the peristaltic motion of the small intestines for AP and a magnetic swimming tail uses the MRI large constant magnetic field in order to swim. In the future other resources in the body may be used by micro systems such as the blood streams high flow rate and pressure or the free ATP molecules in the muscles.

Micro robots for medical purposes still pose great challenges and opportunities for extensive research.

References

1. Zaaroor, M., Kosa, G., Peri-Eran, A., Maharil, I., Shoham, M., Goldsher, D.: Minim. Invas. Neurosurg. **49**, 220 (2006)
2. Kosa, G., Jakab, P., Hata, N., Jolesz, F., Neubach, Z., Shoham, M., Zaaroor, M., Szekely, G.: In: Biomedical Robotics and Biomechatronics, 2008. BioRob 2008. 2nd IEEE RAS & EMBS International Conference, p. 258 (2008)
3. Yesin,K., Vollmers, K., Nelson, B.: Int. J. Robot. Res. **25**, 527 (2006)
4. Berris, M., Shoham, M.: Comp. Aided Surg. **11**, (2006)
5. Iddan, G., Meron, G., Glukhovsky, A., Swain, P.: Nature **405**, 417 (2000)
6. Rey, J.-F., Kuznetsov, K., Vazquez-Ballesteros, E., Gastrointest. Endosc. **63**, AB176 (2006)
7. Wang, J., Li, X.: Medical wireless capsule-type endoscope system has swallowable wireless capsule, portable image recorder, and wireless terminal end connected to medical imaging workstation to exchange data between system and workstation. Chongqing Jinshan Sci & Technology Group Co Ltd (Chon-Non-standard) Chongqing Jinshan Sci & Technology Group (CHON-Non-standard), p. 21 (2005)
8. Moglia, A., Menciassi, A., Schurr, M.O., Dario, P.: Biomed. Microdevices **9**, 235 (2007)
9. Swain, P.: World J. Gastroenterol. **14**, 4142 (2008)
10. Woo, S.H., Jang, J.Y., Jung, E.S., Lee, J.H., Moon, Y.K., Kim, T.W., Won, C.H., Choi, H.C., Cho, J.H.: Electrical stimuli capsule for control moving direction at the small intestine. In: Proceedings of the 24th IASTED international conference on Biomedical engineering, ACTA Press, Innsbruck, Austria (2006)
11. Rao, S.S.C., Kuo, B., McCallum, R.W., Chey, W.D., DiBaise, J.K., Hasler, W.L., Koch, K.L., Lackner, J.M., Miller, C., Saad, R., Semler, J.R. Sitrin, M.D., Wilding, G.E., Parkman, H.P.: Clin. Gastroenterol. Hepatol. **7**, 537 (2009)
12. McKenzie, J.E., Osgood, D.W.: J. Thermal Biol. **29**, 605 (2004)
13. Menciassi, A., Valdastri, P., Harada, K., Dario, P.: In: Biomedical Robotics and Biomechatronics, 2008. BioRob 2008. 2nd IEEE RAS & EMBS International Conference, p. 238 (2008)
14. Dario, P., Valleggi, R., Carrozza, M.C., Montesi, M.C., Cocco, M.: J. Micromech. Microeng. 141 (1992)
15. Ebefors, T., Stemme, G.: In: Gad-El-Hak, M. (ed.) The MEMS Handbook, p. 281. CRC Press, University of Notre Dame (2002)
16. Lim, G., Park, K., Sugihara, M., Minami, K., Esashi, M..: Sens. Actuat. A: Phys. **56**, 113 (1996)
17. Zhang, J., Wei, W., Manolidis, S., Roland, J., Simaan, N.: In: Medical Image Computing and Computer-Assisted Intervention – MICCAI 2008, p. 692 (2008)
18. Fukuda, T., Kawamoto, A., Arai, F., Matsuura, H.: In: Robotics and Automation, 1994. Proceedings, 1994 IEEE International Conference, p. 814 (1994)

19. Watson, B., Friend, J., Yeo, L.: In: Quantum, Nano and Micro Technologies, 2009. ICQNM '09. Third International Conference, p. 81 (2009)
20. Kosa, G., Shoham, M., Zaaroor, M.: IEEE Trans. Robot. **23**, 137 (2007)
21. Guo, S.X., Okuda, Y., Asaka, K.: 2005 IEEE International Conference on Mechatronics and Automations, Conference Proceeding, vols 1–4, P.1604 (2005)
22. Vidal, F., Plesse, C., Palaprat, G., Kheddar, A., Citerin, J., Teyssi, D., Chevrot, C.: Synthetic Met. **156**, 1299 (2006)
23. Nakabo, Y., Mukai, T., Asaka, K.: In: Proceedings of the SPIE - The International Society for Optical Engineering, p. 132. (2004)
24. Kyu-Jin, C., Hawkes, E., Quinn, C., Wood, R.J.: In: Robotics and Automation, 2008. ICRA 2008. IEEE International Conference, p. 706 (2008)
25. Honda, T., Arai, K.I., Ishiyama, K.: Magnetics. IEEE Trans. Magn. **32**, 5085 (1996)
26. Sendoh, M., Ishiyama, K., Arai, K.I.: Magnetics. IEEE Trans. Magn. **39**, 3232 (2003)
27. Dreyfus, R., Baudry, J., Roper, M.L., Fermigier, M., Stone, H.A., Bibette, J.: Nature **437**, 862 (2005)
28. Bell, D.J., Leutenegger, S., Hammar, K.M., Dong, L.X., Nelson, B.J.: In: Robotics and Automation, 2007 IEEE International Conference, p. 1128 (2007)
29. Yesin, K.B., Vollmers, K., Nelson, B.J.: Int. J. Rob. Res. **25**, 527 (2006)
30. Mathieu, J.B., Beaudoin, G., Martel, S.: IEEE Trans. Biomed. Eng. **53**, 292, (2006)
31. Kosa, G., Jakab, P., Hata, N.: In: Joint Annual Meeting ISMRM-ESMRMB 2007, Berlin (2007)
32. Martel, S., Felfoul, O., Mohammadi, M.: In: Biomedical Robotics and Biomechatronics, 2008. BioRob 2008. 2nd IEEE RAS & EMBS International Conference, p. 264 (2008)
33. Behkam, B., Sitti, M.: Appl. Phys. Lett. **90**, 023902 (2007)
34. Steager, E., Kim, C.-B., Patel, J., Bith, S., Naik, C., Reber, L., Kim, M.J.: Appl. Phys. Lett. **90**, 263901 (2007)
35. Gorini, S., Quirini, M., Menciassi, A., Pernorio, G., Stefanini, C., Dario, P.: In: Biomedical Robotics and Biomechatronics, 2006. BioRob 2006. The First IEEE/RAS-EMBS International Conference, p. 443 (2006)
36. Kim, B., Lee, M.G., Lee, Y.P., Kim, Y., Lee, G.: Sens. Actuat. A: Phys. **125**, 429 (2006)
37. Choudhury, N.A., Sampath, S., Shukla, A.K.: Energ. Environ. Sci. **2**, 55 (2009)
38. Dunn-Rankin, D., Leal, E.M., Walther, D.C.: Progr. Energ. Combust. Sci. **31**, 422 (2005)
39. James Larminie, J.L.: In: Electric Vehicle Technology Explained. p. 23 (2003)
40. James Larminie, J.L.: In: Electric Vehicle Technology Explained. p. 69 (2003)
41. James Larminie, J.L.: In: Electric Vehicle Technology Explained. p. 81 (2003)
42. Broussely, M., Archdale, G.: J. Power Sources **136**, 386 (2004)
43. Song, J., Yang, X., Zeng, S.-S., Cai, M.-Z., Zhang, L.-T., Dong, Q.-F., Zheng, M.-S., Wu, S.-T., Wu, Q.-H.: Journal of Micromechanics and Microengineering **9**, 1 (2009)
44. Nathan, M., Golodnitsky, D., Yufit, V., Strauss, E., Ripenbein, T., Shechtman, I., Menkin, S., Peled, E.: J. Microelectromech. Syst. **14**, 879 (2005)
45. Yao, S.-C., Tang, X., Hsieh, C.-C., Alyousef, Y., Vladimer, M., Fedder, G.K., Amon, C.H.: Energy **31**, 639 (2006)
46. Yoo, S.Y., Lee, H.C., Noh, M.D.: In: Control, Automation and Systems, 2008. ICCAS 2008. International Conference, p. 492 (2008)
47. Lewis, D.H., Waypa, J.J., Antonsson, E.K., Lakeman, C.D.E.: Micro-supercapacitor for energy storage device, e.g., battery, comprises predetermined volume of few cubic millimeters, Northrop Grumman Corp, Tpl Inc, California Inst of Technology, (2003)
48. Lenaerts, B., Puers, R. Sens. Actuat. A: Phys. **123–124**, 522 (2005)
49. Kosa, G., Jakab, P., Jolesz, F., Hata, N.: In: Robotics and Automation, 2008. ICRA 2008. IEEE International Conference, p. 2922 (2008)
50. Anton, S.R., Sodano, H.A.: Smart Mater. Struct. **16**, R1 (2007)
51. Renaud, M., Karakaya, K., Sterken, T., Fiorini, P., Van Hoof, C., Puers, R.: Sens. Actuat. A-Phys. **145**, 380 (2008)

52. White, N.M., Harris, N.R., Swee Leong, K., Tudor, M.J.: 2008 2nd Electronics System-integration Technology Conference, p. 589 (2008)
53. Arnold, D.P.: Magnetics. IEEE Trans. **43**, 3940 (2007)
54. Kulkarni, S., Koukharenko, E., Torah, R., Tudor, J., Beeby, S., O'Donnell, T., Roy, S.: Sens. Actuat. A: Phys. **145–146**, 336 (2008)
55. Montane, E., Bota, S.A., Lopez-Sanchez, J., Miribel-Catala, P., Puig-Vidal, M., Samitier, J.: In: Solid-State Circuits Conference, 2001. ESSCIRC 2001. Proceedings of the 27th European, p. 249 (2001)
56. Casanova, R., Saiz, A., Lacort, J., Brufau, J., Arbat, A., Dieguez, A., Miribel, P., Puig-Vidal, M., Samitier, J.: 2005 IEEE/RSJ International Conference on Intelligent Robots and Systems, p. 789 (2005)
57. Casanova, R., Dieguez, A., Arbat, A., Samitier, J.: J. Low Power Electron. **2**, 291 (2006)
58. Casanova, R., Dieguez, A., Arbat, A., Alonso, O., Canals, J., Sanuy, A., Samitier, J.: 2007 50th Midwest Symposium on Circuits and Systems, vols 1–3, p. 695 (2007)
59. Buchegger, T., Ossberger, G., Hochmair, E., Folger, U., Reisenzahn, A., Springer, A.: In: Ultra Wideband Systems, 2004. Joint with Conference on Ultrawideband Systems and Technologies. Joint UWBST & IWUWBS. 2004 International Workshop, p. 356 (2004)
60. Hu, Y.M., Sawan, M., IEEE: In: 47th Midwest Symposium on Circuits and Systems (MWSCAS 2004), Hiroshima, Japan, p. 25 (2004)
61. Kyoung-chul, K., Jinhoon, C., Doyoung, J., Dong-il Dan, C.: In: Intelligent Robots and Systems, 2005. (IROS 2005). 2005 IEEE/RSJ International Conference, p. 1839 (2005)
62. Park, S., Koo, K.I., Bang, S.M., Park, J.Y., Song, S.Y., Cho, D.: J. Micromech. Microeng. **18** (2008)
63. Byun, S., Lim, J.M., Paik, S.J., Lee, A., Koo, K., Park, S., Park, J., Choi, B. D., Seo, J.M., Kim, K., Chung, H., Song, S.Y., Jeon, D., Cho, D.: J. Micromechan. Microeng. **15**, 1279 (2005)
64. Okamura, A.M., Simone, C., O'Leary, M.D.: Biomed. Eng. IEEE Trans. **51**, 1707 (2004)
65. Ishiyama, K., Sendoh, M., Yamazaki, A., Arai, K.I.: Sens. Actuat. A: Phys. **91**, 141 (2001)
66. Nelson, B.J., Dong, L., Arai, F.: In: Springer Handbook of Robotics, p. 411 (2008)
67. Menciassi, A., Scalari, G., Eisinberg, A., Anticoli, C., Francabandiera, P., Carrozza, M.C., Dario, P.: Biomed. Microdevices **3**, 149 (2001)
68. Bhisitkul, R.B., Keller, C.G.: Br. J. Ophthalmol. **89**, 1586 (2005)
69. Niranjan, A., Lunsford, L., Gobbel, G., Kondziolka, D., Maitz, A., Flickinger, J.: Brain Tumor Pathol. **17**, 89 (2000)
70. Li, Y. Shawgo, R.S., Tyler, B., Henderson, P.T., Vogel, J.S., Rosenberg, A., Storm, P.B., Langer, R., Brem, H., Cima, M.J.: J. Controlled Release **100**, 211 (2004)
71. Nisar, A., Aftulpurkar, N., Mahaisavariya, B., Tuantranont, A.: Sens. Actuat. B-Chem. **130**, 917 (2008)
72. Kovacs, G.: Micromachined Transducers Sourcebook. McGraw-Hill, Boston, (1998)
73. Rothstein, R.I.: J. Clin. Gastroenterol. **42**, 594 (2008)
74. Nafis, C., Jensen, V., Beauregard, L., Anderson, P.: In: Cleary, K.R., Galloway, R.L. (eds) Medical Imaging 2006 Conference, San Diego, CA, p. K1410 (2006)
75. Nagaoka, T., Uchiyama, A.: In: Microtechnology in Medicine and Biology, 2005. 3rd IEEE/EMBS Special Topic Conference, p. 130 (2005)
76. Wu, T.Y.T., Swimming and flying in nature. In: Wu, T.Y.T. Brokaw, C.J., Brennen, C. (eds). New York, Plenum Press (1975)
77. Brokaw, C.J.: In: International Workshop on Biofluid Dynamics, Haifa, Israel, p. 1351 (2000)
78. Okada, Y., Takeda, S., Tanaka, Y., Hirokawa, N.: Cell Struct. Func. **30**, 34 (2005)
79. Alt, W., Dembo, M.: In: Conference on Deterministic and Stochastic Modelling of Biological Interaction (DESTOBIO 97), Sofia, Bulgaria, p. 207 (1997)
80. Taylor, G.: Proc. Roy. Soc. Lond. Ser. a-Math. Phys. Sci. **211**, 225 (1952)
81. Purcell, E.M.: Proc. Natl. Acad. Sci. USA. **94**, 11307 (1997)
82. McMillen, T., Holmes, P.: J. Math. Biol. **53**, 843 (2006)
83. Tadmor, E.B., Kosa, G.: J. Microelectromech. Syst. **12**, 899 (2003)

84. Taylor, G.: Proc. Roy. Soc. Lond. Ser. A. Math. Phys. Sci. **211**, 225 (1952)
85. Kosa, G., Shoham, M., Zaaroor, M.: 2006 1st IEEE RAS-EMBS International Conference on Biomedical Robotics and Biomechatronics, vols 1–3, p. 1103 (2006)
86. Cox, R.G.: J. Fluid Mech. Digital Arch. **44**, 791 (1970)
87. Trimmer, W.S.N.: Sens. Actuat. **19**, 267 (1989)
88. Tabib-Azar, M., Garcia-Valenzuela, A.: Sens. Actuat. A: Phys. **48**, 87 (1995)
89. Abbott, J.J., Peyer, K.E., Lagomarsino, M.C., Zhang, L., Dong, L., Kaliakatsos, I.K., Nelson B.J.: How should microrobots swim? The International Journal of Robotics Research **28** (11–12), 1434–1447 (2009)
90. Kok-Leong, C., Bah-Hwee, G.: In: Circuits and Systems, 2006. ISCAS 2006. Proceedings. 2006 IEEE International Symposium, p. 4 (2006)

Chapter 17
Flagellated Bacterial Nanorobots for Medical Interventions in the Human Body

Sylvain Martel

Abstract Enhancing targeting in the smallest blood vessels found in the human microvasculature will most likely require the use of various types of microdevices and nanorobots. As such, biology may play an important role where medical bio-nanorobots including nanorobots propelled in the microvasculature by flagellated bacteria to target deep regions in the human body will become important candidates for such applications. In this chapter, we introduce the concept and show the advantages of integrating biological components and more specifically Magnetotactic Bacteria (MTB) for the development of hybrid nanorobots, i.e., nanorobots made of synthetic and biological nanoscale components, designed to operate efficiently in the human microvascular network. Similarly, the chapter shows the advantage of using Magnetic Resonance Imaging (MRI) as an imaging modality to control and track such medical nanorobots when operating inside the complex human vascular network. The chapter also presents preliminary experimental results suggesting the feasibility of guiding and controlling these nanorobots directly towards specific locations deep inside the human body.

Keywords Bacteria · Nanorobots · Magnetic Resonance Imaging (MRI) · Medical interventions · Tumor targeting

17.1 Introduction and Motivation

Nanorobots capable of operating inside the human body could potentially help various medical cases including but not limited to tumor targeting, arteriosclerosis,

S. Martel (✉)
NanoRobotics Laboratory, Department of Computer and Software Engineering,
Institute of Biomedical Engineering École Polytechnique, de Montréal (EPM),
Station Centre-Ville, Montréal, QC, Canada
e-mail: sylvain.martel@polymtl.ca

J. Rosen et al. (eds.), *Surgical Robotics: Systems Applications and Visions*,
DOI 10.1007/978-1-4419-1126-1_17,

blood clots leading to stroke, accumulation of scar tissue, localized pockets of infection, and many more.

While much speculation has been published on possible far-future applications of nanorobots, i.e., miniature robots based on nanotechnology using advanced materials and manufacturing techniques, little has been published on applying existing engineering technology with biology to implement systems beyond traditional technological limitations allowing such devices to operate in the human vascular network.

There exist several definitions of nanorobots describing different types, from relatively large robotic platform capable of operations at the nanoscale, to the theoretical version of robots with overall dimensions in the nanometer-scale. Theoretical indeed since the implementation of true robots and especially medical robots with overall dimensions of less than approximately 100 nm (being considered as the upper limit in nanotechnology), is well beyond today's technology. Therefore, a new definition for nanorobots that reflects more the reality of what can be achieved has recently emerged, defining such nanorobots has robots with overall dimensions in the micrometer-scale that rely on nanoscale components to enable us to embed particular functionalities. Similarly, medical nanorobots are defined here as microscale robots that exploit nanometer-scale components and phenomena while applying principles of robotics combined with nanomedicine to provide new medical diagnostic and interventional tools. Although there are many potential applications that could take advantage of such nanorobots, tumor targeting for therapeutic purpose has been chosen in this chapter as an important yet challenging application in medicine where such nanorobots could bring significant outcomes.

17.1.1 Main Accessible Regions for Untethered Robots in the Human Body

There are many regions in the human body which can be accessible for untethered robots. Many of these regions can be classified as direct line-of-sight regions, meaning that many visualization and/or tracking techniques (with or without microscopy) used in traditional robotics can feed back information to a controller for servo-control of the untethered robot. This is important since the imaging modality being used may impact and often complicate the control loop implementation by adding significant latencies that would prevent achieving control stability. Some of these regions that could be classified as direct line-of-sight include areas such as inside the mouth, the ears, or the eyes.

Other regions in the human body where direct line-of-sight is not possible include regions such as the digestive track and the vascular network, to name but only two main regions where significant research efforts are underway. In regions such as the digestive track, larger sized robots are more appropriate than microscale robots whereas in the human vasculature, microscale untethered robots become more appropriate. In general, the human vascular network offer more potential for interventions, especially when we consider that it offers close to 100,000 km of routes to access the various regions inside the human body.

The diameter of the various blood vessels in the human vasculature dictates the overall dimensions of the untethered robots. In larger blood vessels such as the arteries, the maximum diameter is a few millimeters. The carotid artery for instance has a diameter of approximately 4–5 mm. Because of the various factors including wall effects which would add drag force to the robots, the diameter of a single robot navigating in such vessels should be approximately no more than half the diameter of the vessels being traveled.

To reach the capillary networks, such robots must travel through the arterioles. Since the diameters of the arterioles may vary from approximately 150 μm down to approximately 50 μm, an overall diameter for each robot of no more than approximately 25 μm would be necessary to reach the capillaries. To reach a target such as a tumor, each robot would have to travel through the capillaries. These capillaries may have a diameter as small as approximately 4 μm, meaning that the maximum diameter of an untethered robot designed to target a tumor should not be larger than approximately 2 μm. Interesting enough is the fact that the same robot does not need to be smaller than 2 μm in diameter unless it must go through the Blood Brain Barrier (BBB).

17.1.2 Embedded Synthetic Versus Biological Propulsion System

Tumoral lesions can be accessed by transiting through anarchic arteriocapillar networks stimulated by tumoral angiogenesis where capillaries located near the tumor could have diameters as small as a red blood cell. Hence, it becomes obvious that the development of self-propelled medical nanorobots relying on an embedded source of propulsion based on a synthetic machine such as envisioned in [1] for instance and capable of providing sufficient thrust force to operate in the human microvasculature, cannot be implemented considering actual technological advances.

As such, existing biological motors such as the molecular motors of bacteria with the attached flagella acting like propellers, becomes an interesting option for nanorobots, especially medical nanorobots operating in the bloodstreams [2, 3]. The flagellated bacteria of type MC-1 for instance is at this time considered to be one of the best candidates for such application for several reasons. First, the diameter of the MC-1 bacterium is $\sim$2 μm being approximately half the diameter of the smallest capillaries found in human. Second, initial tests performed in mice have demonstrated potential for biocompatibility with proper initial response of the immune system. Third, experiments already showed that each bacterium provides thrust force exceeding 4 pico-Newtons (pN) which is at least ten times the thrust force provided by most species of flagellated bacteria. Hence, knowing present technological limits, such flagellated bacterium can be considered as a serious bio-actuator for nanorobots designed to operate in the microvasculature.

17.1.3 Directional Control

Although propulsion is an important factor, directional control is a must when nanorobots must be navigated in complex vascular networks. In [4], the first random motion of an auto-mobile microchip with flagellated bacteria attached was observed. But for targeting specific regions in the human body as in most other applications that could be envisioned with micro-nanorobots propelled by bacteria, steering or displacement control becomes essential. Some level of control of flagellated bacteria has been demonstrated such as chemical stop/resume [5] and phototaxis stop/resume [6] controls. But so far, only a directional control method based on magnetotaxis seems to be suitable for operations in the microvasculature. Although accurate chemotaxis-based steering control performed by computer has not been demonstrated yet, the use of chemicals may in a practical point of view, be extremely difficult if feasible to apply in the human microvasculature. Similarly, phototaxis-based directional control may not be applicable especially when targeting deep in the human body.

But accurate steering control deep in the human body can be achieved with the use of flagellated Magnetotactic Bacteria (MTB). Already, computer steering or control along a pre-programmed path [7] of a flagellated MTB has been demonstrated [8].

Magnetotaxis-based directional control induces a directional torque to a chain of membrane-based nanoparticles (referred to as magnetosomes) embedded in the cell of the MTB. The torque is induced from electrical currents flowing in a special conductor network surrounding the patient. When interfaced to a computer, automatic directional control can be achieved [8].

17.2 Bacterial Mechanical Power, Propulsion, Steering and Tracking

The four fundamental functions that must be embedded in an untethered microscale nanorobot designed to travel in the human microvascular networks are the mechanical power source, the propulsion system, the steering system, and some sort of beacons allowing such nanorobots to be tracked inside the human body.

Interesting enough is the fact all these four fundamental requirements are already embedded in each MC-1 bacterium as depicted in Fig. 17.1.

As depicted in Fig. 17.1, the diameter of the MC-1 magnetotactic bacterium which is spherical in shape, has a diameter of approximately 2 μm, allowing it to operate in the smallest diameter blood vessels found in human.

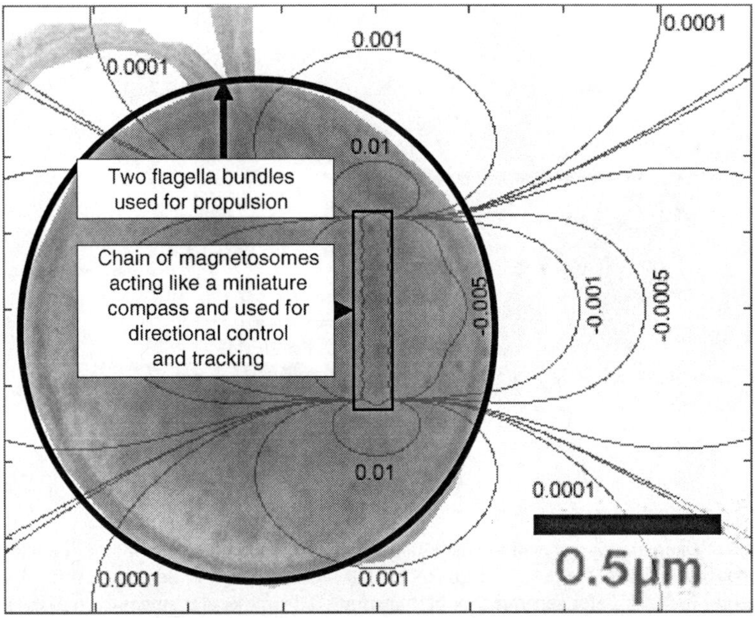

Fig. 17.1 A single MC-1 magnetotactic bacterium imaged using an electron microscope. Propulsion is provided by two flagella bundles while a chain of magnetosomes is used for directional control and tracking. The lines superposed over the image indicate the magnitude of distortion created by the magnetosomes when placed inside an MRI system allowing the bacteria to be tracked deep in the human vasculature where direct line-of-sight is not possible

17.2.1 Bacterial Mechanical Power

The use of bacteria by itself resolves one of the biggest challenges in artificially made micro- or nanorobots since no appropriate technology is available to embed sufficient power within such space constraints. Especially with a microscale robot with overall dimensions of only 2 μm across, the only alternative for engineers and researchers is to induce the power from an external source. An obvious solution which was and remains the most popular to date, is to induce a mechanical force using an external magnetic field typically generated from coils surrounding the patient or the part of the body where the intervention is being performed. Nonetheless, this solution has limitation especially when the distance from the coils and the microscale robots increase to accommodate operation deep in the human body, especially when operating deep in the human torso. In the latter case for instance, overheating of the coils would most likely prevent their usage for such applications. Therefore, an embedded source of propulsion would become more appropriate.

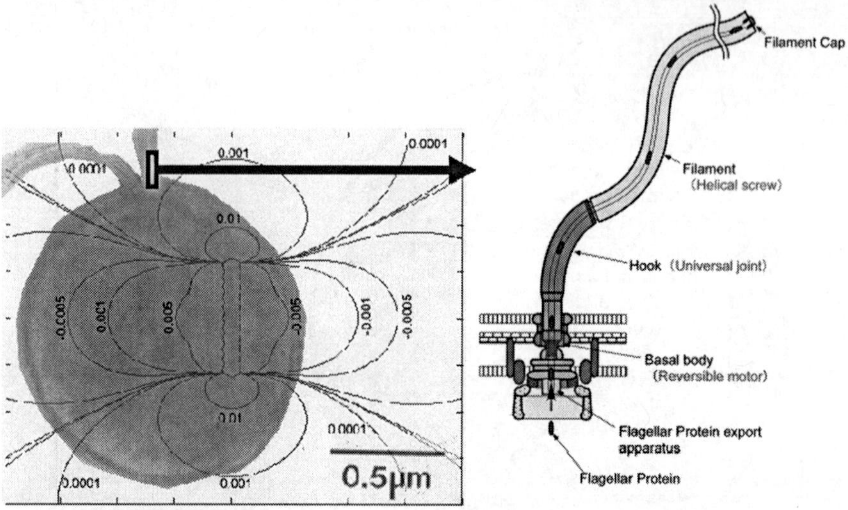

Fig. 17.2 Schematic diagrams of the flagellum connected to the molecular motor of a bacterium. As shown in the figure, the design of this molecular motor is very similar to an artificial version with a rotor inside a stator separated by ball-bearings. This molecular structure consists of three main parts: the basal body, which acts as a reversible rotary motor; the hook, which functions as a universal joint; and the filament, which acts as a helical screw (*OM*: Outer Membrane; PG: Peptidoglycan layer; CM: Cytoplasmic Membrane. (Adapted from Fig. 17.1 in Minamino T. et al. Molecular motors of the bacterial flagella. Curr. Opin. Struct. Biol. **18**, 693–701 (2008))

17.2.2 Bacterial Propulsion System

Each flagellum acting as a propeller is connected to a hook acting as a universal joint that connects to a molecular motor as depicted in Fig. 17.2. The MC-1 bacterium has two bundles of such flagella providing a total thrust force for propulsion between 4.0–4.7 pN > 0.3–0.5 pN for many flagellated bacteria. Each flagellum rotates 360 degrees like a shaft in standard motor and can be reversed for backward motion. The total diameter of each molecular motor is less than 300 nm.

The flagellum acting like a propeller consists of a 20 nm-thick hollow tube. As depicted in the figure, the flagellum next to the outer membrane of the cell has a helical shape with a sharp bend outside. Together, this structure appears like a hook with a shaft running between the hook and the basal body. Similar to the architecture of a potential future artificial version, the shaft then passes through protein rings that act as bearings. For this nanometer-scale propulsion system, counter-clockwise rotations of a polar flagellum will thrust the cell forward while clockwise rotations will result in the cell or the microstructure attached to, to move backward.

Experiments conducted in water and in human blood showed an average velocity for the MC-1 bacteria often exceeding 200 μm/s (compared to an average velocity of approximately 30 μm/s for many other flagellated bacteria). Peak velocities of approximately 300 μm/s, i.e., 150 times its own cell's length per second have also been recorded.

17.2.3 Bacterial Steering System

The direction of motion of most flagellated bacteria is mainly influenced by chemotaxis. With chemotaxis, the swimming direction and the motility of bacteria is influenced by chemical gradients such as nutrient gradients [9–11]. Hence, a chemical approach would first appear to be the appropriate strategy to be used for the directional control of most flagellated bacteria. But although a chemical approach could be suitable in many applications, chemotaxis-based directional control in the human vasculature may be very difficult if suitable to be considered. Furthermore, chemotaxis-based directional control if feasible in the human body may not provide an adequate interface with computers or electronic controllers, which may be an essential requirement for a robotic platform where automatic and accurate motion control along pre-planned paths in the blood vessels must be implemented.

On the other hand, with the right species of flagellated bacteria, such as MTB, a low intensity directional electro-magnetic field capable of penetrating the human body with ease and without harm for the patient can be used while being compatible and easily interfaced with electronic computers and controllers.

For MTB, each cell (body) contains a naturally growth chain of magnetosomes which are membrane-based nanoparticles of a magnetic iron, e.g., iron-oxide for the MC-1 cells. Each nanoparticle has a diameter of only a few tens of nanometers and therefore will show single magnetic domain behavior. This chain acts like a nanometer-sized compass needle and will be oriented with the lines of magnetic field towards an artificial pole generated by computer.

The motion behavior of the MC-1 bacteria can be influenced in many ways including chemotaxis, phototaxis, aerotaxis, and magnetotaxis [12–14], the latter being more appropriate for automatic closed-loop navigation control in the vascular system.

To achieve better computerized control on the swimming direction of the MC-1 cells, the modes which are not or cannot be under the influence of an electronic controller or a software program and which would add uncertainties and errors along a pre-defined swimming path through unpredicted motion behaviors must become negligible. To achieve this, a very low intensity directional magnetic field but slightly higher than the Earth's geomagnetic field of 0.5 G is then applied. When doing so, the directional motion of these self-propelled bacterial steering systems become mainly influenced by magnetotaxis and therefore fully controllable as demonstrated in [8] (Fig. 17.3).

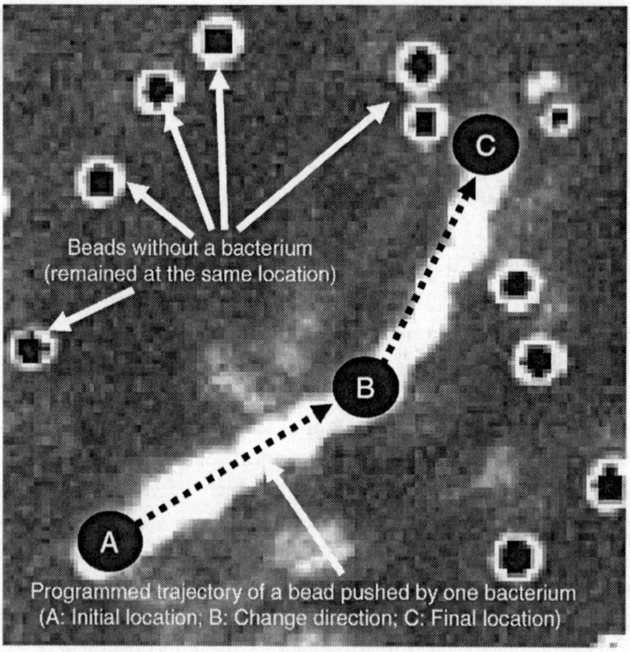

Fig. 17.3 These experimental results captured with an optical microscope, validate the concept of using a single flagellated magnetotactic bacterium as a self-propelled steering system for a microscale robot. Here a single flagellated bacterium was attached using specific antibodies to a 3-μm bead representing the artificial structure or body of a hybrid microrobot. As shown in the upper left section of the image, one can see that such microrobot represented by the bead can be pushed efficiently by a single flagellated bacterium. The lower left section shows the same bead represented by the black circles with numbers along a pre-programmed trajectory and corresponding to a sequence of software instructions, proving that accurate computer-based control of a bacterial propulsion and steering system is possible

17.2.4 Bacterial Tracking System

It does not matter how smart is a nanorobot, for targeting a specific region such as a tumor, any nanorobots will get lost in a maze of close to 100,000 km of blood vessels if a global tracking/navigation system similar to the Global Positioning System (GPS) for humans, is not provided. Therefore, when operating inside the vascular network, since imaging techniques relying on direct line-of-sight is not possible for servo-control purpose, gathering tracking information in order to guide such bacterial nanorobots to a specific targeted location inside the human body using the shortest or any desired paths without getting lost would require an appropriate imaging modality. Magnetic Resonance Imaging (MRI) systems already implemented in most clinics can be used for this purpose. This also adds further advantages especially during the planning phase where soft tissue and the blood vessels can be imaged in 3D.

As mentioned earlier, the magnetosomes embedded in the cell of each MC-1 magnetotactic bacterium are Fe_3O_4 single magnetic domain crystals of a few tens of nanometers in diameter and are used for directional control. Similarly, when placed inside the bore of a clinical scanner, these nanoparticles cause a local distortion of the high intensity DC magnetic field. Such local field distortion caused by each magnetosome can be estimated at a point P of coordinate r (x, y, z) by that of a magnetic dipole as

$$\vec{B}'(P) = \frac{\mu_0}{4\pi}\left(3\frac{(\vec{m}.\vec{r})\vec{r}}{r^5} - \frac{\vec{m}}{r^3}\right). \tag{17.1}$$

In (17.1), $\mu_0 = 4\pi 10^{-7}\,\mathrm{H\,m^{-1}}$ represents the permeability of free space. The dipolar magnetic moment (A m^2) for a magnetosome uniformly magnetized by the high intensity of the MRI scanner is then given by

$$\vec{m} = \frac{4}{3}\pi a^3 \vec{M}_{SAT} \tag{17.2}$$

where the saturation magnetization of the magnetosome and its radius a (m) are taken into consideration. A numerical simulation of a single bacterium magnetic field perturbation plotted over an electron microscopy image assuming 11 aligned magnetosomes, each with a diameter of 70 nm has been plotted in Fig. 17.1. Since the homogeneity level of modern MRI clinical scanner is approximately 5 ppm over a 50 cm diameter spherical volume at 1.5 T, suggest that tracking of bacterial medical nanorobots inside the human body using existing medical imaging platforms is possible. This is confirmed experimentally in Fig. 17.4 where a swarm of MC-1 flagellated bacteria has been imaged with a 1.5 T clinical scanner.

In reality, because of the small overall size of each bacterial nanorobot, delivering sufficient therapeutic agents to a tumor would require more than one nanorobot. As such, directional control of a swarm or agglomeration of flagellated bacteria (or bacterial nanorobots since unlike other flagellated bacteria operating independently of computer commands, they represent fundamental self-propelled entities being controlled by computer in a closed-loop scheme) as shown in Fig. 17.4 will be essential in many instances. Using several bacteria simultaneously not only increases the amount of therapeutics being delivered but makes the tracking and hence the feedback control easier by increasing the signal intensity inside the human body. This is important since although the sensitivity of a clinical MRI system is very high, the limitation in the spatial resolution can be somewhat compensated for with a larger agglomeration of MTB. Furthermore, as depicted in Fig. 17.4, the percentage of bacteria in a specific region can also be evaluated with proper MRI sequences. This is important to evaluate the percentage of MTB that have reached the target. For delivering therapeutics to a tumor for example,

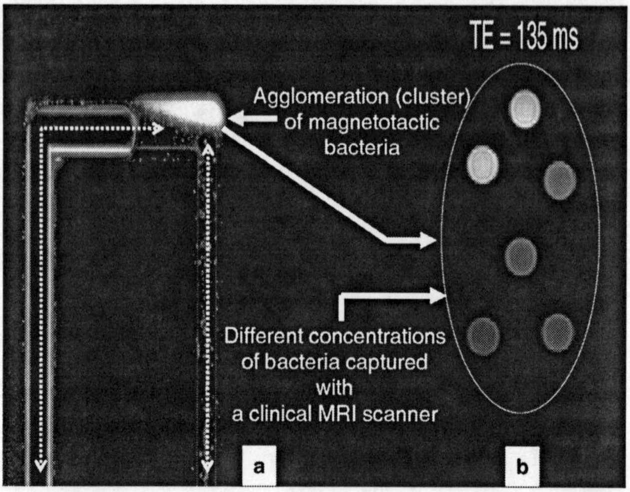

Fig. 17.4 (**a**) The figure shows an example of directional control of an agglomeration of MC-1 magnetotactic bacteria along the dotted path and (**b**), images taken with a clinical MRI system of various concentrations of MTB. The magnetosomes embedded in the cells affect the spin-spin (T2) relaxation times when imaged using fast spin echo sequence. (Images taken with a Siemens Avanto 1.5T clinical scanner using a wrist antenna with sequence parameters: TR/TE = 5,620/ 135 ms, slice thickness of 20 mm, and pixel spacing of 0.254 mm)

determining the percentage of bacterial nanorobots that has reached the target will indicate the amount of secondary toxicity delivered to the systemic blood circulatory system. In turn, this will indicate how many times the operations can be repeated without increasing secondary toxicity in the human body beyond a critical threshold. Similarly, if only one attempt is planned, this will indicate the maximum level of toxicity of the therapeutics that can be carried by the bacterial nanorobots for maximum efficacy at the tumoral lesion.

Figure 17.4a shows an agglomeration of MC-1 bacteria swimming in a tube mimicking a human blood vessel. Feedback directional control was done from tracking information gathered using an optical microscope (Fig. 17.4a). To the right in Fig. 17.4b, one can observe the intensity of the image from various concentration of MTB.

17.3 Envisioned Main Types of Medical Nanorobots

The human vasculature represents a complex environment where the variations in physiological properties represent new difficulties that must be dealt with. For instance, blood flow velocity in the arteries can reach one meter per second whereas in the capillaries it can be as low as 1 mm/s. But in capillaries, because of the size of the blood cells, the medium cannot be considered homogeneous as in the arteries. Therefore, sending a nanorobot to a tumor could be somewhat similar to sending an exploration robot on the surface of planet Mars where different modules must be used to first escape the Earth gravitational force, then travel from the Earth to the orbit of Mars using closed-loop navigation control prior to use a landing module before releasing the exploration robot itself. Similarly, navigating from the release site in larger arteries at the catheterization limit to the tumoral region and passing through vessels such as the arterioles and capillaries, requires various types of microdevices/carriers and/or robots.

17.3.1 Bacterial Nanorobots

Bacterial nanorobots or nanorobots propelled by bacteria are most efficient in Low Reynolds conditions and particularly in the capillary networks. As mentioned earlier, unless designed to travel through the BBB, there is no apparent advantage to implement bacterial nanorobots designed to operate in the human microvasculature with an overall size of less than approximately 2 μm across. Actual designs of nanorobots capable of crossing the BBB are presently based on speculations without real development and experimental data and as such, it is not covered here due to a lack of experimental data and proof-of-concepts. Nonetheless, such devices capable of transiting through the BBB could indeed with the advancement in nanorobotics, become a potential topic in future publications.

The functionality of microscale nanorobots can be enhanced for operations in the human vasculature by taking advantage of the properties offered by nanometer-scale components. The propulsion system with the molecular motors and the flagella, the magnetosomes acting as tracking beacons and steering systems are just a few examples of that.

Because of their overall dimensions, engineers must think differently for the implementation of such microscale nanorobots compared to larger robots. For instance, several parts must rely on biochemistry instead of the more traditional mechanical approaches. As a simple example, a miniature mechanical gripper with recognition sensors for specificity in target recognition that may be connected to an onboard computer can be replaced with the integration of specific antibodies. Specific antibodies can also replace the glue or the screws used to assemble components in the implementation of larger robots. This is shown in Fig. 17.5 where a large bead emulating a potential artificial component is attached to a flagellated bacterium to build a hybrid microscale nanorobot. Other types of antibodies, bacteriophages, ligands, functionalized molecules, and peptides are only a few examples of components that can be used to emulate grippers, implement specificity or to embed target recognition, attach components, anchor the nanorobot, and/or to load therapeutic or other types of agents, to name but a few typical functions.

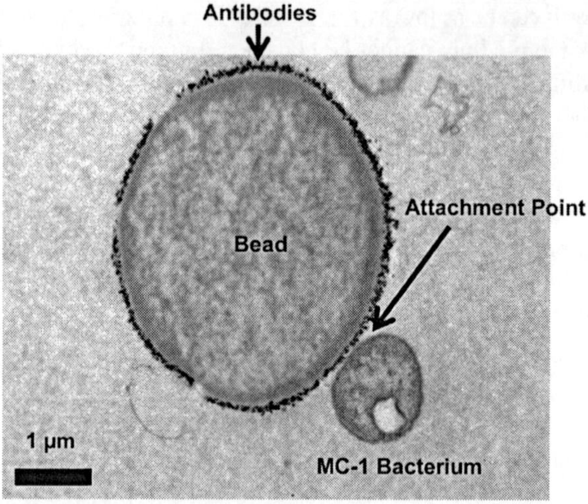

Fig. 17.5 One MC-1 flagellated bacterium being attached reliably to large bead emulating a potential artificial part of a hybrid nanorobot

For instance, loading the bacterial nanorobots can be done in many ways. For example, the MC-1 cell can be loaded with fluorescent cell penetrating peptide which could also be replaced by cytotoxic or radio-active agents. Nanometer-scale biodegradable polymeric nanoparticles loaded with cytotoxic or other agents can also be attached to the cell of the MC-1 bacterium using specific antibodies, again as depicted in the example in Fig. 17.5 where a relatively large artificial structure (here represented by a simple bead) which could be a synthetic part of an hybrid robot, has been attached to a bacterium using specific antibodies.

Bacterial nanorobots have huge advantages compared to an entirely synthetic version. For instance, MC-1-based bacterial nanorobots have superior performance in smaller diameter capillaries over synthetic counterparts such as ferromagnetic-based microrobots propelled by magnetic gradients when operating inside a relatively large animal or an adult size human body.

But the use of non-pathogenic MC-1 bacteria as propulsion system for micro-scale robots has some potential drawbacks that do not exist in synthetic versions. Indeed, initial tests conducted in human blood at 37°C showed that the velocity of the MTB and denoted v_{B37} at a time t (expressed in minutes and within the lifetime of approximately 40 min for the MC-1 bacterium when exposed to the environmental conditions of the human microvasculature) would decrease continuously according to (17.3).

$$v_{B37} = 0.09\,t^2 - 8.10\,t + v_{\mathrm{MTB}}. \tag{17.3}$$

In (17.3), v_{MTB} is the average velocity of the MTB (approximately 188 μm/s depending on cultivation parameters) prior to be exposed to the environmental conditions of the human microvasculature.

Even with such a decrease in velocity when exposed to environmental conditions inside the human body, flagellated MC-1 bacterial nanorobots would remain more effective for at least the time period required to target the tumoral lesion (estimated to be less than approximately 30 min although more tests are required to confirm it) when operating in the human body compared to a synthetic version such as a ferromagnetic microscale robot of similar size with force induced using continuous (50% duty cycle) 500 mT/m gradients being approximately the maximum gradients that can be applied at the human scale (e.g., to accommodate an adult size human torso) inside an upgraded MRI platform.

Effort could be done towards extending the operating life of the bacterium in the human microvasculature using approaches such as the ones relying on RNA. But the lifespan of such bacterium should remain short enough to prevent the bacteria to reproduce, becoming pathogens and a potential threat for the human body. On the other hand, the lifespan and the motility of these bacteria should be sufficient to allow them to reach their targets. Previous studies indicate that 40 min with previously recorded velocities of the MC-1 would most likely be sufficient for such task.

17.3.2 Synthetic Microscale Medical Robots and Carriers

Although bacterial nanorobots may have many advantages especially when we consider their very high efficiency for traveling in the microvasculature including larger capillaries, they become much less effective in larger diameter blood vessels where blood flow is much higher than the blood velocity levels found in the capillary networks. As a simple example to give an idea of their effectiveness, when in an arteriole having a diameter of 100 μm for instance, the bacterial nanorobots will have to deal with an expected average Poiseuille blood flow rate of approximately 0.52 m/s and be able to steer adequately considering an estimated 50 mm between successive blood vessel bifurcations. In this particular environment, an entirely synthetic ferromagnetic nanorobot of approximately 25 μm in diameter and loaded at 50%vol. with proven biocompatible Fe_3O_4 nanoparticles (which has a relatively low saturation magnetization of 0.5 compared to other options), will still perform better. In this case, this synthetic microscale robot will have a velocity of approximately 500 μm/s which is by far higher that the fastest bacterial nanorobot (initial maximum velocity of approximately 300 μm/s) under the same physiological conditions. Although recent experimental results show that the maximum velocity of the MC-1 bacteria can be increased beyond this value, several complementary types of medical nanorobots are still likely to be required considering the very high blood velocity in larger diameter vessels to allow targeting operations deeper in the microvasculature using bacterial nanorobots being carried near the targeting sites using synthetic micro-carriers. So far, recent results and proof-of-concepts suggest that such synthetic untethered micro-carriers and robots will have embedded ferromagnetic or superparamagnetic materials to allow propulsion by magnetic gradients generated by an external source.

It is now known that a ferromagnetic core can be propelled, tracked, and controlled in real-time along a pre-planned path in blood vessels by using magnetic gradients applied in a 3D space. This was shown experimentally *in vivo* in the carotid artery of a living swine [15], an animal model close to human. Equation (17.4) explains the basic principle to induce a propulsion and steering force on ferromagnetic object.

$$\vec{F} = R \cdot V\left(\vec{M} \cdot \nabla\right) \vec{B}. \tag{17.4}$$

The force (17.4) induced on a ferromagnetic core with a volume V within an untethered micro-carriers or robots, depends in great part on the duty cycle R when the propulsion gradients are applied. This duty cycle is limited for two main reasons. First, propulsion cannot be executed when MRI is being performed since the magnetic propulsion gradients would interfere with the imaging gradients used for image slice selection in the bore of the MRI scanner. Second, in particular instances, the coils generating the propulsion gradients may overheat and when this is the case, the duty cycle must be decreased further to allow sufficient time for the coils to cool down to remain within an operational temperature range. All this will contribute to lower the efficiency of the propulsion method especially when the overall dimensions of the ferromagnetic robot become smaller.

Another factor to consider is the magnetization of the core material. Fortunately, when placed in the homogeneous DC magnetic field of 1.5 or 3 T found in modern scanners, it will reach the saturation level (M_{SAT}). Another factor affecting the propulsion force is the maximum gradient that can be applied. Initial estimations and designs suggest that the 40 mT/m of actual clinical MRI scanners could be increased to a maximum level of approximately 500 mT/m with a duty cycle of approximately 50% (to allow for tracking and/or cooling). Preliminary works suggest, although this remains a challenging engineering task at the present time, that this could be feasible within the space constraints inside the bore of a clinical MRI scanner while maintaining sufficient room to place an adult size human.

Nonetheless, since the induced force from magnetic gradients is proportional to the effective volume of magnetic material being embedded in the microscale nanorobot, a given synthetic microscale nanorobot would have to be as large as possible (i.e., with a diameter of approximately half the diameter of the blood vessels being traveled) to gather enough propulsion/steering force to cope with the blood flow in larger diameter vessels. But by doing so, such microscale nanorobot would not be able to transit from larger to smaller diameter blood vessels and will most likely create an undesired embolization.

An approach to resolve this issue is the use of an agglomeration of synthetic microscale robots or carriers. At such a scale and when placed in the high DC magnetic field of an MRI scanner, a magnetic dipole will develop for each micro-scale robots. This will create a dipole-dipole interaction or attraction force between each robot. The objective is to create a strong enough dipole-dipole interaction to increase the effective volume of magnetic material to achieve a higher induced displacement force to cope with the higher flow rate while having a sufficiently low dipole-dipole interaction to allow such agglomeration to change shape when transiting from larger to narrower vessel diameters. Methods such as modifying the characteristics of the surface of each robot can be used to adjust such trade-off in dipole-dipole interaction.

17.4 Combining Synthetic and Bacterial Microscale Nanorobots to Reach Deeper Regions in the Human Vasculature

As stated earlier, preliminary data indicate that a combination of synthetic (artificial) and biological (bacterial) nanorobots would be required for tumor targeting. For instance, the data gathered using a synthetic version indicate that using 50%vol. to allow room for therapeutic loads of material having the highest magnetization saturation *(M_{SAT})* of 2.45 T and an overall size equal to the cell of a single MC-1 bacterium required for operations and navigation in the smallest diameter capillaries under realistic conditions where a diameter of 4–5 μm is expected with 1 mm between bifurcations, and an average Poiseuille blood flow in the order of 0.5 mm/s, will be able to reach a maximum velocity of approximately 4 μm/s. At such velocity, it would be very difficult to operate/navigate within physiological conditions in the angiogenesis networks where the vasculature has narrowed diameter vessels and an impressive number of vessel bifurcations. Looking at one of the most advanced synthetic propulsion system in the form of an artificial flagella built from nanocoils [16] being propelled with the use of a rotating magnetic field where a velocity in a aqueous medium and without flow suggests that it would not be enough for such environment unless the blow flow is completely stopped or at least slowed down significantly. Indeed, the latter achieved a velocity of approximately 4.6 μm/s and as such, it cannot compete with the initial average velocity of approximately 200 μm/s (initial distribution between approximately 30–300 μm/s) of the MC-1 bacteria used as propulsion for the bacterial nanorobots. But as mentioned earlier, the same bacterial nanorobots would be useless in larger diameter vessels where the blood flow is much larger. As such, a combination of synthetic and bacterial nanorobots and/or carriers will be required for many targeting applications performed through the human vasculature. This is shown schematically in Fig. 17.6.

In Fig. 17.6, Large Embolization and Transport (LET) microscale carriers may take various forms. One preferred implementation consists of an entity that is made of several types of material depending on the applications and embedded functionalities required. Within such entity, magnetic nanoparticles are encapsulated with the therapeutic agents to be released if chemo-embolization is supported. For LET, an agglomeration of MTB is also encapsulated. The nanoparticles encapsulated in the polymeric shell can be Fe_3O_4 nanoparticles for proven biocompatibility

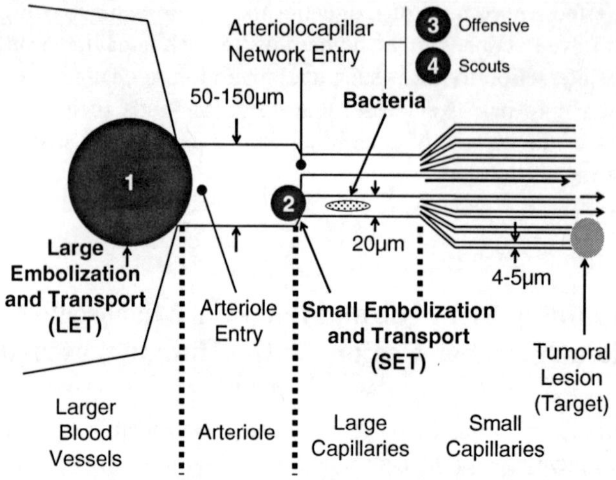

Fig. 17.6 Simple schematic showing the use of synthetic and bacterial microscale robots with their respective operating regions in the human vasculature for tumor targeting

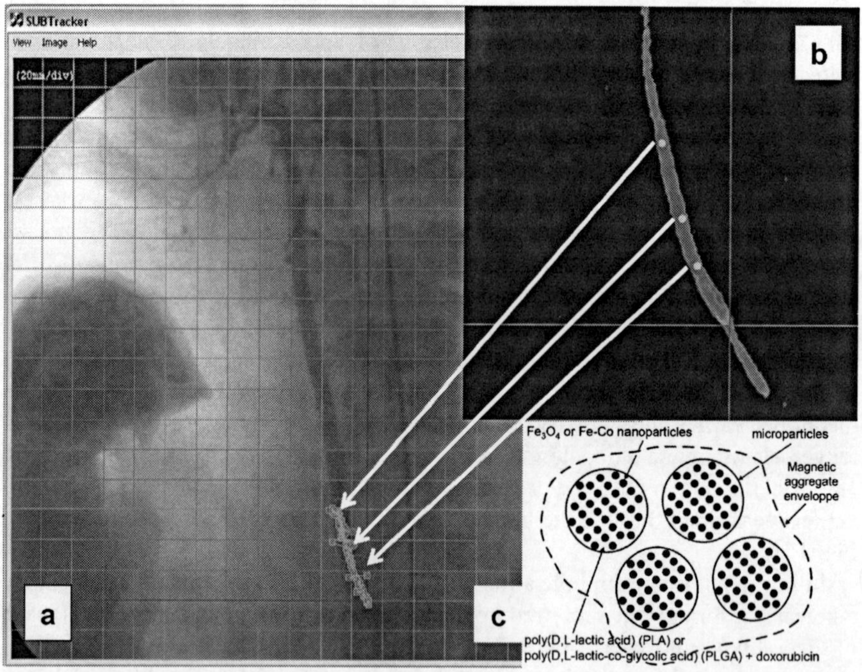

Fig. 17.7 Computer-controlled motion of a 1.5 mm ferromagnetic core in the artery of a living swine using 40 mT/m gradients generated by a clinical MRI scanner; (**a**) corresponding waypoints indicating the planned path being plotted on an acquired imaged of the artery; and (**b**) schematic representation of an agglomeration of magnetic carriers

but relatively low magnetization saturation or FeCo nanoparticles for enhanced propulsion/steering force density through a higher magnetization saturation level. In cases of FeCo, the nanoparticles are often designed to prevent toxic Cobalt ions to come in direct contact with blood. The number of nanoparticles is chosen high enough to provide an effective V for sufficient propulsion/steering force to be induced using magnetic gradients as described in (17.4). This is shown in Fig. 17.7a.

The magnetic nanoparticles are distributed throughout the inner structure to provide in addition to propulsion/steering, local magnetic field distortions (see 17.1) similar to MRI contrast agents that can be exploited for MR-tracking. The overall diameter of the nanoparticles is also chosen in many instances to allow hyperthermia-based computer-triggered release when other release methods such as the ones relying on a time-biodegradable polymer are not used.

Without upgrading typical clinical MRI scanner, the minimum size of each LET carrier is limited to approximately 250 μm across. This size is adequate for navigation in larger blood vessels using the 40 mT/m of each of the three orthogonal gradient coils already implemented in clinical MRI scanner for slice selection during MRI if a blood flow reduction technique such as the use of a balloon catheter is adopted. This will help since the pulsating flow rate in larger arteries may reach a peak velocity of approximately 1 m/s in human. An example of a relatively large ferromagnetic body being navigated in an artery of a living swine (an animal model relatively close to human) is depicted in Fig. 17.7a with the waypoints indicating the planned trajectory in the artery being depicted in Fig. 17.7b. The overall size of the LET carrier would prevent its entrance or navigation in smaller diameter vessels such as the arterioles since embolization would occur before getting deeper in the vasculature. As such, an agglomeration of LET entities (Fig. 17.7c) relying on a proper level of dipole-dipole interaction to aggregate would allow a large quantity of bacterial nanorobots to be transported simultaneously and beyond the entrance of the arterioles. Nonetheless, such LET carriers would create an embolization relatively far from the entrance of capillary network and hence would release such bacterial nanorobots further away from the tumor site. In such a case, the distance separating the tumor and the releasing site would become problematic and may decrease the targeting efficacy especially for targets located deeper in the microvasculature. In order to decrease the distance between the release (embolization) site and the tumor (target) site for instance, Small Embolization and Transport (SET) carriers can be used. The minimum size of each SET carriers would be approximately 40 μm across to create an embolization prior to be brought back to the systemic circulation. Furthermore, such size is also the limit (because of the effective volume of magnetic material that can be embedded) to apply effective steering force for navigation purpose using the highest magnetic gradient amplitude of approximately 500 mT/m that today's technology allows at the human scale (relative to an adult size torso) within the constraints of modern clinical scanners.

Indeed, if the hardware of the MRI scanner is upgraded to provide higher gradient amplitudes, then an agglomeration of SET entities or carriers can be

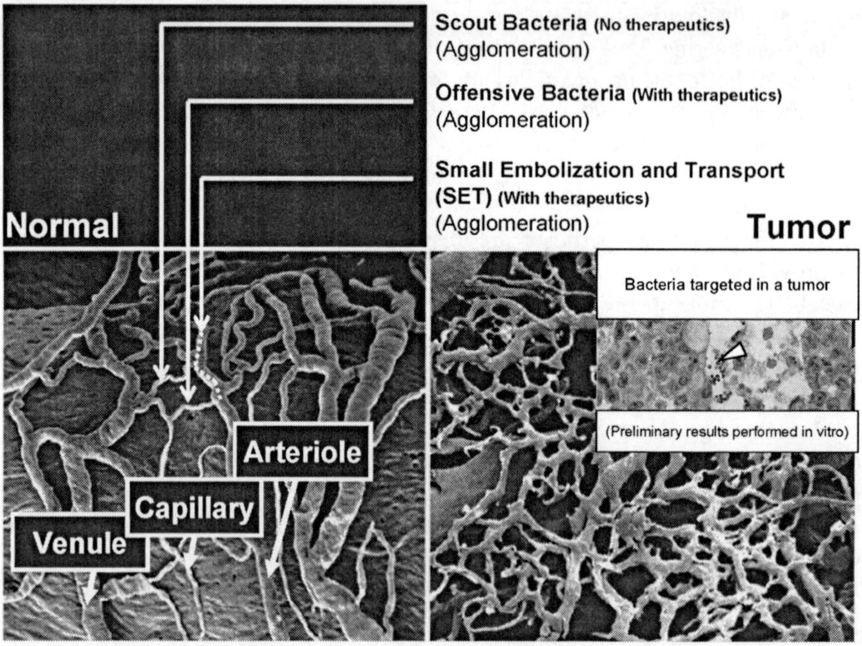

Fig. 17.8 Photographs of the various types of blood vessels with examples of locations accessible to the various types of entities (*left*) and a photograph showing the complexity of a tumoral angiogenesis network (*right*)

envisioned for releasing bacteria closer to the target at the arteriocapillar entry. The smaller number of bacterial nanorobots embedded per SET carriers can be compensated with a larger number of SET carriers. These flagellated bacteria or bacterial nanorobots could be scout bacteria or nanorobots, or offensive bacteria or nanorobots. This is depicted in Fig. 17.8.

Agglomerations of scout bacteria acting like controllable MRI contrast agents can be sent first in order to validate an appropriate path to the target. MRI sequences can be used to estimate the percentage of scout MTB that have reached the target provided that the population is sufficient for the sensitivity and spatial resolution of the scanner. Because of the initial higher risk of missing the target, these scouts unlike the offensive bacteria which transport toxic agents, have no therapeutic or toxic compounds and hence would avoid an increase of secondary toxicity in case they missed their target and reach the systemic circulation through the venules. Once a proper path is determined, offensive bacteria can then be used.

17.5 Conclusion

The complementary use of synthetic and hybrid nanorobots would typically yield better targeting when the target must be reached by navigating through the microvasculature. Although flagellated MTB of type MC-1 with thrust exceeding 4 pN are more efficient when transiting though the capillary networks, they will prove to be useless in larger blood vessels in normal conditions. Therefore, synthetic carriers such as the ones made of polymeric and ferromagnetic materials have been proposed for the transport of these bacteria through the larger blood vessels and towards the release sites at the arteriocapillar network entry if a reduction of the blood flow using instruments such as balloon catheters is not applicable. The displacement behavior of MTB can also be taken into account to avoid the need for traditional feedback control since existing medical imaging modalities do not have a sufficiently high spatial resolution to image small blood vessels such as capillaries. Hence, bacterial nanorobots defined here in its most basic form as MTB navigating under computer control have the capability to enhance targeting in complex microvasculature. Other types of medical nanorobots with different capabilities are under development and were not presented here to maintain the chapter at an introductory level. Indeed, in this chapter, only the fundamental types have been briefly presented as an introduction in this new field of research that may offer promising avenues for the next generation of medical interventions.

References

1. Drexler, K.E.: Nanosystems: molecular machinery, manufacturing, and computation, John Wiley and Sons, New York (1992)
2. Martel, S.: Method and system for controlling micro-objects or micro-particles, U.S. Pat. Appl. 11/145,007 (2005)
3. Martel, S.: Targeted delivery of therapeutic agents with controlled bacterial carriers in the human blood vessels. In: 2nd ASM/IEEE EMBS Conference on Bio, Micro and Nanosystems. San Francisco, USA, 2006
4. Darnton, N., Turner, L., Breuer, K., Berg, H.C.: Moving fluid with bacterial carpet. Biophys. J. **86**, 1863–1870 (2004)
5. Behkam, B., Sitti, M.: Bacterial flagella-based propulsion and on/off motion control of microscale objects. Appl. Phys. Lett. **90**, 023902–023904 (2007)
6. Steager, E., Kim, C-B., Patel, J., Bith, S., Naik, C., Reber, L., Kim, M.J.: Control of microfabricated structures powered by flagellated bacteria using phototaxis. Appl. Phys. Lett. **90**, 263901–263903 (2007)
7. Martel, S.: Controlled bacterial micro-actuation. In: Proceedings of International Conference on Microtechnology in Medicine and Biology (MMB), Okinawa, Japan, 2006
8. Martel, S., Tremblay, C., Ngakeng, S., Langlois, G.: Controlled manipulation and actuation of micro-objects with magnetotactic bacteria. Appl. Phys. Lett. **89**, 233804–233806 (2006)
9. Berg, H.C., Brown, D.A.: "Chemotaxis in Escherichia coli analyzed by three-dimensional tracking. Nature. **239**, 500–504 (1972)
10. Ford, R.M., Phillips, B.R., Quinn, J.A., Lauffenburger, D.A.: Measurement of bacterial random motility and chemotaxis coefficients. I. Stopped-flow diffusion chamber assay. Biotech. Bioeng. **37**(7), 647–660 (1991)
11. Armitage, J.P.: Bacterial motility and chemotaxis. Sci. Prog. **76**, 451–477 (1992)
12. Frankel, R.B. Blakemore, R.P.: Navigational compass in magnetic bacteria. J. Magn. Magn. Mater. **15–18**(3), 1562–1564 (1980)

13. Denham, C., Blakemore, R., Frankel, R.: Bulk magnetic properties of magnetotactic bacteria. IEEE Trans. Magnetism. **16**(5), 1006–1007 (1980)
14. Debarros, H., Esquivel, D.M.S., Farina, M.: Magnetotaxis. Sci. Prog. **74**, 347–359 (1990)
15. Martel, S., Mathieu, J-B., Felfoul, O., Chanu, A., Aboussouan, É., Tamaz, S., Pouponneau, P., Beaudoin, G., Soulez, G., Yahia, L'H., Mankiewicz, M.: Automatic navigation of an untethered device in the artery of a living animal using a conventional clinical magnetic resonance imaging system. Appl. Phys. Lett. **90**(11), 114105–114107 (2007)
16. Bell, D.J., Leutenegger, S., Hammar, K.M., Dong, L.X., Nelson, B.J.: Flagella-like propulsion for microrobots using a nanocoil and a rotating electromagnetic field. In: Proceedings of the IEEE International Conference on Robotics and Automation (ICRA), pp. 1128–1133 (2007)

Part III
Engineering Developments

Chapter 18
Force Feedback and Sensory Substitution for Robot-Assisted Surgery

Allison M. Okamura, Lawton N. Verner, Tomonori Yamamoto, James C. Gwilliam, and Paul G. Griffiths

Abstract It is hypothesized that the lack of haptic (force and tactile) feedback presented to the surgeon is a limiting factor in the performance of teleoperated robot-assisted minimally invasive surgery. This chapter reviews the technical challenges of creating *force* feedback in robot-assisted surgical systems and describes recent results in creating and evaluating the effectiveness of this feedback in mock surgical tasks. In the design of a force-feedback teleoperator, the importance of hardware design choices and their relationship to controller design are emphasized. In addition, the practicality and necessity of force feedback in all degrees of freedom of the teleoperator are considered in the context of surgical tasks and the operating room environment. An alternative to direct force feedback to the surgeon's hands is sensory substitution/augmented reality, in which graphical displays are used to convey information about the forces between the surgical instrument and the patient, or about the mechanical properties of the patient's tissue. Experimental results demonstrate that the effectiveness of direct and graphical force feedback depend on the nature of the surgical task and the experience level of the surgeon.

18.1 Introduction

During traditional open surgery, a surgeon can fluently process his or her own proprioceptive and cutaneous senses to determine the amount of force being applied to a tissue or organ, as well as the state of contact between his or her hands and the patient. Such haptic (force and tactile) information allows for accurate application of forces to and appropriate manipulation of the patient's tissues. However, in traditional and robot-assisted minimally invasive surgery (MIS), sensing and

A.M. Okamura (✉)
Department of Mechanical Engineering,
Laboratory for Computional Sensing and Robotics, Johns Hopkins University,
Baltimore, MD 21218, USA
e-mail: aokamura@jhu.edu

J. Rosen et al. (eds.), *Surgical Robotics: Systems Applications and Visions*,
DOI 10.1007/978-1-4419-1126-1_18, © Springer Science+Business Media, LLC 2011

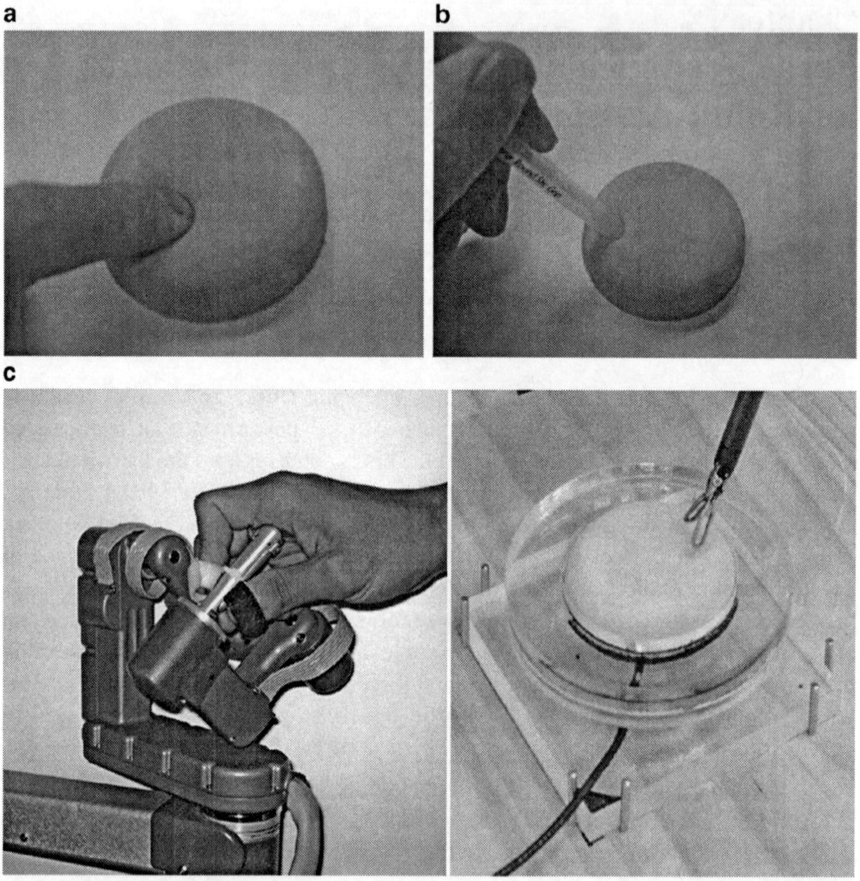

Fig. 18.1 A progressive loss of haptic feedback. (**a**) During open procedures, surgeons have direct contact with tissue, providing excellent haptic feedback. (**b**) Use of a laparoscopic tool (represented here by a simple probe) removes tactile sensation, and also diminishes useful force feedback due to body wall forces. (**c**) Conventional teleoperated robot-assisted minimally invasive surgical systems lack both tactile and force feedback

displaying such haptic information to the surgeon is challenging. While MIS procedures offer patients shorter recovery times and minimized trauma at the site of the incision, the surgeon is required to perform such procedures with significantly degraded or missing haptic feedback (Fig. 18.1). While some force information can be discerned using laparoscopic instruments, almost no force feedback is available when using current clinical robot-assisted minimally invasive surgery (RMIS) systems, such as the da Vinci Surgical System (Intuitive Surgical, Inc., Sunnyvale, CA, USA). Tactile feedback is also limited to the contact between the surgeon's hands and the instrument (in MIS) or a robotic/force feedback device (in RMIS). This chapter focuses on the *force* component of haptics.

18.1.1 The Role of Force Feedback

Force feedback is relevant to surgical task performance in several ways. First, force and movement information can be combined to sense tissue mechanical properties, allowing a surgeon to distinguish diseased and healthy tissue regions. In exploration tasks, such as palpating for cancerous lumps or tumors [24] and identifying calcified regions of cardiac arteries [12], force feedback can help a surgeon identify specific tissue features or regions that are burdensome to identify visually. Second, force feedback can prevent inadvertent trauma to tissue. Without force feedback, a surgeon performing RMIS cannot be sure of the tool-tissue interaction forces. In manipulation tasks such as blunt dissection [43] and suturing [29], force feedback is important to prevent puncture of tissue and vital organs, and to reduce the overall trauma to the tissue. Furthermore, force feedback can help to reduce the likelihood of breaking suture from unintentional large applied forces.

To understand the forces that are used in traditional surgery, MIS, and RMIS without artificial force feedback, our research group studied the difference between forces applied to sutures by attending and resident surgeons during three knot tying exercises: hand ties, instrument ties (using needle drivers), and robot ties (using the da Vinci with no haptic feedback). It was found that the robot ties, but not the instrument ties, differed from hand ties in repeatability of applied force (Fig. 18.2) [5]. This implies that, if RMIS can be imbued with the same kind of kinesthetic feedback provided by hand-held instruments, repeatability would be improved. Although some groups describe force feedback as providing useful "information" for surgery, it is important to note that forces also generate dynamic constraints that inherently change the motion of the surgeon in relation to the tissue, usually resulting in better-controlled interaction forces and manipulation.

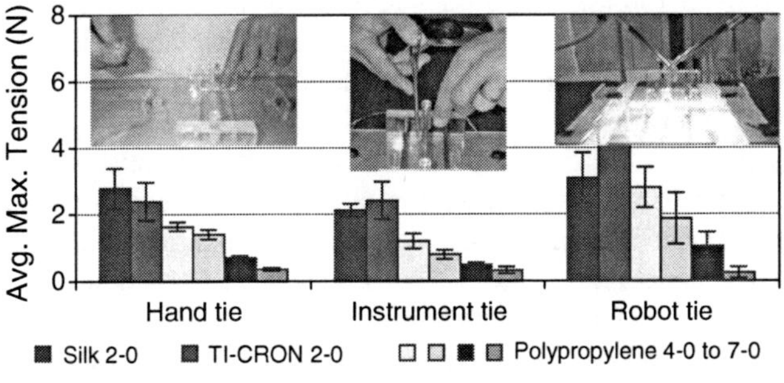

Fig. 18.2 Forces applied to various sutures during a knot tie by an attending surgeon. Instrument tie force levels and standard deviations of the hand tie and instrument tie are similar, while those of the robot tie are different [20], ©Springer 2002

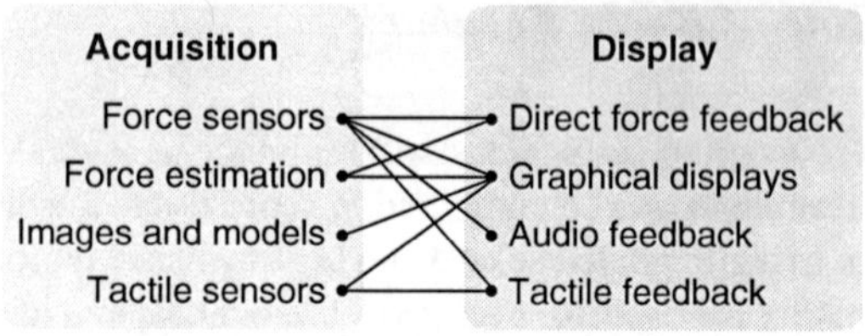

Fig. 18.3 Methods for haptic sensing and display in teleoperated robot-assisted minimally invasive surgery. The lines connecting sensing and display methods indicate combinations that have been explored in commercial or research RMIS systems described in the literature

18.1.2 Methods for Force Feedback

Force feedback for teleoperated robots, including surgical robots, fundamentally requires two capabilities: force sensing (or estimation) and force display. There are numerous technologies that can be used to address each of these capabilities (Fig. 18.3).

18.1.2.1 Force Sensing and Estimation

The inside of a patient's body is a challenging environment in which to sense forces. Force sensors would ideally be placed at or near the tip of the instrument, but this requires biocompatibility, sterilization, and insensitivity to temperature variation. In addition, force/torque sensors can add significant cost to a surgical instrument, especially if measurements must be made in many degrees of freedom and/or the instruments are disposable. To circumvent these challenges, force sensors can be placed on the robot outside the body of the patient. However, this is problematic for most surgical robots because force data would be acquired not only from the delicate interactions between the instruments and patient's tissues, but also from friction, body wall forces, and torques applied to the instrument at the insertion point on the patient. These undesirable forces are often large enough to mask the instrument-tissue interaction forces that should be displayed to the surgeon. Another possible location for force sensors is on the shaft of the instruments. For surgical systems like the da Vinci, which have cable-driven "wrist" degrees of freedom at the tip of the instruments, the internal shaft forces vary widely during manipulation. These internal forces are an order of magnitude higher than the instrument-tissue interaction forces [30]. Calibration to estimate and cancel the internal forces is not practical due to hysteresis and heavy dependence on uncertain starting conditions. Several recent research projects have redesigned the surgical robot and/or instruments in order to enable practical force sensing at instrument tips without compromising dexterity or sterilizability (e.g., [28]).

Using certain teleoperation controllers, forces applied to the patient can be estimated without using force sensors. For example, in patient-side robots designed with low inertia and friction, the difference between the desired and actual position of the patient-side robot is an indication of the level of force applied to the environment. The implementation and analysis of such controllers and their effectiveness as force estimators are described in Sect. 18.2.

18.1.2.2 Force Display

Force data acquired through direct sensing or estimation can be displayed to the user in a number of ways. The most obvious is direct force feedback (DFF) to the surgeon's hands, using a kinesthetic (force-feedback) haptic device. Many teleoperated RMIS systems designed by research groups use three-degree-of-freedom (or higher-degree-of-freedom) commercial devices such as the PHANTOM (SensAble Technologies, Inc., Woburn, MA, USA) [18] and Omega (Force Dimension, Inc., Nyon, Switzerland) [37] as the force-feedback master manipulandum. In contrast, the da Vinci surgical system uses a custom-designed master manipulandum, which is equipped with motors to enable force feedback in six degrees of freedom. (However, force feedback is limited in part due to the challenges described in Sect. 1.2.1.) The consequences of a mismatch between degrees of freedom of motion and degrees of freedom of force feedback in a teleoperator are described in Sect. 3.

While direct force feedback provided through a kinesthetic device is considered the most useful type of haptic feedback for many teleoperation scenarios, force information can also be conveyed to the surgeon through sensory substitution, e.g. visual (graphical) [29], auditory [27], or tactile [19] feedback. As described earlier, such sensory substitution may not be as effective as direct force feedback because it does not generate the natural physical constraints that inherently arise from mechanical interactions. Yet sensory substitution may be significantly more practical in RMIS scenarios because it can be less expensive to implement, is easier to add to existing robot-assisted surgery systems, and eliminates some of the control and stability issues created by direct force feedback (Sect. 2). In addition, sensory substitution may be an effective approach for training surgeons to use RMIS systems. Several methods of graphical displays of force and material property information, as well as their evaluation, are described in Sects. 4 and 5.

18.2 Control Issues in Force-Feedback Teleoperation

The control design problem posed by force feedback teleoperation sets forth stringent requirements for performance, stability, and robustness. This section describes common control architectures used to achieve force feedback, measures of performance in teleoperation, and the passivity criterion for robust stability of teleoperators. Emphasis is placed on the relationship between hardware and control design.

A useful set of control architectures for force-feedback teleoperation is described by the *four-channel* framework proposed by Lawrence [22]. This framework, shown

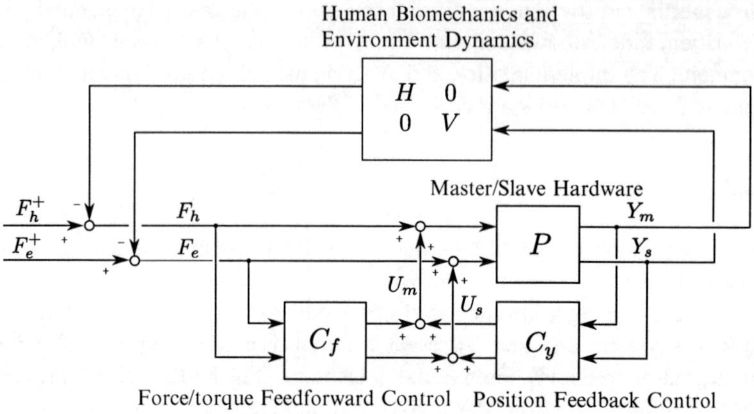

Fig. 18.4 Four-channel control framework for force feedback teleoperation. When C_f and C_y have non-zero off-diagonal terms both force and position measurements are exchanged between the master and slave. Four communication channels are then required to communicate these signals between the master and slave

in Fig. 18.4, describes a linear time-invariant model for teleoperation with various sensor suites, which may include position and force sensing at both the master and patient-side manipulator (also called the "slave"). The hardware model P is a diagonal matrix with transfer functions P_m for the master and P_s for the slave along the diagonal.[1] The force/torque controller C_f and the C_y are 2 ×2 matrices of transfer functions. This is a *four-channel* architecture because four signals that may be exchanged between the master and remote manipulator controller. If all off-diagonal elements of C_f and C_y are non-zero, then two channels are needed to communicate F_h (human-applied force) and Y_m (position of the master) to the remote manipulator, and two more channels are needed to communicate F_e (environment-applied force) and Y_s (position of the slave) back to the master. If delay exists in the communication channels between the master and slave, factors e^{-sT} appear in the off-diagonal elements of C_f and C_y, where T is the one-way transmission delay. Human operator and environment dynamics are modeled in the transfer functions H and V, respectively. While neither the human nor the environment are likely governed by linear time-invariant models, the linearization is frequently useful for local analysis of performance and stability properties.

The applicability of the four-channel framework to surgical teleoperation depends on the extent to which real hardware meets implicit ideal assumptions of the framework. One assumption is that force sensors are located distally such that they read the interaction force with the user or the environment. A second

[1] The four-channel framework describes single-input/single-output hardware P_m and P_s. To extend the framework to multi-degree-of-freedom master and slave devices, one may treat each degree of freedom as an independent scalar feedback loop; however, this assumption ignores interactions between degrees of freedom. Robust, optimal state-space control design such as H_∞ and μ -synthesis [32] provide an alternative integrated multivariable approach.

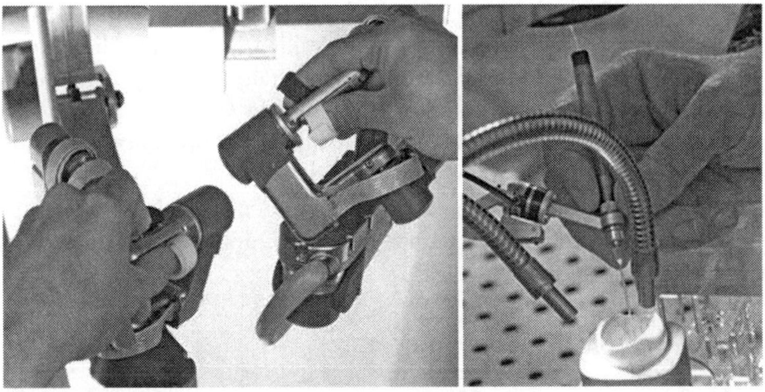

Fig. 18.5 A da Vinci master interface (*left*, ©2010 Intuitive Surgical, Inc.) and a Steady-Hand Robot (*right*, credit: Will Kirk, Johns Hopkins University) are representative of impedance-type and admittance-type master interfaces, respectively

assumption is that forces F_h^+ and F_e^+ and control inputs U_m and U_s affect the plant output through the same dynamics, or in short, for each robot, control and signals may be summed together. These two assumptions are useful idealizations for many teleoperators. However, it is worth noting that practical hardware never meets these idealizations exactly; tool-tips present intervening dynamics between the forces and the sensing element, and the force and the control input do not affect the plant output through the same dynamics when compliance and friction effects are considered. Generally, the second assumption is valid for devices termed *impedance-type*. The master robots of the da Vinci Surgical System, shown in Fig. 18.5, are examples of impedance-type devices. The da Vinci masters move freely when the user manipulates the end-points, as the joints exhibit little friction and the links have low inertia. This is in contrast to the Steady-Hand Robot paradigm [36], in which an admittance-type device is specifically designed to not move unless commanded by the control system. Admittance-type devices require force/torque sensing (visible to the left of the user's hand in Fig. 18.5) and do not fit within the four-channel framework for teleoperation.

There are many possible control designs employing a variety of sensor suites and hardware types. We will present two schemes that are particularly common: a controller based on position sensing only, and a controller in which the environment force is sensed. These two control schemes capture key concepts common to many teleoperator controllers and have both been used in studies of force-feedback surgical teleoperation.

18.2.1 Position-Position Teleoperation

Position sensing alone can be sufficient in certain scenarios to create a virtual coupling between the master and remote robots that achieves force-feedback teleoperation. This coupling may be as simple as a linear virtual spring that stretches as the master and remote robots are pulled in opposite directions. Only position sensors are needed to compute the stretch of the virtual spring. The position sensing control architecture is commonly referred to simply as *position-position* teleoperation to denote the bi-directional exchange of position information between the master and remote manipulator controllers.

A simple but effective position-position controller is modeled after a parallel spring-damper connecting the master and remote manipulator. Let *position coordination error* be defined as

$$E_y \triangleq Y_m - Y_s. \tag{18.1}$$

The control laws for the master and slave, respectively, are then given by

$$U_m = -(k_p + k_d s)Ey \tag{18.2}$$

$$U_s = (k_p + k_d s)Ey, \tag{18.3}$$

where s denotes the Laplace variable (which is equivalent to differentiation in the time domain). This controller may be expressed in the four-channel framework by letting

$$C_y = \begin{bmatrix} -k_p - k_d s & k_p + k_d s \\ k_p + k_d s & -k_p - k_d s \end{bmatrix} \tag{18.4}$$

$$C_f = 0. \tag{18.5}$$

The strength of the coupling is increased as the proportional (k_p) and derivative (k_d) gains are increased, which leads to reduced position coordination error between the master and slave manipulators.

The position-position architecture is well suited for master and remote robots that move freely when the user or environment pushes on the end-effector. The ideal virtual spring-damper described by (18.2) and (18.3) maintains equal and opposite forces on either end. However, hardware dynamics interfere as they detract from this equality. Good hardware design can reduce inherent friction and inertia. Local feedback around the master and remote manipulator may also be used to shape the hardware dynamics. For example, derivative feedback that mimics negative damping can cancel *some* of the inherent hardware damping. Caution is advised, however, as stability margins and robustness seriously degrade as the degree of cancellation increases [11].

18.2.2 Position-Force Teleoperation

Negative attributes of force/torque sensors, notably their cost, fragility, and need for frequent calibration, discourage the addition of force sensing and make position-position teleoperation a generally preferred architecture. However, a force/torque sensor is necessary if either the master or remote robot has large friction or inertia. This is frequently the case with very large or very small robots. Large robots frequently employ gear reducers to obtain greater joint torques from smaller motors. At the same time, gear reduction can greatly increase the apparent inertia and friction when driving the end-effector. Certain gear configurations, such as a worm gear drive, may be completely non-backdrivable. Small robots frequently suffer from large inherent friction relative to the small forces exerted on the end-point.

When inherent friction and inertia of the remote manipulator are large, the tool-tip may be instrumented with force/torque sensing to recover sensitivity to environment forces and torques. At a high level, the master provides position commands to the remote robot and the forces measured at the remote robot are transmitted back to the master. This architecture is called *position-force* teleoperation because the master position and the remote force are exchanged. A simple implementation of position-force control is given by

$$U_m = F_e \tag{18.6}$$

$$U_s = (k_p + k_d s)Ey. \tag{18.7}$$

In terms of the four-channel architecture (Fig. 18.4), the controller is given by

$$C_y = \begin{bmatrix} 0 & 0 \\ k_p + k_d s & -(k_p + k_d s) \end{bmatrix} \tag{18.8}$$

$$C_f = \begin{bmatrix} 0 & 1 \\ 0 & 0 \end{bmatrix}. \tag{18.9}$$

This position-force controller uses proportional-derivative control to make the remote manipulator track the master position. The force signal is commanded to the master interface with unity gain. In practice, the force signal may need to be filtered to remove high frequency noise. Position-force control schemes are susceptible to instability, particularly when communication contributes delay to the feedback system. If significant delay exists, the communicated signals should be transformed into *wave variables* prior to transmission to ameliorate the destabilizing effect. An overview of wave variables is provided by Niemeyer and Slotine [26].

18.2.3 Motion and Force Scaling

Motion scaling between the master and remote tool is a key benefit of teleoperation in minimally invasive surgery, enabling the surgeon to work with greater precision at small scales. With the addition of force feedback, scaling may be applied to both motion and forces. The scaling factors do not appear explicitly in the control laws; rather, they are the product of normalizing the remote robot position and force. Suppose that the un-normalized remote robot dynamics are given by

$$\bar{Y}_s = \bar{P}_s(\bar{U}_s + \bar{F}_e). \tag{18.10}$$

Here $\bar{Y}_s$ is the measured (un-normalized) signal. The normalized position is defined as $Y_s \triangleq \alpha \bar{Y}_s$ where α is the desired ratio for $Y_m/\bar{Y}_s$. Similarly suppose that β is the desired ratio for $-F_h/\bar{F}_e$ and define $F_e \triangleq \beta \bar{F}_e$. The dynamics in terms of normalized positions and forces is

$$\alpha Y_s = \bar{P}_s(\beta U_s + \beta F_e). \tag{18.11}$$

Note that for U_s and F_e to sum together, the normalized control must be defined by $U_s \triangleq \beta \bar{U}_s$. It follows from (18.11) that the normalized remote robot dynamics are $P_s \triangleq \frac{\beta}{\alpha} \bar{P}_s$. The control design is then reduced to teleoperation with unity scaling.

18.2.4 Measures of Teleoperator Performance

The performance objective in force-feedback teleoperation is to provide a stiff, low friction, light weight/inertia tool with which the user can manipulate of explore the remote environment. High stiffness is important for feeling material properties such as surface texture and stiffness. Low inertia and friction are necessary to discern small environment forces. Performance of force-feedback teleoperation may be assessed by a pair of metrics: either position and force coordination, or position coordination and transparency.

Position coordination is measured by the error $E_y \triangleq Y_m - Y_s$. Similarly force coordination may be measured by $E_f \triangleq F_h + F_e$, where perfect coordination is achieved when F_h and F_e are equal and opposite. The conditions under which the error signals E_y and E_f are identically zero are referred to as *ideal kinesthetic coupling* [13, 47]. Measuring performance with position and force coordination error is made complicated by the dependence of these metrics on the user and environment dynamics. Thus, a thorough assessment of performance must consider many possible combinations of user and environment dynamics. An alternate measure of force-feedback performance provides a partial solution to this problem.

Transparency describes how environment dynamics are altered when rendered to the user through the teleoperator. The term *transparency* evokes an analogy to optical transmission through glass: a perfectly transparent pane of glass does not distort an image that passes through it. In the four-channel framework shown in Fig. 18.4, the linear environment dynamics are given by V. The actual dynamics rendered to the user are a function of the environment dynamics V as well as the hardware and controller. Within Fig. 18.4, the rendered dynamics are given by the response from F_h^+ to Y_m without the feedback H from the user:

$$R \triangleq \frac{Y_m}{F_h^+}, \quad (H = 0). \tag{18.12}$$

Transparency is then measured either by the ratio R/V or the error between R and V, called *distortion* [10, 11] and defined as

$$\Theta \triangleq \frac{R - V}{V}. \tag{18.13}$$

Perfect transparency is achieved if distortion Θ is identically zero. In practice, the frequency response of distortion can be made small over some finite bandwidth. The advantage of transparency as a measure of performance is that it does not depend on the user dynamics.

18.2.5 Stability and Passivity

Mechanical contact between the human operator and the master and between the environment and the remote manipulator can dramatically alter the closed-loop system dynamics, potentially yielding instability. A common approach to prevent coupled instability is to design the teleoperator such that its two-port network model is *passive*. (For an in-depth discussion of passivity, see [15] and [17]). In the network modeling framework, the human and environment may be modeled as one-port terminations on the two-port teleoperator model, as shown in Fig. 18.6. Passivity of the two-port network means that the net mechanical work done by the teleoperator on the human and environment does not exceed a finite initial energy. If the human operator and environment dynamics are also passive, the entire coupled system is passive, and thus stable.

Passivity of the teleoperator may be readily determined by a matrix norm condition on the closed-loop dynamics. Let us first describe the teleoperator dynamics, including the hardware and controller, but excluding the human and environment dynamics, by the hybrid matrix:

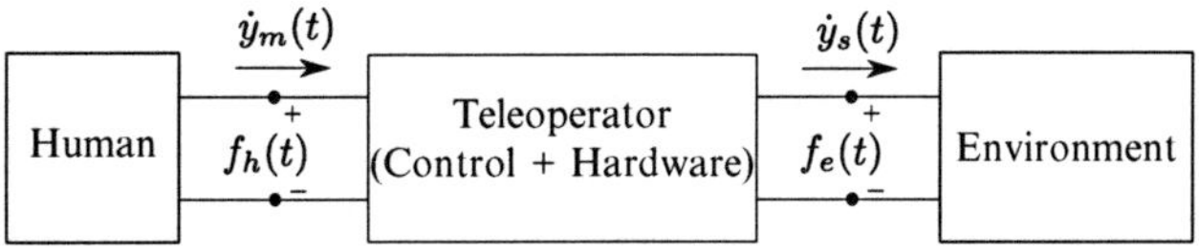

Fig. 18.6 The human and environment may be modeled as one-port terminations on the two-port teleoperator network model

$$\mathcal{H}(s) \triangleq \begin{bmatrix} H_{11}(s) & H_{12}(s) \\ H_{21}(s) & H_{22}(s) \end{bmatrix} \tag{18.14}$$

where

$$\begin{bmatrix} F_h(s) \\ -sY_s(s) \end{bmatrix} = \begin{bmatrix} H_{11}(s) & H_{12}(s) \\ H_{21}(s) & H_{22}(s) \end{bmatrix} \begin{bmatrix} sY_m(s) \\ F_e(s) \end{bmatrix} \tag{18.15}$$

In terms of the hybrid matrix, the *scattering matrix* is defined as [2]

$$S_t(s) \triangleq \begin{bmatrix} 1 & 0 \\ 0 & -1 \end{bmatrix} (\mathcal{H}(s) - I)(\mathcal{H}(s) + I)^{-1}. \tag{18.16}$$

The teleoperator is passive if and only if

$$||S_t(j\omega)||_\infty \leq 1. \tag{18.17}$$

where $||\,.\,||_\infty$ denotes the peak singular value across all frequencies. When this inequality is satisfied, the teleoperator feedback design is stable for all passive human operator and environment dynamics. In practice, such design is overly conservative, since the range of possible dynamic properties of the human and environment is limited. Linear analysis and some limited knowledge of the possible user's dynamics can yield better performance and linear stability guarantees [7].

18.2.6 Hardware-Imposed Limitations on Control Design

We now emphasize the importance of hardware design choices and their relationship to control design, since the performance of a teleoperator is largely determined before any control software is written. The size, weight, linkage design, transmission, and actuators of the master and remote robots determine important properties, including structural rigidity, damping properties, friction, saturation, and back-drivability. These properties guide appropriate sensor selection, control strategy, and ultimately the achievable robust performance and stability of the teleoperator.

Compliance and structural modes of the robot play a key role in the limiting the performance of force-feedback teleoperation. For example, we show in Fig. 18.7 the frequency response from a free-standing first-generation da Vinci master interface[2] (without the user), obtained by applying pseudo-random noise to a single

[2] The tested device differs from the commercial system in its custom mounting, motor amplifiers, data acquisition, and controllers.

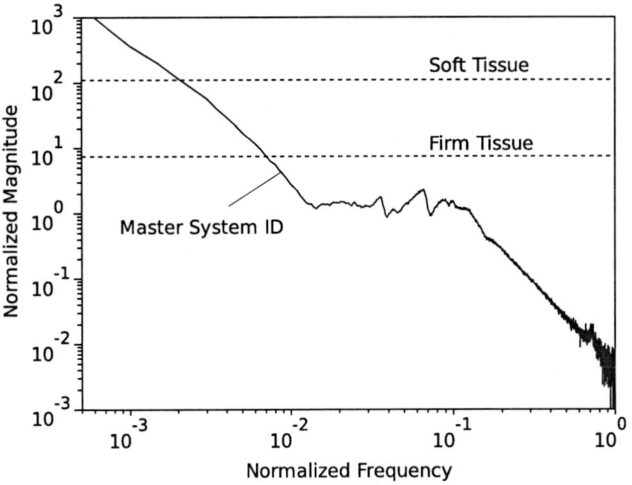

Fig. 18.7 Experimental frequency response of a surgical teleoperator master interface in one degree of freedom. Frequency responses of representative soft and hard tissue stiffnesses are overlaid for reference

joint. The axis scales are normalized to protect proprietary data. This plot exhibits features common to many impedance-type master devices. Below a certain frequency, the linear behavior of the device may be described by a simple mass-damper model. Compliance in cable drives and resonant modes generate multiple resonances and anti-resonances evident at higher frequencies (above 10^{-2} in Fig. 18.7). We remark that some of the resonant modes may be attributed to the distal gimbal assembly and that the characteristics of the higher-frequency modes depend in part on controller used to constrain the assembly. As the frequency response decreases in magnitude at higher frequencies, larger control inputs are required to achieve position coordination. In practice, control authority over the robot endpoint drops off very rapidly above the first anti-resonance.[3] When considered along with limited actuator bandwidth and saturation effects, the resonant modes of the device form practical bandwidth limitations for haptic feedback – limitations which are common to all impedance-type haptic devices.

Saturation and actuator bandwidth are not the only limitations that force the closed-loop bandwidth below the hardware resonances. Good stability robustness is similarly restrictive. As the closed-loop bandwidth is increased, the contribution of resonances to the time-response grows and high-frequency vibrations become noticeable, both visually and haptically. If the first resonant mode is very poorly

[3] While the magnitude of the measured frequency response is relatively flat over the normalized bandwidth of 10^{-2} to 10^{-1} in Fig. 18.7, the ability to move the distal end of the master robot in fact diminishes rapidly above 10^{-2}. The measured response is between a proximal motor and a co-located sensor rather than the robot endpoint. The anti-resonance zeros become poles in the response from motor command to the robot endpoint position.

damped, it may be excited in closed-loop even though the position tracking bandwidth is much slower than the resonance. A notch filter with zeros carefully matched to the resonance can be used to diminish these oscillations. However, this model-based control strategy will be inherently sensitive to the hardware dynamics. Uncertainty in the model, variations in the open-loop dynamics due to the user and environment dynamics, and natural variations in hardware parameters between devices and with age will all degrade the accuracy of cancellation.

Rigidity is an important hardware design consideration, not only for closed-loop bandwidth, but also for steady-state positional accuracy. The robot end-point is typically computed based on measured joint angles and rigid kinematics. However, bending of the robot links under load may be significant and cannot be determined through measurement of joint variables. A compliant drive between the joint and the measured variable (as is frequently the case due to the cost and complexity of instrumenting distal links of the robot) further exacerbates error between the computed robot end-point and the actual end-point. Tavakoli and Howe [35] show that the addition of position sensing on flexible surgical patient-side end-effectors can reduce low-frequency closed-loop flexibility and thus improve position coordination. Absolute positional accuracy is not necessarily important surgical telemanipuation with real-time video. Indeed, clutching allows the surgeon to center the master without moving the remote manipulator. However, absolute positional accuracy becomes important when data acquired from the tool-tip must be registered with itself or with other sensor data.

Other hardware dynamics can limit the ability to achieve high-fidelity haptic feedback at frequencies below the resonant modes of the system. Inertia, damping, and friction may effectively mask small forces applied to the endpoints of the remote and master robot. Filter design can provide partial compensate of these dynamics; however, increased sensitivity to variations in the hardware dynamics is an unavoidable cost. The exact tradeoff has been quantified for linear single-degree-of-freedom haptic rendering under position feedback control [11]. Qualitatively, rendering environments with light dynamics on a heavy device requires hardware compensation. For example, the frequency response of relatively soft tissue stiffness and firm tissue stiffness are overlayed on the frequency response of the master hardware in Fig. 18.7. Frequency by frequency, the softer environment requires more hardware compensation than the hard environment to achieve accurate rendering.

18.3 Partial Force Feedback

Although the utility of force feedback in surgical systems has been demonstrated (as described in Sect. 1), it is difficult and costly to incorporate in clinical practice. Some of these challenges were described from a control systems perspective in the previous section. In addition, force feedback requires additional motors (on the master) and/or sensors (on the patient-side robot) in comparison to a teleoperator with no force feedback. To reduce the need for these additional components, partial force feedback can be used such that force information is sensed and fed back to the user in fewer degrees of freedom than the degrees of freedom in which the surgical system operates. For example, in a robotic system that allows movement in the x, y, and z directions, it may be more cost effective to only provide feedback to the surgeon in one of those directions, as long as the force feedback provided is still relevant to the surgical task at hand.

Although partial force feedback can reduce the cost and complexity of robot-assisted surgical systems, several questions remain regarding the feasibility of partial force feedback in the surgical setting. Partial force feedback creates a sensor/actuator asymmetry in the system, which can potentially be disconcerting to users because of unrealistic force sensations [3] and may alter how a surgical system is used in practice. Also, since force feedback has direct implications on the control strategy of robot-assisted surgical systems, the effect of partial force feedback on the control stability of these systems must be evaluated.

More work to characterize of the importance of force feedback in specific directions of surgical tasks is needed. Saha [30] studied the magnitude and variance of applied force/torque in different degrees of freedom during tasks performed with the da Vinci Surgical System, but the results are inconclusive. Also, systems with sensor/actuator asymmetries cause forces to be displayed in incorrect directions. Work by Barbagli et al. [4] and Tan et al. [33] has shown that human ability to resolve force direction is relatively poor, thus the effect of partial force feedback on the users' understanding of force direction may be minimal.

18.3.1 User Performance with Partial Force Feedback

Results of user studies carried out by our group reveal the effect of partial force feedback on user performance. These studies tested three important hypotheses regarding the viability of partial force feedback:

1. The performance of a system with partial force feedback can approximate the performance of the same system with full force feedback.
2. Applied gripping force cannot be modulated from Cartesian force feedback alone (i.e. gripping force feedback is needed).
3. 3-DOF force only feedback is sufficient in a 6-DOF manipulator (i.e. no feedback in the rotational DOFs needed).

18.3.1.1 Experiment 1: Partial Cartesian Force Feedback

The first study involved a simulated artery dissection task using a 3-DOF research telemanipulator. The goal of the task was to expose the simulated artery from an artificial tissue bed (Fig. 18.8a). The teleoperation system consisted of two PHANTOM haptic interface devices, models Premium 1.0 and 1.5 (SensAble Technologies, Inc., Woburn, MA, USA), and a six-axis force/torque sensor (Nano-17, ATI

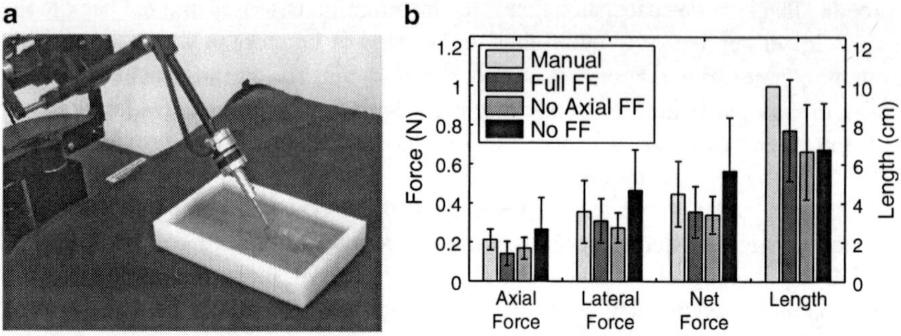

Fig. 18.8 Partial force feedback (FF) experiment. (**a**) Blunt dissection task pictured where operators dissected out an artificial artery from the surrounding tissue bed [31], ©IEEE 2003. (**b**) Average applied force and length of dissection for 10 subjects for a simulated dissection task. The axial and lateral applied forces for the No Axial FF condition approximate the applied forces for the Full FF condition. No significant differences are observed between artery dissection length in the Full FF, No Axial FF, and No FF cases. Manual yielded the longest dissected length, but this was most likely due to increased dexterity and a direct, 3D view of the task [40]

Industrial Automation, Apex, NC, USA). The ATI six-axis force/torque sensor measured forces applied at the tip of the tool to the artificial tissue bed and the teleoperation system was connected via a modified position-force controller.

Ten subjects performed the dissection task under four force feedback conditions: (1) Manual, (2) Full force feedback (Full FF), (3) No tool axis force feedback (No Axial FF), and (4) No force feedback (No FF). Conditions (3) and (4) were created by eliminating force feedback in some or all of the DOFs of the master haptic device. For the manual trials, subjects used a stylus with the force sensor attached between the subjects hand and the tip of tool with direct vision of the task set-up. For the teleoperated trials, subjects used the master robot stylus to control the movements of the tool mounted on the slave; during these trials, subjects watched a magnified digital video camera image on the computer monitor in front of them. The force sensed by the force sensor was fed back to the operator via the operator's haptic device. Each subject was given approximately 15 min of practice time with the surgical model and became familiar with each feedback type prior to being the experiment. Subjects were given two minutes to complete each dissection and the order of the conditions was randomized between subjects.

Results from the simulated blunt dissection task show that performance with the No Axial FF condition is similar to the Full FF condition (Fig. 18.8b). Average applied forces in the axial and lateral directions of the tool for the No Axial FF and Full FF conditions are less than the forces for the No FF condition. Similar results are found for the net force applied during the experiment. Dissection length between the three teleoperated trials was not significant different. ANOVA tests show significant differences in the conditions across all three force metrics ($p < 0.05$). Thus, it is possible for the performance of a system with partial force feedback to approximate the performance of the same system with full force feedback.

18.3.1.2 Experiment 2: Grip Force Feedback

The second study was a soft peg-in-hole placement task completed using a 4-DOF research teleoperation a system. The teleoperation system was similar to the one used in Sect. 18.3.1.1, with an additional custom gripping mechanism attached to both the master and slave haptic devices (Fig. 18.9a). Force sensors on the slave manipulator were used to record the force applied to the peg. Each subject completed the task with four feedback conditions: (1) Full force feedback (Full FF), (2) Translational force feedback only (Translational FF), (3) Gripping force feedback only (Gripping FF), and (4) No force feedback (No FF). Conditions (2), (3), and (4) were created by eliminating force feedback in some or all of the DOFs of the master haptic device. Five subjects repeated each of the four conditions in a random order five times. Subjects were given time to practice with each of the four feedback conditions prior the experiment.

Results from the soft peg-in-hole task show the importance of gripping force feedback (Fig. 18.9b). Applied gripping force at the slave device in the No FF and Translational FF conditions is not significantly different, indicating that translational force feedback does not help users understand the amount of grip force they are applying to the soft peg. Similarly, the applied translational force is not significantly different between the No FF and Gripping FF conditions. Thus, there is no performance crossover between gripping force feedback and translational force feedback. If the amount of gripping force applied to the patient is important due to delicate anatomy, grip force feedback should be provided.

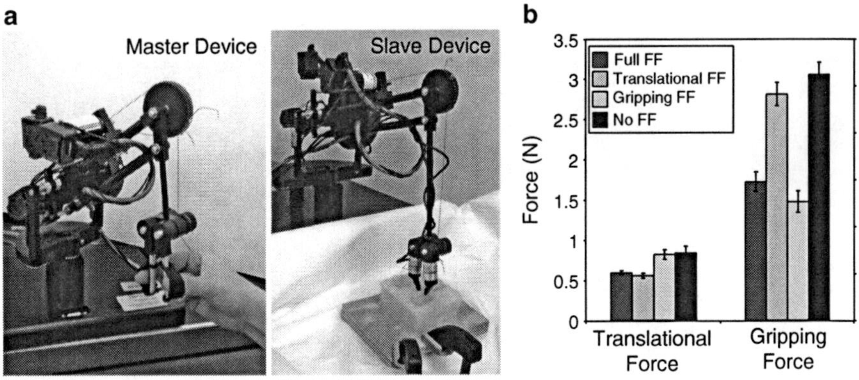

Fig. 18.9 Grip force feedback experiment. (**a**) A 4-DOF telemanipulator, consists of two PHANTOM haptic devices with added gripping mechanism [41], ©IEEE 2007. (**b**) Average applied force for 5 subjects performing the peg-in-hole task. Applied translational forces were only affected by force feedback in the translational DOFs. Similarly, applied grip force was only affected by feedback given in the gripping DOF [40]

18.3.1.3 Experiment 3: 3-DOF vs. 6-DOF Force Feedback

The third user performance study involved force and torque feedback in a 6-degree-of-freedom robot-assisted surgical system. The goal of this study was to determine if 3-DOF force feedback can approximate 6-DOF force and torque feedback in a 6-DOF positioning device. The master telemanipulator of a research version of the da Vinci Surgical System at Johns Hopkins University was used to interact with a virtual environment (master pictured in Fig. 18.1c). Ideally, the experiment would be performed using both the master and patient side manipulators (i.e. telemanipulation). However, few robot-assisted surgical systems have the ability to sense forces and torques in 6 degrees of freedom. Thus, for this experiment, the da Vinci master was used to interface with a virtual environment that can render forces in 6 degrees of freedom.

Six subjects transcribed the word "HAPTICS" using a virtual pen and paper (Fig. 18.10a). The 6-DOF motion of the pen corresponded to the motion of the da Vinci master device. As the tip of the virtual pen penetrated the virtual paper, a force and torque was exerted on the user. The interaction closely parallels the force and torque applied to the hand when writing with a real pen. The force and torque were created by the force of the writing surface applied along the lever arm of the pen. The virtual environment was presented with a 3-dimensional view to the user by the standard stereoviewer of the da Vinci system. The subjects performed the transcription task under four different conditions: (1) No Feedback, (2) Force Only Feedback, (3) Torque Only Feedback, and (4) Force & Torque Feedback. Each user was given unlimited time to practice with each condition.

The results of the experiment show that performance with Force Only Feedback closely approximates the performance with Force & Torque Feedback (Fig. 18.10b). Similarly, Torque Only Feedback provides reasonable performance,

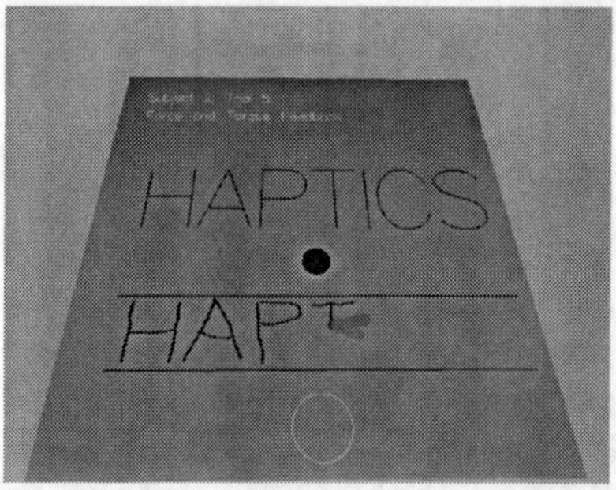

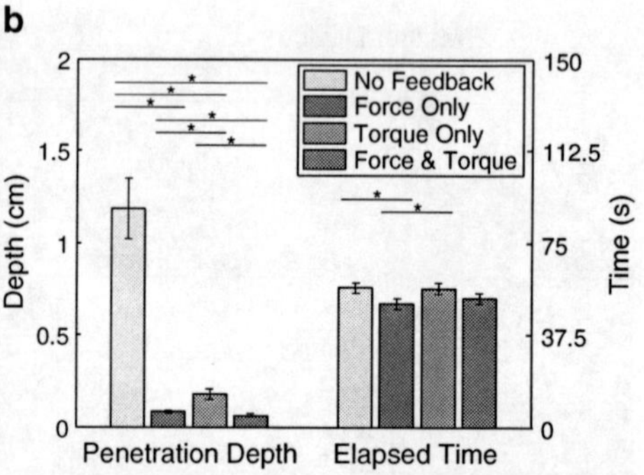

Fig. 18.10 Importance of torque feedback in a 6-DOF system. (**a**) A screen shot of the virtual "paper" that subjects draw on. (**b**) Average depth and elapsed time for the transcribing the word "HAPTICS". Average depth into the virtual wall is significantly increased for the No Feedback condition. The performance of with the Force Only and Force & Torque conditions is similar. Lines with stars above connect conditions that have statistically significantly different means [39], ©IEEE 2009

but it was noted by all of the subjects that this condition felt very disconcerting and unnatural. Results for elapsed time were inconclusive; though there are some significant differences between the conditions (Fig. 18.10b). ANOVA tests show significance differences in the conditions for both time and penetration depth ($p < 0.001$). These results indicate that, for certain tasks with a 6-DOF manipulator, it may be possible results to provide feedback in only 3 DOFs without significantly affecting performance with the robot-assisted surgical system.

18.3.2 Potential Instability with Partial Force Feedback

The results of the studies described in the previous section show that partial force feedback is a promising alternative to full force feedback from the perspective of user performance. However, ensuring the safety of robot-assisted surgical systems is of high priority and therefore the stability of systems with partial force feedback must be demonstrated. As noted in Sect. 2.5, stability of telemanipulators is often guaranteed by passivity conditions applied using the two-port network representation.

In contrast to telemanipulators with full force feedback, telemanipulators with sensor/actuator asymmetries (e.g. partial force feedback) are not guaranteed to be passive. For telemanipulators with sensor/actuator asymmetries, the scattering matrix of the communication block of the telemanipulator without time delay is

$$S_t = \begin{bmatrix} -(I-A)(I+A)^{-1} & 2A(I+A)^{-1} \\ 2(I+A)^{-1} & (I-A)(I+A)^{-1} \end{bmatrix}. \tag{18.18}$$

where A is a singular matrix, called the asymmetry matrix, whose null space corresponds to the degree(s) of freedom without force feedback. It can be shown that, for all singular A, the scattering operator is not bounded from above by 1, or

$$||S_t(j\omega)||_\infty > 1, \tag{18.19}$$

and thus the system is not guaranteed to be passive, even if the underlying control law is passive [40, 42]. Thus, standard "passive" telemanipulator control laws do not ensure stability for telemanipulators with sensor/actuator asymmetries.

Our experimental work and the preceding result regarding the scattering matrix reveal the conservative nature of the passivity criterion when applied to telemanipulators with sensor/actuator asymmetries. While the system may feel unusual and/ or unrealistic, it may not exhibit unstable behaviors, such as the contact instability problem that is common in telemanipulator literature. For example, all of the experiments in Sect. 3.1 were completed without the telemanipulators exhibiting unstable behavior. Further research on control strategies for systems with partial force feedback may make it possible to ensure stability of systems with partial force feedback without the need for passivity requirements.

18.4 Graphical Display of Force Information

Sections 2 and 3 described some of the control challenges associated with force feedback in surgical teleoperation. Such issues can be circumvented by providing graphical displays of force or environment property information, as described in here and in the following section.

During laparoscopic or RMIS procedures, endoscopic cameras are typically used to produce a rich visual display of the surgical site, which a surgeon views on a monitor or through a viewing console. These images can be augmented in real time with overlaid, artificially generated graphics that enhance a surgeon's sensory experience and can display either sensed or estimated forces graphically. This sensory transduction is a form of sensory substitution, referred to as graphical force feedback (GFF), which allows a surgeon to visualize an applied force, rather than feel it directly. Graphical force feedback can take many forms, depending on the nature of the surgical task (e.g., manipulation or exploration). Additionally, the graphics used to augment surgical images should be implemented in such a way to deliver the greatest possible advantage to the surgeon. Considerations include graphic shape, size, location, transparency, and tracking with tool motion. A number of studies have employed graphical force feedback in various forms, as shown in Fig. 18.11.

Since the tip of the surgical instrument is usually in the surgeon's visual field, placement of GFF near the tip in the image should facilitate effective use of the information and require less cognitive processing. Akinbiyi et al. [1] and Reiley et al. [29] augmented the surgeon's visual workspace by overlaying small dots on the image near the robot tooltip during a suturing task. The dots changed colors (from green to yellow to red) as a function of increasing applied force (Fig. 18.11a). For surgeons without significant experience in RMIS, GFF resulted in fewer suture breakages (Fig. 18.12a) and lower peak applied forces (Fig. 18.12b), though the effect was smaller for surgeons with RMIS experience. These results hint that GFF may be especially helpful in training surgeons who are not yet skilled with robot-assisted surgery. In contrast, another study used an external LED display to present applied force [34] (Fig. 18.11b). We propose that augmenting forces graphically on top of the normal visual workspace results in superior performance, as it allows the surgeon to recognize forces without removing visual attention and focus from the point of operation.

Bar graph displays were developed by Mahvash et al. [24] and Gwilliam et al. [12], in order to change both the color and size of the display to correspond with the level of applied force. In the first system, the bar graph did not track the tooltip motion, causing it to occasionally block a portion of the surgeon's field of view in certain configurations in the workspace (Fig. 18.11c). The goal of the second system was to provide force information more intuitively, while avoiding obscuring important visual information, cluttering the surgeon's field of view, or creating an unnecessary distraction. For exploration tasks in which the instrument tip moves in a large workspace, it is important that the graphical display tracks the

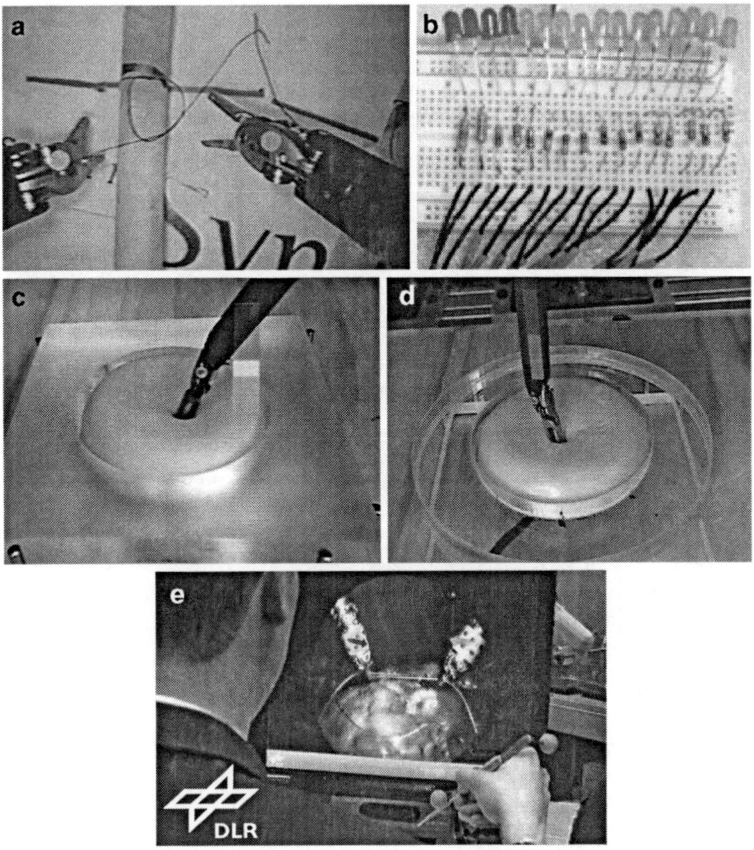

Fig. 18.11 (**a**) Small dots augmented over the surgeon's visual workspace display forces during a suturing task [29]. (**b**) External LED array displaying force levels [34], ©IEEE 2007. (**c**) Bar graph overlaid on visual workspace to display estimated forces during an exploration task [24], ©IEEE 2008. (**d**) Bar graph is aligned with da Vinci tool shaft and tracks its motion with proper disparity and depth perception, while displaying estimated forces [12], ©IEEE 2009. (**e**) 3D colored arrows indicated the force vector applied by the surgical instrument (courtesy DLR, see [37] for details)

tooltip appropriately. Thus, the two-dimensional GFF image for each eye must be rendered with appropriate size, and disparity in the camera frame as the tooltip moves in a 3D workspace. Gwilliam et al. [12] created such a bar graph to display force information near the moving instrument tip. Additionally, it was aligned with the instrument shaft, such that the bar changed size and orientation as the instrument navigated the 3D workspace. This also made the GFF appear with proper disparity and depth perception (Fig. 18.11d). Since the graphical rendering was a bar graph, the surgeon could use both color and size (magnitude of bar graph level) as cues for the applied force. Because of the calibration to the tool shaft, the GFF bar never occluded other objects in the visual workspace that were not already

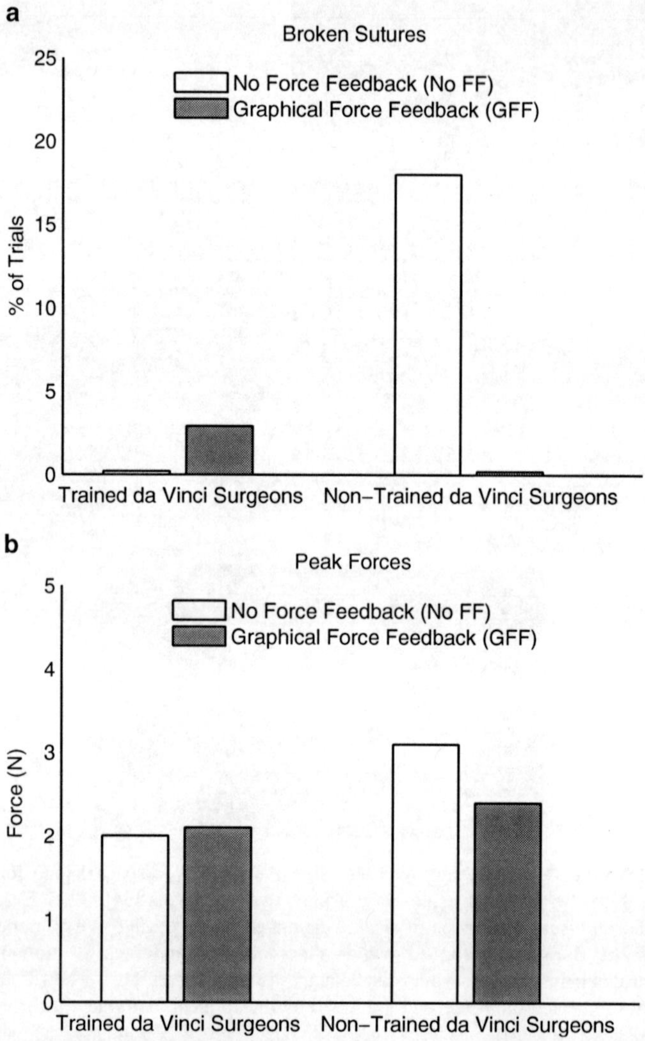

Fig. 18.12 For surgeons with little RMIS experience, graphical force feedback resulted in significantly lower suture breakages (**a**) and lower applied peak forces (**b**) during a knot-tying task [29]

blocked by the tool shaft. An experiment with this system was performed with experienced robot-assisted surgeon subjects. The task was to identify the correct orientation of an artificial calcified artery embedded within an artificial tissue model as part of a surgical palpation task. Figure 18.13 shows that, for this task, graphical force feedback does not significantly improve subject performance relative to no feedback. We conclude that spatially distributed direct force feedback is easier for subjects to integrate than GFF.

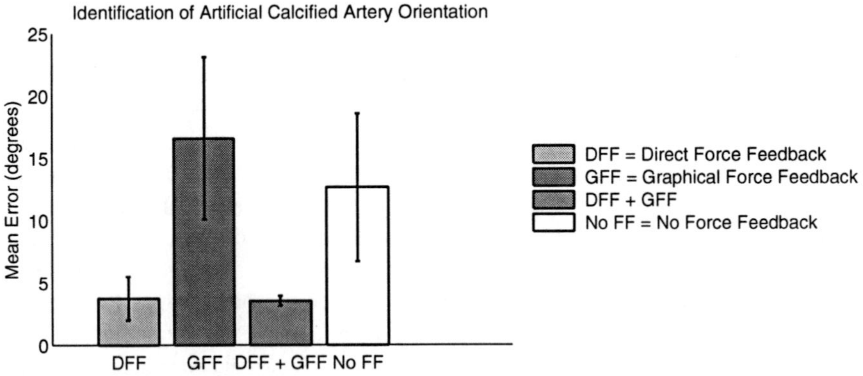

Fig. 18.13 Palpation task performance under various feedback conditions. In contrast to the suture manipulation task shown in Fig. 18.12, graphical feedback in this task does not improve surgeon subject performance over no force feedback [12], ©IEEE 2009

In conclusion, graphical force feedback has produced mixed results. For exploration tasks, such as palpating for tumors in artificial prostate [24] and for calcified arteries in artificial tissue [12], GFF did not significantly improve performance alone, although it did improve performance when paired with direct force feedback. For a manipulation task such as suturing, GFF benefited surgeons whose experience with RMIS systems was minimal, though it did not significantly improve performance of surgeons with considerable RMIS experience [29]. Thus, one's experience with RMIS systems as well as the nature of the surgical task being performed may impact the effectiveness of GFF.

18.5 Graphical Display of Tissue Properties

While force feedback and graphical force feedback provide surgeons with real-time information about applied forces, exploration tasks, such as the palpation scenario described in the previous section, focus on developing a model of tissue properties. In this section, we describe a real-time graphical overlay of stiffness information that is created based on estimated tissue properties during RMIS. The goal of this type of display is to help surgeons identify and characterize tissue abnormalities. For example, surgeons palpate biological tissues to investigate locations of lumps, since malignant tumors are typically stiffer than normal healthy tissues [21, 16]. Although tissue abnormalities can be found preoperatively using medical images such as magnetic resonance imaging (MRI) or computed tomography (CT), such images may not help surgeons localize lumps during actual surgery because of tissue mobility and deformation.

Several studies have explored graphical display of tissue properties. Miller et al. [25] developed a Tactile Imaging System (TIS) that uses a commercial capacitance-based tactile array sensor to measure contact pressure. Based on the measured pressure data from each element of the array, TIS generates a colored tactile map and overlays it on video images in real time. Liu et al. [23] used a rolling indenter with a force/torque sensor to localize hard lumps hidden in soft tissues during RMIS. By keeping the penetration depth constant, a colored force map is created to identify the locations of nodules. Trejos et al. [38] also used a tactile array sensor and developed a Tactile Sensing Instrument (TSI) for localization of tumors during RMIS. They performed a force-controlled autonomous palpation by a robot and generated a pressure map. Xu and Simaan [44] tested force-sensing capabilities of a continuum robot. Similar to Liu et al. and Trejos et al., they also kept a constant penetration depth into the tissues and created a force map to detect embedded hard materials. The stiffness map described in this section differs in that it begins displaying mechanical properties of the tissues in real time as the surgeon is palpating.

18.5.1 Methods for Graphical Display of Tissue Properties

Acquisition of an accurate mathematical model of tissue is challenging since biological tissues are known to be nonlinear and inhomogeneous. Our goal is to use an approximate model of tissue behavior that is sufficiently accurate to give useful discriminatory information, primarily stiffness. Standard online identification techniques can be used to adaptively estimate parameters of a tissue model; however an appropriate form of the model must be assumed a priori. We select the best tissue model through the following offline procedure:

1. Consider several possible mathematical models of the tissue type/organ under consideration.
2. Palpate and record tool-tissue interaction data with real tissue samples. Position, velocity, and acceleration of a tool tip can be calculated from robot kinematics. Force data can be acquired by a force/torque sensor or estimated using a force estimation technique.
3. Postprocess the data to estimate unknown parameters of the model and compare the model accuracy based on force estimation errors (self-validation).
4. Compare the model accuracy based on force estimation errors, using the estimated parameters from self-validation and recorded data (cross-validation).
5. Pick a model that is useful to distinguish hard nodules from healthy soft tissues.

For the parameters of the selected model to be estimated online, there are several estimation techniques available, such as recursive least squares (RLS) and adaptive identification. For detection of lesions in real time, the estimation needs to be so fast that a single palpation may be enough for unknown parameters to converge. Previous results [45] suggest that RLS is a fast, accurate, and simple approach to estimate biological tissue model parameters in real time. In [46], the chosen mathematical model of the artificial tissues is a nonlinear Hunt-Crossley model.

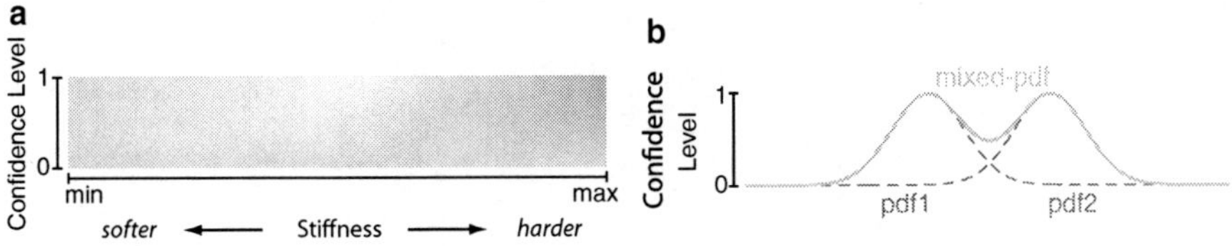

Fig. 18.14 The Hue-Saturation-Luminance representation is used to create the stiffness map. (**a**) Hue is determined by estimated stiffness. (**b**) A weighted probability density function (pdf) is used as confidence level. When two or more pdfs overlap, we take summation of both confidence levels [46], ©IEEE 2009

For estimating the Hunt-Crossley model, two RLS estimators run in parallel as proposed by Diolaiti et al. [8].

Simultaneous to the tissue property estimation, a graphical overlay is created to represent the stiffness at a palpated location. The overlay for a single palpation is a semi-transparent colored circle, which maintains the ability of the surgeon to see the surface of the tissue and laparoscopic instrument tip. With transparency of 50%, the camera images behind the overlay are clearly visible, while the semi-transparent color map is still easy to interpret. A Hue-Saturation-Luminance (HSL) color space is employed to create the color map. The level of stiffness is represented by color (hue) that changes from green to yellow to red as the tissue becomes harder (Fig. 18.14a). The range of a stiffness value can be predetermined based on a preliminary palpation task or updated as new estimation data are added to database. The confidence level of the estimated stiffness is defined by saturation. As a corresponding pixel in the camera image(s) is away from the center of a palpated point, we have less confidence in the estimated stiffness. A weighted probability distribution function of Gaussian distribution is used to calculate confidence level. When multiple probability density functions overlap each other, each confidence level is summed (Fig. 18.14b). This simple interpolation technique creates a nicely blended color map. Luminance is always 0.5 so that the graphical overlay is in color.

18.5.2 Results and Discussion of Graphical Display of Tissue Properties

We tested our approach by using a research version of the da Vinci Surgical System [24] at Johns Hopkins University. An artificial heart tissue made of silicone rubber with an artificial calcified artery embedded was used as a test sample, similar to that used in the palpation experiment described in the previous section. The task was to distinguish the invisible artificial calcified artery from the artificial soft tissue. A sequence of four pictures from a trial are shown in Fig. 18.15. As soon as a palpation tool contacts with the artificial tissue, the estimation algorithm starts estimating unknown parameters of the mathematical tissue model. If a chosen

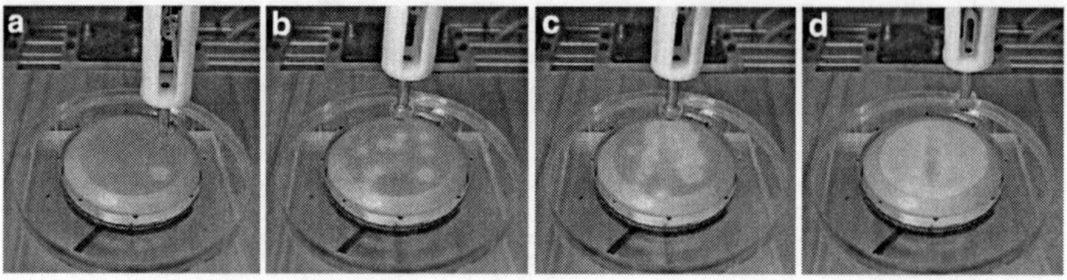

Fig. 18.15 Four representative pictures taken during a palpation experiment. An artificial calcified artery is embedded in the center of an artificial heart tissue vertically. In (**d**), the stiffness map successfully displays the hard artificial artery in the artificial tissue [46], ©IEEE 2009

mathematical model of tissue is accurate enough, estimated parameters converge within one palpation, which generally takes less than one second. When the estimated parameters approach steady-state or, alternatively, parameter updates are small, a semi-transparent colored circle is overlaid on the camera image.

In Fig. 18.15a, there is one green circle overlaid, which indicates the palpated point is relatively soft. Several palpations occurred in Fig. 18.15b, where some of the circles located close to each other are blended smoothly due to the interpolation technique. Fig. 18.15c shows the rough location of the hidden calcified artery. The red region identifies the center of the artificial heart tissue and is surrounded by a green region of "normal" tissue. After the entire surface of the tissue is palpated (Fig. 18.15d), the location of the calcified artery can be easily seen.

To make such graphical overlays appear more natural, tool visualization can be improved by segmentation of the surgical tool from the background in the camera images. In addition, the simple 2D stiffness map shown here should be extended to 3D to accommodate the complex surface geometry of real biological tissues. Such a surface model can be created prior a procedure using pre-operative images, and/or during a procedure using stereo camera images.

18.6 Conclusions

Collectively, results using various forms of force feedback in research systems indicate that such feedback would improve surgeon performance (in some cases, depending on the level of experience of the surgeon) and patient outcomes in RMIS. However, there remains a great deal of work to make force feedback practical and safe for clinical systems. In order for RMIS systems to incorporate force feedback in the most effective manner, such feedback should be considered from the beginning of the design process. Intelligent controller design applied to suboptimal hardware has inherent limitations in transparency and robustness. In addition, it is not yet clear how surgeon experience and training relates to the effectiveness of haptic feedback.

One important area for future work in direct force feedback is the development of systems that incorporate the minimal level and degrees of freedom of force feedback. More studies are necessary to understand how users fill in missing haptic information that is lost when partial force feedback is used. While the phenomenon of visual "fill in" is well-known and documented in the vision literature, the extent to which users can infer force information has not been shown for systems with partial force feedback. Further studies on how people learn to use systems with sensor/actuator asymmetries may also prove to be insightful. Such studies could help to understand how people adapt to reduced force information, which could be a useful for designers considering which degrees of freedom of force feedback are necessary in a surgical system.

While indirect feedback (or sensory substitution) of force information is promising, further research is clearly necessary to establish the appropriateness of such feedback for manipulation (e.g. suturing) versus exploration (e.g. palpation) tasks. Other future work may include identifying more intuitive ways of graphically displaying force information. To date, only a limited number of graphical representations have been realized, and there are likely other representations appropriate for specific surgical tasks or settings.

As shown in Sect. 5, acquired tissue models can be used to generate compelling graphical displays for exploratory procedures. Future work could incorporate such models in teleoperation controllers to improve transparency for direct force feedback. While most teleoperation controllers assume no knowledge of the environment [22, 14], an estimated tissue model may be also useful to create enhanced sensitivity [6, 9] in direct force feedback systems.

Acknowledgements This work was supported in part by Johns Hopkins University, National Science Foundation grants 0347464, 9731478, and 0722943, and National Institutes of Health grant EB002004. The authors thank Dr. David Yuh, Dr. Li-Ming Su, Dr. Mohsen Mahvash, Carol Reiley, Balazs Vagvolgyi, Masaya Kitagawa and Wagahta Semere for their contributions to this work, and Intuitive Surgical, Inc. for access to surgical robotics hardware.

References

1. Akinbiyi, T., Reiley, C.E., Saha, S., Burschka, D., Hasser, C.J., Yuh, D.D., Okamura, A.M.: Dynamic augmented reality for sensory substitution in robot-assisted surgical systems. In: 28th Annual International Conference of the IEEE Engineering in Medicine and Biology Society, pp. 567–570 (2006)
2. Anderson, R.J., Spong, M.W.: Bilateral control of teleoperators with time delay. IEEE Trans. Automat. Contr. **34**(5), 494–501 (1989)
3. Barbagli, F., Salisbury, K.: The effect of sensor/actuator asymmetries in haptic interfaces. In: Proceedings of 11th Symposium on Haptic Interfaces for Virtual Environments and Teleoperator Systems, pp. 140–147 (2003)
4. Barbagli, F., Salisbury, K., Ho, C., Spence, C., Tan, H.: Haptic discrimination of force direction and the influence of visual information. ACM Trans. Appl. Percept. **3**(2), 125–135 (2006)

5. Bethea, B.T., Okamura, A.M., Kitagawa, M., Fitton, T.P., Cattaneo, S.M., Gott, V.L., Baumgartner, W.A., Yuh, D.D.: Application of haptic feedback to robotic surgery. J. Laparoendosc. Adv. Surg. Tech. **14**(3), 191–195 (2004)
6. Cavusoglu, M.C., Sherman, A., Tendick, F.: Design of bilateral teleoperation controllers for haptic exploration and telemanipulation of soft environments. IEEE Trans. Rob. Autom. **18** (4), 641–647 (2002)
7. Daniel, R.W., McAree, P.R.: Fundamental limits of performance for force reflecting teleoperation. Int. J. Rob. Res. **17**(8), 811–830 (1998)
8. Diolaiti, N., Melchiorri, C., Stramigioli, S.: Contact impedance estimation for robotic systems. IEEE Trans. Robot. **21**(5), 925–935 (2005)
9. Gersem, G.D., Brussel, H.V., Tendick, F.: Reliable and enhanced stiffness perception in soft-tissue telemanipulation. Int. J. Rob. Res. **24**(10), 805–822 (2005)
10. Griffiths, P.G., Gillespie, R.B.: Characterizing teleoperator behavior for feedback design and performance analysis. In: Symposium on Haptic Interfaces for Virtual Environment and Teleoperator Systems, pp. 273–280 (2008)
11. Griffiths, P.G., Gillespie, R.B., Freudenberg, J.S.: A fundamental tradeoff between performance and sensitivity in haptic rendering. IEEE Trans. Robot. **24**(3), 537–548 (2008)
12. Gwilliam, J., Mahvash, M., Vagvolgyi, B., Vacharat, A., Yuh, D.D., Okamura, A.M.: Effects of haptic and graphical force feedback for teleoperated palpation. In: IEEE International Conference on Robotics and Automation, pp. 677–682 (2009)
13. Hannaford, B.: A design framework for teleoperators with kinesthetic feedback. IEEE Trans. Rob. Autom. **5**(4), 426–434 (1989)
14. Hashtrudi-Zaad, K., Salcudean, S.E.: Analysis of control architectures for teleoperation systems with impedance/admittance master and slave manipulators. Int. J. Rob. Res. **20**(6), 419–445 (2001)
15. Haykin, S.S.: Active Network Theory. Addison-Wesley, London (1970)
16. Hoyt, K., Castaneda, B., Zhang, M., Nigwekar, P., di Sant'agnese, P.A., Joseph, J.V., Strang, J., Rubens, D.J., Parker, K.J.: Tissue elasticity properties as biomarkers for prostate cancer. Cancer Biomark. **4**(4–5), 213–225 (2008)
17. Khalil, H.K.: Nonlinear Systems, 3rd edn. Prentice Hall, Upper Saddle River, NJ (2002)
18. Kim, K., Cavusoglu, M.C., Chung, W.K.: Quantitative comparison of bilateral teleoperation systems using μ-synthesis. IEEE Trans. Robot. **23**(4), 776–789 (2007)
19. King, C.H., Culjat, M.O., Franco, M.L., Lewis, C.E., Dutson, E.P., Grundfest, W.S., Bisley, J.W.: Tactile feedback induces reduced grasping force in robot-assisted surgery. IEEE Trans. Haptics **2** (2), 103–110 (2009)
20. Kitagawa, M., Bethea, B.T., Gott, V.L., Okamura, A.M.: Analysis of suture manipulation forces for teleoperation with force feedback. In: T. Dohi, R. Kikinis (eds.) Medical Image Computing and Computer Assisted Intervention. Lecture Notes in Computer Science, vol. 2488, pp. 155–162. Springer, Berlin (2002)
21. Krouskop, T.A., Wheeler, T.M., Kallel, F., Garra, B.S., Hall, T.: Elastic moduli of breast and prostate tissues under compression. Ultrasonic imaging **20**(4), 260–274 (1998)
22. Lawrence, D.: Stability and transparency in bilateral teleoperation. IEEE Trans. Rob. Autom. **9**(5), 624–637 (1993)
23. Liu, H., Noonan, D.P., Challacombe, B.J., Dasgupta, P., Seneviratne, L.D., Althoefer, K.: Rolling Mechanical Imaging for Tissue Abnormality Localization During Minimally Invasive Surgery. IEEE Trans. Biomed. Eng. **57**(2), 404–14 (2010)
24. Mahvash, M., Gwilliam, J., Agarwal, R., Vagvolgyi, B., Su, L.M., Yuh, D.D., Okamura, A. M.: Force-feedback surgical teleoperator: Controller design and palpation experiments. In: 16th Symposium on Haptic Interfaces for Virtual Environments and Teleoperator Systems, pp. 465–471 (2008)
25. Miller, A.P., Peine, W.J., Son, J.S., Hammoud, Z.T.: Tactile imaging system for localizing lung nodules during video assisted thoracoscopic surgery. In: IEEE International Conference on Robotics and Automation, pp. 2996–3001 (2007)

26. Niemeyer, G., Slotine, J.E.: Telemanipulation with time delays. Int. J. Rob. Res. **23**(9), 873–890 (2004)
27. Okamura, A.M.: Methods for haptic feedback in teleoperated robot-assisted surgery. Ind. Rob. **31**(6), 499–508 (2004)
28. Ortmaier, T., Deml, B., Kuebler, B., Passig, G., Reintsema, D., Seibold, U.: Robot assisted force feedback surgery. In: Ferre, M., Buss, M., Aracil, R., Melchiorri, C., Balaguer, C. (eds.) Advances in Telerobotics, Springer Tracts on Advanced Robotics STAR 31, pp. 361–379. Springer, Berlin (2007)
29. Reiley, C.E., Akinbiyi, T., Burschka, D., Chang, D.C., Okamura, A.M., Yuh, D.D.: Effects of visual force feedback on robot-assisted surgical task performance. J. Thorac. Cardiovasc. Surg. **135**(1), 196–202 (2008)
30. Saha, S.: Appropriate degrees of freedom of force sensing in robot-assisted minimally invasive surgery. M.S. thesis, Department of Biomedical Engineering, The Johns Hopkins University (2006)
31. Semere, W., Kitagawa, M., Okamura, A.M.: Teleoperation with sensor/actuator asymmetry: Task performance with partial force feedback. In: Proceedings of 12th Symposium on Haptic Interfaces for Virtual Environments and Teleoperator Systems, pp. 121–127 (2004)
32. Skogestad, S., Postlethwaite, I.: Multivariable Feedback Control: Analysis and Design. Wiley, New York (1997)
33. Tan, H., Barbagli, F., Salisbury, K., Ho, C., Spence, C.: Force-direction discrimination is not influenced by reference force direction. Haptics-e, Electron J. Haptics Res. (www.haptics-e.org) **4**(1) (2006)
34. Tavakoli, M., Aziminejad, A., Patel, R., Moallem, M.: High-fidelity bilateral teleoperation systems and the effect of multimodal haptics. IEEE Trans. Syst. Man Cybern. B Cybern. **37** (6), 1512–1528 (2007)
35. Tavakoli, M., Howe, R.D.: Haptic effects of surgical teleoperator flexibility. Int. J. Rob. Res. **28**(10), 1289–1301 (2004)
36. Taylor, R., Jensen, P., Whitcomb, L., Barnes, A., Kumar, R., Stoianovici, D., Gupta, P., Wang, Z., Dejuan, E., Kavoussi, L.: Steady-hand robotic system for microsurgical augmentation. Int. J. Rob. Res. **18**(12), 1201–1210 (1999)
37. Tobergte, A., Konietschke, R., Hirzinger, G.: Planning and control of a teleoperation system for research in minimally invasive robotic surgery. In: IEEE International Conference on Robotics and Automation, pp. 4225–4232 (2009)
38. Trejos, A.L., Jayender, J., Perri, M.T., Naish, M.D., Patel, R.V., Malthaner, R.A.: Robot-assisted Tactile Sensing for Minimally Invasive Tumor Localization. Int. J. Rob. Res. **28**(9), 1118–1133 (2009)
39. Verner, L.N., Okamura, A.M.: Force & torque feedback vs force only feedback. In: Third Joint Eurohaptics Conference and Symposium on Haptic Interfaces for Virtual Environment and Teleoperator Systems (World Haptics), pp. 406–410 (2009)
40. Verner, L.N.: Sensor/actuator asymmetries in telemanipulators. Ph.D. in Mechanical Engineering, The Johns Hopkins University (2009)
41. Verner, L.N., Okamura, A.M.: Effects of translational and gripping force feedback are decoupled in a 4-degree-of-freedom telemanipulator. In: Second Joint Eurohaptics Conference and Symposium on Haptic Interfaces for Virtual Environment and Teleoperator Systems (World Haptics), pp. 286–291 (2007)
42. Verner, L.N., Okamura, A.M.: Telemanipulators with sensor/actuator asymmetries fail the robustness criterion. In: Proceedings of 16th Symposium on Haptic Interfaces for Virtual Environments and Teleoperator Systems, pp. 267–271 (2008)
43. Wagner, C.R., Stylopoulos, N., Jackson, P.G., Howe, R.D.: The benefit of force feedback in surgery: Examination of blunt dissection. Presence: Teleoperators and Virtual Environments **16**(3), 252–262 (2007)
44. Xu, K., Simaan, N.: An investigation of the intrinsic force sensing capabilities of continuum robots. IEEE Trans. Robot. **24**(3), 576–587 (2008)

45. Yamamoto, T., Bernhardt, M., Peer, A., Buss, M., Okamura, A.M.: Multi-estimator technique for environment parameter estimation during telemanipulation. In: IEEE International Conference on Biomedical Robotics and Biomechatronics, pp. 217–223 (2008)
46. Yamamoto, T., Vagvolgyi, B., Balaji, K., Whitcomb, L., Okamura, A.M.: Tissue property estimation and graphical display for teleoperated robot-assisted surgery. In: IEEE International Conference on Robotics and Automation, pp. 4239–4245 (2009)
47. Yokokohji, Y., Yoshikawa, T.: Bilateral control of master-slave manipulators for ideal kinesthetic coupling–formulation and experiment. IEEE Trans. Rob. Autom. **10**(5), 605–20 (1994)

Chapter 19
Tactile Feedback in Surgical Robotics

Martin O. Culjat, James W. Bisley, Chih-Hung King, Christopher Wottawa, Richard E. Fan, Erik P. Dutson, and Warren S. Grundfest

Abstract While commercial surgical robotic systems have provided improvements to minimally invasive surgery, such as 3D stereoscopic visualization, improved range of motion, and increased precision, they have been designed with only limited haptic feedback. A number of robotic surgery systems are currently under development with integrated kinesthetic feedback systems, providing a sense of resistance to the hands or arms of the user. However, the application of tactile feedback systems has been limited to date. The challenges and potential benefits associated with the development of tactile feedback systems to surgical robotics are discussed. A tactile feedback system, featuring piezoresistive force sensors and pneumatic silicone-based balloon actuators, is presented. Initial tests with the system mounted on a commercial robotic surgical system have indicated that tactile feedback may potentially reduce grip forces applied to tissues and sutures during robotic surgery, while also providing high spatial and tactile resolution.

Keywords Tactile feedback · Haptic feedback · Pneumatic · Actuator · Piezoresistive · Sensor · PDMS · Silicone · Mechanoreceptor · da Vinci Surgical System · Grip force · Robotic surgery · Surgical robotics · Balloon · Membrane

19.1 Introduction

Robotic minimally invasive surgery offers advantages such as improved range of motion, enhanced visualization, and higher precision compared to standard laparoscopic techniques. However, the surgeon sits remotely from the patient, therefore

M.O. Culjat (✉)
Center for Advanced Surgical and Interventional Technology (CASIT), University of California, Los Angeles, Los Angeles, CA 90095, USA;
Department of Surgery, UCLA, Los Angeles, CA 90095, USA;
Department of Bioengineering, UCLA, Los Angeles, CA 90095, USA
e-mail: mculjat@mednet.ucla.edu

J. Rosen et al. (eds.), *Surgical Robotics: Systems Applications and Visions*,
DOI 10.1007/978-1-4419-1126-1_19, © Springer Science+Business Media, LLC 2011

limiting haptic information interchange between the surgeon and the patient. The near absence of haptic feedback in commercial robotic surgery systems requires surgeons to rely almost exclusively on visual cues. Visual information is sufficient for many procedures; however the addition of haptic information may enable surgeons to feel tissue characteristics, appropriately tension sutures, and identify pathologic conditions, possibly decreasing the learning curve associated with the adoption of robotic surgery and also enabling its expansion to other minimally invasive procedures.

Haptic feedback refers to the restoration of both tactile and kinesthetic information [1]. Tactile information is useful in providing touch, or cutaneous, sensation to the user, while kinesthetic, or force, feedback provides a sense of position and movement of the robotic end-effectors relative to tissues in the body. The da Vinci surgical robotic system (Intuitive Surgical), a commercially available translational master-slave system (Fig. 19.1), has minimal kinesthetic feedback that limits tool collision and over-extension of joints, but does not prevent damage to tissues or sutures. The ZEUS (Computer Motion), an earlier translational master-slave system, did not feature kinesthetic feedback. Other commercial systems that provide some form of kinesthetic feedback include the NeuroMate (Integrated Surgical Systems, Inc.) for guidance of neurosurgical tools; the RoboDoc (Integrated Surgical Systems, Inc.), CASPAR (Ortomaquet), and HipNav (CASurgica) systems for bone-milling during hip and knee replacement surgeries; the MAKO RIO (MAKO Surgical Corp.) for guidance during minimally invasive knee replacement surgery; the Sensei Robotic Catheter System (Hansen Medical) for cardiac catheter placement; and the Acrobat Sculptor (Acrobat Company Ltd.) for precision bone-drilling during knee replacement surgery. The AESOP (Computer

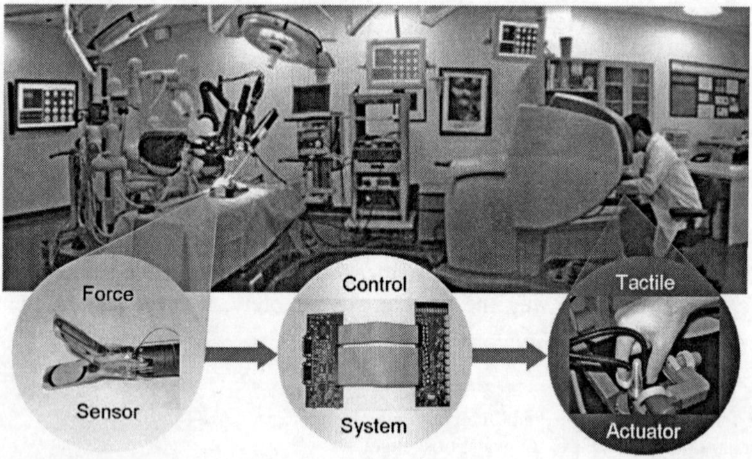

Fig. 19.1 Tactile feedback system concept, with force sensors transmitting information to actuators at the surgeon's fingers. Surgeon is seated at the master console of the da Vinci Surgical System (*left*), and the robotic end-effectors are positioned above the patient cart (*left*)

Motion) and EndoAssist (Armstrong Healthcare) systems, used for control of camera orientation, and the PathFinder (Armstrong Healthcare Ltd), used to actively align tools using pre-operative imaging, do not have haptic feedback. A number of robotic surgical systems being developed at academic institutions are equipped with kinesthetic feedback, such as the UW RAVEN II, the MIT Black Falcon, the JHU Steady-Hand, UC Berkley/UCSF Robotic Telesurgical Workstation, NASA-JPL RAMS system, U Calgary NeuroArm and the UCLA LapRobot [2–7].

Although some surgical robotic systems provide kinesthetic feedback, no commercially available systems provide tactile feedback. Two tactile feedback systems have previously been developed for robotic teleoperation, but both were designed for mounting on specialized mechanical systems rather than robotic surgical systems [8, 9]. Similarly, many tactile feedback system components, both sensors and actuators, have been fabricated for specialized mechanical or robotic applications [9–15]. While promising, these designs are either not intended for surgical robotics or have limited applicability to surgical robotic systems due to size, complexity, or limitations with respect to tactile output.

In many surgical robotic systems, such as those intended for passive positioning, tactile feedback may not provide substantial benefit. Tactile feedback may be more applicable to active master-slave systems, in which the sense of touch can improve precision and control during surgical manipulation. The full effect of the application of augmentative tactile feedback to surgical robotics has not been quantified, primarily because tactile feedback has not yet been applied to surgical robotic systems [16, 17].

This chapter focuses on tactile feedback as it relates to surgical robotics. The importance of tactile feedback is discussed, and a tactile feedback system is described that features silicone-based pneumatic balloon actuators and piezoelectric force sensors, paired with a control system. This system has been fitted directly onto the da Vinci surgical robotic system, allowing the forces applied at the robotic end-effectors to be felt on the fingers of surgeons or other system operators (Fig. 19.1). The system is the first complete tactile feedback system known to be applied to a commercial robotic surgical system. Preliminary system tests with human subjects are described that examine the potential benefit and hindrance of the system in the performance of simple surgical tasks.

19.2 Tactile Perception

Before describing the approach to designing a tactile feedback system, it is worth mentioning how the body collects and processes haptic information. Kinesthetic information, which comes from proprioceptive nerve afferents, contains information about the current musculoskeletal configuration of the body. Golgi tendon organs convey information about muscle tension and drive reflexes that keep muscles from being damaged from excessive contraction. Muscle spindles are stretch receptors that drive reflexes to allow muscles to automatically adapt when environmental forces change. Joint receptors are only active at the extreme limits of the joint's movement. In addition to kinesthetic information, fine tactile information, which originates from four main classes of nerve afferents, contains information about local interactions between the skin and the objects it touches. The four afferents are divided into two classes, based on their physiological responses to skin indentation: fast adapting (FA) afferents respond only when the skin initially contacts a stimulus; whereas slowly adapting (SA) afferents respond both to the initial contact and then continue to be active throughout the period of contact. Within the two classes, the afferents are further characterized by the average area of skin on which contact induces afferent activity. The type I afferents (SAI and FAI) respond to indentation over small regions of skin. Type II afferents respond to stimuli presented over larger areas, with SAII afferents preferring skin stretch and FAII afferents responding best to high frequency vibration, such as that induced when the finger rubs over a fine texture. The responses of these four afferent neurons are driven by the mechanical deformation of the skin, and this deformation is usually caused by local force from a stimulus indenting into the skin. It is important to note that both proprioceptive receptors and tactile receptors respond to force, but it is the location and magnitude of the force that differentially excites these receptors.

Although proprioception no doubt plays an integral role in motor control, the fine control of grip force appears to be driven by a reflex mediated by the responses of FAI afferents. When an object is held between the thumb and forefinger, it is held with just enough force to keep the object in place. Too little force results in the object being dropped or slipping, too much force may damage the object and is inefficient, as it may result in more rapid muscle fatigue. The grip force is usually determined by sensory-motor memory, but it is fine tuned by reflexes that monitor slippage of the object. Recordings from single human afferents have found that the responses from muscle spindles cannot account for this reflex and full reliance on these afferents, induced by anesthetizing the nerves innervating the fingertips, produces a lack of controlled changes in grip force [18–19]. This would indicate that the tactile afferents innervating the fingertip make up the sensory leg of the reflex, and single human afferent recordings have confirmed that the responses of FAI afferents can explain such changes in grip force [20].

Forces at the fingertips in fine manipulation are also not measured by proprioceptive receptors, but by the receptors connected to these tactile afferents. Thus, for fine manipulation in surgery, it is more appropriate to feed this tactile signal back than proprioceptive information. However, it is worth noting that muscle spindles are clearly important for perturbances in digit (or limb) positioning, but when forces are this strong, they have likely passed the point of being potentially detrimental to the tissue being manipulated. Thus, tactile feedback is the more pertinent haptic mechanism for fine manipulation, which is directly applicable to direct grasping of tissues and suture needles.

These principles were used to guide the design of the tactile feedback system. Knowing that it is the FAI afferent that plays a major role in the grip force control reflex, and that the FAI afferent has a small receptive field and responds to indentation of the skin, the tactile actuator was designed such that it exerts pressure onto the skin in a controlled manner. Translating forces from surgical tools directly to the skin of the finger avoids issues related to sensory substitution. These issues include the learning curve associated with the translation of a sense through a different sensing modality. This added step may even require active processing by the surgeon, potentially attenuating the benefits of real-time sensory feedback with time delays and added stress. By providing direct tactile feedback, it was possible to tap into existing grip force control reflexes. In order to do this, the actuator must stimulate the finger in exactly the same way as the object being manipulated. FAI afferents are known to be particularly sensitive to vibrations between 30–50 Hz. However, we felt that the use of a vibratory stimulus would be unsuitable because the time-locked response it elicits would likely swamp the small burst in activity normally seen when a change in force occurs. Thus, we would not be able to tap into the reflex; this would result in a less efficient feedback system. The ideal actuator would also stimulate the SAI and SAII afferents so the surgeon would be able to get a sense of the local force and size of the object being manipulated. SAI afferents are known to give the most accurate representation of the pattern of skin indentation, which correlates to the shape and force of the stimulus and direct stimulation of SAII afferents conveys further information about force [21]. If tactile stimuli can be provided that innervates the FAI, SAI and SAII afferents on a surgeon's fingertips that correspond to detected forces on a robotic grasper, detailed information about the shape size and grip force of a grasped object can be conveyed. Since two of these tactile afferents have small receptive fields, it is possible for an appropriately designed array to transmit information about all three of these variables.

19.3 Actuation in Tactile Feedback

There are various challenges involved with mounting actuators on surgical robotic systems, including limited mounting space, movements of the robotic system, timing, and synchrony between sensing and actuation. Additional requirements are that it must have a light weight, compact size, durability, and negligible hysteresis over thousands of actuation cycles. An ideal actuator is scalable and adaptable, has broad application to multiple fields, and remains viable even as robotic systems are redesigned. Both the Zeus and the da Vinci are finger-controlled systems with limited mounting space, and the master controls are subject to various rotational and translational movements. In order to provide sufficient information to the operator while occupying a small footprint, the individual actuator array elements must be small (0.5–5.0 mm) and allow for narrow element-to-element spacing. In addition, designs must incorporate durable cabling that does not interfere with system operation.

Previous psychophysical, surgical, and haptic feedback studies provide further guidelines regarding tactile actuator design, especially when stimuli are provided to the finger. The temporal and spatial resolutions of the fingertip in response to pin stimuli are approximately 5.5 ms and 2 mm, respectively [22]. For circular or hemispherical stimuli to the fingers, Braille dots serve as a better model than pins. In Standard American Braille, Braille dots are 1.45 mm in diameter, 0.48 mm in height, spaced 0.89 mm apart, and grouped in 3.79 × 6.13 mm cells [23]. The slowly adapting (SA) mechanoreceptors of the finger are sensitive to spherical stimuli and encode Standard American Braille with high efficiency [24].

Grasping frequencies in minimally invasive surgery (MIS) typically do not exceed 3 Hz, so 5–10 Hz operation is sufficient for local procedures [25]. However, for application to tele-robotic surgery or tele-mentoring, frequencies in the 50 Hz range (20 ms cycle) may be desired to minimize the impact of latency between movements. During the first transatlantic telerobotic surgery demonstration, an experimentally verified safe latency period between movement and video images over long distance was found to be within a 300 ms [26]. The minimum detectable force change on the finger pad, ΔF, depends on the applied force. For tapping stimuli, $\Delta F \approx F/2.5$, where F is the applied force [22]. Tactile displays in the literature have had element diameters and spacing as small as 1.0 mm, and have operated in the 6 mN to 3 N force range [27, 28].

Several tactile feedback schemes have been developed as non-invasive means to provide sensory feedback, including motor driven actuation, vibrotactile displays, piezoelectric actuators, shape memory alloys, rheological fluids, and pneumatically driven actuators. These systems function by applying either static or dynamic mechanical deformations along the skin's surface, triggering the sensory receptors in the skin. Many of these tactile systems are effective for a variety of applications; however adaptation effects, low force output, slow response time or bulky mechanical apparatus limits some of these actuators. These characteristics are detailed in Table 19.1.

Vibrotactile actuators, which are among the most popular tactile actuation schemes, have been shown to function with the greatest spatial resolution at 250 Hz [40]. Since this optimized vibrotactile scheme operates in the range of hundreds of hertz, they tend to innervate the fast adapting Pacinian corpuscle sensory receptors (FA II) [41]. While vibrotactile actuators have been shown to operate with very high spatial resolution, it has also been demonstrated that the stimulation of the fast adapting sensory receptors has a deleterious effect on the long-term perception of these actuators [30]. Therefore, despite the many successes of vibrotactile actuators, it does not appear to be a clinically viable choice for extended use in surgical robotics [42].

Pneumatic balloon actuators are well suited for surgical robotics, since they address many of the design constraints described above. They have the advantage of large force output, large deflection, rapid response time, and low mass [43]. Actuation frequency and intensity can be electronically specified to ameliorate adaptation limitations that are prevalent with vibrotactile actuators. This concept has been explored previously for different applications, including aerodynamics control movements and tactile feedback to the fingertip [44–46]; however these designs were not intended for surgical robotics.

Table 19.1 Tactile actuation technologies

Technology	Operating principle	Advantages	Disadvantages	Ref.
Motor-driven	Step-motor rotation translated into linear movement via lead-screw	Precise control of actuation via step-motor, success relaying complex objects and shapes	High response time difficult to provide real-time feedback, large mechanical apparatus	[29]
Vibrotactile	Vibrating motors typically with tunable frequency	High spatial resolution	Susceptible to sensory adaptation effects,	[30]
Piezoelectric	Bimorph plate array of elements that mechanical deform driven by voltage	Can integrate many elements using MEMS processing,	Similar to vibrotactile functionality	[31]
Shape memory alloys	Mechanical deflection induced by thermal expansion	Large force output	High response time, hysteresis	[32, 33]
Rheological fluids	Electrical or magnetic field inputs alter the viscosity of target fluid to allow passive tactile displays	Can be developed for both kinesthetic and tactile feedback.	Large input requirements (voltage or magnetic field), only passive tactile display (in response to finger scanning or rolling)	[34, 35]
Pneumatically driven	Pneumatic source provides vertical actuation on either pin or membrane to provide tactile feedback	Large force output, low response time	Bulky mechanical apparatus	[36–39]

19.4 Pressure Sensing in Tactile Feedback

Pressure sensing systems in robotic surgery are primarily limited by the small mounting surface, the need for serrated edges, large grasping forces encountered on robotic instruments. Forces felt by surgical tools are typically in the 0–5 N range, but have been reported as high as 40 N, and over an average duration of 2–3 s [25, 47, 48]. Recording or transmitting force data can require wire traces along the shaft or a wireless transmitter. In either case, the sensor needs to be powered, also requiring wiring. Sensing systems must be biocompatible, autoclavable, and be capable of withstanding a wet biological environment as well as temperature variations between the body and operating room. Tool-mounted sensor systems must not interfere with the tool's basic functionality (such as the grasping action of a grasper). Therefore ideal sensing systems would have a low-profile, be highly durable to different environments, and could be integrated directly onto or into the instrument.

Many force sensing technologies are strong candidates for force sensing in tactile feedback systems. Some of these technologies include capacitive sensors, magnetic coil sensors, magnetoelastic sensors, optical sensors, piezoelectric sensors, piezoresistive sensors, strain gauges, spring-loaded whisker sensors, and ultrasonic sensors. Some of the advantages and disadvantages of each modality are described in Table19.2. Of these technologies, capacitive force sensors and piezoresistive force sensors are among the most promising for miniaturization and mounting onto surgical robotic end-effectors. Capacitive force sensors have high sensitivity, low temperature dependence, and long-term stability, but are often beset

Table 19.2 Force sensing technologies

Technology	Operating principle	Advantages	Disadvantages	Ref.
Capacitive	Force → change in capacitance	High sensitivity, low temperature dependence, long term stability, ideal for measuring dynamic loads	Static loads difficult to measure, hysteresis, nonlinearity, complicated circuitry	[49, 50]
Magnetic coil	Force → probe displacement → change in magnetic field	Simple design, wide dynamic range, large detection range	Bulky size, mechanical failure under shear loads	[51]
Magnetoelastic	Force → deformation of magnetoelastic material → change in magnetic field	Wide dynamic range, linear response, low hysteresis, low temperature dependence	Susceptible to external magnetic fields, small output power	[52]
Optical	Force → change in light intensity			[53]
Piezoelectric	Force → deformation → change in electric field	Linearity, good at dynamic measurements	Limited in the ability to measure static loads	[54, 55]
Piezoresistive	Force → deformation → change in resistance	Common, simple design, good static and dynamic response	Nonlinearity, hysteresis, creep	[56]
Ultrasonic	Force → change in density and speed of sound in material → change in acoustic wavelength	High linearity, high frequency response, and high sensitivity	Complex electronic interface, limited number of sensing element, hysteresis effect due to material property	[57, 58]
Whisker	Force → deformation of spring-loaded whisker → change in displacement angle of whisker	High sensitivity, simple electronic interface.	Low spatial resolution, bulky size, difficult for microfabrication	[59]

by hysteresis, have complex control circuitry, and are limited in their ability to measure static forces. Piezoresistive force sensors are among the most common force sensing mechanisms, and have the ability to accurately detect both static and dynamic stimuli with simpler electronics, but are also often beset by hysteresis, creep, and signal nonlinearity. Both capacitive and piezoresistive sensors have been developed using sufficiently small MEMS packaging.

Several research groups have developed sensing systems to evaluate tool-tissue force interactions during minimally invasive surgery. Howe et al. [33] developed an 8×8 capacitive sensor array on a probe that measures the distribution of pressure across the tissue contact [50]. Matsumoto et al. [58] developed an ultrasonic sensing probe for stiffness detection of soft tissue, capable of detecting the presence of a gallstone in the gallbladder or cholecystic duct during laparoscopic surgery [58]. Bicchi et al. [60] used an optical sensor to measure contact area spread rate and then correlated this with grasping force [60]. Dargahi et al. [54] developed a 1×4 piezoelectric sensor array using thin film PVDF on a custom made laparoscopic grasper [54]. Kattavenos et. al [61] fabricated a 1×8 piezoresistive array using resistive paste, thin polyester film, and screen-printing technologies [61]. Schostek et. al. [62] integrated a sensor array (32 hexagonal elements) into a customized 10 mm grasper tip. Forces exerted on the silicone rubber layer resulted in a change of the contact area between the conductive polymer sheet and the electrode surface, and thus in a change in electrical resistance between the copper foil and the respective electrode [62]. Mirbagheri et al. [63] developed a four-element force sensor using an instrumented membrane [63].

19.5 Tactile Feedback System Design

A tactile feedback system was developed at the UCLA Center for Advanced Surgical and Interventional Technology (CASIT) in order to address the need for haptics in surgical robotics and to explore its benefits. The tactile feedback system features piezoresistive force sensors, silicone-based balloon actuators, and a pneumatic control system. Pneumatic balloon-based actuation was selected due its ability to provide high force output and be easily integrated and piezoresistive sensing was selected due to thin profile and monotonic response.

19.5.1 Pneumatic Balloon Actuator Design

The actuators are composed of a macromolded polydimethylsiloxane (PDMS) substrate housing a 3×2 array of pneumatic channels and a thin film silicone membrane (Fig. 19.2a). Vertical pneumatic channels within the PDMS substrate connect to horizontal channels that lead to a series of polyurethane tubes, connected with silicone adhesive sealant. The thin film silicone membrane at the output of the

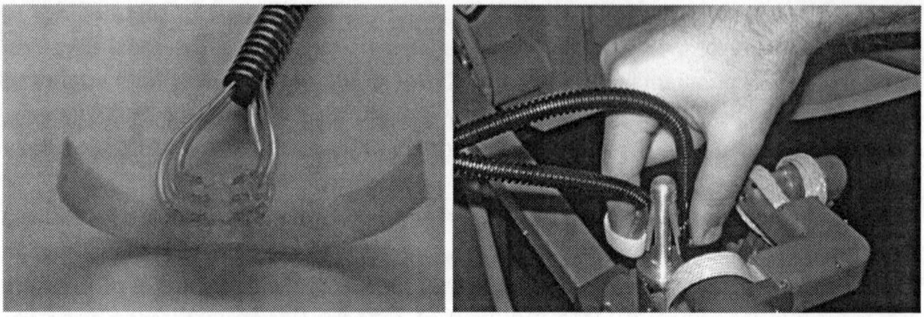

Fig. 19.2 Actuator array (*left*) and mounting on the da Vinci master controls (*right*). (© 2008 IEEE), reprinted with permission

vertical channels form an array of hemispherical balloons that provide pneumatic pressure stimuli to the fingers. Velcro straps were attached to the rear of the PDMS substrate with silicone adhesive sealant for rapid mounting directly onto the da Vinci master controls. The existing Velcro straps were removed from the da Vinci and replaced with the Velcro straps of the actuator array by feeding them through a pair of metal brackets on each master control. The actuator array body fits tightly against the metal master control surface at each finger contact. The Velcro straps were fastened around each finger, and the pneumatic tubing was fed upwards such that it does not interfere with movements of the da Vinci master controls (Fig. 19.2b). Two tactile actuators arrays were developed for each master control, such that a separate actuator array interacts with each finger (left/right index fingers, left/right thumbs).

Mechanical tests and human psychophysical tests were performed to determine the most effective actuator architecture; it was determined that 3 mm diameter balloons, with 1.5 mm element spacing and 300 μm membrane thickness, provided the highest level of performance both mechanically and perceptually [64, 65]. This configuration allowed a maximum of six tactile balloon elements, to be placed on the 1.8 cm × 1.0 cm size mounting area of the da Vinci master controls. Vertical balloon deflection was found to be directly proportional to input pressure. The relationship is monotonic, which allows deterministic and controllable deflection output. Fatigue tests were performed and found the balloon actuators to have negligible hysteresis over at least 150,000 inflation-deflation cycles, therefore ensuring consistent and reliable performance during surgical procedures [64].

19.5.2 Piezoresistive Sensor Design

Commercially available piezoresistive force sensors (Tekscan FlexiForce) were selected for mounting onto the robotic end-effectors due to their thin-film profile (208 μm), small diameter (10 mm), appropriate force range (0–110 N), high linearity, and good static and dynamic response. The sensors were diced to a

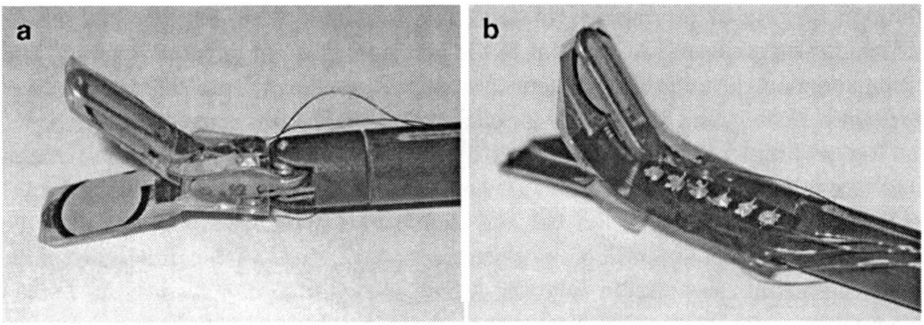

Fig. 19.3 Cadiere graspers with single-element sensors (*left*) and 3 × 2 array (*right*) (© 2008 IEEE), reprinted with permission

width of 5 mm and mounted onto Cadiere tools, which had had their serrations machined down to create a flat surface (Fig. 19.3) [66]. Silver conductive strips (2 mm × 10 mm) taken from the FlexiForce sensors served as the electrodes of the sensors. Thin copper wire was connected to the electrodes using silver epoxy. An inverting amplification circuit was used to translate the change in resistance of the sensor to a change in voltage.

The sensors were able to operate both as a single element sensor and as a 3 × 2 array, with the single element sensor used for feasibility studies and the array developed to examine spatial resolution (Fig. 19.3). The array was constructed by incorporating a 3 × 2 element array of gold electrode pads into the sensing pad. The gold sensing electrodes, traces, and interconnect electrodes were deposited on 127 μm thick polyimide film (Dupont Kapton HN type) using a microfabrication process [67]. The electrode pads were each 1.75 mm × 1.75 mm and separated by 1.5 mm.

The sensors were characterized using an Instron mechanical loading system (Instron 5544), which was programmed to perform 20 loading cycles at the speed of 1 cycle/min over a force range of 0–25 N [66, 67]. Both the single-element sensors and sensor array elements demonstrated a linear resistance decrease with increasing force. The linear response of the mounted piezoresistive sensors was essential to simplify the translation of the input forces from the sensors into proportional output pressures to the balloon actuators. As designed, the force sensing robotic end effectors were sufficient for laboratory testing, but not for clinical use.

19.5.3 Pneumatic and Electronic Control System

A control system was developed that converts the forces detected at the grasper tips to pressures at the surgeon's finger tips via the pneumatic balloon actuators. The control system has both pneumatic and electronic subsystems.

The pneumatic subsystem uses a parallel arrangement of on/off solenoid valves to route different pressures to each actuator. These pre-set pressure levels are achieved through the use of pressure regulators. The solenoid valve arrangement selects which pre-set pressure level is sent to the actuator. Press-fit pneumatic tubing and fittings connect the individual system components, one-way check valves are used to prevent backflow, and an external air canister acts as the pressure source.

The solenoid valves are controlled by the electronic subsystem. A peripheral interface controller (PIC) measures sensor resistances, determines the desired level of balloon inflation for each channel, and outputs a set of digital control signals to the solenoid valves. Signal conditioning electronics convert the variable resistances of the force sensors into measurable voltages. A built-in analog-to-digital converter (ADC) reads these values into the PIC. Software on the PIC divides the voltage range into five distinct regions, using calibration data from surgeons, with buffer regions to limit level transition jitter. In previous experiments, it was determined that five discrete levels were the maximum number that could accurately be detected by the finger [65].

System performance was primarily limited by the response of the pneumatic system, which had a combined filling and exhausting time that was under 50 ms and a frequency response up to 20 Hz. Prior temporal human perceptual tests with the tactile feedback system indicated that the minimum threshold for accurate detection of two time-separated actuations was 62.4 ms, or 16 Hz [65]. Therefore the maximum frequency of the system exceeds the maximum frequency that can be perceived by the finger. For reference, grasping frequencies in local minimally invasive surgery procedures do not typically exceed 3 Hz, and the experimentally verified safe latency period between movement and video images for long distance telesurgery was determined to be within 300 ms [68, 69].

19.6 System Testing

Three system evaluation studies were performed to determine the viability of tactile feedback in surgical robotics, specifically for the master-slave da Vinci surgical system. The first study explored the effect of tactile feedback on grasping, and found that grip force was significantly decreased when tactile feedback was present. A second study examined the potential hindrance of the tactile feedback system to normal operation, and determined that there was no negative effect on the performance of simple tasks. The third study investigated array-based tactile feedback, and determined that spatial information could improve discrimination of objects held within the grasper jaws.

19.6.1 Grip Force

Two grasping studies were performed to examine the potential benefit of tactile feedback in robotic surgery, using the tactile feedback system with the mounted single-element force sensors [66, 70]. The tactile feedback system provided five discrete pressure levels to the finger, corresponding to five grip force ranges at the

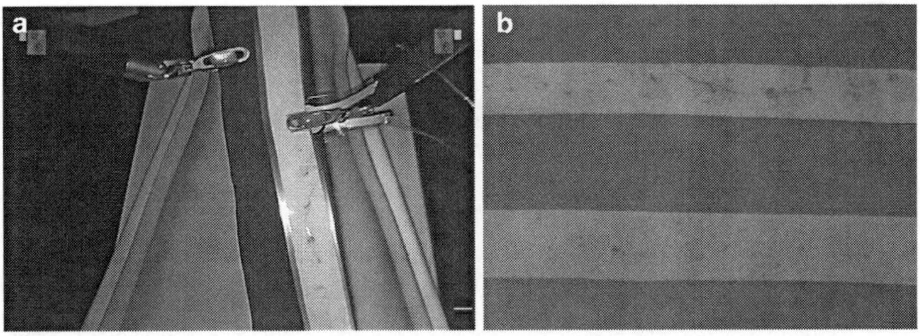

Fig. 19.4 (**a**) Grasping of the tissue phantom with the da Vinci robotic system. *The left grasper* passed the neoprene phantom; *the right grasper* (with mounted sensors) passed the neoprene phantom with attached pressure-indicating film. Marks appeared on the film at grasping locations; color intensity of the marks was proportional to the grasping force. (**b**) Pressure-indicating film resulting from a grasping trial without tactile feedback (*top*) and with tactile feedback (*bottom*). These results indicate that less force was applied to the film with tactile feedback. (© 2008 IEEE), reprinted with permission

surgical end-effectors. In the first of the two studies, grip force was observed qualitatively by asking subjects to grasp a pressure-indicating film, and observing the film following grasping with tactile feedback and without tactile feedback [66]. The film (Fuji Prescale Film LLLW) was attached to a rectangular neoprene substrate (Dupont Corp), with the film and substrate together serving as a soft tissue phantom. When pressure was applied to the pressure-indicating film, the rupture of microcapsules within the film generated a red hue on the white film surface, with increased pressure corresponding to increasing color intensity. Four subjects were asked to perform two trials each, in which they passed the phantom from one end to the other and back (Fig. 19.4a). One trial was performed with the tactile feedback system active, and other with the system inactivated, with the order of trials reversed for each subject. The intent of the study was to determine whether grip force was decreased when the tactile feedback system was activated. All four phantoms had fewer and lower intensity red spots when tactile feedback was used (Fig. 19.4b), therefore indicating that grip force may be lower when the subjects used tactile feedback.

A follow-up study was performed to quantify the effects of tactile feedback on grip force [70]. Twenty subjects (16 novices, 4 experts) were asked to perform a single-hand peg transfer task using the da Vinci system with the tactile feedback system in place, and the subjects' performance was analyzed. Each subject used his/her dominant hand to transfer rubber pieces between two parallel rows of pegs (Fig. 19.5a). The peg transfer test was composed of three consecutive blocks of trials; in each block, a subject performed 18 peg transfers. In the first block, the subject was tested with no tactile feedback; in the second block, the subject was tested with the tactile feedback system activated; and in the third block, the subject was again tested without tactile feedback. The peg transfer task was adapted from the standard peg transfer test in the

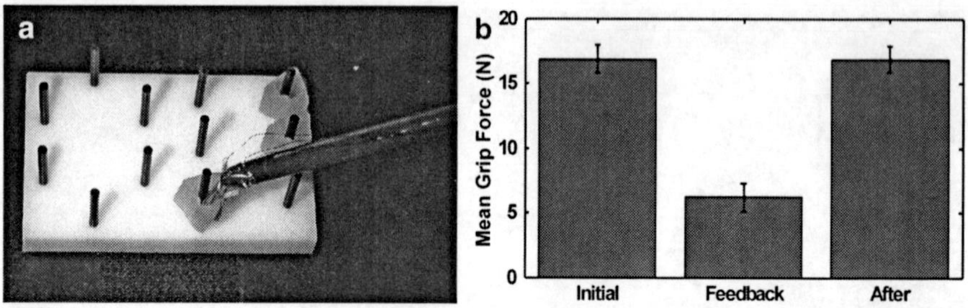

Fig. 19.5 **(a)** Single-hand peg transfer task showing grasping and transfer of a rubber piece using the da Vinci robotic system. **(b)** The mean grip forces exerted by an expert subject during each block of the peg transfer test. Error bars indicate standard error of the mean. (© 2009 IEEE), reprinted with permission

Fundamentals of Laparoscopic Surgery (FLS) education module developed by the Society of American Gastrointestinal Endoscopic Surgeons (SAGES).

Addition of tactile feedback reduced grip force by more than a factor of two. The overall mean grip force for a representative expert subject during each block is shown in Fig. 19.5b. The overall mean grip force during the initial block, in which there was no tactile feedback, was substantially and significantly greater than during the block with feedback ($p << 0.01$, t-test). Since the subject maintained control of the peg throughout the duration of the feedback block, these data suggest that grip force in the absence of feedback is significantly higher, even in a skilled surgeon used to using the da Vinci system. The mean grip force in the block after the feedback was switched off was similar to the initial block, and significantly more than the mean grip force in the block with feedback ($p << 0.01$, t-test). This result indicates that after the tactile feedback was removed, the subject's grip force rapidly returned to the level in the initial block. Similar results were seen in all 20 subjects [70].

19.6.2 System Operability

A study was performed to examine whether the addition of the mounted actuators and pneumatic system, while inactive, impeded system performance [71]. Seven subjects performed four trials each of a Fundamentals of Laparoscopic Surgery peg transfer task; two trials were performed with the inactive actuators and pneumatic system attached to the master controls and master console of da Vinci system, and two trials were performed using the baseline da Vinci system without the actuators or pneumatics attached to the system. All peg transfer trials were timed, and the subjects were asked to respond to a questionnaire regarding the potential impact of the actuators and pneumatic system on task performance.

The results of the peg transfer test found no significant difference in task performance ($p = 0.078$, Wilcoxon Sign Rank test) with and without the inactivated actuators and pneumatics attached to the robotic system. Survey responses from the study indicated that the mounted, inactive actuators and pneumatics did not hinder task performance, ergonomics, or finger or arm movements. The PDMS/silicone actuators themselves did not present added difficulties in grasping; in fact, some of the subjects' responses indicated that the silicone surface of the actuators improved their grip of the robotic master controls.

19.6.3 Array-Based Tactile Feedback

A third study using the mounted 3 × 2 element force sensor array was performed to determine whether the array-based tactile feedback system could accurately transmit tactile information spatially [70]. Forces were applied to individual force sensing elements on the robotic end-effectors, and transmitted to the corresponding balloon elements at the fingers of the robotic operator. Pressure stimuli was provided to six subjects by applying forces to individual sensor elements, to rows and columns of sensor elements simultaneously, and to rows and columns of elements sequentially. The subjects were asked to identify the location and/or sequence of the stimuli.

The spatial perception tests using the system with the 3 × 2 element force sensor array found that there was a high accuracy in detection of single elements (>96%), in rows and columns of elements excited simultaneously (100%), and in rows and columns of elements excited sequentially (100%). The high accuracy in the detection of array-based tactile information suggests that spatial information can accurately be transmitted from the grasper surface to the finger pad, potentially enabling discrimination of variations in tissue properties and slippage of tissues or sutures within the grasper jaws.

19.7 Conclusion

Tactile feedback has not previously been applied to surgical robotics, and still has not been demonstrated for clinical use, primarily because of the challenges associated with the development and integration of proper sensing and actuation technologies. The work described here demonstrates that tactile feedback can be supplied accurately and effectively using thin film piezoresistive force sensors and compact silicone pneumatic balloon actuators with minimal impact to task performance. However, for in-vivo use, force sensing technologies must be better integrated into robotic end-effectors, such that the grasper serrations can remain. The studies described here have also shown that tactile feedback has quantifiable benefits to surgical task performance, and that the array-based tactile feedback approach is viable. Therefore, further development and in-vivo testing of tactile feedback array systems should be pursued for surgical robotics.

Reduction of grip force was a direct outcome of the addition of tactile feedback during surgical task performance. There are many other potential benefits to tactile feedback that are yet to be characterized, such as the sensing of tissue properties, force distributions, and dynamic force changes. The spatial perception tests with the force sensing array have demonstrated that spatial information can be accurately processed from the finger when gripping the robotic master controls. The combination of pressure magnitude and spatial information may provide a surgeon with the full spectrum of tactile information needed during surgery, including grip force, tissue stiffness, tissue conformation, finger positioning, and tissue or suture slippage. Additional quantitative studies must be carried through to examine the magnitude of these effects, and to determine whether these effects extend from the dry laboratory to the in-vivo environment.

The tactile feedback system that was developed for robotic surgery was designed such that it is modular, it is scalable, and its force input and pressure output ranges can be modulated. The system therefore can be rapidly adapted for industrial robotic applications, as well as for future robotic surgery or robotic microsurgery systems, and can be integrated with existing force feedback systems. The system has already been adapted to laparoscopic tools and to provide dynamic Braille information to the blind [72, 73]. The system has also been redesigned and integrated into a lower-limb prosthetic system that transmits tactile information from the base of a lower-limb prosthesis to the upper residual limb of an amputee in order to improve gait and rehabilitation [74]. Other potential applications of the tactile feedback technology include upper limb rehabilitation, virtual reality, gaming, aeronautic feedback, and device packaging.

Acknowledgments The authors would like to thank Dr. E. Carmack Holmes and Mrs. Cheryl Hein for their support, and Mr. Miguel Franco, Ms. Adrienne Higa, and Dr. Catherine Lewis for their hard work and dedication to this project. The authors most gratefully appreciate funding provided for this work by the Telemedicine and Advanced Technology Research Center (TATRC)/Department of Defense under award number W81XWH-05-2-0024. Additional funding for James W. Bisley is provided by an Alfred P. Sloan Research Fellowship and a Klingenstein Fellowship Award.

References

1. Burdea, G.C.: Force and Touch Feedback for Virtual Reality. pp. 3–4. Wiley, New York ((1996)).
2. Lum, M.J.H., Friedman, D.C.W., King, H.H.I., Donlin, R., Sankaranarayanan, G., Broderick, T.J., Sinanan, M.N., Rosen, J., Hannaford, B.: Teleoperation of a surgical robot via airborne wireless radio and transatlantic internet links. In: Laugier, C., Siegwart, R. (eds.) Field and Service Robotics. Springer, Berlin (2008)
3. Madhani, A.J., Niemeyer, G., Salisbury, J.K.: The black falcon: a teleoperated surgical instrument for minimally invasive surgery. In: Proceedings of Intelligent Robots and Systems (1998)

4. Taylor, R., Jensen, P., Whitcomb, L., Barnes, A., Kumar, R., Stoianovici, D., Gupta, P., Wang, Z., deJuan, E., Kavoussi, L.: A steady-hand robotic system for microsurgical augmentation. Int. J. Robot. Res. **18**, 1201–1210 (1999)
5. Cavusoglu, M.C., Tendick, F., Cohn, M., Sastry, S.S.: A Laparoscopic Telesurgical Workstation. IEEE T. Robo. Autom. **15**(4), 728–739 (1999)
6. Das, H., Zak, H., Johnson, J., Crouch, J., Frambach, D.: Evaluation of a telerobotic system to assist surgeons in microsurgery. Comput. Aided Surg. **4**(1), 15–25 (1999)
7. McBeth, P., Louw, D., Rizun, P., Sutherland, G.: Robotics in neurosurgery. Am. J. Surg. **188** (4), 68–75 (2004)
8. Kitagawa, M., Dokko, D., Okamura, A., Yuh, D.D.: Effect of sensory substitution on suture-manipulation forces for robotic surgical systems. J. Thorac. Cardiovasc. Surg. **129**(1), 151–158 (2005)
9. Kontarinis, D.A., Son, J.S., Peine, W., Howe, R.D.: A tactile shape sensing and display system for teleoperated manipulation. Proc. IEEE Int. Conf. Robot. Autom. **1**, 641–646 (1995)
10. Moy, G.: Bidigital teletaction system design and performance. Ph.D. Dissertation,University of California at Berkeley 2002
11. Hayward, V., Cruz-Hernandez, M.: Tactile display device using distributed lateral skin stretch. In: Proceedings Haptic Interfaces for Virtual Environment and Teleoperator Systems Symposium, pp. 1309–1314 (2000)
12. Fukuda, T., Morita, H., Arai, F., Ishihara, H., Matsuura, H.: Microresonator using electromagnetic actuator for tactile display. In: Proceedings 1997 International Symposium on Micromechatronics and Human Science, pp. 143–148 (1997)
13. Wagner, C.R., Lederman, S.J., Howe, R.D.: A tactile shape display using RC servomotors. In: Proceedings HAPTICS, pp. 354–355 (2002)
14. Moy, G., Wagner, C., Fearing, R.S.: A compliant tactile display for teletaction. Proc. IEEE Int. Conf. Robot. Autom. **4**, 3409–3415 (2000)
15. Caldwell, D.G., Tsagarakis, N., Giesler, C.: An integrated tactile/shear feedback array for stimulation of finger mechanoreceptor. Proc. IEEE Int. Conf. Robot. Autom. **1**, 287–292 (1999)
16. Okamura, A.M.: Haptic feedback in robot-assisted minimally invasive surgery. Curr. Opin. Urol. **19**, 102–107 (2009)
17. van der Meijden, O.A.J., Schijven, M.P.: The value of haptic feedback in conventional and robot-assisted minimal invasive surgery and virtual reality training: a current review. Surg. Endosc. 23, 1180–1190 (2009)
18. Macefield, V.G., Johansson, R.S.: Control of grip force during re-straint of an object held between finger and thumb: responses of muscle and joint afferents from the digits. Exp. Brain Res. **108**, 172–184 (1996)
19. Johansson, R.S., Hager, C., Backstrom, L.: Somatosensory control of precision grip during unpredictable pulling loads. III. Impairments during digital anesthesia. Exp. Brain Res. **89**, 204–213 (1992)
20. Macefield, V.G., Hager-Ross, C., Johansson, R.S.: Control of grip force during restraint of an object held between finger and thumb: responses of cu-taneous afferents from the digits. Exp. Brain Res. **108**, 155–171 (1996)
21. Goodwin, A.W., Macefield, V.G., Bisley, J.W.: Encoding of object curvature by tactile afferents from human fingers. J. Neurophysiol. **78**, 2881–2888 (1997)
22. Schiff, W., Foulke, E.: Tactual Perception: A Sourcebook. Cambridge University Press, Cambridge, UK (1982)
23. Tiresias: (2008) Braille cell dimensions. http://www.tiresias.org/reports/braile_cell.htm
24. Phillips, J.R., Johansson, R.S., Johnson, K.O.: Representation of Braille characters in human nerve fibers. Exp. Brain Res. **81**, 589–592 (1990)
25. Brown, J.D., Rosen, J., Chang, L., Sinanan, M.N., Hannaford, B.: Quantifying surgeon grasping mechanics in laparoscopy using the blue DRAGON system. Stud. Health Technol. Inform. **98**, 34–36 (2004)

26. Marescaux, J., Leroy, J., Gagner, M., Rubino, F., Mutter, D., Vix, M., Butner, S.E., Smith, M.K.: Transatlantic robot-assisted telesurgery. Nature **413**(6854), 379–380 (2001)
27. Fukuda, T., Morita, H., Arai, F., Ishihara, H., Matsuura, H.: Micro resonator using electromagnetic actuator for tactile display. In: Proceedings 1997 International Symposium on Micromechatronics and Human Science, pp. 143–148 (1997)
28. Caldwell, D.G., Tsagarakis, N., Giesler, C.: An integrated tactile/shear feedback array for stimulation of finger mechanoreceptor. Proc. IEEE Int. Conf. Robot. Autom. **1**, 287–292 (1999)
29. Shinohara, M., Shimizu, Y., Mochizuki, A.: Three-dimensional tactile display for the blind. IEEE Trans. Neural Syst. Rehabil. Eng. **6**(3), 249–256 (1998)
30. Kaczmarek, K.A., Webster, J.G., Bach-y-Rita, P., Tompkins, W.J.: Electrotactile and vibrotactile displays for sensory substitution systems. IEEE Trans. Biomed. Eng. **38**(1), 1–16 (1991)
31. Summers, I.R., Chanter, C.M.: A broadband tactile array on the fingertip. J. Acoust. Soc. Am. **112**, 2118–2126 (2002)
32. Haga, Y., Mizushima, M., Matsunaga, T., Esashi, M.: Medical and welfare applications of shape memory alloy microcoil actuators. Smart Mater. Struct. **14**(5), S266–S272 (2005)
33. Howe, R.D.Kontarinis, D.A. Peine, W.J.: Shape memory alloy actuator controller design for tactile displays. In: Proceedings of the 34th IEEE Conference on Decision and Control, vol. 4, pp. 3540–3544 (1995)
34. Taylor, P.M., Hosseini-Sianaki, A., Varley, C.J.: An electrorheological fluid-based tactile array for virtual environments. Proc. IEEE Int. Conf. Robot. Autom. **1**, 18–23 (1996)
35. Bicchi, A., Scilingo, E.P., Sgambelluri, N., De Rossi, D.: Haptic interfaces based on magnetorheological fluids. In: Proceedings of Eurohaptics, pp. 6–11 (2002)
36. Cohn, M.B., Lam, M., Fearing, R.S.: Tactile feedback for teleoperation. SPIE Telemanipulator Technol. **1833**, 240–255 (1992)
37. Asamura, N., Yokoyama, N., Shinoda, H.: Selectively stimulating skin receptors for tactile display. IEEE Comput Graph. Appl. **18**(6), 32–37 (1998)
38. Sato, K., Igarashi, E., Kimura, M.: Development of non-constrained master arm with tactile feedback device. In: ICAR, Fifth International Conference on Advanced Robotics 'Robots in Unstructured Environments', pp. 334–338 (1991)
39. Moy, G., Wagner, C., Fearing, R.S.: A Compliant Tactile Display for Teletaction. In: ICRA 2000 IEEE International Conference on Robotic Automation, pp. 3409–3415 (2000)
40. Rogers, C.H.: Choice of stimulator frequency for tactile arrays. IEEE Trans. Man Mach. Syst. **MMS-11**, 5–11 (1970)
41. Phillips, J.R., Johnson, K.O.: Neural Mechanisms of scanned and stationary touch. J. Acoust. Soc. Am. **77**, 220–224 (1985)
42. Petzold, B., Zaeh, M.F., Faerber, B., Deml, B., Egermeier, H., Schilp, J., Clarke, S.: A Study on Visual, Auditory, and Haptic Feedback for Assembly Tasks. Presence **13**(1), 16–21 (2004)
43. King, C.H., Franco, M., Culjat, M.O., Bisley, J.W., Dutson, E., Grundfest, W.S.: Fabrication and characterization of a balloon actuator array for haptic feedback in robotic surgery. ASME J. Medical Devices **2**, 041066-1-041066-7 (2008)
44. Grosjean, C., Lee, G.B., Hong, W., Tai, Y.C., Ho, C.M.: Micro Balloon Actuators for Aerodynamic Control. In: Proceedings of IEEE Micro Electro Mechanical Systems, pp. 166–171 (1998)
45. Yuan, G., Wu, X., Yoon, Y.K., Allen, M.G.: Kinematically-Stabilized Microbubble Actuator Arrays, Micro Electro Mechanical Systems. MEMS 2005. 18th IEEE International Conference on 30 Jan.-3 Feb. 2005, pp. 411–414 (2005)
46. Moy, G., Wagner, C., Fearing, R.S.: A compliant tactile display for teletaction. In: ICRA 2000 IEEE Int.Conf. Rob. Automat, pp. 3409–3415 (2000)
47. Brown, J.D., Rosen, J., Kim, Y.S., Chang, L., Sinanan, M.N., Hannaford, B.: In-vivo and in-situ compressive properties of porcine abdominal soft tissues. Stud. Health Technol. Inform. **94**, 26–32 (2003)

48. Okamura, A., Simone, C., O'Leary, M.: Force modeling for needle insertion into soft tissue. IEEE Trans. Biomed. Eng. **51**(10), 1707–1715.(2004).
49. Gray, B.L., Fearing, R.S.: A surface micromahined microtactile sensor array. IEEE Int. Conf. Robot. Autom. **1**, 1–6 (1996)
50. Howe, R.D., Peine, W.J., Kontarinis, D.A., Son, J.S.: Remote palpation technology for surgical applications. IEEE Eng. Med. Biol. Mag. **14**(3), 318–23 (1995)
51. Sato, N., Heginbotham, W.B., Pugh, A.: A method for three dimensional part identification by tactile transducer. Proc. 7th Int. Symp. Indust. Robots **26**(5), 311–317 (1987)
52. Luo, R.C., Wang, F., Liu, Y.: An imaging tactile sensor with magnetostrictive transduction. Robot Sensors: Tactile Non-Vision **2**, 199–205 (1986)
53. Begej, S.: Planar and finger-shaped optical tactile sensors for roboticapplications. IEEE J. Robot. Autom. **4**(5), 472–484 (1988)
54. Dargahi, J., Parameswaran, M., Payandeh, S.: A micromachined piezoelectric tactile sensor for an endoscopicgrasper-theory, fabrication and experiments. J. Microelectromech. Syst. **9**(3), 329–335 (2000)
55. Krishna, G.M., Rajanna, K.: Tactile sensor based on piezoelectric resonance. IEEE Sens. J. **4**(5), 691–697 (2004)
56. Beebe, D.J., Hsieh, A.S., Denton, D.D., Radwin, R.G.: A silicon force sensor for robotics and medicine. Sens. Actuators **50**, 55–65 (1995)
57. Shinoda, H., Ando, S.: A tactile sensor with 5-D deformation sensing element. IEEE Int. Conf. Robot. Autom. **1**, 7–12 (1996)
58. Matsumoto, S., Ooshima, R., Kobayashi, K., Kawabe, N., Shiraishi, T., Mizuno, Y., Suzuki, H., Umemoto, S.: A tactile sensor for laparoscopic cholecystectomy. Surg. Endo. **11**(9), 939–941 (1997)
59. Russell, R.A.: Using tactile whiskers to measure surface contours. IEEE Int. Conf. Robot. Autom. **1**, 1295–1299 (1992)
60. Bicchi, A., Scilingo, E.P., De Rossi, D.: Haptic discrimination of softness in teleoperation: the role of the contact area spread rate. IEEE Trans. Rob. Autom. **16**, 496–504 (2000)
61. Kattavenos, N., Lawrenson, B., Frank, T.G., Pridham, M.S., Keatch, R.P., Cuschieri, A.: Force-sensitive tactile sensor for minimal access surgery. Minim. Invasive Ther. Allied Technol. **1**, 42–46 (2004)
62. Schostek, S., Ho, N., Kalanovic, D., Schurr, M.O.: Artificial tactile sensing in minimally invasive surgery – a new technical approach. Minim. Invasive Ther. Allied Technol. **15**(5), 296–304 (2006)
63. Mirbagheri, A., Dargahi, J., Narajian, S., Ghomshe, F.T.: Design, Fabrication, and Testing of a Membrane Piezoelectric Tactile Sensor with Four Sensing Elements. Am. J. Appl. Sci. **4**(9), 645–652 (2007)
64. King, C.H., Franco, M.L., Culjat, M.O., Higa, A.T., Bisley, J.W., Dutson, E., Grundfest, W.S.: Fabrication and characterization of a balloon actuator array for haptic feedback in robotic surgery. ASME J. Med. Devices **2**, 041006-1-041006-7 (2008)
65. King, C.H., Culjat, M.O., Franco, M.L., Bisley, J.W., Dutson, E., Grundfest, W.S.: Optimization of pneumatic balloon tactile display for robotic surgery based on human perception. IEEE Trans. Biomed. Eng. **55**(11), 2593–2600 (2008)
66. Culjat, M.O., King, C.H., Franco, M.L., Lewis, C.E., Bisley, J.W., Dutson, E.P., Grundfest, W.S.: A Tactile feedback system for robotic surgery. Proc. IEEE Eng. Med. Biol. Soc. **1**, 1930–1934 (2008)
67. King, C.H., Culjat, M.O., Franco, M.L., Bisley, J.W., Carman, G.P., Dutson, E.P., Grundfest, W.S.: A multi-element tactile feedback system for robot-assisted minimally invasive surgery. IEEE Trans. Haptics **2**(1), 52–56 (2009)
68. Brown, J.D., Rosen, J., Chang, L., Sinanan, M.N., Hannaford, B.: Quantifying surgeon grasping mechanics in laparoscopy using the blue DRAGON system. Stud. Health Technol. Inform. **98**, 34–36 (2004)

69. Marescaux, J., Leroy, J., Gagner, M., Rubino, F., Mutter, D., Vix, M., Butner, S.E., Smith, M.K.: Transatlantic robot-assisted telesurgery. Nature **413**(6854), 379–380 (2001)
70. King, C.H., Culjat, M.O., Franco, M.L., Lewis, C.E., Dutson, E.P., Grundfest, W.S., Bisley, J.W.: Tactile feedback induces reduced grasping force in robotic surgery. IEEE Trans. Haptics (2009)
71. Franco, M.L., King, C.H., Culjat, M.O., Lewis, C.E., Bisley, J.W., Holmes, E.C., Grundfest, W.S., Dutson, E.P.: An integrated pneumatic tactile feedback actuator array for robotic surgery. Int. J. Med. Robot. and Comput. Assist. Surg. **5**(1), 13–19 (2009)
72. Wottawa, C., Fan, R.E., Lewis, C.E., Jordan, B., Culjat, M.O., Grundfest, W.S., Dutson, E.P.: Laparoscopic grasper with integrated tactile feedback system, Proceedings of the 2009 ICME/IEEE International Conference in Complex Medical Engineering. 9–11 April (2009), Tempe, AZ, 1–5 (2009)
73. Fan, R.E., Feinman, A., Wottawa, C., King, C.H., Franco, M.L., Dutson, E.P., Grundfest, W.S., Culjat, M.O.: Characterization of a pneumatic balloon actuator for use in refreshable Braille displays. Stud. Health Technol. Inform. **142**, 94–96 (2009)
74. Fan, R.E., Culjat, M.O., King, C.H., Franco, M.L., Boryk, R., Bisley, J.W., Dutson, E., Grundfest, W.S.: A haptic feedback system for lower-limb prostheses. IEEE Trans. Neural Syst. Rehabil. Eng. **16**(3), 270–277 (2008)

Chapter 20
Robotic Techniques for Minimally Invasive Tumor Localization

Michael D. Naish, Rajni V. Patel, Ana Luisa Trejos, Melissa T. Perri, and Richard A. Malthaner

Abstract The challenges imposed by Minimally Invasive Surgery (MIS) have been the subject of significant research in the last decade. In the case of cancer surgery, a significant limitation is the inability to effectively palpate the target tissue to localize tumor nodules for treatment or removal. Current clinical technologies are still limited and tumor localization efforts often result in the need to increase the size of the incision to allow finger access for direct palpation. New methods of MIS tumor localization under investigation involve restoring the sense of touch, or haptic feedback. The two most commonly investigated modes of haptic perception include kinesthetic and tactile sensing, each with its own advantages and disadvantages. Work in this area includes the development of customized instruments with embedded sensors that aim to solve the problem of limited haptic feedback in MIS. This chapter provides a review of the work to date in the use of kinesthetic and tactile sensing information in MIS for tissue palpation, with the goal of highlighting the benefits and limitations of each mode when used to locate hidden tumors during MIS.

Keywords Tumor localization · Palpation · Minimally invasive surgery (MIS) · Minimally invasive therapy · Cancer treatment · Haptics · Force feedback · Kinaesthetic feedback · Tactile sensing · Instrument design · Palpation depth · Palpation velocity · Palpation force · Tactile pressure maps

---

M.D. Naish (✉)
Department of Mechanical & Materials Engineering, Department of Electrical & Computer Engineering, The University of Western Ontario, London, Ontario, Canada N6A 5B9;
Lawson Health Research Institute (LHRI), Canadian Surgical Technologies & Advanced Robotics (CSTAR), 339 Windermere Road London, Ontario, Canada N6A 5A5
e-mail: mnaish@uwo.ca; michael.naish@lawsonresearch.com

J. Rosen et al. (eds.), *Surgical Robotics: Systems Applications and Visions*,
DOI 10.1007/978-1-4419-1126-1_20,

20.1 Introduction

Cancer is the second leading cause of death in North America for both men and women [1]. For early stages of cancer, surgery is usually the treatment of choice [2], traditionally performed through a large 12-cm incision that allows direct access to the organ of interest. Although preoperative images provide a general idea of the tumor location, intraoperative localization of the cancerous nodule is also required and is commonly performed by manually palpating the organ since nodules are typically stiffer than the surrounding healthy tissue [3].

With the advancement and integration of technology into medicine, minimally invasive surgery (MIS) has become increasingly prevalent over traditional open surgical procedures. MIS is characterized by the use of long, slender instruments to perform procedures through small incisions, 5–12 mm in diameter. However, with the physical constraints presented in MIS, direct palpation of the organ by the surgeon is no longer possible. As a result, the surgeon is limited to using visual or limited tactile cues to determine the position of the tumor.

Current clinical technologies adopted for MIS tumor localization procedures include preoperative Computed Tomography (CT) scans, and the intra-operative use of ultrasound or endoscopic graspers. CT is one conventional method used to diagnose cancer; however, due to tissue shift between preoperative imaging and the operative procedure, the CT image is often an unreliable resource during surgery [4]. This is further aggravated in the case of the lung, since the lung of interest must be collapsed prior to the operative procedure. Intraoperatively, diagnostic ultrasound can be used for tumor localization, and is usually favored due to its real-time non-invasive, portability and versatility. However, ultrasound technologies require specialized personnel to be present in the operating room to interpret the ultrasound images. Furthermore, its use in the lung is impeded by the presence of residual air that often results in poor quality and distorted images [5]. Finally, the use of endoscopic instruments for tissue palpation is affected by forces exerted by the tissue at the instrument entry port [6], thereby corrupting the force feedback to the surgeon.

In the case that these methods fail to locate a tumor intraoperatively, the size of the incision must be increased and the ribs spread apart to allow finger access for direct palpation. Therefore, there is a clear need for an alternative method for locating tumors during MIS while preserving all of the associated benefits of MIS.

20.1.1 Role of Kinesthetic and Tactile Feedback in Tumor Localization

New methods of tumor localization during MIS have been investigated that involve the re-creation of haptic cues, or the "sense of touch," from tissue–instrument interactions at the surgeon–instrument interface.

Haptic information can be divided into two categories: kinesthetic and tactile [7]. Kinesthetic information relates to the movement and bulk forces acting in the joints of an arm, either human or robotic, and at the point of contact. The contour and stiffness of an object can be determined through kinesthetic information via a force/torque sensor. In contrast to kinesthetic information, tactile information includes the sensation of surface textures, or distributed pressures acting across the contacting surface. In this category of haptic information, multiple contact pressures or forces can be concurrently measured using a tightly packed array of sensors. Between the two modes of haptic information, tactile feedback is inherently more complicated than kinesthetic feedback since it often requires an array of sensing elements to determine pressures over a small area. The benefit of tactile feedback, however, is that significantly more information can potentially be collected while palpating tissue.

In MIS procedures, the absence of haptic feedback can be regarded as a safety concern potentially leading to accidental puncture of vessels, or severe bulk tissue damage through the application of high forces. Due to these risks, both modes of haptic feedback have been the focus of active research. This research involves the development of customized instruments with embedded sensors that aim to solve the problems of limited haptic feedback in MIS.

20.1.2 Instrument Design Goals

When designing an instrument specifically for use in MIS procedures, several design challenges are encountered. The design constraints include the following: the sensor size is limited to about 10 mm in diameter in order to fit through the small port incisions; the sensor must be able to withstand temperature variations without affecting its sensing accuracy; it must be sterilizable using at least one commonly used sterilization method; and the sensor must be biocompatible or operate with a protective sheath. Other considerations could require that the instrument be easy and intuitive to use (similar to current conventional surgical devices, in order to reduce the amount of training required); provide real-time information to the user; and have the capability of providing a warning when excessive forces are being applied to the sensor or palpation surface.

The integration of tactile or kinesthetic sensors into MIS instruments aims to increase patient safety, reduce the operation time, and avoid the need to convert from a minimally invasive to an open procedure. The following sections provide a review of the work to date in the use of kinesthetic and tactile sensing information in MIS for tissue palpation. The goal is to outline the benefits and limitations of each mode when used to locate hidden tumors during MIS.

20.2 Kinesthetic Methods

The simplest mode of haptic perception is provided by kinesthetic information, referring to the bulk forces, position, or movement occurring at the joints of an arm when applying forces at the end-effector. These force measurements can be acquired through the use of simple force/torque sensors in order to establish the contour and stiffness of nodules during tumor localization. The following section outlines the state of the art in tumor localization techniques using kinesthetic feedback.

20.2.1 Prior Art

For medical diagnosis, instrument design, and improved simulations, it is important to be able to measure tissue compliance during palpation in order to accurately portray kinesthetic feedback. Indentation testing of in vivo and ex vivo human and porcine abdominal organs was conducted in [8]. While healthy tissues demonstrated similar compliance, a noticeable difference was observed for diseased tissues. These results suggest that an indentation instrument could be used as diagnostic palpation device during MIS.

The simplest palpation device capable of providing kinesthetic feedback is a rod that palpates axially, fitted with a sensor capable of registering the amount of force applied during palpation. One such device that measures tissue elasticity using a force-sensing resistor is presented in [9]. Similar technology was employed in [10] to localize breast tumors. With the aid of finite element analysis, the authors were able to determine the pressure applied during palpation on breast tissue containing an underlying tumor. An indirect measurement approach is used in [11], where an infrared cut filter and image processing algorithms are used in combination with an endoscopic camera to detect forces on the tip of a laparoscopic instrument. While promising, the system must be calibrated for each operating environment, has a relatively slow 5 Hz update rate, and suffers from noise. An axial probe that employs a distal wheel to roll over tissue is presented in [12]. Palpation force is measured using a force-sensing resistor. By rolling the probe over a tissue in several passes, a two-dimensional image of the underlying tissue stiffness can be created.

With probe-type palpation devices, the tip shape and approach method can yield significantly varying results [13]. To address this, a number research groups have developed grasping type designs. In [14], a six degree-of-freedom (DOF) force/torque sensor based on a strain gauge-instrumented Stewart platform is incorporated into a grasper system. The sensor is suitable for MIS, can be placed at the distal end of the instrument to directly measure tool–tissue interaction forces, and has a force resolution of 0.25 N in the z direction and 0.05 N in the x and y directions. A three-DOF sensor with 0.04 N resolution based on fiber-optic sensing is presented in [15]. Strain gauges, sealed in silicone to permit operation in a wet saline environment, were mounted to the back of the jaws of a Babcock grasper in [3]. Qualitative changes in grasping force as small as "a few grams" were relayed to the user via an array of light emitting diodes (LEDs) near the instrument handle. An alternate configuration is described in [16], where a number of strain gauges and a single-axis load cell have been integrated into a custom endoscopic instrument. This system was used to explore user performance during soft tissue discrimination and lump localization [17, 18]. A customized instrument with interchangeable handles and tips that can measure all five DOF available during MIS is presented in [19]. RMS error below 0.07 N was reported for forces in the x and y directions, with 0.35 N for actuation (grasping) and 1.5 N·mm for torsion about the instrument shaft; axial (z) measurements were affected by significant noise. A refined version of the instrument [20] improves slightly upon the measurement of the x, y, and actuation forces; RMS error for torsion was reduced to 0.07 N·mm and for z to 0.14 N.

While sensing at the distal end is preferable because it avoids friction and fulcrum effects of the entry point, it is easier to measure outside the body. A Babcock grasper modified with strain gauges near the handle is presented in [21]. The grasping force and grasper position were indicated, along with a measure of compliance, which could be used to differentiate between objects of various stiffnesses. Another instrumented grasper, described in [22], includes a computerized endoscopic surgical grasper that utilizes shafts and tips from existing surgical tools. It performs an automatic palpation consisting of three cycles of a 1 Hz sinusoidal displacement of the grasper. Experimental results indicate that the tool can distinguish different mechanical properties of tissues by relating motor torque to grasper displacement. In comparative testing, it outperformed a standard laparoscopic grasper but could not discriminate stiffness as well as a gloved hand, the standard approach in open surgery [23, 24]. A similar approach is used in [25], in which tissue stiffness is determined by measuring the amount of current applied to the motor of a motorized grasper. This sensing method is affected by the friction present within the moving components of the grasper, leading to inaccuracies and nonlinearities.

By combining two piezoresistive sensors for measuring lateral forces at the grasper and a thin film force sensor for normal force sensing, tri-directional force sensing may be achieved [26, 27]. The device enables both palpation in an axial direction by closing the grasper jaws and the use of shear force measurements to detect abnormalities by sliding the jaws over a tissue sample. An automatic palpation function was used during validation to ensure consistent palpation. In [28], an ionic polymer metal composite (IPMC) sensor was designed to determine the mechanical properties of soft tissue for open brain tumor resection procedures. Research in [29] used a polyvinylidene fluoride (PVDF) sensor to determine and record mechanical properties during a non-invasive mammogram procedure.

Other research has focused on using pulsated air jets as a means to locate tumors in an internal organ. A non-contact active sensing system is described in [30] to visually represent the dynamic behavior of internal organs for pulmonary tumor localization during VATS procedures. The system is composed of a nozzle to supply pulsated air jets to an internal organ, a strobe system for visualizing the dynamic behavior of tissue, and a camera for capturing strobe images.

The ability to localize tumors using kinesthetic feedback alone depends on a series of requirements and unknowns:

1. What is the ideal palpation method: to palpate to a particular depth and measure the resulting force or to palpate to a particular force and measure depth?
2. How many times can tissue be palpated before the properties of the tissue change and alter the results?

3. If measurements are made to a particular depth, how deeply should the tissue be palpated to best characterize underlying tissue stiffness?
4. In order to reliably measure palpation depth, how can the first point of tissue contact be reliably established?
5. If a robot is used for palpation, what control method produces better results: a constant velocity or accelerating approach?
6. Considering the viscoelastic response of tissue, what forces should be measured during palpation and are the peak or settled forces better suited to the task of identifying the presence of a tumor? This also determines how long the tissue should be palpated for.

The first of these questions was examined both analytically and experimentally for viscoelastic materials in [31]. This study found that palpating to a specific depth from the tissue surface and measuring the force response was more sensitive to stiffness deviations than palpating to a predetermined force and measuring depth. The second question relates to an effect known as pre-conditioning, where soft tissue response approaches a steady-state stiffness and hysteresis over repetitive strain cycles. In [32], in vivo and in situ tissue did not tend to reach a pre-conditioned state within ten cycles.

A series of experiments were performed with the goal of answering the remaining questions. A simple probe, instrumented to provide kinesthetic feedback, was applied to the task of tumor localization in lung tissue. The following sections provide a description of the experiments as well as the results obtained. A final analysis is presented in order to answer the questions above and provide guidelines for future work in the area.

20.2.2 Robotic System and Control

Consistently palpating tissue following a selected palpation method while accurately measuring force requires the ability to accurately control the position and motion of the palpating probe. To achieve this, a robotic system was selected. The benefits of using robotic systems lie in their ability to perform repetitive tasks with great accuracy, consistency and control. For tumor localization, this translates into the ability to control the velocity, location and depth of palpation with increased precision and better repeatability.

For the experiments outlined herein, a Mitsubishi PA10-7C robot was used. An aluminum rod 50 cm in length with a hemispherical tip 9 mm in diameter was mounted on a Gamma force/torque (F/T) sensor (ATI Industrial Automation Inc.). The sensor was directly attached to the mounting plate on the robot. The robot was programmed to move the probe in a vertical position straight down toward the surface of the lung. A picture of the experimental setup is shown in Fig. 20.1.

Considering the variability in tissue thickness between samples and within each sample, a method was developed to determine when the probe made first contact

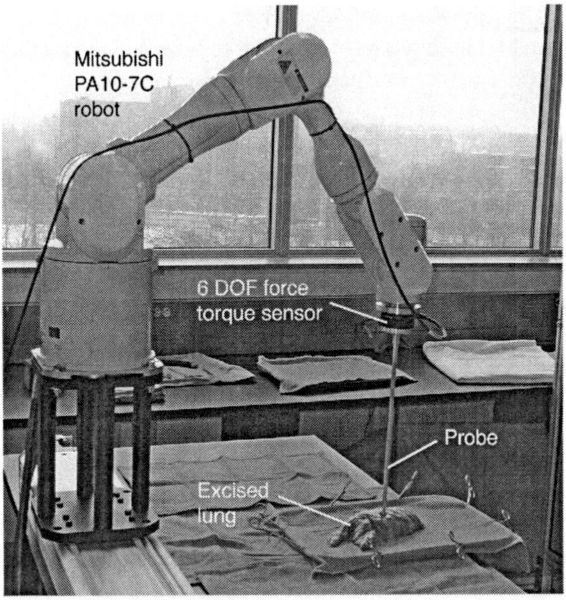

Fig. 20.1 Experimental setup showing the Mitsubishi PA10-7C robot with the palpation probe mounted at the end-effector through a force/torque sensor

with the tissue. The probe was positioned above the tissue and then moved down in 0.5 mm increments until a threshold force was detected in the axial direction.

Due to the variability in the tissue surface, the forces acting on the probe were not always in line with the axis of the probe. Therefore, the magnitude of the resulting forces was computed from the force data for the three orthogonal directions. To determine the value of the threshold force that establishes the first point of contact, a series of trials were performed to establish the minimum value that would not produce false triggers caused by inertial forces acting on the probe. A value of 0.04 N was shown to be ideal. After establishing first contact with the tissue, the robot was held stationary while recording a series of measurements (17 measurements were recorded at 250 Hz) to establish the base force value. Using this approach to ensure consistency between separate palpations, a variety of different palpation methods were assessed. As discussed later in this section, the depth to which the tissue was palpated and the palpation velocity were varied in an effort to determine the ideal method for probe-based tumor localization.

20.2.3 Experimental Evaluation

The feasibility of using kinesthetic feedback alone to establish the presence of a tumor within soft tissue can only be established through an appropriate model of the diseased tissue. The following sections outline the selection and handling of tissues for testing, the method used to create artificial tumors and the methodology used to assess different approaches to palpation with a rigid probe.

20.2.3.1 Preparation of a Tissue Model

For the series of experiments presented in this section, lung was selected as the target organ. Current practice for MIS lung tumor resection involves deflating the lung by occluding the bronchus. Over 30–45 min, the residual gases in the alveoli are absorbed by blood circulating through the lungs. Unfortunately, obtaining collapsed lungs for experimental purposes is often not possible. According to research reported in [33], uncollapsed lung tissue will be equal in stiffness, or slightly stiffer, than intra-operative lung. Thus, detecting a tumor in uncollapsed ex vivo lung tissue will be of similar difficulty as the intra-operative case, if not more so.

Ex vivo porcine lungs were collected from local abattoirs. When harvested, the right and left lungs were left connected by an intact portion of the bronchi. Artificial tumors were embedded into the left lung, while the right served as a control in all experiments. This approach served to mitigate the effects of tissue autolysis [34], temperature cycles and hydration that could obscure the effect of the artificial tumor. Furthermore, all experiments were completed within 60 hours of harvest. The lungs were sealed individually in plastic bags containing saline solution and refrigerated when not undergoing testing.

20.2.3.2 Manufacturing of Artificial Tumors

After testing a variety of methods, including placing small volumes of material into incisions or on a mesh placed between the halves of bisected tissue, it was discovered that any method that introduced major incisions affected the force-deformation behavior of the tissue [13]. To preserve the tissue characteristics, material was instead injected. A preparation of Sigma Gelrite Gellan Gum (agar), prepared in a 30:1 ratio of water to agar by weight, boiled and then injected into cold tissue was determined to provide the most suitable tumor model. This ratio results in artificial tumors that have a stiffness within the range of those found clinically. In qualitative terms, tumor stiffness generally varies from that of a grape to that of a rock.

To produce a tumor with a diameter of approximately 1 cm, a volume of 1.5 ml of agar was injected. The additional volume accounted for losses of liquid through the injection site – no evidence of absorption of agar into the surrounding tissue was observed. Examples of artificial tumors, excised from ex vivo porcine lung after testing, are shown in Fig. 20.2.

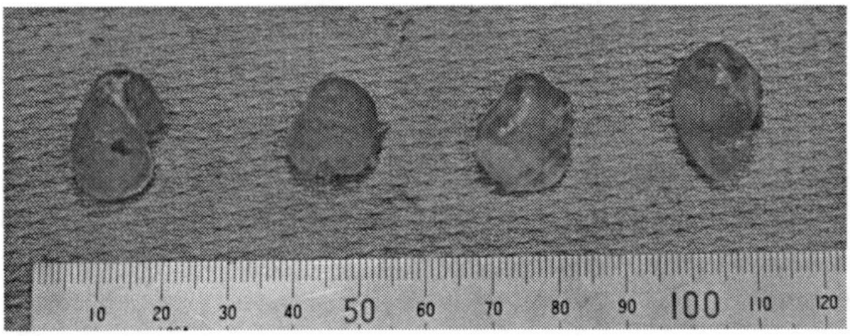

Fig. 20.2 Sample artificial tumors excised from ex vivo lung

20.2.3.3 Methods

Tests were conducted on paired lungs, as described in Sect. 20.2.3.1. To avoid preconditioning as much as possible, test sites were selected to be at least 30 mm apart, minimizing the influence of adjacent sites. The testing grid was designed to sample only the lower lobe and to avoid the main bronchial branches. Each test site underwent only two palpations: an initial sample was collected for the intact tissue, followed by a second sample after a tumor had been injected.

Each test began by locating the lung surface as described in Sect. 20.2.2. Once contact was verified, the probe was advanced into the tissue to a pre-programmed depth. Depths of 5, 7, and 9 mm were selected after preliminary testing revealed that the artificial tumors would occasionally crack when palpated beyond 10 mm. Upon reaching the target depth, the probe was held stationary for the remainder of the sample period. Continuous force measurements were recorded in three orthogonal directions for 10 s at a sample rate of 250 Hz. At the end of the test, the probe was retracted and moved to the next test site. After testing was completed for each lung, each palpation site was marked with permanent marker.

Once all initial tests were completed for a specific depth, artificial tumors were injected into each palpation site of the left lung, to approximately the mid-thickness of the lung (the right-side control lung was left intact). The palpation tests were then repeated. Due to variability in the tumor formation process, each tumor was located using manual palpation and aligned with the probe prior to testing. The test sites were numbered such that the measurements from before and after the tumor was injected could be paired for each location.

To account for the effect that palpation velocity might have on the ability to detect underlying tumors, the tests were performed using two velocity control methods. The variable-velocity approach was realized by setting the maximum velocity of the robot to 40 mm/s. With the relatively short translations required for palpation, the robot did not achieve this velocity, but rather accelerated towards it. As a result, the variable-velocity approach resulted in higher peak velocities as the palpation depth was increased. In contrast, the constant-velocity approach set the maximum translation velocity of the robot to 2 mm/s. This velocity was reached for all depths.

20.2.4 Effect of Palpation Depth and Velocity

The detection of tumors is based on the change in observed force when palpating to a fixed depth. Thus, the change in force feedback, $\Delta F = F_2 - F_1$, reflects the degree to which a tumor is likely present, where F_1 and F_2 are the forces measured during the first and second palpations respectively. The greater the force increase, the better the response to the presence of underlying tumors can be considered to be. For the control lungs, the expected force increase was 0 N. For the experiments presented here, F_1 represents the force recorded from the initial palpation of either the test or control lung and F_2 is the measurement record with the tumor present, or the second test of the control lung.

To account for any changes that could be caused by factors other than the presence of a tumor, the test values were adjusted using the upper quartile measurement from the associated control. This effectively reduces the observed force increase. Following this correction, the differing palpation methods were assessed by grouping the ΔF values into sets according to depth and the control approach used. These results are presented in Tables 20.1 and 20.2. A Wilcoxon paired-sample test found the p values to be less than 10^{-4} in all cases.

20.2.4.1 Depth

All of the selected palpation depths produced an observable force increase when palpating tumors. The best depth is then the one that produces the largest consistent force increase. Ideally, the deviation from the median value is small and there are no negative differences. Negative differences are problematic as they indicate a

Table 20.1 Lower quartile (LQ), median, and upper quartile (UQ) of force difference, ΔF, observed when palpating tumors using a variable-velocity approach

	Peak force (N)			Settled force (N)		
Palpation depth (mm)	LQ	Median	UQ	LQ	Median	UQ
5	0.0675	0.1092	0.1657	0.0587	0.0871	0.1294
7	0.1466	0.2295	0.4643	0.0703	0.1609	0.2908
9	0.0425	0.3995	0.6853	0.0301	0.1666	0.4311

Table 20.2 Lower quartile (LQ), median, and upper quartile (UQ) of force difference, ΔF, observed when palpating tumors using a constant-velocity approach

	Peak force (N)			Settled force (N)		
Palpation depth (mm)	LQ	Median	UQ	LQ	Median	UQ
5	0.0521	0.0946	0.1458	0.0434	0.0764	0.1303
7	0.1011	0.1529	0.1923	0.0724	0.1246	0.1568
9	0.0560	0.1445	0.2481	0.0576	0.1300	0.2225

Table 20.3 Percentage of observed negative force differences, indicating rate of false negatives

	Variable-velocity approach		Constant-velocity approach	
Palpation depth (mm)	Peak (%)	Settled (%)	Peak (%)	Settled (%)
5	10	10	8	8
7	3	5	3	3
9	21	21	3	3

false negative—a case where a tumor is undetected. The numbers of negative differences observed are presented in Table 20.3.

On average, the force difference and variance observed for the 5 mm palpation tests were lower than the 7 and 9 mm tests and in over 8% of the cases, the difference was negative, resulting in a false negative interpretation. In contrast, the false negative rates for both 7 and 9 mm constant-velocity tests were 3%. Note that, in clinical practice, the surgeon would likely perform multiple palpations, mitigating the impact of false-negative results.

The 7 mm constant-velocity tests tended to perform best in terms of statistical confidence in the observed force increase, tight data spread and the number of false-negative results, leading to a more accurate conclusion regarding the presence of an underlying tumor. However, the performance of the 9 mm constant-velocity approach was very similar to that of the 7 mm, with the exception of a wider data spread. Overall, the 7 and 9 mm constant-velocity approaches were clearly superior to the other methods tested.

20.2.4.2 Velocity

Comparing the peak and settled values of the variable- and constant-velocity approach methods in Tables 20.1 and 20.2, it is apparent that the slower constant-velocity approach exhibits less of a difference between the peak and settled force values. This implies that the constant-velocity approach permits reasonably accurate comparisons between measurements collected at any time after the desired palpation depth has been reached. Figure 20.3 illustrates the nearly constant stress resulting from the constant-velocity method. This may be advantageous when considered in a clinical setting, where maintaining a constant strain for an extended period may not be possible. When considering the faster, variable-velocity approach, the pronounced peak and exponential stress decay imply that, to make comparisons between two tests, the time at which the response is analyzed is critical.

20.2.5 Applicability to the Design of Tactile Instruments

Table 20.4 presents the raw peak values that were recorded when palpating tissue after a tumor had been injected. These represent the maximum expected forces during probe-based kinesthetic lung tumor localization. Note that, unlike the

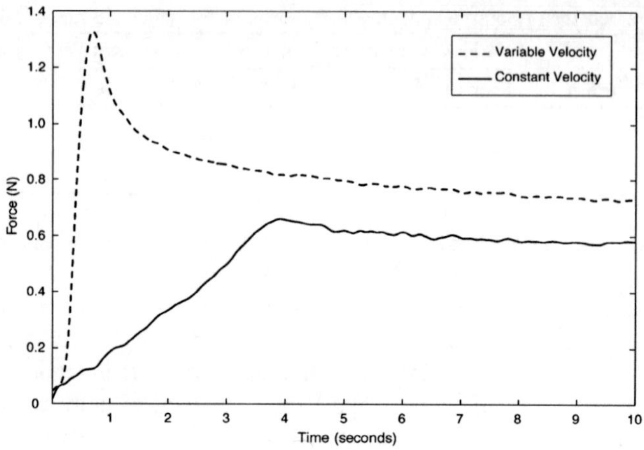

Fig. 20.3 Sample force response of variable-velocity and constant-velocity palpation to a depth of 9 mm

Table 20.4 Lower quartile (LQ), median, and upper quartile (UQ) of peak forces observed in the presence of a tumor

	Variable-velocity approach (N)			Constant-velocity approach (N)		
Palpation depth (mm)	LQ	Median	UQ	LQ	Median	UQ
5	0.2405	0.2862	0.3414	0.2320	0.3025	0.3416
7	0.4333	0.5260	0.7396	0.4061	0.4607	0.5096
9	0.5154	0.9057	1.1401	0.4562	0.5987	0.6858

values in Tables 20.1 and 20.2, these do not represent differences, nor are they adjusted by the controls.

Overall, peak measurements lie in the range of <2 and <1 N for the variable-velocity and constant-velocity tests, respectively. Thus, depending on the control method, it would be reasonable to define the full scale range of a sensor system as 0–2 or 0–1 N.

The required resolution can be determined from the corrected ΔF values given in Tables 20.1 and 20.2. The 7 mm constant-velocity approach will be used for this purpose, due to its superior performance. Considering the upper and lower quartile bounds from both the peak and settled analysis, a significant portion of the ΔF values are in the range 0.07–0.16 N. To be able to detect the majority of underlying tumors, a kinesthetic sensing system must be able to resolve forces in this range. A resolution of 0.07 N would result in a false-negative error rate of 25%, based on a single palpation. To avoid quantization error, the resolution of the sensor should be one order of magnitude less than this value: 0.007 N, or approximately 0.01 N.

20.2.6 Conclusion

The experiments performed as part of this study aimed to determine the ideal palpation method when using kinesthetic information for tumor localization. Through a review of the literature, it was determined that, ideally, a probe palpates to a particular depth and a sensor measures the resulting force on the probe. Other research supports the idea of pre-conditioning tissue, showing that the number of palpation cycles must be kept below ten in order to minimize the effect of tissue change caused by palpation. Following this reasoning, a series of experiments were conducted that showed the ideal palpation depth to be 7 mm, with good results also possible with a 9 mm palpation depth, at the expense of higher forces and possible tissue damage. The experiments also showed that constant-velocity tests tended to perform the best in terms of statistical significance of the force increase caused by the presence of a tumor, limited data spread, and the number of false negative results observed. Furthermore, as the constant-velocity approach shows a nearly constant stress curve as a function of time, taking the settled measurement over the peak measurement is not significantly different.

Through these experiments, the importance of clearly identifying when the probe first makes contact with the tissue was determined, in order to ensure that the palpation depth is the same at each palpation point. The ideal method for establishing the first point of contact was to slowly advance the probe towards the tissue until a consistent value of 0.04 N was detected by the force sensor.

The results of these experiments indicate that sensors for minimally invasive tumor localization should have a sensing range of 0–2 N and a minimum resolution of approximately 0.01 N. This information is not only valuable for the development of kinesthetic sensors for tumor localization, but also for tactile sensors, as outlined in the following section.

20.3 Tactile Approaches

Although kinesthetic information is useful in minimally invasive tumor localization techniques, there are limitations with providing the surgeon with solely kinesthetic feedback. Most importantly, the use of kinesthetic feedback for tumor localization requires absolute control of the palpation depth relative to the tissue surface, which is not always feasible in actual clinical settings.

An alternative technology involves the use of tactile sensors, usually formed by an array of pressure sensors that can map relative pressure differences on the contacting surface. The resulting pressure map provides information on the relative difference in the stiffness of the tissue immediately below the sensor, regardless of the applied force or depth of palpation. To achieve this, many sensing elements must be embedded onto a small sensing area, taking into account manufacturability and the complexity of integration with minimally invasive instruments. Despite this difficult task, most researchers value the advantage of relaying more information about the tissue over the added design complexity. Therefore, the use of tactile sensors with different feedback methods to aid in tumor localization has become an active research area.

20.3.1 Prior Art

A variety of tactile instruments have been developed to measure tool-tissue interaction forces and relative tissue stiffness. Technologies employed for tactile sensors include piezoelectric, piezoresistive, capacitance-based, optical fiber and polymer-based sensors. These have been integrated into passive and active devices for MIS. Manual instruments generally rely on the user to properly position the sensor and apply the correct level of contact force, while active devices automate the sensing process to ensure consistent operation. Examples of these devices are summarized below:

A simple four-element one-dimensional array sensor for a grasper-type minimally invasive instrument is developed in [35]. The sensor is constructed from layers of PVDF as the sensing medium for a tactile sensor with four sensing elements. While the sensor is compact and inexpensive, the one-dimensional configuration is uncommon. Instead, most tactile devices employ two-dimensional arrays that limit the amount of motion required to build a topographical pressure map of the underlying tissue.

A number of elastomer-based tactile sensors are reviewed in [36]. In one type, carbon or silver was embedded within rubber, such that compression of the rubber caused a local increase in the concentration of the embedded media, thereby increasing the conductivity. The resulting sensor had a spatial resolution of a few millimeters. In general, these sensors suffer from noise, hysteresis, creep, cross-talk, and limited dynamic range.

Piezoelectric and piezoresistive sensor technologies have been incorporated into a variety of devices for tumor localization in organs. For example, a device consisting of a modified ultrasound scaler and a piezoelectric sensor used as a sensing element for brain tumor localization is proposed in [37]. When tested, this device could determine the difference between healthy tissue and hard occlusions in gelatin models. However, due to the nature of the design, the instrument does not meet MIS design constraints. The design of an 8×8 array of piezoresistive-based sensors incorporated on forceps for MIS examination of the bowel for tumors was proposed in [38]. The resulting prototype was slightly too large for traditional MIS. A similar design was proposed in [39] for non-invasive mammogram examinations. The instrument consisted of a 64-element tactile sensor array using piezoelectric technology; however, it also did not meet MIS size constraints.

A high-sensitivity capacitance-based sensor is described in [40]. A variety of different arrays may be formed from perpendicular copper strips separated by thin pieces of silicone dielectric. Devices based on this technology include a modified conventional laparoscopic grasper with a 60-element capacitive array sensor [41, 42]. This device was able to determine the hardness and size of rubber balls in water-filled porcine intestine. Using similar technology, another device was proposed in [43] for pulmonary tumor localization during video-assisted thoracoscopic surgery (VATS). When tested on ex vivo porcine lungs, the system was capable of outputting streaming video images overlaid with color-contour maps representing the pressure distribution of the tissue. Limited haptic feedback and difficulties in articulating the end-effector of the probe were reported to be problematic.

A three-layer hexagonal array comprised of spherical conductor electrodes and thin conductive films is presented in [44]. Signals from the tactile array are multiplexed at the sensor, minimizing the required number of signal wires. This device has been incorporated into the jaw of a custom laparoscopic grasper by completely sealing it in silicone rubber. The sensor provides an operating range of 0–40 N and a 1.4 mm spatial resolution.

In [45], pairs of LEDs and photodetectors are mounted within a special casing. Deformation of the casing changes the internal optical reflectance of each pair, permitting contact forces to be estimated. This sensor is simple and inexpensive; however it is currently too large for MIS applications. Using optical fiber technology, a force-sensitive wheeled probe is proposed in [46] to generate a color-contour image representing the stiffness distribution of tissue. However, the successful use of this system is entirely dependent on the accurate movement of the wheeled probe across the surface of the tissue at an equal palpation depth, which is not realistic in actual clinical applications.

A robotic system consisting of an anthropomorphic finger with a tactile sensor array in the fingertip is proposed in [47] to autonomously palpate for a patient's arterial pulse at the wrist. This system is the only automated system, apart from the work presented herein, to use tactile sensing for diagnostic purposes. To detect the presence of a 19 mm acrylic ball embedded in rubber, a master–slave robotic system equipped with tactile sensing capabilities was evaluated by [48]. When compared to direct manipulation of the tactile sensor, the results showed that the performance of the system was greatly dependent on how well the exploration force could be controlled by the user.

20.3.2 Tactile Instrument Design

To address the limitation of poor tactile sensing capabilities during MIS, a device, termed the Tactile Sensing Instrument (TSI), was designed to meet the sensor

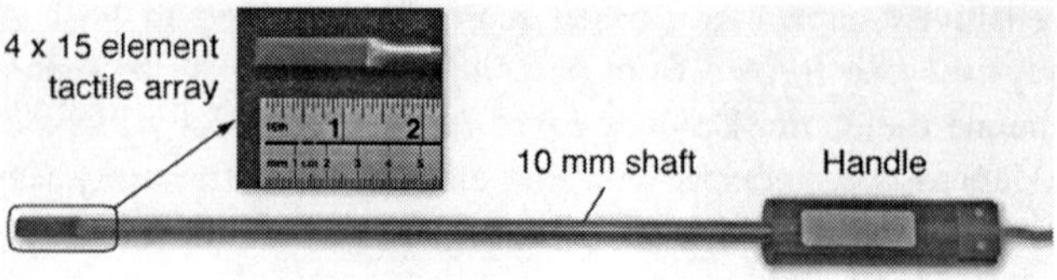

Fig. 20.4 The Tactile Sensing Instrument (TSI)

Table 20.5 Specifications of the TSI

Probe shaft length	385 mm	Spatial resolution	2 mm
Probe shaft diameter	10 mm	ADC resolution	12-bit
Active area of sensor	10 × 35 mm	Pressure range of sensor	0–14,000 kPa
Number of sensor elements	60	Temperature range of sensor	−40 to 200 C

performance conditions determined through the experiments presented in the previous section [49]. The TSI, developed by a research team at Canadian Surgical Technologies and Advanced Robotics (CSTAR, London, Ontario), uses a commercially available pressure sensor that was incorporated on a metal probe that meets the design constraints for MIS procedures (see Fig. 20.4 and Table 20.5). The instrument has been designed using only biocompatible materials that could withstand various sterilization procedures. To protect the sensor during use, a disposable laparoscopic latex sleeve can be placed over both the sensor and the shaft of the probe.

The pressure sensor on the TSI uses capacitance-based technology, marketed as the Industrial TactArray sensor from Pressure Profile Systems (PPS, Los Angeles, CA). This sensor can offer excellent sensitivity and repeatability while maintaining satisfactory results in resolution, temperature, and design flexibility [50]. The PPS Industrial TactArray sensor was custom-built using only biocompatible materials and adhesives to meet the size constraints of MIS procedures. The resulting sensor has a sensing area 10 × 35 mm formed by 60 sensing elements arranged in 15 rows and 4 columns. In the sensor array, two layers of electrodes are oriented orthogonally to each other with each overlapping area between the row and column electrodes forming a distinct capacitor [51]. In between the capacitor plates is a compressible dielectric matrix that effectively acts as a spring between the electrodes. When a force is applied to the top mobile capacitor plate, the decrease in distance between the two capacitor plates generates an increase in the output voltage. Once pressure is no longer applied to the sensor, the spring-like dielectric matrix allows the capacitor plates to return to their resting position.

Preliminary experiments performed with the TSI [52] showed a marginal improvement over palpating with a surgical grasper in a minimally invasive box trainer. This led to the notion that robotic manipulation of the instrument could improve the results, as presented in the following section.

20.3.3 Robotic System and Control

20.3.3.1 Experimental Setup

Experiments using the TSI were performed using a seven-DOF Mitsubishi PA10-7C robot. The laboratory setup is shown in Fig. 20.5. To measure the forces exerted by the robot end-effector on the tissue, an ATI Gamma six-DOF force/torque sensor (F/T Sensor B) was fixed to the wrist of the robot.

20.3.3.2 Robot Control

The control system was designed to allow for force-controlled tissue palpation. An Augmented Hybrid Impedance Control (AHIC) scheme [53–55] was implemented on the PA10-7C robot. The AHIC scheme allowed for the control of palpation force and position of the robot's end-effector in Cartesian space. The task space in AHIC was divided into two modes. The first mode allowed for force control in the direction of palpation (z direction). The second mode controlled the position and orientation of the end-effector of the robot in the remaining orthogonal directions (x, y directions). A simple interface allowed a user to indicate the desired tissue area to palpate by entering the direction of palpation and the number of points to palpate. After the required input information was provided by the user, robot palpation occurred in a completely autonomous manner.

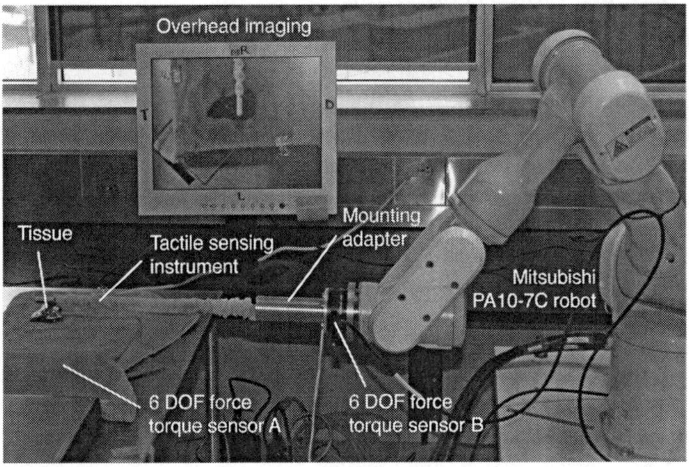

Fig. 20.5 Robotic experimental setup for tissue palpation with the TSI

The robot control strategy was implemented in two settings: force control and position control. In the force-control setting, the robot was designed to approach the tissue (in the *z* direction) under force control until the ATI force/torque sensor (F/T Sensor B) located on the robot end-effector registered a pre-determined threshold force. A threshold force of 4 N was determined through preliminary experiments using the PA10-7C robot, as well as from other experiments that found that the average maximum force applied by surgeons when manually palpating tissue was 4.4 N [56, 57]. Upon reaching the threshold force of 4 N, the client computer recorded the robot end-effector coordinates (also corresponding to the tissue surface coordinates) and the readings from the TSI, indicating the force profile of the contact made by the robot onto the tissue. After data logging of the desired information was complete, the robot autonomously moved to the next palpation point by moving both up (*z* direction) and 3 mm sideways in either the *x* or *y* direction, depending on the user's input. This process continued until all user instructions were executed and the entire tissue surface was palpated.

Similarly, the position-control setting ensures that the position of the robot end-effector is controlled in all Cartesian directions. In this mode, the desired trajectory in the *z* direction was designed to ensure that the robot made contact with the surface of the tissue and moved below the surface under position control. To determine how deep below the surface the robot end-effector should go, readings from F/T Sensor B were constantly polled to detect when the contact at the end-effector reached the pre-determined threshold force of 4 N. When the threshold force of 4 N was reached, the robot position and TSI data were recorded and the robot autonomously moved to the next palpation point. This process continued until the entire area defined by the user had been palpated.

20.3.4 Experimental Evaluation

A study was designed to assess the feasibility of using the TSI under robotic control to reliably locate underlying tumors while reducing collateral tissue trauma. The performance of humans and a robot using the TSI to locate tumor phantoms embedded into ex vivo bovine livers was compared. Two experimental setups were used for comparison; both setups incorporated the use of the TSI.

20.3.4.1 Tissue Preparation

To perform these experiments, ex vivo bovine liver slices were used as tissue. To simulate the tumors, 5 and 10 mm diameter spheres were prepared using thermo-plastic adhesive with thin metal wires embedded within them. This ensured that the tumors were visible in radiographic images later used to measure accuracy. The spheres were pressed into the underside of the liver. Each liver sample had the possibility of containing zero to two tumors, determined a priori through a block randomization process. For each of the palpation methods, nine ex vivo livers were prepared with small tumors and nine with large tumors.

20.3.4.2 Manual Control

For the manual control experiments, four participants were recruited. The layout of the manual setup, shown in Fig. 20.6, consisted of an ATI Gamma six-DOF force/torque sensor (F/T Sensor A), a 0 degree endoscope with standard resolution camera (Stryker Endoscopy, Inc.), a monitor to provide the endoscope display, an adjacent monitor to display the tactile map, and the TSI. The TSI was used by the participant to palpate tissue resting on a platform that incorporated the ATI Gamma force/torque sensor to register the forces being applied to the tissue during palpation. To mimic the indirect vision conditions present during MIS, the tissue samples were shielded from the view of the participant during tissue palpation, such that the working field was only visible via the monitor.

The tactile display uses the commercially available PPS Sapphire® Visualization software, designed to display the results from the tactile sensor in a meaningful way to the user. This real-time pressure profiling system converts the measured voltage values from the capacitive sensor to pressure measurements, and displays these results in a color-contour map of pressure distributions. The visualization software uses the

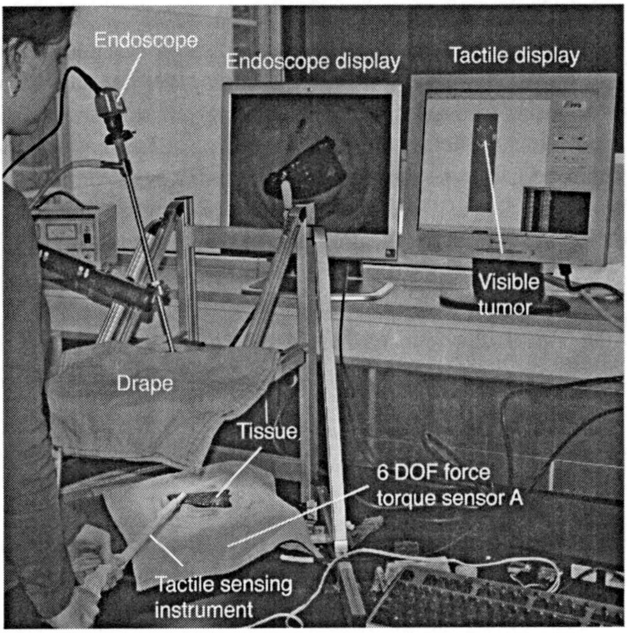

Fig. 20.6 Experimental setup for manual testing of TSI

visual color spectrum to indicate the levels of localized pressure intensity experienced by the probe. Pink areas of the contour map signify areas on the sensor that are experiencing maximal pressures and blue areas of the contour map signify areas that are experiencing small or no contact pressures. Pressures lying within this dynamic range are linearly represented by the visual color spectrum, listed in increasing pressure as blue, green, yellow, orange, red, and pink. In these experiments, since the tumors are stiffer than the surrounding liver tissue, a typical color-contour map displayed when a tumor is located would show a pink region of localized high pressure (representing the tumor) surrounded by a blue region of low pressure (representing the surrounding tissue), shown in Fig. 20.6, thereby clearly distinguishing a tumor from the surrounding tissue.

20.3.4.3 Manual Tests

A total of eighteen trials were completed by each of the four subjects to locate artificial tumors embedded in liver (nine livers with 10 mm tumors and nine with 5 mm tumors). The livers used in each trial were randomly assigned to the subjects, ensuring that the each would palpate the same number of tumors as the robot.

The task completion time and the palpation force, as indicated by F/T Sensor A shown in Figs. 20.5 and 20.6, were recorded. The location of the tumors found by the participants were marked using a plastic instrument marker and marking pins. To assess the success of the manual control method in locating tumors, all livers were imaged using a fluoroscopic radiographic machine after palpation. The images were later assessed by an additional four volunteers to determine the detection sensitivity by determining if the tumors located by the participant were actually a true positive or false positive result. Similarly, instrument accuracy was determined by identifying the true positive and true negative results after palpation. It was decided that the best way of determining whether the participant had in fact correctly identified a tumor, without introducing a proctor bias into the results, was to see if the area of the plastic instrument marker and the tumor in the radiographic images intersected, thereby indicating the correct localization of a tumor by the participant.

20.3.5 Data Visualization and Interpretation

For the tests performed by the PA10-7C robot using both force and position control methods, a software program was custom-designed to integrate the tactile pressure map with the robot position information for the palpated tissues. This was achieved by post-processing the data obtained by the client computer and plotting the data on a 2D topographical map representing the surface of the liver, with palpation force information overlaid directly on this map. This graph results in an indication of tumors located by the robot during palpation, with tumors presented as red areas in the color map, and blue areas representing healthy palpated tissue (see Fig. 20.7). The analysis and assessment of the 2D plot was performed by four volunteers

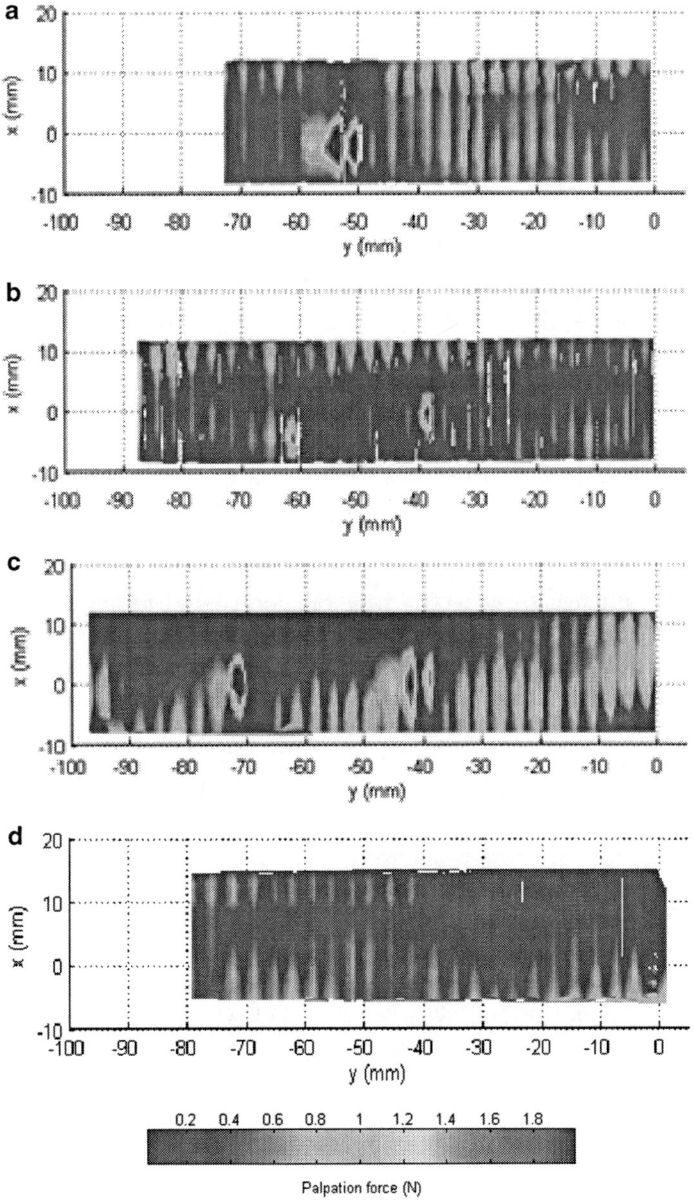

Fig. 20.7 Sample pressure maps for the robot palpation experiments showing (**a**) one large tumor, (**b**) two small tumors, (**c**) two large tumors, and (**d**) no tumors. Visible tumors are indicated by arrows

who were blinded to the number of tumors present and the control method used in each trial to create an unbiased record of the location and number of tumors that were located by the robot.

20.3.6 Discussion of Results

The results of the experiments were evaluated in terms of the maximum palpation force, the task completion time, and the accuracy and sensitivity of the system for both small and large tumors. The maximum palpation force for all three methods was calculated by the mean of the maximum force applied in the eighteen trials with each of the methods. It was shown that there was a significant difference ($p < 0.001$) in the forces applied by the TSI using human control methods (8.13 ± 3.2 N) in comparison to both robotic control methods (5.17 ± 0.63 N for force control; 5.24 ± 0.9 N for position control). It was also observed that in the human trials, the application of forces greater than 6 N for extended periods of time caused visible, irreversible damage to the palpated tissue. Both robotic control methods on average stayed below these damaging forces. Between the two robotic control methods, however, there is no significant difference between the forces applied on the tissue ($p = 0.784$). For performance, the force-control method ($p = 0.180$ small to large) showed marginally better performance for smaller tumors over the position-control method ($p = 0.537$ small to large). However, the average times for the two robotic methods are significantly different ($p < 0.001$) with position control having a quicker task completion time (100.6 ± 14.9 s) than the force-control method (142.9 ± 31.5 s). Task completion time for the human was 197.2 ± 126.4 s. This can only be used as a reference and cannot be directly compared to the robot trials since task completion time for the human control methods also includes the time it took to assess the locations of tumors from the tactile display. For the robot control methods, the assessment of tumor location was executed during offline analysis of the topographical maps.

The measures of accuracy and sensitivity can be used to assess the effectiveness of each control method in determining tumor presence. Force control, position control, and manual control achieved 92, 90, and 59% accuracy, respectively. Force control (94%) was also better with respect to sensitivity when compared to position control (86%), and manual control (81%) methods. Overall, both force and position robot control methods were better in all of the measures when compared to manual control.

There is no clear indication whether the robot force-control method or position-control method is better for tissue palpation purposes. There was an insignificant difference in the amount of force applied during palpation for both methods. Although position control reduced task completion time by about 40%, force control provided better accuracy and sensitivity measures, which are significant indicators that a greater proportion of tumors were correctly identified. Regardless of the robot control method used, the experimental results demonstrated that the detection of tumors using tactile sensing is highly dependent on how consistently the forces on the tactile sensing area are applied, and that robotic assistance can be of great benefit when trying to localize tumors in VATS.

20.3.7 Factors Influencing Performance

The primary difference between robot and human tissue palpation is the ability to use the information from the robot to create a complete contiguous force map of the entire palpated surface. This can be achieved since the robot has the ability to apply a consistent amount of force when palpating tissue at all points and can move systematically over the entire surface of the tissue. It is also possible to record the exact position of the robot end-effector during palpation. The creation of a topographical force map is the equivalent to having one large tactile sensor that covers the entire specimen and applies an ideal force to the entire surface of the tissue. On the other hand, when palpation is performed manually, there is no method to determine the amount of force being applied to one area of the tissue, particularly when compared to the amount of force applied to other areas of the tissue. Due to the sensitivity of the tactile sensor, a particular feature on the tissue might be highlighted solely because a higher palpation force was applied to that area, or conversely, a tumor might not be detected only because inadequate levels of palpation force were applied to that area. Manual control also cannot guarantee that the surface of the tactile sensor and tissue are parallel, thereby creating the potential for inaccurate results simply due to the contact angle between the instrument and the tissue becoming oblique without the intention of the user. Even with surgical experience and knowledge of acceptable pressures for tissue palpation, humans cannot always control the amount of force being applied to the tissue during palpation. Humans have a natural tendency to apply increased force to areas of palpated tissue that indicate the possible presence of a tumor. Hence, there were significant increases in task completion times and applied forces, sometimes leading to irreversible tissue damage. This highlights the advantage of using a robot. Not only can the robot be designed to administer a consistent force onto the tissue, it can also be designed to always restrict the applied forces to lie within safe limits, thereby reducing the possibility of severe tissue damage.

20.4 Concluding Remarks and Future Directions

This chapter has presented a review of the current state of the art in the development of kinesthetic and tactile sensing techniques for tissue palpation during tumor localization with the goal of highlighting the merits of each approach for minimally invasive tumor localization. Haptic feedback has been shown to significantly assist in localizing tumors in minimally invasive procedures by re-establishing the ability to palpate and detect differences in tissue stiffness. Robotic control of the haptic sensors has been shown to increase localization accuracy while applying force levels that ensure adequate sensitivity and minimize tissue damage.

Kinesthetic sensing alone is capable of identifying the presence of an underlying tumor in tissue; however, determining the first point of contact with tissue and very accurate control of the palpation depth are critical. This level of control and accuracy can only be achieved through the use of robotic manipulators or mechatronic devices. Although most sensors currently available are not small enough to enter the patient's body through the small incisions characteristic of MIS procedures, several research groups have been successful in designing sensors adequate for these applications. The true advantage of using kinesthetic sensors lies in the simplicity of their design, compared to the more complex tactile sensors. Through experimental evaluation using kinesthetic techniques, it has been established that sensors for MIS tumor localization have an ideal range of 0–2 N with a 0.01 N resolution. This information is also applicable to the development of MIS tactile sensors.

Unlike the kinesthetic methods, tactile sensing approaches are less sensitive to the way the tissue is approached and how deeply the instrument palpates the tissue. Instead, the success of tactile sensing depends upon ensuring that the angle of approach is such that even contact of all of the sensing elements is made with the target tissue. Additionally, to ensure that underlying regions of increased stiffness are adequately detected, a minimum contact force threshold must be met, while simultaneously ensuring that an upper force threshold is not exceeded in order to prevent tissue damage. A force of 4 N was found to be optimal during the experimental evaluation presented herein.

The advantages of using haptics-enabled robotic assistance for the detection of underlying tumors in MIS lie in the ability to generate consistent and complete pressure maps of the target tissue. Through a systematic palpation approach, combined with control of the palpation force, accurate tumor localization and detection of small tumors (under 10 mm in diameter) can be realized, while minimizing tissue damage.

Despite these advances, approved sensors for use in MIS procedures are not currently available. The challenge lies in developing devices that meet the size constraints of MIS, in addition to sterilization and biocompatibility requirements. The design of minimally invasive sensing instruments is further limited by the presence of wires and cables required to interface with the sensors. The bulk of these wires becomes all the more problematic in the development of articulating instruments that are capable of applying even palpation forces for a range of access conditions and tissue orientations. From a usability perspective, it is in the development of these articulating instruments that the future of tumor localization devices is headed.

An additional limitation of current tactile sensors is their coarse spatial resolution, constrained by how densely the sensing elements can be packed and the number of wires required to transfer pressure signals. Development of sensors capable of submillimeter spatial resolution, while still meeting the force range and resolution requirements of MIS, would facilitate the detection of millimeter-scale tumors. Local signal processing and data encoding at the sensor location could serve to significantly reduce the number of signal wires that must pass through the distal end of the instrument. This would create more available space to allow for more complex articulation mechanisms and/or enable more compact instruments.

It is expected that as haptics-enabled robotics-assisted minimally invasive tumor localization techniques become further refined and find their way into common clinical practice, the treatment of cancer will be enhanced, leading to improved patient care and offering the potential for new surgical and therapeutic techniques. The ability to localize increasingly small tumors is expected to lead to a reduced number of cancer deaths and an increased quality of life.

Acknowledgements The authors would like to acknowledge the contributions of Greig McCreery, who designed the TSI presented in Sect. 20.3.2 and who, along with Chris Kong, performed the kinesthetic tumor localization experiments described in Sect. 20.2. The AHIC-based position and force control strategies for the PA10-7C robot were developed by Dr. J. Jayender. The authors would also like to acknowledge his contributions to the implementation of robot-assisted tissue palpation, used for the experimental work described in Sects. 20.3.3 and 20.3.4.

References

1. American Cancer Society (2007) Global cancer facts and figures 2007. http://www.cancer.org/downloads/STT/Global_Facts_and_Figures_2007_rev2.pdf. Accessed 12 Nov 2009
2. Jeremic, B., Bamberg, M.: External beam radiation therapy for bronchial stump recurrence of non-small-cell lung cancer after complete resection. Radiother. Oncol. **64**, 251–257 (2002)
3. Dargahi, J., Najarian, S.: An integrated force–position tactile sensor for improving diagnostic and therapeutic endoscopic surgery. Biomed. Mater. Eng. **14**,151–166 (2004)
4. Kaneko, M., Toya, C., Okajima, M.: Active strobe imager for visualizing dynamic behavior of tumours. In: Proceedings of IEEE International Conference on Robotics and Automation, pp. 3009–3014. Rome (2007).
5. Hornblower, VDM., Yu, E., Fenster, A., et al.: 3D thoracoscopic ultrasound volume measurement validation in an ex vivo and in vivo porcine model of lung tumors. Phys. Med. Biol. **52**, 91–106 (2007)
6. Trejos, AL., Patel, RV., Naish, MD.: Force sensing and its applications in minimally invasive surgery and therapy: a survey. Proceedings of the Institution of Mechanical Engineers, Part C: Journal of Mechanical Engineering Science Special Issue on Robots and Devices in Surgery **224**(7), 1435–1454 (2010)
7. Ottermo, MV., Ovstedal, M., Lango, T., et al.: The role of tactile feedback in laparoscopic surgery. Surg. Laparosc. Endosc. Percutan. Techn. **16**(6), 390–400 (2006).
8. Carter, FJ., Frank, TG., Davies, PJ., et al.: Measurements and modeling of the compliance of human and porcine organs. Med. Image. Anal. **5**, 231–236 (2001).
9. Roham, H., Najarian, S., Hosseini, SM., et al.: Design and fabrication of a new tactile probe for measuring the modulus of elasticity of soft tissues. Sens. Rev. **27**, 317–323 (2007).
10. Hosseini, SM., Najarian, S., Motaghinasab, S., et al.: Prediction of tumor existence in the virtual soft tissue by using tactile tumor detector. Am. J. Appl. Sci. **5**(5), 483–489 (2008).
11. Takashima, K., Yoshinaka, K., Okazaki, T., et al.: An endoscopic tactile sensor for low invasive surgery. Sens. Actuators A Phys. **119**(2), 372–383 (2005).
12. Liu, H., Noonan, DP., Althoefer K., et al.: Rolling mechanical imaging: a novel approach for soft tissue modeling and identification during minimally invasive surgery. In: Proceedings of IEEE International Conference on Robotics and Automation, pp. 845–850. Pasadena, 2008

13. McCreery, GL., Trejos, AL., Patel, RV., et al.: Evaluation of force feedback requirements for minimally invasive lung tumour localization. In: IEEE/RSJ International Conference on Intelligent Robots and Systems, pp. 883–888. San Diego, 2007
14. Seibold, U., Kubler, B., Hirzinger, G.: Prototype of instrument for minimally invasive surgery with six-axis force sensing capability. In: Proceedings of IEEE International Conference on Robotics and Automation, pp. 498–503. Barcelona, 2005
15. Peirs, J., Clijnen, J., Reynaerts, D., et al.: A micro optical force sensor for force feedback during minimally invasive robotic surgery. Sens. Actuators A Phys. **115**: 447–755 (2004)
16. Tavakoli, M., Patel, RV., Moallem, M.: Haptic interaction in robot-assisted endoscopic surgery: a sensorized end-effector. Int. J. Med. Robot. Comput. Assist. Surg. **1**(2), 53–63 (2005)
17. Tavakoli, M., Aziminejad, A., Patel, RV., et al.: Multi-sensory force/deformation cues for stiffness characterization in soft-tissue palpation. In: Proceedings of IEEE International Conference in Engineering in Medicine and Biology Society, pp. 847–840. New York City, 2006
18. Tavakoli, M., Aziminejad, A., Patel, RV., et al.: Tool/tissue interaction feedback modalities in robot-assisted lump localization. In: Proceedings of IEEE International Conference in Engineering in Medicine and Biology Society, pp. 3854–3857. New York City, 2006
19. Trejos, AL., Patel, RV., Naish, MD., et al.: (2009) A sensorized instrument for skills assessment and training in minimally invasive surgery. ASME J. Med. Devices **3**(4), 041002-1–041002-12 (2009)
20. Trejos, AL., Lyle, AL., Escoto, A., et al.: Force/position-based modular system for minimally invasive surgery. In: IEEE International Conference Robotics and Automation, pp. 3660–3665. Anchorage, Alaska, 2009
21. Bicchi, A., Canepa, G., De Rossi, D., et al.: A sensorized minimally invasive surgery tool for detecting tissutal elastic properties. In: IEEE International Conference Robotics and Automation, pp. 884–888. Minneapolis, 1996.
22. Hannaford, B., Trujillo, J., Sinanan, M., et al.: Computerized endoscopic surgical grasper. In: Med Meet Virtual Real, San Diego, 1998
23. MacFarlane, M., Rosen, J., Hannaford, B., et al.: Force-feedback grasper helps restore sense of touch in minimally invasive surgery. J. Gastrointest. Surg. **3**(3), 278–285 (1999)
24. Rosen, J., Hannaford, B.: Force controlled and teleoperated endoscopic grasper for minimally invasive surgery – experimental performance evaluation. IEEE Trans. Biomed. Eng. **46**(10), 1212–1221 (1999)
25. Tholey, G., Desai, JP., Castellanos, AE.: Force feedback plays a significant role in minimally invasive surgery. Ann. Surg. **241**(1), 102–109 (2005)
26. Tholey, G., Pillarisetti, A., Desai JP.: On-site three dimensional force sensing capability in a laparoscopic grasper. Ind. Robot. Int. J. **31**(6), 509–518 (2004)
27. Tholey, G., Pillarisetti, A., Green, W., et al.: Design, development, and testing of an automated laparoscopic grasper with 3-D force measurement capability. In: International Symposium on Medical Simulation, pp. 38–48. Cambridge, 2004
28. Bonomo, C., Brunetto, P., Fortuna, L., et al.: A tactile sensor for biomedical applications based on IPMCs. IEEE Sens. J. **8**(8), 1486–1493 (2004)
29. Dargahi, J., Najarian, S., Mirjalili, V., et al.: Modeling and testing of a sensor capable of determining the stiffness of biological tissues. Can. J. Electr. Comput. Eng. **32**(1), 45–51 (2007)
30. Kaneko, M., Toya, C., Okajima, M.: Active strobe imager for visualizing dynamic behavior of tumours. In: Proceedings of IEEE International Conference on Robotics and Automation, pp. 3009–3014. Roma, 2007.
31. Yen, PL.: Palpation sensitivity analysis of exploring hard objects under soft tissue. In: Proceedings of IEEE/ASME International Conference on Advanced Intelligent Mechatronics, pp. 1102–1106. Kobe, Japan, 2003.

32. Brown, JD., Rosen, J., Kim, YS., et al.: In-vivo and in-situ compressive properties of porcine abdominal soft tissues. In: Westwood, JD., Hoffman, HM., Mogel, GT., et al., (eds.). Medicine Meets Virtual Reality 11, pp. 26–32. IOS Press, Amsterdam, The Netherlands (2003)
33. Lai-Fook, SJ., Wilson, TA., Hyatt, RE., et al.: Elastic constants of inflated lobes of dog lungs. J. Appl. Physiol. **40**(4), 508–513 (1976)
34. Ottensmeyer, MP., Kerdok, AE., Howe RD., et al.: The effects of testing environment on the viscoelastic properties of soft tissues. In: Proceedings of 2nd International Symposium on Medical Simulation, pp. 67–76. Springer, Cambridge, MA (2004)
35. Dargahi, J., Najarian, S., Narjarian, K.: Development and three-dimensional modeling of a biological-tissue grasper tool equipped with a tactile sensor. Can. J. Elect. Comput. Eng. **30**, 225–230 (2005)
36. Eltaib, MEH., Hewit, JR.: Tactile sensing technology for minimal access surgery—a review. Mechatronics. **13**,1163–1177 (2003)
37. Hemsel, T., Stroop, R., Olivia Uribe, D., et al.: Resonant vibrating sensors for tactile tissue differentiation. J. Sound Vib. **308**(3–5), 441–446 (2007)
38. Kattavenos, N., Lawrenson, B., Frank, TG., et al.: Force-sensitive tactile sensor for minimal access surgery. Min. Invasive Ther. Allied Technol. **13**(1), 42–46 (2004)
39. Murayama, Y., Haruta, M., Hatakeyama, Y., et al.: Development of new instrument for examination of stiffness in the breast using haptic sensor technology. Sens. Actuators A Phys. 143(2), 430–438 (2008)
40. Pawluk, DTV., Son, JS., Wellman, PS., et al.: A distributed pressure sensor for biomechanical measurements. Trans. ASME. **120**, 302–305 (1998)
41. Ottermo, MV., Øvstedal, M., Langø, T., et al.: The role of tactile feedback in laparoscopic surgery. Surg. Laparosc. Endosc. Percutan. Techn. **16**(6), 390–400 (2006).
42. Ottermo, MV., Stabdahl, Ø., Johansen, TA.: Palpation instrument for augmented minimally invasive surgery. In: Proceedings of IEEE/RSJ International Conference on Intelligent Robotic Systems, pp. 3960–3964. Sandal, 2004
43. Miller, AP., Peine, WJ., Son, JS., et al.: Tactile imaging system for localizing lung nodules during video assisted thoracoscopic surgery. In: Proceedings of IEEE International Conference on Robotics and Automation, pp. 2996–3001. Roma, 2007
44. Schostek, S., Ho, CN., Kalanovic, D., et al.: Artificial tactile sensing in minimally invasive surgery—a new surgical approach. Min. Invasive Ther. Allied Technol. **15**, 296–304 (2006)
45. Dollar, AM., Wagner, CR., Howe, RD.: Embedded sensors for biomimetic robotics via shape deposition manufacturing. In: Proceedings of the First IEEE RAS/EMBS International Conference on Biomedical Robotics and Biomechatronics, pp. 763–768. Pisa, Italy, 2006.
46. Noonan, DP., Liu, H., Zweiri, YH. et al.: A dual-function wheeled probe for tissue visoelastic property identification during minimally invasive surgery. In: Proceedings of IEEE International Conference on Robotics and Automation, pp. 2629–2634. Roma, 2007.
47. Dario, P., Bergamasco, M.: An advanced robot system for automated diagnostic tasks through palpation. IEEE Trans. Biomed. Eng. **35**(2), 118–126 (2007)
48. Feller, RL., Lau, CKL., Wagner, CR., et al.: The effect of force feedback on remote palpation, In: Proceedings of IEEE International Conference on Robotics and Automation, pp. 782–788. New Orleans, 2004
49. McCreery, GL., Trejos, AL., Naish, MD., et al.: Feasibility of locating tumours in lung via kinaesthetic feedback. Int. J. Med. Robot. Comput. Assist. Surg. **4**(1), 58–68 (2008)
50. Pressure Profile Systems (2007) Capacitive sensing. Pressure Profile Systems, Inc. http://www.pressureprofile.com/technology-capacitive.php. Accessed 24 August 2009
51. Pawluk, DTV., Son, JS., Wellman, PS., et al.: A distributed pressure sensor for biomechanical measurements. ASME J. Biomech. Eng. **102**(2), 302–305 (1998)
52. Perri, MT., Bottoni, DA., Trejos, AL., et al.: A new tactile imaging device to aid with localizing lung tumours during throacoscopic surgery. Int. J. Comput. Assist. Radiol. Surg. **3**, S257–S258 (2008)

53. Patel, RV., Talebi, HA., Jayender, J.,et al.: A robust position and force control strategy for 7-DOF redundant manipulators. IEEE/ASME Trans. Mechatron. **14**(5), 575–589 (2009)
54. Patel, RV., Shadpey, F.: Control of redundant robot manipulators: theory and experiments. Lecture Notes in Control and Information Sciences, vol. 315. Springer-Verlag, Heidelberg (2005)
55. Jayender, J.: Haptics enabled robot-assisted active catheter insertion. PhD Thesis, The University of Western Ontario, London, Ontario (2007)
56. Trejos, AL., Jayender, J., Perri, MT., et al.: Experimental evaluation of robot-assisted tactile sensing for minimally invasive surgery. In: Proceedings of 2nd IEEE RAS/EMBS International Conference on Biomedical Robotics and Biomechatronics, pp. 971–976. Arizona, 2008
57. Trejos, AL., Jayender, J., Perri, MT., et al.: Robot-assisted tactile sensing for minimally invasive tumour localization. Int. J. Robot. Res. Special Issue on Medical Robotics **28**(9), 1118–1133 (2009)

Chapter 21
Motion Tracking for Beating Heart Surgery

Rogério Richa, Antônio P. L. Bó, and Philippe Poignet

21.1 Introduction

The past decades have seen the notable development of minimally invasive surgery (MIS), in which the surgical gesture is performed through small incisions in the patient's body. The benefits of this modality of surgery for patients are numerous, shortening convalescence, reducing trauma and surgery costs. However, several difficulties are imposed to the surgeon, such as decreased mobility, reduced visibility, uncomfortable working posture and the loss of tactile feedback. In this context, robotic assistance aims to aid surgeons to overcome such difficulties, making the surgical act more intuitive and safer. Consequently, commercially available surgical platforms such as the daVinci^{TM} (Intuitive Surgical) quickly became popular.

However, current robotic platforms display limitations, specially regarding the absence of tactile feedback and the absence of active physiological motion compensation, impeding their use in a broader range of surgical scenarios. For instance, active physiological motion compensation in cardiac surgery imposes challenges on several levels of the robot's design. The ultimate goal is the creation of a virtual stable operating site, where the surgeon is given the impression to be working on a motionless environment. The potential improvements of precision and repeatability of surgical gestures when working on a stabilized environment have been revealed in recent studies [30].

In the domain of cardiac MIS, heartbeat and respiration represent two important sources of disturbances. Even though miniature versions of heart stabilizers have been conceived for the MIS scenario, residual motion is still considerable [15] and has to be manually canceled by the surgeon, complicating the realization of the delicate surgical tasks. For actively compensating the residual heart motion, the heart must be accurately tracked. In the literature, different kinds of sensors have been employed for estimating the heart motion, ranging from contact sensor systems [3], sonometric systems [7], laser range finding systems [13], miniature mobile robots [25] and vision-

R. Richa (✉)
LIRMM 161 Rue Add, 34392 Montpellier Cedex 5, France
e-mail: philippe.poignet@lirmm.rf

J. Rosen et al. (eds.), *Surgical Robotics: Systems Applications and Visions*,
DOI 10.1007/978-1-4419-1126-1_21, © Springer Science+Business Media, LLC 2011

based systems [21]. From a practical point of view, vision-based techniques are preferred since they rely only on the visual feedback provided by the laparoscope and avoid the insertion of additional sensors in the limited surgical workspace. In this chapter we discuss the problem of tracking the beating heart, present the main difficulties involved in this task and give a summery of recent advances in the field. In addition, we describe a visual tracking method for estimating the beating heart motion specially tailored for the specific challenges of MIS.

21.2 Tracking the Heart Surface

Heart surgery has critical accuracy constraints, since fine cardiac structures are manipulated. If we consider a common task such as performing suture points on vessel 2 mm diameter wide, motion estimation accuracy under 20 μm is required. It is important to remark that the estimation accuracy depends fundamentally on the tracking algorithms but also largely on hardware setup and specifications.

The challenges involved in tracking the beating heart begin with the acquisition system. Image quality (resolution, lens deformation), synchronization and acquisition speed are essentially hardware problems. Ginhoux et al. [11] pointed out that the heart motion has very fast transients and with a slow acquisition rate, information loss due to aliasing is not negligible. They also suggested an acquisition speed not smaller than 100 Hz for compensating the heart motion. This observation has direct consequences on the design of a tracking algorithm.

From the computer vision point of view, the problems encountered when tracking the heart can be grouped in three categories: illumination issues, appearance changes and lack of visual information. These categories are presented below in detail. With current commercially available acquisition hardware, motion blur is an important problem but since it is a hardware-related problem, it is not cited below. For our investigations, we use in vivo images of a total endoscopic coronary bypass grafting captured by a daVinci^{TM} platform as a realistic clinical scenario. Extracts of this image sequence are presented in Fig. 21.1.

- *Illumination issues* – The wet-like appearance of soft-tissue gives rise to specular reflections. These specular reflections work as occluders and considerably reduce

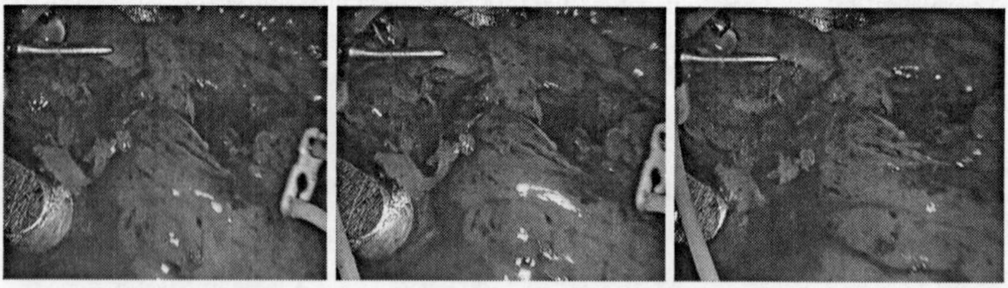

Fig. 21.1 Images acquired by the left camera of a stereoendoscope during a total endoscopic coronary bypass grafting procedure using a daVinci^{TM} surgical platform

the available visual information used by the tracking algorithm. Another important source of disturbance for the tracking algorithm are lighting changes. Due to the restricted workspace in MIS, the light source illuminates unevenly the operating site. Therefore, the brightness constancy assumption on which various tracking methods are based upon is violated, complicating the visual tracking task.

- *Appearance changes* – Liquids and smoke which are present at the operating site often disturb visual tracking. It is also expected that as the surgeon manipulates the heart its appearance will significantly change. Due to the unpredictable nature of the changes, their modeling is very difficult.
- *Lack of visual information* – Certain regions of the heart surface do not offer a set of identifiable and stable features or texture information, from which we can estimate motion. Since certain regions of the heart surface do not offer enough image gradient information and the use of artificial markers is not practical, visual tracking is challenging.

21.3 State of the Art

The first visual-based system for heart motion compensation was proposed by Nakamura et al. [21], who introduced the concept synchronizing the surgical tools with the heart motion (dubbed *heartbeat synchronization*). In his first prototype, the 2D motion of artificial markers fixed on the heart surface were estimated on the image plane by a vision system using high-speed cameras. Artificial markers were also employed in different works to facilitate the visual tracking task but for the medical community the fixation of markers is not practical. This issue motivated the use of natural structures on the heart surface to estimate the heart motion.

The first work that explored natural landmarks on the heart surface for estimating the heart motion was Ortmaier et al. [23], where a reduced affine tracking method was proposed for tracking the 2D displacement of salient features on the heart surface. Only with the recent advent of the stereo endoscope, research began to explore stereo vision techniques for 3D tracking the heart motion. Stoyanov et al. [29] presented a modified Lucas–Kanade feature tracker for tracking salient features on the heart surface. Feature-based techniques display performance issues related to the deformable nature of the heart surface and the complexity of the operating field. Addressing these problems, Mountney et al. [20] recently proposed an on-line learning based feature tracking method that improves tracking performance when facing drift, occlusions and tissue deformation. Another interesting study by Mountney et al. [19] evaluates existing feature descriptors in computer vision in the specific context of deformable tissue tracking. Texture information has also been investigated in Noce et al. [22] for increasing feature discriminability.

Another class of methods for tracking the heart surface are region-based methods. This class of methods, also called direct methods, are based on the key assumption that the heart surface is smooth, continuous and is sufficiently textured. If this assumption stands, a parametric function can be used to describe the heart surface deformation and tracking is formulated as an iterative registration problem. Traditionally, region-based methods have been applied for estimating depth in the intraoperative field rather than 3D motion tracking. Lau et al. [14] proposed a B-Spline parametric model to obtain depth from the disparity between stereo images of the heart. A similar approach by Stoyanov et al. [28] employed a piecewise bilinear map for modeling the heart surface. Due to the complex nature of the heart surface, a suitable choice for a deformation model is complicated.

An alternative to parametric interpolating functions are physically based models, as proposed by Bogatyrenko et al. [5]. Nevertheless, convincing experimental results are still required to demonstrate the validity of such models. A novel and very promising approach was recently proposed by Lo et al. [17], where the fusion of multiple visual cues such as Shape-from-Shading (SFS) and stereo correspondence using a Markov Random Fields (MRF) Bayesian belief propagation is proposed for increasing reconstruction accuracy.

An analysis of currently proposed techniques suggests that region-based techniques cope better with the challenges of tracking soft-tissue such as large deformations and illumination variations. In the next section, a 3D tracking method based on a Radial Basis Function called Thin-Plate Spline (TPS) is described in detail. The TPS modeling enables dense motion estimation and copes better with textureless regions on the heart surface.

21.4 Motion Tracking Using a TPS Deformable Model

In this section, we present the TPS for modeling the heart surface deformation. This tracking method was originally proposed in [24]. First, we give a brief review of the TPS transformations following the parameterization proposed by Lim et al. [16]. Secondly, we describe the extension for tracking in 3D using a stereo camera system and how the model parameters can be estimated using efficient minimization techniques.

21.4.1 The Thin-Plate Spline Warping

The Thin-Plate Spline is a radial basis function (RBF) that specifies an approximation function $f : \Re^2 \rightarrow \Re$ which minimizes the bending energy:

$$E_f = \int\int_{\Re^2} (f_{xx}^2 + 2f_{xy}^2 + f_{yy}^2)dxdy \qquad (21.1)$$

The function f of a point $\mathbf{x} = (x, y)$ in a plane is defined by the Thin Plate Spline basis function $U(s) = s^2 log(s^2)$, a $(n+3)$ parameter vector $\mathbf{t} = (w_1, \ldots, w_n, r_1, r_2, r_3)^T$ and a set of n control points $\mathbf{c}_i = (\check{x}_i, \check{y}_i)$, such that:

$$f(\mathbf{x}) = r_1 + r_2 x + r_3 y + \sum\nolimits_{i=1}^{n} w_i U(||\mathbf{c}_i - \mathbf{x}||) \tag{21.2}$$

Also, side conditions must be considered in order for the function $f(\mathbf{x})$ to have square-integrable second derivatives:

$$\sum\nolimits_{i=1}^{n} w_i = \sum\nolimits_{i=1}^{n} w_i \check{x}_i = \sum\nolimits_{i=1}^{n} w_i \check{y}_i = 0 \tag{21.3}$$

To write a $\Re^2 \rightarrow \Re^2$ mapping m we stack two RBFs f^x and f^y sharing their control points:

$$m(\mathbf{x}) = \begin{bmatrix} f^x \\ f^y \end{bmatrix} = \begin{bmatrix} r_2^x & r_3^x & r_1^x \\ r_2^y & r_3^y & r_1^y \end{bmatrix} \begin{bmatrix} x \\ y \\ 1 \end{bmatrix} + \sum\nolimits_{i=1}^{n} \begin{bmatrix} w_i^x \\ w_i^y \end{bmatrix} U(||\mathbf{c}_j - \mathbf{x}||) \tag{21.4}$$

The mapping m can be used to describe the positions of the pixel coordinates of a reference image on the current image. Bookstein [6] proposed a linear system to calculate the parameter vectors $\mathbf{t}^x$ and $\mathbf{t}^y$ of each Cartesian coordinate that define f^x and f^y based on the mapped positions of the control points $\mathbf{c}$ on the original image. This linear system can be written by stacking the mapped values $\mathbf{c}' = (\check{x}', \check{y}') = m(\mathbf{c})$ of each control point $\mathbf{c}$ into a matrix $\check{\mathbf{P}} = [\check{\mathbf{x}}', \check{\mathbf{y}}']$:

$$\begin{bmatrix} \mathbf{L} & \mathbf{P} \\ \mathbf{P}^T & \mathbf{O} \end{bmatrix} [\mathbf{t}^x \quad \mathbf{t}^y] = \begin{bmatrix} \check{\mathbf{P}} \\ \mathbf{0} \end{bmatrix} \tag{21.5}$$

where $L_{ij} = U(\| \mathbf{c}_j - \mathbf{c}_i \|)$, $\mathbf{P}$ are the stacked coordinates of the control points $(1, \check{x}, \check{y})$ on the original plane, $\mathbf{O}$ and $\mathbf{0}$ are 3×3 and 3×2 zero matrices respectively. The mapped control points $\mathbf{c}'$ are called *control point correspondences*. Denoting the leftmost matrix as $\mathbf{K}$, we can invert the system to solve for $\mathbf{t}^x$ and $\mathbf{t}^y$:

$$(\mathbf{t}^x \quad \mathbf{t}^y) = \mathbf{K}^{-1} \begin{bmatrix} \check{\mathbf{P}} \\ \mathbf{0} \end{bmatrix} \tag{21.6}$$

Finally, since we are considering the correspondence between two images, we can calculate the transformed coordinates $\mathbf{x}' = (x', y')$ of the reference image as a function of control point correspondences $\mathbf{c}'$ (see Fig. 21.2a). For simplification purposes, we denote the $(n+3) \times n$ sub-matrix of $\mathbf{K}^{-1}$ as $\mathbf{K}_*$. This eliminates the need for adding zeros in the rightmost matrix in the above equation. Let $\mathbf{P}'$ be the q stacked transformed pixel coordinates $\mathbf{x}'_j$ such that $\mathbf{P}' = [\mathbf{x}'_1, \mathbf{x}'_2, \ldots, \mathbf{x}'_q]^T$:

$$\mathbf{P}' = [\mathbf{V} \quad \mathbf{W}] \mathbf{K}_* \check{\mathbf{P}} \tag{21.7}$$

where $V_{ji} = U(\| \mathbf{x}_j - \mathbf{c}_i \|)$ and $\mathbf{W}_j = (1, x_j, y_j)$. The matrix $[\mathbf{V}\ \mathbf{W}]$ is denoted as $\mathbf{M}$. If the position of the control points in the reference plane does not change, the matrices $\mathbf{M}$ and $\mathbf{K}_*$ can be pre-computed.

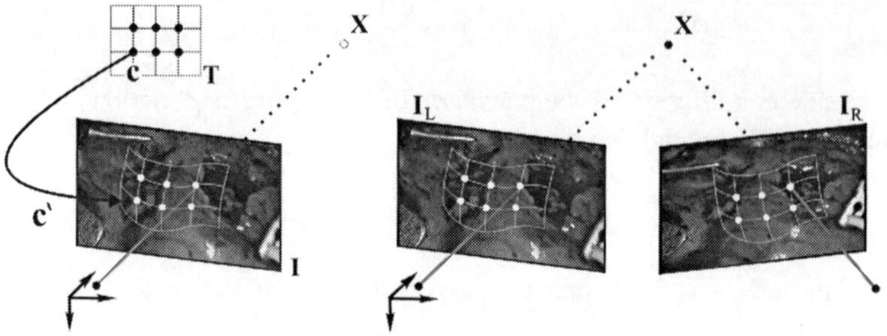

Fig. 21.2 (**a**) The control points mapped on the current image are projections of 3D points in space onto the image. (**b**) If we seek the alignment of the same reference image on both cameras of the stereo pair, the TPS control points are necessarily the projections of the same 3D points in space onto both images. If we consider a calibrated system, these 3D points can be estimated from both images

From the formulation above, we can define the function $w(\mathbf{x}_i, \check{\mathbf{P}})$ that warps the i-th pixel $\mathbf{x}_i$ of the reference image based on control point correspondences $[\check{\mathbf{x}}, \check{\mathbf{y}}]$ stored in the matrix $\check{\mathbf{P}}$.

$$w(\mathbf{x}_i, \check{\mathbf{P}}) = [x'_i \;\; y'_i] = [\mathbf{M}_i \mathbf{K}_* \check{\mathbf{x}} | \mathbf{M}_i \mathbf{K}_* \check{\mathbf{y}}] \tag{21.8}$$

where $\mathbf{M}_i$ is the i-th row of the matrix $\mathbf{M}$.

21.4.2 Extension to 3D Tracking

In the formulation given above, the TPS warping defines a mapping between each pixel position on $\mathbf{T}$ to the current image $\mathbf{I}$ of the surface, as defined in (21.7). Nevertheless the standard TPS warping is intrinsically affine and cannot capture projective deformations with a finite number of control points. Addressing this problem, Malis [18] proposed the parameterization of the projective depths of the tracked surface in order to represent projective deformations.

Following the TPS formulation adopted in this paper, the mapping m proposed in (21.4) can be extended to model surface depth by adding the necessary degrees of freedom with an additional TPS function f^z:

$$m(\mathbf{x}) = \begin{bmatrix} f^x \\ f^y \\ f^z \end{bmatrix} = \begin{bmatrix} r_2^x & r_3^x & r_1^x \\ r_2^y & r_3^y & r_1^y \\ r_2^z & r_3^z & r_1^z \end{bmatrix} \begin{bmatrix} x \\ y \\ 1 \end{bmatrix} + \sum_{i=1}^{n} \begin{bmatrix} w_i^x \\ w_i^y \\ w_i^z \end{bmatrix} U(||\mathbf{c}_j - \mathbf{x}||) \tag{21.9}$$

The transformed pixel coordinates $\mathbf{x}' = (sx', sy', s)$ from the original image (now in projective coordinates) can be calculated in the same fashion as (21.7):

$$\mathbf{P}' = \mathbf{M}\mathbf{K}_* \left[\check{\mathbf{P}} \quad \check{\mathbf{z}} \right] \tag{21.10}$$

where the additional column $\check{\mathbf{z}}$ represents the stacked depth positions for each control point, in pixels. We then propose the expansion of this framework for the stereo case, by first making the observation that, by definition, the position of the control points correspondences $\mathbf{c}'$ are the projections of 3D points onto the image plane.

$$\begin{bmatrix} \check{\mathbf{P}} & \check{\mathbf{z}} \end{bmatrix} = (\mathbf{C} \begin{bmatrix} \mathbf{X} \\ \mathbf{1}^T \end{bmatrix})^T = \begin{bmatrix} \mathbf{X} & \mathbf{1} \end{bmatrix} \mathbf{C}^T \tag{21.11}$$

where $\mathbf{C}$ is the 3 $\times$4 camera calibration matrix, $\mathbf{X}$ the n $\times$3 matrix of stacked 3D cartesian coordinates of the points that map to $\mathbf{c}'$ on the image and $\mathbf{1}$ a n column vector of ones. Following this interpretation, the (21.10) can be rewritten as a function of $\mathbf{X}$.

$$\mathbf{P}' = \mathbf{M}\mathbf{K}_* \begin{bmatrix} \mathbf{X} & \mathbf{1} \end{bmatrix} \mathbf{C}^T \tag{21.12}$$

In result, a linear relationship can be established between the reference image and its pixel's 3D coordinates. Assuming the world coordinate frame centered on the camera, the camera matrix $\mathbf{C}$ is composed uniquely by its intrinsic parameters $\mathbf{C}^*$ and the stacked 3D coordinates $\mathbf{X}'$ of every pixel from the reference image $\mathbf{T}$ (in the camera coordinate frame) can be found by inverting (21.12):

$$\mathbf{X}' = (\mathbf{C}^{*-1} \mathbf{P}'^T)^T = \mathbf{C}^{*-T} \mathbf{P}' \tag{21.13}$$

To fully constrain the estimation of the transformed pixel coordinates on (21.10) and consequently their 3D coordinates (21.13), we consider a calibrated stereo rig as illustrated in Fig. 21.2b, with the world coordinate frame arbitrarily centered on the left camera. The tracking problem then consists on minimizing the alignment error between the reference image $\mathbf{T}$ in both left and right cameras $\mathbf{I}_l$ and $\mathbf{I}_r$. Redefining (21.8) for the 3D case, the new warping function $w_{3D}(\mathbf{x}, \mathbf{h}, \mathbf{C})$ is defined by the camera calibration matrix $\mathbf{C}$ and a $3n$ parameter vector $\mathbf{h} = (\mathbf{h}^{xT}, \mathbf{h}^{yT}, \mathbf{h}^{zT})^T$ composed by the stacked columns of the matrix $\mathbf{X} = [\mathbf{h}^x, \mathbf{h}^y, \mathbf{h}^z]$:

$$w_{3D}(\mathbf{x}_i, \mathbf{h}, \mathbf{C}) = [sx'_i sy'_i s_i] = \mathbf{M}_i \mathbf{K}_* \begin{bmatrix} \mathbf{h}^x & \mathbf{h}^y & \mathbf{h}^z & \mathbf{1} \end{bmatrix} \mathbf{C}^T \tag{21.14}$$

where $\mathbf{M}_i$ is the i-th row of the matrix $\mathbf{M}$ corresponding to $\mathbf{x}_i$.

21.4.3 Efficient Warp Estimation

Unlike the 2D case where the problem consists in estimating the control point correspondences between the reference and current images, in the stereo case we estimate the 3D coordinates of the points in space that map to control point positions that minimize the alignment error between the reference image and both stereo images simultaneously. The concept is illustrated in Fig. 21.2b. Letting $\mathbf{C}_l$ and $\mathbf{C}_r$ be the camera calibration matrices for the left and right cameras respectively and $\mathbf{I}_l$ and $\mathbf{I}_r$ their respective images, this problem can be formulated as:

$$\min_{\mathbf{h}} \epsilon = \sum_{\mathbf{x}\in A} [[\,\mathbf{I}_l(w_{3D}(\mathbf{x},\mathbf{h},\mathbf{C}_l)) - \mathbf{T}(\mathbf{x})\,]^2 + [\,\mathbf{I}_r(w_{3D}(\mathbf{x},\mathbf{h},\mathbf{C}_r)) - \mathbf{T}(\mathbf{x})\,]^2] \tag{21.15}$$

where A is the set of the template coordinates and $\mathbf{I}(w_{3D}(\mathbf{x},\mathbf{h},\mathbf{C}))$ is the current left or right image transformed by the warping function $w_{3D}(\mathbf{x},\mathbf{h},\mathbf{C})$. Notice that we minimize the alignment error of the same reference image in both images from the stereo pair.

In the literature, the above minimization problem is traditionally solved using Gauss–Newton or Leverberg–Marquardt techniques. In our study, we apply the efficient second-order minimization (ESM) algorithm, proposed by Benhimane and Malis [4]. The ESM displays a significant improvement over the traditional template-based tracking techniques, since it has a faster convergence rate and larger convergence basin.

Let $\mathbf{h}_a$ be the current warping parameters, $\mathbf{h}_o$ the parameters at startup and $\delta\mathbf{h}$ the increment such that $\mathbf{h}+\delta\mathbf{h}$ minimizes (21.15). Since we consider images from both stereo pairs $\mathbf{I}_l$ and $\mathbf{I}_r$, the increment $\delta\mathbf{h}$ can be calculated by stacking the left and right image Jacobians and error images:

$$\delta\mathbf{h} = -2\begin{bmatrix}\mathbf{J}(\mathbf{T},\mathbf{h}_o,\mathbf{x}) + \mathbf{J}(\mathbf{I}_l,\mathbf{h}_a,\mathbf{x})\\ \mathbf{J}(\mathbf{T},\mathbf{h}_o,\mathbf{x}) + \mathbf{J}(\mathbf{I}_r,\mathbf{h}_a,\mathbf{x})\end{bmatrix}^{+}\begin{bmatrix}\delta\mathbf{I}_l\\ \delta\mathbf{I}_r\end{bmatrix} \tag{21.16}$$

where $\delta\mathbf{I}$ represents the stacked error image $(\mathbf{I}(w(\mathbf{x},\mathbf{h}_a)) - \mathbf{T}(\mathbf{x}))$ for all $\mathbf{x}\in A$. The Jacobian matrix $\mathbf{J}$ is calculated by deriving the warping function (21.14) with respect to each parameter h_j of $\mathbf{h}_j$, with $j\in\{\,1,\ldots,3n\}$. The i-line of $\mathbf{J}$, corresponding to the pixel $\mathbf{x}_i$ is given as:

$$\mathbf{J}_i(\mathbf{I},\mathbf{h},\mathbf{x}_i) = \left[\frac{\partial\mathbf{I}(w(\mathbf{x}_i,\mathbf{h}))}{\partial h_1} \cdots \frac{\partial\mathbf{I}(w(\mathbf{x}_i,\mathbf{h}))}{\partial h_{3n}}\right] \tag{21.17}$$

Equation (21.16) is iterated to achieve higher precision. The whole procedure could also be performed at multiple scales to aid the convergence. When tracking through an image sequence, we make the assumption that the inter-frame surface deformation is sufficiently small. Hence, for every new frame, the previous surface parameters $\mathbf{h}_a$ can be used as initial guess in (21.16) for the current parameters.

In this framework for 3D tracking, no explicit matching between the stereo camera images is performed. Consequently no intermediate steps such as rectification are needed and tracking accuracy is increased.

21.4.4 Illumination Compensation

Due to the illumination conditions in MIS, lighting variations on the heart surface are not negligible. The lighting on the heart surface is uneven and its glossy nature gives rise to specular reflections that reduce the observable surface. For tracking to be robust, these phenomena must be taken into account. In this section we incorporate in our tracking method an illumination compensation step which is adapted for beating heart imaging conditions.

In the specific context of soft-tissue tracking, Gröeger [12] was the first to treat illumination changes, based on a simple mean intensity normalization. Other commonly used methods, such as the affine compensation can also significantly improve tracking robustness but only global illumination changes are modeled. Recently, Silveira and Malis [26] proposed an efficient method for modeling arbitrary illumination changes which does not require any knowledge of the reflectance properties of the target surface or the characteristics of the illumination source. Here we apply this technique for modeling the illumination variations on the heart surface.

The illumination compensation can be formulated as the problem of finding the elementwise multiplicative lighting variation $\hat{\mathbf{I}}$ for each pixel $\mathbf{x}_i$ of the current image $\mathbf{I}$ and a global bias b such that $\mathbf{I}'$ matches the closest the illumination conditions of the reference image $\mathbf{T}$:

$$\mathbf{I}'(\mathbf{x}_i) = \hat{\mathbf{I}}(\mathbf{x}_i)\mathbf{I}(\mathbf{x}_i) + b \tag{21.18}$$

The element $\hat{\mathbf{I}}$ can be modeled as a parametric surface $\hat{\mathbf{I}} = f(\mathbf{x}, \mathbf{u})$, where $\mathbf{x}$ are pixel coordinates and $\mathbf{u}$ are the surface parameters which vary with time. As suggested in [26], the surface $\hat{\mathbf{I}}$ can be modeled as a Thin-Plate Spline surface:

$$\hat{\mathbf{I}}(\mathbf{x}, \mathbf{u}) \approx u_{m+1} + \mathbf{x}[u_{m+2}, u_{m+3}]^T + \sum_{k=1}^{m} u_k U(||\mathbf{x} - \mathbf{q}_k||) \tag{21.19}$$

where the control points $\mathbf{q}$ are evenly spaced on the image to best capture the illumination changes. The only assumption made when using a TPS surface is that the lighting changes are continuous over the surface, which is a weak assumption. Furthermore, in our application the illumination transformations of the left and right cameras of the stereo pair are treated independently.

In our study, we develop the illumination parametrization based on the TPS using the same formulation as for the deformation warping presented in Sect. 21.4.1. The TPS surface of parameters $\mathbf{u}$ in (21.19) can be reparameterized as a function of $\mathbf{y}$, which are the stacked values of $\hat{\mathbf{I}}$ at each control point $\mathbf{q}$. Therefore, matrices $\mathbf{K}^{il}$ and $\mathbf{M}^{il}$ can be deduced similarly to (21.6) and (21.7) and $\hat{\mathbf{I}}$ can be calculated as:

$$\hat{\mathbf{I}} = \mathbf{M}^{il}\mathbf{K}^{il}_{*}\mathbf{y} \tag{21.20}$$

For estimating the illumination compensation parameters, we use the same optimization framework based on the ESM. Similarly to (21.16), an increment $\delta\mathbf{h}' = [\mathbf{h}, \mathbf{y}_l, \mathbf{y}_r, b_l, b_r]$ can be defined, such that $\delta\mathbf{h}'$ incorporates the warping and illumination parameters. The minimization problem in (21.15) can be redefined as:

$$\min_{\mathbf{h},\mathbf{y}^l,b^l,\mathbf{y}^r,b^r} \epsilon = \sum_{\mathbf{x}\in A} \left[\left[\hat{\mathbf{I}}(\mathbf{x},\mathbf{y}^l)\; \mathbf{I}_l(w_{3D}(\mathbf{x},\mathbf{h})) + b^l - \mathbf{T}(\mathbf{x})\right]^2 + \right. \tag{21.21}$$

$$\left. + \left[\hat{\mathbf{I}}(\mathbf{x},\mathbf{y}^r)\; \mathbf{I}_r(w_{3D}(\mathbf{x},\mathbf{h})) + b^r - \mathbf{T}(\mathbf{x})\right]^2\right] \tag{21.22}$$

After the incorporation of the illumination compensation parameters, new Jacobian matrices are required. In this case, they are composed of 3 sub-matrices:

$$\mathbf{J}'(\mathbf{I},\mathbf{h},\mathbf{y},b,\mathbf{x}) = \left[\mathbf{J}^w(\mathbf{I},\mathbf{h},\mathbf{y},\mathbf{x}), \mathbf{J}^y(\mathbf{I},\mathbf{x}), \mathbf{J}^b(\mathbf{I},b,\mathbf{x})\right] \tag{21.23}$$

The matrix $\mathbf{J}^w(\mathbf{I},\mathbf{h},\mathbf{y},\mathbf{x})$ is the Jacobian matrix corresponding to the warping parameters. It can be computed by replacing $\mathbf{I}$ by $\mathbf{I}'$ in (21.17):

$$\mathbf{J}^w(\mathbf{I},\mathbf{h},\mathbf{y},\mathbf{x}_i) = \frac{\partial \mathbf{I}'(w(\mathbf{x}_i,\mathbf{h}))}{\partial \mathbf{h}} = \hat{\mathbf{I}}(\mathbf{x}_i,\mathbf{y})\frac{\partial \mathbf{I}(w(\mathbf{x}_i,\mathbf{h}))}{\partial \mathbf{h}} \tag{21.24}$$

The second sub-matrix, $\mathbf{J}^y(\mathbf{I},\mathbf{x})$ corresponds to the TPS surface that models $\hat{\mathbf{I}}$. Notice that it no longer depends on the illumination parameters $\mathbf{y}$:

$$\mathbf{J}^y(\mathbf{I},\mathbf{x}_i) = \mathbf{I}(\mathbf{x}_i) M^{il} K^{il}_{*} \tag{21.25}$$

The final third sub-matrix $\mathbf{J}^b(\mathbf{I},b,\mathbf{x})$ is a column of ones, is the result of deriving (21.18) with respect to b. An interesting aspect about this illumination compensation technique is that little extra computational cost is needed, since only a few extra parameters are estimated per iteration. More details about this technique can be found in [26].

21.4.4.1 Specular Highlights Detection

Specular reflections on the heart surface considerably affects tracking performance because they act as occluders, reducing the available visual information. Most works propose to detect these effects based on simple thresholding of the intensity level, which is a practical solution once little computational effort is required.

The specular highlights detection step consists on finding saturated zones on the image by thresholding the intensity level for generating a specularity map. Based on the specularity map, the pixels corresponding to specular highlights on the images are removed from the estimation of the illumination and warping parameters during the optimization process (21.22). Figure 21.3 shows some example results of the specular reflection detection.

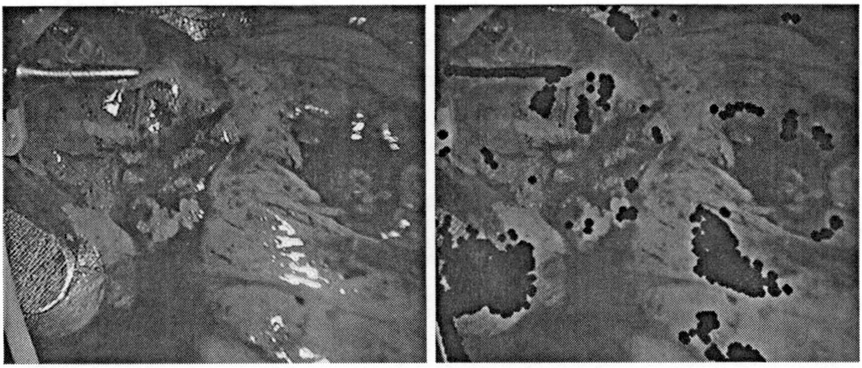

Fig. 21.3 The proposed specular detection method. The pixels on the original image (*left*) affected by specular reflections are painted in black (*right*). The specular reflections are detected and removed from the computation of the warping parameters

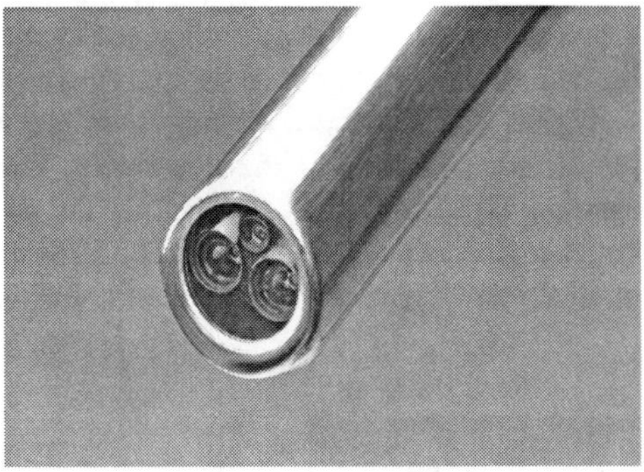

Fig. 21.4 The stereo endoscope that integrates the daVinci platform. Photo courtesy of Intuitive Surgical, Inc., 2008

21.4.5 Tracking Results

An *in-vivo* analysis of the performance of the proposed algorithm was performed using the image sequence of a totally endoscopic coronary artery bypass graft (TECAB) from the daVinci^{TM} (Intuitive Surgical, CA) surgical platform previously presented in section 2. Since the real 3D structure of the heart surface could not be measured, we evaluate the results based on the coherence of the retrieved 3D shape and motion of the heart surface.

21.4.5.1 Experimental Setup

The cameras of the stereo endoscope (Figure 21.4) were synchronized using a proprietary FPGA device designed by Intuitive Surgical and the stereo cameras were calibrated before the procedure using a planar calibration object [33]. A mechanical stabilizer was positioned on the patient's heart previous to the acquisition. Thesequence consists in 320x288 color images of the heart captured at 50 Hz but for generalization purposes the image sequence is converted to 8-bit grayscale. Even though the distance between camera centers is small, accurate 3D reconstruction using the stereoendoscope is still possible since the system has low calibration error and the distance to the operating field is small (less than 80mm).

21.4.5.2 Tracking performance

In our experiments, we apply the proposed method to track a 80x80 pixel image of a region on the heart surface using 6 control points (Fig. 5 (left)). The reference image of the heart surface is manually selected by the surgeon and defines the region of interest to be virtually stabilized. The left camera image is arbitrarily used for defining the reference image. The control points are manually placed on regions with enough texture. To avoid the aperture problem, the number of control points is kept small. The factors that limit the spatial resolution of the tracking algorithm are the amplitude of the surface motion and available texture information. Some of these limitations can be overcome by applying a coarse-to-fine estimation procedure.

Figure 21.7 illustrates the target surface viewed by both left and right cameras for different instants of the heartcycle. The world frame is centered at the left camera. The 3D displacement of a point of interest situated in the center of the target region is plotted in figure 21.5. The cardiac and respiratory motion for this interest point has an amplitude ranging from [9.89,5.13] mm, [0.92, -1.30] mm and

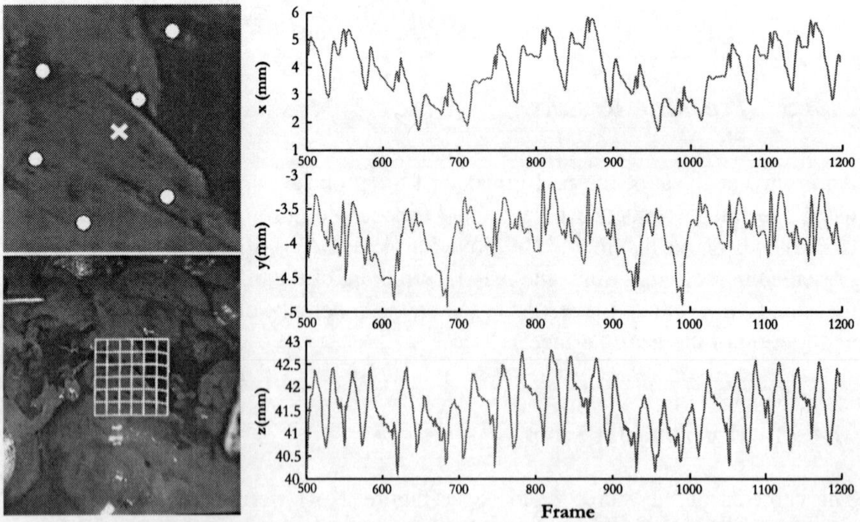

Fig. 21.5 (Top-left) The position of control points on the reference image (marked as white circles). A point of interest in the center of the target region is marked by a cross. (Bottom-left) The target region on the heart surface. (Right) The retrieved 3D coordinates of a point of interest on the heart surface

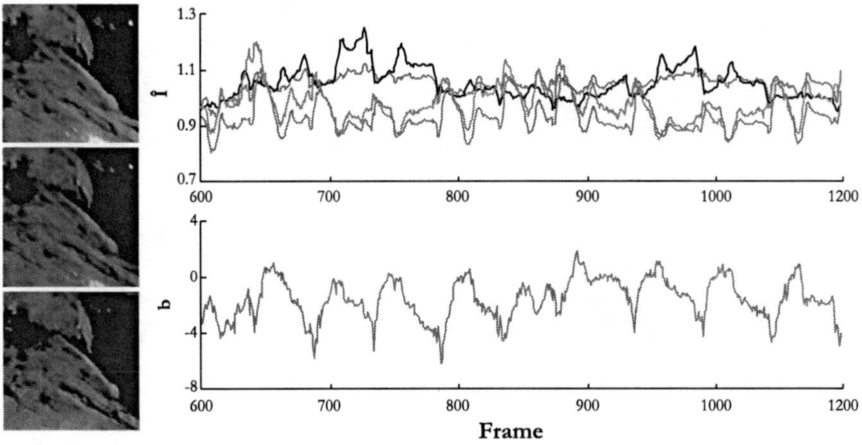

Fig. 21.6 Illustration of the compensated back-warped images at frame 220 and the illumination compensation parameters $\mathbf{y}^l$; b^l of the left stereo image. (Left) From top to bottom, the original, compensated back-warped images and the reference template. (Top-right) The evolution of the values of **I** at the four control points that model the surface. (Bottom-right) The global bias b

[43.02, 36.94] mm for the xyz coordinates respectively. This corresponds approximately to the amplitude verified by the experiment performed by Stoyanov [29] on the same image sequence. From the experimental results, we verify that the TPS model successfully copes with the real heart surface deformation and the retrieved 3D heart motion illustrates the high accuracy of the tracking algorithm.

Back-warped images from the tracking sequence illustrated in Fig. 21.6 (left) present significant illumination variations. In our experiments, we found that a TPS surface with 4 control points was sufficient to model and compensate the illumination variations. Figure 21.6 (left) illustrates the visual result of the illumination compensation and Fig. 21.6 (right) shows the evolution of $\hat{\mathbf{I}}_l$ and b_l at the four control points in time. One can easily notice the correlation between the illumination parameters variation and the heart motion, as expected. The amplitude of $\hat{\mathbf{I}}_l$ varies of $\pm 20\%$ in average, indicating a considerable lighting variation. The same is observed for b_l, that models the global intensity shift, that has an average amplitude of 5 gray levels (over 256 gray levels). The average number of iterations for the tracking algorithm to converge (including the illumination parameters for both left and right images) is 8 iterations, suggesting that the warping and illumination parameters converge rapidly even though interframe motion at 50 Hz is relatively large.

21.4.6 Real-Time Implementation on a Graphics Processor Unit

The visual tracking algorithm must extract the heart motion on-line for an accurate synchronization with the robotic tools. The first implementation of the tracking algorithm proposed in the previous sections was done using Intel Performance Primitives (IPP) and its performance was far below the desired tracking framerate

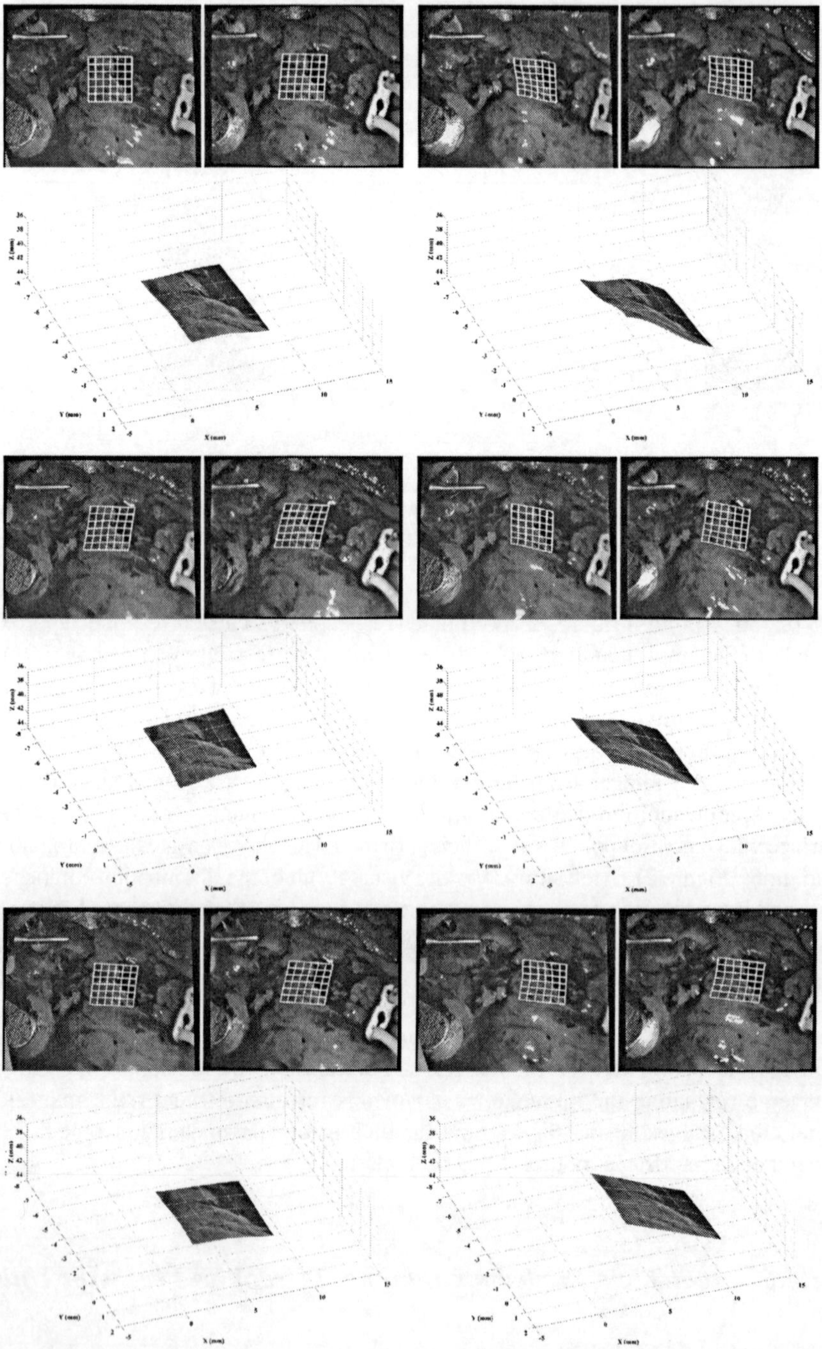

Fig. 21.7 For different instants of the heart cycle, we illustrate the target region tracked on the left and right images of the stereo pair and the corresponding TPS surface approximation. The TPS model successfully copes with the real tissue deformation. The world coordinate frame is centered in the left camera

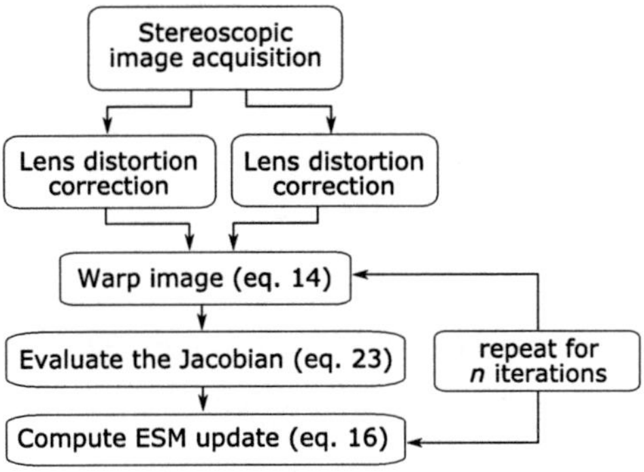

Fig. 21.8 A schematic overview of the tracking algorithm

($\approx$ 5 Hz). The computational requirements of the application motivated us to explore the computational power of recently released graphics processor units (GPU).

The market for real-time high-definition 3D graphics has given birth to high-parallel, multi-core processors with great computation power. In our work, we seek to exploit this computational power for executing the beating heart tracking algorithm. Since GPUs are specially well-suited to address problems that can be expressed as data-parallel computations, the tracking algorithm was adapted in order to take advantage of the hardware efficient processing. For our experiments, we used a NVidia GTX280 (Santa Clara, EUA) graphics card programmed using CUDA (a programming extension to C).

An overview of the tracking algorithm described earlier is given in Fig. 21.8. To illustrate the performance gain using the GPU, we have summarized the gains for each step in the real-time tracking algorithm.

- **Step 2** The lens distortion on the endoscopic images is considerable and its correction is done by remapping the pixels using a pre-computed lookup table. The computation of the interpolated gray-values using bilinear interpolation is hardware accelerated on the graphics card. Therefore, the gain with respect to the CPU implementation where the intensity value each pixel is computed sequentially is considerable.
- **Step 3** For each camera, the mapped position of each pixel from the template is given by a matrix-vector multiplication (see (21.12)). The computational speed-up is obtained by computing each line of the matrix-vector multiplication on separate threads, which is far more efficient than the CPU implementation. Finally, the computation of the interpolated gray-values using bilinear interpolation is hardware accelerated.
- **Step 4** For mounting the Jacobian matrix, each line of the matrix is handled by a separate thread. The low latency access to the graphics card internal memory allows for a significant speed-up compared to the CPU implementation. Secondly, the computation of the gradient of the warped images by convolution is highly efficient on the graphics card, since the convolution with a kernel can be efficiently parallelized on the graphics card.
- **Step 5** The computation of the ESM update is performed both on the CPU and the GPU. For the computation of the pseudo-inverse required by the

Table 21.1 Computational times for the GPU implementation

Step	Required time (ms)
2	0.25
3	0.6
4	0.6
5	0.8
Total	2.25 (per iteration)

minimization, the Moore-Penrose pseudo-inverse is used. Since the computation of a matrix inverse is required in the process and since its implementation on a GPU is still an open problem, we opted to transfer the matrices to the CPU and use LAPACK C library functions to perform the inverse. As expected, this is the most time-costly operation required by the tracking algorithm. The computation of the Hessian matrix $\mathbf{J}^T\mathbf{J}$, where $\mathbf{J}$ is the Jacobian matrix detailed in Sect. 21.4.2 as well as the multiplication of the Jacobian transpose with the error image is performed on the GPU. However, no significant speed-up is obtained with respect to the CPU implementation. This is due specially to the shape of the Jacobian matrix, which does not allow efficient accesses to the graphics card memory and the parallel computation using a larger number of threads.

The time required for each step is detailed in Table 21.1. The most costly operation is the Hessian computation and the inverse. It is important to note that the number of iterations required for the tracking algorithm to converge depends on the inter-frame motion, which is directly related to the image acquisition speed. Therefore, if we increase the acquisition speed, the computational requirement decreases. Considering a total processing time of 2.25 ms per iteration in addition to the time required to acquire the images from the camera to the graphics card memory, tracking at speeds above 80 fps can be achieved. It is important to remark that the time required by the memory transfers between the CPU and the GPU can be neglected in our problem.

21.5 Predicting the Beating Heart Motion

In the previous sections, a visual tracking method for estimating the 3D motion of the heart surface was proposed. Tracking is solved as an iterative registration method: at every frame, the previous estimated tracking parameters are used to initialize the minimization procedure. Furthermore, no regularization or constraint is applied to the TPS model deformation. Thus, even though this formulation gives full freedom for the deformable model to adapt to complex shapes, it does not take into account the quasi-periodicity of the beating heart motion.

Clearly, valuable information can be extracted from the beating heart motion dynamics for improving the accuracy and robustness of visual tracking. More specifically, if the heart dynamics can be modeled, the future heart motion can be anticipated and tracking disturbances such as occlusions (e.g. surgical instruments, smoke, blood, specular reflections, etc) or tracking failures can be bridged.

In this section, the problem of modeling and estimating the beating heart motion for predicting its future behavior in the context of visual tracking is studied. We propose to estimate the future heart motion using a predictive Extended Kalman Filter (EKF) based on a time-varying dual Fourier series for modeling the quasi-periodic beating heart motion. The performance of the proposed heart motion prediction method is assessed through several experiments using in vivo data. The resulting improvements in the tracking robustness facing tracking failures and occlusions are presented and discussed. Preliminary results have been published at [?]. Given the tracking framework proposed above, we seek to model the 3D motion of the points in space that project to the TPS control points on both stereo cameras.

21.5.1 Background

The heart motion modeling and prediction is also useful in several levels of a surgical robotic assistant design. Notably, error feedback alone is insufficient for controlling a robotic actuator with a sufficiently low tracking error [7]. In the literature, various paradigms for predicting future positions of a point of interest (POI) on the heart surface have been proposed. In Franke et al. [10], a generalized linear predictor was designed for providing future position estimations of a POI in a robot tracking task. Similarly, Bebek et al. [2] uses a copy of the previous heartbeat cycle synchronized with an electrocardiogram (ECG) signal for predicting the following heartbeat cycle. For tracking features on the beating heart using vision, Ortmaier et al. [23] uses embedded vectors of previous heart cycles for increasing tracking resilience. A thorough investigation of the heart motion was performed by Cuvillon [8], who proposed a motion prediction algorithm based on a Linear Parameter Variant model that is a function of the ECG and respiratory signals.

An alternative paradigm also found in the literature for describing the quasi-periodic heart motion is the Fourier series model and different approaches exist for estimating its coefficients. In Ginhoux et al. [11] a known and steady cardiac rhythm is assumed, whereas Thakral et al. [31] proposed the estimation of a non-stationary Fourier series coefficients using Recursive Least Squares. A similar approach has also been proposed by Yuen et al. [32] for tracking the Mitral valve annulus motion on a single axis. However, in the methods above, the respiratory and cardiac motions that comprise the heartbeat are treated separately or only the cardiac motion is modeled. This is an obvious drawback since these two sources of motion are coupled [8]. Alternative approaches have also been proposed, such as the membrane model presented in Bader et al. [1] and the motion model based on a combination of several basis functions proposed by Duindam et al [9], but experimentations using real heart motion are not presented.

In our previous work, we introduced an estimation framework based on the Extended Kalman Filter (EKF), which offers the advantage of explicitly modeling the sthocastic uncertainties associated with estimating the heart motion, offering also the possibility of fusing external signals such as the ECG for improved prediction quality. However, the previous formulation was not capable of predicting the heart motion for larger prediction horizons, as needed in case of occlusions by surgical instruments. In this section, we present an improved motion model based on a time-varying dual Fourier series that explicitly models the cardiac and respiratory components that comprise the heart motion. Details of the design of the Extended Kalman filter for estimating the series parameters are given. Secondly, we illustrate the contributions of the proposed prediction method in visual tracking, enabling us to successfully bridge tracking disturbances and automatically reestablish tracking in case of occlusions. For designing and testing the motion prediction scheme, we use in vivo porcine beating heart motion data.

21.5.2 Beating Heart Motion Dynamics

Here we present the key characteristics of the heart motion. High-speed images are necessary to properly capture the heart dynamics [11] and since the acquisition speed in commercial stereo endoscopes is limited, we used two high speed 1M75 DALSA cameras attached to Storz endoscopes mounted with a small baseline to simulate a real stereo endoscope. The system was calibrated using standard techniques [33]. Figure 21.9 displays the y trajectory of a POI on the surface of a porcine heart, acquired at 83.3 Hz using artificial markers attached to the heart surface. The acquisition setup is illustrated in Fig. 21.9. The heart was imaged for 60 s in an open chest configuration with no mechanical stabilization, which yields a heart motion amplitude considerably larger than in the MIS scenario. A frequency analysis shows that the dominant frequencies are situated between 0 and 2 Hz, with significant energy up to 5 Hz. These observations match similar experimental data reported in [2]. Also from the FFT plot, the two dominant frequencies associated to the respiratory and cardiac motions can be easily detected.

21.5.3 Non-Stationary Dual Fourier Series Model

The heart motion can be considered as the sum of the respiratory and cardiac motions, which can be represented as a dual non-stationary Fourier series. Given the 3D coordinates $\boldsymbol{d} = [{}^{x}d\,{}^{y}d\,{}^{z}d]$ of a POI on the heart surface, the motion dynamics d of each Cartesian coordinate at a given instant t can be parameterized as:

$$\begin{aligned} d(t) = {} & \sum_{h=1}^{H_r}\left[a_h \sin\left(h\sum_{k=0}^{t}\omega_r(k)\right) + b_h \cos\left(h\sum_{k=0}^{t}\omega_r(k)\right)\right] + \\ & c_r + \sum_{h=1}^{H_c}\left[d_h \sin\left(h\sum_{k=0}^{t}\omega_c(k)\right) + e_h \cos\left(h\sum_{k=0}^{t}\omega_c(k)\right)\right], \end{aligned} \tag{21.26}$$

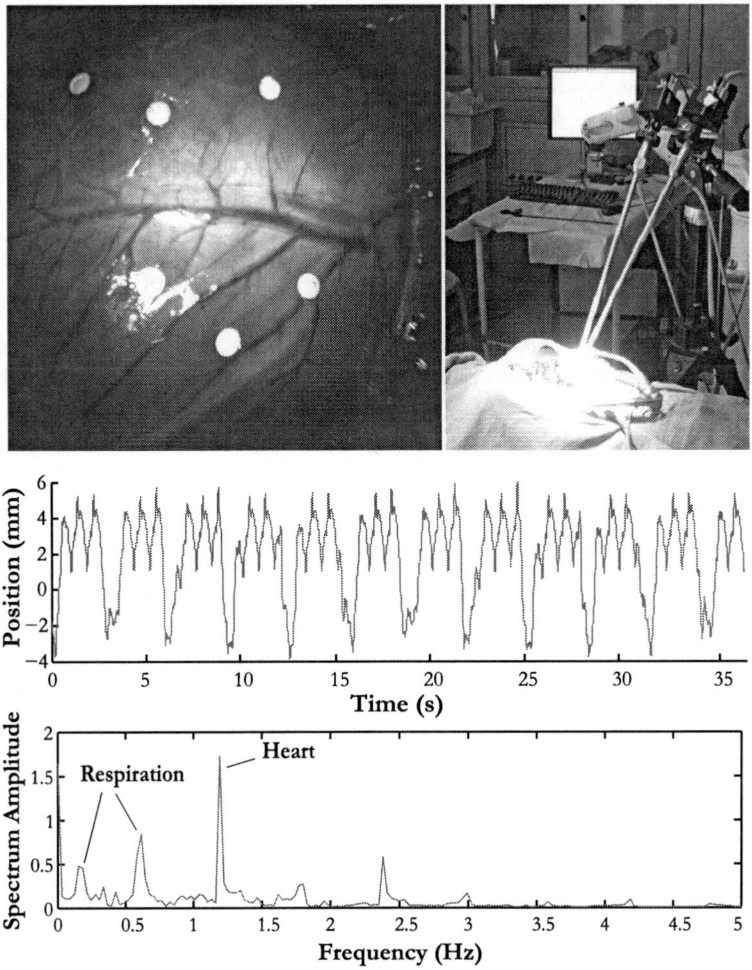

Fig. 21.9 (Top left) Markers used to retrieve the beating heart motion (Top right) In vivo experiment setup (Middle) y motion of a POI on the heart (Bottom) FFT spectrum of the y coordinate motion. The main peaks related to the respiratory and cardiac motions are highlighted

where H_r and H_c are the number of harmonics for modeling the respiratory and cardiac components respectively, ω_r and ω_c are the respiratory and cardiac frequencies and the vector $\mathbf{p} = [a_1, ..., a_{H_r}, b_1, ..., b_{H_r}, c_r, d_1, ..., d_{H_c}, e_1, ..., e_{H_c}]$ are its Fourier coefficients. Note that the number of harmonics H_r and H_c among the x y z directions may vary due to differences in their motion complexity. Finally, a POI has $n = ({}^xH_r + {}^xH_c) + ({}^yH_r + {}^yH_c) + ({}^zH_r + {}^zH_c)$ parameters plus the respiratory and cardiac frequencies, which are shared among coordinates. At a given instant t, the computation of q-step future position estimates using (21.26) is straightforward, considering a stationary system within the prediction horizon.

21.5.4 The Extended Kalman Filter

In our formulation, we employ the Kalman Filter (KF) for the recursive estimation of the Fourier series parameters. The KF offers several advantages, such as explicit modeling of the uncertainties associated with the proposed motion model and position measures. Since estimating the Fourier series parameters is a nonlinear problem, the EKF is used [27].

The EKF state vector **x** for estimating the trajectory of p POIs is composed $(p \cdot n+2)$ parameters, where n is the number of parameters of the Fourier series mentioned in the previous section. It is composed of the Fourier parameters $[\mathbf{p}_x, \mathbf{p}_y, \mathbf{p}_z]$ for the Cartesian coordinates of all estimated POIs and the cardiac and respiratory frequencies. When initializing the filter, no a priori knowledge of the signal is needed and all state vector values are set to zero except for the two frequencies, for which initial values (extracted from ECG and ventilation machine) are normally available in practice.

In the KF, estimation is divided between the prediction and correction phases. All parameters are modeled as random walk processes. In the prediction phase, the a priori estimate of the filter's state $\mathbf{x}^-$ at an instant k is given by the state values from the previous instant *k-1*. This implies that the error covariance matrix **P** is propagated according to:

$$\mathbf{P}^-(k) = \mathbf{P}(k-1) + \mathbf{Q} \tag{21.27}$$

where **Q** is the process covariance matrix.

It is important to remark that an interesting feature that can be incorporated into the filter formulation is motion correlation between POIs, since close POIs tend to have similar trajectories. This notion is incorporated in the process noise covariance matrix **Q** by giving small values to the non-diagonal elements of the matrix. When initializing the filter, we use a process variance related to the respiratory components $\sigma_r = 10^{-5}$ (with a variance $\sigma_c = 10^{-6}$ for the offset c_r), the cardiac components $\sigma_c = 10^{-5}$ and the frequencies variance $\sigma_w = 10^{-8}$. To consider the correlation between POIs, the non-diagonal elements of **Q** are set to 10^{-10}.

In the correction phase, the available measurements from the visual tracking algorithm **y** are used to update the initial estimates $\mathbf{x}^-$, yielding the a posteriori estimate $\mathbf{x}^+$. The updated state is computed as:

$$\mathbf{x}^+(k) = \mathbf{x}^-(k) + \underbrace{\mathbf{P}^-(k)\mathbf{C}^T(\mathbf{C}\mathbf{P}^-(k)\mathbf{C}^T + \mathbf{R})^{-1}}_{\mathbf{K}(k)}(\mathbf{y}(k) - \mathbf{x}^-(k)) \tag{21.28}$$

where **R** is a diagonal matrix containing the measurement error covariance (the diagonal elements are set to 10^{-1}) and **C** is a Jacobian matrix whose lines contain the derivatives of each motion component (as in 21.26) with respect to the Fourier series parameters. The matrix **K**(k) indicated above in 21.28 is the Kalman gain matrix. As an example, if we were to estimate only the x component of one POI, the matrix $\mathbf{C}(k)$ would be a $1 \times ({}^{x}H_r + {}^{x}H_c + 2)$ matrix given as:

$$\mathbf{C}(k) = \left.\frac{\partial d(t, x_k)}{\partial \mathbf{x}_k}\right|_{\mathbf{x}_k = \hat{\mathbf{x}}_{k|k-1}} \tag{21.29}$$

The updated error covariance matrix is given by:

$$\mathbf{P}^{+}(k) = (\mathbf{I} - \mathbf{K}(k)\mathbf{C})\mathbf{P}^{-}(k) \tag{21.30}$$

where $\mathbf{I}$ is the identity matrix.

21.6 Experimental Results

21.6.1 Predictive Filter Performance

For evaluating the performance of the proposed prediction method, we use the experimental data presented in Sect. 21.5.2. As stated in Sect. 21.5.3, predictions are computed using (21.26), considering a stationary system within the prediction horizon. This idea is illustrated in Fig. 21.10, featuring the estimated Fourier series at an instant t_0. Since motion is estimated on-line with all parameters initialized with zeros, 1.5 respiratory cycle ($\sim$400 samples) is needed for the model's parameters to converge. The number of harmonics for modeling respiration H_r for the x y z directions was set to 3, while for H_c, 5 harmonics were sufficient.

For investigating the prediction performance in time, we evaluate the prediction error at every instant for 15, 83 and 250-step prediction horizons (0.18, 1 and 3 s respectively). The error is calculated as the Euclidean distance $\| \| \boldsymbol{d} - \boldsymbol{p} \| \|$ between the predicted $\boldsymbol{d}$ and true $\boldsymbol{p}$ 3D positions of the POI. The root mean square and peak prediction errors at every motion sample for a given prediction horizon are measured. The prediction errors are plotted in Fig. 21.11 and quantified in

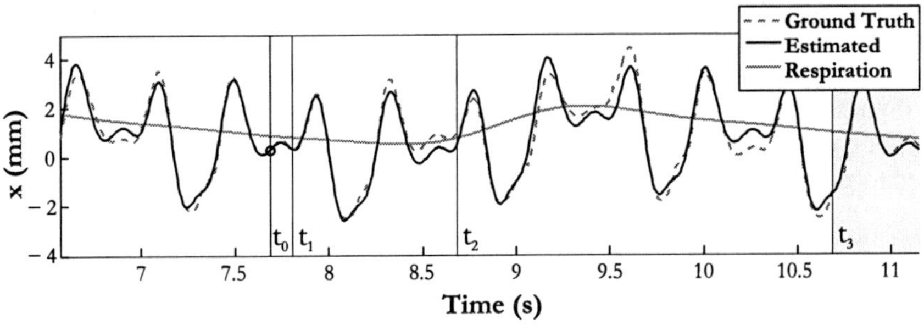

Fig. 21.10 Ground truth and estimated Fourier series at an instant t0. For evaluating prediction performance, we measure the prediction error for a 15, 83 and 250-step horizon prediction (0.18, 1 second and 3 seconds respectively)

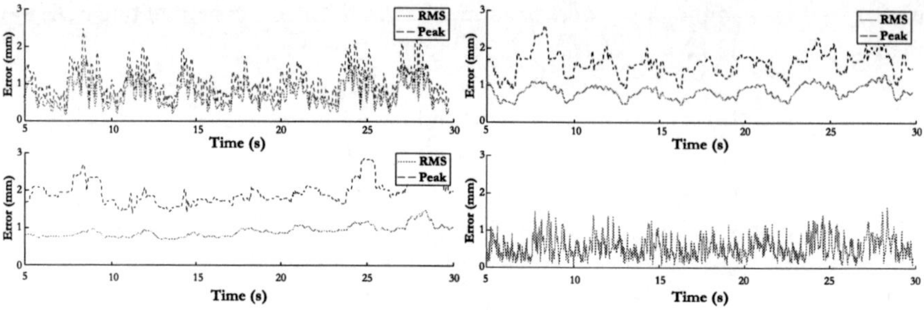

Fig. 21.11 From top to bottom, the RMS and peak prediction errors for a 15-step, 83-step, 250-step prediction horizon and the identification error, respectively [48]

Table 21.2 Predictive filter performance on in-vivo data

Horizon (s)	Average RMS error (mm)	Average peak error (mm)
0.18	0.8076	1.1829
1	0.8785	1.6764
3	1.0209	2.0496

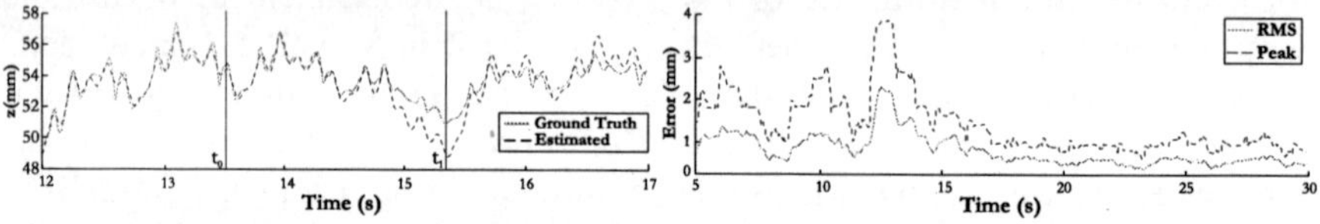

Fig. 21.12 (Top) The original heart motion is slowly damped from t_0 to t_1 to 50% of its original amplitude. (Bottom) The prediction error plot for a 1 second prediction horizon indicates the fast filter adaptation to the amplitude change

Table 21.2. The identification error, which is the difference between the true position and the position estimated by the filter at a given instant is also plotted in Fig. 21.11.

The prediction filter must be robust to changes in the heart rhythm. In order to investigate the filter's behavior to amplitude changes, we slowly damp the original heart motion signal with respect to its mean to 50% of its original value in all coordinates and analyze the behavior of the filter and the evolution of the prediction errors for a 1 s prediction horizon. Results are plotted in Fig. 21.12. Cardiac arrhythmia may also cause irregular heart motion and in order to analyze the filter's performance under such circumstances, we have simulated a disturbance in the heartcycle in all coordinates as illustrated in Fig. 21.13. The effects on the prediction error for a 1 s prediction horizon is plotted in Fig. 21.13.

21.6.2 Improvements in Visual Tracking

For evaluating the improvements in visual tracking, we use another set of in vivo data consisting of images of a porcine beating heart acquired using the same experimental setup presented in Sect. 21.5.2.

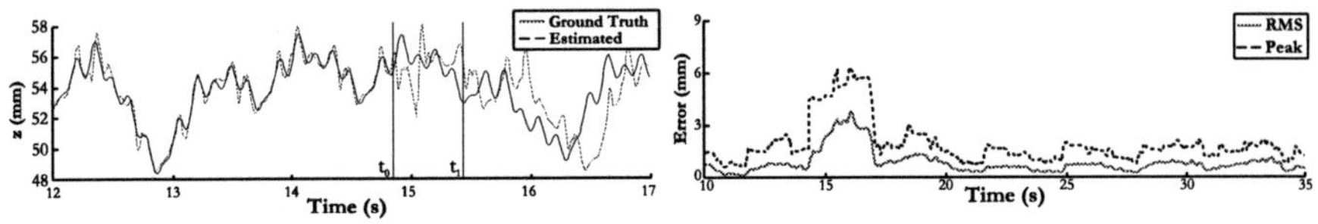

Fig. 21.13 (Top) A signal disturbance that resembles an arrhythmia is induced from t_0 to t_1. (Bottom) The prediction error plots for a 1 second horizon indicate high prediction errors during the disturbance but the prediction quality is quickly reestablished past the disturbance

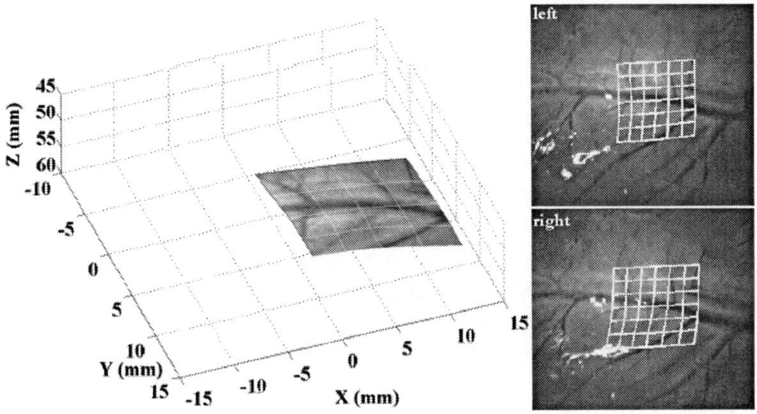

Fig. 21.14 (Left) The white mesh represents the TPS surface that models a selected ROI on the heart. (Right) The 3D shape of a ROI on the heart surface.

Recalling Fig. 21.2, we note that the parameter vector **h** is composed of 3D points that projects to the TPS control points on both stereo cameras and therefore, each point can be considered as a POI in the prediction framework presented before. Figure 21.14 displays a 128 ×128 pixel ROI on the heart tracked using the proposed method, using 6 control points for modeling the heart surface deformation.

21.6.2.1 Specular Reflections

Specular reflections are the direct reflection of the illumination source on the glossy, wet-like heart surface. Such reflections saturate the affected pixels, disturbing considerably the visual tracking task. Figure 21.16 illustrates a case of tracking under such phenomenon. Although the tracking method automatically removes the affected regions from the estimation of the warping parameters, certain control points of the TPS mesh whose support region on the image is more severely affected by this phenomenon may be poorly estimated. In addition, depending on the duration of the perturbation, the estimation of the warping parameters may get stuck in local minima and diverge.

In the motion prediction context, specular reflections can be considered as occluders since no texture information is available from the affected areas. If the area affected by specular reflections goes beyond a critical level, tracking is suspended. In this context, the predicted heart motion can be used to bridge such disturbances, which normally last for approximately 0.12 s (15 frames in the acquisition speed used for the experiments).

21.6.2.2 Occlusions

The proposed motion prediction method also allows us to tackle the problem of occlusion by surgical tools. Surgical instruments eventually occlude the operating site for considerably longer periods of time and the proposed prediction scheme offers a solution for automatic tracking reinitialization in such cases.

In Fig. 21.15, the result of a simulated 3-s occlusion is presented, displaying the successful tracking reestablishment after the event. For simulating the occlusion, the correction step of the Kalman filter is suspended at an arbitrary instant t_0 and the predicted heart motion at t_1 3 s later was used to reestablish tracking. For visualizing the accuracy of the predicted heart motion, tracking results (the motion plots of one TPS control point) are also presented throughout the whole sequence for comparison purposes. An important remark after a visual inspection of the results is that although the predicted motion is accurate enough to restart tracking, it cannot be used for motion compensation since its error is too large.

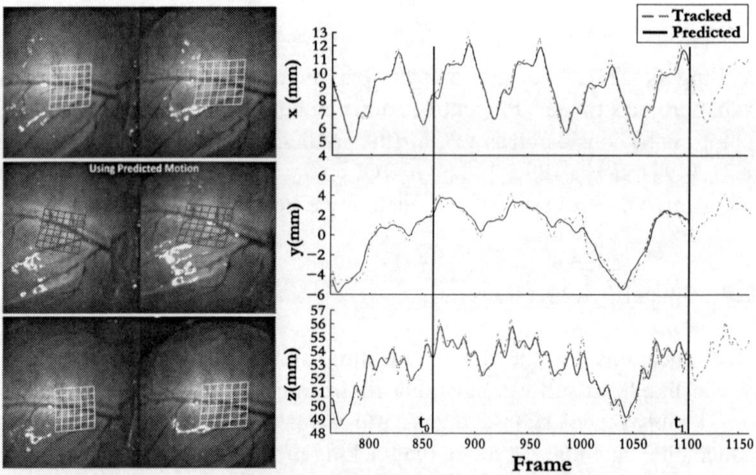

Fig. 21.15 (*Left Top*) The mesh illustrates the tracked region of interest on both left and right endoscopic images. (*Left Middle*) During the simulated occlusion, tracking is suspended amd the predicted trajectory of the several POI that comprise the mesh is displayed as the mesh in red. (*Left Bottom*) Tracking is successfully reestablished past the occlusion. (*Right*) The predicted and tracked heart motion at an instant t_0 for the x component of a POI.

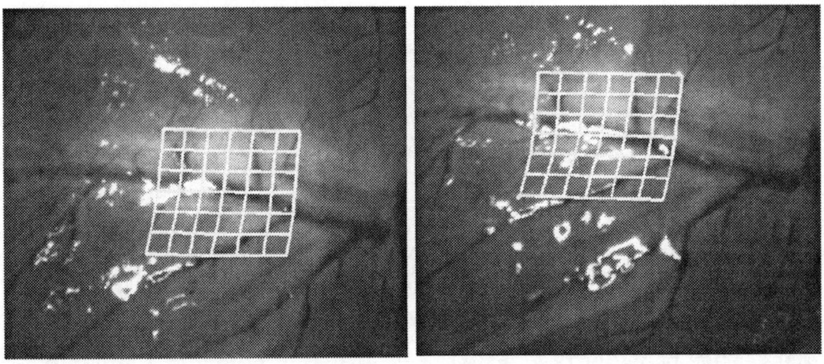

Fig. 21.16 The endoscopic illumination source reflects on the wet-like surface of the heart, giving rise to specular reflections that considerably disturb visual tracking.

21.6.3 Discussion of the Predictive Filter Performance

The prediction errors presented in Fig. 21.11 reveal the good performance of the predictive filter, successfully acquiring the true heart dynamics. In addition, the error values given in Table 21.2 indicate that the RMS and peak errors do not increase significantly when the prediction horizon is extended, suggesting the proposed motion model fits properly the heart motion. If the prediction horizon is further expanded, the prediction quality gradually degrades, due specially to small cardiac frequency variations.

The capacity of the filter to adapt to amplitude variations was properly displayed by the extreme simulation presented in Fig. 21.12. From the error plots, we can verify that the filter takes approximately 1 s to re-adapt to the signal changes, later stabilizing in a lower level since the signal's amplitude was reduced by 50%.

Next, the simulated arrhythmic heart behavior presented in Fig. 21.13 is the most challenging event for the prediction filter. In fact, if we analyze the filter parameters in detail we learn that the abnormal heart behavior is interpreted as a drastic phase change. Although the error plots show a considerable large error during the event, the filter successfully copes with the disturbance, converging quickly past the event. It is also important to remark that although the simulated arrhythmia generates a major disturbance in the predictive filter, such abnormal heart behaviors can be easily detected from the heart electric activity and in practice they do not represent a critical problem.

Moreover, the relatively high identification error indicates that the heart motion model describes the coarse heart trajectory. An increase of the number of harmonics does not lower this error, which suggests a relation to the natural variability of the heart motion.

The performance increase obtained with a Kalman filtering framework has been analyzed in previous works [32]. However, the direct comparison with experimental results of similar techniques proposed in the literature is not possible since the used experimental database highly influences the prediction errors, as clearly demonstrated when analyzing the amplitude changes in the input signal shown in the experimental section. This is due to the non linearity of the RMS error measure and the fact that the error amplitude is dependent on the amplitude of the heart motion.

The heart motion prediction can be further improved using the proposed formulation by exploring the ECG and respiratory signals directly in the filter design, as suggested in Sect. 21.5.4. For instance, the heart electric activity precedes the mechanical motion and the ECG waves may help predict more accurately the heartbeat contraction and relaxation cycles. The EKF framework adopted in this paper provides an elegant framework for fusing different sources of information in a straightforward fashion. Furthermore, the ECG can also be used for detecting abnormal cardiac behavior (e.g. arrhythmia). Another aspect of Kalman filtering is the adequate choice for the filter's covariance matrices. Although the values proposed in Sect. 21.5.4 were not "fine" tuned for the experimental database, an adaptive update of the filter's uncertainty parameters could better estimate the heart motion, hence producing better future estimates.

21.7 Conclusion

In the previous sections, we have described a performing tracking method for estimating the beating heart motion. The proposed tracking model based on the TPS mapping in association with the motion prediction scheme attests the feasibility of a visual motion compensation system for robotized surgery. The tracking algorithm is robust to illumination variations, specular reflections and occlusions by surgical instruments. In addition, real-time performance has been achieved using a GPU. Experiments conducted on in vivo data attest the good performance of the approach.

There are several routes for improving the current design of the tracking algorithm. Although region-based techniques such as the TPS based tracker proposed in this paper provide dense reconstruction with respect to feature-based tracking, they still provide limited reconstruction quality due to the parametric models used to approximate the heart surface. A possible solution is the use of physically inspired models supported by pre-operative data for modeling more accurately the heart dynamics. An alternative for increasing tracking accuracy is the fusion of multiple visual tracking techniques (such as feature-based and direct tracking methods, shape from shading, etc), compensating the downsides of one technique with the upsides of others. Another important aspect of the problem of designing a visual motion compensation system which also must be taken into consideration are the hardware requirements, briefly discussed in Sect. 21.2. High resolution and acquisition rate are essential for accurate motion estimation.

Currently, our works focus on the incorporation of the motion tracking methods in a robotized prototype surgical platform. We are also evaluating the performance of different predictive filters for increasing the heart motion prediction.

References

1. Bader, T., Wiedermann, A., Roberts, K., Hanebeck, U.D.: Model-based motion estimation of elastic surfaces for minimally invasive cardiac surgery. In: Proceedings of IEEE Conference on Intelligent Robots and Systems (IROS '07), pp. 871–876. San Diego, USA, (2007)
2. Bebek, O., Cavusoglu, M.C.: Intelligent control algorithms for robotic-assisted beating heart surgery. IEEE Trans. Robot. **23**(3), 468–480 (2007)
3. Bebek, O., Cavusoglu, M.C.: Whisker sensor design for three dimensional position measurement in robotic assisted beating heart surgery. In: Proceedings of IEEE Conference on Intelligent Robots and Systems (IROS '07), pp. 225–231. San Diego, USA, (2007)
4. Benhimane, S., Malis, E.: Real-time image-based tracking of planes using efficient second-order minimization. In: Proceedings of IEEE Conference on Intelligent Robots and Systems (IROS '04), pp. 943–948. Sendai, Japan (2004)
5. Bogatyrenko, E., Hanebeck, U.D., Szabó, G.: Heart surface motion estimation framework for robotic surgery employing meshless methods. In: Proceedings of IEEE Conference on Intelligent Robots and Systems (IROS '09), pp. 67–74. St. Louis, USA, (2009)
6. Bookstein, F.L.: Principal warps: Thin-plate splines and the decomposition of deformations. IEEE Trans. Pattern Anal. Mach. Intell. (PAMI) **11**(6), 567–585 (1989)
7. Cavusoglu, M.C., Rotella, J., Newman, W.S., Choi, S., Ustin, J., Sastry, S.S.: Control algorithms for active relative motion cancelling for robotic assisted off-pump coronary artery bypass graft surgery. In: Proceedings of the 12th International Conference on Advanced Robotics (ICAR), pp. 431–436. Seattle, USA, (2005)
8. Cuvillon, L., de Mathelin, M., Forgione, A.: Towards robotized beating heart TECABG: Assessment of the heart dynamics using high-speed vision. Comput. Aided Surg. **11**(5), 267–277 (2006)
9. Duindam, V., Sastry, S.: Geometric motion estimation and control for robotic-assisted beating-heart surgery. In: Proceedings of IEEE Conference on Intelligent Robots and Systems (IROS '07), pp. 871–876. San Diego, USA, (2007)
10. Franke, T.J., Bebek, O., Cavusoglu, M.C.: Prediction of heartbeat motion with a generalized adaptive filter. In: Proceedings of IEEE Conference on Robotics and Automation (ICRA '08), pp. 2916–2921. Pasadena, USA, (2008)
11. Ginhoux, R., Gangloff, J., de Mathelin, M., Soler, L., Sanchez, M.A., Marescaux, J.: Active filtering of physiological motion in robotized surgery using predictive control. IEEE Trans. Robot. **21**(1), 67–79 (2003)
12. Gröger, M., Ortmaier, T., Sepp, W., Gerd, H.: Tracking local motion on the beating heart. In: Imaging, S.M. (ed.) Visualization, Image-Guided Procedures and Display, vol. 4681, pp. 233–241. SPIE, San Diego (2002)
13. Hayashibe, M., Suzuki, N., Nakamura, Y.: Laser-scan endoscope system for intraoperative geometry aqusition and surgical robot safety management. Med. Image Anal. (10), 509–519 (2006)
14. Lau, W., Ramey, N.A., Corso, J.J., Thakor, N.V., Hager, G.D.: Stereo-based endoscopic tracking of cardiac surface deformation. In: Medical Image Computing and Computer-Assisted Intervention (MICCAI '04), Saint Malo, France. Lecture Notes in Computer Science (LNCS), vol. 3217, pp. 494–501. Springer, Heidelberg (2004)
15. Lemma, M., Mangini, A., Redaelli, A., Acocella, F.: Do cardiac stabilizers really stabilize? experimental quantitative analysis of mechanical stabilization. Interact. CardioVasc. Thorac. Surg. **4**, 222–226 (2005)
16. Lim, J., Yang, M.: A direct method for modeling non-rigid motion with thin plate spline. In: Proceedings of IEEE Conference on Computer Vision and Pattern Recognition (CVPR '05), pp. 1196–1202. Washington, USA, (2005)
17. Lo, B., Chung, A.J., Stoyanov, D., Mylonas, G., Yang, G.Z.: Real-time intra-operative 3D tissue deformation recovery. In: Proceedings of IEEE International Symposium on Biomedical Imaging (ISBI '08), pp. 1387–1390, Paris, France, (2008)

18. Malis, E.: An efficient unified approach to direct visual tracking of rigid and deformable surfaces. In: Proceedings of IEEE Conference on Intelligent Robots and Systems (IROS '07), pp. 2729–2734, San Diego, USA, (2007)
19. Mountney, P., Lo, B., Thiemjarus, S., Stoyanov, D., Yang, G.Z.: A probabilistic framework for tracking deformable soft tissue in minimally invasive surgery. In: Medical Image Computing and Computer-Assisted Intervention (MICCAI '07), Brisbane, Australia. Lecture Notes in Computer Science (LNCS), vol. 4792, pp. 34–41. Springer, Berlin (2007)
20. Mountney, P., Yang, G.Z.: Soft tissue tracking for minimally invasive surgery: Learning local deformation online. In: Medical Image Computing and Computer-Assisted Intervention (MICCAI '08), New York, USA. Lecture Notes in Computer Science (LNCS), vol. 5242, pp. 364–372. Springer, Berlin (2008)
21. Nakamura, Y., Kishi, K., Kawakami, H.: Heartbeat synchronization for robotic cardiac surgery. In: Proceedings of IEEE International Conference on Robotics and Automation (ICRA '01), vol. 2, pp. 2014–2019, Seoul, Korea, (2001)
22. Noce, A., Triboulet, J., Poignet, P.: Efficient tracking of the heart using texture. In: Proceedings of IEEE International Conference of the Engineering in Medicine and Biology Society (EMBS '07), pp. 4480–4483, Lyon, France, (2007)
23. Ortmaier, T., Groger, M., Boehm, D.H., Falk, V., Hirzinger, G.: Motion estimation in beating heart surgery. IEEE Trans. Biomed. Eng. **52**(10), 1729–1740 (2005)
24. Richa, R., B\'o, A.P.L., Poignet, P.: Beating heart motion prediction for robust visual tracking. In: Proc. IEEE Conf. Robot. Automat. (ICRA '10). Anchorage, USA (2010)
25. Richa, R., Poignet, P., Liu, C.: Efficient 3D tracking for motion compensation in beating heart surgery. In: Medical Image Computing and Computer-Assisted Intervention (MICCAI '08), New York, USA. Lecture Notes in Computer Science (LNCS), pp. 684–691. Springer, Heidelberg (2008)
26. Riviere, C.N., Gangloff, J., de Mathelin, M.: Robotic compensation of biological motion to enhance surgical accuracy. Proc. IEEE, **94**(9), 1705–1716 (2006)
27. Silveira, G., Malis, E.: Real-time visual tracking under arbitrary illumination changes. In: Proceedings of IEEE Conference on Computer Vision and Pattern Recognition (CVPR '07), pp. 1–6, Minneapolis, USA, (2007)
28. Simon, D.: Optimal State Estimation: Kalman, "H_∞" and Nonlinear Approaches. Wiley Inc., New York (2006)
29. Stoyanov, D., Darzi, A., Yang, G.Z.: A practical approach towards accurate dense 3D depth recovery for robotic laparoscopic surgery. Comput. Aided Surg. **4**(10), 199–208 (2005)
30. Stoyanov, D., Mylonas, G.P., Deligianni, F., Darzi, A., Yang, G.Z.: Soft-tissue motion tracking and structure estimation for robotic assisted mis procedures. In: Medical Image Computing and Computer-Assisted Intervention (MICCAI '05), Palm Springs, USA. Lecture Notes in Computer Science (LNCS), vol. 3750, pp. 139–146. Springer, Heidelberg (2005)
31. Stoyanov, D., Yang, G.Z.: Stabilization of image motion for robotic assisted beating heart surgery. In: Medical Image Computing and Computer-Assisted Intervention (MICCAI '07), Brisbane, Australia. Lecture Notes in Computer Science (LNCS), vol. 4791, pp. 417–424. Springer, Berlin (2007)
32. Thakral, A., Wallace, J., Tomlin, D., Seth, N., Thakor, N.V.: Surgical motion adaptive robotic technology (s.m.a.r.t): Taking the motion out of physiological motion. In: Medical Image Computing and Computer-Assisted Intervention (MICCAI '01), Utrecht, The Netherlands. Lecture Notes in Computer Science (LNCS), pp. 317–325. Springer, Berlin (2001)
33. Yuen, S.G., Novotny, P.M., Howe, R.D.: Quasiperiodic predictive filtering for robot-assisted beating heart surgery. In: Proceedings of IEEE Conference on Robotics and Automation (ICRA '08), pp. 1050–1057, Pasadena, USA, (2008)
34. Zhang, Z.: A flexible new technique for camera calibration. IEEE Trans. Pattern Anal. Mach. Intell. (PAMI) **22**(11), 1330–1334 (2000)

Chapter 22
Towards the Development of a Robotic System for Beating Heart Surgery

Özkan Bebek and M. Cenk Çavuşoğlu

Abstract The use of intelligent robotic tools promises an alternative and superior way of performing off-pump coronary artery bypass graft (CABG) surgery. In the robotic-assisted surgical paradigm proposed, the conventional surgical tools are replaced with robotic instruments, which are under direct control of the surgeon through teleoperation. The robotic tools actively cancel the relative motion between the surgical instruments and the point-of-interest on the beating heart, in contrast to traditional off-pump CABG where the heart is passively constrained to dampen the beating motion. As a result, the surgeon operates on the heart as if it were stationary. We call the proposed algorithm "Active Relative Motion Cancelling" (ARMC) to emphasize the active cancellation. This chapter will provide a review of our research towards developing robotic tools for off-pump CABG surgery. First, we will explain the algorithm we have developed to achieve effective motion cancellation. Second, we will explain the necessary sensory system for the beating heart surgery and the developed whisker sensors to detect three-dimensional heart motion. Third, we will explain the millirobotic gripper developed for minimal invasive surgery. Finally, we will outlay the overall system design for robotic-assisted beating heart surgery.

22.1 Model-Based Active Relative Motion Canceling

In this section of the chapter, we will first discuss the model-based intelligent ARMC algorithm we have developed to achieve effective motion cancellation. The developed algorithm relies on the quasiperiodic nature of the heart motion, using feedforward predictive motion control with estimation of heart motion. The developed model-based algorithm employs biological signals, such as electrocardiogram, in estimation of heart motion, and uses this for integrated arrhythmia

M.C. Çavuşoğlu (✉)
Department of Electrical Engineering and Computer Sciences, Case Western Reserve University, 308 Glennan Building, Cleveland, OH, USA
e-mail: cavusoglu@case.edu

J. Rosen et al. (eds.), *Surgical Robotics: Systems Applications and Visions*,
DOI 10.1007/978-1-4419-1126-1_22, © Springer Science+Business Media, LLC 2011

detection and handling, to provide safety during the operation. One of the goals of this research was to improve the tracking performance so that the degree of necessary tracking can be achieved. To this end, the tracking performance research has primarily been focused on developing estimation methods for use with a receding horizon model predictive controller. Next in this section, a heart motion prediction method based on adaptive filter techniques will be explained. A recursive least squares based adaptive filter algorithm was used to parameterize a linear system to predict the heart motion. Finally, we will discuss a generalized adaptive filter to further improve the prediction by using parameterized linear predictors for points throughout the prediction horizon independently. The feasibility and effectiveness of the approach will be demonstrated.

22.1.1 Robotic-Assisted Beating Heart Surgery

Improving the treatment for coronary heart disease (CHD) should be a priority in terms of developing relevant treatment options as the statistics of the Centers for Disease Control and Prevention (CDC) indicate heart disease as a leading cause of death [1, Table 7]. In the medical field, intelligent robotic tools reshape the surgical procedures by providing shorter operation times and lower costs. This technology also promises an enhanced way of performing off-pump coronary artery bypass graft (CABG) surgery. In the robotic-assisted off-pump CABG surgery, the surgeon operates on the beating heart using intelligent robotic instruments. Robotic tools actively cancel the relative motion between the surgical instruments and the point of interest (POI) on the beating heart, dynamically stabilizing the heart for the operation. This algorithm is called Active Relative Motion Canceling (ARMC).

Although off-pump CABG surgery is in a nascent stage and only applicable to limited cases, it is preferred over on-pump CABG surgery because of the significant complications resulting from the use of cardio-pulmonary bypass machine, which include long term cognitive loss [34], and increased hospitalization time and cost [38]. On the other hand off-pump grafting technology is crude and only applicable to a small portion of the cases because of the technological limitations: inadequacy with all but the largest diameter target vessels, ineffectiveness with the coronary arteries on the side and the back of the heart, and its limitation to small number of bypasses. Off-pump procedures represent only 15–20% of all CABG surgeries, at best [32]. Manual tracking of the complex heartbeat motion cannot be achieved by a human without phase and amplitude errors [16]. Use of robotics technology will overcome limitations as it promises an alternative and superior way of performing off-pump CABG surgery.

Therefore, it is aimed to develop telerobotic tools to actively track and cancel the relative motion between the surgical instruments and the heart by Active Relative Motion Canceling (ARMC) algorithms, which will allow CABG surgeries to be performed on a stabilized view of the beating heart with the technical convenience of on-pump procedures. Towards this direction, electrocardiogram (ECG) is utilized as a biological signal in the estimation of the heart motion for an effective motion canceling in the model-based intelligent ARMC algorithm.

22.1.2 System Concept for Robotic Telesurgical System for Off-Pump CABG Surgery

Robotic-assisted surgery concept replaces conventional surgical tools with robotic instruments which are under direct control of the surgeon through teleoperation, as shown in Fig. 22.1. The surgeon views the surgical scene on a video display with images provided by a camera mounted on a robotic arm that follows the heart motion, showing a stabilized view. The robotic surgical instruments also track the heart motion, canceling the relative motion between the surgical site on the heart and the surgical instruments. As a result, the surgeon operates on the heart as if it were stationary, while the robotic system actively compensates for the relative motion of the heart. This is in contrast to traditional off-pump CABG surgery where the heart is passively constrained to dampen the beating motion. The proposed control algorithm is called "Active Relative Motion Canceling (ARMC)" to emphasize this difference. Since this method does not rely on passively constraining the heart, it is possible to operate on the side and back surfaces of the heart as well as the front surface using millimeter scale robotic manipulators that can fit into spaces the surgeon cannot reach.

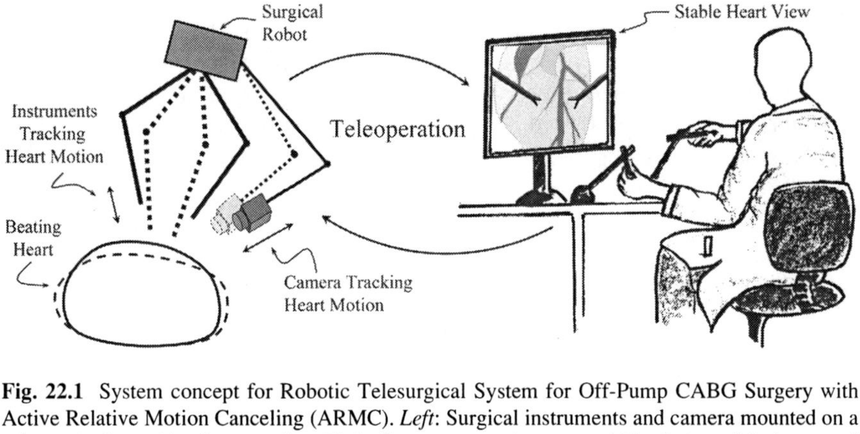

Fig. 22.1 System concept for Robotic Telesurgical System for Off-Pump CABG Surgery with Active Relative Motion Canceling (ARMC). *Left*: Surgical instruments and camera mounted on a robot actively tracking heart motion. *Right*: Surgeon operating on a stabilized view of the heart, and teleoperatively controlling robotic surgical instruments to perform the surgery (© IEEE 2007), reprinted with permission

22.1.3 Motion Canceling in Medical Interventions

The earlier studies in the literature on canceling biological motion in robotic-assisted medical interventions focus on canceling respiratory motion. Sharma et al. [42] and Schweikard et al. [41] studied the compensation of the breathing motion in order to reduce the applied radiation dose to irradiate tumors. Both studies concluded that motion compensation was achievable. Riviere et al. [39] looked at the cancelation of respiratory motion during percutaneous needle insertion. Their results showed that an adaptive controller was able to model and predict the breathing motion. Trejos et al. [46] conducted a feasibility study on the ability to perform tasks on motion-canceled targets, and demonstrated that tasks could be performed better using motion canceling.

Madhani and Salisbury [30] developed a 6-DOF telesurgical robot design for general minimally invasive surgery, which was later adapted by Intuitive Surgical Inc., Palo Alto, CA, for their commercial system, called daVinci. Computer Motion Inc., Goleta, CA,[1] developed a 5-DOF telesurgical robotic system, called Zeus, with scaled motions for microsurgery and cardiac surgery. Both of these systems are currently in use for cardiothoracic surgery applications. These systems are designed to enable dexterous minimally invasive cardiac surgery, and they are neither intended nor suitable for off-pump CABG surgery with active relative motion canceling, due to their size, bandwidth, and lack of motion tracking capabilities. These systems can only perform on-pump or off-pump CABG surgery using passive stabilizers, therefore have the same limitations as conventional tools described above.

Gilhuly et al. [20] tested suturing on a beating heart model using optical stabilization through strobing. They found that participants' reaction times were too slow to adjust to the changing light conditions, and concluded that stabilization methods should not rely on surgeons' reaction times.

Nakamura et al. [33] performed experiments to track the heart motion with a 4-DOF robot using a vision system to measure heart motion. The tracking error due to the camera feedback system was relatively large (error on the order of few millimeters in the normal direction) to perform beating heart surgery. There are also other studies in the literature on measuring heart motion. Thakor et al. [45] used a laser range finder system to measure one-dimensional motion of a rat's heart. Groeger et al. [22] used a two-camera computer vision system to measure local motion of heart and performed analysis of measured trajectories, and Koransky et al. [28] studied the stabilization of coronary artery motion afforded by passive cardiac stabilizers using 3-D digital sonomicrometer.

Ortmaier et al. [35, 36] used ECG signal in visual measurement of heart motion using a camera system for estimation of the motion when the surgical tools occluded the view. They reported significant correlations between heart surface trajectory and ECG signals, which implies these inputs can be used

[1] Computer Motion Inc. was acquired by Intuitive Surgical Inc., and does not exist anymore.

interchangeably. Therefore, these two independent components were considered as inputs to the estimation algorithm. In their study, heart motion estimation was not based on a heart motion model and it was completely dependent on previously recorded position data. Actual tracking of the heart motion using a robotic system was planned as future work.

More recently, in a pair of independent parallel studies by Ginhoux et al. [21] and Rotella [40], motion canceling through prediction of future heart motion was demonstrated. In both studies, model predictive controllers were used to get higher precision tracking. In the former, a high-speed camera was used to measure heart motion. Their results indicated a tracking error variance on the order of 6–7 pixels (approximately 1.50–1.75 mm calculated from the 40 pixel/cm resolution reported in [21]) in each direction of a 3-DOF tracking task. Although it yielded better results than earlier studies using vision systems, the error was still too large to perform heart surgery, as operation targets to be manipulated using the robotic systems in a CABG surgery are blood vessels with 2 mm or less in diameter. In Rotella's study [40], using a 1-DOF test bed system, accuracy very close to the desired error specifications for heart surgery was achieved, and it was concluded that there still was a need for better prediction of heart motion.

A heart model was proposed by Cuvillon et al. [14], based on the extraction of the respiration motion from the heartbeat motion using the QRS wave form of the ECG and lung airflow information as sensory inputs. They concluded that heartbeat motion is not the product of two independent components, rather the heartbeat motion is modulated by the lung volume.

Duindam and Sastry [15] proposed a method to separate 3-D quasiperiodic heart motion data into its two periodic components using ECG and respiratory information. Future heart surface motion was estimated using the separated periodic components; and an explicit model based controller was proposed to asymptotically cancel the relative motion between surgical tools and a region on the heart surface.

22.1.4 Model-Based Active Relative Motion Cancelation: Motivation and Methodology

The control algorithm is the core of the robotic tools for tracking heart motion during coronary artery bypass graft (CABG) surgery. The tools need to track and manipulate a fast moving target with very high precision. During free beating, individual points on the heart move as much as 7–10 mm. Although the dominant mode of heart motion is on the order of 1–2 Hz, measured motion of individual points on the heart during normal beating exhibit significant energy at frequencies up to 26 Hz. The coronary arteries that are operated on during CABG surgery range from 2 mm in diameter down to smaller than 0.5 mm, which means the system needs to have a tracking precision in the order of 100 μm. This corresponds to a less than 1% dynamic tracking error up to a bandwidth of about 20–30 Hz.

The specifications for tracking heart motion are very demanding. These stringent requirements could not be achieved using traditional algorithms in earlier attempts reported in the literature [33, 45], as they rely solely on feedback signal from measurement of heart motion using external sensors, and do not use any physiological model of the heart motion.

Using a basic model of heart motion can significantly improve tracking performance since heart motion is quasiperiodic [21]. It is also possible to use the information from the biological signals, such as ECG activity, and aortic, atrial and ventricular blood pressures, to control the robotic tools tracking the heart motion.

The proposed control architecture is shown in Fig. 22.2. In this architecture, the control algorithm utilizes the biological signals in a model-based predictive control fashion. Using biological signals in the control algorithm improves the performance of the system since these signals are products of physiological processes which causally precede the heart motion. Therefore a heart motion model can be formed by combining motion data and biological signal data.

22.1.5 Intelligent Control Algorithms for Model-Based ARMC

In the Model-Based ARMC Algorithm architecture, shown in Fig. 22.2, the control algorithm uses information from multiple sources: mechanical motion sensors which measure the heart motion, and sensors measuring biological signals. The control algorithm identifies the salient features of the biological signals and uses this information to predict the feedforward reference signal.

The control algorithm also handles the changes in the heart motion, including adapting to slow variations in heart rhythm during the course of the surgery, as well as handling occasional arrhythmias which may have natural causes or may be due to the manipulation of the heart during surgery.

In this control scheme, the two dominant modes of the motion of POI are separated by using a pair of complementary filters. The control path for tracking of the heartbeat component of the motion has significantly more demanding requirements in terms of the bandwidth of the motion that needs to be tracked. That is why a more sophisticated feedforward algorithm is employed for this part. Respiratory motion has significantly lower frequency, and it is canceled by a purely feedback based controller. In the proposed architecture (Fig. 22.2), the robot motion control signal is computed by combining these two parts. The feedforward part is calculated with the signal provided by the heart motion model, and the feedback signal is calculated with the direct measurements of heartbeat and respiratory motions. The feedforward controller was designed using the model predictive control [19] and optimal control [11, 3] methodology of modern control theory, as described in Sect. 22.1.6.

The confidence level reported by the heart motion model is used as a safety switching signal to turn off the feedforward component of the controller if an arrhythmia is detected, and switch to a further fail-safe mode if necessary. This

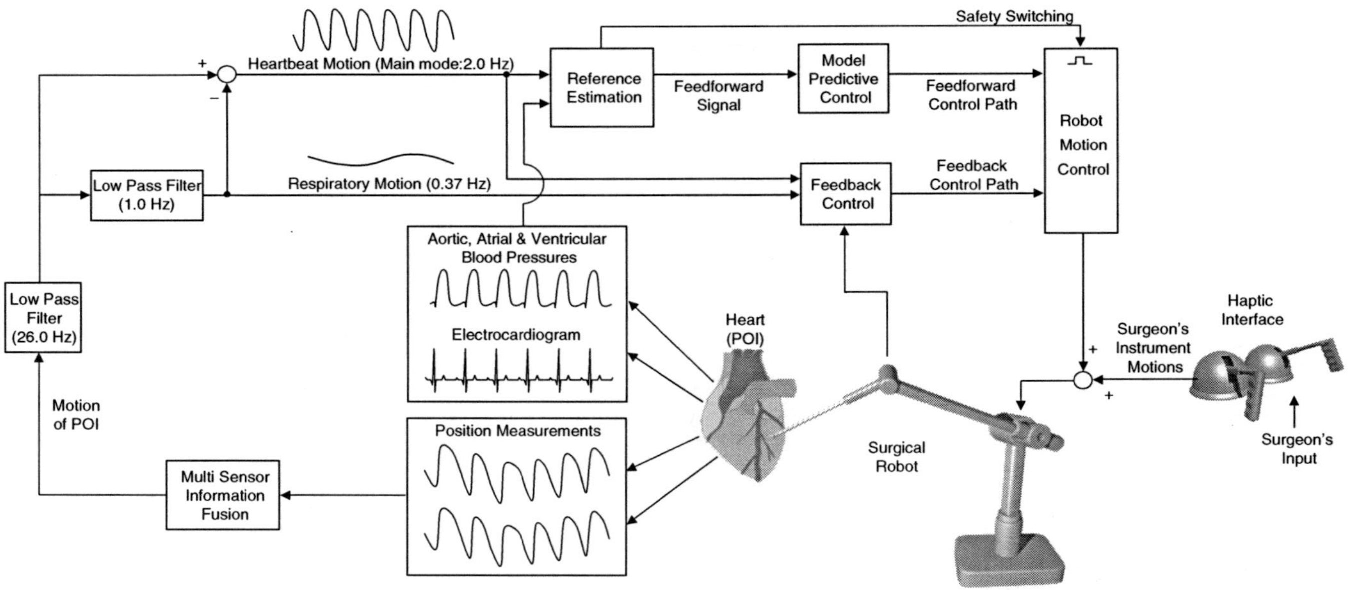

Fig. 22.2 Proposed control architecture for designing Intelligent Control Algorithms for Active Relative Motion Canceling on the beating heart surgery (© IEEE 2007), reprinted with permission

confidence level will also be used to adaptively weigh the amount of feedforward and feedback components used in the final control signal. These safety features will be an important component of the final system. Therefore, the best design strategies for developing feedforward motion control was aimed.

In Fig. 22.3, a finite-state model for the cardiac cycle is shown. The model involves primary states of the heart's physiological activity. Transitions between the states are depicted using the states of the mitral and aortic valves of heart and P, R and T waves of the ECG. During the ECG wave form detection process, QRS complex is detected and used in substitute to R wave. Any out of sequence or abnormal states in the cycle can be identified as irregularity. Using this model, rhythm abnormalities and arrhythmias can be spotted and system can be switched to a safer mode of operation.

Although, some of the system concepts in the literature are similar to this scheme at the most basic level, there are significant differences including their lack of intelligent model-based predictive control using biological signals, and multi-sensor fusion with complementary and redundant sensors, which form the core of our proposed architecture. The system by Nakamura et al. [33] used purely position feedback obtained from a two-camera computer vision system. Neither biological signals were used in the system, nor was a feedforward control component present. The system by Ginhoux et al. [21] utilized a feedforward control algorithm based on model predictive control and adaptive observers; however, it did not utilize any biological signals. Ortmaier et al. [36] utilized ECG using a "model free" method, i.e., without using a heart model in the process.

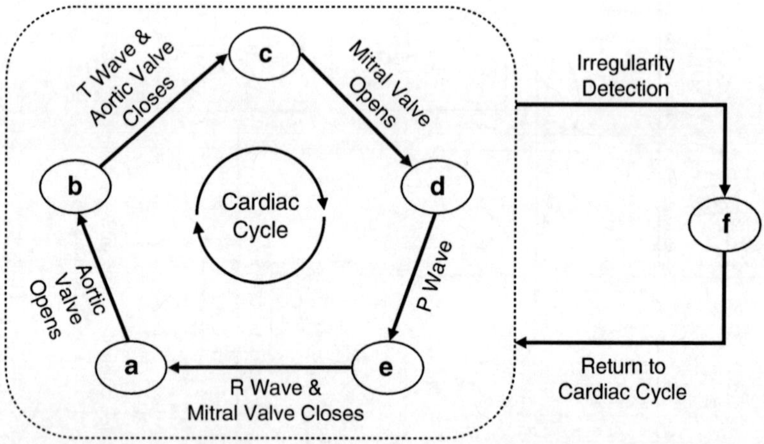

Fig. 22.3 State model of the beating heart. Transition between the states are depicted using ECG waves and the motion of the heart valves, which can be inferred from blood pressure measurements. States forming the cardiac cycle are: (**a**) Isovolumic contraction; (**b**) Ejection; (**c**) Isovolumic relaxation; (**d**) Ventricular filling; (**e**) Atrial Systole; (**f**) Irregularity in the Cardiac Cycle (© IEEE 2007), reprinted with permission

With the architecture proposed, the degree of awareness is increased by utilizing a heart motion model in reference signal estimation. Inclusion of biological signals in a model-based predictive control algorithm increases the estimation quality, and such a scheme provides better safety with more precise detection of anomalies and switching to a safer mode of tracking.

22.1.6 Control Algorithms

The control algorithm is the core of the robotic tools for tracking heart motion during coronary artery bypass graft (CABG) surgery. The robotic tools should have high precision to satisfy the tracking requirements [more than 97% motion cancelation (details in Sect. 22.1.5)]. During free beating, individual points on the heart move as much as 10 mm. Although the dominant mode of heart motion is in the order of 1–2 Hz, measured motion of individual points on the heart during normal beating exhibit significant energy at frequencies up to 26 Hz.

The heart motion is quasiperiodic and previous beats can be used as a feedforward signal during the control of the robotic tool for ARMC. Rotella [40] compared a model-based predictive controller, using the estimation of the heart motion, with feedback based controllers on a 1-DOF robotic test-bed system. The model-based predictive controller outperformed the feedback based controllers both in terms of the RMS error and the control action applied. In [5], Bebek extended the comparison of model-based predictive controller and feedback based controllers to 3-D case with a 3-DOF robotic test-bed system. The results show that model-based predictive controller is more robust and effective than the traditional controllers in tracking the heart motion.

A key component of the ARMC algorithm, when a predictive controller is used, is estimation of the reference motion of the heart which is provided to the feedforward path. If the feedforward controller has high enough precision to perform the necessary tracking, the tracking problem can be reduced to predicting the estimated reference signal effectively.

Ginhoux et al. [21] used an adaptive observer, which identifies the Fourier components of the past motion at the base heart rate frequency and its several harmonics to estimate the future motion. This approach assumes that the heartbeat rate stays constant. Ortmaier et al. [36] estimated the heart motion by matching the current heart position and ECG signals of sufficient length with recorded past signals, assuming that with similar inputs, heart would create outputs similar to the ones detected in the past.

In Sects. 22.1.6.1 and 22.1.6.3, reference signal estimation schemes used for the ARMC algorithm are described. Section 22.1.6.2 explains the Electrocardiogram (ECG) and ECG wave form detection methodology used in this study. Finally, the control problem and its solution are given in Sects. 22.1.6.4 and 22.1.6.5 respectively.

22.1.6.1 Reference Signal Estimation

A simple prediction scheme that assumes constant heartbeat rate can be used for reference signal estimation. Heartbeat is a quasiperiodic motion with small variations in every beating cycle. If the past heartbeat motion cycle is known, it can be used as an estimate reference signal for the next cycle. Any measured heart position value can be approximated forward one cycle as long as the heartbeat period for that cycle is known. In this case, a constant heartbeat period (0.5 s) was used to store one period length of the heartbeat signal. The motion of the heart from the previous cycle was used as a prediction of the next cycle. The stored beating cycle was used as the approximate future reference beating signal in the ARMC algorithm.

Using the last heartbeat cycle exactly as the future reference would result in large errors due to the quasiperiodic characteristics of the heart motion and other irregularities of the signal. Therefore, instead of using the past beating cycle directly, the reference signal was processed online. To achieve this, any position offset between the starting point of the past cycle and the starting point of the next cycle (i.e., current position in time) were lined up by subtracting the difference. But the added offset was gradually decreased over a constant length of time (hereafter this length will be referred to as *horizon*, T) using a high order error correction function. This calculation was carried out T steps ahead. So, only some percentage of the current error was added to the future signals, and no error was added to the signals T steps ahead. In Fig. 22.4, the actual and the estimated motions can be seen as the control executes. This maintained the continuity of the signal estimate and converges it onto the actual signal within the horizon ahead.

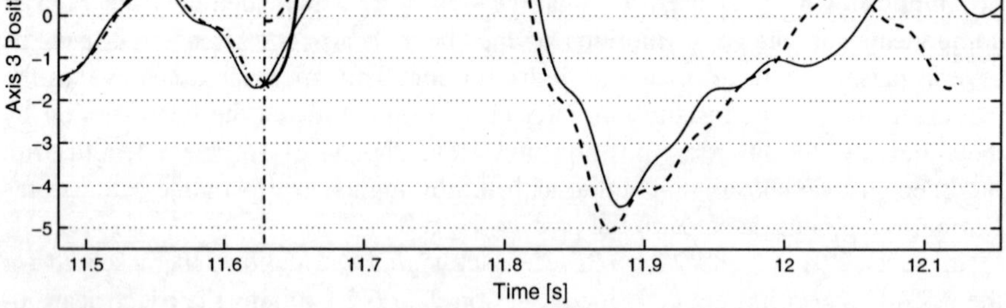

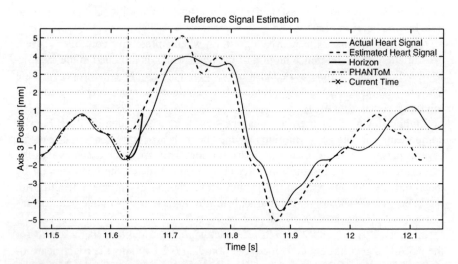

Fig. 22.4 Reference Signal Estimation during control action. Observe the horizon signal where the offset between the current position and estimated signal is added gradually starting from current time to horizon steps ahead (© IEEE 2007), reprinted with permission

22.1.6.2 Electrocardiogram as the Biological Signal

The human body acts as a giant conductor of electrical currents. Connecting electrical *leads* to any two points on the body may be used to register an electrocardiogram (ECG). Thus, ECG contains records for the electrical activity of the heart. The ECG of heart forms a series of waves and complexes that have been labeled in alphabetical order, the P wave, the QRS complex, the T wave and the U wave [2]. Depolarization of the atria produces the P wave; depolarization of the ventricles produces the QRS complex; and repolarization of the ventricles causes the T wave. The significance of the U wave is uncertain [31]. Each of these electrical stimulations results in a mechanical muscle twitch. This is called the electrical excitation-mechanical contraction coupling of the heart [9, 10]. Thus, the identification of such waves and complexes can help determine the timing of the heart muscle contractions.

Using ECG in the control algorithm can improve the performance of the position estimation because these wave forms are results of physiological processes that causally precede the heart motion and also because ECG is significantly correlated with heartbeat motion [36].

The difficulty in detection of the ECG wave forms arises from the diversity of complex wave forms and the noise and artifacts accompanying the ECG signals. In this work, the significant ECG wave forms and points, such as P, QRS and T, were detected in hard real time[2] by an algorithm adapted from Bahoura et al. [4]. This one was selected among other available algorithms because it employs signal localization both in time and frequency using wavelet analysis, characterization of the local regularity of the signal and separation of the ECG waves from serious noise, artifacts, and baseline drifts in real time.

Bahoura et al. [4] evaluated the original algorithm in real time with the MIT-BIH Arrythmia Database [37]. This database contains 48 half-hour excerpts of two-channel ambulatory ECG recordings. The recordings were digitized at 360 samples per second per channel with 11-bit resolution over a 10 mV range. Two or more cardiologists independently annotated each record; disagreements were resolved to obtain the computer-readable reference annotations for each beat included with the database. Using the database, Bahoura et al. [4] reported a 0.29% false detection rate (135 false positive beats and 184 false negative beats out of 109,809 beats), showing the algorithms capability in detecting QRS complexes. Constant detection parameters rather than adaptive ones were used, and this produced a 1.49% false detection rate using the same database (408 false positive beats and 709 false negative beats out of 75,010 healthy beats).

[2] In hard real time no corrections are allowed to be performed to the past data after the operation deadline expires.

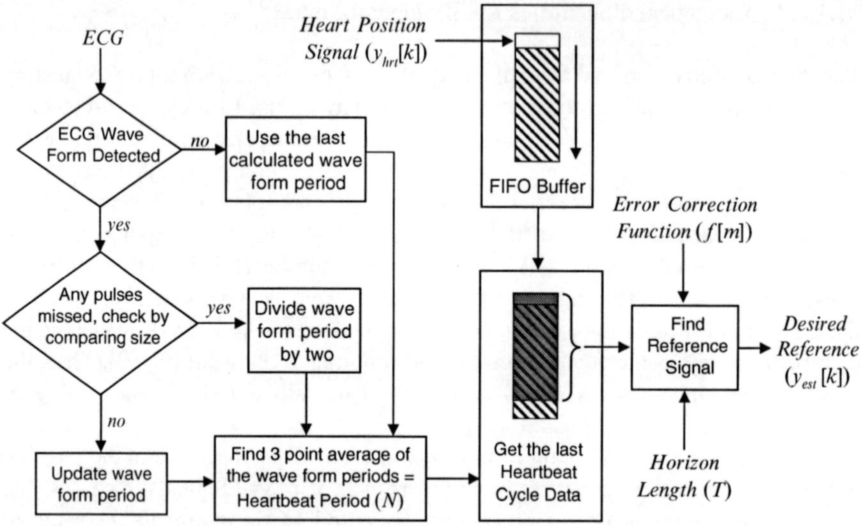

Fig. 22.5 Simplified finite state model of the Reference Signal Estimation algorithm using ECG. Detected ECG Wave forms were used in the estimation of *Reference*, with the buffered *Past Heart Position Data* (© IEEE 2007), reprinted with permission

With this method, QRS-T-P waves were detected in real time for the collected 56 s ECG data[3] with 100% QRS complex and T wave detection rates, and 97.3% P wave detection rate. Detected signals were used to estimate the Reference Signal as described in Sect. 22.1.6.3 (Fig. 22.5).

22.1.6.3 Reference Signal Estimation Using Biological Signals

Although the position offset between the previous and current beating cycles can be eliminated gradually using the technique given in Sect. 22.1.6.1, the error due to changes in heartbeat period remains. Because heartbeat is a quasiperiodic motion with small period variations in every beating cycle, these period changes could result in large offsets in the estimated signal, and can cause jumps during the tracking.

As mentioned earlier in the Sect. 22.1.6.2, ECG signal is very suitable for period-to-period synchronization. In this reference signal estimation scheme, rather than using a constant heartbeat period, a variable period that was calculated using ECG was used. QRS, P, and T waves were used as check points for detecting heartbeat period. In Fig. 22.5, the block diagram for reference signal estimation using ECG is illustrated.

[3] The ECG signal employed in this research was collected with the analog data acquisition part of the sonomicrometry system used. The ECG data were recorded simultaneously with the collection of the heart motion data at the same sampling rate of 257 Hz.

Here, past heart position data were stored on the fly into a FIFO buffer which was 1,300 elements long (i.e. 650 ms of data; also note that average heartbeat period is about 500 ms long). The most recently stored part of the heart position buffer, in the length of updated heartbeat period using ECG, was used in the estimation.

The current heartbeat period was calculated by averaging the periods of the three ECG wave forms. The period was updated continuously as new wave forms were detected. If detection of any ECG wave form was missed, the period of the missed signal was doubled to compensate for the missing signal. Some upper and lower period boundaries were imposed in order to eliminate any misses by the detection algorithm.

In Fig. 22.6 the estimated signals just before and after the detection of a new wave form are shown. In Fig. 22.6b, observe that after the T wave was detected, the past heartbeat period time mark was shifted back in time as a result of the increase in the heartbeat period. In the example shown with Fig. 22.6a, b, RMS estimation error for one heartbeat period ahead decreased from 0.887 mm to 0.456 mm after the shift. With the use of ECG in ARMC algorithm, heartbeat period in the estimation of reference signal can be adjusted online.

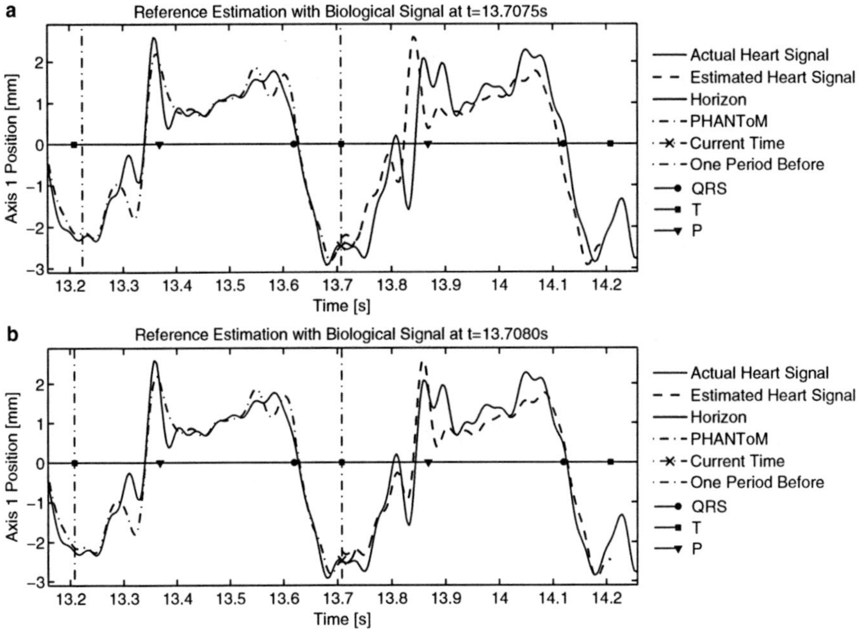

Fig. 22.6 Reference Estimation with Biological Signal. (**a**) *Just before T Wave was detected:* Estimated Heart Signal did not fit well with the Actual Heart Signal. T Wave detection was shown with *filled square* markers. (**b**) *T Wave has been detected:* Heart Period and Estimated Signal were adjusted. Observe that the beginning of the previous heartbeat period marker [·—·—·](t ≈13.22 s) was shifted back in time (t ≈13.21 s) as a result of the increase in the heartbeat period. Accordingly, Estimated Heart Signal was changed to adjust with the new period. RMS estimation error was decreased from 0.887 mm to 0.456 mm with the shift (© IEEE 2007), reprinted with permission

22.1.6.4 Control Problem

Having the estimated trajectory of the next cycle in hand, the following control problem arises: Tracking of heart motion where there is some knowledge of the future motion. Then, this optimal tracking problem can be stated as follows.

Suppose the dynamics of the robotic surgical manipulator is given by an n-dimensional linear system having state equations

$$x[k+1] = \mathbf{\Phi} x[k] + \mathbf{\Gamma} u[k] \tag{22.1}$$

$$y[k] = \mathbf{H} x[k] \; . \tag{22.2}$$

Here, if the dimensions of $\mathbf{\Phi}$, $\mathbf{\Gamma}$, and $\mathbf{H}$ are $n \times n$, $n \times m$ and $l \times n$ respectively; then $x[k] \in \mathbb{R}^n$ denotes the system state at time k where $x[k_0]$ is given for some time k_0 such that $k_0 \leq k$; $u[k] \in \mathbb{R}^m$ denotes system control at time k; and the $y[k] \in \mathbb{R}^l$ denotes the system output at time k where l entries of y are linearly independent, or equivalently, the matrix $\mathbf{H}$ has rank l. Suppose we are also given an l-vector $y_{est}[k]$ for all k in the range $k_0 \leq k \leq k_0{+}T$ for some times k_0 and T. The optimal tracking problem is then to find the optimal control u for the system (1)–(2), such that the output y tracks the signal y_{est}, minimizing the index (3)

$$J[k] = \sum_{k=k_0}^{k_0+T} \left((x[k] - x_{est}[k])^T \mathbf{Q} (x[k] - x_{est}[k]) + u^T[k] \mathbf{R} u[k] \right) \tag{22.3}$$

$$x_{est} = \mathbf{L} y_{est} \tag{22.4}$$

where $\mathbf{Q}$ is a non-negative definite symmetric matrix and $\mathbf{R}$ is a positive definite symmetric matrix, and, $\mathbf{L}$ and $\mathbf{Q}$ are

$$\mathbf{L} = \mathbf{H}^T (\mathbf{H}\mathbf{H}^T)^{-1} \tag{22.5}$$

$$\mathbf{Q} = (\mathbf{I} - \mathbf{L}\mathbf{H})^T \mathbf{Q}_1 (\mathbf{I} - \mathbf{L}\mathbf{H}) + \mathbf{H}^T \mathbf{Q}_2 \mathbf{H}. \tag{22.6}$$

where $\mathbf{Q}_1$ and $\mathbf{Q}_2$ are non-negative definite symmetric matrices.

22.1.6.5 Receding Horizon Model Predictive Control

Solution to the control problem given in Sect. 22.1.6.4 was derived in [5] using the method given in [11]. An optimal tracking system can be derived using regulator theory. Such controller consist of a standard optimal feedback regulator involving the backwards solution of a Riccati equation, and an external signal (feedforward) that results from the backwards solution of a linear differential equation. Unmeasurable states can be replaced by state estimates, under observability of $\mathbf{\Phi}$ and $\mathbf{H}$. The solution to the control problem is

$$u[k] = -(\mathbf{\Gamma}^T\mathbf{S}[k+1]\mathbf{\Gamma} + \mathbf{R})^{-1}\mathbf{\Gamma}^T(\mathbf{S}[k+1]\mathbf{\Phi}x[k] + \mathbf{M}[k+1]) \tag{22.7}$$

where **S** and **M** are given by the iterative equations

$$\mathbf{S}[k] = \mathbf{\Phi}^T(\mathbf{S}[k+1] - \mathbf{S}[k+1]\mathbf{\Gamma}(\mathbf{\Gamma}^T\mathbf{S}[k+1]\mathbf{\Gamma} + \mathbf{R})^{-1}\mathbf{\Gamma}^T\mathbf{S}[k+1])\mathbf{\Phi} + \mathbf{Q} \tag{22.8}$$

$$\mathbf{M}[k] = (\mathbf{\Phi}^T + \mathbf{K}^T[k]\mathbf{\Gamma}^T)\mathbf{M}[k+1] - \mathbf{QL}y_{est}[k] \tag{22.9}$$

and **K** is

$$\mathbf{K}[k] = -(\mathbf{\Gamma}^T\mathbf{S}[k+1]\mathbf{\Gamma} + \mathbf{R})^{-1}\mathbf{\Gamma}^T\mathbf{S}[k+1]\mathbf{\Phi} \tag{22.10}$$

The resulting control algorithm is composed of feedback and feedforward parts which are identified, respectively, as follows:

$$u_{fb}[k] = -(\mathbf{\Gamma}^T\mathbf{S}[k+1]\mathbf{\Gamma} + \mathbf{R})^{-1}\mathbf{\Gamma}^T\mathbf{S}[k+1]\mathbf{\Phi}x[k] \tag{22.11}$$

$$u_{ff}[k] = -(\mathbf{\Gamma}^T\mathbf{S}[k+1]\mathbf{\Gamma} + \mathbf{R})^{-1}\mathbf{\Gamma}^T\mathbf{M}[k+1] \tag{22.12}$$

such that

$$u[k] = u_{fb}[k] + u_{ff}[k] \tag{22.13}$$

Parameters **S** and **M** are calculated iteratively backwards with final conditions **S** $[T] = \mathbf{Q}$ and $\mathbf{M}[T] = 0$. The iterations are carried out for *horizon*, T, times. Every iteration corresponds to one control cycle set of gains. In effect, calculating T iterations is like calculating time varying gains up to T steps ahead even though only the gain for the current time is used. This type of control is also known as Receding Horizon Control [11], and in this framework, we call the control defined in (7) as the Receding Horizon Model Predictive Control (RHMPC). With every new control cycle, a new point on the desired signal is used and an old point is dropped in the gain calculation. The calculation is then repeated at every control cycle. The prediction horizon recedes as time progresses such that the furthermost point ahead of the horizon is considered to be moving one step for every control cycle.

22.1.6.6 Simulation and Experimental Results

Simulations and experiments were carried out for the estimation algorithms with RHMPCs as presented in the previous section.

In order to find a baseline performance of the estimation algorithms, a RHMPC with known future reference signal was also tested. Knowing the future reference signal for the RHMPC algorithm provides close to perfect tracking. However, using the future reference signal in heart tracking is not feasible as this makes the algorithm acasual. In this case, it was used to show the base line performance.

This RHMPC can handle time-varying systems and weighting matrices. For the applications used herein, constant weighting matrices, $\mathbf{Q}_1, \mathbf{Q}_2$, and $\mathbf{R}$, and a constant horizon value, T, were used along with constant state-space models. The only true time-varying gain matrix within the algorithm was $\mathbf{M}$, which was calculated from the heartbeat data. As a result, feedforward control term (12) was time-varying, and $\mathbf{M}$ was calculated iteratively on the fly every control cycle.

Feedback term (11) was not dependent on any time-varying values. Consequently, calculated gains were constant for a given horizon. Once the horizon value was set, there was no need to calculate the feedback gains in every control cycle.

Although, using the past heart cycle as an estimate of future reference signals would cause large errors in extended estimates, it was not a deterministic issue in this approach, since the horizon used in the RHMPC algorithm (25 ms) was relatively short compared to the heartbeat period ($\approx$ 500 ms).

The robot was made to follow the combined motion of heartbeat and breathing as described in Sect. 22.1.4. Separating the respiratory motion enabled better heart motion estimation. In terms of control performance, controlling the respiratory motion separately did not affect the heart tracking accuracy when the results of the combined motion tracking were compared with the pure heartbeat motion tracking results. This validates our earlier observation that heartbeat motion tracking will be the bottleneck in motion tracking and the breathing motion can be easily tracked using a pure feedback controller.

22.1.6.7 Test Bed System

In order to develop and test the algorithms, a hardware test bed system, PHANToM Premium 1.5A, was used and modeled. For detailed derivation of the mathematical modeling of the PHANToM robot, see [12]. The PHANToM robot possesses characteristics similar to an actual surgery robot. Its lightweight links, low inertia design and low friction actuation system allows sufficient motion and speed abilities for tracking the heartbeat signal. In the experimental setup, the control algorithms were executed on a PC equipped with a 2.33 GHz Dual-Core Intel Xeon 5140 processor running MATLAB xPC Target v3.3 real-time kernel with a sampling time of 0.5 ms.

22.1.6.8 Experimental Results

In both simulations and experiments, the same methods and reference data were used. Some slight differences in parameters were observed due to the mathematical modeling of the robot. To validate the algorithms effectiveness, first 10 s of the 56-s long data was used to tune the control parameters. Then, the experiments were carried out using the same prerecorded 56-s long heart motion data. The robot was made to follow the combined motion of heartbeat and breathing. The system was run using prerecorded data points in place of online measurements.

Table 22.1 End-effector simulation and experimental results

	RMS (Max) Position error [mm]		RMS control effort [Nmm]	
End-effector tracking results	Simulation	PHANToM	Simulation	PHANToM
Receding horizon MPC with exact reference information	0.295 (1.732)	0.277 (2.066)	14.8	46.9
Receding horizon MPC with reference signal estimation	0.726 (3.826)	0.906 (5.958)	17.6	66.5
Receding horizon MPC with reference signal estimation using ECG	0.533 (3.066)	0.682 (4.921)	16.3	55.9

Summary of the end-effector RMS position error, max position error (in parenthesis) and RMS control effort values for the control algorithms used with 56-s data. Heartbeat and breathing motions were filtered offline. Some of theexperimental results are underlined to point out the effect of biological signal onthe estimation.

Matrix weighting parameters of the optimal index were tuned to minimize RMS tracking error. Parameters were selected in order to accentuate the states and hence regulate more quickly, with higher control efforts. Tuning was performed to avoid the high frequency resonances so that no vibration would be reflected to the structure.

For each case, experiments on PHANToM robot were repeated ten times. The reported RMS errors are calculated from the difference between the prerecorded target points and the actual end effector positions. It was noted that the deviations between the trials are very small. Among these results, the maximum values for the *End-effector RMS and Maximum Position Errors* in 3D and *RMS Control Effort* are summarized in Table 22.1 to project the worst cases.

It is believed that the maximum error values are affected from the noise in the data collected by sonomicrometric sensor. Although high-frequency parts of the raw data were filtered out, relatively low "high frequency" components stayed intact. It is unlikely that the POI on the heart is capable of moving 5 mm in a few milliseconds. The measured data has velocity peaks that are over 13 times faster than the maximum LAD velocity measurements reported by Shechter et al. [43]. Heavy filtering should have been performed to delete the high frequency motions, but they were kept, as currently we do not have an independent set of sensor measurements (such as from a vision sensor) that would validate this conjecture. This also gives a conservative measurement of the performance of the system.

22.1.6.9 Discussion of the Results

If we compare the results of the algorithms with each other, as expected, the RHMPC with Reference Signal Estimation Using Biological Signals algorithm outperformed the RHMPC with Reference Signal Estimation algorithm. Results proved that by using ECG signal in the motion estimation, heart position tracking was not only improved but also became more robust. The system was more responsive to sudden changes in the heart motion with the addition of ECG signal, accordingly the variance of the error distribution decreased by half. One-way ANOVA was used to test the statistical significance of the results and they were found to be significantly different ($F(1,38) = 6809$, $p < 0.001$).

22.1.7 Adaptive Filtering-Based Motion Predictor

In this section, a heart motion prediction method based on adaptive filter techniques is studied. A recursive least squares based adaptive filter algorithm used to parameterize a linear system to predict the heart motion [17].

A new prediction technique using adaptive filters is designed and used in the controller described in Sect. 22.1.6. Since the new linear predictor is parameterized by a least squares algorithm, the predictor is inherently robust to noise. The predictor only uses observations close to and including the present making it less susceptible to differences between heart periods than the algorithm of 22.1.6.1. Where as Ginhoux et al. [21] formulated prediction for periodic POI motion, no assumptions are made towards periodicity of the system a priori, rather the predictor is unconstrained so that it can best mimic the motion of the POI.

Fundamentally, the predictor is a mapping from the samples in its memory to the best estimate of the next observation. The estimate for the next value is such to minimize the error between the prediction and the actual position of the POI as it will appear in the future. An Mth order predictor has memory of the past $M - 1$ observations as well as having access to the current observation. It uses these M observations to generate the next expected observation. To generate the observation after that one, what was just estimated is treated as if it were actually observed and is added to the set of observations. The next observation predicted from this set is the next value in the prediction horizon. Proceeding inductively by this method, any number of future estimates can be made, stopping when all the predictions for the horizon have been made.

In order to generate predictions in this way, the one step prediction function must be known. Abstractly, the motion of the POI is represented as a continuous time dynamic system. An analogous discrete system must be created for estimation. However, the state space of the heart is not known – not even its dimension. To simplify prediction, the heart model must use a finite, and rather low order, state vector. The discrete transition function maps from the approximate model's state space onto the same space. The transition function for the discrete, approximate model is, in general, nonlinear. As is, this would be very difficult to parameterize; so the transition function will be assumed to be linear. This assumption is justified by the nature of the signal that is being predicted.

The adaptive predictor consists of two principle parts: a linear filter and an adaption algorithm. The input-output relation of the adaptive filter is determined by the linear filter. Prediction error is used to update the filter weights and therefore prediction error is minimized adaptively.

If the adaptive algorithm is able to forget the past, just as it was able to converge to a stationary signal, it can track a signal with changing statistics [23]. In the special case that the statistics change slowly relative to the algorithm's ability to adapt, then the filter can track the ideal time-varying solution. Further, if the statistics change slowly relative to the length of the prediction horizon as well as the length of the state vector, then the adaptive filter can be considered to be locally linear time-invariant. The two afore mentioned conditions are the case with modeling the heart motion during most normal situations.

Recursive least square is the adaptive algorithm used to update the filter weights. The adaption algorithm uses an exponential window to weight past observations so that more recent observations carry more weight. Due to the exponential windowing, the adaptive predictor is able to track the heart signal even if the statistics of the heart signal change slowly with time.

22.1.7.1 Simulation and Experimental Results

The controller described in Sect. 22.1.6 was modified to include the new prediction algorithm, and the proposed algorithm was tested on the 3-DOF robotic test bed using the same experimental procedure given in Sect. 22.1.6.8.

As can be seen from Table 22.2, in the simulation the estimator out performed the exact heart signal. This is likely due a combination of two factors. First, the simulation model is a linearized, reduced order model of the actual hardware. Second, the estimator has a robustness characteristic that makes its output less noisy than the actual heart data. The combination of these two factors yields good results in the linear case. However, when the experiment is performed on the hardware, the effects of the nonlinearities are seen when the performance of the estimator-driven controller decreases. It should be noted that though the simulation provides valuable insight to the effectiveness of the controller, it is the experimental trials that are the best indicator of performance.

Table 22.2 End-effector simulation and experimental results

	RMS position error [mm]		RMS control effort [Nmm]	
End-effector tracking results	Simulation	PHANToM	Simulation	PHANToM
Receding horizon MPC with exact reference information	0.283	0.287	14.9	48.4
Receding horizon MPC with one step adaptive filter estimation	0.258	0.499	15.4	54.5

Summary of the end-effector RMS position error and RMS control effort values forthe control algorithms used with 56-s data.

22.1.8 Generalization of the Adaptive Filtering-Based Motion PredictionMethod

In this section, we will discuss a generalized adaptive filter to further improve the prediction by using parameterized linear predictors for points throughout the prediction horizon independently [18].

The generalized method does not assume that the horizon can be generated through recursive implementation of a one-step predictor; instead, estimators for samples throughout the horizon are independently parameterized. In this way, there is no presumed linear dynamics governing the POI motion, in contrast to the recursive application of a one-step predictor which is a linear time-invariant model. A predictor that has horizon estimates related by a linear system is a special case of a model which presumes no such dependencies and therefore the former predictor is more general.

In the recursive least squares estimation scheme (details were given in Sect. 22.1.7), two assumptions were made. First, predictor is linear and therefore can be represented by matrix multiplication. Second, that estimates further in the prediction horizon can be generated by recursively applying the single step predictor. In order to have an online, adaptive method for determining predictor, the recursive least squares (RLS) algorithm was employed. The updating of predictor was done through an adaptive filter which used the collection of observations from the previous sample as input and the current observation as the desired output [23]. The estimator is able to adapt to slowly changing heart behavior since it is updated at every time step.

The linear one step predictor method approximates the heart dynamics as being a linear discrete time system and leads to sub-ideal predictions, as the POI motion has nonlinear dynamics. However the assumption of a linear system relation between consecutive time samples is abandoned in the generalized prediction method. Instead, a linear estimator for each point in the horizon is independently estimated. With this generalization, nonlinear dynamics throughout the prediction horizon can be better predicted.

22.1.8.1 Simulation and Experimental Results

The proposed estimation algorithm was tested on the same PHANToM Premium 1.5A haptic device, which is a 3-DOF robotic test-bed system. The controller from Sect. 22.1.6 was modified to include the new prediction algorithm. The same experimental procedure given in Sect. 22.1.6.8 was used to test the algorithm.

The end-effector RMS position errors in millimeters along with maximum end-effector position errors are reported in Table 22.3. The three axes mean of the RMS control efforts are also tabulated. Tracking results with the generalized adaptive filter estimation is shown in Fig. 22.7. In the figure magnitude of the end effector position error superimposed with the reference signal for the x-axis is shown.

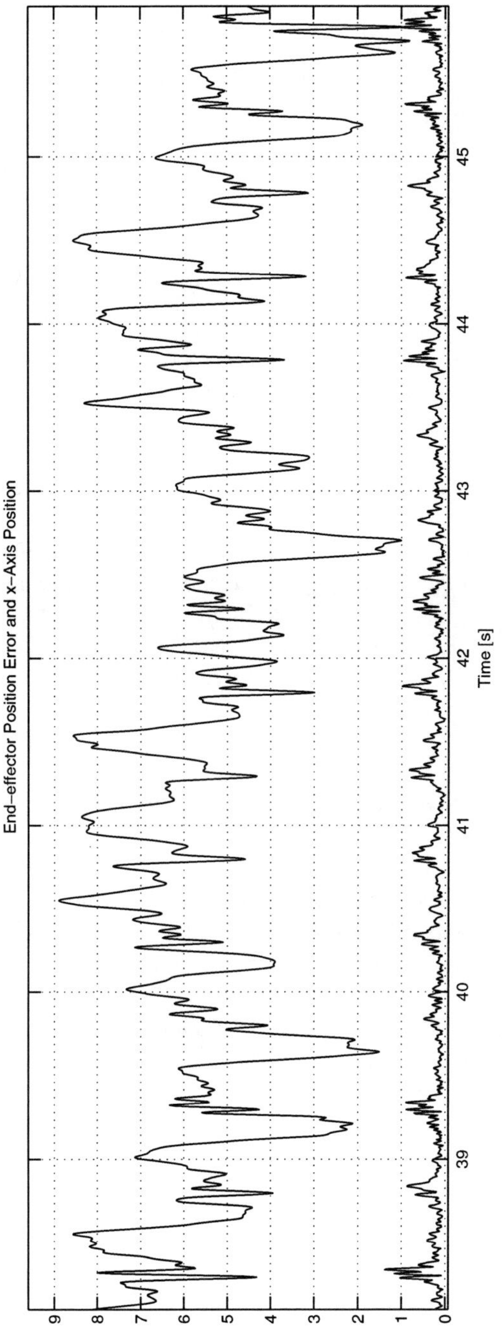

Fig. 22.7 Tracking results with Receding Horizon Model Predictive Controller with Generalized Adaptive Filter Estimation. Magnitude of the end effector error (*below*) superimposed with the reference signal for the x-axis (© IEEE 2008), reprinted with permission

Table 22.3 End-effector simulation and experimental results

	RMS (Max) position error [mm]		RMS control effort [Nmm]	
End-effector tracking results	Simulation	PHANToM	Simulation	PHANToM
Receding horizon MPC with exact reference information	0.295 (1.732)	0.312 (1.993)	14.8	45.2
Receding horizon MPC with generalized adaptive filter estimation	0.295 (1.680)	0.305 (1.813)	14.6	45.6

Summary of the end-effector RMS position error, max position error and RMScontrol effort values for the control algorithms used with 56-s heart motiondata.

As explained in Sect. 22.1.6.8, the maximum error values are affected from the noise in the data collected by Sonomicrometry sensor as it is unlikely that the POI on the heart is capable of moving 5 mm in a few milliseconds. The data has been kept as-is without applying any filtering to eliminate these jumps in the sensor measurement data as currently we do not have an independent set of sensor measurements (such as from a vision sensor) that would confirm this conjecture.

As can be seen from Table 22.3, in the simulation the estimator and the exact heart signal performed almost equally. The maximum error and control effort were slightly smaller with the estimated horizon. In the experiments, the controller with estimator outperformed the controller with exact heart signal reference. The maximum error and control effort were also slightly smaller, similar to the simulation results. However, the control effort of the new predictor was slightly larger in the hardware trials, indicating that though the tracking performance increased, it did so at a tradeoff.

Ginhoux et al. [21] used motion canceling through prediction of future heart motion using high-speed visual servoing with a model predictive controller. Their results indicated a tracking error variance on the order of 6–7 pixels (approximately 1.5–1.75 mm calculated from the 40 pixel/cm resolution reported in [21]) in each direction of a 3-DOF tracking task. Although it yielded better results than earlier studies using vision systems, the error was still very large to perform heart surgery.

Bebek et al. used the past heartbeat cycle motion data, synchronized with the ECG data, in their estimation algorithms. They achieved 0.682 mm RMS end-effector position error on a 3-DOF robotic test-bed system [6].

Franke et al. used a recursive least squares based adaptive filter algorithm for parameterizing a linear system to predict the heart motion. The predictor was used with the model predictive controller presented by Bebek et al. [6]. They reported 0.449 mm RMS end-effector position tracking results in [17].

The generalized predictor proposed in this paper represent the best results reported in the literature. These results show that the model predictive controller with the proposed generalized estimator and the exact reference data performed equally well, which indicates that the main cause of error is no longer the prediction but the performance limitations of the robot and controller. It is important to note that the results also need to be validated in vivo, which was the case in [21].

22.2 Sensing Systems

In this section of the chapter, we will present the design of a whisker-like three-dimensional7 position sensor [7]. This flexible, high precision, high bandwidth contact sensor designed to measure biological motion of soft tissue has low stiffness and prevents damage on the tissue during its contact. Two different designs, one for measuring large displacements, the other for small displacements will be described.

22.2.1 Whisker Sensor Design

The scope of this work is to create a miniature whisker sensor to measure the position of point of interest on the tissue or skin during medical interventions. Physically a whisker sensor is a long thin, and flexible extension used to detect the surrounding objects as well as their position, orientation and profiles. Design limitations include size constraints to make the tool usable in minimally invasive operations. A typical heartbeat motion is in the order of 1–2 Hz with 12 mm maximum peak displacement [6]. The resolution of the sensor needs to be in the range of 50 μm in order to track the beating heart using the control algorithm described in Sect. 22.1.6.

Two whisker sensor designs are proposed to be used in two different scenarios.

Design 1 employs a linear position sensor connected to two flexible cantilever beams that are attached orthogonally with a ridged joint. The one dimensional linear motion along the normal dimension of the tip is measured with the linear position sensor and the two dimensional lateral motion of the tip is measured with strain gauge sensors placed on the beams by separating the motion into its two orthogonal components (Fig. 22.8). These kind of beam designs are used in flexure joint mechanisms [29]. The design shown in Fig. 22.8 can be attached to the robotic manipulator base to provide continuous contact. Even though the surgical tools are not in close proximity to the heart the sensor is capable of measuring the biological motion. The operation range of the sensor is adjusted to fit the heart motion, 12 mm peak to peak max displacement [13].

Design 2 employs a cross shaped flexible structure at the back of the linear sensor, which allows the lateral motion on the tip to be measured by the strain in the legs of the cross structure (Fig. 22.9). A similar cross-shaped structure design was used by Berkelman [8] to measure force/torque values of the sensor tip. One major difference is the higher stiffness of their design, which was intended for force sensing. In the second design, a smaller linear position sensor with a spring loaded coil is used, since a smaller operation range of 5 mm in each direction is aimed. Cross shaped whisker sensor design is manufactured in relatively smaller dimensions and it is planned to be used with the system in a slightly different way as a result of its smaller size. The sensor will be attached to the surgical tool to

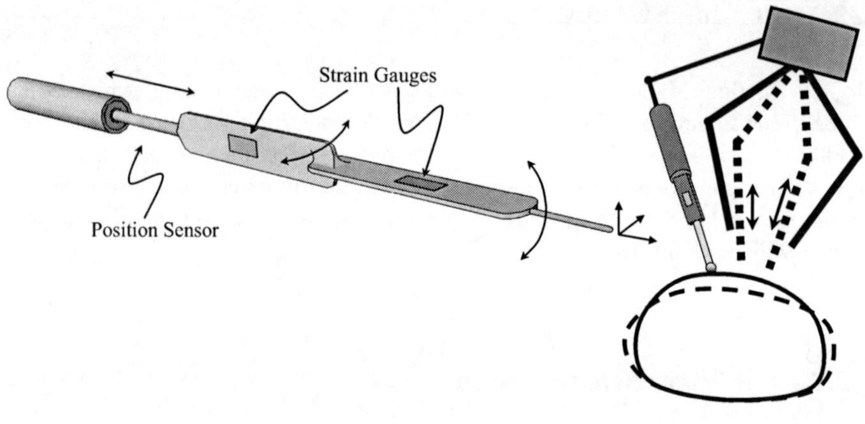

Fig. 22.8 Whisker Sensor Design 1. *Left*: One linear position sensor and two orthogonally placed flexure beams with strain gauges are used to measure the three-dimensional position of the sensor tip. *Right*: Sensor is attached to the manipulator base to provide continuous contact even when the surgical tools are not in close proximity, and to measure the heart position (© IEEE 2008), reprinted with permission

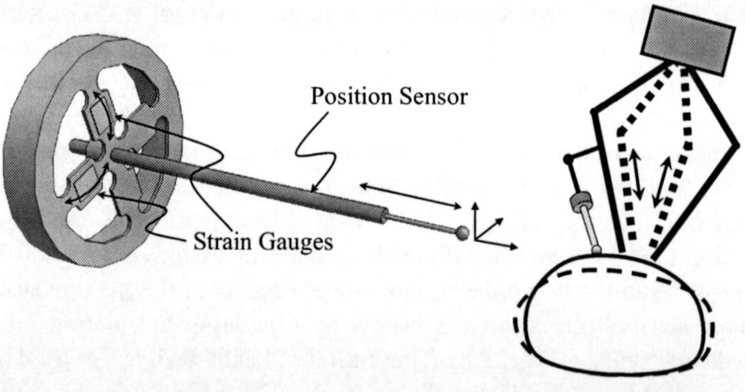

Fig. 22.9 Whisker Sensor Design 2. *Left*: One linear position sensor and a cross (×) shaped flexible structure with strain gauges are used to measure the three-dimensional position of the sensor tip. *Right*: Sensor is attached to the robot arm to measure the displacement between the heart and the surgical tools (© IEEE 2008), reprinted with permission

measure the displacement between heart and surgical tools. The spring coiled position sensor will provide continuous contact with the tissue and give measurements with respect to the whisker base. The position of the point of interest will be estimated by the combination of the robot kinematics and the sensor measurements. This will bring more dexterity to the system, since the sensor base moves with the surgical tool.

Both of the proposed whisker sensor designs use a one axis linear position sensing element (i.e., a Linear Variable Displacement Transducer) and a two axes flexure strain gauge position sensor. The reason for using linear position sensors to

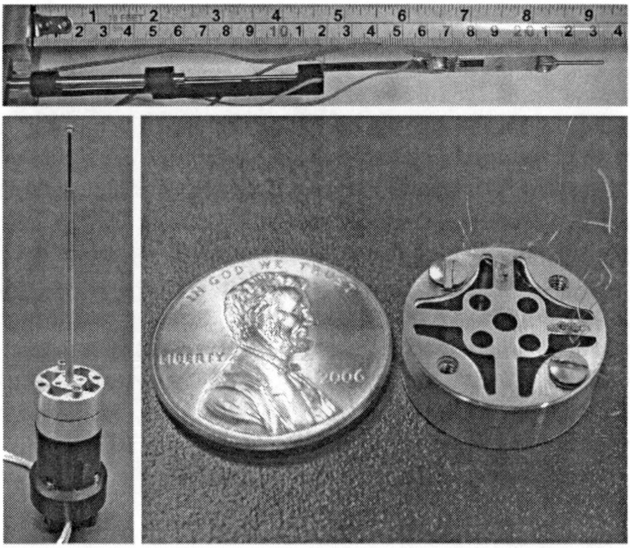

Fig. 22.10 *Top* – Whisker Sensor Design 1 prototype is shown. Overall length of the sensor when the linear stage is fully retracted and extended are 213 and 239 mm, respectively. Its largest diameter is 12.5 mm. *Bottom Left* – Whisker Sensor Design 2 is shown. Overall length of the sensor when the linear stage is uncoiled is 60.0 mm. Its largest diameter is 15.3 mm. *Bottom Right* – Flexure part of the Design 2 shown next to a cent. (© IEEE 2008), reprinted with permission

measure the motion in the normal direction of the sensor is to provide low stiffness. The positions in the lateral axes are to be measured with strain gauges attached to flexure beams. As mentioned earlier, similar geometrical designs are used in flexural joint mechanism designs [29]. Flexural joints are preferred because of the absence of friction and backlash. A drawback of the flexural elements is their limited deflection, which needs to be considered during the design.

Note that, due to the constraints of minimally invasive surgery, both of these designs will to be fitted inside a narrow cylindrical volume. The sensor design shown in Fig. 22.8 is relatively bigger in size with respect the one shown in Fig. 22.9 since the linear transducer needs to support the flexure beams holding the strain sensors. This necessity for support requires a structurally stronger therefore bigger linear sensor. However, smaller linear sensors can be used in the design shown in Fig. 22.9.

The estimated ideal resolution of the sensors can be calculated using mechanics of the flexure beams and the resolution of the DAC system. The calculated resolution of the Design 1 is $(x, y, z) = (10.1, 4.7, 5.7)\ \mu m$. The resolution difference in the x and y axes are due to the relative placement of the strain gauges to sensor tip and the length of flexures. Resolution of Design 2 is $(x, y, z) = (0.9, 0.9, 4.5)\ \mu m$, note that resolution of x and y axes are same due to symmetry of the design. This resolution estimate is valid for the ideal case and does not include the effects of noise and unmodeled nonlinear effects.

22.2.2 Whisker Sensor Prototypes

The prototypes of the designs are shown in Fig. 22.10. Design 1 can be attached to a robotic manipulator base to provide continuous contact. Even though the surgical tools are not in close proximity to the heart, this sensor is capable of measuring the biological motion. The operation range of the sensor is adjusted to fit the heart motion, which is 12 mm peak to peak max displacement. Design 2 has a smaller operation range of 5 mm in each direction. The sensor is manufactured in relatively smaller dimensions. This sensor will be attached to the surgical tool to measure the displacement between heart and surgical tools. The spring coiled position sensor will provide continuous contact with the tissue and give measurements with respect to the whisker base. This will bring more dexterity to the system. The manufactured prototypes showed that the use of proposed whisker sensors are promising and able to effectively measure dynamic motion at a bandwidth of 10 Hz.

22.3 Design of Novel Actuators

We developed a millirobotic gripper with integrated actuation for minimal invasive surgery [26, 24, 27]. The diameter and length of the tool is restricted in order to give maneuverability to the minimally invasive robots and to crate less damage to the tissue during the incision. The design is aimed to have a high gripping force and small dimensions than existing designs. In this section working details of the millirobotic gripper will be given.

22.3.1 End Effectors in Minimally Invasive Surgery

Commercially available telemanipulation systems, that are currently in use for minimally invasive surgery (MIS), are designed in such a way that the actuators are placed outside the human body to enable high gripping forces at the end effectors while maintaining small instrument sizes. Due to the present actuator technologies available for the end effector and power transmission for orienting the end effector, it is difficult to design a multi DOF wrist for orienting the end effector of a small diameter (e.g. 5 mm) robotic instrument for MIS [44].

If the robot has a multi DOF wrist then the number of DOF for the end effector can be increased, which allows the surgeon to perform the surgery with more dexterity. A conflicting objective is to reduce the diameter of the multi DOF robotic surgical instruments to allow their use in variety of applications, including minimally invasive pediatric, neonatal and fetal surgery.

If we can integrate a local actuating system with the end effector itself, this would simplify the wrist design, facilitating construction of wrists with higher DOF at smaller diameters, as we can eliminate the transmission of the mechanical power through the wrist to the end effector. Developing a device for local actuation is therefore an important enabling technology for designing new tools for MIS.

22.3.2 Design of the Hybrid Actuator

Developed novel hybrid actuator is driven by a miniature brushless DC motor and shape memory alloy (SMA) actuator. The two stages of the hybrid actuator are connected in series, and it works as a two-phase actuator. Hybrid design is used to overcome the present limitations of current MIS tools. The hybrid actuator is used to actuate a 5-mm diameter endoscopic needle driver (adapted from a manual endoscopic needle driver manufactured by Ethicon Endo-Surgery, Inc.) that will be used as an end-effector in a robotic telesurgical system for minimally-invasive cardiothoracic surgery. The built actuator, shown in Fig. 22.11, is 5.14 mm in diameter (including the casing) and 40 mm in length and is used to actuate 20 mm long needle driver assembly, while generating a force of 24 N, resulting in a gripping force of 8 N. The total stroke length of the actuator is 1 mm, which results in a 45 ° opening of the needle driver jaw with a gap of 8 mm in between the jaws.

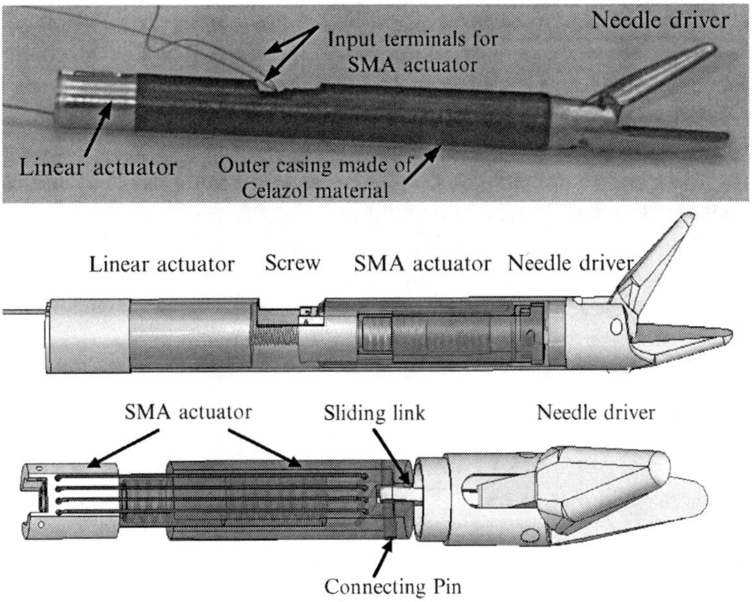

Fig. 22.11 The second generation design prototype. *Top* – Photograph of the prototype. *Middle* – CAD model of the design. *Bottom* – CAD model of the assembly of SMA actuator and the needle driver (© IEEE 2007), reprinted with permission

22.3.3 *Operation of the Hybrid Actuator*

Opening action of the gripper jaws: First the DC motor is actuated. The total time taken by the DC motor to close the gripper is 0.4 s. After the gripper closes, the SMA actuator is actuated by switching an electro mechanical relay. The SMA actuator is actuated while the gripper needs to hold the needle. The amount of gripping force applied on a needle located at approximately 7.5 mm (3/4 of the jaw length) from the gripper joint is 8 N.
Closing action of the gripper jaws: The DC motor is not actuated immediately after switching OFF the SMA actuator. It is rather actuated after a time delay of 1 s. The time delay is necessary to make sure that the SMA actuator has relaxed and is not applying a high axial force on the linear actuator. At the end, the DC motor is actuated to open the gripper jaws.

The total time required for one complete cycle excluding the duration while holding the needle is 1.8 s. This cycle time includes the time required for cooling down of the SMA. A video of the tests of the system is available for view at [25].

22.4 Conclusions and Future Work

In this final section of the chapter, we will outline the development progress of the overall system and the future technologies that will be required. We will explain the proposed system design to perform off-pump coronary artery bypass graft surgery.

This chapter provides a review of our research towards developing robotic tools for off-pump CABG surgery. In Sect. 22.1.6.3, the use of biological signals in the model-based intelligent Active Relative Motion Canceling (ARMC) algorithm to achieve better motion canceling was presented. In order to reduce the tracking error resulting from heart rate variations, ECG wave forms were detected and used to adjust heartbeat period during the tracking.

The estimation algorithms were implemented and tested on a realistic 3 degrees-of-freedom robotic test-bed system, with root-mean-square tracking errors (305 μm) more than 5.5 times better than the best results reported in the literature.

Experimental results showed that using ECG signal in ARMC algorithm improved the reference signal estimation. It is important to note that, for patients with severe rhythm abnormalities, the detection of the ECG waveforms present a challenge for the proposed method. Biological signals other than ECG that can be used to assist the tracking of heart motion include aortic, atrial and ventricular blood pressures. Similar to the ECG signal, these blood pressures are significant indicators of the heart motion as they can be used to predict when the heart valves will be opening and closing, which in turn helps us determine the distinct phases of the heart cycle. The blood pressure signals also give additional independent information, which can be used in conjunction with ECG signal to improve noise robustness and to reliably detect unexpected rhythm abnormalities and arrhythmias, which will be a challenging part for the realization of the ARMC algorithm.

In this study we used position data from a sonomicrometer. An important part of this robotic system is the development of sensing systems that will be appropriate to use in a tight control loop for active tracking of the heart. These sensing components will track the heart motion, monitor biological signals, and provide force feedback. Multi sensor fusion with complementary and redundant sensors will be used for superior performance and safety. The whisker sensor introduced here is a high sensitivity contact sensor. The advantage of whisker sensor is that it will directly give the relative motion of the heart with respect to the robotic manipulator. Merging the sensor data from multiple position sources would increase the accuracy of motion detection and improve tracking results. Complementary sensors that are planned to be used to collect heart motion data include sonomicrometric sensors, whisker sensors, a multi-camera vision system and inertial sensors. Also in addition to whisker sensors, adding more mechanical sensors that measure heart motion would improve the measurement precision.

In Sect. 22.3, we presented a novel millirobotic gripper with integrated actuation. Designing and prototyping of milli- and micro-scale dexterous robotic instruments are necessary to perform minimally invasive coronary artery surgery (also called limited access coronary artery surgery). Development of these capable tools would enhance the quality of the operations by shortening the operation times and reduce the restrictions faced during the robotic-assisted interventions.

Our future work includes the development of intelligent control algorithms for model based ARMC using additional biological signals in estimation of heart motion. Biological signals are significant indicators of the heart motion and they give additional independent information of the heart behavior. Therefore, they can be used to improve noise robustness. We will also design and build prototypes of milli- and micro-scale dexterous robotic instruments and sensors to be used in the slave manipulator.

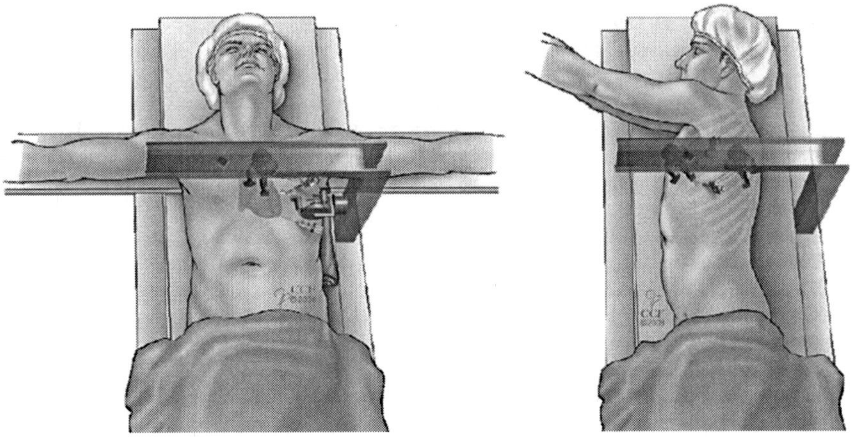

Fig. 22.12 Illustration of robotic assisted off-pump coronary artery bypass graft surgery. *Left*: Antero-lateral approach. Patient positioned in supine position. *Right*: Postero-lateral surgical approach. Patient placed in a full left posterolateral thoracotomy position

Proposed robotic platform for off-pump CABG will have table top mount 6-DOF robotic arms at the patient side for interventions (see Fig. 22.12). The slave manipulators will have two main stages: A 3-DOF, high bandwidth, lightweight, small footprint macro stage that will accomplish the active relative motion cancellation during off-pump CABG surgery; and a micro stage with 3 DOF wrist with integrated micro-sensing and actuation. This robotic system will integrate the already developed millirobotic gripper and sensing systems. With the integrated actuation, the system will allow the surgeon to perform the surgery with more dexterity.

Acknowledgements The authors would like to thank Dr. Mark Ratcliffe for his help during collection of the heart data; Dr. Hung-I Kuo and Engin Pehlivanoğlu for their help during the bonding of the strain gauges to the whisker sensor flexure beams.

This work was supported in part by National Science Foundation under grants CISE IIS-0222743, EIA-0329811, and CNS-0423253, US DoC under grant TOP-39-60-04003, and Case Western Reserve University with a Support of Undergraduate Research and Creative Endeavors (SOURCE) award.

References

1. Deaths: Preliminary data for 2004 (2006). URL http://www.cdc.gov/nchs/fastats/heart.htm
2. American Heart Association (2007). URL http://www.americanheart.org
3. Anderson, B.D., Moore, J.B.: Optimal Control Linear Quadratic Methods. Prentice Hall, Englewood Cliffs, NJ (1990)
4. Bahoura, M., Hassani, M., Hubin, M.: DSP implementation of wavelet transform for real wave forms detection and heart rate analysis. Comput. Methods Programs Biomed. **52**, 35–44 (1997)
5. Özkan Bebek: Robotic-assisted beating heart surgery. Ph.D. Dissertation, Case Western Reserve University, Cleveland, OH, USA (2008). URL http://rave.ohiolink.edu/etdc/view?acc_num=case1201289393
6. Bebek, O., Cavusoglu, M.C.: Intelligent control algorithms for robotic assisted beating heart surgery. IEEE Trans. Robot. **23**(3), 468–480 (2007)
7. Bebek, O., Cavusoglu, M.C.: Whisker-like position sensor for measuring physiological motion. IEEE/ASME Trans. Mech. **13**(5), 538–547 (2008)
8. Berkelman, P.J., Whitcomb, L.L., Taylor, R.H., Jensen, P.: A miniature microsurgical instrument tip force sensor for enhanced force feedback during robot-assisted manipulation. IEEE Trans. Robot. Automat. **19**(5), 917–922 (2003)
9. Berne, R.M., Levy, M.N.: Principles of Physiology, 3rd edn. Mosby Inc, St. Louis (2000)
10. Berne, R.M., Levy, M.N.: Cardiovascular Physiology, 8th edn. Mosby Inc, St. Louis (2001)
11. Camacho, E., Bordons, C.: Model Predictive Control. Springer, Berlin (1999)
12. Cavusoglu, M.C., Feygin, D., Tendik, F.: A critical study of the mechanical and electrical properties of the PHANToM haptic interface and improvement for high performance control. Presence **11**(6), 555–568 (2002)
13. Cavusoglu, M.C., Rotella, J., Newman, W.S., Choi, S., Ustin, J., Sastry, S.S.: Control algorithms for active relative motion cancelling for robotic assisted off-pump coronary artery bypass graft surgery. In: Proc. of the 12th International Conference on Advanced Robotics (ICAR), pp. 431–436. Seattle, WA, USA (2005)
14. Cuvillon, L., Gangloff, J., DeMathelin, M., Forgione, A.: Toward robotized beating heart TECABG: Assessment of the heart dynamics using high-speed vision. In: Proc. of Medical

Image Computing and Computer-Assisted Intervention (MICCAI), vol. 2, pp. 551–558. Palm Springs, USA (2005)
15. Duindam, V., Sastry, S.: Geometric motion estimation and control for robotic-assisted beating-heart surgery. In: Proc. IEEE/RSJ Int. Conf. Intell. Robot. Sys. (IROS), pp. 871–876. San Diego, CA, USA (2007)
16. Falk, V.: Manual control and tracking-a human factor analysis relevant for beating heart surgery. Ann. Thorac. Surg. **74**, 624–628 (2002)
17. Franke, T., Bebek, O., Cavusoglu, M.C.: Improved prediction of heart motion using an adaptive filter for robot assisted beating heart surgery. In: Proc. IEEE/RSJ Int. Conf. Intell. Robot. Sys. (IROS), pp. 509–515. San Diego, CA, USA (2007)
18. Franke, T., Bebek, O., Cavusoglu, M.C.: Prediction of heartbeat motion with a generalized adaptive filter. In: Proc. Int. Conf. Robot. Autom. (ICRA), pp. 2916–2921. Pasadena, CA, USA (2008)
19. Garcia, C.E., Prett, D.M., Morari, M.: Model predictive control: Theory and practice-a survey. Automatica **25**(3), 335–348 (1989)
20. Gilhuly, T.J., Salcudean, S.E., Lichtenstein, S.V.: Evaluating optical stabilization of the beating heart. IEEE Eng. Med. Biol. Mag. **22**(4), 133–140 (2003)
21. Ginhoux, R., Gangloff, J.A., DeMathelin, M.F., Soler, L., Leroy, J., Sanchez, M.M.A., Marescaux, J.: Active filtering of physiological motion in robotized surgery using predictive control. IEEE Trans. Robot. **21**(1), 67–79 (2005)
22. Groeger, M., Ortmaier, T., Sepp, W., Hirzinger, G.: Tracking local motion on the beating heart. In: Proc. of the SPIE Medical Imaging Conference, vol. 4681 of SPIE, pp. 233–241. San Diego, CA, USA (2002)
23. Haykin, S.: Adaptive Filter Theory, 4th edn. Prentice Hall, New Jersey (2001)
24. Kode, V.R.C.: Design and characterization of a novel hybrid actuator using shape memory alloy and D.C. motor for minimally invasive surgery applications. M.S. thesis, Case Western Reserve University, Cleveland, OH, USA (2006)
25. Kode, V.R.C., Cavusoglu, M.C.: Millirobotic tools for minimal invasive surgery: Video of hybrid gripper tests. Video (2006). URL http://robotics.case.edu/research.html#medicalrobotics
26. Kode, V.R.C., Cavusoglu, M.C.: Design and characterization of a novel hybrid actuator using shape memory alloy and dc micro-motor for minimally invasive surgery applications. IEEE/ASME Trans. Mech. **12**(4), 455–464 (2007)
27. Kode, V.R.C., Cavusoglu, M.C., Tabib-Azar, M.: Design and characterization of a novel hybrid actuator using shape memory alloy and DC motor for minimally invasive surgery applications. In: In Proc. IEEE Int. Conf. Mech. Autom. (ICMA), pp. 416–420. Niagara Falls, Ontario, Canada (2005)
28. Koransky, M.L., Tavana, M.L., Yamaguchi, A., Robbins, R.: Quantification of mechanical stabilization for the performance of offpump coronary artery surgery. In: Proc. of the Meeting of the International Society for Minimally Invasive Cardiac Surgery (ISMICS). Munich, Germany (2001). (Abstract)
29. Koster, M.P.: Flexural joints in mechanisms. In: Proceedings of the ASME Dynamic Systems and Control Division, vol. 2, pp. 855–859 (2000)
30. Madhani, A.J., Niemeyer, G., Salisbury, J.K.: The black falcon: a teleoperated surgical instrument for minimally invasive surgery. In: Proc. IEEE/RSJ Int. Conf. Intell. Robot. Sys. (IROS), vol. 2, pp. 936–944. Victoria, BC, Canada (1998)
31. Malmivuo, J., Plonsey, R.: Bioelectromagnetism - Principles and Applications of Bioelectric and Biomagnetic Fields. Oxford University Press, New York (1995)
32. Mark B. Ratcliffe, M.: Personal Communication Chief of Surgery, San Francisco VA Medical Center, Professor in Residence
33. Nakamura, Y., Kishi, K., Kawakami, H.: Heartbeat synchronization for robotic cardiac surgery. In: Proc. Int. Conf. Robot. Autom. (ICRA), vol. 2, pp. 2014–2019. Seoul, Korea (2001)

34. Newman, M.F., Kirchner, J.L., Phillips-Bute, B., Gaver, V., Grocott, H., Jones, R.H., et al.: Longitudinal assessment of neurocognitive function after coronary-artery bypass surgery. N. Engl. J. Med. **344**(6), 395–402 (2001)
35. Ortmaier, T.: Motion compensation in minimally invasive robotic surgery. Doctoral thesis, Technical University of Munich, Germany (2003)
36. Ortmaier, T., Groeger, M., Boehm, D.H., Falk, V., Hirzinger, G.: Motion estimation in beating heart surgery. IEEE Trans. Biomed. Eng. **52**(10), 1729–1740 (2005)
37. PhysioNet: MIT-BIH Arrhythmia Database (1980). URL http://www.physionet.org/physiobank/database/mitdb/
38. Puskas, J.D., Wright, C.E., Ronson, R.S., Brown, W.M., Gott, J.P., Guyton, R.A.: Off-pump multi-vessel coronary bypass via sternotomy is safe and effective. Ann. Thorac. Surg. **66**(3), 1068–1072 (1998)
39. Riviere, C., Thakral, A., Iordachita, I.I., Mitroi, G., Stoianovici, D.: Predicting respiratory motion for active canceling during percutaneous needle insertion. In: Proc. Int. Conf. IEEE EMBS (EMBC), pp. 3477–3480. Istanbul, Turkey (2001)
40. Rotella, J.: Predictive tracking of quasi periodic signals for active relative motion cancellation in robotic assisted coronary artery bypass graft surgery. M.S. thesis, Case Western Reserve University, Cleveland, OH, USA (2004)
41. Schweikard, A., Glosser, G., Bodduluri, M., Murphy, M., Adler, J.: Robotic motion compensation for respiratory movement during radiosurgery. Comput. Aided Surg. **5**(4), 263–77 (2000)
42. Sharma, K., Newman, W.S., Weinhous, M., Glosser, G., Macklis, R.: Experimental evaluation of a robotic image-directed radiation therapy system. In: Proc. Int. Conf. Robot. Autom. (ICRA), vol. 3, pp. 2913–2918. San Francisco, CA, USA (2000)
43. Shechter, G., Resar, J.R., McVeigh, E.R.: Displacement and velocity of the coronary arteries: Cardiac and respiratory motion. IEEE Trans. Med. Imag. **25**(3), 369–375 (2006)
44. Taylor, R.H., Stoianovici, D.: Medical robotics in computer integrated surgery. IEEE Trans. Robot. Automat. **19**(5), 765–781 (2003)
45. Thakral, A., Wallace, J., Tomlin, D., Seth, N., Thakor, N.V.: Surgical motion adaptive robotic technology (S.M.A.R.T.): Taking the motion out of physiological motion. In: Proc. of 4th International Conference on Medical Image Computing and Computer-Assisted Intervention (MICCAI), pp. 317–325. Utrecht, The Netherlands (2001)
46. Trejos, A.L., Salcudean, S.E., Sassani, F., Lichtenstein, S.: On the feasibility of a moving support for surgery on the beating heart. In: Proc. of Medical Image Computing and Computer-Assisted Interventions (MICCAI), pp. 1088–1097. Cambridge, UK (1999)

Chapter 23
Robotic Needle Steering: Design, Modeling, Planning, and Image Guidance

Noah J. Cowan, Ken Goldberg, Gregory S. Chirikjian, Gabor Fichtinger, Ron Alterovitz, Kyle B. Reed, Vinutha Kallem, Wooram Park, Sarthak Misra, and Allison M. Okamura

Abstract This chapter describes how advances in needle design, modeling, planning, and image guidance make it possible to steer flexible needles from outside the body to reach specified anatomical targets not accessible using traditional needle insertion methods. Steering can be achieved using a variety of mechanisms, including tip-based steering, lateral manipulation, and applying forces to the tissue as the needle is inserted. Models of these steering mechanisms can predict needle trajectory based on steering commands, motivating new preoperative path planning algorithms. These planning algorithms can be integrated with emerging needle imaging technology to achieve intraoperative closed-loop guidance and control of steerable needles.

23.1 Introduction

From biopsies to brachytherapy, needle-based interventions already comprise a substantial fraction of minimally invasive medical procedures. The small diameter of a needle enables it to access subsurface targets while inflicting minimal tissue damage and, once in place, the needle's lumen provides a conduit through which to deliver a wide variety of therapies, such as drugs, radioactive seeds, and thermal ablation. In addition to therapeutic delivery, needles are also commonly used for diagnostic procedures, such as biopsy. As biosensors, manipulators, ablation tools, and other "end-effector" technologies continue to get smaller, applications for needle-based interventions will also expand. This chapter reviews the state-of-the-art in steerable needle technologies, including device design, modeling, path planning, and image-guided control.

N.J. Cowan (✉)
Department of Mechanical Engineering, Johns Hopkins University,
Baltimore, MD 21218, USA
e-mail: ncowan@jhu.edu

J. Rosen et al. (eds.), *Surgical Robotics: Systems Applications and Visions*,
DOI 10.1007/978-1-4419-1126-1_23,

Targeting accuracy is crucial for needle-based procedures. For example, poor placement during biopses leads to false negatives. Inaccurate seed placement during brachytherapy destroys healthy instead of cancerous tissue, sometimes with catastrophic outcomes [12]. Robotic needle placement under image guidance promises to improve substantially targeting accuracy – and therefore clinical outcomes – of such procedures. Toward this end, exciting progress has been made engineering needle-placement robots for prostate biopsy and brachytherapy under a variety of imaging modalities, including ultrasound [26], magnetic resonance imaging [38, 64], and multi-imaging scenarios [47]. These robots represent a substantial advance for procedures that require multiple insertions, for example in thermal tumor ablation, because dosimetry and target planning can be updated from one insertion to the next based on intraoperative images. These general image-guided needle aiming systems work in an iterative fashion in which intraoperative imaging is used between insertions to update a plan of subsequent insertions (for example to optimize dosimetry), leaving the physician in the loop to adjust the plan and/or control the invasive (insertion) degree of freedom under image feedback.

These image-guided robotic systems are clinically viable and promise to substantially enhance targeting accuracy in needle-based interventions. However, to date these systems require minimal tissue and needle deformation, and substantial effort is committed to preventing such deformation [47] because unmodeled deflections of the needle or tissue during insertion, if not compensated, will lead to gross targeting inaccuracy. Recently, needle steering researchers have begun taking the next critical step of harnessing and amplifying such deformations as mechanisms for steering a needle to a subsurface target; in this chapter we specifically focus on these recent efforts to steer needles under image feedback once they are inside the tissue using a wide variety of mechanisms, all of which involve deflecting the needle, tissue, or both as depicted in Fig. 23.1.

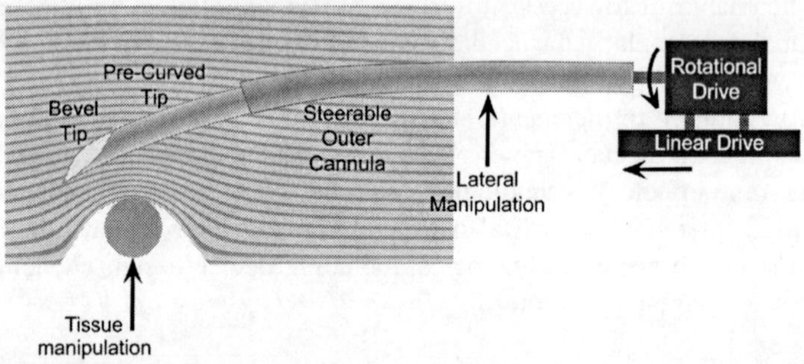

Fig. 23.1 This chapter focuses on subsurface needle steering, wherein a computer-integrated system can actively modify the trajectory through some combination of steering mechanisms. A needle can be steered to a target using several different methods: generating forces at the needle tip using an asymmetric tip [60, 70, 71], lateral manipulation [28], and pushing on the tissue to move the target into the needle's path [40]. A steerable cannula can be used to provide dexterity prior to (and possibly during) insertion (cf. [62] and references therein)

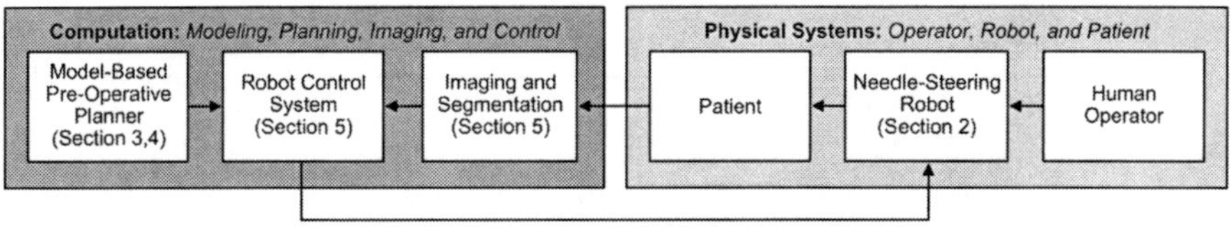

Fig. 23.2 A successful robotically controlled needle-steering system must be comprised of a combination of computational algorithms and physical systems

This chapter describes needle steering approaches in which needles are manipulated from outside the tissue in order to change the path of the needle tip inside tissue. Alternatively, active elements could be invoked to bend the needle once inside tissue, but to our knowledge this approach has not been extensively studied from a computer-integrated surgery perspective. The advantage of passive needle steering approaches is that all the electromechanical mechanisms remain outside the patient, enabling the use of thinner needles, larger actuators, and a clearer path to clinical application.

Figure 23.2 shows the various computational and physical systems needed to achieve robot-assisted needle steering, and provides a graphical outline for this chapter. Section 23.2 provides a taxonomy of needle-steering mechanisms and robots, and Sect. 23.3 reviews the models (both phenomenological and mechanics-based) that describe these steering mechanisms. Sections 23.4 and 23.5 describe a rich variety of robotic planning, imaging, and control literature that has emerged as a consequence of these new technologies. Finally, concluding remarks are provided in Sect. 23.6.

23.2 Steering Approaches and Devices

This section reviews several methods for steering needles inside tissue (Fig. 23.1), and describes example robotic devices that have been used to achieve needle steering (Fig. 23.3). Ultimately, a combination of the needle steering approaches described here – needle flexibility, bevel asymmetry and shape, pre-bent elements, tissue manipulation, and needle base actuation – will likely lead to systems with superior steering capability over any one method alone.

23.2.1 Tip-Steerable Needles

Conventional needles used in percutaneous therapy and biopsy can be classified as symmetric (e.g. conical or triangular prismatic) or asymmetric (e.g. beveled), as shown in Fig. 23.4. It has been shown that inserting needles with asymmetric tips results in larger lateral (bending) forces than needles with symmetric tips [50]. These lateral bending forces result in deviation of the needle from a straight line path, even if the tissue does not deform. Physicians often spin asymmetric-tipped

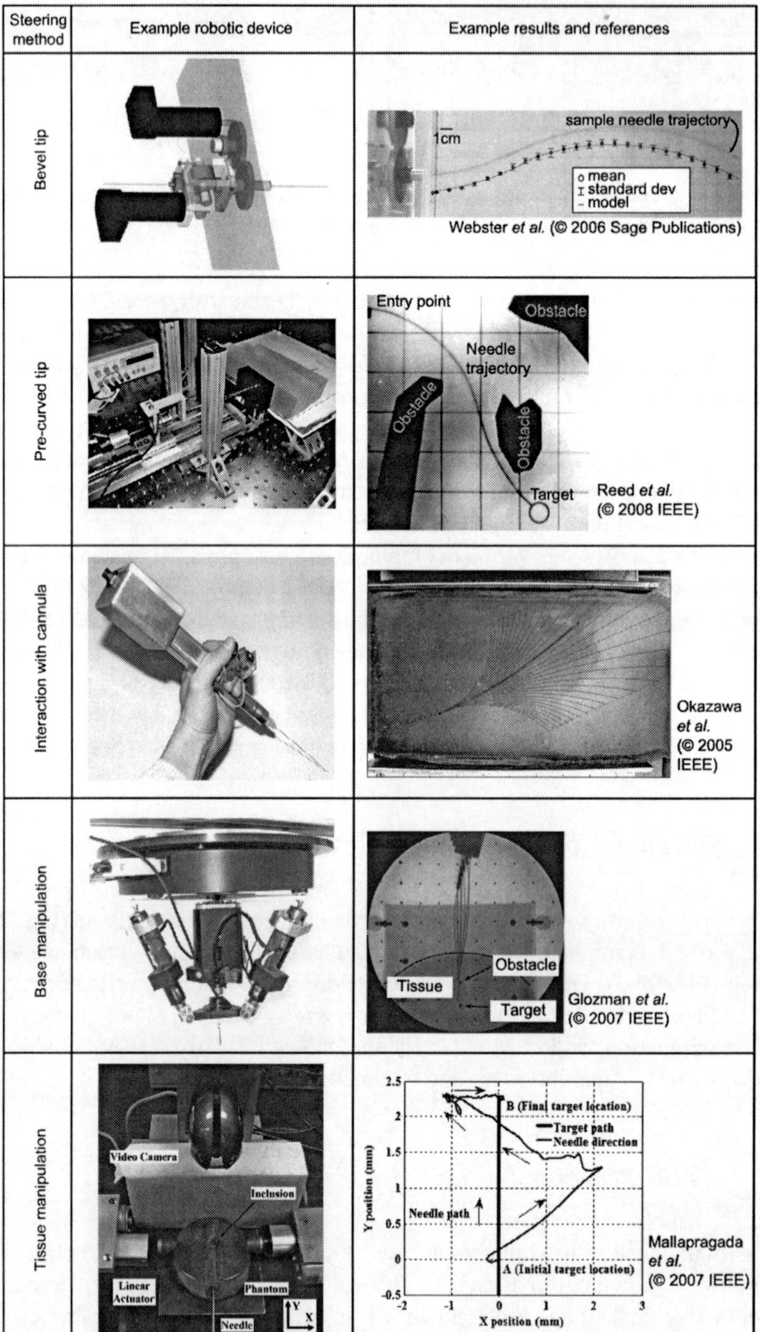

Fig. 23.3 Steering methods, example robotic devices, and example results from needle steering systems in the literature, including Webster et al. [70], Reed et al. [60], Okazawa et al. [52], Glozman et al. [28], and Mallapragada et al. [40]. All figures reprinted with permission

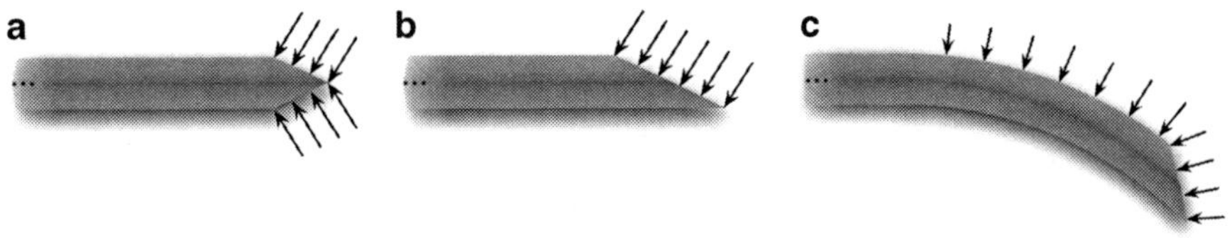

Fig. 23.4 Needle tips: (**a**) a symmetric conical tip, (**b**) an asymmetric bevel tip, (**c**) an asymmetric pre-bent/curved tip. Tip-based steering relies on an asymmetric design such as (**b**) or (**c**)

needles by hand in order to reduce needle bending during insertion, and engineers have developed devices to enhance this effect by "drilling" the needle to reduce friction and cutting forces [74]. The use of symmetric-tip needles or drilling of asymmetric-tip needles does not guarantee that a target can be reached. In both cases, needles can deviate slightly from a straight-line path due to tissue deformation or inhomogeneity, with no way to correct for this error after insertion. Also, these methods assume that there exists a straight-line path between the insertion point of the needle and the target.

In contrast, some needle steering techniques intentionally use the asymmetry of the needle tip to cause needle bending inside tissue. This can be used to enhance targeting accuracy by redirecting the path of the needle when it deviates from a desired trajectory. In addition, needle steering can allow a needle to go around obstacles or sensitive tissues to acquire targets that are inaccessible by straight-line paths. Physicians who perform targeted needle insertion currently use a number of ad-hoc methods to approximate steering, such as rotating the bevel tip of a needle, causing it to deflect slightly as inserted, or externally manipulating the tissue to guide the needle in a desired direction. However, without computer assistance, these manual needle steering techniques require the physician to have excellent 3D spatial reasoning, extensive experience, and precise coordination with high-resolution real-time image feedback.

The simplest type of asymmetric tip is a bevel tip. Bevel-tip needles are commonplace because they are straightforward to manufacture and they can be used to (slightly) direct the flow of therapeutic drugs. Bevel-tip needle steering arises from a combination of needle insertion, which causes the needle naturally to follow a curved path due to asymmetric tip forces (Fig. 23.1), and spinning the needle about its axis, which changes the direction of subsequent bending [69]. The needle spin speed can be "duty cycled" to vary the curvature of the needle path [41], although the maximum curvature is always limited by the combined mechanical properties of the needle and tissue. In addition, "airfoil" tips can be added to increase the area of a bevel tip and increase the curvature of the needle path [25]. It is important to note that needles steered in this fashion can only steer when cutting a new path. When the tissue does not deform, the entire needle will follow the tip path [70]. When a needle is removed (by simply pulling on the needle base), it follows the same path as insertion but in the opposite direction. The bevel-tip needle steering method is most effective when the needle is highly flexible (structurally having low stiffness) compared to the medium in which it is being steered. Thus, the superelastic (and biocompatible) material Nitinol has been used in some bevel-tip needle steering studies. Models for bevel-tip needle steering are discussed in Sect. 23.3.

In order to insert needles for bevel-tip steering, specialized devices are required. Automated flexible needle insertion is challenging because needles tend to buckle if not supported outside the tissue. Humans are not able to insert a needle with a precise velocity, and they may inadvertently apply lateral forces or torque about the needle axis. Webster et al. [70] developed two different robotic devices for steering needles using tip asymmetry. Each device is able to control insertion velocity and the rotation (spin about the needle axis) velocity. The first device is based on a friction drive concept, which has advantages of compactness and simplicity. However, major drawbacks to this design include slippage in the insertion degree of freedom (DOF), a slight spin of the needle during insertion due to imperfect alignment of the friction drive, and difficulty in measuring insertion force and spin torque. The second device involves driving the needle from its base (the distal end) while using a telescoping support sheath to prevent the needle from buckling. A needle rotation module is attached to the translational stage to spin the needle and enable steering. Although this device is larger than the first, it provides more control over needle insertion parameters, and also enables straightforward integration of force/torque sensing, making it ideal for laboratory experiments.

A needle with a curve or pre-bend near the tip achieves a smaller radius of curvature than a bevel tip alone [63, 59, 72], but can be controlled much like a bevel-tip needle [59]. The smaller radius comes from the larger asymmetry at the tip of a pre-bent needle, which creates a larger force perpendicular to the insertion direction during an insertion. Several studies have demonstrated that the radius of curvature of pre-bent [63] and curved [72] needles varies with the length and angle of the asymmetry. For long pre-curved needles, the radius of curvature approaches the radius of curvature of the needle at the tip [72]. Although using pre-bent needles allows greater dexterity, a pre-bent needle might detrimentally affect a medical procedure; for example, a pre-bent needle tip can potentially cut tissue when the needle base is rotated while not simultaneously being inserted, placing constraints on planning and control algorithms.

The curvature of a needle as it is inserted into tissue can also be modulated by changing the curvature of the needle tip. One method uses small wires inside the needle to pull the tip in the desired direction. Another method varies the tip curvature by placing a curved needle inside a stiff straight outer cannula [51]. Extending the needle so the curved section protrudes from the cannula provides an asymmetric surface that causes the needle-cannula system to bend during insertion. The amount of needle protrusion can be controlled directly and dictates the radius of curvature. For example, if the needle is entirely inside the cannula, the needle will travel in a roughly straight line. Once the cannula tip is in position, the needle can be withdrawn completely, allowing the lumen of the cannula to be used for a medical procedure. This method requires control of three DOFs: the insertion distance of both needle and cannula, and the rotation of the inner needle.

A generalization of the concentric cannula-needle system is an "active cannula" or "concentric tube" robot [71, 62, 61, 23], in which any number of concentric flexible tubes can interact with each other to change the three-dimensional (3D) shape of the device. Rotating and inserting/retracting each of the individual tubes allows control of the device tip within a large set of configurations. These concentric-tube devices do not depend directly on needle-tissue interaction, but can be used as steerable needles.

23.2.2 Lateral Manipulation

An alternative method of steering the needle involves moving the base of the needle perpendicular to the insertion axis [18, 27]. The perpendicular motions cause the entire needle shaft to move inside the tissue where the needle acts, much like a beam resting on a compliant fulcrum. Once the needle is inserted sufficiently far inside the tissue, motion of the needle base orthogonal to needle shaft direction causes the tip to move in roughly the opposite direction. However, there is substantial path dependence, making it challenging to develop closed-form models (Sect. 23.3.4).

Maneuvering a needle using lateral manipulation may require Cartesian motions and rotations. The only DOF not required is the rotation of the needle around the insertion axis, which is one of the two required inputs to control a tip-steered needle, so lateral manipulation may allow added maneuverability to a tip-steered needle.

Lateral manipulation can achieve large changes in the needle path near the surface, but the effect decreases as the needle is inserted further into the tissue. The needle must transmit all the force from the base to the tip and, as the needle is inserted further, more tissue can resist the force and the moment arm increases. To generate the same change in path throughout the insertion, the force at the base must increase, but the tissue can only withstand so much force before tearing. Tip-steered needles, however, are approximately depth independent, since the dominant steering force is generated at the tip of the needle. Lateral manipulation and tip-steered needles can be used together for additional control over the needle throughout the entire insertion.

23.2.3 Tissue Manipulation

In addition to manipulating the needle in order to acquire targets in soft tissue, it is also possible to manipulate the *tissue* in order to move targets into the path of the needle or push obstacles and sensitive tissues out of the path of the needle. Physicians already perform such tissue manipulation by hand, and recent work has provided insight regarding robotic control to achieve the same effects. Robotic tissue manipulation systems could improve both the accuracy of target acquisition and the accessibility of targets, and be combined with the other needle steering approaches described above.

Mallapragada et al. [39] developed a method for real-time tumor manipulation, in which a robotic controller takes as input real-time medical images of a tumor and outputs an appropriate external force to move the tumor to a desired position. During needle insertion (in an approximately straight line path), blunt robotic end-effectors push on the tissue to move the tumor onto the needle path (Fig. 23.3). In simulations, Torabi et al. [66] considered a more complex tissue manipulation problem, in which robots are used to both move obstacles out of the way of the path of the needle and the target onto the path of the tissue. A two-dimensional mass-spring simulation demonstrated the effectiveness of the planner/controller combination in reducing targeting errors and shifting obstacles.

23.3 Modeling

The design of needle steering planners and most types of controllers requires a model of needle-tissue interaction that predicts needle or needle-tip motions given inputs at the needle base. This section describes several phenomenological models that capture needle-tissue behavior sufficient to inform planning and control design, as well as ongoing efforts to create more accurate mechanics-based models.

23.3.1 Nonholonomic Steering

A bevel-tip needle inserted into homogenous tissue will follow a stereotyped path. Webster et al. [69] demonstrated that the kinematics of a bevel tip needle can be modeled as a non-holonomic system with a constant steering constraint. According to this model, the needle tip advances forward in a curved path, but cannot translate when embedded in tissue. The kinematic model is similar to the motion of a unicycle or bicycle with the handlebars locked in one position. The wheels of a bicycle cannot instantaneously move sideways, yet the bicycle can attain any desired pose in the plane through a more complex sequence of motions. Whereas bicycle steering occurs in plane, needle steering occurs in 3D space.

Webster et al. performed experiments and statistical analysis verifying that the-nonholonomic model fits a limited battery of insertions and found that the two-parameter bicycle model described the needle behavior better than a single-parameter unicycle model, although the unicycle model's simplicity and reasonable accuracy has made it a good choice for control systems design [37, 35, 36, 34]. Many of the models, planning algorithms, and control systems described throughout the remainder of this chapter build upon these nonholonomic models of needle motion.

The kinematic model can be mathematically expressed as follows. Attach a reference frame to the needle tip with the local z-axis denoting the tangent to the needle shaft and x-axis denoting the axis orthogonal to the direction of infinitesimal motion induced by the bevel (i.e. the needle bends in the instantaneous y-z plane). The nonholonomic kinematic model for the evolution of the frame at the needle tip was developed based on a unicycle model in [53, 69] as

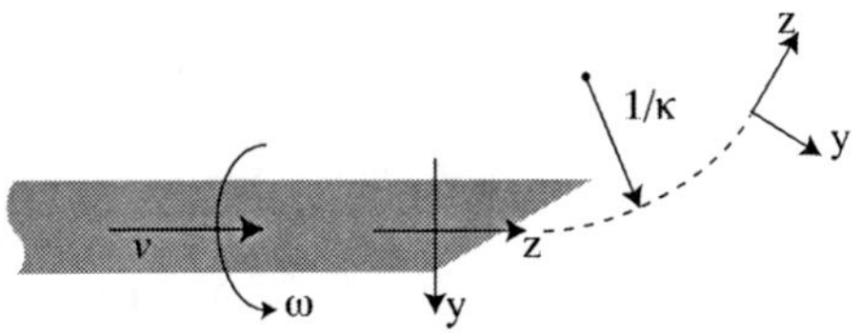

Fig. 23.5 The definition of parameters and frames in the nonholonomic needle model [70, 54] (Reprinted with permission from [56], © 2010 Sage Publications)

$$\xi(t) = (g^{-1}(t)\dot{g}(t))^{\vee} = \left[\kappa v(t)\ \ 0\ \ \omega(t)\ \ 0\ \ 0\ \ v(t)\right]^T, \tag{23.1}$$

where $g(t)$ is the element of the Euclidean motion group, SE(3) and ξ is the element of see (23.3), which is the Lie algebra associated with SE(3). Here, $g(t)$ is the 6-DOF pose of the frame attached to the needle tip in 3D space and $\xi(t) \in \mathbf{R}^6$ in denotes the 6D translational and rotational velocity of the frame. The control inputs, $\omega(t)$ and $v(t)$, are the rotation and insertion velocities, respectively, and κ is the curvature of the needle curve. The frames and parameters for the needle are shown in Fig. 23.5.

23.3.2 Stochastic Modeling

Although the kinematic model for needle steering describes the motion of the needle, there is inherently variation between insertions. If everything were certain, and if this model were exact, the motion, $g(t)$, could be obtained by simply integrating the ordinary differential equation in (23.1). However, a needle that is repeatedly inserted into a medium, such as a gelatin used to simulate tissue [69], will demonstrate an ensemble of slightly different trajectories.

A simple stochastic model [53, 54] is obtained by adding noise to the two input parameters in the ideal model:

$$\omega(t) = \omega_0(t) + \lambda_1 w_1(t) \quad \text{and} \quad v(t) = v_0(t) + \lambda_2 w_2(t),$$

where $\omega_0(t)$ and $v_0(t)$ are what the inputs would be in the ideal case, $w_1(t)$ and $w_2(t)$ are uncorrelated unit Gaussian white noises, and λ_1 and λ_2 are constants. Thus, the nonholonomic needle model with noise is

$$(g^{-1}(t)\dot{g}(t))^{\vee} dt = [\kappa v_0(t)\quad 0\quad \omega_0(t)\quad 0\quad 0\quad v_0(t)]^T dt + \begin{bmatrix} 0 & 0 & \lambda_1 & 0 & 0 & 0 \\ \kappa\lambda_2 & 0 & 0 & 0 & 0 & \lambda_2 \end{bmatrix}^T \begin{bmatrix} dW_1 \\ dW_2 \end{bmatrix},$$

where $dW_i = W_i(t + dt) - W_i(t) = w_i(t)dt$ are the non-differentiable increments of a Wiener process $W_i(t)$. This noise model is a stochastic differential equation (SDE) on SE(3). As shorthand, we write this as

$$(g^{-1}(t)\dot{g}(t))^{\vee} dt = \mathbf{h}(t)dt + H d\mathbf{W}(t).$$

23.3.3 Torsional Modeling

In order to change the direction of curvature of a tip-steered needle, the base of the needle must be rotated. As the needle rotates inside the tissue, friction opposes the needle's rotation and can cause the angle at the tip to lag behind the angle at the base (Fig. 23.6). Some artificial tissues exert enough friction to cause over a 30° difference between the base and tip angles for an insertion distance of 10 cm [60]. These large angle misalignments are thought to account for some of the reduced performance in the image-guided controllers discussed in Sect. 23.5.3. Although the torques applied during a prostrate brachytherapy are not significant enough to cause any torsion windup in the typical steel needles used for percutaneous procedures [57], the torques are likely to cause a significant discrepancy in the flexible needles required for needle steering [60]. Unfortunately, there is a tradeoff that arises due to the flexibility of the needle; increased flexibility enhances steering, but also increases the amount of torsion windup when rotating the needle.

State-of-the-art imaging is unable to accurately measure the tip angle of the small needles used in percutaneous procedures, but the angle lag at the tip of the needle can be estimated using a force sensor at the base of a bevel-tip needle [1, 58]. One method to overcome torsion estimates the angle lag from the measured torque and rotates the needle several times in alternating directions to orient the entire needle shaft to the desired orientation [58]. However, this method only works when the needle is not being inserted during rotation.

When the needle is being simultaneously rotated and inserted through the tissue, the effects of stiction are not present since the needle is continuously sliding past the tissue. In this case, the needle-tissue interaction can be modeled as viscous damping and a modal analysis can determine the dynamics of the needle tip, and a parsimoneous finite-dimensional model can be obtained using modal analysis [60]. The estimated tip position and measured base angle can then be used in a controller to increase the base-tip convergence time and decrease the positioning error.

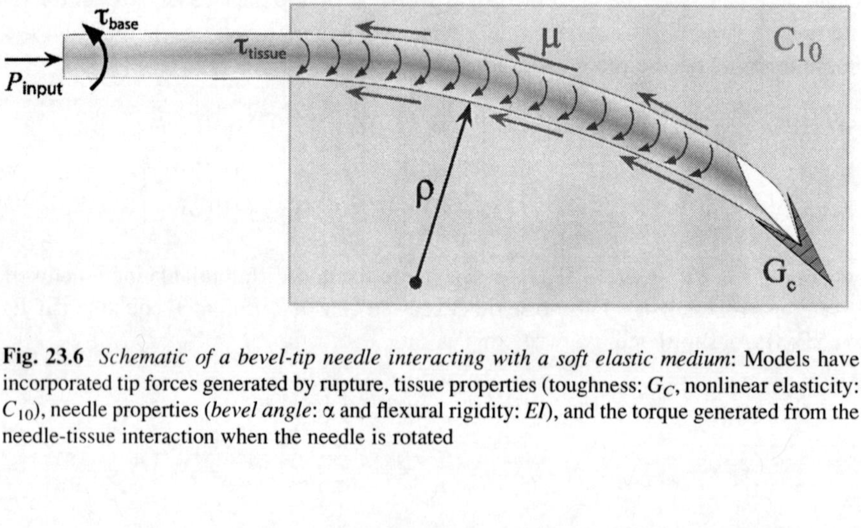

Fig. 23.6 *Schematic of a bevel-tip needle interacting with a soft elastic medium*: Models have incorporated tip forces generated by rupture, tissue properties (toughness: G_C, nonlinear elasticity: C_{10}), needle properties (*bevel angle*: α and flexural rigidity: EI), and the torque generated from the needle-tissue interaction when the needle is rotated

23.3.4 "Tissue Jacobian" Approaches

Changing the insertion direction of a needle by manipulating the base of the needle outside the tissue requires an understanding of how the flexible needle will interact with soft tissue. Two models relate the motions at the base of the needle to motions at the tip of the needle. In one method, the inverse kinematics of the needle are used to determine the path [27]. The kinematics are derived from modeling the soft tissue as springs with stiffness coefficients that vary along the length of the needle. The needle is modeled as a linear beam.

Another model involves numerically calculating the Jacobian for the tissue deformation and needle deflection [18]. Given the velocity of the base, this model determines the tip velocities. A needle path is computed based on potential fields: a repulsive field drives the needle away from obstacles and an attractor field drives the needle toward the desired target.

23.3.5 Toward Fundamental Mechanics-Based Models

Several research groups have developed physics-based needle and soft tissue interaction models [7, 15, 17, 30, 31, 48]. A general survey of surgical tool and tissue interaction models, which describes both physics- and non-physics-based interaction models, is provided in [42]. As described in Sect. 23.3.1, Webster et al. [69] presented a phenomenological nonholonomic model for steering flexible needles with bevel tips. The parameters for their model were fit using experimental data, but this model is not informed by the fundamental mechanical interaction of a needle with an elastic medium. For improved planning and control, as well as the optimization of needle design for particular medical applications, an ideal model would relate needle tip forces to the amount of needle deflection based on the fundamental principles of continuum and fracture mechanics.

Mechanics-based needle-tissue interaction models aim to relate the needle's radius of curvature to the material and geometric properties of the tissue and needle. The radius of curvature of a bevel-tipped needle is a function of several parameters (Fig. 23.6): the needle's Young's modulus (E), second moment of inertia (I), and bevel-tip angle (α); the tissue's nonlinear (hyperelastic) material property (C_{10}), rupture toughness (G_c), and coefficient of friction (μ); and the input insertion force from the robot controller (P_{input}).

Misra et al. [43] investigated the sensitivity of the tip forces to the tissue rupture toughness, linear and nonlinear tissue elasticity, and needle bevel-tip angle. In order to find the forces acting at the needle tip, they measured the rupture toughness and nonlinear material elasticity parameters of several soft tissue simulant gels and chicken tissue. These physical parameters were incorporated into a finite element model that included both contact and cohesive zone models to simulate tissue cleavage. The model showed that the tip forces were sensitive to the rupture toughness.

In addition, Misra et al. [44, 45, 46] developed an energy-based formulation incorporating tissue-specific parameters such as rupture toughness, nonlinear material elasticity, interaction stiffness, and needle geometric and material properties. This mechanics-based model was guided by microscopic and macroscopic experiments. The functional form for the deflection of the needle in an elastic medium was initially assumed and the Rayleigh-Ritz approach was used to evaluate the coefficients of the deflection equation. The Rayleigh-Ritz method is a variational method in which the minimum of a potential defined by the sum of the total energy and work done by the system are calculated. The system potential, Λ, of a needle interacting with an elastic medium, is given by

$$\Lambda = \underbrace{(N_E + S_E)}_{\text{energy}} + \underbrace{(-W_Q - W_P - W_R)}_{\text{work}} + \underbrace{(P_{\text{input}} l_i)}_{\text{input work}}, \tag{23.2}$$

where N_E and S_E are the energies associated with needle bending and needle-tissue interaction, respectively, and W_Q and W_P are the work due to transverse and axial bevel tip loads, respectively, and W_R is the work done to rupture the tissue. Explicit expressions for each of the terms in (23.2) are provided in [45]. Simulation results follow similar trends (deflection and radius of curvature) to those observed in experimental studies of a robot-driven needle interacting with different kinds of gels. These results contribute to a mechanics-based model of robotic needle steering, extending previous work on kinematic models.

23.4 Needle Path and Motion Planning

Directing steerable needles to specific targets while avoiding anatomical obstacles requires planning paths through the patient's anatomy. For steerable needles, this planning is often beyond the capabilities of human intuition due to the complex kinematics discussed in Sect. 23.3 and the effects of tissue deformation, tissue inhomogeneities, and other causes of motion uncertainty. In order to harness the full potential of steerable needles, efficient computational methods can help physicians plan paths and actions.

When steerable needles are used with image guidance, the physician can specify the target to be reached, feasible needle insertion locations, and the locations of anatomical obstacles, including those that cannot be passed through such as bones as well as sensitive anatomical structures that ought to be avoided such as blood vessels or nerves. Using patient-specific information about such anatomical structures, a motion planning algorithm determines a sequence of actions (such as insertions and bevel direction changes for bevel-tip needles) so that the needle tip reaches the specified target while avoiding the clinician-specified obstacles. Planning can be used purely preoperatively to generate a plan that is then followed by the robot or physician during the procedure. Planning can be also used intraoperatively by updating the plan in real time based on intraoperative images and other sensor feedback.

23.4.1 3D Path Planning with Obstacles

Motion planning algorithms have been developed to compute optimal trajectories for bevel-tip steerable needles in 3D environments with obstacles. Using the model of Webster et al. [69], Duindam et al. [21] computed piece-wise helical motions of the needle tip. The method optimizes a cost function that numerically quantifies the planning objective, including penalties for deviation from the target location, large control actions, and obstacle penetration. The algorithm uses a suitable discretization of the control space to quickly compute a needle path with (locally) minimal cost. In a second algorithm, Duindam *et al.* rely on an explicit expression of the inverse kinematics of the needle to generate a range of valid needle paths from start to target, from which the best solution can be selected [22]. Although this algorithm generally does not compute a (locally) optimal solution, it does not require iteration to converge to a solution and is hence much faster than the first algorithm. Depending on the required balance between speed and optimality, either algorithm can be advantageous. Xu et al. present a sampling-based motion planning technique based on the Rapidly-exploring Random Trees (RRTs) method [73]. The planner quickly builds a tree to search the configuration space using random sampling of the control space. Recently, Hauser et al. explored the use of a model predictive control strategy that chooses a needle twist rate such that the predicted helical trajectory minimizes the distance to the target, which can be used both for preoperative planning and intraoperative control [29].

23.4.2 Planning for Deformable Tissues

Inserting needles into soft tissues causes the surrounding tissues to displace and deform. Ignoring these deformations can result in substantial placement error. For example, while performing prostate brachytherapy cancer treatment, an experienced physician implanting radioactive seeds in 20 patients achieved an average placement error of 0.63 cm, a substantial error of over 15% of average prostate diameter [65] (Fig. 23.7).

Computer simulations that model soft tissue deformations can assist in preoperative planning by enabling clinicians a priori to optimize paths for needle insertion procedures [5]. Building on their prior work on simulation of rigid needles into deformable tissue [10, 9, 7], Alterovitz et al. developed a simulation of bevel-tip steerable needles in 2D [6] and Chentanez et al. developed a 3D simulation [13]. These simulations model the coupling between a steerable needle and deformable tissue using the finite element method (FEM) – a mathematical method based on continuum mechanics for modeling the deformations and motions of solids and fluids. The simulations model patient-specific anatomy using a mesh composed of triangular (2D) or tetrahedral (3D) elements. As the needle moves, the simulations model needle friction and cutting forces, as described in the models in Sect. 23.3. The simulations use novel re-meshing to ensure conformity of the mesh to the

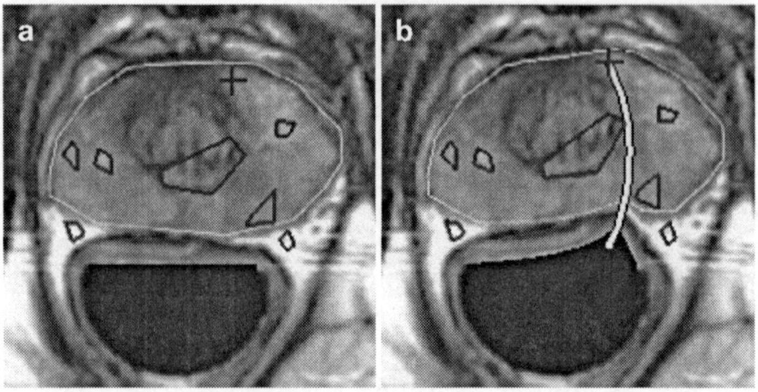

Fig. 23.7 A needle steering planner that considers 2D tissue deformation [6] (© 2005 IEEE), reprinted with permission. The magnetic resonance images show a tumor target (*cross*) in the prostate with obstacles that preclude a straight-line trajectory. The images show (**a**) the initial configuration and (**b**) a planned path for a bevel-tip steerable needle deployed from a transrectal probe. This locally optimal plan compensates for tissue deformations, avoids obstacles, and minimizes insertion distance

curvilinear needle path. Achieving a computationally efficient simulation is challenging; the FEM computation in [13] is parallelized over multiple cores of an 8-core 3.0 GHz PC and achieve a 25 Hz frame rate for a prostate mesh composed of 13,375 tetrahedra.

To help physicians anticipate and correct for the effects of tissue deformations, Alterovitz et al. developed a planner for bevel tip steerable needles that uses the simulation to compensate for predicted tissue deformations and to minimize placement error [6]. To compute the optimal initial insertion location and orientation, the planner formulates the planning problem as an optimization problem. The planner minimizes the distance the needle is inserted subject to the constraints that the needle tip reaches the target, the needle path does not intersect any obstacles, and the control inputs are within feasible ranges. The planner uses the simulation to predict the path of the needle when evaluating the objective function and constraints, and it employs a penalty method to convert the nonlinear, constrained optimization problem into a sequence of unconstrained problems that can be solved quickly. The method computes a solution in just a couple of minutes on a standard processor.

As discussed in Sect. 23.2, some needle steering approaches leverage tissue deformation in order to generate curved paths through tissue. DiMaio and Salcudean introduced simulation and planning for flexible symmetric-tip needles in 2D deformable tissue by controlling motion of the needle base [18]. Their Jacobian-based planner relied on a quasi-static FEM simulation to estimate the needle and tissue deformations. This simulation was designed for offline planning and does not achieve frame rates needed for interactive simulation or global optimization. Glozman and Shoham accelerate this approach by approximating the tissue using springs to compute local, but not global, deformations, enabling a fast planning algorithm based on inverse kinematics [27].

23.4.3 Planning Under Motion Uncertainty

Although detailed models are available for predicting the motion of steerable needles, a steerable needle may deflect from its expected path due to tissue inhomogeneities, transitions between tissue layers, local tissue deformations, patient variability, and uncertainty in needle/tissue parameters. Medical imaging can be used to measure the needle's current position and orientation, but this measurement by itself provides no information about the effect of future deflections on procedure outcome (Fig. 23.8).

Alterovitz et al. have developed planners that explicitly consider uncertainty in needle motion in order to maximize the probability of avoiding collisions and successfully reaching the target [3, 4, 11]. The Stochastic Motion Roadmap (SMR) framework efficiently samples the state space, builds a "roadmap" through the tissues that encodes the system's motion uncertainty, formulates the planning problem as a Markov Decision Process (MDP), and determines a solution using dynamic programming to maximize the probability of successfully reaching the target. This framework was applied to compute steerable needle paths around obstacles to targets in tissues imaged using 2D slices. Explicitly accounting for uncertainty can lead to significantly different motion plans compared to traditional shortest paths, such as longer paths with greater clearance from obstacles in order to increase the probability of success.

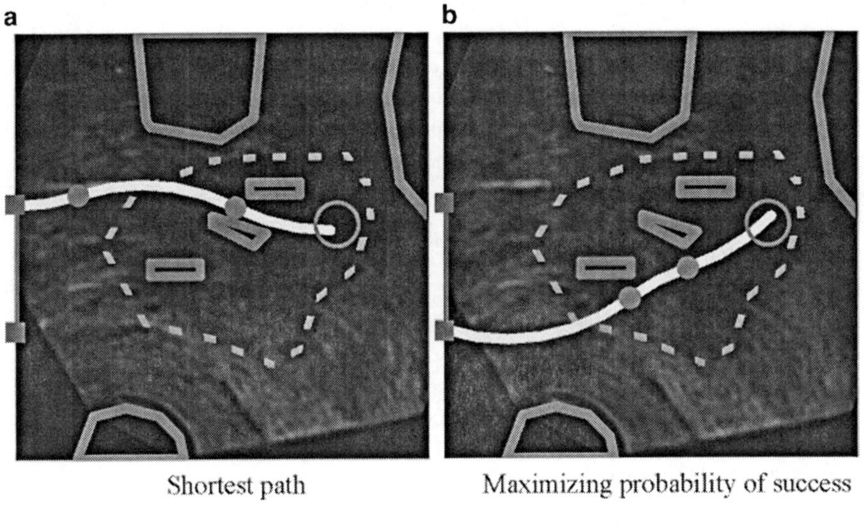

Fig. 23.8 The motion planner computes a sequence of insertions (*curved lines*) and direction changes (*dots*) to steer the needle from a start region at the left to the target (*circle*) while avoiding obstacles (*grey outlines*) [4] (© 2008 Sage Publication), reprinted with permission. The planner computes (**a**) the shortest path, which passes close to obstacles, and (**b**) a better path generated by explicitly considering uncertainty in the planning stage, which increases the probability of successfully avoiding obstacles while reaching the target

Reed et al. integrated this planner into an image-guided robotic needle steering system that includes a robotic device that can control the needle in artificial tissue and a low-level image-guided feedback controller to maintain the needle on a 2D plane [60]. The needle successfully reached targets in artificial tissues and the system experimentally demonstrated that the planner is robust to initial positioning errors of 2 cm.

The SMR framework described above transforms the continuous workspace into a discrete roadmap that encodes actions, motions, and uncertainty. An alternative approach considers the ensemble of needle trajectories obtained by repeated insertion with the same control inputs. The trajectories will be slightly different from each other due to uncertainty that may exist in the control mechanism and the interaction between the needle and the tissue. Park et al. [53, 54] developed such a path planning method for needle steering that actively utilizes this stochastic behavior of the flexible needles. This algorithm is an adapted version of the path-of-probability (POP) algorithm in [24]. A similar trajectory planning method can also be found in [40].

In the POP algorithm, the whole trajectory is obtained by serially pasting together several intermediate paths. Based on the stochastic behavior of the flexible needle, the probability density function of the needle tip pose can be estimated and evaluated. The intermediate steps are determined so as to maximize the probability that the needle tip hits the target pose.

Figure 23.9 shows the concept of the POP algorithm. The planning goal is to find a needle path that starts at $g_0 \in SE(3)$ and ends at $g_{goal} \in SE(3)$ using M intermediate steps. The homogeneous transformation matrix, $g_i \in SE(3)$ $(i = 1, 2, \ldots, M)$, represents the position and rotation of the ith frame with respect to $(i-1)^{th}$ frame. Suppose that the $(i-1)$ intermediate steps $(g_1, g_2, \cdots, g_{i-1} \in SE(3))$ have already been determined. The intermediate step, g_i, is determined to maximize the probability that the remaining needle insertion reaches the goal. The shaded ellipses depict the probability density function that represents the probability of the needle tip pose after the remaining $(M-i)$ steps. In other words, after the remaining $(M-i)$ steps, the final pose will be placed in the dark area with higher probability than the bright area. Comparing the two simplified cases in Fig. 23.9, if the previous intermediate steps $(g_1, g_2, \cdots, g_{i-1})$ are the same for both cases, g_i

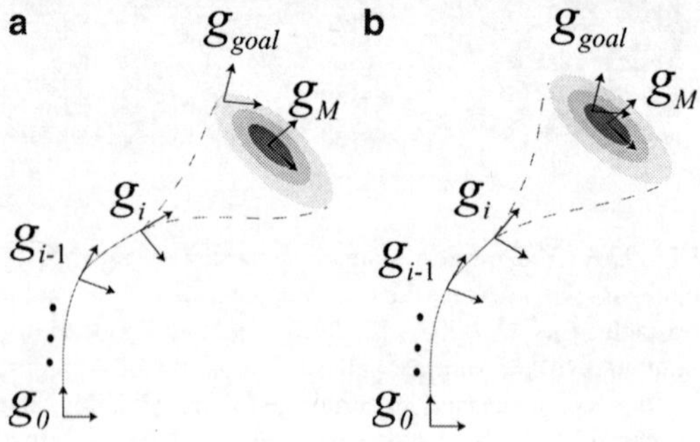

Fig. 23.9 The path-of-probability algorithm at the ith step [56] (© 2010 Sage Publications), reprinted with permission. (**a**) An intermediate step, g_i, resulting in low probability of reaching the goal. (**b**) An intermediate step, g_i, resulting in high probability of reaching the goal

shown in Fig. 23.9b is a better choice, because it guarantees with higher probability that the final pose reaches the goal pose.

Computing the probability density function plays a crucial role in the POP algorithm. The probability density function can be obtained using the stochastic model for the flexible needle stochastic differential equations (SDE) as reviewed in Sect. 23.3.2. The Fokker-Planck equation corresponding to the SDE defines a function representing the probability density of the needle tip pose. Rapid evaluation of the probability function is required for fast path planning. Specifically, the probability density is estimated by a Gaussian function [54, 56], and the mean and covariance are estimated using error propagation techniques developed for the motion groups [67, 68].

23.5 Image Guidance

23.5.1 Needle Localization in Medical Images

The problem of needle localization in images might seem straightforward, yet practical implementations have seldom appeared. Usually, a cascade of basic image filters (such as thresholding, edge detection, image smoothing and noise removal filters) are combined with more sophisticated feature detections routines, such as a variant of Hough transform. Significant literature exists on the theory, use, and extension of Hough transform; a succinct summary and background reading are given in [19]. Many localization methods entail two steps: first, points or fragments of the needle are extracted from the images and then a 3D geometric model (straight line, polynomial, etc.) is fit to the fragments, typically in a least-squares optimization scheme. The two steps can be combined in a probabilistic framework, where points of high probability of belonging to the needle are fitted on a 3D geometrical model. This approach is especially suitable when the quality of images (resolution, dynamic range, etc.) is poor, such as in ultrasound images. In this section, we survey the most popular needle localization methods used with various imaging modalities, namely fluoroscopy, computed tomography (CT), magnetic resonance imaging (MRI), and ultrasound (US).

Fluoroscopy. Metal needles, being of high density, tend to be visible in X-ray images such as those obtained from CT and fluoroscopy. In fluoroscopy, a single projection image is insufficient for reconstruction of the needle in 3D. Two images are sufficient to reconstruct a straight needle, while three or more images and some amount of prior knowledge about the curve are necessary for 3D reconstruction of a curved needle. For needles that lie in a plane, polynomial models are preferable because polynomials are invariant to perspective projection. For example, Jain et al. used a combination of 0th-, 1st-, and 2nd-degree polynomials to fit image points on a 3D model with sub-millimeter and sub-degree accuracy [32]. When a needle is driven out of plane, spatial reconstruction becomes more demanding and requires more images and/or a more elaborate 3D model for the needle. A seemingly innocuous and often underrated problem in fluoroscopy is that the device must be precisely calibrated, including the relative pose of the fluoroscopy images [32].

CT. Although CT can produce a 3D volume, needle insertion is often performed in a single 2D plane, with the CT gantry tilted in order to show the needle in the 2D image. Newer CT scanners provide short acquisition time with reasonably low dose, convenient for intermittent observation of the needle. Many CT scanners also provide continuous beam mode, yielding a single CT image of low resolution at high frame rate ($\approx$ 10 fps). There is a trade-off between image quality (resolution and dynamic range), frame rate, and X-ray dose. Modern CT scanners can also produce multiple slices (i.e. thin 3D volume) and high-end scanners even provide multiple slices in continuous beam mode.

A universal problem of any X-ray imaging modality (fluoroscopy and CT included) is that for safety reasons image acquisition cannot be triggered by the surgical navigation software and images are acquired under the command of a human operator. This process is time consuming, cumbersome and error prone. The available alternative is using continuous X-ray, exposing the patient and physician to excessive radiation.

MRI. For needle localization, the one major advantage of MRI over X-ray imaging is the absence of harmful radiation. In practice, there is typically a compromise on both spatial resolution and acquisition time: MR images used in surgical guidance tend to be of much lower resolution than diagnostic images, and the acquisition is usually not real-time. A further disadvantage of MRI is that metal needles create a large signal void in the image. Further, the signal void does not coincide with the true position of the needle, and the displacement between the two depends on the configuration of the needle, the B0 field and the gradient field [16]. It is not uncommon for a 1 mm diameter needle to leave a 5 mm signal void in the image; hiding both the needle and the surrounding anatomy.

Ultrasound. Ultrasound (US) is an attractive needle guidance modality, due to its low cost, widespread availability, and safety. US imaging is an operator-dependent manual process. It also causes some degree of tissue deformation and dislocation as the transducer makes contact with the tissue scanned. US images tend to be noisy, due to reflections, reverberations, shadows, air pockets, and biological speckle, which makes needle localization challenging. Some needle localization methods use 2D images [19, 14, 52], while others compound a 3D volume from a tracked sweep of 2D images [20, 2]. For completeness, we note that, due to current limitations on voxel resolution and transfer speed, 3D US probes have not been practical for image-based needle guidance. Novotni et al. tracked laparoscopy instruments (which are larger than needles), but this requires a research agreement with the vendor of the ultrasound machine [49].

To localize straight needles in 2D ultrasound, Ding et al. introduced a sophisticated derivative of the Hough transform [19]. Cheung et al. proposed an enhancement algorithm that maximizes the received reflections by steering the ultrasound beam to be precisely perpendicular to the needle [14]. Surface-coated needles are available commercially, to enhance ultrasonic visibility of the needle, which in turn increases friction during insertion and thus may not be appropriate for needle steering. Okazawa et al. localized bent needles in a 2D image plane by warping an initial guess straight line into a 2D parametric curve fitting on probable needle points [52]. This method works well for conventional needles, but it breaks under excessive curvature often observed with elastic needles and catheters. Ding et al. constructed a 3D volume from a sweep of tracked 2D images, cropped the volume sensibly and created several orthogonal projection images. They segmented the needle in the projections with the Hough transform and then reconstructed the needle from its 2D projections as a straight line. Aboofazeli et al. recently localized curved non-planar needles in 3D space [2]. They pre-filtered a compounded 3D US volume and produced series of 2D images by ray casting. In the projected images, the needle was segmented with the Hough transform and fitted onto a polynomial model. From the series of 2D polynomial curves, they reconstructed a surface that contains the needle. This 3D surface was smoothed and the needle was detected on the surface using the Hough transform followed by a polynomial curve fitting. The end result was a continuous 3D curve consisting of polynomial patches.

Localization of the needle tip has been a major challenge, especially in 2D US, where it is difficult to determine whether the needle tip is inside or outside the plane of imaging. The non-uniform thickness of the US beam adds further to the localization error. When using bevel-tip needles, the physician often rotates the needle to create a visible, fluctuating artifact at the needle tip. Harmat et al. created mechanical vibrations on the needle tip and measured the resulting Doppler effects [28]. Their prototype robustly detected the needle tip, but it did not seem to provide sufficient accuracy for localizing the needle tip for controlled insertion.

23.5.2 State Estimation of Unmeasured Degrees of Freedom

As described above, except in MR images, researchers have had reasonable success in localizing needles, but estimating the full 6-DOF pose of the needle tip directly from medical images, including rotation about the needle axis, remains elusive. However, this rotation information is necessary for control and planning purposes. To overcome this, Kallem et al. designed dynamical observers (analogous to a Kalman filter) based on kinematic models of needle steering that can be used to estimate full 6-DOF needle tip pose from a sequence of 3D position measurements [33]. They showed that the rotation of the needle tip may be inferred from the measurements of the needle tip position over time and developed model-based asymptotic observers that exploit the task-induced reduction to estimate the full needle pose.

Needle steering is highly nonlinear, which makes the estimation and control problem coupled, unlike in linear systems. Building on the nonholonomic model of Webster et al. (see Sect. 23.3.1), Kallem and Cowan [37, 36] exploit the fact that, to drive the needle to a desired 2D plane (y-z plane without any loss of generality), only three of the six degrees of freedom need to be considered. Using this reduction, they first developed an observer to estimate the x position, the pitch of the needle tip, and the roll of the needle from just x position measurements. In [33] a linear model to represent the dynamics of the other three states (y, z positions and yaw of the needle) is created by state immersion into a finite higher dimensional manifold; based on this, Luenberger observers for this smaller system are designed. This two-stage coupled observer estimates the complete needle orientation and also filtered the noisy position measurements. For other tasks, similar controller-observer pairs need to be developed to estimate needle orientation.

23.5.3 Image-Guided Control of Needle Steering

As described in Sect. 23.3, considerable progress has been made developing "plant models" for manipulating a needle from outside the patient. These models enable development of model-based feedback controllers to steer the needle inside the tissue. Glozman and Shoham [27] developed an image-guidance strategy for flexible needles without a bevel tip. First they plan a needle path that avoids obstacles in the workspace. Then at every time step they invert a virtual spring model to obtain the translation and orientation of the needle base (the inputs) in order to drive the needle back to the planned path in one step.

Kallem and Cowan [37, 36] took a systems-theoretic perspective to develop feedback-based controller-observer pairs for tip-steerable needles. A tip-steerable needle has been modeled as a 6-DOF nonholonomic system (1) with two inputs and nonholonomy degree four. Furthermore, when the needle is pulled out of the tissue, no cutting forces are generated and thus the needle follows the same path as during insertion into the tissue. These constraints imply that asymptotic controllers do not exist for certain tasks, such as driving the needle tip to reach a desired pose in 6 DOF or following a circular path whose radius is the natural radius of curvature of the needle inside the tissue. To overcome these challenges, the approach taken in [37, 36] is to develop low-level, asymptotic controllers that only control a subset of the degrees of freedom. These controllers are designed to cooperate with the higher-level 2D planners from Alterovitz et al. [8, 11]. These planners, which rely on the needle staying within a specified 2D plane, construct a sequence of circular arcs of the natural radius of the needle that can be achieved via alternating insertions and 180° rotations of the needle shaft. In effect, the low-level 2D plane-following controller designed by Kallem et al., described below, ensures that the needle remains close to a desired 2D plane, on top of which Alterovitz et al.'s planner can operate.

Kallem and Cowan [36, 37] developed a feedback-based estimator-controller pair to drive the needle to a desired plane, and subsequently generalized this to other subspace trackers [35]. The feedback signal used is the needle-tip position. For this task, they showed that considering a three-state system is sufficient, which simplified the estimation and control design needed to achieve the task. This controller has been successfully tested in simulations and in artificial tissue. Figure 23.10

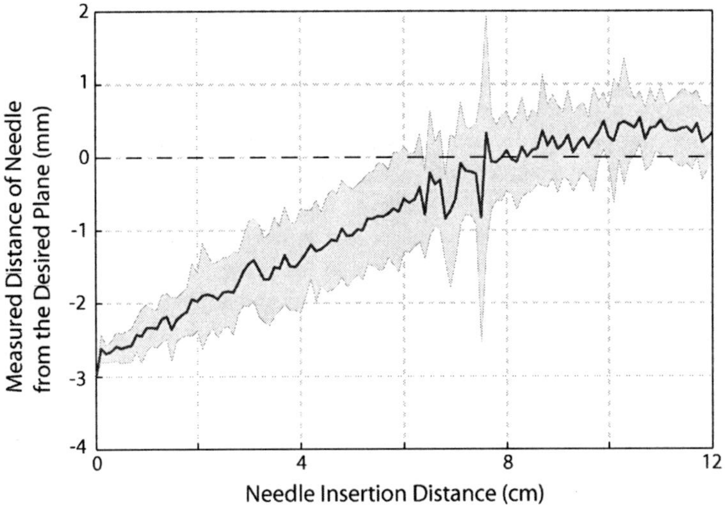

Fig. 23.10 Nine experimental trials were used to validate an image-guided controller [38] (© 2009 IEEE), reprinted with permission. The mean distance of the needle-tip from the desired plane of all trials is plotted against the insertion distance of the needle into the tissue (*solid bold line; gray region* indicates mean ± standard deviation). All trials control approach the desired 2D plane and stay within the noise levels of the position measurements of approximately 1 mm

shows successful experimental results of a needle being driven to a desired plane when inserted into artificial tissue.

Reed et al. [59] integrated the full 6-DOF asymptotic observer and the planar controller with the 2D planner of Alterovitz et al. [11] and the torsional compensator of Reed [58]. Figure 23.3 *Second Row, Right Column* shows the path taken by the unified system to reach a target inside the tissue. The goal is to reach the circular target while avoiding the polygon obstacles in the workspace. The planar controllers act every 1 mm of needle insertion into the tissue to drive the needle to the desired 2D plane and the planner acts at 1 cm insertion intervals. With integrated planning and control, the needle successfully reaches the target (as shown in the pre-curved tip example in Fig. 23.3 [59]).

23.6 Conclusions

This chapter provides an overview of the technological and algorithmic state-of-the-art in needle steering. As can be seen from this chapter, numerous components are required to enable needle steering. Ultimately, the clinical success of needle steering depends on uniting these pieces and reducing them to practice in a driving application to create a fully integrated clinical needle steering system. As shown in Fig. 23.2, such a system includes a set of computational and physical components – including the robotic device and steering mechanism, modeling, planning, imaging, and control – each of which is addressed in Sects. 23.2–23.5.

A potential first driving application for needle steering is transperineal prostate brachytherapy, a treatment that involves implantation of radioactive seeds by needles into the prostate in order to kill cancer with radiation. Literature shows that reducing surgical trauma of the prostate reduces the severity of edema, thereby improving implant dosimetry and reducing toxicity. Current manual needle placement can involve multiple reinsertions and adjustments of the needle before it reaches a target, causing excessive trauma to the prostate. We hypothesize robotic needle steering will eliminate needle reinsertions and adjustments, and thus lead to reduction of surgical trauma. Efforts are underway by some of the authors of this chapter to create a clinically viable needle steering system for prostate brachytherapy. Along the way, we expect there to be continued advances in devices, models, planning, sensing, and control that will lead to advances in needle steering, as well as robotics in general.

Acknowledgements The authors thank Dr. Purang Abolmaesumi and Meysam Torabi for their detailed feedback on this chapter. This work was supported in part by the National Institutes of Health under Grants R21-EB003452, R01-EB006435, and F32-CA124138.

References

1. Abolhassani, N., Patel, R.V., Ayazi, F.: Minimization of needle deflection in robot-assisted percutaneous therapy. Int. J. Med. Robot. and Comp. Assist. Surg. **3**, 140–148 (2007)
2. Aboofazeli, M., Abolmaesumi, P., Mousavi, P., Fichtinger, G.: A new scheme for curved needle segmentation in three-dimensional ultrasound images. In: Proc. IEEE Int. Symp. on Biomedical Imaging, pp. 1067–1070. Boston, MA (2009)
3. Alterovitz, R., Branicky, M., Goldberg, K.: Constant-curvature motion planning under uncertainty with applications in image-guided medical needle steering. In: Akella, S., Amato, N.M., Huang, W.H., Mishra, B. (eds.) Algorithmic Foundations of Robotics. Springer Tracts in Advanced Robotics, vol. 47, pp. 319–334. Springer, Berlin (2008)
4. Alterovitz, R., Branicky, M., Goldberg, K.: Motion planning under uncertainty for image-guided medical needle steering. Int. J. Robot. Res. **27**(11–12), 1361–1374 (2008)
5. Alterovitz, R., Goldberg, K.: Motion Planning in Medicine: Optimization and Simulation Algorithms for Image-Guided Procedures. Springer Tracts in Advanced Robotics, vol. 50. Springer, Berlin (2008)
6. Alterovitz, R., Goldberg, K., Okamura, A.M.: Planning for steerable bevel-tip needle insertion through 2D soft tissue with obstacles. In: Proc. IEEE Int. Conf. Robot. and Autom., pp. 1652–1657. Barcelona, Spain (2005)
7. Alterovitz, R., Goldberg, K.Y., Pouliot, J., Hsu, I.C.: Sensorless motion planning for medical needle insertion in deformable tissues. IEEE Trans. Inf. Technol. Biomed. **13**(2), 217–225 (2009)
8. Alterovitz, R., Lim, A., Goldberg, K., Chirikjian, G.S., Okamura, A.M.: Steering flexible needles under Markov motion uncertainty. In: Proc. IEEE/RSJ Int. Conf. on Intell. Robots and Syst., pp. 1570–1575 (2005)
9. Alterovitz, R., Pouliot, J., Taschereau, R., Hsu, I.C., Goldberg, K.: Needle insertion and radioactive seed implantation in human tissues: Simulation and sensitivity analysis. In: Proc. IEEE Int. Conf. Robot. and Autom., vol. 2, pp. 1793–1799. Taipei, Taiwan (2003)

10. Alterovitz, R., Pouliot, J., Taschereau, R., Hsu, I.C., Goldberg, K.: Simulating needle insertion and radioactive seed implantation for prostate brachytherapy. In: Westwood, J.D., Hoffman, H.M., Mogel, G.T., Phillips, R., Robb, R.A., Stredney, D. (eds.) Medicine Meets Virtual Reality, pp. 19–25. IOS Press, Newport Beach, CA (2003)
11. Alterovitz, R., Siméon, T., Goldberg, K.: The Stochastic Motion Roadmap: A sampling framework for planning with Markov motion uncertainty. In: Burgard, W., Brock, O., Stachniss, C. (eds.) Proc. Robotics: Science and Systems, pp. 246–253. MIT Press, Cambridge, MA (2008)
12. Bogdanich, W.: At V.A. hospital, a rogue cancer unit. The New York Times (2009)
13. Chentanez, N., Alterovitz, R., Ritchie, D., Cho, L., Hauser, K.K., Goldberg, K., Shewchuk, J.R., O'Brien, J.F.: Interactive simulation of surgical needle insertion and steering. ACM Transactions on Graphics (Proc. SIGGRAPH). **28**(3), 88:1–10 (2009)
14. Cheung, S., Rohling, R.: Enhancement of needle visibility in ultrasound-guided percutaneous procedures. Ultrasound Med. Biol. **30**(5), 617–624 (2004)
15. Crouch, J.R., Schneider, C.M., Wainer, J., Okamura, A.M.: A velocity-dependent model for needle insertion in soft tissue. In: Medical Image Computing and Computer Assisted Intervention. Lecture Notes in Computer Science, vol. 3750, pp. 624–632. Springer, Berlin (2005)
16. DiMaio, S.P., Kacher, D.F., Ellis, R.E., Fichtinger, G., Hata, N., Zientara, G.P., Panych, L.P., Kikinis, R., Jolesz, F.A.: Needle artifact localization in 3T MR images. Stud. Health Technol. Inform. **119**, 120–125 (2006)
17. DiMaio, S.P., Salcudean, S.E.: Needle insertion modeling and simulation. IEEE Trans. Robot. Autom. **19**(5), 864–875 (2003)
18. DiMaio, S.P., Salcudean, S.E.: Needle steering and motion planning in soft tissues. IEEE Trans. Biomed. Eng. **52**(6), 965–974 (2005)
19. Ding, M., Fenster, A.: A real-time biopsy needle segmentation technique using hough transform. Med. Phys. **30**(8), 2222–2233 (2003)
20. Ding, M., Fenster, A.: Projection-based needle segmentation in 3D ultrasound images. Comput. Aided Surg. **9**(5), 193–201 (2004)
21. Duindam, V., Alterovitz, R., Sastry, S., Goldberg, K.: Screw-based motion planning for bevel-tip flexible needles in 3D environments with obstacles. In: Proc. IEEE Int. Conf. Robot. and Autom., pp. 2483–2488 (2008)
22. Duindam, V., Xu, J., Alterovitz, R., Sastry, S., Goldberg, K.: Three-dimensional motion planning algorithms for steerable needles using inverse kinematics. Int. J. Robot. Res. **29**(7), 789–800 (2010)
23. Dupont, P.E., Lock, J.L., Itkowitz, B., Butler, E.: Design and control of concentric-tube robots. IEEE Trans. Robot. **26**(2), 209–225 (2010)
24. Ebert-Uphoff, I., Chirikjian, G.S.: Inverse kinematics of discretely actuated hyper-redundant manipulators using workspace densities. In: Proc. IEEE Int. Conf. Robot. and Autom., pp. 139–145 (1996)
25. Engh, J., Podnar, G., Kondziolka, D., Riviere, C.: Toward effective needle steering in brain tissue. In: Proc. IEEE Int. Conf. Eng. Med. Biol. Soc., pp. 559–562 (2006)
26. Fichtinger, G., Fiene, J., Kennedy, C.W., Kronreif, G., Iordachita, I., Song, D.Y., Burdette, E.C., Kazanzides, P.: Robotic assistance for ultrasound guided prostate brachytherapy. In: Medical Image Computing and Computer Assisted Intervention. Lecture Notes in Computer Science, pp. 119–127. Springer, Brisbane, Australia (2007)
27. Glozman, D., Shoham, M.: Image-guided robotic flexible needle steering. IEEE Trans. Robot. **23**(3), 459–467 (2007)
28. Harmat, A., Rohling, R.N., Salcudean, S.E.: Needle tip localization using stylet vibration. Ultrasound Med. Biol. **32**(9), 1339–1348 (2006)
29. Hauser, K., Alterovitz, R., Chentanez, N., Okamura, A., Goldberg, K.: Feedback control for steering needles through 3D deformable tissue using helical paths. In: Proc. Robotics: Science and Systems. Seattle, USA (2009)

30. Heverly, M., Dupont, P., Triedman, J.: Trajectory optimization for dynamic needle insertion. In: Proc. IEEE Int. Conf. Robot. and Autom., vol. 1, pp. 1646–1651. Barcelona, Spain (2005)
31. Hing, J.T., Brooks, A.D., Desai, J.P.: Reality-based needle insertion simulation for haptic feedback in prostate brachytherapy. In: Proc. IEEE Int. Conf. Robot. and Autom., vol. 1, pp. 619–624. Orlando, USA (2006)
32. Jain, A.K., Mustafa, T., Zhou, Y., Burdette, C., Chirikjian, G.S., Fichtinger, G.: Ftrac–a robust fluoroscope tracking fiducial. Med. Phys. **32**(10), 3185–3198 (2005)
33. Kallem, V.: Vision-based control on lie groups with application to needle steering. Ph.D. thesis, Johns Hopkins University (2008)
34. Kallem, V., Chang, D.E., Cowan, N.J.: Task-induced symmetry and reduction in kinematic systems with application to needle steering. In: Proc. IEEE/RSJ Int. Conf. on Intell. Robots and Syst., pp. 3302–3308. San Diego, CA (2007)
35. Kallem, V., Chang, D.E., Cowan, N.J.: Task-induced symmetry and reduction with application to needle steering. IEEE Trans. Automat. Contr. **55**(3), 664–673 (2010)
36. Kallem, V., Cowan, N.J.: Image-guided control of flexible bevel-tip needles. In: Proc. IEEE Int. Conf. Robot. and Autom., pp. 3015–3020. Rome, Italy (2007)
37. Kallem, V., Cowan, N.J.: Image guidance of flexible tip-steerable needles. IEEE Trans. Robot. **25**, 191–196 (2009)
38. Krieger, A., Susil, R.C., Ménard, C., Coleman, J.A., Fichtinger, G., Atalar, E., Whitcomb, L.L.: Design of a novel MRI compatible manipulator for image guided prostate interventions. IEEE Trans. Biomed. Eng. **52**(2), 306–313 (2005)
39. Mallapragada, V.G., Sarkar, N., Podder, T.K.: Robot-assisted real-time tumor manipulation for breast biopsy. IEEE Trans. Robot. **25**(2), 316–324 (2009)
40. Mason, R., Burdick, J.: Trajectory planning using reachable-state density functions. In: Proc. IEEE Int. Conf. Robot. and Autom., pp. 273–280 (2002)
41. Minhas, D.S., Engh, J.A., Fenske, M.M., Riviere, C.N.: Modeling of needle steering via duty-cycled spinning. Proc. IEEE Int. Conf. Eng. Med. Biol. Soc. **2007**, 2756–2759 (2007)
42. Misra, S., Ramesh, K.T., Okamura, A.M.: Modeling of tool-tissue interactions for computer-based surgical simulation: A literature review. Presence: Teleoperators & Virtual Environments **17**(5), 463–491 (2008)
43. Misra, S., Reed, K.B., Douglas, A.S., Ramesh, K.T., Okamura, A.M.: Needle-tissue interaction forces for bevel-tip steerable needles. In: Proc. IEEE/RASJ Int. Conf. on Biomed. Robotics and Biomechatronics, pp. 224–231. Scottsdale, USA (2008)
44. Misra, S., Reed, K.B., Ramesh, K.T., Okamura, A.M.: Observations of needle-tissue interactions. In: Proc. IEEE Int. Conf. Eng. Med. Biol. Soc., pp. 262–265. Minneapolis, USA (2009)
45. Misra, S., Reed, K.B., Schafer, B.W., Ramesh, K.T., Okamura, A.M.: Observations and models for needle-tissue interactions. In: Proc. IEEE Int. Conf. Robot. and Autom., pp. 2687–2692. Kobe, Japan (2009)
46. Misra, S., Reed, K.B., Schafer, B.W., Ramesh, K.T., Okamura, A.M.: Mechanics of flexible needles robotically steered through soft tissue. Int. J. Robot. Res. (2010). URL http://ijr.sagepub.com/content/early/2010/06/02/0278364910369714.short?rss=1&ssource=mfc
47. Mozer, P.C., Partin, A.W., Stoianovici, D.: Robotic image-guided needle interventions of the prostate. Rev. Urol. **11**(1), 7–15 (2009)
48. Nienhuys, H.W., van der Stappen, F.A.: A computational technique for interactive needle insertions in 3D nonlinear material. In: Proc. IEEE Int. Conf. Robot. and Autom., vol. 2, pp. 2061–2067. New Orleans, USA (2004)
49. Novotny, P.M., Stoll, J.A., Vasilyev, N.V., del Nido, P.J., Dupont, P.E., Zickler, T.E., Howe, R.D.: GPU based real-time instrument tracking with three-dimensional ultrasound. Med. Image Anal. **11**(5), 458–464 (2007)
50. Okamura, A.M., Simone, C., O'Leary, M.D.: Force modeling for needle insertion into soft tissue. IEEE Trans. Biomed. Eng. **51**(10), 1707–1716 (2004)
51. Okazawa, S., Ebrahimi, R., Chuang, J., Salcudean, S.E., Rohling, R.: Hand-held steerable needle device. IEEE ASME Trans. Mechatron. **10**(3), 285–296 (2005)

52. Okazawa, S.H., Ebrahimi, R., Chuang, J., Rohling, R.N., Salcudean, S.E.: Methods for segmenting curved needles in ultrasound images. Med. Image Anal. **10**(3), 330–342 (2006)
53. Park, W., Kim, J.S., Zhou, Y., Cowan, N.J., Okamura, A.M., Chirikjian, G.S.: Diffusion-based motion planning for a nonholonomic flexible needle model. In: Proc. IEEE Int. Conf. Robot. and Autom., pp. 4600–4605. Barcelona, Spain (2005)
54. Park, W., Liu, Y., Zhou, Y., Moses, M., Chirikjian, G.S.: Kinematic state estimation and motion planning for stochastic nonholonomic systems using the exponential map. Robotica **26**, 419–434 (2008)
55. Park, W., Wang, Y., Chirikjian, G.S.: The path-of-probability algorithm for steering and feedback control of flexible needles. Int. J. Robot. Res. **29**(7), 813830 (2010)
56. Park, W., Wang, Y., Chirikjian, G.S.: Path planning for flexible needles using second order error propagation. In: Chirikjian, G.S., Choset, H., Morales, M., Murphey, T. (eds.) Algorithmic Foundations of Robotics VIII. Springer Tracts in Advanced Robotics, pp. 583–598. Springer, Berlin (2010)
57. Podder, T., Clark, D., Sherman, J., Fuller, D., Messing, E., Rubens, D., Strang, J., Liao, L., Ng, W.S., Yu, Y.: In vivo motion and force measurement of surgical needle intervention during prostate brachytherapy. Med. Phys. **33**(8), 2915–2922 (2006)
58. Reed, K.B.: Compensating for torsion windup in steerable needles. In: Proc. IEEE/RASJ Int. Conf. on Biomed. Robotics and Biomechatronics, pp. 936–941. Scottsdale, AR, USA (2008)
59. Reed, K.B., Kallem, V., Alterovitz, R., Goldberg, K., Okamura, A.M., Cowan, N.J.: Integrated planning and image-guided control for planar needle-steering. In: Proc. IEEE/RASJ Int. Conf. on Biomed. Robotics and Biomechatronics, pp. 819–824. Scottsdale, AR, USA (2008)
60. Reed, K.B., Okamura, A.M., Cowan, N.J.: Modeling and control of needles with torsional friction. IEEE Trans. Biomed. Eng. **56**(12), 2905–2916 (2009)
61. Rucker, D.C., Webster, R.J. III, Chirikjian, G.S., Cowan, N.J.: Equilibrium conformations of concentric-tube continuum robots. Int. J. Robot. Res. (2010). In press (published online April 1, 2010)
62. Sears, P., Dupont, P.: A steerable needle technology using curved concentric tubes. In: Proc. IEEE/RSJ Int. Conf. on Intell. Robots and Syst., pp. 2850–2856 (2006)
63. Sitzman, B.T., Uncles, D.R.: The effects of needle type, gauge, and tip bend on spinal needle deflection. Anesth. Analg. **82**(2), 297–301 (1996)
64. Susil, R.C., Ménard, C., Krieger, A., Coleman, J.A., Camphausen, K., Choyke, P., Fichtinger, G., Whitcomb, L.L., Coleman, C.N., Atalar, E.: Transrectal prostate biopsy and fiducial marker placement in a standard 1.5T magnetic resonance imaging scanner. J. Urol. **175**(1), 113–120 (2006)
65. Taschereau, R., Pouliot, J., Roy, J., Tremblay, D.: Seed misplacement and stabilizing needles in transperineal permanent prostate implants. Radiother. Oncol. **55**(1), 59–63 (2000)
66. Torabi, M., Hauser, K., Alterovitz, R., Duindam, V., Goldberg, K.: Guiding medical needles using single-point tissue manipulation. In: Proc. IEEE Int. Conf. Robot. and Autom., pp. 2705–2710. Kobe, Japan (2009)
67. Wang, Y., Chirikjian, G.S.: Error propagation on the Euclidean group with applications to manipulator kinematics. IEEE Trans. Robot. **22**(4), 591–602 (2006)
68. Wang, Y., Chirikjian, G.S.: Nonparametric second-order theory of error propagation on motion groups. Int. J. Robot. Res. **27**, 1258–1273 (2008)
69. Webster, R.J. III, Kim, J.S., Cowan, N.J., Chirikjian, G.S., Okamura, A.M.: Nonholonomic modeling of needle steering. Int. J. Robot. Res. **25**(5–6), 509–525 (2006)
70. Webster, R.J. III, Memisevic, J., Okamura, A.M.: Design considerations for robotic needle steering. In: Proc. IEEE Int. Conf. Robot. and Autom., vol. 1, pp. 3588–3594. Barcelona, Spain (2005)
71. Webster, R.J. III, Romano, J.M., Cowan, N.J.: Mechanics of precurved-tube continuum robots. IEEE Trans. Robot. **25**, 67–78 (2009)

72. Wedlick, T., Okamura, A.: Characterization of pre-curved needles for steering in tissue. In: Proc. IEEE Int. Conf. Eng. Med. Biol. Soc., pp. 1200–1203 (2009)
73. Xu, J., Duindam, V., Alterovitz, R., Goldberg, K.: Motion planning for steerable needles in 3D environments with obstacles using rapidly-exploring random trees and backchaining. In: Proc. IEEE Int. Conf. Automation Sci. and Eng., pp. 41–46 (2008)
74. Yan, K.G., Ng, W.S., Ling, K.V., Yu, Y., Podder, T.: High frequency translational oscillation and rotational drilling of the needle in reducing target movement. In: IEEE Int. Symp. Comp. Intell. in Robot. and Autom., pp. 163–168 (2005)

Chapter 24
Macro and Micro Soft-Tissue Biomechanics and Tissue Damage: Application in Surgical Robotics

Jacob Rosen, Jeff Brown, Smita De, and Blake Hannaford

Abstract Accurate knowledge of biomechanical characteristics of tissues is essential for developing realistic computer-based surgical simulators incorporating haptic feedback, as well as for the design of surgical robots and tools. Most past and current biomechanical research is focused on soft and hard anatomical structures that are subject to physiological loading while testing the organs in situ. Internal organs are different in that respect since they are not subject to extensive loads as part of their regular physiological function. However, during surgery, a different set of loading conditions are imposed on these organs as a result of the interaction with the surgical tools. The focus of the current study was to obtain the structural biomechanical properties (engineering stress-strain and stress relaxation) of seven abdominal organs, including bladder, gallbladder, large and small intestines, liver, spleen, and stomach, using a porcine animal model. The organs were tested in vivo, in situ, and ex corpus (the latter two conditions being postmortem) under cyclical and step strain compressions using a motorized endoscopic grasper and a universal-testing machine. The tissues were tested with the same loading conditions commonly applied by surgeons during minimally invasive surgical procedures. Phenomenological models were developed for the various organs, testing conditions, and experimental devices. A property database—unique to the literature—has been created that contains the average elastic and relaxation model parameters measured for these tissues in vivo and postmortem. The results quantitatively indicate the significant differences between tissue properties measured in vivo and postmortem. A quantitative understanding of how the unconditioned tissue properties and model parameters are influenced by time postmortem and loading condition has been obtained. The results provide the material property foundations for developing science-based haptic surgical simulators, as well as surgical tools for manual and robotic systems.

J. Rosen (✉)
Department of Computer Engineering, Jack Baskin School of Engineering,
University of California Santa Cruz, 1156 High street, Santa Cruz, CA 95064-1099, USA
e-mail: rosen@ucsc.edu

J. Rosen et al. (eds.), *Surgical Robotics: Systems Applications and Visions*,
DOI 10.1007/978-1-4419-1126-1_24, © Springer Science+Business Media, LLC 2011

Keywords Soft tissue · Biomechanics · Internal organs · Surgical robotics · Surgical simulation · Haptics · Surgical tools

24.1 Introduction

New technologies have fundamentally changed the practice of surgery. Having recently introduced minimally invasive (laparoscopic) techniques, surgery is now poised to take another big step by incorporating surgical robotic systems into practice. These robotic devices [1] are only in their first generation of development but promise to significantly improve surgical dexterity in small and remote body cavities. Along with surgical robots, surgical simulators are being introduced into the curriculum for surgical training. To perform or simulate manipulation of soft tissues, both surgical robots and surgical simulators must be engineered with knowledge of the biomechanical properties of the tissues most relevant to the clinical application. To date, there have been little biomechanical data available, and current simulators and robots have largely been engineered to accomplish acceptable "handling" characteristics, as determined by expert surgeon consultants. While the initial pioneering surgical robots from Intuitive Surgical and Computer Motion, Inc. (now merged) have achieved FDA approval and some commercial success without detailed biomechanical data, as this field matures, the need for precise instrument design based on quantitative evaluation of tissue biomechanical properties will increase. Accurate models of clinically relevant tissues will allow designers to predict manipulation forces and torques required. The first step to understanding the consequences of tissue stress is a better understanding of the biomechanics of the tissues.

Surgical training has been affected by many factors such as statutory limitation of work hours, patient safety concerns, and a growing regulatory push for credentialing of surgical trainees. Formal curriculum development with specific milestones and significant improvement in computer-based surgical simulation as a training tool have also augmented the surgical armamentarium. However, initial simulation efforts did not focus on the accuracy with which they render deformation forces and displacements of the tissues and few provided any haptic feedback. As the next generation of simulators are developed, biomechanical data are essential for making this feedback accurate. The consequences of inaccurate tissue deformation modeling on clinical performance after simulation training has not been formally studied, but it is reasonable to imagine that students accustomed to inaccurate forces or displacements from simulation training might be at greater risk of tissue injury when applying their clinical skills in the actual operating room.

With few exceptions, most of the existing literature on the biomechanics of internal organ tissue comes from measurements taken from non-living tissue. Often the tissue has been frozen and thawed for convenient laboratory use. Physiologic changes in living tissue certainly influence the mechanical properties of soft tissues in-vivo. Another issue is the effect of fluid within the tissue. For example, after several similar loading cycles, the non-linear stiffness and hysteresis of soft tissues typically stabilizes – a phenomenon known as conditioning [2]. Most researchers "precondition" their tissue samples to obtain consistent results by cycling them 10–20 times before collecting data. This process runs counter to the normal conditions found in surgery since surgeons do not precondition tissues before manipulating them. "First squeeze" behavior of tissues has not been widely reported.

In 1967, Fung published classic work on rabbit mesentery in uni-axial tension [3]. Yamada in 1973 reported results of tests on esophagus, stomach, small and large intestines, liver, and gallbladder [4]. Much of this work was done with animal organs in-vitro, but some data was presented from human cadavers. Most of the data were expressed as tissue tension with an emphasis on measurement of failure levels.

A large literature describes testing abdominal organs in relation to blunt impact injury, especially in the context of automobile accidents. Yoganandan et al. (2001) and Rouhana (1993) reviewed many of these studies [5, 6]. More detailed measurements of specific organs include shear measurements of liver [7–9], and distension of intestine (relation between pressure and volume) [10, 11].

In the context of laparoscopic surgery, Carter et al. [12] measured the uniaxial force required to puncture pig and sheep livers with a scalpel as well as the displacement of the tissue when puncture occurred. Other studies by these researchers [13, 14] used a bench-top device in ex-corpus testing of pig and sheep liver and spleen. They also performed in-vivo measurements reviewed below. Tamura et al. [15] studied porcine liver, spleen, and kidney in-vitro by compression loading of rectangular-shaped samples. Elastic and stress relaxation properties were examined, but the nature of the studies – single-point displacement of small fragments of tissue – limit application to clinical conditions.

In an effort to improve the physiological accuracy of ex-corpus testing, some studies have used perfusion of the excised organ. Davies et al. [14] tested artificially perfused spleen, while Melvin et al. (1973) [16] placed intact kidney and liver into a uni-axial compression testing machine while still perfused by the body. The emphasis was on measurement of tissue failure (as low as 293kPa for liver). Other interesting in-vivo results have been obtained in research on prosthetics [17–19]. Zheng et al. (1999) used a combination of load cell and ultrasound to measure compressive properties non-invasively [17, 18]. However, this method requires a rigid backing, such as bone.

Brouwer et al. [20] developed several instruments for tensile and compressive testing of porcine tissues in-vitro and in-vivo. One of these devices contained two grippers whose separation was controlled by a lead-screw and stepper motor. Ottensmeyer and Salisbury [21] developed the TeMpEST 1-D device which applies high frequency, low amplitude compressive displacements to the surface of an organ. In-vivo testing with this device showed a relationship between elastic modulus and frequency. Carter et al. [13] used a similar hand-held indentation device and recorded the only published in-vivo data obtained from living human subjects. Maximum applied strain was 60kPa. Interestingly, diseased liver was at least twice as stiff as normal, which supports the clinical surgical impression. Kalanovic et al. [22] developed a rotary shear device (ROSA-2), which used a 6mm right cylindrical contact surface that rotates relative to a fixed outer ring. Slippage was prevented by a needle array or cyanoacrylate adhesive. Calculated material parameters agreed in the range of 0–10 Hz with those found with the TeMpEST 1-D.

In a departure from prior studies with specialized stress-strain measurement devices, Bicchi et al. [23] applied sensors to standard surgical tools, in this case adding force and position sensors to measure jaw force and angle in endoscopic surgical pliers. Morimoto et al. [24] instrumented a laparoscopic Babcock grasper with a six-axis force/torque sensor to record forces and torques applied during animal procedures. Their device successfully isolated tool-tissue interaction forces from forces arising from the abdominal wall and port. Brouwer et al. [20] used a six-axis force/torque sensor mounted to a modified grasper to measure the forces and torques applied to the tool while driving a needle through porcine abdominal tissues. Greenish et al. [25] instrumented scissors to collect in-situ data during cutting of skin, abdominal wall, muscle, and tendon tissues from sheep and rats.

Building on this experience, our group has developed a series of devices for measurement of tool-tissue interactions during surgery. We have developed a laparoscopic tissue grasper with six-axis force/torque and grasp force-sensing capability embedded in an articulated mechanism for measurement of motion in five axes (the "Blue DRAGON" system) [26–28]. The devices described above were passive, or human-powered, and were used for measurement of activity during simulated surgical procedures. For example, two of the Blue DRAGON devices were used on the left and right hand tools in experiments recording a database from 30 surgeons performing portions of a laparoscopic cholecystectomy (gallbladder removal) and gastric fundoplication (antireflux surgery) in pigs [83]. We have also developed and evaluated several motorized and teleoperated graspers, including the Force-Reflecting Endoscopic Grasper (FREG) [36]. Active, or motorized, tools facilitate the application of controlled displacements or forces to tissue under computer control. The FREG was used [29] to test several porcine liver, spleen, stomach, small and large intestine, and lung specimens in-vivo and measure their force-displacement response to stresses up to 100kPa and compressive strains up to 60%. The measured force-deformation responses could be fit with an exponential function, resulting in two coefficients that could differentiate the tissues.

Based on data collected with the Blue DRAGON system, the Motorized Endoscopic Grasper (MEG) was designed to reproduce the maximum grasping forces and velocities observed during clinical surgical tissue manipulation and acquire more extensive and reliable compressive data from abdominal organs [30, 31]. Full characterization of a non-linear, fluid-perfused, non-isotropic and non-homogeneous material such as the major internal organs is a complex endeavor. To name just one difficulty, proper modeling of bulk materials requires knowledge from tri-axial testing that can only come from tissue biomechanical studies that are not similar to surgical conditions. Although in general it will not be possible to fully characterize these materials with the uni-axial compressive tests our instrument can perform, we must begin to measure at least basic in-vivo properties.

The emphasis in this paper is on the devices and methodology for collection of tissue performance data rather than tissue modeling. For clarity, a few curves were fit to the data and have been included. A more complete description of tissue models derived from these data are provided in a companion paper [32].

There is substantial literature on mathematical models for the response of soft tissue to mechanical testing. Fung [2] noted that many tissues seem to follow an exponential relationship between stress and strain. Soft tissues also exhibit hysteresis between loading and unloading. The loading and unloading curves are generally different, and we will concentrate on the loading curve only (pseudo-elasticity). For example, Brouwer et al. [20] fit their data to Fung's exponential curve. A similar procedure was used by Rosen et al. [29] and Tamura et al. [15].

There are many approaches for modeling the time-dependent response of soft tissues, including Quasi-Linear Viscoelasticity (QLV) [2, 33, 34], bi-phasic models [35–41], and even tri-phasic theory [42] involving solid, fluid, and ionic concentration state variables. While there is much potential to apply sophisticated time-dependent models to our data, at this point we will limit ourselves to simply fitting our data with first order exponential time functions.

As indicated in this literature review, biomechanical properties were studied at the macro scale level in selected well controlled experimental conditions, however little is known about the types of stresses that can be safely applied using surgical instruments while limiting tissue damage and potentially injury. In earlier work with the porcine animal model, we measured relationships between acute indicators of tissue injury and average surgical grasping stress [47]. In these experiments, tissue damage was observed even at low average grasping forces, suggesting that the observed tissue damage might correspond to the spatial stress distribution between the grasper jaws instead of average stress.

It is evident from the literature that four things are lacking for modeling tissues in the context of surgery: (1) an understanding of how surgeons interact with tissues (i.e., to establish the relevant scale of stress and strain), (2) compression testing, (3) in-vivo data, and (4) human data (5) an understanding of tissue damage a the cellular level as well as the relationship between the stress developed at the tissue and acute tissue injury. Most studies have tested tissues in-vitro in tension using excised animal specimens (often after freezing and thawing).

24.2 Methods

24.2.1 Macro Scale Biomechanics

24.2.1.1 Definitions

In this study, in-vivo will refer to testing done inside an intact live specimen, with the organ in its normal position. In-situ will refer to testing the same organs after the animal has died, but with the organs still in the body proper. In-vitro refers to testing done outside the body, using tissue samples that have been excised from the bulk organ. Finally, *ex-corpus* will refer to intact, non-living organs removed from the body, and possibly stored before testing some time postmortem.

24.2.1.2 Tools

Two types of tools were used to acquire the biomechanical properties of internal organs in-vivo, in-situ and *ex-corpus*: (1) a custom-made motorized endoscopic grasper (MEG), used in all conditions; and (2) a servohydraulic universal testing material testing system by MTS Corporation (Eden Prairie, Minnesota), used for testing tissue only *ex*-corpus only.

24.2.1.3 Motorized Endoscopic Grasper

The motorized endoscopic grasper is the second generation of Force-Reflecting Endoscopic Grasper (FREG) [29] that was originally designed as a 1 degree-of-freedom (DOF), bi-lateral teleoperated system, but was also capable of applying in-vivo computer controlled sequences of compressive force via a flat-coil actuated endoscopic grasper (slave element). As such, it was used to test several porcine abdominal tissues in-vivo to measure their stress-strain response but could only apply approximately 8N compressive force that was estimated by measuring the current to the flat-coil actuator. Following these research efforts the Motorized Endoscopic Grasper (MEG) was designed to further examine the compressive properties of porcine abdominal organs [30, 31]. The engineering specifications of the MEG were based on data collected from previous experiments using the Blue DRAGON surgical tool tracking system [43]. These data were examined in order to determine the forces, deformations, and timing of compressive loads applied on tissues.

The MEG uses a brushed DC motor (RE25, 10W, Maxon Precision Motors – Fall River, MA) with a 19:1 planetary gearhead (GP26, Maxon Precision Motors – Fall River, MA) to drive a Babcock grasper (#33510 BL, Karl Storz – Germany) – Fig. 24.1. The motor is attached to a capstan that drives a cable and partial pulley. The pulley is attached to a cam joint that converts the rotational motion of the motor and pulley to a linear translation of the grasper shaft, which opens and closes the jaws. A 500-count digital encoder (HEDL55, Hewlette-Packard – Palo Alto, CA), attached to the motor, measures angular position. The mechanism's overall effective gearing ratio is approximately 190:1, including the planetary gearhead ratio (19:1) and the partial pulley-capstan gearing ratio (10:1), increasing the 29 mNm of continuous torque generated by the motor to 5.51Nm applied by the partial pulley. A wide variety of standard Karl Storz laparoscopic instruments can be attached to the base plate mount, but a Babcock grasper (Fig. 24.1c) was selected as the primary loading device due to its special geometry. Range of motion for the Babcock jaws is 54.3 deg,

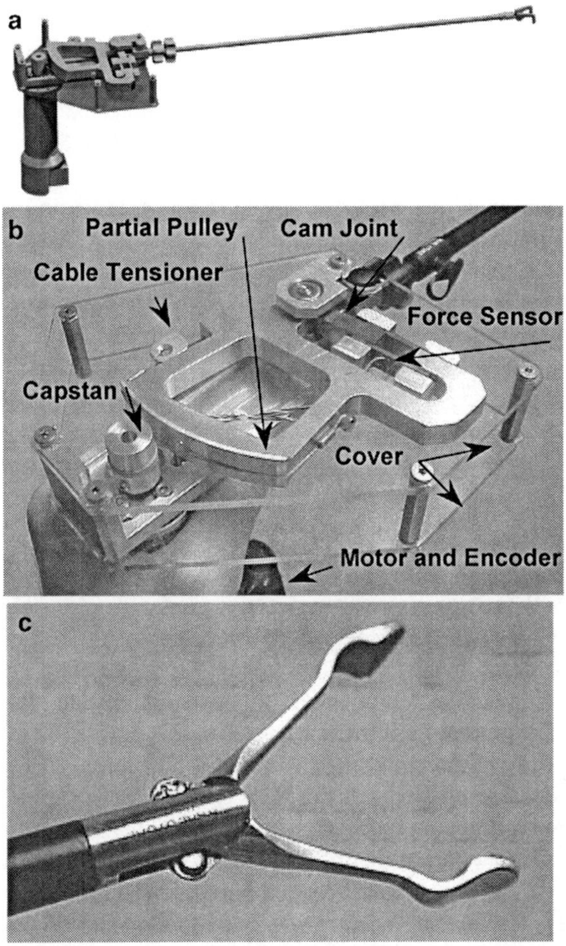

Fig. 24.1 The Motorized Endoscopic Grasper (MEG): (**a**) rendered CAD drawing of MEG (protective top cover not shown), (**b**) close-up photograph of the MEG's drive mechanism, (**c**) close-up photograph of the MEG's Babcock grasper end effector

or 184 deg at the capstan. Resolution of jaw angle is approximately 1.13×10^{-2} deg per encoder count (5.5×10^{-3}mm at the jaws' grasping surfaces). At full opening, the two grasping surfaces are 26.3mm apart.

A double-beam planar force sensor (FR1010, 40lb, Futek – Irvine, CA) is mounted in the partial pulley, measuring force applied to the end effector. The signals are amplified with a Futek signal conditioning unit (model JM-2). The resolution of force signals following a 16-bit A/D conversion is 0.6 mN. A noise level of up to 50mN, including the quantization noise, was observed, which represents 0.0 25% of the sensor's full scale. The maximum continuous motor torque of 29 mNm is equivalent to 26.5 N of grasping force by the Babcock grasper's jaws, after transmission through the mechanism, which is greater than the average force applied by surgeons during typical surgical tasks [43]. Based on the Babcock grasper's jaw dimensions, the application of 26.5 N is equivalent to a compressive stress of 470 kPa. The MEG is hand-held and weighs 0.7 kg. It is inserted into the body through standard 10mm endoscopic "ports" used for passing videoendoscopic instruments into the body without losing the gas pressure in the abdomen.

Computer control of the MEG is provided via a PC using a proportional-derivative (PD) position controller implemented in Simulink (Mathworks – Natick, MA) and dSPACE (Novi, MI) user interface software (ControlDesk) and hardware (DS1102). Current is supplied to the motor via a voltage-controlled current supply (escap ELD-3503, Portescap – Hauppauge, NY) controlled by the output from the dSPACE board (D/A 16-bit). The control loop runs at 1kHz. The MEG was calibrated to address the nonlinear relationship between the position of and the force applied by the distal tool tips with respect to the sensors located on the proximal end of the tool (defined analytically in [29]), as well as to compensate for mechanism compliance and backlash.

24.2.1.4 MTS Setup

The testing system by MTS Corporation is a standard servo-hydraulic universal-testing machine often used in material testing in the field of biomechanics. The custom-built frame was used with a model 252 valve. Maximum closed-loop velocity of the ram using this valve is 500 mm/s.

The experimental setup used with the MTS machine for tissue testing is shown in Fig. 24.2. The top and bottom indenters were identical 7 mm diameter right circular cylinders providing a contact area of 38.5 mm^2, compared to the MEG's contact area of 56.4 mm^2. The top indenter screwed into the MTS ram (the moving portion of the machine). The bottom indenter was fixed to the tension/compression force sensor (44.5N tension/compression unit, Sensotec model #31/1426–04). The force sensing resolution was 21.7m N. A noise level of up to 9m N including the quantization noise, was observed, which represents 0.019% of the sensor's full scale. The force sensor rested in a stainless steel base plate that was affixed to the MTS frame. The top of the base plate and the top of the bottom indenter were aligned. The organ rested on the base plate and the bottom indenter. The opening was just large enough to accommodate the force sensor but not allow the tissue to droop significantly. Additionally, the base plate had two grooves, one vertical slot for routing the force sensor's wire and the other a horizontal one around the entire base for cinching down a very thin plastic sheet with a rubber band. This plastic sheet protected the force sensor from fluids present during testing. Despite the presence of this sheet and the fact that the effective top of the force sensor and the rest of the plate were level, it was assumed that the force sensor would measure the majority of the applied pressure, since the film was very thin and flexible and there was a relatively large gap surrounding the force sensor indenter (Fig. 24.2).

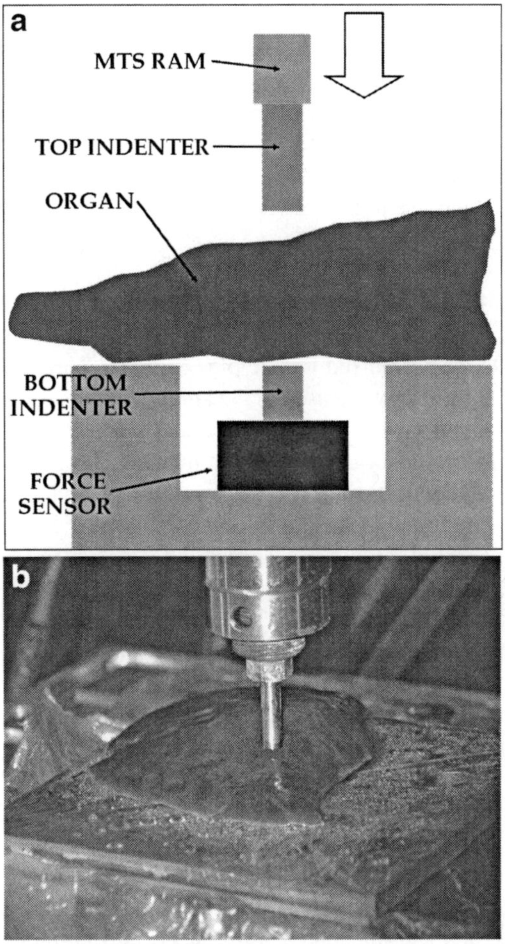

Fig. 24.2 MTS experimental testing machine setup: (**a**) schematic overview of the system, (**b**) the setup with a liver ex-corpus

The MTS ram was operated in a position-control mode using TestStar II software and hardware. Axial position was sensed using a linear displacement transducer (LVDT) mounted on the hydraulic ram in the frame's crosshead (model 244.11). Analog signal conditioning was performed in hardware before passing to the PCI-based, 12-bit analog/digital conversion board (PCI-6071-E, National Instruments). The axial position resolution was 0.0074 mm in a preset $\pm$15.24 mm range. Data were sampled at 1 KHz or faster.

One may note that the sensors on the MTS and MEG aimed to measure the end effector position (and therefore the tissue thickness) and the forces applied on the tissue are located at different places along their respective kinematic chains, starting at the actuator and ending at the end effector. However, given the kinematics and the dynamics of each chain, the sensors' readings were mapped from their locations to the devices' end effectors. Locating the MEG's position and force sensors proximal to the end effector without altering the end effector itself was motivated by the desire to use standard surgical instruments in a typical surgical environment. Placing a sensor on the end effector that could survive the in-vivo environment while not significantly altering the tool's geometry and/or ability to be used in-vivo would be extremely difficult. Moreover, since the endoscopic tool remained unchanged it is possible to remove it completely from the MEG, sterilize it, and use it in a survival procedure.

24.2.1.5 Experimental Protocol and Loading Conditions

Three-month-old female pigs (porcine Yorkshire cross) with an average weight of 37 ($\pm$5) kg were used as the animal models for the experimental protocol. The same animal model is used for training laparoscopic surgeons due to its similar internal abdominal organ anatomy to humans. Seven internal organs (liver, spleen, bladder, gallbladder, small and large intestines, and stomach) taken from 14 different pigs were tested in various testing conditions (in-vivo, in-situ, and *ex-corpus*). The MEG was used for testing all seven organs of six animals, whereas the MTS machine was used on four organs (liver, spleen, small intestine, and stomach) from three animals. The MEG was used in all conditions, while the MTS was obviously used for only *ex-corpus* testing. (The remaining animals were tested with some mix of condition and organ.) The in-vivo and in-situ experiments were recorded visually using the endoscopic camera, synced with force-deformation data, and recorded on digital video for off-line analysis and archival.

In-vivo tests were performed on a sedated and anesthetized animal as per standard veterinary protocols and typical for a laparoscopic training procedure at the University of Washington Center for Videoendoscopic Surgery, an AALAC-accredited facility. The abdomen was insufflated with CO_2 to a pressure of 11–12 mmHg, as typical in porcine MIS procedures. Three laparoscopic ports (10mm in diameter) were placed into the abdomen, which allowed access to all the organs to be tested as well as visualization of the tool tip by the endoscopic camera. In-situ tests were conducted under the same experimental conditions on the euthanized animal immediately postmortem. *Ex-corpus* testing was performed at the UW Applied Biomechanics Laboratory. For the organ harvesting, blood vessels to the organs were cut, and blood was free to drain and clot. Hollow organs were stapled and then cut to ensure any contents remained intact. The organs were kept moist with 0.9% saline solution and stored in an ice chest with ice packs. The ex-corpus testing took place in a climate-controlled room; the temperature was held at 22.7°C with a humidity of 22% during all the tests. During the ex-corpus tests, the tissues were constantly kept moist with sprays of saline solution; the organs were never frozen.

Cyclic and step strains were used as the two loading conditions for testing the various soft tissues. In addition to these two loading conditions, the tissues were tested to failure, defined by a tissue fracture, by both devices *ex-corpus*. The loading characteristics used as part of the experimental protocol were defined based on a detailed analysis of the grasping action in laparoscopic surgery, as measured by the Blue DRAGON system [43]. Moreover, since laparoscopy, by definition, is performed in-vivo, collecting load-response data under similar conditions is paramount to reflecting the nature of these biological materials as presented to the surgeon clinically. Emulating surgical conditions as part of the experimental protocol guaranteed that models that were developed based on the collected data reflected the appropriate nature of these biomaterials for future applications, such as haptic virtual reality surgical simulators. This concept manifested itself profoundly in the experimental protocol design and execution.

One of the major deviations from a more common soft tissue biomechanical testing protocol was in regard to tissue preconditioning. Due to the viscous nature of soft tissues, their deformation response changes with each successive loading cycle [2]. A stable behavior can develop after several loading cycles, at which point the tissue has been "conditioned," and its hysteresis loop is minimized. Conditioning a tissue before testing (referred to as "preconditioning") often takes 10–20 cycles, depending on the tissue and the loading condition [2]. Since tissues are not preconditioned before being manipulated in surgery, first-cycle behavior is of great interest, as is steady-state behavior and the number of cycles to reach conditioning. No preconditioning was performed during this study. A new site (location on the organ) was used for each test regime to ensure the natural (unconditioned) state of the tissue was measured.

Initial tissue thickness was determined by the distance between the tool tips (or indenters) at the point of first contact. Each subsequent cycle used this same value, whether or not the tissue was actually in contact at this distance. This was done to observe any depressions left in the tissue after the previous compression.

The first type of load applied was a cyclic position (strain) waveform, in order to examine the tissues' elastic stress-strain response. The constant velocity (triangle-shaped) strain signal was the cyclic loading profile of choice for the following reasons: (1) it allows controlled strain rate, (2) it facilitates tool-tip contact detection based on deviation from nominal velocity, (3) it has been used in previous studies. The second type of load applied was a single position (strain) step, in order to examine the stress-relaxation properties of the tissues. A viscous material exhibits an exponential decrease in the measured stress within the material while the strain is held constant. Analysis of measurements made with the Blue DRAGON [43] indicated that maximum grasp time during various surgical tasks was 66.27 s. The average maximum grasp time was 13.37 ± 11.42 s, the mean grasp time was 2.29 ± 1.65 s, and 95% of each subject's grasps were held for less than 8.86 ± 7.06 s. Based on these results, a short hold time (10s or less) could be used for loading the tissues. However, it is useful for modeling purposes to examine the relaxation over a longer period of time, in order to better characterize the behavior. For practical purposes, the step strain was held for 60 s at three different strain levels (in different tests), targeted between 42 and 60% strain. During the step strain tests, the MEG end effector was commanded to close as rapidly as mechanically possible. It is important to note that the entire organ under study remained intact throughout the experimental protocol. Although the compressive loads were applied uni-axially on the various organs, the surrounding tissues of the organs themselves define the boundary conditions. These boundary conditions are fundamentally different from the boundary conditions of a sample of tissue removed (excised) from an organ. With such a sample either free boundary conditions or confined boundary conditions within a fixed geometry can be used. Setting such controlled boundary conditions is a common practice in material testing; however, keeping the organ intact better reflects the boundary conditions encountered during real surgery. These testing conditions imply that the results reported in this study refer to both structural *and* material properties of tissues, not just to the material properties. In addition to the loading and boundary condition, the testing location on the organs were limited to the organs' peripheries for both the MEG and the MTS. These testing locations were selected due to the fact that the Babcock jaws of the MEG were less than 3 cm long; it was impossible to test the interior bulk of the larger organs like liver and stomach with the MEG.

24.2.1.6 Data Analysis: Phenomenological Models

Two fundamental approaches exist for developing models of soft tissue mechanical behavior: (1) constitutive, physical law-based models, such as strain energy function models; and (2) phenomenological models based on curve-fitting experimental data. The former approach leads to easier extraction of physical meaning of the parameters but may not have perfect fits with the acquired data. The latter approach has little or no physical relevance but may achieve excellent fits to the acquired data with potentially less computationally intensive functions. Due to the empirical emphasis of this study, a phenomenological modeling approach was used. In order to evaluate which of these methods should be selected, a series of candidate curves were defined and evaluated for their ability to fit a significant portion of the dataset accurately and consistently. The measures of fit that were examined were the mean, median, and standard deviation of both R^2 (regression coefficient) and RMSE (root mean squared error).

Elastic Models

Eight functions were chosen to model the elastic characteristics of the tissue. In these equations, the engineering (nominal) stress (σ) is defined to be the ratio of compression force (F) applied on the tissue to the contact area (A) – (24.1a). The engineering strain (ε) is defined as the difference between the initial thickness of the tissue (l_0) under no load and the actual thickness under the compression load (l) normalized with respect to the initial thickness. Each model assumes zero compressive stress (σ) at zero strain (ε), and a positive stress at positive strain. Theoretically, compressive strain must be less than unity (1), since a value of 1.0 indicates the material has been totally compressed.

$$\sigma = \frac{F}{A} \tag{24.1a}$$

$$\varepsilon = \frac{l_0 - l}{l_0} \tag{24.1b}$$

The first function (24.2) to be examined is a basic exponential function, referred to as EXP. Various forms of this equation have been used by several researchers [2, 15, 18, 20, 29]. α and β are coefficients determined by curve-fitting the experimental data.

$$\sigma = \beta(e^{\alpha \in} - 1) \tag{24.2}$$

The second function (24.3) is an expansion of EXP, introducing a linear term and increasing the order of strain to ε^2. This equation was developed for this study and is referred to as EXP2. Again, γ is a coefficient obtained by curve-fitting experimental data.

$$\sigma = \beta(e^{\alpha \in^2} - 1) + \gamma \in \tag{24.3}$$

The third function (24.4) incorporates the inverse of strain and is referred to as INV. This equation introduces a vertical asymptote in the stress-strain relation. This asymptote must lie between $\varepsilon = 0$ and $\varepsilon = 1$. There may be some physical relevance to the value of this strain asymptote: it may reflect the amount of fluid within the tissue that cannot be exuded, or the point at which the tissue becomes incompressible.

$$\sigma = \beta(\frac{1}{1 - \alpha \in} - 1) \tag{24.4}$$

The fourth function (24.5) is a uni-axial form of a Blatz-Ko model and is referred to as BLATZ. This equation was previously used to model the kidney and liver under compression loading [44].

$$\sigma = \frac{-\gamma}{\alpha + 1}((1 - \in)e^{(\alpha((1-\in)^2 - 1)} - \frac{1}{(1 - \in)^2} e^{\alpha(\frac{1}{1-\in} - 1)} \tag{24.5}$$

The final functions (described by (24.6)) are polynomials with increasing order from second (i=2) to fifth (i=5). They are referred to as POLY2 through POLY5.

$$\sigma = \sum_{i=1}^{n} c_i \varepsilon^i \tag{24.6}$$

The derivative of a stress-strain function with respect to strain defines the material stiffness, or tangent modulus. A linearly elastic material's stiffness would be a constant, or Young's modulus. The derivative of an exponential stress-strain relationship is a function of its strain (e.g., the derivative of (24.3) with respect to strain results in (24.7)). The overall stiffness indicators defined for EXP2 are $\beta\times\alpha$ and $\beta\times\alpha+\gamma$, which serve as useful scalars for roughly approximating overall stiffness of a material and allowing quick comparisons between materials.

$$\frac{d\sigma}{d\in} = 2\alpha(\beta e^{\alpha\in^2})\in+\gamma \tag{24.7}$$

Stress Relaxation Model

Three functions were selected to model the stress-relaxation data. The first function (24.8) is a logarithmic function with two time constants [2, 15] that is referred to as RLOG:

$$\sigma(t) = -A\,In\,(t) + B \tag{24.8}$$

where

$$A = \frac{c}{1 + c\,In(\tau_2) - c\,In(\tau_1)}$$

$$B = A\left(\frac{1}{c} - \gamma + In(\tau_2)\right)$$

and γ is the Euler constant (γ=0.5772). Curve-fitting experimental data results in τ_1 and τ_2 (time constants) and c.

The second stress-relaxation function (24.9) is a decaying exponential function with a single time constant [2, 18, 45, 46] that is referred to as REXP1:

$$\sigma(t) = 1 - a + ae^{\frac{-t}{\tau}} \tag{24.9}$$

with *a* being a curve-fit coefficient.

The third equation (24.10) is a decaying exponential raised to a power, with a single time constant. This function is referred to as REXP2.

$$\sigma(t) = \exp\left(\left(\frac{-t}{\tau}\right)^{\beta}\right) \tag{24.10}$$

24.2.2 *Micro Scale Biomechanics*

Compression stresses at magnitudes between 0 and 250 kPa were applied in vivo to porcine abdominal organs using a motorized endoscopic grasper. Test tissues were harvested after 3h, and tissue injury was measured from histological sections based on cell death, fibrin deposition, and neutrophil infiltration. Based on preliminary FE models [47], it was determined that the central portion of the compression site would have a uniform stress level. The central uniform region was chosen as the site of histological damage measurement in order to reduce variance from spatial stress variation.

24.3 Results

24.3.1 *Macro Scale Biomechanics*

24.3.1.1 Elastic Testing

Compression stress-strain experimental data plots of various internal organs are depicted in Fig. 24.3 and the associated elastic phenomenological model (EXP, EXP2, and INV) curve fits are plotted in Fig. 24.4. Example organ response data, as well as the phenomenological models and their fit are plotted for the liver in Fig. 24.5. The average of the individual EXP2 model parameters across all conditions based on the MEG and MTS measurements in-vivo and *ex-corpus* are summarized in Table 24.1.

As indicated in Fig. 24.3, there is a major change in the stress-strain curve between the first and fifth loading cycles. Moreover, Fig. 24.3 depicts the spectrum of stress-strain characteristics bounded by the two extreme experimental conditions: (1) first cycle compression in-vivo – a typical loading condition during surgery (Fig. 24.3a), and (2) near-preconditioned fifth compression cycle *ex-corpus* – a loading condition more typical to biomechanical characterization analysis of soft tissue (Fig. 24.3b).

In general, it appeared that a tissue's stiffness increased with subsequent loading cycles for the first 7–10 loading cycles, at which point the stress-strain behavior reached a steady-state phase, indicating the point at which the tissue likely became conditioned. Note the marked difference in shape of the stress-strain curve between first and fifth loading cycles in spleen (Figs. 24.3 and 24.4). This behavior was noted visually during spleen testing by the fact that the MEG jaws tended to leave a deep impression in the organ after the first loading cycle; the tissue did not recover to its initial thickness after the first loading cycle. The spleen also appeared to have a nearly constant stiffness on first compression but became more exponential on subsequent cycles. The hollow organs, particularly small intestine, tended to have two distinct parts to their stress-strain curves, separated by an abrupt change in stiffness. The first part represents moving of the walls and compression of the

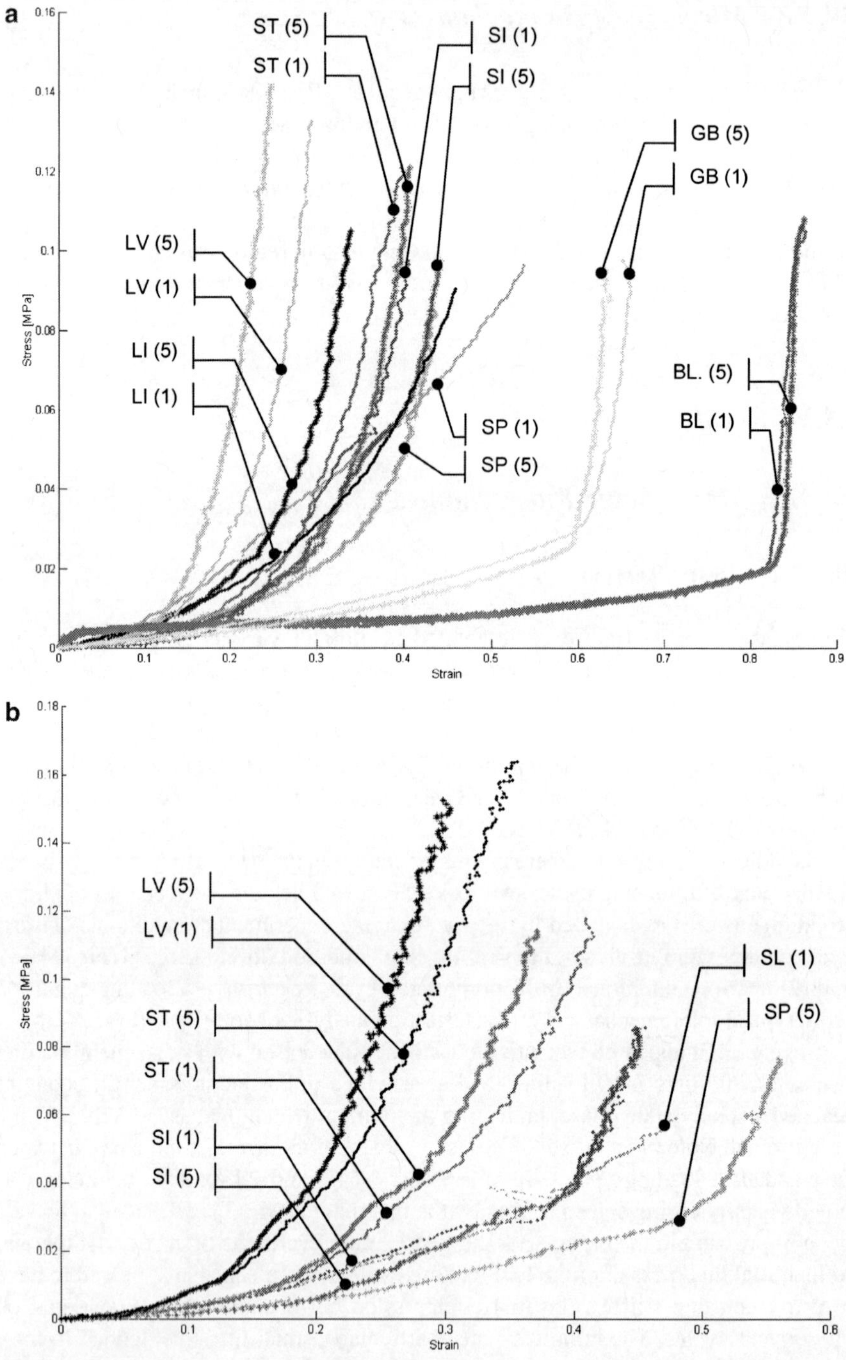

Fig. 24.3 Stress-strain curves for all organs under study, as measured with the MEG at 5.4 mm/s loading velocity (first and fifth cycles shown): (**a**) in-vivo, (**b**) ex-corpus. Organs' legends: *BL* bladder, *GL* gallbladder, *LI* large intestine, *LV* liver, *SI* small intestine, *SP* spleen, *ST* stomach. The loading cycle number (1 or 5) is defined in the brackets

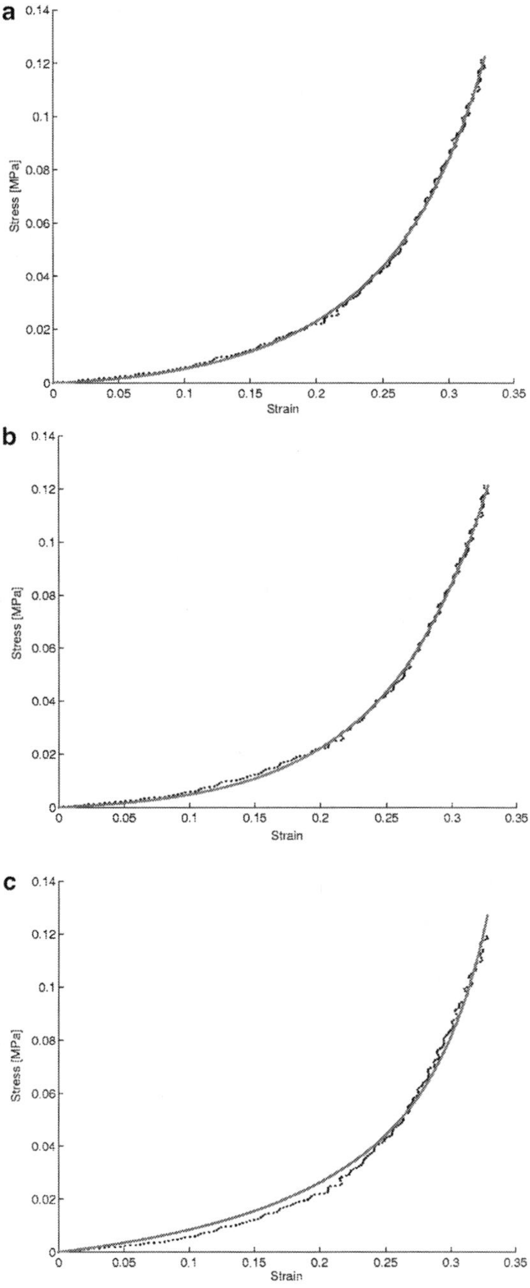

Fig. 24.5 Measured data and phenomenological models of liver tissue under compression loading. The same in-vivo data measured by the MEG was fit with various models. The measures of fit for these models are: (**a**) EXP2, R2 = 0.9989, RMSE = 1.5048E3; (**b**) EXP, R2 = 0.9984, RMSE = 1.5166E3; (**c**) INV, R2 = 0.9931, RMSE = 3.0291E3

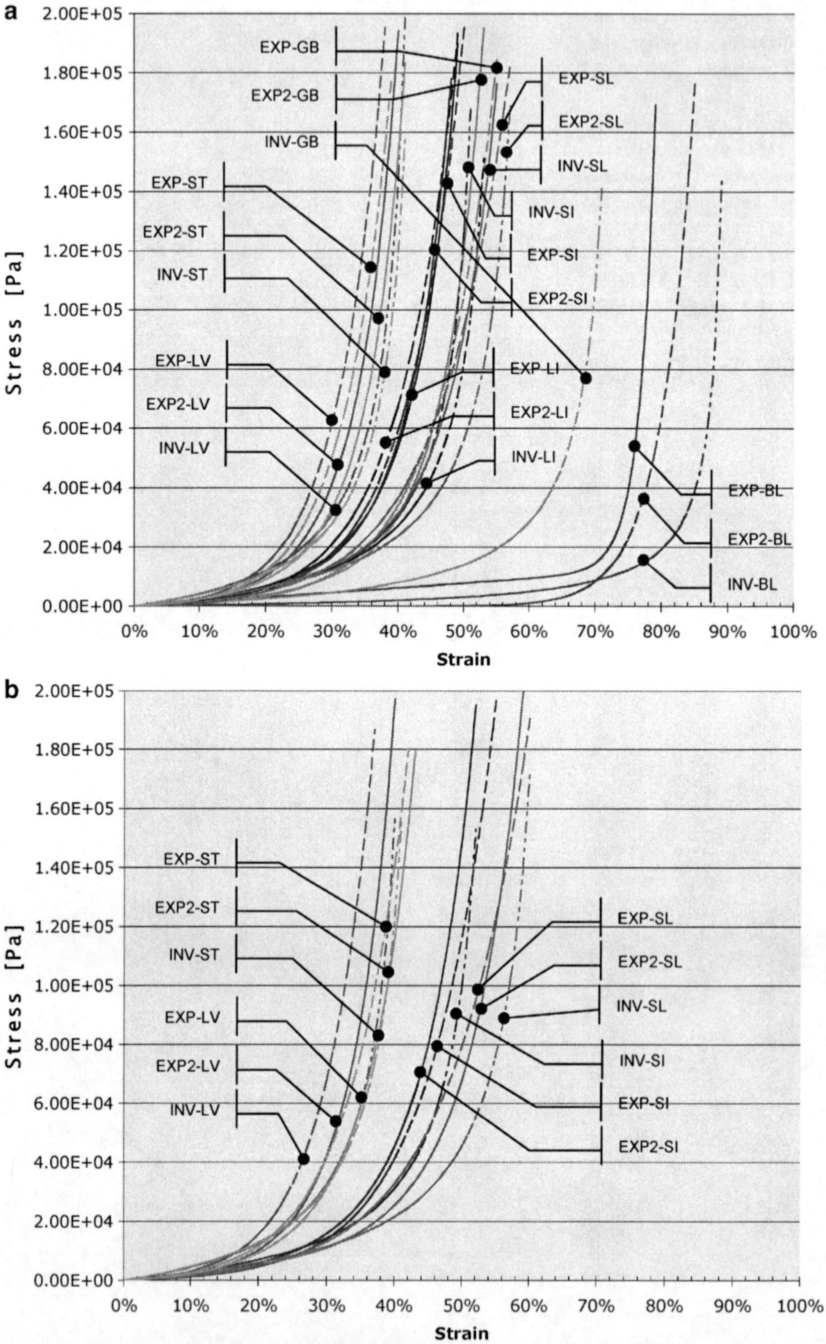

Fig. 24.4 Stress-strain curves for all organs with average curve-fit parameters across all conditions: (**a**) in-vivo data measured by the MEG, (**b**) ex-corpus data measured by the MEG, (**c**) ex-corpus data measured by the MTS. Organ legend: *BL* bladder, *GL* gallbladder, *LI* large intestine, *LV* liver, *SI* small intestine, *SP* spleen, *ST* stomach. See text for the definitions of the functions EXP, EXP2, INV

Table 24.1 Mean values of the EXP2 model parameters (α, β, γ) for each organ, in-vivo and ex-corpus, as tested by the MEG and MTS, across all animals, loading velocities, and cycle number

Device:	MEG			MEG			MTS		
Condition:	In-vivo			Ex-corpus			Ex-corpus		
Parameters: Organ	β (Pa)	α	γ (Pa)	β (Pa)	α	γ (Pa)	β (Pa)	α	γ (Pa)
Bladder	0.0041	27.98	15,439.2	N/A	N/A	N/A	N/A	N/A	N/A
Gallbladder	2,304.5	15.75	9,622.2	N/A	N/A	N/A	N/A	N/A	N/A
Large intestine	3,849.7	16.14	16,544.1	N/A	N/A	N/A	N/A	N/A	N/A
Liver	7,377.1	20.63	3,289.4	7,972.1	20.29	781.0	8,449.8	26.26	1,679.4
Small intestine	3,857.3	16.60	11,273.8	6,166.5	12.81	7,967.5	1,745.9	13.60	2,580.9
Spleen	3,364.4	12.94	19,853.1	3,798.8	11.31	14,440.4	2,764.9	11.85	13,103.8
Stomach	4,934.9	21.51	11,105.9	8,107.0	16.91	6,483.8	2,247.6	21.22	6,803.3

contents (solid, air, or liquid). The second part occurs when the two walls of the organ contact each other. This portion, then, can be considered the actual deformation behavior of the *tissue* and should appear similar to the responses obtained by the other (solid) organs. One could argue the entire curve represents the clinically relevant behavior of the *organ*.

Large intestine response to loading was different then the small intestine, which could be attributed to its thicker walls and generally larger shape (Fig. 24.3a). However, because it contained stool, it tended to show drastically different biomechanical behavior between the first and subsequent squeezes as the contents were compressed and moved about. Small intestine tended not to have as much volume of contents as did the large intestine.

Two other hollow organs that show different behavior from the other organs, bladder and gallbladder, were fluid-filled. Therefore, their initial response was simply from the stretching of the membranous walls – more like tensile testing than compression. When the walls finally came together, because they were so thin, the jaws were essentially touching and the sudden change in stiffness to nearly rigid was observed (Fig. 24.3a). *Ex-corpus* results were generally similar to those seen in-vivo (Fig. 24.3b). For example, small intestine still had the two-part shape, and first-load cycle of spleen tended to be different from subsequent cycles. Ranges of stress and strain appeared to be similar, as well. One key difference was the amount of internal compression variability. Aside from the difference between first and second loading cycles, the stress-strain behavior reached a consistent response more quickly. This may indicate a more rapid onset of tissue conditioning, or it could be less influence from in-vivo factors such as ventilator motion and tissue re-perfusion.

24.3.1.2 Stress-Relaxation Testing

Experimental data of normalized stress-relaxation under compression loading are depicted in Fig. 24.6a for the liver. The stress was normalized with respect to

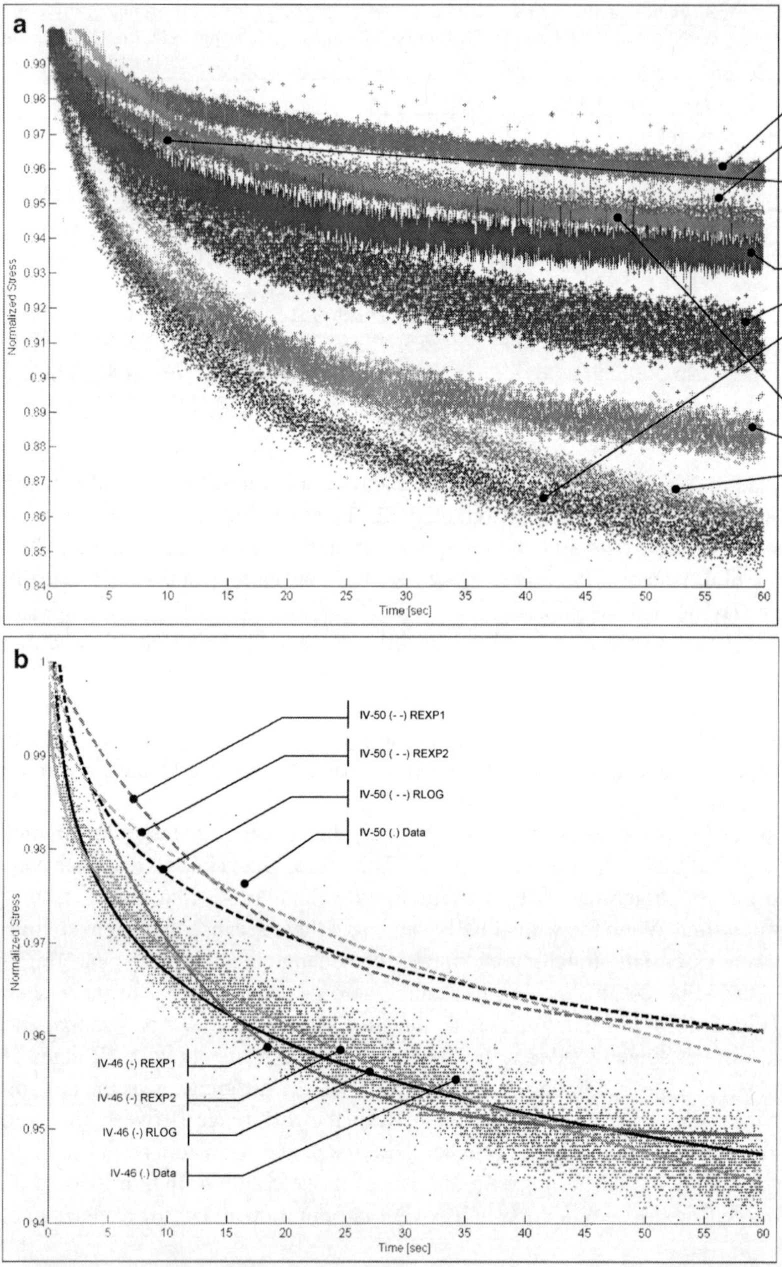

Fig. 24.6 Normalized stress-relaxation curves as a function of time for one liver tested with the MEG: (**a**) three different testing conditions (*IV* in-vivo, *IS* in-situ, *EC* ex-corpus) and strain levels (indicated in the legends as a two-digit numeral [% strain]; (**b**) measured data and phenomenological models of two strain levels. Their measures of fit: 46% strain [REXP1 (R2 = 0.8948, RMSE = 0.0042), REXP2 (R2 = 0.9261, RMSE = 0.0030), RLOG (R2 = 0.9084, RMSE = 0.0034)], and strain 50% [REXP1 (R2 = 0.9387, RMSE = 0.0026), REXP2 (R2 = 0.9526, RMSE = 0.0021), RLOG (R2 = 0.9140, RMSE = 0.0028)]

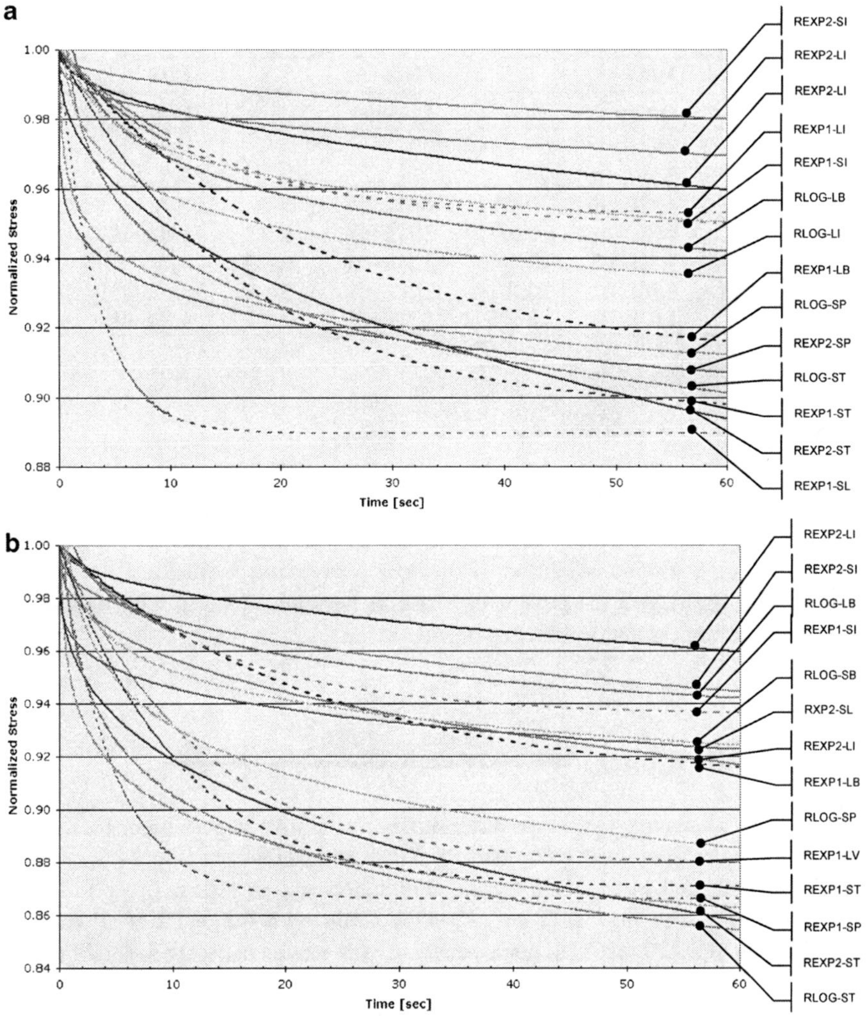

Fig. 24.7 Average normalized stress-relaxation curves for internal organs, based on mean values of REXP1, REXP2, and RLOG models: (**a**) in-vivo, (**b**) ex-corpus. Organ legend: *BL* bladder, *GL* gallbladder, *LI* large intestine, *LV* liver, *SI* small intestine, *SP* spleen, *ST* stomach. See text for the definitions of the functions REXP1, REXP2, RLOG

the maximal value of the stress that applied during the loading phase. The associated phenomenological models (REXP1, REXP2, and RLOG) curve-fit functions are plotted in Figs. 24.6b and 24.7. The average of the individual REXP2 (the overall best fitting model) parameters across all conditions based on MEG and MTS measurements in-vivo and *ex-corpus* are summarized in Table 24.2.

The stress-relaxation data acquired from liver in-vivo and *ex-corpus* for various step strain levels are depicted in Fig. 24.6. The maximum value of the total decrease

Table 24.2 Mean values of the REXP2 model parameters (τ, β) for each organ, in-vivo and ex-corpus, as tested by the MEG and MTS across all animals, loading velocities, and cycle number

Device:	MEG		MEG		MTS	
Condition:	In vivo		Ex-corpus		Ex-corpus	
Parameter: Organ	τ(s)	β	τ(s)	β	τ(s)	β
Large intestine	4.72E+04	0.479	N/A	N/A	N/A	N/A
Liver	4.95E+06	0.307	3.71E+04	0.381	1.40E+00	0.233
Small intestine	7.87E+05	0.412	1.13E+05	0.380	N/A	N/A
Spleen	6.70E+07	0.167	1.10E+07	0.208	8.84E-01	0.188
Stomach	1.03E+04	0.425	1.73E+04	0.331	4.59E-01	0.189

in stress was about 4–6% over the 60s test in-vivo, while the in-situ and *ex-corpus* maximum total decreases were 6–14%. The data indicate three general trends: (1) greater percent decreases in stress in the in-situ and postmortem conditions compared to the in-vivo condition, (2) greater decrease in normalized stress with less applied strain, and (3) greater decrease in normalized stress with increasing time postmortem (in-situ versus *ex-corpus*.

24.3.1.3 Failure: Liver

One benefit of testing tissues postmortem is the ability to test them to failure. Failure for liver tissue was examined for MEG and MTS tests (Fig. 24.8). Tissue failure is indicated in Fig. 24.8 by an abrupt decrease in stress. Liver failed at 35–60% strain with the MEG and 30–43% strain with the MTS at stresses of 160–280kPa and 220–420kPa, respectively. These results compare favorably with previously collected data reporting ultimate strain for liver at 43.8±4.0% (range: 39.0–49.1%) and an ultimate stress of 162.5 ± 27.5kPa (range: 127.1–192.7kPa), when loaded at 5mm/s [15]. It is important to mention the difference in the boundary conditions between the two studies: in the study by Tamura et al. [15], rectangular samples were used rather than intact organs, as in this study. Some differences are therefore to be expected, but the orders of magnitude are similar, suggesting good agreement for both MEG and MTS results.

It was observed that failure mode was different for the MEG and MTS devices. The MEG, with its rounded and smooth jaw edges, tended to crush the internal structure of the liver, the parenchyma, a condition known as liver fracture. No damage to the outer capsule was visible, other than a depression. The indenter on the MTS machine, however, tended to tear the capsule before fracturing. This was likely due to the indenter's sharp edges and the sloping of the organ surface (Fig. 24.2).

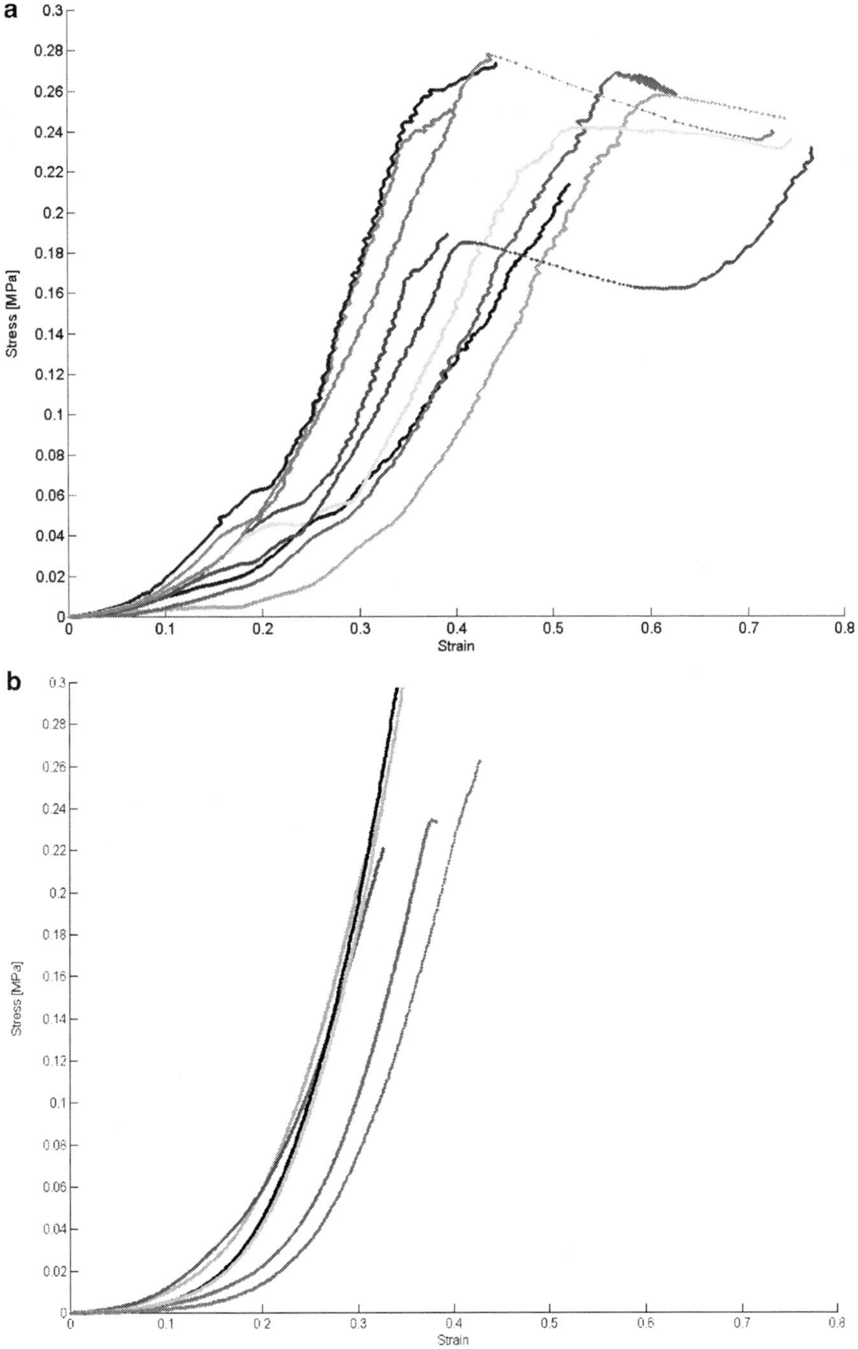

Fig. 24.8 Ex-corpus stress-strain characteristics of the liver under compression loading to failure: (**a**) MEG, (**b**) MTS

Table 24.3 The best fit of phenomenological models to the in-vivo experimental data acquired from various internal organs by the MEG under the two compression loading conditions (elastic stress-strain and stress relaxation) across all conditions

Organ	Data type	Model
Bladder	Elastic	EXP2
Gallbladder	Elastic	INV
Large intestine	Elastic	EXP2
	Relaxation	REXP2
Liver	Elastic	EXP2
	Relaxation	RLOG (REXP2)
Small intestine	Elastic	EXP2
	Relaxation	REXP2
Spleen	Elastic	EXP2
	Relaxation	RLOG (REXP2)
Stomach	Elastic	EXP2
	Relaxation	REXP2 (REXP2)

Models in parentheses are based on data acquired by the MTS system (ex-corpus only)

24.3.1.4 Phenomenological Model Fit

Ranking the phenomenological models based on measures of fit (mean, median, and standard deviation of both R2 and RMSE) separately and summing the ranks identified the best fitting model for each organ, summarized in Table 24.3. The phenomenological model parameters were identified for each set of acquired data (per organ, testing condition, cycle number, etc.). One may note that that the hollow organs appeared to be fit best by REXP2, while the solid organs were fit best by RLOG.

24.3.1.5 Statistical Analysis of phenomenological Model Parameters

One-way ANOVAs were performed for each factor-measure combination, with a probability value of 95% (α=0.05). In Figs. 24.9 and 24.10, each measure is plotted against the levels for each factor (such as organ or compression cycle). The diamonds represent the mean for a given level (e.g., liver is a level of the factor organ), and the horizontal bars indicate the standard deviation. The black dots are the individual data points. The right-hand side of the plots depict the results from post hoc Tukey-Kramer HSD (Honestly Significant Difference) analysis, as performed in the statistical software JMP (Cary, NC). This statistical test finds which pairs of levels have significantly different means, which is represented graphically by the circles: the center of each circle lies at the mean with the radius of the circle encompassing the region of confidence. If two circles overlap, then their means may not be significantly different and vice versa. The circles simply serve as a means for rapidly visually identifying significantly different groups.

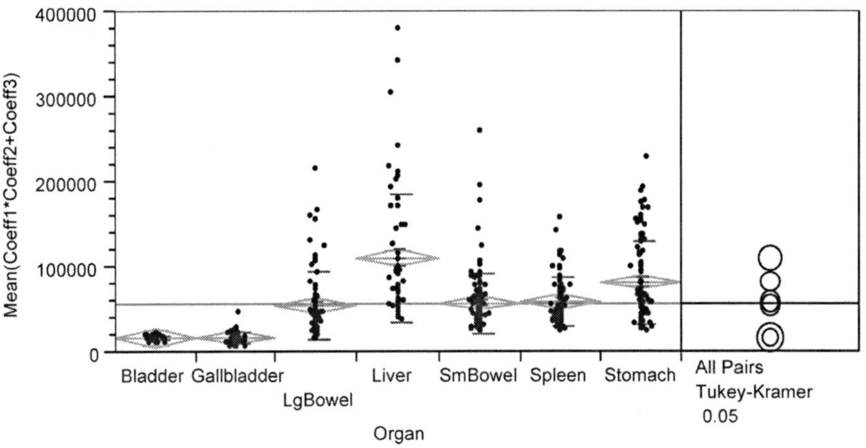

Fig. 24.9 The stiffness indicator scalar b × a + g of the EXP2 phenomenological model plotted for various organs for measured elastic data. The right-hand side of the plot depicts the results from post hoc Tukey-Kramer HSD analysis. The radius of the circle represents the region of confidence (95%)

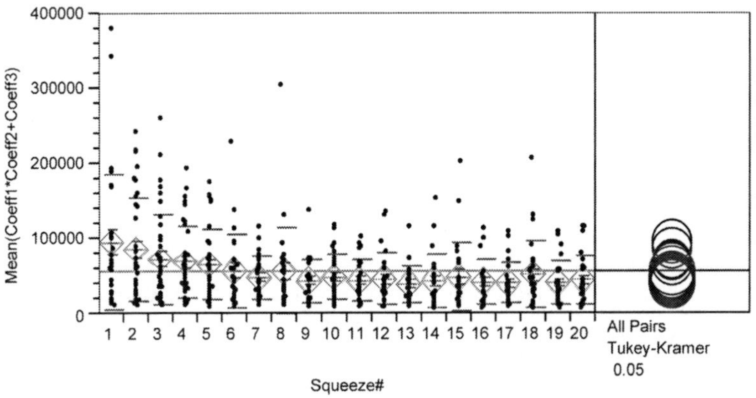

Fig. 24.10 The stiffness indicator scalar b ? a + g of the EXP2 phenomenological model plotted as a function of loading cycle for measured elastic data. The right-hand side of the plot depicts the results from post hoc Tukey-Kramer HSD analysis. The radius of the circle represents the region of confidence (95%)

Using the general stiffness indicator scalar $\beta \times \alpha + \gamma$ derived from (24.7) as a single indicator of the phenomenological model, a significant difference ($p<0.0001$) was found between the organs, indicating a significant difference in "stiffness" between the most of the organs. Only 4 of the possible 21 organ pairs were not found to be significantly different: spleen and small intestine, spleen and large intestine, small intestine and large intestine, and bladder and gallbladder (Fig. 24.9).

It is interesting to note that small and large intestine were not significantly different from each other using the overall stiffness measure ($\beta\times\alpha+\gamma$). Only when looking solely at the γ term does one find a significant difference. This would indicate that the overall behavior of the intestines is similar, especially at higher strains, but their behavior is significantly different at low strains.

There was a significant difference ($p<0.0001$) found between loading cycle with respect to stiffness indicator scalar $\beta\times\alpha+\gamma$ (Fig. 24.10). The stiffness indicator scalar for the first loading cycle was significantly greater than the seventh loading cycle and cycles 9–20. Moreover, the stiffness indicator scalar of the second loading cycle was greater than that from the 13th, 16th, 17th, and 19th loading cycles. These results indicate that the stiffness indicator scalar in the first six loading cycles is generally larger than latter loading cycles. A stable condition appears to be reached after 7–9 loading cycles.

Statistical analysis of the models' parameters indicated several significant differences as the function of the testing conditions (in-vivo, in-situ, and *ex-corpus*).

24.3.2 Micro Scale Biomechanics

Histological analyses and tests showed that the sites of compression injury in the porcine liver exhibited early signs of hepatic necrosis in hematoxylin and eosin (H&E) stained sections. Figure 24.11 shows a plot of percent necrosis versus average applied stress based on 48 liver tissue samples from nine animal experiments. The results indicate a graded acute injury response to compression stress in the range of stresses typical to MIS.

Figure 24.12a is a composite of several microphotographs showing an H&E stained section of liver from the aforementioned animal after an applied compression stress of 200kPa. A finite element models (FEM) corresponding to the histological sections experiments was developed and depicted in Fig. 24.12b. The tissue was assumed to be linear, isotropic, and homogeneous. The Young's modulus equal to 300kPa was chosen based on previous measurements in relevant stress ranges [Sect. 3.1]. Poisson's ratio was set at 0.4, reflecting the nearly incompressible nature of soft tissues [48–50]. The two-dimensional FEM model was used to simulate the center plane of the grasping site, which was a plane of symmetry of the three-dimensional geometry and analogous to the histological sections. The mesh consisted of six-noded triangular elements.

An overlay of the computed von Mises stress contour lines of the FE model and the H&E section is presented in Fig.24.12c. Most evident when comparing the histological section and FEM is the correlation between the high stress concentrations at the corners of the compression site in the model and the sites of hemorrhage in the tissue. The colored lines indicate the different stress levels in regular increments. The areas in which histology indicates early necrosis were colored manually in a solid color, with a different color used for each contour band. Any

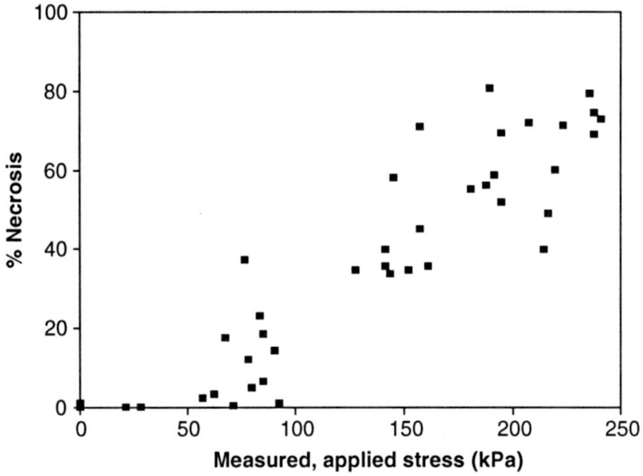

Fig. 24.11 Plot of necrosis in section images as a function of applied stress in the liver. Each data point is the average of four measurements

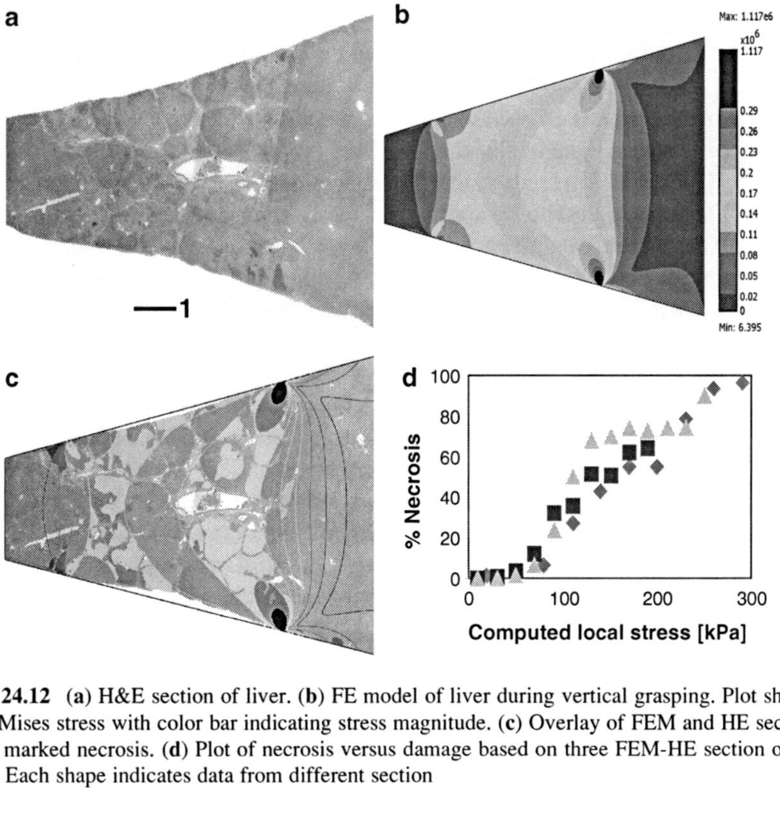

Fig. 24.12 (**a**) H&E section of liver. (**b**) FE model of liver during vertical grasping. Plot shows von Mises stress with color bar indicating stress magnitude. (**c**) Overlay of FEM and HE section with marked necrosis. (**d**) Plot of necrosis versus damage based on three FEM-HE section overlays. Each shape indicates data from different section

large space void of hepatocytes, such as that formed by a large blood vessel, was subtracted when calculating percent damage. This analysis was repeated using two different histological sections from two additional animals (not shown). Figure 24.12d is a plot of percent necrosis by level of stress, as indicated by the contour bands, for the three analyzed sections.

It was expected that the central region of each compression site with the uniform stress to be uniformly damaged. However, upon closer inspection, there was clear spatial variation in necrosis within the lobules. This was seen in almost all histological sections.

In light of these results, a small scale FEM was developed to explore if the variations in damage within the compression region, which is theoretically under uniform stress is due to the microstructure of liver. The micro scale FEM represent 2.94mm $\times$ 2.2mm section of a (Fig. 24.13a). The pressure applied on this H&E stained section was 90kPa (compression – grasping). A magnified region from the center of the section is shown in Fig. 24.13b with six lobules from the center of the compression site chosen for the FEM outlined in blue. The identified regions of necrosis are outlined in green, illustrating the typical irregular patterns of damage within lobules.

The mesh of the microscopic model is seen in Fig. 24.13c and utilized six-node triangular elements. Similar to the global model, the local model was two-dimensional and assumed linearity and isotropy. Boundary conditions, or displacement inputs, for the local model were based on the x- and y- displacements calculated in the global model for a box corresponding to the position and outer dimensions of the local model. The Young's moduli of the stroma and hepatocytes were based on previous measurements of stress-strain characteristics of in vivo bulk liver. The hepatic tissue was assigned a linear approximation for Young's modulus of 160kPa [Sect. 3.1]. This is lower than the Young's modulus used for the first model (Fig. 24.12) because tissues have a lower linearly approximated stiffness at lower stresses. The stroma was given a higher Young's modulus of 1.6MPa to reflect higher values for collagen, a component of the stroma [51]. The Poisson's ratio for hepatocytes was kept at 0.4, but the value for lobular walls was reduced to 0.2, again to reflect lower Poisson's ratio values found in the literature for collagen [51].

One clear observation from these plots was that this model of microscopic tissue heterogeneity produced only small variations in stress and strain distributions within each of the lobules. Figure 24.13d shows the von Mises stress contours when the Young's modulus of the stroma was assumed to be one order of magnitude stiffer than the hepatic tissue. A direct comparison of the simulation results in the form of von Mises stress (Fig. 24.13d) or other aforementioned types of stress to histological results (Fig. 24.13b) did not show a match between higher areas of stress and areas of necrosis. This mismatch may be explained by the composite nature of the soft tissue and the small scale in which the stiff structure shielded the stress from the softer structures.

The Young's modulus used to characterize the stroma was simply an estimate since there have been no separate measurements on the two specific material types. Therefore, a parametric analysis (where the simulation is run several times while altering the value of one variable over a range) was employed to alter the Young's modulus of the stroma compared to the hepatocytes. This allowed to both confirm that the model was

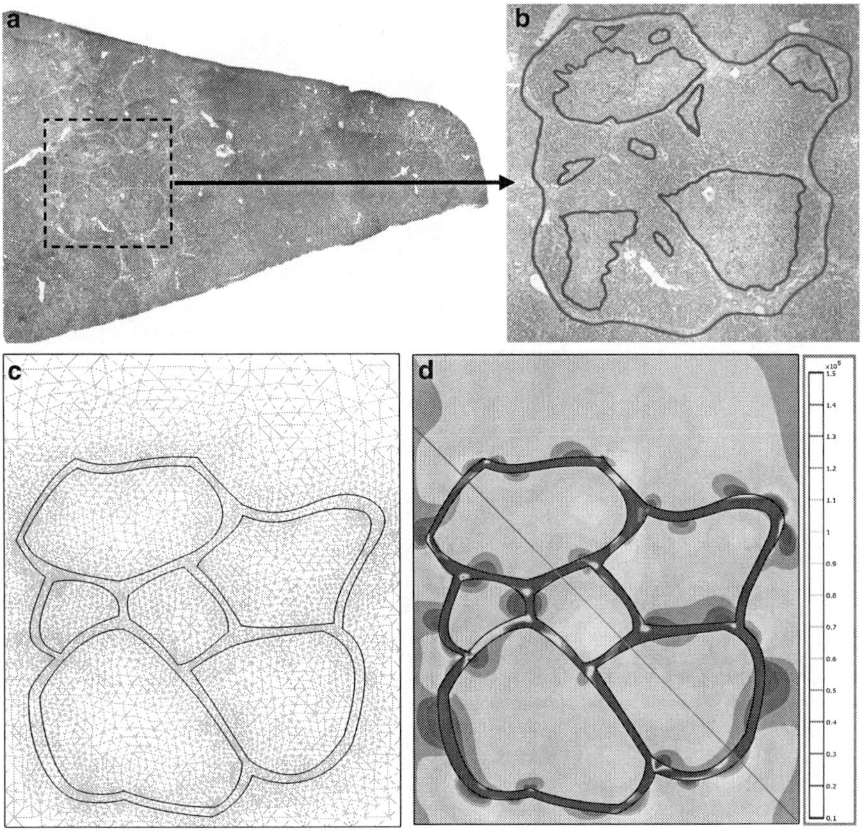

Fig. 24.13 (**a**) H&E stained section of liver after 90 kPa applied compression stress. Box approximates boundaries for group of lobules shown in (**b**). (**b**) *Blue line* delineates lobules used to create microscopic FE model. *Green* indicates regions indicating early necrosis. (**c**) Mesh of FE lobule model (a finer mesh was used for the final analysis). (**d**) von Mises stress plot with contour band values given along the *right* (*red* = higher stress; *blue* = lower stress)

stable and to determine if observations regarding stress variations changed greatly with different Young's moduli. A plot of von Mises stress along a diagonal cross-sectional line through the lobule model (indicated by the red line in Fig. 24.13d) at various levels of Young's moduli is given in Fig. 24.14, with "spikes" or "dips" representing the connective tissue between lobules. The hepatocyte material property had a constant Young's modulus of 160kPa, while the connective tissue Young's modulus varied between 16kPa and 1.6MPa. Results from the parametric study suggested that variation of Young's modulus between the stroma and hepatocytes affected the magnitude of stress contours, but only small variation existed *within* lobules with spikes resulting at the stromal boundaries. This is further exemplified by the line E = 1.6 e5 in that small "spikes" still result when the only difference in the two materials is in the Poisson's ratio.

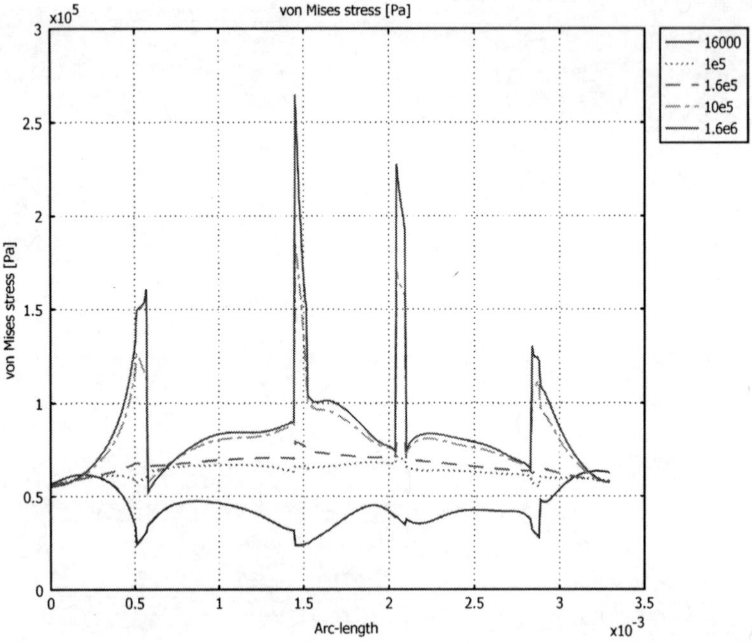

Fig. 24.14 Plot of von Mises stress through diagonal (*red line* Fig. 24.13d) for parametric analysis altering stromal Young's modulus over range indicated by legend

Initial inspection of these results suggested that inclusion of heterogeneity in a finite element model resulted in stress variations that might be able to explain the irregular damage patterns observed in experimental sections. However, closer scrutiny showed that the stress and strain variations within the lobules, as exemplified by von Mises stress plots in Fig. 24.14, were quite minimal regardless of the assigned Young's moduli. The maximum variation within a lobule generally appeared to be 20–30 kPa. This range was essentially "flat" compared to the entire stress profile. A comparable analysis for a different histological section from another animal produced similar results, suggesting that either stress is not directly correlated to damage at this microscopic level or that FE modeling as a method for predicting damage at this level most-likely requires inclusion of more complex properties, both biological and structural.

24.4 Conclusions and Discussion

Structural biomechanical properties (stress-strain and stress-relaxation) of seven abdominal organs (bladder, gallbladder, large and small intestine, liver, spleen, and stomach) have been obtained using a porcine animal model. The organs were tested in-vivo, in-situ, and *ex-corpus* under compressive loadings using a novel device, the Motorized Endoscopic Grasper (MEG), and a standard universal material testing system (MTS). The tissues were tested with the same loading conditions commonly applied by surgeons during minimally invasive surgical procedures. phenomenological models were developed for the various organs, testing conditions, and experimental devices. The results indicate significant quantitative differences between tissue properties measured in-vivo and postmortem conditions that will be of value for developing performance criteria for the next generation of surgical robots and simulators.

One of the most difficult aspects of any testing of biological materials is the large degree of variability (difference between animals, heterogeneity of the organs, strain history-dependence, strain rate-dependence, etc.). This particular study compounded this problem by testing bulk organs in-vivo and without preconditioning. Testing tissues in-vitro, using specimens of known shape under very controlled loading and boundary conditions, can usually lead to results with lower variability, particularly if the tissues are preconditioned. Testing in-vivo also introduces potential sources of noise, such as movement artifacts from beating heart and respiration, varying rates of tissue re-perfusion, etc. Unfortunately, this variability may mask effects from other factors. Some of this might have been quantified by repeated testing of the same site, but the fact that the tissues exhibit strain history-dependence makes this impractical: the sites would have to be allowed to fully recover to their natural state before subsequent testing, requiring the animal to be anesthetized for extended amounts of time. While this variability makes finding statistical significance in the data difficult, for the scope of surgical simulation, it is worthwhile to determine ranges of tissue properties.

With this information, simulators can realistically change the organs' virtual mechanical behavior so that the virtual liver operated on in one session would be different from the next. Providing realistic force magnitudes identical to those felt by surgeons when grasping organs during actual surgery is the first step towards more realistic and scientifically-based surgical simulators incorporating haptic feedback. In addition, surgical instruments and surgical robot manufacturers can use this information for optimizing their products to provide sufficient grasping traction while minimizing trauma. This could decrease costs and improve patient outcome.

The goodness of fit measures of the phenomenological models to the experimental data are based on residual error. In the case of the elastic tests, residual error is typically highest at large strains, where small changes in strain cause rapid increases in stress. Therefore, the best fitting curves are often the ones that fit best in the large strain region (the steepest part of the curve) but may or may not fit as well at lower strains. Study of the stress-strain database shows that nearly any set of data can be fit well by a sufficiently high-order equation. However, this becomes unwieldy and physically irrelevant. Due to the large number of parameters in POLY4 and POLY5 and the fact that the functions are not monotonically increasing, these models are not the model of choice for internal organ soft tissues, despite their good measures of fit. Moreover, the functions POLY2 and POLY3 and BLATZ lacked sufficient goodness of fit. The INV and EXP2 models provided better results than EXP, which is a curve commonly used by soft tissue studies. The EXP model may be better suited for tensile experiments, where there is no vertical asymptote before failure. Due to the nature of compression, strain varies from 0 to 1 and can never reach unity (1). For bulk materials that have not failed, there will always be a strain asymptote between 0 and 1. INV provides this number explicitly by its β term: the asymptote occurs at $\varepsilon=1/\beta$. This may shed some physical insight into the nature of the tissues. Perhaps this value of β represents the thickness of the fluid within the tissue that cannot be exuded, thus leading to an incompressible state. While EXP2 does not provide this physical information and has three parameters instead of two, it overwhelmingly is the best fitting of all the exponential-type functions and the best fitting of all functions under study.

Fitting models to stress-relaxation tests are highly dependent on the duration of the test. Extrapolation beyond the testing may lead to inaccurate results. Only the REXP1 model, of the three models examined, has a stress asymptote (of value 1-*a*), which is usually what is observed in tissue. Soft tissues are generally considered viscoelastic, which means there is some elastic component and a viscous component. After infinite time in compression, little stress is developed in the viscous component, and only the elastic component will remain, which is a finite, nonzero value. Models such as REXP2 and RLOG lack the asymptotic behavior as contained in REXP1. Therefore, extrapolating data based on these two models may predict non-physical behavior in which the stress continually decreases as a function of time, even beyond a value of zero – a physical impossibility. Despite this, REXP2 was overwhelmingly the best fit model to the data.

Analyzing the models' parameters of all the tissues under study that were tested with the MEG across the various conditions (in-vivo, in-situ, and ex-corpus) indicated the following characteristics. Given the elastic model EXP2 ((24.3) and (24.7)), the parameter g decreased significantly ($p < 0.0068$) as a function of the time postmortem. The parameter g represents the linear portion of the stress-strain curve, which dominates the stresses generated at low strains. Therefore, the results indicate that lower stresses were developed for small strains postmortem as opposed to in-vivo. The stiffness indicators $b \times a$ and $b \times a+g$ were significantly increasing ($p < 0.0001$) as a function of the time postmortem. The results of the stress-relaxation tests indicated that the tissue recovery between successive periodic step stains was greater for longer rest periods and for in-vivo. These phenomena can be explained in part by the higher perfusion of pressurized fluids within the tissues in-vivo, which may also contribute to the greater relaxation of the tissue postmortem than in-vivo. Despite the variability in the data, this study is a first step towards characterizing the highly complex behavior of abdominal soft tissues in their in-vivo state. The MEG is a useful and effective device capable of measuring compressive structural properties of abdominal tissues under in-vivo and surgically realistic conditions.

A full experimental characterization of a non-linear, fluid-perfused, non-isotropic material such as the major internal organs in-vivo is a complex endeavor. Proper modeling of bulk materials requires knowledge from tri-axial testing that can only come from tissue biomechanical studies that are not similar to surgical conditions.

The aim of this experimental protocol is to characterize the tissues' response to typical loading conditions in minimally invasive surgery. In that respect, the results reported in this study represent only one axis (dimension) of the tissue's tri-axial response. However, it should be emphasized that given the inherent dependencies between the three dimensions, the two unloaded dimensions are reflected in the dimension under study here. In addition, the dimension under study is the very same dimension that the surgeon is exposed to as he or she palpates the tissue with standard surgical tools. Moreover, one may note that one underlying assumption of the elastic model was that the compression stresses are zero at zero strain. This initial condition limits the reported elastic model to incorporate the soft tissues' residual stresses due to hydration and natural internal boundary conditions which in turn limits the model to accurately predict the tissues' stress response to small strains. This limitation is diminished for large strains, which are what surgeons typically apply during tissue manipulation.

Better understanding of the tool-tissue interface in MIS can lead to development of safer and more effective surgical instruments, and this may allow to overcome some of the limitations of novel MIS devices. Comparison of FE models corresponding to tissue sections subjected to compressive stress in vivo produced damage-stress relationships (Fig. 24.12d) similar to that obtained from analysis of multiple samples from multiple animals (Fig. 24.11), suggesting that FEM can predict tissue damage at a macroscopic level (centimeter scale). There are several implications of this result. First, FEM could be used for surgical instrument design by modifying tools to apply favorable stress distributions to tissue to reduce the potential for injury. Second, surgical simulators that utilize FEM modeling could provide feedback to trainees regarding tissue damage based on computed stress levels and extended data analogous to Fig. 24.12d. Finally, tissue damage could be minimized during a procedure through advanced treatment planning or improved control algorithms in surgical robots.

The liver has a highly complex, yet variable system of blood vessels and ducts as well as a dual blood supply (hepatic and portal). In addition, there are functional differences within the organ that may results in variation in oxygenation or metabolic burden. Incorporation of such biological and structural intricacies into an FE model could help provide a more complete understanding of tissue during surgical grasping.

Stress computed by homogeneous FEM of surgical grasping of liver correlated with damage seen in experimental tissues at a macroscopic level. This relationship was observed both by taking multiple tissue samples from multiple animals as well as by comparing single histological sections to their corresponding computed stress profiles. Microscopically, we did not see a similar correlation, which suggests that incorporating three dimensions or other anatomical and physiological effects in microscopic simulation models may be required to better predict tissue damage at that scale.

References

1. Madhani, A.J., Niemeyer, G., Salisbury, J.K. Jr.: The Black Falcon: a teleoperated surgical instrument for minimally invasive surgery. In: IEEE/RSJ International Conference on Intelligent Robots and Systems, New York, NY, 1998, vol. 2, pp. 936–944
2. Fung, Y.C.: Biomechanics: Mechanical Properties of Living Tissues, 2nd edn. Springer Verlag, New York (1993)
3. Fung, Y.C.: Elasticity of soft tissues in simple elongation. Am. J. Physiol. **213**(6), 1532–1544 (1967)
4. Yamada, H.: Strength of Biological Materials. Robert E. Krieger Publishing, Huntington, NY (1973)
5. Yoganandan, N., Pintar, F.A., Maltese, M.R.: Biomechanics of abdominal injuries. Crit. Rev. Biomed. Eng. **29**(2), 173–246 (2001)
6. Rouhana, S.W.: Biomechanics of abdominal trauma. In: Nahum, A.M., Melvin, J.W. (eds.) Accidental Injury: Biomechanics and Prevention, pp. 391–428. Springer-Verlag, New York (1993)
7. Liu, Z., Bilston, L.: On the viscoelastic character of liver tissue: experiments and modelling of the linear behaviour. Biorheology **37**(3), 191 (2000)
8. Arbogast, K.B., Thibault, K.L., Pinheiro, B.S., Winey, K.I., Margulies, S.S.: A high-frequency shear device for testing soft biological tissues. J. Biomech. **30**(7), 757–759 (1997)
9. Dokos, S., LeGrice, I.J., Smaill, B.H., Kar, J., Young, A.A.: A triaxial-measurement shear-test device for soft biological tissues. Trans. ASME J. Biomech. Eng. **122**(5), 471–478 (2000)
10. Gao, C.W., Gregersen, H.: Biomechanical and morphological properties in rat large intestine. J. Biomech. **33**(9), 1089 (2000)
11. Gregersen, H., Emery, J.L., McCulloch, A.D.: History-dependent mechanical behavior of guinea-pig small intestine. Ann. Biomed. Eng. **26**(5), 850 (1998)
12. Carter, F.J., Frank, T.G., Davies, P.J., Cuschieri, A.: Puncture forces of solid organ surfaces. Surg. Endosc. **14**(9), 783–786 (2000)
13. Carter, F.J., Frank, T.G., Davies, P.J., McLean, D., Cuschieri, A.: Measurements and modelling of the compliance of human and porcine organs. Med. Image Anal. **5**(4), 231–236 (2001)
14. Davies, P.J., Carter, F.J., Cuschieri, A.: Mathematical modelling for keyhole surgery simulations: a biomechanical model for spleen tissue. IMA J. Appl. Math. **67**, 41–67 (2002)
15. Tamura, A., Omori, K., Miki, K., Lee, J.B., Yang, K.H., King, A.I.: Mechanical characterization of porcine abdominal organs. In: Proceedings of the 46th Stapp Car Crash Conference, 2002, vol. 46, pp. 55–69
16. Melvin, J.W., Stalnaker, R.L., Roberts, V.L., Trollope, M.L.: Impact injury mechanisms in abdominal organs. In: Proceedings of the 17th Stapp Car Crash Conference, 1973, pp. 115–126
17. Zheng, Y.P., Mak, A.F.T., Lue, B.: Objective assessment of limb tissue elasticity: development of a manual indentation procedure. J. Rehabil. Res. Dev. **36**(2) (1999)
18. Zheng, Y.P., Mak, A.F.T.: Extraction of quasi-linear viscoelastic parameters for lower limb soft tissues from manual indentation experiment. J. Biomech. Eng. **121**(3), 330–339 (1999)
19. Pathak, A.P., Silver, T.M.B., Thierfelder, C.A., Prieto, T.E.: A rate-controlled indentor for in vivo analysis of residual limb tissues. IEEE Trans. Rehabil. Eng. **6**(1), 12–20 (1998)
20. Brouwer, I., Ustin, J., Bentley, L., Sherman, A., Dhruv, N., Tendick, F.: Measuring in vivo animal soft tissue properties for haptic modeling in surgical simulation. Stud. Health Technol. Inform. **81**, 69–74 (2001)
21. Ottensmeyer, M.P., Salisbury, J.: In-vivo mechanical tissue property measurement for improved simulations. Proc. SPIE **4037**, 286–293 (2000)
22. Kalanovic, D., Ottensmeyer, M.P., Gross, J., Buess, G., Dawson, S.L.: Independent testing of soft tissue viscoelasticity using indentation and rotary shear deformations. In: Medicine Meets Virtual Reality, Newport Beach, CA, 22–25 January. Stud. Health Technol. Inform. **94**, 137–143 (2003)

23. Bicchi, A., Canepa, G., De, R.D., Iacconi, P., Scillingo, E.P.: A sensor-based minimally invasive surgery tool for detecting tissue elastic properties. In: Proceedings 1996 IEEE International Conference on Robotics and Automation, New York, NY, 1996, vol. 1, pp. 884–888
24. Morimoto, A.K., Foral, R.D., Kuhlman, J.L., Zucker, K.A., Curet, M.J., Bocklage, T., MacFarlane, T.I., Kory, L.: Force sensor for laparoscopic Babcock. In: Medicine Meets Virtual Reality, 1997, pp. 354–361
25. Greenish, S., Haywar0, V., Chial, V., Okamura, A., Steffen, T.: Measurement, analysis, and display of haptic signals during surgical cutting. Presence Teleop. Virt. Environ. **11**(6), 626–651 (2002)
26. Brown, J.D., Rosen, J., Longnion, J., Sinanan, M., Hannaford, B.: Design and performance of a surgical tool tracking system for minimally invasive surgery. In: ASME International Mechanical Engineering Congress and Exposition, Advances in Bioengineering, New York, 11–16 November 2001, vol. 51, pp. 169–170
27. Rosen, J., Brown, J.D., Barreca, M., Chang, L., Hannaford, B., Sinanan, M.: The Blue DRAGON – a system for monitoring the kinematics and the dynamics of endoscopic tools in minimally invasive surgery for objective laparoscopic skill assessment. In: Medicine Meets Virtual Reality, Newport Beach, CA, 23–26 January 2002. Stud. Health Technol. Inform. **85**, 412–418 (2002)
28. Rosen, J., Brown, J.D., Barreca, M., Chang, L., Sinanan, M., Hannaford, B.: The Blue-DRAGON – a system for measuring the kinematics and the dynamics of minimally invasive surgical instruments in-vivo. In: 2002 IEEE International Conference on Robotics and Automation, Washington, DC, 2002, vol. 2, pp. 1876–1881
29. Rosen, J., Hannaford, B., MacFarlane, M.P., Sinanan, M.N.: Force controlled and teleoperated endoscopic grasper for minimally invasive surgery – experimental performance evaluation. IEEE Trans. Biomed. Eng. **46**(10), 1212–1221 (1999)
30. Brown, J.D., Rosen, J., Moreyra, M., Sinanan, M., Hannaford, B.: Computer-controlled motorized endoscopic grasper for in vivo measurement of soft tissue biomechanical characteristics. In: Medicine Meets Virtual Reality, Newport Beach, CA, 23–26 January. Stud. Health Technol. Inform. **85**, 71–73 (2002)
31. Brown, J.D., Rosen, J., Kim, Y.S., Chang, L., Sinanan, M.N., Hannaford, B.: In-vivo and in-situ compressive properties of porcine abdominal soft tissues. In: Medicine Meets Virtual Reality, Newport Beach, CA, 22–25 January. Stud. Health Technol. Inform. **94**, 26–32 (2003)
32. Brown, J.D., Rosen, J., Sinanan, M.N., Hannaford, B.: In-vivo and postmortem compressive properties of porcine abdominal organs. In: MICCAI 2003, Montreal, Canada. Lecture Notes in Computer Science, 2003, vol. 2878, pp. 238–245
33. Mkandawire, C., Ledoux, W., Sangeorzan, B., Ching, R.: A quasi-linear viscoelastic model of foot-ankle ligaments. In: 25th Annual Meeting of the American Society of Biomechanics, University of California-San Diego, San Diego, CA, 8–11 August 2001
34. Woo, S.L., Simon, B.R., Kuei, S.C., Akeson, W.H.: Quasi-linear viscoelastic properties of normal articular cartilage. J. Biomech. Eng. **102**(2), 85–90 (1980)
35. Mow, V.C., Kuei, S.C., Armstrong, C.G.: Biphasic creep and stress relaxation of articular cartilage in compression: theory and experiments. Trans. ASME J. Biomech. Eng. **102**(1), 73–84 (1980)
36. Ateshian, G.A., Warden, W.H., Kim, J.J., Grelsamer, R.P., Maw, V.C.: Finite deformation biphasic material properties of bovine articular cartilage from confined compression experiments. J. Biomech. **30**(11/12), 1157–1164 (1997)
37. DiSilvestro, M.R., Suh, J.K.: A cross-validation of the biphasic poroviscoelastic model of articular cartilage in unconfined compression, indentation, and confined compression. J. Biomech. **34**(4), 519–525 (2001)
38. DiSilvestro, M.R., Qiliang, Z., Marcy, W., Jurvelin, J.S., Jun, K.F.S.: Biphasic poroviscoelastic simulation of the unconfined compression of articular cartilage: I-Simultaneous

prediction of reaction force and lateral displacement. Trans. ASME. J. Biomech. Eng. **123**(2), 191–197 (2001)
39. DiSilvestro, M.R., Qiliang, Z., Jun, K.F.S.: Biphasic poroviscoelastic simulation of the unconfined compression of articular cartilage: II-Effect of variable strain rates. Trans. ASME. J. Biomech. Eng. **123**(2), 198–200 (2001)
40. Fortin, M., Hat, J., and Hmann, M.D.: Unconfined compression of articular cartilage: nonlinear behavior and comparison with a fibril-reinforced biphasic model. J. Biomech. Eng. **122**(2), 189–195 (2000)
41. Suh, J.K., Spilker, R.L.: Indentation analysis of biphasic articular cartilage: nonlinear phenomena under finite deformation. J. Biomech. Eng. **116**(1), 1–9 (1994)
42. Lai, W.M., Hou, J.S., Mow, V.C.: A triphasic theory for the swelling and deformation behaviors behaviors of articular cartilage. Trans. ASME. J. Biomech. Eng. **113**(3), 245–258 (1991)
43. Brown, J.R.: J.; Chang, L.; Sinanan, M.N.; Hannaford, B., Quantifying surgeon grasping mechanics in laparoscopy using the Blue DRAGON system. Medicine Meets Virtual Reality. Stud. Health Technol. Inform. **98**, 34–36 (2004)
44. Farshad, M., Barbezat, M., Flueler, P., Schmidlin, F., Graber, P., Niederer, P.: Material characterization of the pig kidney in relation with the biomechanical analysis of renal trauma. J. Biomech. **32**(4), 417–425 (1999)
45. Wang, J., Brienza, D.M., Bertocci, G., Karg, P.: Stress relaxation properties of buttock soft tissues: in vivo indentation test. In: Proceedings of the RESNA 2001 Annual Conference, RESNA, Reno, Nevada, 22–26 June 2001, pp. 391–393
46. Simon, B.R., Coats, R.S., Woo, S.L.: Relaxation and creep quasilinear viscoelastic models for normal articular cartilage. J. Biomech. Eng. **106**(2), 159–164 (1984)
47. De, S., et al.: Assessment of tissue damage due to Mechanical Stresses, Int. J. Rob. Res. **26**, 1159 (2007)
48. Niculescu, G., Foran, D.J., Nosher, J.: Non-rigid registration of the liver in consecutive CT studies for assessment of tumor response to radiofrequency ablation. Conf. Proc. IEEE Eng. Med. Biol. Soc. **1**, 856 (2007)
49. Brock, K.K., Sharpe, M.B., Dawson, L.A., Kim, S.M., Jaffray, D.A.: Accuracy of finite element model-based multi-organ deformable image registration. Med. Phys. **32**, 1647 (2005)
50. Chui, C., Kobayashi, E., Chen, X., Hisada, T., Sakuma, I.: Transversely isotropic properties of porcine liver tissue: experiments and constitutive modelling. Med. Biol. Eng. Comput. **42**, 787 (2004)
51. Sasaki, N., Odajima, S.: Stress–strain curve and Young's modulus of collagen molecules as determined by the X-ray diffraction technique. J. Biomech. **29**, 655–658 (1996)

Chapter 25
Objective Assessment of Surgical Skills

Jacob Rosen, Mika Sinanan, and Blake Hannaford

Abstract Minimally invasive surgery (MIS) involves a multi-dimensional series of tasks requiring a synthesis between visual information and the kinematics and dynamics of the surgical tools. Analysis of these sources of information is a key step in mastering MIS but may also be used to define objective criteria for characterizing surgical performance. The BlueDRAGON is a new system for acquiring the kinematics and the dynamics of two endoscopic tools synchronized with the visual view of the surgical scene. It includes passive mechanisms equipped with position and force torque sensors for measuring the position and the orientation (P/O) of two endoscopic tools along with the force and torque (F/T) applied on them by the surgeon's hands. The analogy between Minimally Invasive Surgery (MIS) and human language inspires the decomposition of a surgical task into its primary elements in which tool/tissue interactions are considered as "words" that have versions pronunciations defined by the F/T signatures applied on the tissues and P/O of the surgical tools. The frequency of different elements or "words" and their sequential associations or "grammar" both hold critical information about the process of the procedure. Modeling these sequential element expressions using a multi finite states model (Markov model – MM) reveals the structure of the surgical task and is utilized as one of the key steps in objectively assessing surgical performance. The surgical task is modeled by a fully connected, 30 state Markov model representing the two surgical tools where each state corresponds to a fundamental tool/tissue interaction based on the tool kinematics and associated with unique F/T signatures. In addition to the MM objective analysis, a scoring protocol was used by an expert surgeon to subjectively assess the subjects' technical performance. The experimental protocol includes seven MIS tasks performed on an animal

J. Rosen (✉)
Department of Computer Engineering, Jack Baskin School of Engineering,
University of California Santa Cruz, 1156 High Street, Santa Cruz, CA 95064, USA
e-mail: rosen@ucsc.edu

J. Rosen et al. (eds.), *Surgical Robotics: Systems Applications and Visions*,
DOI 10.1007/978-1-4419-1126-1_25, © Springer Science+Business Media, LLC 2011

model (pig) by 30 surgeons at different levels of training including expert surgeons. Analysis of these data shows that the major differences between trainees at different skill levels were: (a) the types of tool/tissue interactions being used, (b) the transitions between tool/tissue interactions being applied by each hand, (c) time spent while performing each tool/tissue interaction, (d) the overall completion time, and (e) the variable F/T magnitudes being applied by the subjects through the endoscopic tools. An objective learning curve was defined based on measuring quantitative statistical distance (similarity) between MM of experts and MM of residents at different levels of training. The objective learning curve (e.g. statistical distance between MM) was similar to that of the subjective performance analysis. The MM proved to be a powerful and compact mathematical model for decomposing a complex task such as laparoscopic suturing. Systems like surgical robots or virtual reality simulators in which the kinematics and the dynamics of the surgical tool are inherently measured may benefit from incorporation of the proposed methodology for analysis of efficacy and objective evaluation of surgical skills during training.

Keywords Dynamics · Human Machine Interface · Haptics · Kinematics · Manipulation · Markov Model · Minimally Invasive · Simulation · Surgery · Surgical Skill Assessment · Soft Tissue · Surgical Tool · Robotics · Vector Quantization

25.1 Introduction

Evaluation of procedural skills in surgery can be performed utilizing three different modalities: during actual open or minimally invasive clinical procedures; in physical or virtual reality simulators with or without haptic feedback; and during interaction with surgical robotic systems (Fig. 25.1). In each of these interactions,

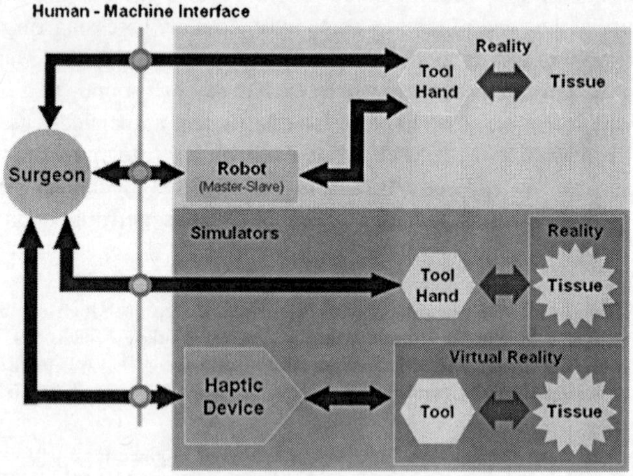

Fig. 25.1 Modalities for performing surgery

the surgeon is separated from the treated tissue or medium by an instrument, an interface that is at least mechanical, but may be a combination of mechanical and virtual representation of the anatomy (simulator). The intermediate modality in all of these options can be considered interchangeable.

During open or minimally invasive surgical (MIS) procedures, the surgeon interacts with the patient's tissue either directly with his/her hands or through the mediations of tools. Surgical robotics enables the surgeon to operate in a teleoperation mode with or without force feedback using a master/slave system configuration. In this mode of operation, visualization is obtained from either an external camera or an endoscopic camera. Incorporating force feedback, allows the surgeon to feel through the master console the forces being applied on the tissue by the surgical robot, the slave, as he/she interacts with it from the master console. For training in a simulated virtual environment, the surgical tools, the robot – slave, and the anatomical structures are replaced with virtual counterparts. The surgeon interacts with specially-designed input devices, haptic devices when force feedback is incorporated, that emulate surgical tools, or with the master console of the robotic system itself, and perform surgical procedures in virtual reality.

One element that all these modalities have in common is the human–machine interface in which visual, kinematic, dynamic, and haptic information are shared between the surgeon and the various modalities. This interface, rich with multi-dimensional data, is a valuable source of objective information that can be used to objectively assess technical surgical and medical skill within the general framework of surgical and medical ability. Algorithms that are developed for objective assessment of skill are independent of the modality being used, and therefore, the same algorithms can be incorporated into any of these technologies.

Advances in surgical instrumentation have expanded the use of minimally invasive surgical (MIS) techniques over the last decade. Using a miniature video camera and instruments inserted through small incisions, operations previously performed through large incisions are now completed with MIS techniques leading to a much shorter recovery time and decreased risk of surgical site infections. However from surgeon's perspective, this new technology requires a new set of skills. The new human–machine interface, the associated loss of 3-D vision, and degraded haptic sensation introduce new challenges. Moreover, the use of this technology has also presented a new dilemma – namely the training of individuals to perform surgical procedures that require a new set of skills. This is especially problematic in the field of MIS where the teacher is one step removed from the actual conduct of the operation.

Developing objective methodology for surgical competence and performance are paramount to superior surgical training. Moreover, alternatives to the traditional apprenticeship model of surgical training are necessary in today's emphasis on cost containment and professional competency and patient safety. There is a need to demonstrate continuing competency among practicing surgeons as well as confirming competency in trainees early on, before surgical trainees are thrust into the role of primary assistant or surgeon in the operating room. Inherent difficulties in evaluating clinical competence for physicians and physicians-in-training have spawned the wide use of various assessment techniques including Objective Structured Clinical Examinations (OSCE), oral examinations, standardized patient examinations, and simulation technology. While successful evaluation of cognitive skills using these methods have been reported, objective evaluation of procedural skills remains difficult. As the medical profession is faced with demands for greater accountability and patient safety, there is a critical need for the development of consistent and reliable methods for objective evaluation of clinician performance during procedures.

Objective assessment of surgical competence during MIS procedures, defined as caring out the surgical procedure in a minimally invasive surgical setup, is a multidimensional problem. MIS performance is comprised of physiological constraints (stress fatigue) equipment constraints (camera rotation and port location), team constraints (nurses) and MIS ability. Ability when referred to surgery, is defined as the natural state or condition of being capable; innate aptitude (prior to training) which an individual brings for performing a surgical task [1]. MIS ability, by itself, includes cognitive factors (knowledge and judgment) and technical factors (psychomotor ability, visio-spatial ability and perceptual ability). By definition, fundamental psychometric abilities are fixed at birth or early childhood and show little or no learning effect [2]. However training enables the subject to perform as close as possible to his or her inherent psychometric abilities.

The methodology for assessing surgical skill as a subset of surgical ability, is gradually shifting from subjective scoring of an expert which may be a variably biased opinion using vague criteria, towards a more objective, quantitative analysis. This shift is enabled by using instrumented tools [3–7], measurements of the surgeon's arm kinematics [8], gaze patterns [9], physical simulators [10], a variety of virtual reality simulators with and without haptics [1, 11–32], and robotic systems. Regardless of the modality being used or the clinical procedure being studied, task deconstruction or decomposition is an essential component of a rigorous objective skills-assessment methodology. By exposing and analyzing the internal hierarchy of tasks a broader understanding of procedures is achieved while providing objective means for quantifying training and skills acquisition.

Task decomposition is associated with defining the prime elements of the process. In surgery, a procedure is traditionally and methodologically divided into steps, stages, or phases with well-defined intermediate goals. Additional hierarchical decomposition is based upon identifying tasks or subtasks [33] composed of sequence of and actions or states [3–7]. In addition, other measurable parameters such as workspace [34] completion time, tool position, and forces and torques were studied individually [3–7]. Selecting low-level elements of the task decomposition allows one to associate these elements with quantifiable and measurable parameters. The definition of these states, along with measurable, quantitative data, are the foundation for modeling and examining surgical tasks as a process.

In the current study, an analogy between (MIS) and the human language inspires the decomposition of a surgical task into its primary elements. Modeling the sequential element expressions using a multi states model (Markov model) reveals the internal structure of the surgical task, and this is utilized as one of the key steps in objectively assessing surgical performance. Markov Modeling (MM) and its subset – Hidden Markov Modeling (HMM) were extensively developed in the area of speech recognition [35] and further used in a broad spectrum of other fields, e.g. human operator modeling, robotics, and teleoperation [36–40], gesture recognition and facial expressions [41, 42] DNA and protein modeling [43], and surgical tools in MIS setup [5, 44]. These studies indicate that MMs and HMMs provide adequate models to characterize humans operating in complex interactive tasks with machines among other applications.

The aim of the study was to develop a system of acquiring data in a real MIS setup using an animal model and a methodology for decomposing two-handed surgical tasks using Markov models (MM) based on the kinematics and the dynamics of the surgical tools. Measuring the statistical similarity between the models representing subjects at different levels of their surgical training enables an objective assessment of surgical skills.

25.2 Tools and Methods

A novel system named the BlueDRAGON was designed, constructed and used for acquiring the kinematics (position and orientation) and the dynamics (force and torque) of two endoscopic tools during MIS procedures in real-time. The data were acquired during a surgical task performed by 30 subjects at different levels of surgical training followed by objective and subjective surgical skill analysis based on task decomposition. The novel objective methodology was based upon a multi-state Markov model whereas the subjective methodology utilized a standard scoring system for analyzing the videotapes of the surgical scene recorded during the experiment. The following subsections describe the system and the methodologies that were used in the current study.

25.2.1 Tools: The BlueDRAGON System

The BlueDRAGON is a system for acquiring the kinematics and the dynamics of two endoscopic tools along with the visual view of the surgical scene while performing a MIS procedure (Fig. 25.2). The system includes two four-bar passive mechanisms attached to endoscopic tools [4]. The endoscopic tool in minimal invasive surgery is inserted into the body through a port located for example in the abdominal wall. The tool is rotated around a pivot point within the port that is inaccessible for sensors aimed to measure the tool's rotation. The four bar mechanism is one of several mechanisms that allows mapping of the tool's rotation around the port's pivot point. This mapping is enabled by aligning a specific point on the mechanism, where all its rotation axes are intersecting, with the pivot point of the endoscopic tool (Fig. 25.2b). The tool's positions and orientations, with respect to the port, are then tracked by sensors that are incorporated into the mechanism's joints. Moreover,

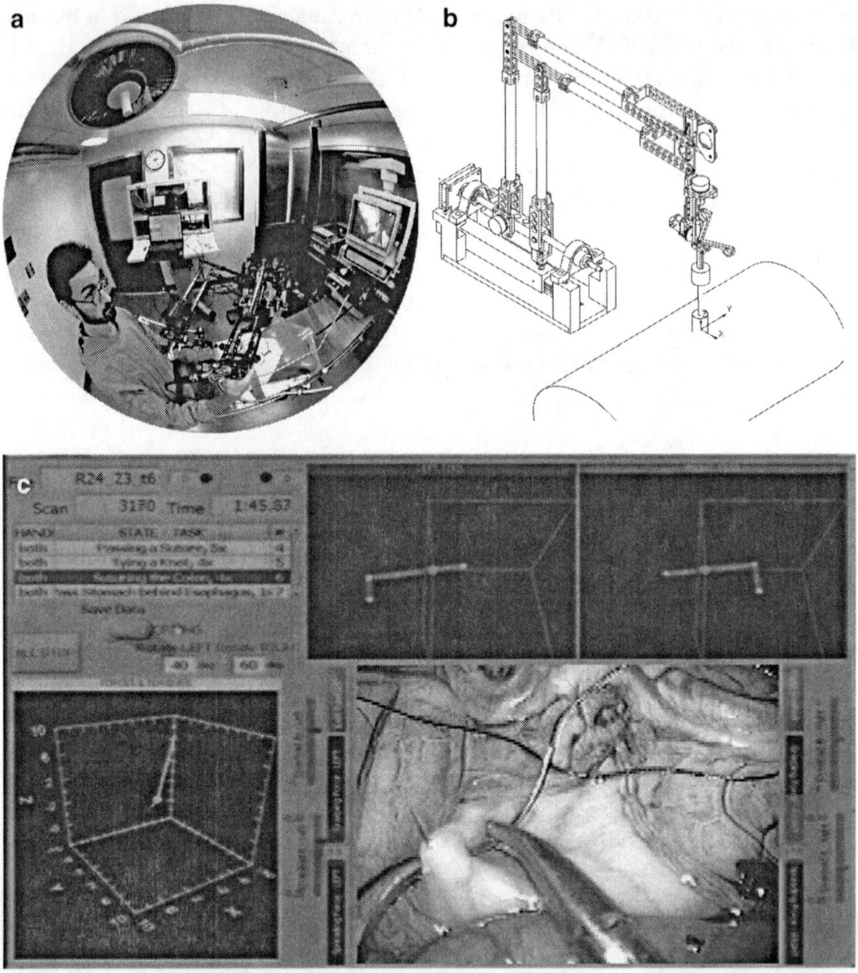

Fig. 25.2 The BlueDRAGON system (**a**) The system integrated into a minimally invasive surgery operating room (**b**) CAD drawing of the BlueDRAGON four bar mechanism and its coordinate system properly aliened with the MIS port. (**c**) Graphical user interface (GUI) incorporating visual view of the surgical seen acquired by the endoscopes video camera (*bottom right*) and real-time information measured by the BlueDRAGONs. On the top right side of the GUI, a virtual representation of the two endoscopic tools are shown along with vectors representing the instantaneous velocities. On the *bottom left* a three dimensional representation of the forces and torque vectors are presented. Surrounding the endoscopic image are *bars* representing the grasping/spreading forces applied on the handle and transmitted to the tool tip via the tool's internal mechanism, along with virtual binary LED indicating contact between the tool tips and the tissues

the mechanism's axes alignment with the pivot point in the port prevents the application of additional moments applied on the skin and internal tissues that may result from misalignment and the fact that an external mechanism is used and attached to the tools. On the other hand, this setup makes the mechanism totally transparent to the moments that are generated intentionally by using the tools.

Substantial effort was made, during the design process, to minimize the weight and the inertia of the mechanism. This was accomplished by using carbon fibers tubes for the links, and by optimizing the shapes of the links for minimizing the mass distribution. The mass of the mechanism's moving parts is 1.36 kg and its maximal moment of inertia relative to the X-axis (I_{xx}) depicted in Fig. 25.2b is 0.157 kg m^2. Moreover, the gravitational forces applied on the surgeon's hand when the mechanism is placed away from its neutral position are compensated by an optimized spring connecting the base with the first two coupled links.

The two mechanisms are equipped with three classes of sensors: (a) position sensors (potentiometers – Midori America Corp.) are incorporated into four of the mechanisms' joints for measuring the position, the orientation and the translation of the two instrumented endoscopic tools attached to them. In addition, two linear potentiometers (Penny & Giles Controls Ltd.) that are attached to the tools' handles are used for measuring the endoscopic handle and tool tip angles; (b) three-axis force/torque (F/T) sensors with holes drilled at their center (ATI-Mini sensor) are inserted and clamped to the proximal end of the endoscopic tools' shafts. In addition, double beam force sensors (Futak) were inserted into the tools' handles for measuring the grasping forces at the hand/tool interface; and (c) contact sensors, based on RC circuit, provided binary indication of any tool-tip/tissue contact.

Data measured by the BlueDRAGON sensors are acquired using two 12-bit USB A/D cards (National Instruments) sampling the 26 channels (four rotations, one translation, one tissue contact, and seven channels of forces and torques from each instrumented grasper) at 30 Hz. In addition to the data acquisition, the synchronized view of the surgical scene is incorporated into a graphical user interface displaying the data in real-time (Fig. 25.2c).

25.2.2 Experimental Protocol

The experimental protocol included 30 surgeons at different levels of expertise from surgeons in training to surgical attendings skilled in laparoscopic surgery. There were five subjects in each group representing the 5 years of surgical training, (5 × R1, R2, R3, R4, R5 – where the numeral denotes year of training) and five expert surgeons. Each subject was given instruction on how to perform an intracorporeal knot through a standard multimedia presentation. The multimedia presentation included a written description of the task along with a video clip of the surgical scene and audio explanation of the task. Subjects were then given a maximum of 15 min to complete this task in a swine model. This complex, integrative task includes many of the elements of advanced MIS techniques.

In addition to the surgical task, each subject performed 15 predefined tool/tissue and tool/needle-suture interactions (Table 25.1). The kinematics (the position/

Table 25.1 Definitions of the 15 states based on spherical coordinate system with an origin at the port

Type	No.	State name	State acronym	Tissue contact	Position/Orientation					Force/Torque						
					ω_x	ω_y	ω_z	v_r	ω_g	F_x	F_y	F_z	T_x	T_y	T_z	F_g
I	1	Idle	ID	–	$\pm\varepsilon_{\omega_x}$	$\pm\varepsilon_{\omega_y}$	$\pm\varepsilon_{\omega_z}$	$\pm\varepsilon_{v_r}$	$\pm\varepsilon_{\omega_g}$	$\pm\varepsilon_{F_x}$	$\pm\varepsilon_{F_y}$	$\pm\varepsilon_{F_z}$	$\pm\varepsilon_{T_x}$	$\pm\varepsilon_{T_y}$	$\pm\varepsilon_{T_z}$	$\pm\varepsilon_{F_g}$
	2	Closing Handle (Grasping/ Cutting)	CL	+		$\pm\varepsilon_{\omega_y}$	$\pm\varepsilon_{\omega_z}$	$\pm\varepsilon_{v_r}$	$\omega_g<\varepsilon_{\omega_g}$ $\pm\varepsilon_{\omega_x}$	$\pm\varepsilon_{F_x}$	$\pm\varepsilon_{F_y}$	$\pm\varepsilon_{F_z}$	$\pm\varepsilon_{T_x}$	$\pm\varepsilon_{T_y}$	$\pm\varepsilon_{T_z}$	$F_g>\varepsilon_{F_g}$
	3	Opening Handle (Spreading)	OP	+	$\pm\varepsilon_{\omega_x}$	$\pm\varepsilon_{\omega_y}$	$\pm\varepsilon_{\omega_z}$	$\pm\varepsilon_{v_r}$	$\pm\varepsilon_{T_z}$ $\omega_g>\varepsilon_{\omega_g}$	$\pm\varepsilon_{F_x}$	$\pm\varepsilon_{F_y}$	$\pm\varepsilon_{F_z}$	$\pm\varepsilon_{T_x}$	$\pm\varepsilon_{T_y}$		$F_g<-\varepsilon_{F_g}$
	4	Pushing	PS	+	$\pm\varepsilon_{\omega_x}$	$\pm\varepsilon_{\omega_y}$	$\pm\varepsilon_{\omega_z}$	$v_r<-\varepsilon_{v_r}$	$\pm\varepsilon_{\omega_g}$	$\pm\varepsilon_{F_x}$	$\pm\varepsilon_{F_y}$	$F_z>\varepsilon_{F_z}$	$\pm\varepsilon_{T_x}$	$\pm\varepsilon_{T_y}$	$\pm\varepsilon_{T_z}$	$\pm\varepsilon_{F_g}$
	5	Rotating (Sweeping)	RT	+	$\omega_x>\|\varepsilon\|$	$\omega_x>\|\varepsilon\|$	$\pm\varepsilon_{\omega_z}$	$\pm\varepsilon_{v_r}$	$\pm\varepsilon_{\omega_g}$	$F_x>\|\varepsilon_{F_x}\|$	$F_y>\|\varepsilon_{F_y}\|$	$\pm\varepsilon_{F_z}$	$T_x>\|\varepsilon_{T_x}\|$	$T_y><\varepsilon_{T_y}\|$	$\pm\varepsilon_{T_z}$	$\pm\varepsilon_{F_g}$
II	6	Closing – Pulling	CL–PL	+	$\pm\varepsilon_{\omega_x}$	$\pm\varepsilon_{\omega_y}$	$\pm\varepsilon_{\omega_z}$	$v_r>\varepsilon_{v_r}$	$\pm\varepsilon_{\omega_y}$ $\omega_g<\varepsilon_{\omega_g}$	$\pm\varepsilon_{F_x}$	$\pm\varepsilon_{F_y}$	$F_z<-\varepsilon_{F_z}$	$\pm\varepsilon_{T_x}$	$\pm\varepsilon_{T_y}$	$\pm\varepsilon_{T_z}$	$F_g>\varepsilon_{F_g}$
	7	Closing – Pulling	CL–PS	+	$\pm\varepsilon_{\omega_x}$		$\pm\varepsilon_{\omega_z}$	$v_r<-\varepsilon_{v_r}$	$\omega_g<\varepsilon_{\omega_g}$	$\pm\varepsilon_{F_x}$	$\pm\varepsilon_{F_y}$	$F_z>\varepsilon_{F_z}$	$\pm\varepsilon_{T_x}$	$\pm\varepsilon_{T_y}$	$\pm\varepsilon_{T_z}$	$F_g>\varepsilon_{F_g}$
	8	Closing – Pulling	CL–RT	+	$\omega_x>\|\varepsilon_{\omega_x}\|$	$\omega_y>\|\varepsilon_{\omega_y}\|$	$\pm\varepsilon_{\omega_z}$	$\pm\varepsilon_{v_r}$	$\omega_g<\varepsilon_{\omega_g}$	$F_x>\|\varepsilon_{F_x}\|$	$F_y>\|\varepsilon_{F_y}\|$	$\pm\varepsilon_{F_z}$	$\pm\varepsilon_{T_x}$	$\pm\varepsilon_{T_y}$	$\pm\varepsilon_{T_z}$	$F_g>\varepsilon_{F_g}$
	9	Pushing – Opening	PS–OP	+	$\pm\varepsilon_{\omega_x}$	$\pm\varepsilon_{\omega_y}$	$\pm\varepsilon_{\omega_z}$	$v_r<$	$\omega_g>\varepsilon_{\omega g}$	$\pm\varepsilon_{F_x}$	$\pm\varepsilon_{F_y}$	$F_z<-\varepsilon_{F_z}$	$\pm\varepsilon_{T_x}$	$\pm\varepsilon_{T_y}$	$\pm\varepsilon_{T_z}$	$F_g<-\varepsilon_{F_g}$
	10	Pushing – Rotating	PS–RT	+	$\omega\vert\varepsilon_x$	$\omega\vert\varepsilon_y$	$\pm\varepsilon\omega_z$	$v_r<\varepsilon v_r$	$\omega_x>\|\varepsilon_{\omega_x}\|$ $\pm\varepsilon_{\omega_g}$	$F_x>\|\varepsilon_{F_x}\|$	$F_y>\|\varepsilon_{F_y}\|$	$F_z>\varepsilon_{F_z}$	$\pm\varepsilon_{T_x}$	$\pm\varepsilon_{T_y}$	$\pm\varepsilon_{T_z}$	$\pm\varepsilon_{F_g}$
	11	Rotating – Opening	RT–OP	+					$\omega_x>\|\varepsilon_{\omega_x}\|$	$F_x>\|\varepsilon_{F_x}\|$	$F_y>\|\varepsilon_{F_y}\|$	$\pm\varepsilon_{F_z}$	$T_x>\|\varepsilon_{T_x}\|$	$T_y>\|\varepsilon_{T_y}\|$	$\pm\varepsilon_{T_z}$	$F_g<-\varepsilon_{F_g}$
III	12	Closing – Pulling – Rotating	CL–PL–RT	+	$\omega_x>\|\varepsilon_{\omega_x}\|$	$\omega_y>\|\varepsilon_{\omega_y}\|$	$\pm\varepsilon_{\omega_z}$	$v_r<-\varepsilon_{v_r}$	$\omega_g<\varepsilon_{\omega_g}$	$F_x>\|\varepsilon_{F_x}\|$	$F_y>\|\varepsilon_{F_y}\|$	$F_z<-\varepsilon_{F_z}$			$\pm\varepsilon_{T_z}$	$F_g>\varepsilon_{F_g}$
	13	Closing – Pushing – Rotating	CL–PS–RT	+	$\omega_x>\|\varepsilon_{\omega_x}\|$	$\omega_y>\|\varepsilon_{\omega_y}\|$	$\pm\varepsilon_{\omega_z}$	$v_r<-\varepsilon_{v_r}$	$\omega_g<\varepsilon_{\omega_g}$	$F_x>\|\varepsilon_{F_x}\|$	$F_y>\|\varepsilon_{F_y}\|$	$F_z\|\varepsilon_{F_z}$	$T_x>\|\varepsilon_{T_x}\|$	$T_y>\|\varepsilon_{T_y}\|$	$\pm\varepsilon_{T_z}$	$F_g>\varepsilon_{F_g}$
	14	Pushing – Rotating – Opening	PS–RT–OP	+	$\omega_x>\|\varepsilon_{\omega_x}\|$	$\omega_y>\|\varepsilon_{\omega_y}\|$	$\pm\varepsilon_{\omega_z}$	$v_r<-\varepsilon_{v_r}$	$\pm\varepsilon_{T_x}$ $\omega_g>\varepsilon_{\omega_g}$	$F_x>\|\varepsilon_{F_x}\|$	$F_y>\|\varepsilon_{F_y}\|$	$F_z>\varepsilon_{F_z}$		$\pm\varepsilon_{T_y}$	$\pm\varepsilon_{T_z}$	$F_g<-\varepsilon_{F_g}$
II	15	Closing Handle – Spinning	CL–SP	+	$\pm\varepsilon_{\omega_z}$	$\pm\varepsilon_{\omega_y}$	$\|\omega_z\|>\varepsilon_{\omega_z}$	$\pm\varepsilon_{v_r}$	$\omega_g<\varepsilon_{\omega_g}$	$\pm\varepsilon_{F_x}$	$\pm\varepsilon_{F_y}$	$\pm\varepsilon_{F_z}$	$\pm\varepsilon_{T_x}$	$\pm\varepsilon_{T_y}$	$T_z>\|\varepsilon_{T_z}\|$	$F_g>\varepsilon_{F_g}$

Each state is characterized by a unique set of angular/linear velocities, forces and torques and associated with a specific tool/tissue or tool/object interaction. A non-zero threshold value is defined for each parameter by ε. The states' definitions are independent from the tool tip being used e.g. the state defined as Closing Handle might be associated with grasping or cutting if a grasper or scissors are being used respectively orientation – P/O of the tools in space with respect to the port) and the dynamics (forces and torque F/T applied by the surgeons on the tools) of the left and right endoscopic tools along with the visual view of the surgical scene were acquired by a passive mechanism that is part of the BlueDRAGON. The aim of this experimental segment was to study the F/T and velocity signatures associated with each interaction that were further used as the model observations associated with each state of the model. All animal procedures were performed in an AALAC-accredited surgical research facility under an approved protocol from the institutional animal care committee of the University of Washington.

25.2.3 Objective Analysis: MIS Task Decomposition and Markov Model

25.2.3.1 Surgery as a Language: The Analogy and The State Definitions

The objective methodology for assessing skill while performing a procedure is inspired by the analogy between the human language and surgery. Further analysis of this concept indicates that these two domains share similar taxonomy and internal etymological structure that allows a mathematical description of the process by using quantitative models. Such models can be further used to objectively assess skill level by revealing the internal structure and dynamics of the process. This analogy is enhanced by the fact that in both the human language and in surgery, an idea can be expressed and a procedure can be preformed in several different ways while retaining the same cognitive meaning or outcome. This fact suggests that a stochastic approach might describe the surgical or medical examination processes incorporating the inherent variability better then a determinist approach.

Table 25.2 summarizes the analogy between the two entities, human language and surgery, along with the corresponding modeling elements in a hierarchal fashion. The critical step in creating such an analogy is to identify the prime elements. In the human language, the prime element is the '*word*' which is analogous to a '*tool/tissue interaction*' in surgery. This prime element is modeled by a '*state*' in the model. As in a spoken language, words have different '*pronunciations*' and yet preserve their meaning. In surgery, various '*force/torque*

Table 25.2 The analogy between the human language and surgery as manifested it self in a similar taxonomy and sub structures along with the corresponding element of the finite state Markov model

Language	Surgery	Model
Book	Operation/Procedure	Multiple Models
Chapter	Step of the Operation	Model
Word	Tool/Tissue Interaction	State
Pronunciation	Force Torque Velocity magnitude	Observation

magnitudes' can be applied on the tissues and still be classified under the same tool/tissue interaction category. These various force/torque magnitudes are simulated by the '*observations*' in the model. In a similar fashion to the human language in which a sequence of words are comprised into a sentence, and sentences create a book '*chapter*', a sequence to tool/tissue interactions form a step of an operation in which an intermediate and specific outcome can be completed. Each step of the operation is represented by a single model. '*Multiple models*' can be further describing a multi-step '*surgical operation*' that is analogous to a '*book*'. One may note that the sub-structures like a sentence and a section were omitted in the current analogy; however, identifying the corresponding elements in surgical procedure may increase the resolution of the model.

Analyzing the degrees of freedom (DOF) of a tool in MIS indicates that the due to the introduction of the port through which the surgeon inserts tools into the body cavity, two DOF of the tool are restricted. The six DOF of a typical open surgical tool is reduced to only four DOF in a minimally invasive setup (Fig. 25.3). These four DOF include rotation along the three orthogonal axes (x,y,z) and translation along the long axis of the tool's shaft (z). A fifth DOF is defined as the tool-tip jaws angle, which is mechanically linked to the tool's handle, when a grasper or a scissor is used. Additional one or two degrees of freedom can be obtained by adding a wrist joint to the MIS tool. The wrist joint has been incorporated into commercially available surgical robots in order to enhance the dexterity of the tool within the body cavity.

Surgeons, while performing MIS procedures, utilize various combinations of the tools' DOF while manipulating them during the interaction with the tissues or other items in the surgical scene (needle, suture, staple etc.) in order to achieve the desire outcome. Quantitative analysis of the tool's position and ordination during surgical procedures revealed 15 different combinations of the tool's five DOF, while interacting with the tissues and other objects. These 15 DOF combinations will be further referred to, and modeled as states (Table 25.1). The 15 states can be grouped into three types, based on the number of movements or DOF utilized simultaneously. The fundamental maneuvers are defined as Type I. The 'idle' state was defined as moving the tool in space (body cavity) without touching any internal organ, tissue, or any other item in the scene. The forces and torques developed in this state represent the interaction with the port and the abdominal wall, in addition to the gravitational and inertial forces. In the 'grasping' and 'spreading' states, compression and tension were applied on the tissue through the tool tip by closing and opening the grasper's handle, respectively. In the 'pushing' state, the tissue was compressed by moving the tool along the Z axis. 'Sweeping' consisted of placing the tool in one position while rotating it around the X and/or Y axes or in any combination of these two axes (port frame). The rest of the tool/tissue interactions in Types II and III were combinations of the fundamental ones defined as Type I. The only one exception was state 15 that was observed only in tasks involved suturing when the surgeon grasps the needle and rotates it around the shaft's long axis to insert it into the tissue. Such a rotation was never observed whenever direct tissue interaction was involved.

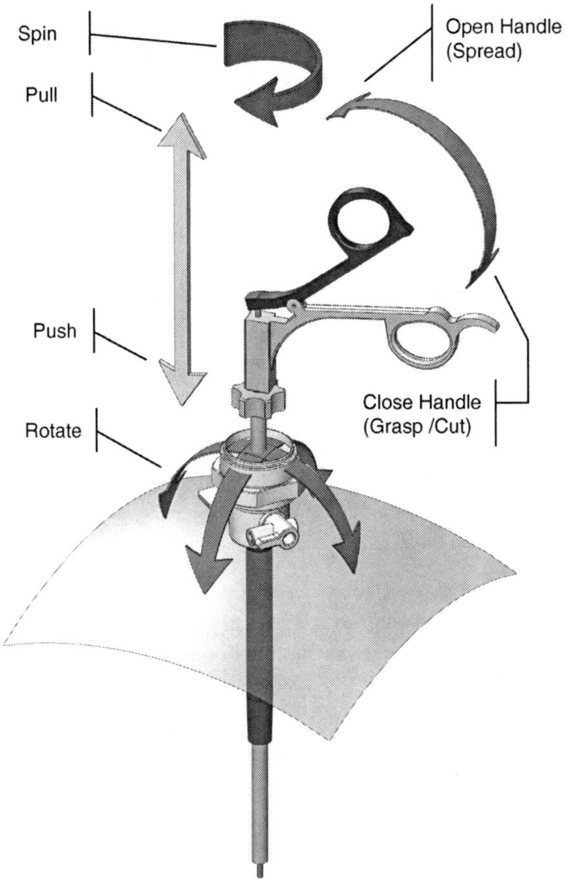

Fig. 25.3 Definition of the five degrees of freedom – DOF (marked by *arrows*) of a typical MIS endoscopic tool. Note that two DOF were separated into two distinct actions (Open/Close handle and Pull/Push), and the other two were lumped into one action (Rotate) for representing the tool tip tissue interactions (omitted in the illustration). The terminology associated with the various DOF corresponds with the model state definitions (Table 25.1)

25.2.3.2 Vector Quantization (VQ)

Each one of the 15 states was associated with a unique set of forces, torques angular and linear velocities, as indicated in Table 25.2. Following the language analogy, in the same way as a word is correlated to a state may be pronounced differently and still retains the same meaning, the tool might be in a specific state while infinite combinations of force, torque angular and linear velocities may be used. A significant data reduction was achieved by using a clustering analysis in a search for discrete number of high concentration cluster centers in the database for each one of the 15 states. As part of this process, the continuous 12 dimensional vectors were transformed into one dimensional vector of 150 symbols (ten symbols for each state).

The data reduction was performed in three phases. During the first phase a subset of the database was created appending all the 12 dimensional vectors associated with each state measured by the left and the right tools and preformed by all the subjects (see Sect. 25.2.2 for details). The 12 dimensional subset of the database $(\dot{\theta}_x, \dot{\theta}_y, \dot{\theta}_z, \dot{\theta}_g, F_x, F_y, F_z, T_x, T_y, T_z, F_g)$ was transformed into a 9 dimensional vector $(\dot{\theta}_{xy}, \dot{\theta}_z, \dot{\theta}_g, F_{xy}, F_z, T_{xy}, T_z, F_g)$ by calculating the magnitude of the angular velocity, and the forces and torques in the XY plane. This process canceled out differences between surgeons due to variations in position relative to the animal and allowed use of the same clusters for the left and the right tools.

As part of the second phase, a K-means vector quantization algorithm [48] was used to identify ten cluster centers associated with each state. Given M patterns $\bar{X}_1, \bar{X}_2, \bar{X}_M$ contained in the pattern space $\bar{S}$, the process of clustering can be formally stated as seeking the regions $\bar{S}_1, \bar{S}_2, \bar{S}_K$ such that every data vector $\bar{X}_i$ $(i = 1, 2M)$ falls into one of these regions and no $\bar{X}_i$ is associated in two regions, i.e.

$$\begin{aligned} &\bar{S}_1 \cup \bar{S}_2 \cup \bar{S}_3 \ldots \cup \bar{S}_K = \bar{S} \quad \text{(a)} \\ &\bar{S}_i \cap \bar{S}_j = 0 \quad \forall i \neq j \quad \text{(b)} \end{aligned} \tag{25.1}$$

The K-means algorithm, is based on minimization of the sum of squared distances from all points in a cluster domain to the cluster center,

$$\min \sum_{X \in S_j(k)} (\bar{X} - \bar{Z}_j)^2 \tag{25.2}$$

where $S_j(k)$ was the cluster domain for cluster centers $\bar{Z}_j$ at the kth iteration, and $\bar{X}$ was a point in the cluster domain.

The pattern spaces $\bar{S}$ in the current study were composed from the F/T applied on the surgical tool by the surgeon along with the tool's angular and linear velocities for different states. A typical data vector $\bar{X}_i$, was a 9 dimensional vector defined as $\{\dot{\theta}_{xy}, \dot{\theta}_z, \dot{\theta}_g, F_{xy}, F_z, T_{xy}, T_z, F_g\}$. The cluster regions $\bar{S}_i$ represented by the cluster centers $\bar{Z}_j$, defined typical signatures or codeword (pronunciations in the human language realm) associated with a specific state (e.g. PS, PL, GR etc.). The number of clusters identified in each type of state was based upon the squared error distortion criterion (25.3). As the number of clusters increased, the distortion decreased exponentially. Following this behavior, the number of clusters was constantly increased until the squared error distortion gradient as a function of k decreased below a threshold of 1% that results in ten cluster centers for each state.

$$d(\bar{X}, \bar{Z}) = \|\bar{X} - \bar{Z}_j\|^2 = \sum_{i=1}^{k} (\bar{X} - \bar{Z}_i)^2 \tag{25.3}$$

In the third phase, the ten cluster centers $\bar{Z}_j$ for each state (Table 25.2) forming a codebook of 150 discrete symbols were then used to encode the entire database of the actual surgical tasks converting the continuous multi-dimensional data into a one-dimensional vector of finite symbols. This step of the data analysis was essential for using the discrete version of Markov Model.

25.2.3.3 Markov Model (MM)

The final step of the data analysis was to develop a model that represents the process of performing MIS along with the methodology for objectively evaluating surgical skill. The Markov Model was found to be a very compact statistical method to summarize a relatively complex task such as a step or a task of a MIS procedure. Moreover, the skill level was incorporated into the MM by developing different MMs based on data acquired for different levels of expertise starting from a first year residents up to a level of expert surgeons.

The modeling approach underling the methodology for decomposing and statistically representing a surgical task is based on a fully connected, symmetric 30 states MM where the left and the right tools are represented by 15 states each (Fig. 25.4). In view of this model, any MIS task may be described as a series of states. In each state, the surgeon is applying a specific force/torque/velocity signature, out of ten signatures that are associated with that state, on the tissue or on any other item in the surgical scene by using the tool. The surgeon may stay within same state for specific time duration using different signatures associated with that state and then perform a transition to another state. The surgeon may utilize any of the 15 states by using the left and the right tools independently. However, the states representing the tool/tissue or tool/object interactions of the left and the right tools are mathematically and functionally linked.

The MM is defined by the compact notation in (25.4). Each Markov sub-model representing the left and the right tool is defined by λ_L and λ_R (25.4). The sub model is defined by: (a) The number of states –N whereas individual states are denoted as $S = \{s_1, s_1, \ldots s_N\}$, and the state at time t as q_t

(b) The number of distinct (discrete) observation symbol – M whereas individual symbols are denoted as $V = \{v_1, v_1, \ldots v_M\}$

(c) The state transition probability distribution matrix indicating the probability of the transition from state $q_t = s_i$ at time t to state $q_{t+1} = s_j$ at time $t+1$ - $A = \{a_{ij}\}$, where $a_{ij} = P[q_{t+1} = s_j | q_t = s_i]$ $1 \leqslant i, j \leqslant N$.

Note that $A = \{a_{ij}\}$ is a non-symmetric matrix ($a_{ij} \neq a_{ji}$) since the probability of performing a transition from state i to state j using each one of the tools is different from the probability of performing a transition from state j to state i.

(d) The observation symbol probability distribution matrix indicating the probability of using the symbol v_k while staying at state s_j at time t – $B = \{b_j(k)\}$, where for state j $b_j(k) = P[v_k \text{at } t | q_t = s_j] 1 \leqslant j \leqslant N, 1 \leqslant k \leqslant M$.

(e) The initial state distribution vector indicating the probability of starting the process with state s_i at time $t = 1 - \pi$ where $\pi_i = P[q_1 = s_i] 1 \leqslant i \leqslant N$.

The two sub models are linked to each other by the left-right interstate transition probability matrix or the cooperation matrix indicating the probability for staying

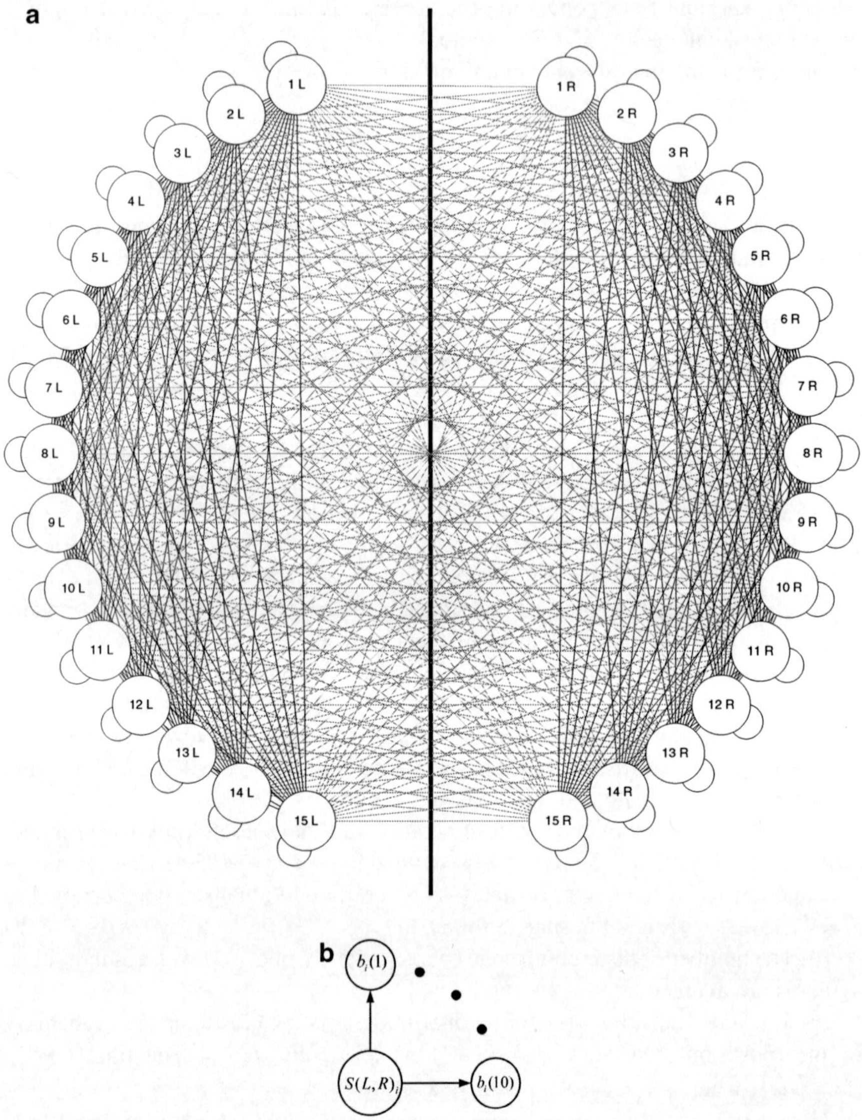

Fig. 25.4 Finite State Diagrams (FSD) – (**a**) Fully connected FSD for decomposing MIS. The tool/tissue and tool/object interactions of the left and the right endoscopic tools are represented by the 15 fully connected sub-models. *Circles* represent states whereas *lines* represent transitions between states. Each *line*, that does not cross the *center-line*, represents probability value defined in the state transition probability distribution matrix$A = \{a_{ij}\}$. Each *line* that crosses the *center-line*, represents probability for a specific combination of the left and the right tools and defined by the interstate transition probability distribution matrix, or the cooperation matrix$C = \{c_{lr}\}$. Note that since the probability of performing a transition from state *i* to state *j* by each one of the tools is different from probability of performing a transition from state *j* to state *i*, these two probabilities should have been represented by two parallel lines connecting state *i* to state *j* and representing the

in states s_l with the left tool s_r with the right tool at time t –$C = \{c_{lr}\}$, where $c_{lr} = P[q_{tL} = s_l \cup q_{tR} = s_r]$ $1 \leqslant l, r \leqslant N$

Note that $C = \{c_{lr}\}$ is a non-symmetric matrix $c_{lr} \neq c_{rl}$ since it representing the combination of using two states simultaneously by the left and the right tools.

The probability of observing the state transition $Q = \{q_1, q_2, \ldots q_T\}$ and the associated observation sequence $O = \{o_1, o_2, \ldots o_T\}$, given the two Markov sub models (25.4) and interstate transition probability matrix, is defined by (25.5)

$$\lambda_L = (A_L, B_L, \pi_L) \quad \lambda_R = (A_R, B_R, \pi_R) \tag{25.4}$$

$$P(Q, O|\lambda_L, \lambda_R, C) = \pi_{q_L}\pi_{q_R}\prod_{t=0}^{T} a_{q_tq_{t+1}L}b_{q_tL}(o_t)a_{q_tq_{t+1}R}b_{q_tR}(o_t)c_{q_{tL}q_{tR}} \tag{25.5}$$

Since probabilities by definition have numerical value in the range of 0–1, for a relatively short time duration, the probability calculated by (25.5) converges exponentially to zero; and therefore exceeds the precision range of essentially any machine. Hence, by using a logarithmic transformation, the resulting values of (25.5) in the range of [0 1] are mapped by (25.6) into $[-\infty\ 1]$.

$$Log(P(Q, O|\lambda_L, \lambda_R, C)) = Log(\pi_{q_L}) + Log(\pi_{q_R}) + \sum_{t=1}^{T} Log(a_{q_tq_{t+1}L}) + Log(b_{q_tL}(o_t)) +; Log(a_{q_tq_{t+1}R}) + Log(b_{q_tR}(o_t)) + Log(c_{q_tLq_tR}) \tag{25.6}$$

Due to the nature of the process associated with surgery in which the procedure, by definition, always starts at the idle state (state 1), the initial state distribution vector is defined as follows:

$$\begin{aligned} \pi_{1L} &= \pi_{1R} = 1 \\ \pi_{iL} &= \pi_{iR} = 0 \quad 2 \leqslant i \leqslant N \end{aligned} \tag{25.7}$$

Once the MMs were defined for specific subjects with specific skill levels, it became possible to calculate the statistical distance factors between them. These statistical distance factors are considered to be an objective criterion for

two potential transitions. However for simplifying the graphical representation of$A = \{a_{ij}\}$ only one line is plotted between state *i* to state *j*. (**b**) Each state out of the 15 states of the left and the right tool$b(L, R)_i$ is associated with the ten force/torque/velocity signature or discrete observation$b_i(1)\ldots b_i(10)$. Each *line*, that connects the state with a specific observation represents probability value defined in observation symbol probability distribution matrix$B = \{b_j(k)\}$. The sub-structure appeared in (b) that is associated with each state was omitted for simplifying the diagram in (a) evaluating skill level if for example the statistical distance factor between a trainee (indicated by index R) and an expert (indicated by index E) is being calculated. This distance indicates how similar is the performance of two subjects under study. Given two MMs λ_{E} (Expert) and λ_{R} (Novice) the nonsymmetrical statistical distances between them are defined as $D_1(\lambda_{\mathrm{R}}, \lambda_{\mathrm{E}})$ and $D_2(\lambda_{\mathrm{E}}, \lambda_{\mathrm{R}})$. The natural expression of the symmetrical statistical distance version D_{ER} is defined by (25.8).

$$D_{ER} = \frac{D_1(O_E, Q_E, O_R, Q_R, \lambda_E) + D_2(O_E, Q_E, O_R, Q_R, \lambda_R)}{2}$$
$$= \left(\frac{\log P(O_R, Q_R|\lambda_E)}{\log P(O_E, Q_E|\lambda_E)} + \frac{\log P(O_R, Q_R|\lambda_R)}{\log P(O_E, Q_E|\lambda_R)}\right) \Big/ 2 \tag{25.8}$$

Setting an expert level as the reference level of performance, the symmetrical statistical distance of a model representing a given subject from a given expert (D_{ER}) is normalized with respect to the average distance between the models representing all the experts associated with the expert group ($\bar{D}_{EE}$) and expressed in (25.9). The normalized distance $\|D_{ER}\|$ represents how far (statistically) is the performance of a subject, given his or her model, from the performance of the average expert.

$$\|D_{ER}\| = \frac{D_{ER}}{\bar{D}_{EE}} = \frac{D_{ER}}{\frac{1}{l}\sum\limits_{u=1;v=1}^{u=l;v=l} D_{E_u E_v}} \quad for\ u \neq v \tag{25.9}$$

For the purpose of calculating the normalized learning curve, the 20 distances between all the expert subjects was first calculated $D_{E_u E_v}$ – (for five subjects in the expert group –$u = v = 1...5 - l = 20$) using (25.8). The denominator of (25.9) was then calculated. Once the reference level of expertise was determined, the statistical distances between each one of the 25 subjects, grouped into five levels of training (R1, R2, R3, R4, R5), and each one of the experts was calculated (five distances for each individual, 25 distances for each group of skill level and 125 distances for the entire data base) using (25.8). The average statistical distance and its variance defines the learning curve of a particular task.

25.2.3.4 Complimentary Objective Indexes

In addition to the Markov models and the statistical similarity analysis, two other objective indexes of performance were measured and calculated, including the task completion time and the overall length (L) of the path of the left and the right tool tips. Where d_L, d_R are the distances between two consecutive tool tip positions

$P_L(t-1), P_R(t-1)$ and $P_L(t), P_R(t)$ as a function of time of the left and the right tools respectively.

$$L = \sum_{t=1}^{T} d_L(P_L(t-1), P_L(t)) + d_R(P_R(t-1), P_R(t)) \tag{25.10}$$

25.2.4 Subjective Analysis: Scoring

The subjective performance analysis was based on an off-line unbiased expert surgeon review (blinded to the subject and training level of each individual) of digital videos recorded during the experiment. The review utilized a scoring system of four equally weighted criteria: (a) overall performance (b) economy of movement (c) tissue handling (d) number of errors including drop needle, drop suture, lose suture loop, breaking suture, needle injury to adjacent tissue, inability to puncture bowel with needle. Criteria (a), (b), and (c) included five levels. The final scores were normalized to the averaged experts scoring.

25.3 Results

25.3.1 Force and Torque Position and Orientation

Typical raw data of forces and torques (F/T) and tool tip position were plotted using three dimensional graphs. The graphs show the kinematics and dynamics of the left and right endoscopic tools as measured by the Blue DRAGON while performing MIS intracorporeal knot by junior trainee (R1 – Fig. 25.5a, c) and expert surgeon (E – Fig. 25.5b, d). The F/T vectors can be depicted as arrows with origins located at the port, changing their lengths and orientations as a function of time and as a result of the F/T applied by the surgeon's hand on the tool. In a similar fashion, the traces of the tool tips with respect to the ports were plotted in Fig. 25.5c, d as their positions changed during the surgical procedure.

The forces along the Z axis (in/out of the port) were higher compared to the forces in the XY plane. On the other hand, torques developed by rotating the tool around the Z axis were extremely low compared to the torques generated while rotating the tool along the X and Y axis while sweeping the tissue or performing lateral retraction. Similar trends in terms of the F/T magnitude ratios between the X, Y, and Z axes were found in the data measured in other MIS tasks.

These raw data demonstrate the complexity of the surgical task and the multi-dimensional data associated with it. This complexity can be resolved in part by decomposing the surgical task into its primary elements enabling profound understanding of the MIS task.

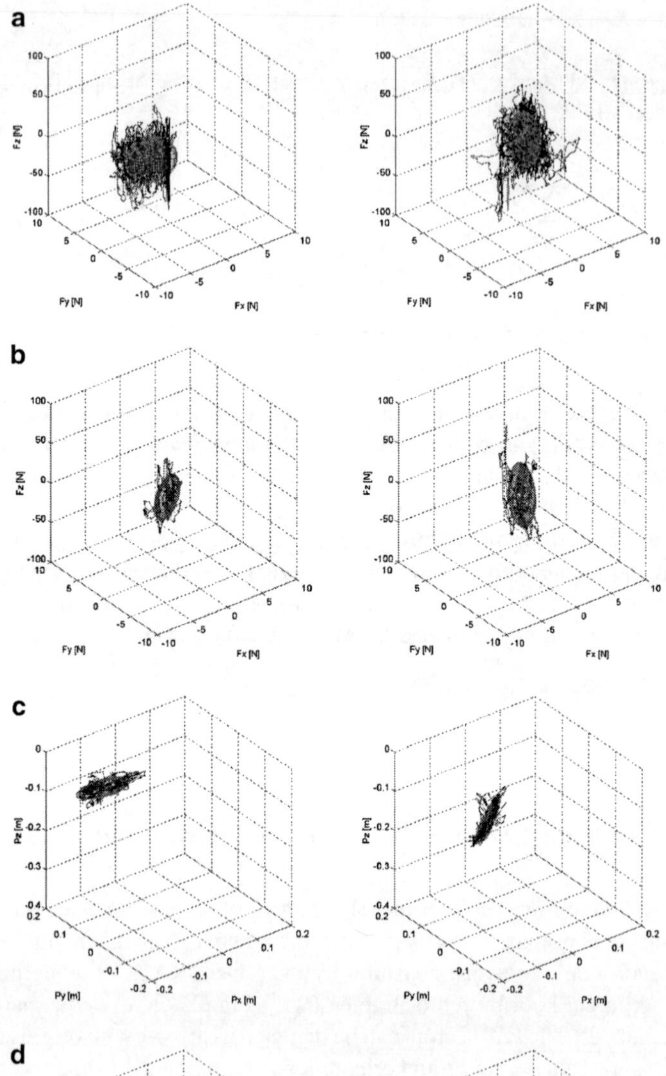

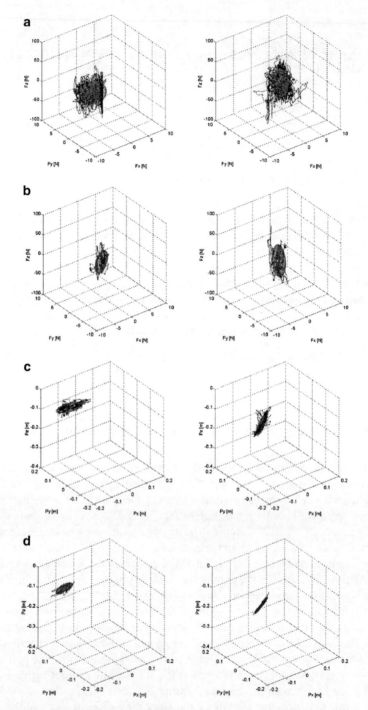

Fig. 25.5 Kinematics and dynamic data from left and the right endoscopic tools measured by the Blue DRAGON while performing MIS suturing and knot tying by a trainee surgeon (**a**, **c**) and an expert surgeon (**b**, **d**) – (**a**, **b**) Forces; (**b**, **c**) tool tip position. The ellipsoids contain 95% of the data points

25.3.2 Cluster Centers and Markov Models

A cluster analysis using the K-means algorithm was performed to define typical cluster centers in the database. These were further used as code-words in the MM analysis. A total of 150 cluster centers were identified, ten clusters centers for each type of tool/tissue/object interaction as defined in Table 25.1. Figure 25.6 depicts the ten cluster centers associated with each one of the 15 states identified in the data. For example, Fig. 25.6 (13) represents ten cluster centers associated with the state defined by Grasping-Pushing-Sweeping (Table 25.1 – State No. 13). Grasping-Pushing-Sweeping is a superposition of three actions. The surgeon grasps a tissue or an object which is identified by the positive grasping force (F_g) acting on the tool's jaws and the negative angular velocity of the handle (ω_g) indicating that the handle is being closed. At the same time the grasped tissue or object is pushed into the port indicated by positive value of the force (F_z) acting along the long shaft of the tool and negative linear velocity (V_r) representing the fact the tool is moved into the port. Simultaneously, sweeping the tissue to the side manifested by the force and the torque in the XY plane (F_{xy}, T_{xy}) that are generated due to the deflection of the abdominal wall, the lateral force applied on the tool by the tissue or object being swept along with the lateral angular velocity (ω_{xy}) indicating the rotation of the tool around the pivot point inside the port.

Both static, quasi-static and dynamic tool/tissue or tool/object interactions are represented by the various cluster centers. Even in static conditions, the forces and torques provide a unique and un-ambivalent signature that can be associated with each one of the 15 states.

The 150 cluster centers (Table 25.1 and Fig. 25.6) form a code-book that is used to encode the entire database of the actual surgical procedure converting the continuous multi-dimensional data into a one-dimensional vector of finite symbols. It should be noted that since each state is associated with a unique set of ten cluster centers and vice versa, a specific cluster center was associated with only one state, and as a by-product of the encoding process, the states were also identified.

25.3.3 Objective and Subjective Indexes of Performance

Given the encoded data, the MM for each subject was calculated defining the probabilities for performing certain tool transitions ([$\boldsymbol{A}$] matrix), the probability of combining two states ([$\boldsymbol{C}$] matrix), and the probability of using the various signatures in each state ([$\boldsymbol{B}$] matrix) – Fig. 25.7. The highest probability values in the [A] matrix usually appeared along the diagonal. These results indicate that a transition associated with staying at the same state is more likely to occur rather than a transition to any one of the other 15 potential states. In minimally invasive surgical suturing, the default transition between any state is to the grasping state (state number 2) as indicated by the high probability values along the second column of the [$\boldsymbol{A}$] matrix. Probability of using one out of the 150 cluster centers defined in Fig. 25.7 is graphically represented by the [$\boldsymbol{B}$] matrix. Each line of the [$\boldsymbol{B}$] matrix is

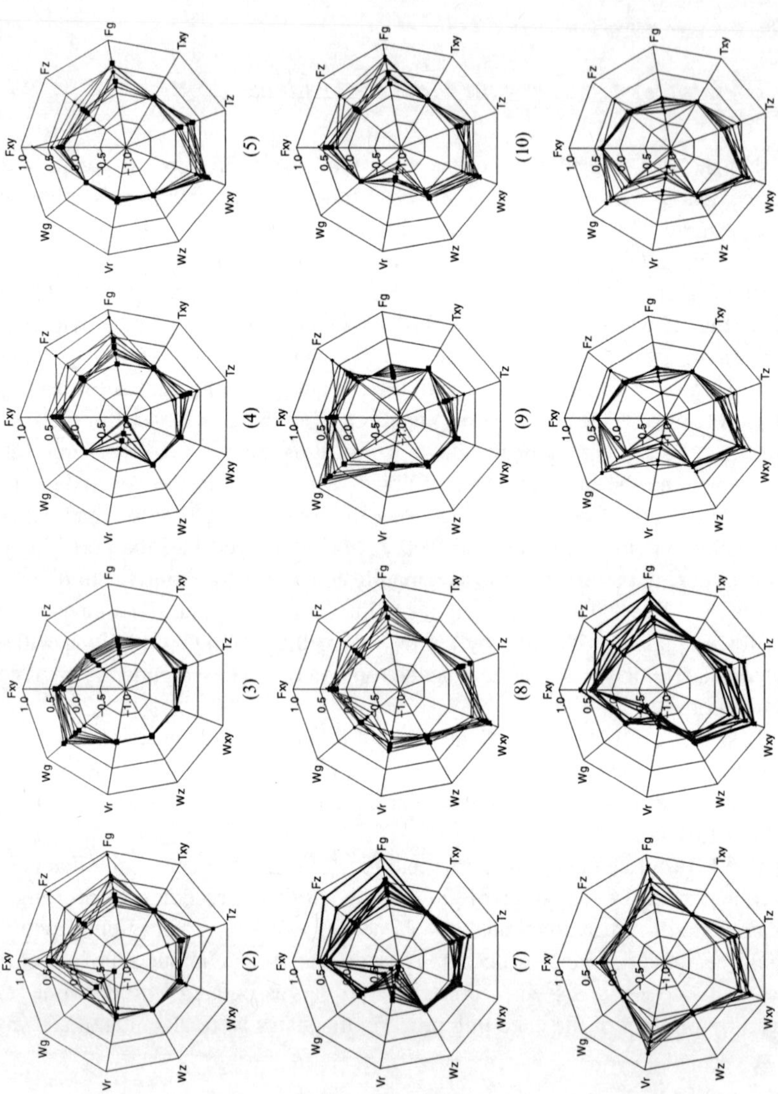

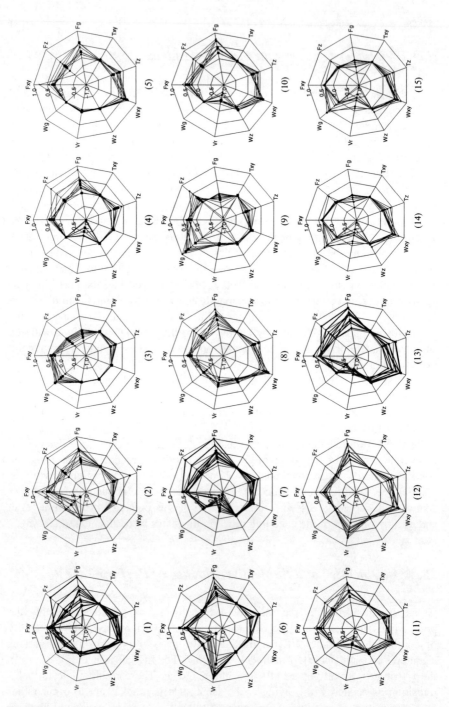

Fig. 25.6 Cluster centers definition – Ten signatures of forces torques linear and angular velocities associated with the 15 types of states (tool/tissue or tool/object interaction) defined by Table 25.1. In these graphs each one of the ten polar lines represent one cluster. The clusters were normalized to a range of [−1 1] using the following min/max values: $\omega_{xy} = 0.593$[r/s], $\omega_Z = 2.310$ [r/s], $V_r = 0.059$[m/s], $\omega_g = 0.532$[r/s], $F_{xy} = 5.069$[N], $F_Z = 152.536$[N], $F_g = 33.669$[N], $T_{xy} = 9.792$[Nm], $T_Z = 0.017$[Nm]. The numbers correspond to the 15 states as defined by Table 25.1

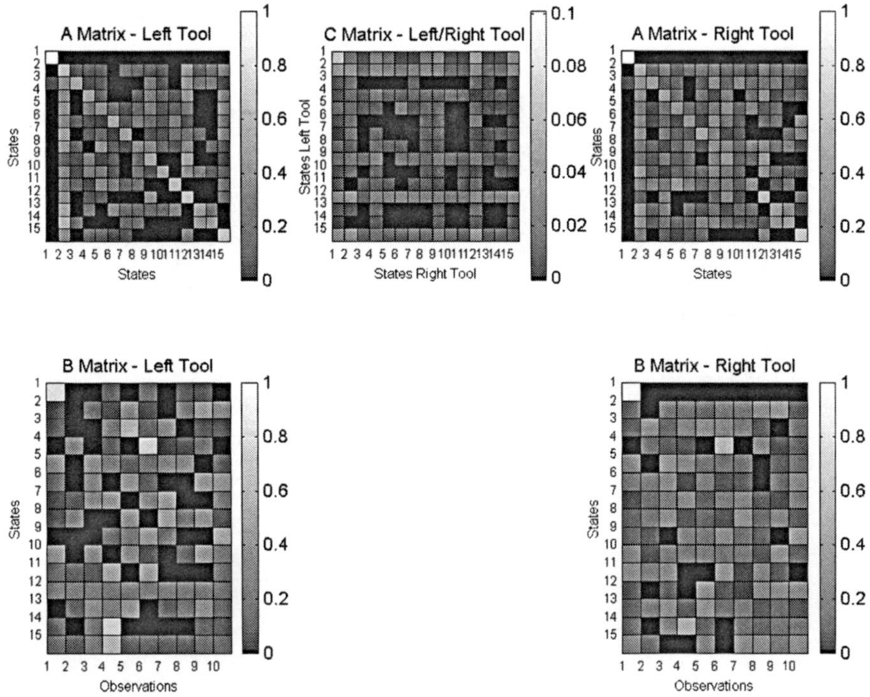

Fig. 25.7 A typical Markov Model where the matrices [*A*], [*B*], [*C*], are represented as color-coded probabilistic maps

associated with one of the ten states. The clusters were ranked according to the mechanical power. The left and the right tool used different distribution of the clusters. Whereas with the left tool the most frequent clusters that were used are related to mid-range power with the right tool the cluster usage is more evenly distributed among the different power levels. The collaboration matrix [*C*] indicates that the most frequently used state with both the left and the right tools are idle (state 1), grasping (state 2) grasping pulling and sweeping (state 12) and grasping rotating (state 15) with the left tool. Once one of the tools utilizes one of these states, the probability of using any of the states by the other tool is equally distributed between the states which is indicated by the bright horizontal stripes in the graphical representation of the [*C*] matrix.

The Idle state (state 1) in which no tool/tissue interaction is performed, was mainly used by both expert and novice surgeons, to move from one operative state to the another. However, the expert surgeons used the idle state only as a transition state while the novices spent a significant amount of time in this state planning the next tool/tissue or tool/object interaction. However, in case of surgical suturing and knot tying, the grasping state (state 2) dominated the transition phases since grasping state maintained the operative state in which both the suture and the needle were held by the two surgical tools.

Figure 25.8a–c represent the normalized MM-based statistical distance as a function of the training level, the normalized completion time, and the normalized path length of the two tool tips respectively. The subjective normalized scoring is

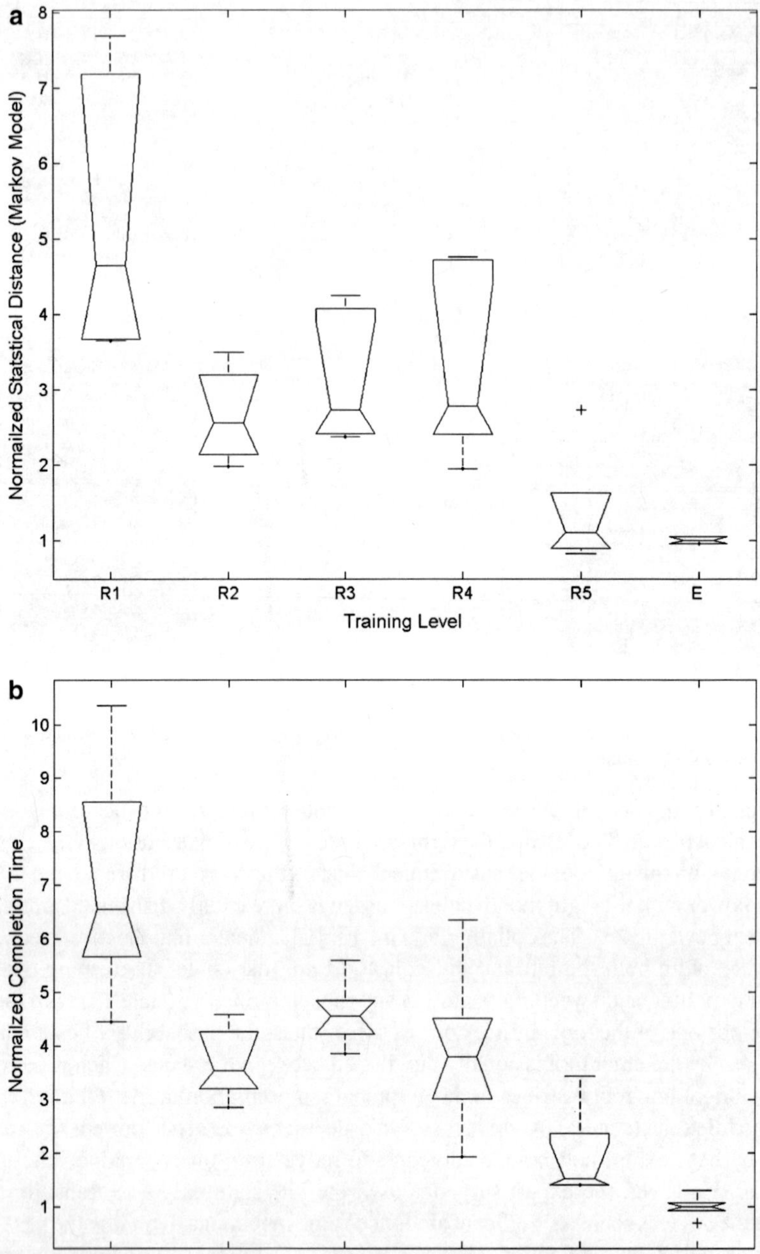

Fig. 25.8 Objective and subjective assessment indexes of minimally invasive suturing learning cure. The objective performance indexes are based on: (**a**) Markov model normalized statistical distance, (**b**) normalized completion time, and

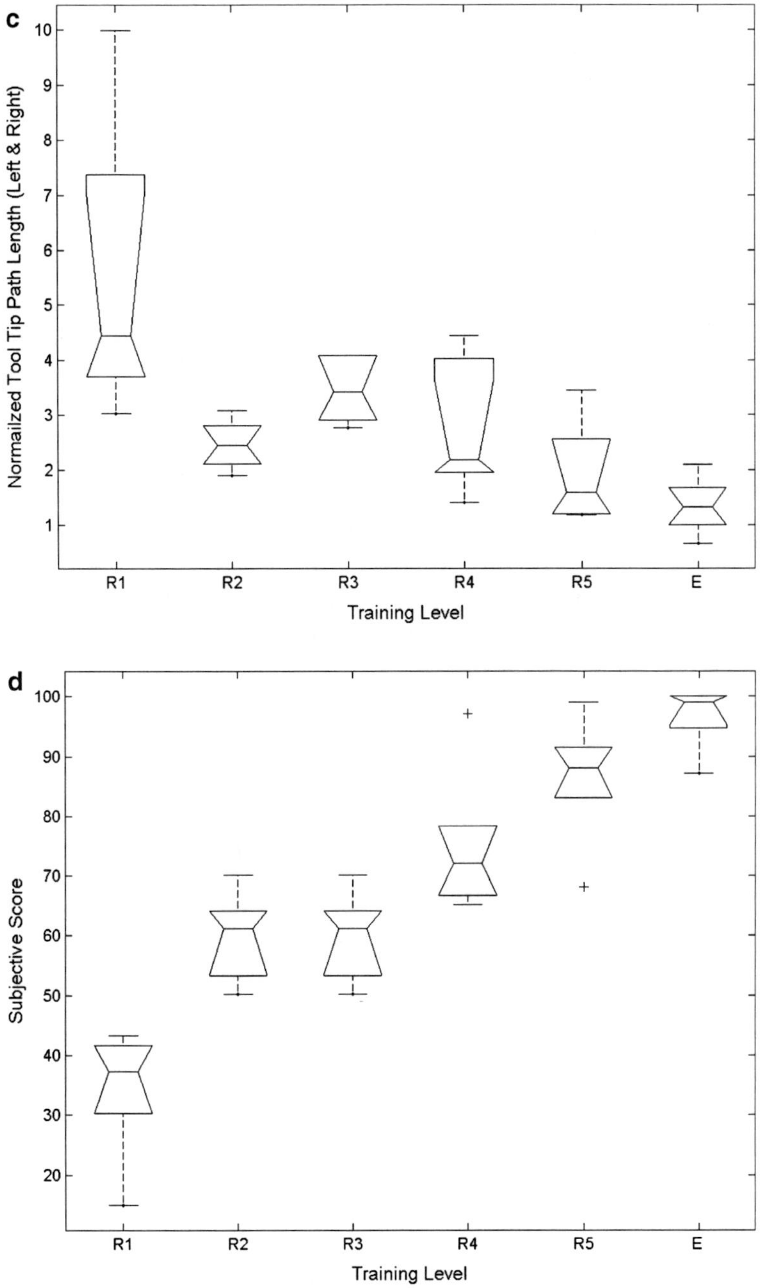

Fig. 25.8 (continued) (**c**) normalized path length of the two tool tips. The average task completion time of the expert group is 98 s and the total path length of the two tools is 3.832 m. The subjective performance index is based on visual scoring by an expert surgeon normalized with respect to experts' performance (**d**)

depicted in Fig. 25.8d. The data demonstrate that substantial suturing skills are acquired during the first year of the residency training. The learning curves do not indicate any significant improvement during the second and the third years of training. The rapid improvement of the first year is followed by lower gradient of the learning curve as the trainees progress toward the expert level. However, the MM based statistical distance along with the completion time criteria show yet another gradient in the learning curve that occurs during the fourth year of the residency training followed by slow conversion to expert performance. Similar trends in the learning curve are also demonstrated by the subjective assessment. One of the subjects in the R2 group outperformed his peers in his own group and some subjects in a more advanced groups (R3, R4). Although, statistically insignificant, the performance slightly altered the overall trend of the learning curves as defined by the different criteria.

A correlation analysis was performed between the means of the objective normalized MM based statistical distance and the subjective normalized scoring. The correlation factor R^2 was found to be 0.86. This result established the linkage between objective and subjective methodologies for assessing surgical skill (Fig. 25.9).

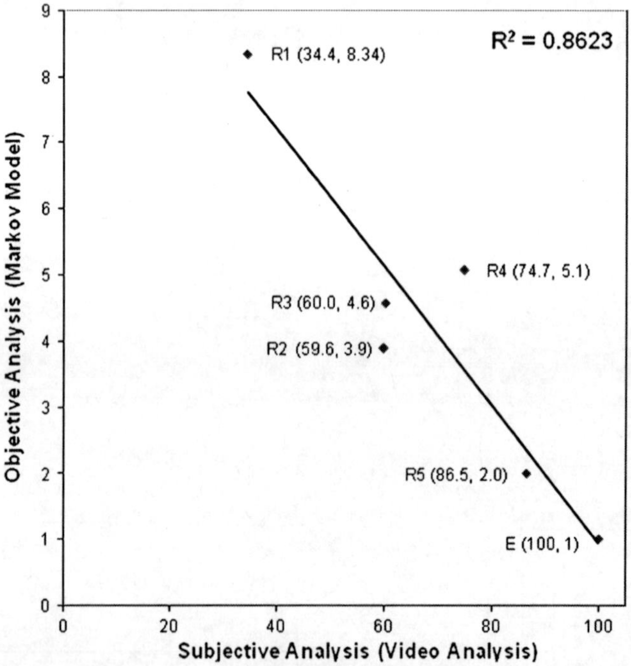

Fig. 25.9 Linear correlation between the normalized mean performance obtained by a subjective video analysis and objective analysis using Markov models and the statistical distance between models of trainees (R1–R5) and experts (E). The notations R1, R2, R3, R4, R5 represent the various residence groups where the number denotes year of training and E indicate expert surgeons. The *values* in the *brackets* represent the normalized mean scores using the subjective and the objective methodologies respectively

Detailed analysis of the MM shows that major differences between surgeons at different skill levels were: (a) the types of tool/tissue/object interactions being used, (b) the transitions between tool/tissue/object interactions being applied by each hand, (c) time spent while performing each tool/tissue/object interaction, (d) the overall completion time, (e) the various F/T/velocity magnitudes being applied by the subjects through the endoscopic tools, and (f) two-handed collaboration. Moreover, the F/T associated with each state showed that the F/T magnitudes were found to be task-dependent. High F/T magnitudes were applied by novices compared to experts during tissue manipulation, and vice versa during tissue dissection. High efficiency of surgical performance was demonstrated by the expert surgeons and expressed by shorter tool tip displacements, shorter periods of time spent in the 'idle' state, and sufficient application of F/T on the tissue to safely accomplish the task.

25.4 Discussion

Minimally invasive surgery, regardless of the performance modality, is a complex task that requires synthesis between visual and kinesthetic information. Analyzing MIS in terms of these two sources of information is a key step towards developing objective criteria for training surgeons and evaluating the performance in different modalities including real surgery, master/slave robotic systems or virtual reality simulators with haptic technology.

Following two steps of data reduction, data that was collected by the Blue DRAGON were further used to develop models representing MIS as a process. In any data reduction there is always a compromise between decreasing the input dimensionality while retaining sufficient information to characterize and model the process under study. Utilizing the VQ algorithm the 13 dimensional stream of acquired data were quantized into 150 symbols with nine dimensions each.

The data quantization included two substeps. In the first steps the cluster centers were identified. As part of the second step the entire database was encoded based on the cluster centers defined in the first step. Every data point needs to meet two criteria in order to be associated with one of the 150 cluster centers defined in the first step. The first criterion is to have the minimal geometrical distance to one of the cluster centers. Once the data point was associated with a specific cluster center it is by definition associated with a specific state out the 15 defined. Based on expert knowledge of surgery, Table 25.1 defines the 15 states and unique sets of individual vector components. The second criterion is that given the candidate state and the data vector, the direction of each component in the vector must match the one defined by the table for the selected state. It was indicated during the data processing that these two criteria were always met suggesting that the data quantization process is very robust in it nature. Following the encoding process a 2 dimensional input (one dimension for each tool) was utilized to form a 30 state fully connected Markov model. The coded data with their close association to the measured data, as well as the Markov model using these codes as its observations distributed among its states, retain sufficient multi model information in a compact mathematical formulation for modeling the process of surgery at different levels.

MIS is recognized both qualitatively and quantitatively as multidimensional process. As such, studying one parameter e.g. completion time, tool-tip paths, or force/torque magnitudes reveals only one aspect of the process. Only a model that truly describes MIS as a process is capable of exposing the process internal dynamics and provides wide spectrum information about it. At the high level, a tremendous amount of information is encapsulated into a single objective indicator of surgical skill level and expressed as the statistical distance between the surgical performance of a particular subject under study from a surgical performance of an expert. As part of an alternative approach a combined score could be calculated by studying each parameter individually (e.g. force, torque, velocity, tool path, completion time etc.), assigning a weight to each one of these parameters, which is a subjective process by itself, and combining them into a single score. The assumption underlying this approach is that a collection of aspects associated with surgery may be used to assess the overall process. However this alternative approach ignores the internal dynamics of the process that is more likely to be revealed by a model such as the Markov model. In addition, as opposed to analyzing individual parameters, studying the low levels of the model provides profound insight into the process of MIS in a way that allows one to offer constructive feedback for a trainee regarding performance aspects like the appropriate application of F/T, economy of motion, and two handed manipulation.

The appropriate application of F/T on the tissue has a significant impact on the surgical performance efficiency and outcome of surgery. Previous results indicated that the F/T magnitudes are task dependent [3–7]. Experts applied high F/T magnitudes on the tissues during tissue dissection as apposed to low F/T magnitudes applied on the tissues by trainees that were trying to avoid irreversible damage. An inverse relationship regarding the F/T magnitudes was observed during tissue manipulation in which high F/T magnitudes applied on the tissue by trainees exposed them to acute damage. It is important to point out that these differences were observed in particular states (e.g all the states including grasping for tissue manipulation and all the state that involved spreading for tissue dissection). Due to the inherent variance in the data even multidimensional ANOVA failed to identify this phenomena once the F/T magnitudes are removed from the context of the multi state model. Given the nature of surgical task, the Markov model [B] Matrix, encompassing information regarding the frequency in which the F/T magnitudes were applied, may be used to assess whether the appropriate magnitudes F/T were applied for each particular state. For obvious reasons, tissue damage is correlated with surgical outcome, and linked to the magnitudes and the directions in which F/T were applied on the tissues. As such, tissue damage boundaries may be incorporated into the [B] matrix for each particular state. Given the surgical task, this additional information may refine the contractive feedback to the trainee and the objective assessment of the performance.

The economy of motion and the two hand collaboration may be further assessed by retrieving the information encapsulated into the [A], and [C] matrices. The amount of information incorporated into these two data structures is well beyond the information provided by a single indicator such as tool-tip path length, or completion time for the purpose of formulating constructive feedback to the trainee. Given a surgical task, utilizing the appropriate sets of states and state transitions are skill dependent. This information is encompassed in the [A] matrix indicating that states that were in use and the state transitions that were performed. Moreover, the ability to refine the time domain analysis using the multi state Markov model indicated, as was observed in previous studies, that the 'idle' state is utilized as a transition state by expert surgeons whereas a significant amount of time is spent in that state by trainees [3–7]. In addition, coordinated movements of the two tools is yet another indication of high skill leveling MIS. At a lower skill level the dominant hand is more active than the non-dominant hand as opposed to a high skill level in which the two tools are utilized equally. The collaborated [C] matrix encapsulates this information and quantifies the level of collaboration between the to tools.

In conclusion, the MM model provides insight into the process of performing MIS. This information can be translated into a constructive feedback to the trainee as indicated by the model three matrices [A], [B] and [C]. Moreover, the capability of running the model in real-time and its inherent memory allows a senior surgeon supervising the surgery or an artificial intelligent expert system incorporated into a surgical robot or a simulator provides an immediate constructive feedback during the process as previously described.

A useful analogy of the proposed methodology for decomposing the surgical task is the human spoken language. Based on this analogy, the basic states – tool/tissue interactions are equivalent to '*words*' of the MIS '*language*' and the 15 states form the MIS '*dictionary*' or set of all available words. In the same way that a single word can be pronounced differently by various people, the same tool/tissue or tool/object interaction can be performed differently by different surgeons. Differences in F/T magnitudes account for this different '*pronunciation*', yet different pronunciation of a '*word*' have the same meaning, or outcome, as in the realm of surgery. The cluster analysis was used to identify the typical F/T and velocities associated with each one of the tool/tissue and tool/object interactions in the surgery '*dictionary*', or using the language analogy, to characterize different pronunciations of a '*word*'. Utilizing the '*dictionary*' of surgery, the MM was then used to define the process of each task or step of the surgical procedure, or in other words, '*dictating chapters*' of the surgical '*story*'. This analogy is reinforced by an important finding in the field of Phonology suggesting that all human languages use a limited repertoire of about 40–50 sounds defined as phones [45] e.g. the DARPA phonetic alphabet, ARPAbet used in American English or the International Phonetic Alphabet (IPA). The proposed methodology retains its power by decomposing the surgical task to its fundamental elements – tool/tissue and tool/object interactions. These elements are inherent in MIS no matter which modality is being used.

One may note that although the notations and the model architecture of the proposed Markov model (MM) and the hidden Markov model (HMM) approach are similar, there are several fundamental differences between them. Strictly speaking, the proposed MM is a *white box* model in which each state has a physical meaning describing a particular interaction between the tools and tissue or other objects in the surgical scene like sutures and needles. However, the HMM is a *black box* model in which the states are abstract and are not related to a specific physical interaction. Moreover, in the proposed white box model, each state has a unique set of observations that characterize only the specific state. By definition, once the discrete observation is matched with a vector quantization code-word the state is also defined. States in the HMM share the same observations, however different observation distributions differentiate between them. The topology of the proposed MM suggests a hybrid approach between the two previously described models. It adds to the classic Markov model another layer of complexity by introducing the observation elements for each state. The model also provides insight into the process by linking the states to physical and meaningful interactions. This unique quality adds to the classic notation of the introduction of the collaboration matrix [$\boldsymbol{C}$]. This matrix is not present in either the MM or the HMM. The [$\boldsymbol{C}$] matrix was introduced as a way to link between the models representing the left and right hand tools since surgery is a two-handed task.

Quantifying the advantages and the disadvantages of each modeling approach (MM or HMM) is still a subject for active research. Whereas the strength of the MM is expressed by providing physical meaning to the process being modeled, development of HMM holds the promise for more compact model topology which avoids any expert knowledge incorporated into the model. Regardless of the type of the model, defining the scope of the model and its fundamental elements, the state and the observation are subjects of extensive research. In the current study the entire surgical task is modeled by a fully connected model topology were each tool/tissue/object interaction is modeled as a state. In a different approach, using a state of the art methodology in speech recognition in which each phenomenon is represented by a model with abstract states, each tool/object interaction is modeled by entire model using more generalized definitions for these interactions e.g. place position, insert remove [46, 47]. This approach may require additional model with a predetermined overall structure that will represent the overall process.

The scope of the proposed model is limited to objectively assess technical factors of surgical ability. Cognitive factors per se cannot be assessed by the model unless a specific action is taken as a result of a decision making process. In any case, the model is incapable of tracing the process back to its cognitive origin. In addition, the underlying assumption made by using a model is that there is a standard technique with insignificant variations by which expert surgeons perform a surgical task. Any significant variation of the surgical performance, regardless of the surgical outcome, is penalized by the model and associated with low scores. If such a surgical performance variation from the standard surgical technique is associated with a better outcome for the patient the model is incapable of detecting it.

Decomposing MIS and analyzing it using MM is one approach for developing objective criteria for surgical performance. The availability of validated objective measures of surgical performance and competency is considered critical for training surgeons and evaluating their performance. Systems like surgical robots or virtual reality simulators that inherently measure the kinematics and the dynamics of the surgical tools may benefit from inclusion of the proposed methodology. Using this information in real-time during the course of learning as feedback to the trainee surgeons or as an artificial intelligent background layer, may increase performance efficiency in MIS and improve patient safety and outcome.

References

1. Satava, R.: Metrics for Objective Assessment of Surgical Skills Workshop – Developing Quantitative Measurements through Surgical Simulation. Scottsdale, Arizona (2001)
2. Gallagher, A.G., Satava, R.M.: Virtual reality as a metric for the assessment of laparoscopic psychomotor skills. Learning curves and reliability measures. Surg Endosc. **16**(12), 1746–1752 (2002)
3. Rosen, J., Hannaford, B., Richards C., Sinanan, M.: Markov modeling of minimally invasive surgery based on tool/tissue interaction and force/torque signatures for evaluating surgical skills. IEEE Trans. Biomed. Eng. **48**(5), 579–591 (2001)
4. Richards, C., Rosen, J., Hannaford, B., MacFarlane, M., Pellegrini, C., Sinanan, M.: Skills evaluation in minimally invasive surgery using force/torque signatures. Surg. Endosc. **14**(9), 791–798 (2000)
5. Rosen J., Solazzo, M., Hannaford, B., Sinanan, M.: Objective evaluation of laparoscopic skills based on haptic information and tool/tissue interactions. Comput. Aided Surg. **7**(1), 49–61 (2002)
6. Rosen J., Brown, J.D., Barreca, M., Chang, L., Hannaford, B., Sinanan, M.: The blue DRAGON, a system for monitoring the kinematics and the dynamics of endoscopic tools in minimally invasive surgery for objective laparoscopic skill assessment. In: Proceedings of MMVR 2002. IOS Press, Amsterdam (2002)
7. Rosen, J., Brown, J.D., Chang, L., Barreca, M., Sinanan, M., Hannaford, B.: The blue DRAGON – a system for measuring the kinematics and the dynamics of minimally invasive surgical tools in-vivo. In: Proceedings of the 2002 IEEE International Conference on Robotics & Automation, Washington DC, USA, 11–15 May 2002
8. McBeth, P.B., Hodgson, A.J., Nagy, A.G., Qayumi, K.: Quantitative methodology of evaluating surgeon performance in laparoscopic surgery. In: Proceedings of MMVR 2002. IOS Press, Amsterdam (2002)
9. Ibbotson, J.A., MacKenzie, C.L., Cao, C.G.L., Lomax, A.J.: Gaze patterns in laparoscopic surgery. In: Westwood, J.D., Hoffman, H.M., Robb, R.A.,Stredney, D. (eds.) Medicine Meets Virtual Reality, vol. 7, pp. 154–160. IOS Press, Washington, DC (1999)
10. Pugh, C.M., Youngblood, P.: Development and validation of assessment measures for a newly developed physical examination simulator. J. Am. Med. Inform. Assoc. **9**(5), 448–460 (2002)
11. Noar, M.: Endoscopy simulation: a brave new world? Endoscopy **23**, 147–149 (1991)
12. Satava, R.: Virtual reality surgical simulator. Surg. Endosc. **7**, 203–205 (1993)
13. Ota, D., Loftin, B., Saito, T., Lea, R., Keller, J.: Virtual reality in surgical education. Comput. Biol. Med. **25**, 127–137 (1995)
14. Berkely, J., Weghorst, S., Gladstone, H., Raugi, G., Ganter, M., Fast Finite Element Modeling for Surgical Simulation, Proceedings. Stud. Health Technol. Inform. **62**, 55–61 (1999)
15. Tseng, C.S., Lee, Y.Y., Chan, Y.P., Wu, S.S., Chiu, A.W.: A PC-based surgical simulator for laparoscopic surgery. Stud. Health Technol. Inform. **50**, 155–160 (1998)
16. Delingette, H., Cotin, S., Ayache, N.: Efficient linear elastic models of soft tissues for real-time surgery simulation. Stud. Health Technol. Inform. **62**, 100–101 (1999)
17. Basdogan, C., Ho, C.H., Srinivasan, M.A.: Simulation of tissue cutting and bleeding for laparoscopic surgery using auxiliary surfaces. Stud. Health Technol. Inform. **62**, 38–44 (1999)

18. Acosta, E., Temkin, B., Krummel, T.M., Heinrichs, W.L.: G2H–graphics-to-haptic virtual environment development tool for PC's. Stud. Health Technol. Inform. **70**, 1–3 (2000)
19. Akatsuka, Y., Shibasaki, T., Saito, A., Kosaka, A., Matsuzaki, H., Asano, T., Furuhashi, Y.: Navigation system for neurosurgery with PC platform. Stud. Health Technol. Inform. **70**, 10–16 (2000)
20. Berkley, J., Oppenheimer, P., Weghorst, S., Berg, D., Raugi, G., Haynor, D., Ganter, M., Brooking, C., Turkiyyah, G.: Creating fast finite element models from medical images. Stud. Health Technol. Inform. **70**, 26–32 (2000)
21. el-Khalili, N., Brodlie, K., Kessel, D.: WebSTer: a web-based surgical training system. Stud. Health Technol. Inform. **70**, 69–75 (2000)
22. Friedl, R., Preisack, M., Schefer, M., Klas, W., Tremper, J., Rose, T., Bay, J., Albers, J., Engels, P., Guilliard, P., Vahl, C.F., Hannekum, A.: CardioOp: an integrated approach to teleteaching in cardiac surgery. Stud. Health Technol. Inform. **70**, 76–82 (2000)
23. Gobbetti, E., Tuveri, M., Zanetti, G., Zorcolo, A.: Catheter insertion simulation with co-registered direct volume rendering and haptic feedback. Stud. Health Technol. Inform. **70**, 96–98 (2000)
24. Gorman, P., Krummel, T., Webster, R., Smith, M., Hutchens, D.: A prototype haptic lumbar puncture simulator. Stud. Health Technol. Inform. **70**, 106–109 (2000)
25. Anne-Claire, J., Denis, Q., Patrick, D., Christophe, C., Philippe, M., Sylvain, K., Carmen, G.: S.P.I.C. pedagogical simulator for gynecologic laparoscopy. Stud. Health Technol. Inform. **70**, 139–145 (2000)
26. Tasto, J.L., Verstreken, K., Brown, J.M., Bauer, J.J.: PreOp endoscopy simulator: from bronchoscopy to ureteroscopy. Stud. Health Technol. Inform. **70**, 334–349 (2000)
27. Wiet, G.J., Stredney, D., Sessanna, D., Bryan, J.A., Welling, D.B., Schmalbrock, P.: Virtual temporal bone dissection: an interactive surgical simulator. Otolaryngol. Head Neck Surg. **127** (1), 79–83 (2002)
28. John, N.W., Thacker, N., Pokric, M., Jackson, A., Zanetti, G., Gobbetti, E., Giachetti, A., Stone, R.J., Campos, J., Emmen, A., Schwerdtner, A., Neri, E., Franceschini, S.S., Rubio, F.: An integrated simulator for surgery of the petrous bone. Stud. Health Technol. Inform. **81**, 218–224 (2001)
29. Bielser, D., Gross. M.H.: Open surgery simulation. Stud. Health Technol. Inform. **81**, 57–63 (2001)
30. Berg, D., Berkley, J., Weghorst, S., Raugi, G., Turkiyyah, G., Ganter, M., Quintanilla, F., Oppenheimer, P.: Issues in validation of a dermatologic surgery simulator. Stud. Health Technol. Inform. **81**, 60–65 (2001)
31. Manyak, M.J., Santangelo, K., Hahn, J., Kaufman, R., Carleton, T., Hua, X.C., Walsh, R.J.: Virtual reality surgical simulation for lower urinary tract endoscopy and procedures. J. Endourol. **16**(3), 185–190 (2002)
32. Basdogan, C., Ho, C., Srinivasan, M.A.: Virtual environments for medical training: graphical and haptic simulation of common bile duct exploration (PDF). IEEE/ASME Trans. Mechatron. (special issue on Haptic Displays and Applications) **6**(3), 267–285 (2001)
33. Cao, C.G.L., MacKenzie, C.L., Ibbotson, J.A., Turner, L.J., Blair, N.P., Nagy, A.G.: Hierarchical decomposition of laparoscopic procedures. In: Westwood, J.D., Hoffman, H.M., Robb, R.A., Stredney, D. (eds.) Medicine Meets Virtual Reality, vol. 7, pp. 83–89. IOS Press, Washington, DC (1999)
34. M.C., Villanueva, I., Tendick, F.: Workspace analysis of robotic manipulator for teleoperated suturing task. In: Proceeding of IEEE/IROS, Maui Hawaii, USA (2001)
35. Rabiner, L.R.: A tutorial on hidden Markov models and selected application in speech recognition. Proc. IEEE **77**(2) (1989)
36. Hannaford, B., Lee, P.: Hidden Markov model of force torque information in telemanipulation. Int. J. Robot. Res. **10**(5), 528–539 (1991)
37. Pook, P., Ballard, D.H.: Recognizing teleoperated manipulations. In: Proc. IEEE Robotics and Automation, vol. 2, pp. 578–585, Atlanta, GA, May 1993

38. Nechyba, M.C., Xu, Y.: Stochastic similarity for validating human control strategy models. IEEE Trans. Robot. Autom. **14**(3), 437–451 (1998)
39. Yang, J., Xu, Y., Chen, C.-S.: Human action learning via hidden Markov model. IEEE Trans. Syst. Man Cybern. A (Syst. Hum.) **27**(1), 34–44 (1997)
40. Itabashi, K., Hirana, K., Suzuki, T., Okuma, S., Fujiwara, F.: Modeling and realization of the peg-in-hole task based on hidden Markov model. In: Proc. IEEE Intl. Conf. on Robotics and Automation, Leuven, Belgium, pp. 1142, May 1998
41. Wachsmuth I, Frohlich M. (eds.) Gesture and Sign Language in Human–Computer Interaction, International Gesture Workshop Proceedings, xi+308 pp. Springer, Berlin, Germany (1998)
42. Lien, J.J., Kanade, T., Cohn, J.F., Li, C.C.: Automated facial expression recognition based on FACS action units. In: Proceedings Third IEEE International Conference on Automatic Face and Gesture Recognition (Cat. No.98EX107), Nara, Japan, pp. 390–395, 14–16 April 1998
43. Baldi, P., Brunak S.: Bioinformatics. MIT Press, Cambridge, MA (1998)
44. Murphy, T.E., Vignes, C.M., Yuh, D.D., Okamura, A.M.: Automatic Motion Recognition and Skill Evaluation for Dynamic Tasks. In: Eurohaptics (2003)
45. Russell, S.J., Norvig, P.: Artificial Intelligence – A Modern Approach, 2nd edn. Pearson Education, Inc., Upper Saddle River, NJ (2003)
46. Li, M., Okamura, A.M.: Recognition of operator motions for real-time assistance using virtual fixtures. In: 11th International Symposium on Haptic Interfaces for Virtual Environment and Teleoperator Systems, IEEE Virtual Reality, pp. 125–131 (2003)
47. Hundtofte, C.S., Hager, G.D., Okamura, A.M.: Building a task language for segmentation and recognition of user input to cooperative manipulation systems. In: 10th International Symposium on Haptic Interfaces for Virtual Environment and Teleoperator Systems, pp. 225–230 (2002)
48. Bow, S.T.: Pattern Recognition. Marcel, Dekker Inc., New York (1984)

Part IV
Clinical Applications/Overviews

Chapter 26
Telesurgery: Translation Technology to Clinical Practice

Mehran Anvari

Abstract The ability to extend the physical reach of a surgeon to treat a patient surgically in another locality was one of the many promises which came with the introduction of Robotic and Computer Assisted Technology into the field of surgery in the late 1970s and early 1980s. In fact, it was the possibility of using a robot as surgeons' hands and eyes at a distance which led to some of the major grants from DARPA, NASA and NIH for the development of the prototypes of the da Vinci and the Zeus Systems which revolutionized the practice of Robotic and Computer Assisted Surgery in the late 1990s.

The primary incentive of these agencies for making such investments was to develop a system to allow them to provide emergency surgical care to the remote operatives. Others saw parallel uses in enhancing quality of surgical care which can be provided to settlements in remote parts of the world or at times of major disasters. And yet another use of telesurgery was an application for practical knowledge translation and a means for an expert surgeon to effectively achieve tele-presence during telementoring of another surgeon with acquisition of new surgical skills. The ability for two surgeons to collaborate across distances during a surgical act was seen as the ultimate achievement in knowledge translation in surgery.

It was these promises which sparked the efforts of many surgeons, engineers, and inventors who dedicated a significant portion of their lives into enhancing the field of Robotic Telesurgery.

M. Anvari (✉)
Department of Surgery, McMaster Institute for Surgical Innovation, Invention and Education, Faculty of Health Sciences, McMaster University, Hamilton, ON, Canada
and
St. Joseph's Healthcare, 50 Charlton Ave, East Room 805, Hamilton, ON L8N 4C6, Canada
e-mail: anvari@mcmaster.ca

J. Rosen et al. (eds.), *Surgical Robotics: Systems Applications and Visions*,
DOI 10.1007/978-1-4419-1126-1_26,

26.1 Early Experience

Telemedicine is defined as the use of medical information exchanged from one site to another via electronic communications for the health and education of the patient or healthcare provider and for the purpose of improving patient care [1]. This technology has been in practice for decades, however recent advancements in telecommunications and computer technology have allowed for the development of more efficient complex applications. In the 1950s, the University of Nebraska became one of the first institutions to employ telemedicine when they used closed circuit television as a means of supporting patient care in remote areas of the state. A decade later, the same institute connected to another state hospital more than 100 mi. distant to establish hospital to hospital telecommunication [1, 2]. In 1967, Massachusetts General Hospital used a microwave connection to Logan Airport in Boston to provide emergency medical care for travelers [2]. Around the same time Dr. Michael DeBakey used a broadband satellite connection to perform real-time surgical teaching between Europe and the United States. Despite these early efforts, telemedicine did not infiltrate main stream medicine until the development and commercialization of remotely controlled robotic surgical devices in the 1990s at which time telementoring and telesurgery became widespread [3].

In 1994 Dr. Louis Kavoussi successfully performed three telementored surgeries in which the instructing surgeon was in the adjacent room [4]. Kavoussi used a surgical robotic system called the Automated Endoscopic System for Optical positioning (AESOP). AESOP is the first FDA approved procedure specific robotic surgical system. It is a voice controlled robotic arm used to hold and position a laparoscopic camera. During this clinical study, Kavoussi et al. observed operating times similar to those reported during non-telepresence surgery [4]. Since this study, AESOP has been shown to be significantly steadier than a human camera holder and has been validated in multiple urologic, gynecologic and general surgical laparoscopic procedures. It eliminates the need for an extra person in the operating suite and is a viable solution in situations where there is a shortage of personnel [5]. Although this early robotic system opened the door to telementoring, it is not without its limitations. Kavoussi reported that for this system to work more efficiently, high bandwidth telecommunications are required to transmit real time video and audio data effectively [4]. In addition, this system cannot control the camera as rapidly or with as little conscious instruction from the surgeon as a well-trained human assistant who can often anticipate the next surgical step [5, 6]. One study reported that the continuous talking of the surgeon during the operation was distracting and bothersome [6]. Cost and the potential to interfere with the surgical training of medical students and residents are also limitations. Despite these limitations, this system has been demonstrated to effectively replace a human assistant without effecting surgical outcomes or significantly increasing operating time [5]. AESOP served as the foundation for the development of many other robotic surgical systems.

26.2 The Lindbergh Project

The first successful demonstration of two handed telesurgery was performed by Professor Jacques Marescaux and his team from IRCAD in September, 2001. During this one hour demonstration which was named the Lindburgh operation, Professor Marescaux, using a teleoperable Zeus system, (Zeus TS, Computer Motion, CA) successfully performed a laparoscopic cholecystectomy from New York on a patient located in Strasbourg, France [3]. This operation was performed under test conditions at a tertiary hospital with multiple experienced surgeons present to intervene if necessary. The surgeons used the Zeus robotic system to perform this surgery. Zeus is a master-slave telerobotic surgical system using two main components: the master console and the slave robotic arms. The surgeon sits in the master's console which uses a computer system to translate and digitize the surgeons' movements. There are three robotic arms, one is a voice controlled camera, essentially an AESOP, and the other two arms hold surgical instruments. These robotic arms, which can be located hundreds of miles away, translate the input into actual instrument manipulations or camera movements [5, 7]. Zeus TS robotic system addressed many of the limitations experienced with the early systems including high resolution video, a steady camera platform, and instruments with multiple degrees of freedom [5]. This operation used ATM connectivity between New York and France. ATM network nodes are interconnected through a high speed terrestrial fiber-optic network that transports data through virtual connections dedicated per customer. This type of connection is very high quality with low transport delay and low packet loss ratio and is 99.99% reliable in terms of network outage, but come at a significant price which makes its everyday clinical use impractical. A backup line was available in case of line congestion. For this operation, the mean total dime delay was 155 ms and there were no unexpected telecommunications or robotic mishaps [7, 8].

26.3 Canadian Telesurgical Network

On the footsteps of the Lindburgh experiment, in February 2003, the world's first clinical telesurgical service was established in Ontario, Canada, between a teaching hospital and a rural community hospital located over 400 km away. Through regular commercially used IP (internet protocol) telecommunication lines an experienced laparoscopic surgeon located at the Centre for Minimal Access Surgery at St. Joseph's Healthcare in Hamilton Ontario Canada was able to perform telerobotic surgery and/or telerobotic assistance and mentoring to a local surgeon at North Bay General Hospital in North Bay Ontario [9]. This service demonstrated the efficacy of telesurgery in a rural setting allowing community surgeons to offer better quality and more advanced laparoscopic procedures to patients in a rural community [10].

The robotic system used for this service was a Zeus TS microjoint system. This system is comprised of three arms which were set up by the local surgeon in North Bay and controlled by the remote surgeon in Hamilton (Fig. 26.1). A two-way audio system, large video displays and digital cameras were used to give the remote surgeon adequate communication with the local surgeon and visualization of the operative field and OR setting (Fig. 26.2). Prior to the start of the service, the local laparoscopic surgeons and nursing team in North Bay were trained to use the robotic arms and instrumentation. There was also a technician present to ensure proper set up of the robotic system [3, 9].

The linkage was established using an IP/VPN network which operated at a bandwidth of 15 Mbps. This network provided quality of service, security and privacy and included an active backup line which could be used immediately if the first line failed. Signals used during telesurgical applications take priority of any other traffic on the network as they operate at the highest priority of Quality of Service to ensure transmission at the most rapid rate possible [9].

The first surgeries performed using this telesurgical system were two laparoscopic Nissen Fundoplications. A subsequent 20 remote procedures were performed including laparoscopic right hemicolectomies, laparoscopic anterior resections, laparoscopic sigmoid resections and laparoscopic hernia repairs [9]. For the first several surgeries in this series, the remote surgeon acted as the primary surgeon while the local surgeon assisted. Gradually, the local surgeon performed more advanced procedures under the instruction of the remote telementor. At the completion of this series, the local surgeon was able to perform each procedure as the primary surgeon, while the remote surgeon assisted and mentored. The methods used allowed either the remote or local surgeon to perform all aspects of the procedures and switch roles between primary and assisting surgeon seamlessly. Each of these surgeries were performed with no major complications and no

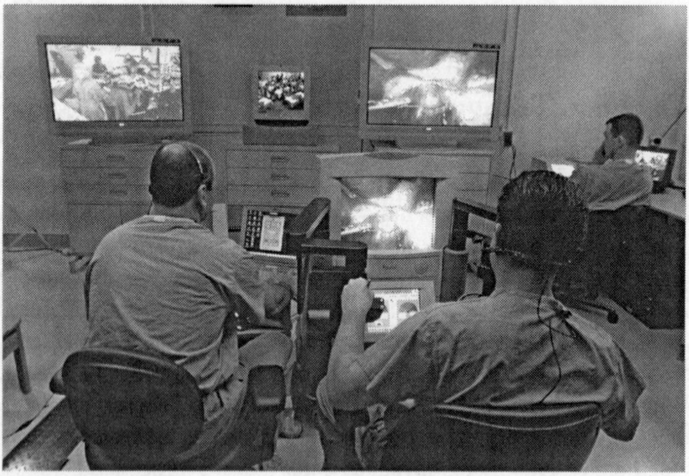

Fig. 26.1 Remote operative console at St. Joseph's Healthcare in Hamilton

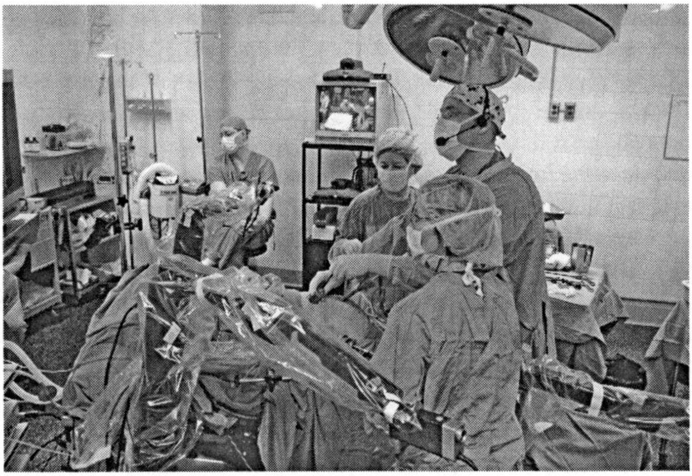

Fig. 26.2 Operating room in North Bay during remote telesurgical procedure

conversions to open. The recovery of all patients was uneventful and clinical outcomes were comparable to those of procedures performed at tertiary care centres [3]. Although an overall latency of 135–140 ms was reported, the telerobotic surgeon was able to adapt without difficulty to perform procedures including fine dissection, identification of key structures and suturing [9].

A number of key lessons were learnt from this first experience with clinical application of telesurgery. These included:

1. The patients are willing to accept the application of telerobotic surgery and the involvement of a remote expert surgeon in their care as long as they understand the potential clinical benefits and are keen to avoid unwanted travel.
2. The telepresence environment created using the high quality video and audio communication between the remote surgeon and a local surgeon is very important in creating a seamless and rich interaction very similar to the interaction between two surgeons assisting in the same operating room.
3. Commercially available land based communication lines can be utilized effectively to provide secure and reliable telelinkage required for remote telesurgery.
4. The distances where telesurgery is applied is less important when land lines are used as the latency introduced into the system by the codec devices which compress and decompress the video signals is far greater than the latency due to signal travel between sites.

26.4 Mode of Communications

The most effective mode of telecommunication between the surgeon console and the robotic slave arms by the side of patient is broad band IP (VPN) connectivity with Quality of Service (QoS) assuring rapid transit of information between the two sites.

The speed of signal transmission with such land lines allows for latencies of less than 150 ms across North America, or between Europe and North America. At such latencies the surgeon is able to complete the most complex of surgical tasks without significant error and in reasonable time. Earlier research has shown that once the latency rises above 200 ms the surgeon has to slow down significantly to complete the task and once the latency rises above 500 ms it becomes difficult to accomplish the difficult two handed surgical tasks [10, 11]. Thus, at time delays experienced when using Geo-orbital satellites (750 ms), it is not possible to complete traditional tele-surgery. During NEEMO 9 we planned surgeries at 2 second time delay to mimic the latency experienced between earth and moon base and we were only able to complete the most rudimentary surgical tasks. At such high latency connections a semi-autonomous robotic procedure may be more appropriate.

Another important facet of telecommunication required for effective telesurgery, is redundancy, to ensure availability of a second telecommunication line should one line go down. Furthermore, encryption of data prior to transfer and decryption at the receiving site will allow for the necessary privacy and security against possible hackers when using public IP lines.

26.5 Managing Surgical Responsibility, Emergencies and Outcomes

Telesurgery can provide for an effective collaboration between two surgeons while performing a complex surgery even if the two are separated by hundreds or thousands of miles. One of the most important elements for successful telesurgery is effective communication between the remote surgeon and the patient-side surgical and nursing team. Another factor is preparation prior to surgery; this includes pre-surgical discussion and choreography of the case and clear plan of how to deal with potential mishaps or complications which may arise. The local surgeon and his team have to be able to deal with possible complications, and to complete or temporize the situation in case of complete loss of communication. Presence of redundancy in the system should mitigate such risks, but in the case of robotic failure, the tele-surgeon can still rely on the ability to tele-mentor the local surgeon in completing the operation.

The outcomes expected in telesurgery should be equivalent to results the patient should anticipate if operated on by the expert tele-surgeon in his/her institution. In our series, this was achieved without ever having to rely on our contingency plans.

26.6 Regulations and Policies

The primary challenges to wide adoption of telesurgery are the limitations posed by the licensing requirements and medico-legal coverage across state lines or even jurisdictions. Clearly, as a surgeon actively involved in surgical care of a patient, the tele-surgeon requires licensing, medico-legal coverage and appropriate privileges at the hospital where the patient care is provided, even if the surgeon never physically practices there.

In Canada, where there is national medical protective coverage from CMPA, medico-legal considerations are considerably less than the USA where most surgeons have state-wide or regional malpractice insurance, limiting their ability to provide care via telemedicine technologies such as telerobotic intervention to patients outside their practice jurisdiction.

Future expansion of all telemedical applications including telesurgery requires for development of national or even transborder licensing and medico-legal coverage.

26.7 Future Developments

The research and limited clinical experience with remote telerobotic surgery has suggested that it will have a significant future in surgical/interventional knowledge translation as well as provision of emergency care to remote patients. The concept of teleoperability has been shown to add minimal risk to robotic intervention and undoubtedly, future surgical robots will all have such capabilities built into their design. There are still a number of unanswered questions with respect to impact of latency on haptics which need to be further discerned and improved. However, before telerobotic surgery can find clinical application in every day practice there are a number of less technical obstacles which need to be dealt with. These include the issue of clinical jurisdiction over the patient, medico-legal and licensing requirements for telesurgical acts, as well as appropriate remuneration for a telesurgical act. While many of these issues appear mundane as compared to the engineering sophistication which has gone into the creation of telerobotic platforms, they are currently the primary reason why we have not seen a wider application of remote telerobotic surgery. There is no doubt that as the clinical sphere of telemedical applications increases many of these issues will be addressed. With this optimistic view, the author strongly believes that remote telerobotic assisting and surgery will be common place in the next decade.

Acknowledgments The author would like to acknowledge Karen Barlow for her assistance with preparation of the manuscript.

References

1. Senapati, S., Advincula, A.: Telemedicine and robotics: Paving the way to globalization of surgery. Int. J. Gynecol. Obstet. **91**, 210–216 (2005)
2. Merrell, R.C.: Telemedicine development: setting the record straight. In: Latifi, R. (ed.) Establishing Telemedicine In Developing Countries: From Inception To Implementation. IOS Press, Amsterdam (2004)

3. Anvari, M.: Telementoring and remote telepresence surgery. In: Faust, R.A. (ed.) Robotics in Surgery: History, Current and Future Applications. Nova Science publishers, New York (2007)
4. Kavoussi, L.R., Moore, R.G., Partin, A.W., Bender, J.S., Zenilman, M.E., Satava, R.M.: Telerobotic assisted laparoscopic surgery: initial laboratory and clinical experience. Urology **44**(1), 15–19 (1994)
5. Ewing, D.R., Pigazzi, A., Wang, Y., Ballantyne, G.H.: Robots in the operating room-the history. Semin. Laparosc. Surg. **11**(2), 63–71 (2004)
6. Kraft, B.M., Jäger, C., Kraft, K., Leibl, B.J., Bittner, R.: The AESOP robot system in laparoscopic surgery: increased risk or advantage for surgeon and patient? Surg. Endosc. **18** (8), 1216–1223 (2004)
7. Marescaux, J., Leroy, J., Rubino, F., Vix, M., Simone, M., Mutter, D.: Transcontinental robot assisted remote telesurgery: feasibility and potential applications. Ann. Surg. **235**, 487–492 (2002)
8. Marescaux, J., Leroy, J., Gagner, M., Rubino, F., Mutter, D., Vix, M., Butner, S., Smith, M.: Transatlantic robot-assisted telesurgery. Nature **413**, 379–380 (2001)
9. Anvari, M., McKinley, C., Stein, H.: Establishment of the world's first telerobotic remote surgical service for provision of advanced laparoscopic surgery in a rural community. Ann. Surg. **241**(3), 460–464 (2005)
10. Anvari, M.: Remote telepresence surgery: The Canadian experience. Surg. Endosc. **21**(4), 537–541 (2007)
11. Anvari, M., Broderick, T., Stein, H., Chapman, T., Ghodoussi, M., Birch, D.W., McKinley, C., Trudeau, P., Dutta, S., Goldsmith, C.H.: The impact of latency on surgical precision and task completion during robotic-assisted remote telepresence surgery. Comp. Aided Surg. **10** (2), 93–99 (2005)

Chapter 27
History of Robots in Orthopedics

Michael Conditt

27.1 Industry

The concept of machines performing tasks normally done by humans was first introduced in 1921 by Czechoslovakian playwright Karel Capek. His play "Rossum's Universal Robots" was a satirical piece intended to protest the growth of technology in Western civilization. However, much to his dismay the play had the opposite effect. Public fascination with robots increased and to this day is still a fascination of modern society. The word robot is derived from the Czechoslovakian word "robata" which is defined as forced labor or servitude. According to Merriam-Webster's Dictionary its simplest definition is "a device that automatically performs complicated often repetitive tasks" [1]. The term robotics was first introduced by Isaac Asimov in 1938 in his short story "Runaround" for *Super Science Stories* Magazine. This was followed by a series of short stories that were later collected and published as "I Robot" in 1942. He used the term robotics to describe three laws governing robot behavior which later became the inspiration for the 2004 movie "I Robot". In society today robots are used to perform highly specific and precise tasks that are impossible to perform or difficult to perform reproducibly by humans. They are utilized in manufacturing, exploration of space and the deep sea, and work in hazardous environments to name a few examples [2].

The first commercial robot was the Unimate built in 1958 by George Charles Devol and Joseph F. Engelberger for General Motors to be used for spot welding and extracting die casting. It was first utilized on the factory floor in 1961 and within 7 years was implemented in multiple factories. Industrial robots continued to evolve in the 1960s with the Rancho Arm being the first artificial robotic arm to be controlled by a computer. It contained six joints which gave it the flexibility of a human arm and was utilized in the assistance of the handicapped. The sixties also saw the opening of artificial intelligence research laboratories at M.I.T, Stanford

M. Conditt (✉)
MAKO Surgical Corp.,
2555 Davie Road, Fort Lauderdale, FL 33317, USA
e-mail: mconditt@makosurgical.com

J. Rosen et al. (eds.), *Surgical Robotics: Systems Applications and Visions*,
DOI 10.1007/978-1-4419-1126-1_27, © Springer Science+Business Media, LLC 2011

Research Institute, Stanford University, and the University of Edinburgh. The 1970s saw incorporation of minicomputers to control robots and designs with touch and pressure sensors. During the 1980s the robot industry started to grow rapidly with a new robot company entering the market every month [3, 4].

27.2 Medicine

In 1985 robots found their way into the operating room with the PUMA 560 performing neurological biopsies. The PUMA 560 was not originally designed to assist in surgical procedures but was first an industrial robot. Its effectiveness in precise and accurate navigation of surgical tools was utilized by Kwoh et al. to excise brain tumors which had previously been inoperable [5]. The robot was essentially used as a positioning fixture similar to that of the stereotactic frame. Once the robot was positioned correctly it was locked in position and power was removed. The preliminary results were promising and compared to conventional stereotactic frames the robot could position itself automatically and accurately. Despite initial positive outcomes it was still an industrial machine not originally designed to be used in close contact with people. Secondary to this, the companies that marketed the Puma 560 did not continue to allow its use in the operating room on the basis that it was unsafe [5].

One year after the inaugural robotic assisted surgery of the Puma 560 IBM and the University of California, Davis began a joint research project to develop a robot for total hip arthroplasty. Its role in the surgery was limited to reaming the proximal femur in preparation for prosthetic implant insertion. The first clinical test took place on a dog in May of 1990 and by November of the same year Integrated Surgical Systems, Inc. was formed to develop the technology for human clinical use. The robot was modified with a Sanko-Seiki control system to increase its safety profile. The modified Sanko-Seiki robot system would later be known as Robodoc which made its human operating room debut in 1991. Robodoc was used in total hip arthroplasty (THA) to ream the femur in preparation for femoral stem implant [2].

27.3 Other Technology

27.3.1 Virtual Reality and Telepresence

Also developing during the eighties and early nineties was virtual reality and the concept of telepresence. Scott Fisher, PhD of National Air and Space Administration (NASA) Ames Research Center and Joe Rosen, MD of Stanford University were developing virtual reality environments with a future vision of surgeon controlled remote robotic arms. They collaborated with Phil Green PhD of the Stanford Research Institute for assistance in robotic arm development which resulted in a surgical telemanipulator designed for hand surgery. After general surgeon Richard

Savata MD joined the NASA-Ames team, the clinical use of this technology in laparoscopic surgery was realized and became the focus of future developments. The Pentagon's Defense Advanced Research Projects Agency (DARPA) became aware of this technology and was interested in its application on the battlefield. With their funding the first voice activated robotic camera for laparoscopic surgery was developed. It was called AESOP (Automated Endoscopic System for Optimal Positioning) and it replaced the need for a surgical assistant to handle the camera [2].

27.3.2 Haptics

Another technology that blossomed from DARPA funding was haptics or force and tactile feedback which had been lacking in the robotic arms to date. Akhil Madhani was a graduate student at MIT that had seen DARPA's vision of telesurgery. His advisor at the time, Dr. Ken Salisbury, encouraged him to pursue a thesis on teleoperated surgical instrumentation with force feedback sensors. With help from Dr. Salisbury and Dr. Gunter Niemeyer, a post-doctoral fellow at MIT and good friend of Madhani, the project took 3 years to complete. Dr. Salisbury provided ideas and guidance while Dr. Niemeyer is credited with all the software programming. Their work developed into the "Black Falcon" the first teleoperated surgical instrument for minimally invasive surgery. The commercial possibilities with the merger of these two technologies became evident and companies wanting to harness this potential eventually formed. Intuitive Surgical, Inc (Sunnyvale, CA) hired Madhani and Dr. Salisbury as consultants and eventually developed the da Vinci telerobotic surgical system in 1997 while Computer Motion Inc developed the Zeus in 1999 which was similar to the da Vinci but differed in the surgeon's console. Computer Motion was eventually purchased by Intuitive surgical which currently markets the da Vinci alone. The da Vinci has been approved for use in mitral valve repair, prostatectomy, Nissen fundoplication gastric bypass surgery, hysterectomy and uterine myomectomy [2, 6]. There are over 300 sites in the United States that are currently using the da Vinci surgical system and the number of surgeries that it is being utilized for is growing. The use of robots has grown from 1,500 cases in the year 2000 to an estimated 20,000 cases in 2004 [7].

27.4 Clinical Applications

27.4.1 General Surgery

Though the da Vinci was originally designed for microvascular surgery its potential in general surgery has been realized and is currently applied in a wide variety of general surgery procedures. As mentioned earlier, Richard Savata MD who joined the NASA-Ames group recognized the potential that telepresence technology had in the realm of laparoscopic surgery. Initial clinical trials done by Marescaux et al. proved the robotic assisted cholecystectomies are safe and feasible while enhancing the skill of the surgeon to perform technically precise suturing and dissection [8]. In a large community hospital Giulianotti et al. have displayed the versatility of the da Vinci in the operating room. Between October 2000 and November 2002, 193 patients underwent a minimally invasive robotic procedure. A total of 207 robotic surgical operations were performed. The most frequently performed procedure was cholecystectomy for gallstones ($n = 66$ cases, 52 single and 14 associated with another operation). This procedure was chosen for the initial learning curve. In the first 26 single procedures the mean operative time was 96.5 vs 66.7 min for the second 26 single procedures. A retrospective comparison between robotic and laparoscopic cholecystectomies showed that after 20 robotic cholecystectomies, the mean operative time (66.7 min) was similar to laparoscopic cholecystectomies (65 min). The conversion rate was higher with the laparoscopic technique (3.5% vs 1.9%). The mean postoperative stay and morbidity rate were similar for the robotic (2 days; 1.9%) and laparoscopic (3 days; 2.2%) groups. Fundoplication for gastroesophageal reflux disease was the second most frequently performed procedure ($n = 51$). The mean operative time was 110 min for robotic fundoplication and 120 min for laparoscopic fundoplication. The conversion rate was similar in both groups. The morbidity rate and mean postoperative hospital stay for robotic and laparoscopic procedures were 4.8% and 4 days and 11.4% and 6 days respectively. Other surgeries performed included esophageal, gastric, pancreatic, hepatobiliary, colorectal, thoracic, adrenalectomy, splenectomy and a few various others such as one fallopian tube anastomosis and one aneurysmectomy of the renal artery. In 193 patients, four procedures required conversion to open surgery and three were converted to conventional laparoscopy. The total reoperation rate was 3.1% and the total perioperative morbidity rate was 9.3%. The total postoperative mortality rate was 1.5%. The authors concluded that robotic assisted surgery is feasible, safe, and easily managed in a big community hospital. The learning curve is relatively short, the robot is safe and sturdy, and setup time decreases as the experience of the operative team increases [9].

27.4.2 Neurosurgery

Neurosurgery, where the use of robots in the operating room began, has evolved in its utilization of robots from the simple stereotactic passive robot used by Kwoh to numerous passive and active robots used today. Neurosurgery is one of the best suited specialties to apply medical robotics because the brain is firmly contained in a rigid structure allowing for fixation of devices and because neurosurgery requires the most precision of the surgical specialties. Robots are used for soft tissue cutting and ablation, bone drilling and bolting, resection of tumors, in conjunction with microscopes, and telepresence to name a few. Neurosurgeons and engineers continue to push the envelope and broaden the horizons of robotic technology.

Garnette R. Sutherland, MD from the University of Calgary and colleagues have designed and manufactured an image-guided robotic system that is compatible with magnetic resonance imaging. Because intraoperative imaging interrupts surgery and prolongs operating room time the need to reduce the number of imaging studies was apparent. The robotic system is void of magnetic and conductive materials and can be integrated with intraoperative MRI to maintain surgical rhythm by avoiding interruptions for imaging [10].

One of the most significant advances in neurosurgery, though not thought of as surgery in the traditional sense, is radiosurgery. A robotic radiosurgery system such as the Cyberknife (Accuray Inc) is a non-invasive alternative to surgery for the treatment of tumors. It delivers beams of high dose radiation to tumors with extreme accuracy [11]. It was developed at Stanford by John Adler, MD where the first patient was treated in 1994. It became FDA approved in 2001 and is commonly used for treatment of arteriovenous malformations, acoustic neuromas, meningiomas, schwannomas, trigeminal neuralgia, and pituitary adenomas but can be used anywhere on the body. Stanford's current clinical trials include brain metastasis, trigeminal neuralgia, and lung cancer [12]. Cho et al. recently reported on 26 patients with pituitary adenomas receiving stereotactic radiosurgery with follow up periods ranging from 7 to 47 months. The tumor control rate was 92.3% and hormonal function was improved in all of nine functioning adenomas. They concluded that CyberKnife is safe and effective in selected patients with pituitary adenomas [13]. As of 2008 more than 40,000 patients have undergone CyberKnife therapy. It is yet another example of the use of robots to make a procedure more precise and safe for the patient.

27.4.3 Cardiothoracic Surgery

Cardiothoracic surgery has been slow in its incorporation of robotic assistance relative to other surgical specialties because of the anatomical, physiological, and microsurgical requirements involved. However, the da Vinci robotic system was initially designed for microvascular surgery and is currently finding its niche in the cardiothoracic world. Current robotic applications include mitral valve repair, atrial septal defect closure, and totally endoscopic coronary artery bypass grafting (TECAB). Robotic assistance allows for difficult procedures such as multivessel coronary bypass to be performed through a small incision without the morbidity of sternotomy. Subramanian et al. evaluated the feasibility of this minimally invasive approach with positive outcomes. Thirty patients had undergone off-pump minimally invasive multivessel coronary bypass. Internal mammary arteries were harvested with robotic telemanipulation using three ports. Twenty-three patients had anterior thoracotomy approach and seven had transabdominal approach. There was no mortality in hospital or on 30-day follow-up. Twenty-nine patients were extubated on the operating table. Within 24 h of surgery 50% ($n = 15$) of patients were discharged, 10% ($n = 3$) were discharged in 24–36 h, 17% ($n = 5$)

were discharged in 48–72 h, and two patients stayed more than 3 days in the hospital. They concluded that the majority of the patients can be safely discharged within 36 h of robotic assisted multivessel coronary bypass making the ultimate goal of same day hospital discharge more reality than fantasy [14].

Grossi et al. found positive results as well for mitral valve reconstruction. They compared intermediate-term results of minimally invasive versus sternotomy approaches for mitral reconstruction. One hundred patients underwent primary mitral reconstruction through a minimally invasive right anterior thoracotomy. Outcomes were compared with those of 100 previous patients undergoing primary mitral repair who were operated on with the standard sternotomy approach. They found that after 1 year follow-up both echocardiographic results and New York Heart Association functional improvements were comparable with results achieved with the standard sternotomy approach. The minimally invasive approach for mitral valve reconstruction provided equally durable results with more advantages for patients and according to the authors should be more widely adopted [15].

27.4.4 Urology

Urology, much like neurosurgery, was there at the beginning of the robotics in medicine movement. As mentioned earlier, from the work of Davies came the PROBOT which used a robotic frame to guide a rotating blade to complete a transurethral resection of the prostate. Currently, the da Vinci is the most popular robotic surgical system in the specialty of Urology. From 2000 to 2004 urology accounted for the largest single specialty increase in procedures with over 8,000 robotic prostatectomies performed in 2004. Other applications include cystectomy, pyeloplasty, radical and simple nephrectomy, live-donor nephrectomy, adrenalectomy, sural nerve grafting, ureteric reimplantation, colposuspension and renal transplant [16].

Ficarra et al. performed a systematic review of robot-assisted laparoscopic radical prostatectomy (RALP) and found that it was a feasible procedure, with limited blood loss, favorable complication rates, and short hospital stays. They concluded that the literature showed that RALP has a short learning curve and promising functional outcome, especially with regard to continence recovery [17].

27.4.5 Gynecology

Though relatively late in its acceptance of robotic assistance gynecology is starting to incorporate and appreciate the benefits of robotic surgery. The da Vinci system was approved for gynecologic procedures by the Food and Drug Administration in 2005. It is currently used for hysterectomy, myomectomy, tubal reanastomosis, ovarian transpositions, oncology procedures, and pelvic reconstruction. Hysterectomy is one of the most promising applications of minimally invasive techniques. Reynolds and Advincula reported on 16 consecutive laparoscopic hysterectomies and noted that robotic technology was advantageous in the patients with scarred or obliterated surgical planes since 13 of the 16 had prior pelvic surgery and required intraoperative management of pelvic adhesions to complete the hysterectomy. They also reported the same advantage in six patients undergoing hysterectomy who had a scarred or obliterated anterior cul de sac secondary to prior cesarean deliveries [18].

Robotics has also found a role in reproductive surgery as a means to reverse the effects of a previous procedure. Patients are usually given grim prospects when consulted about the chances of reversing a tubal ligation in the future. Degueldre et al.. has reported some positive results in tubal reanastomosis. In 2000, 16 tubes were successfully reanastomosed with five of the eight patients demonstrating at least unilateral patency. Two pregnancies were reported within 4 months of surgery. The authors noted that the operating time compared favorably with that required to perform open microsurgery [18].

27.4.6 Orthopedic Surgery

Orthopedics, of all the surgical specialties, is probably the most compatible with robotic systems because of its rigid surgical medium. Nearly 20% of all medical robots are used in orthopedics [19]. With the exception of neurosurgery, which relies on the cranium more as a stabilizing and referencing tool than as the subject of the procedure, bone is the most important aspect of the orthopedic procedure. Bone provides excellent contrast in X-ray/CT images and has high rigidity. These two qualities allow for preoperative and intraoperative models that stay, under most circumstances, true to the patient's anatomy. The rigidity of bone also permits the robot to register itself to preoperative and intraoperative plans as well to its current location in the operating room relative to the patient. The soft tissue is not as important as in other surgical specialties but is important in a secondary sense. For example, in total knee arthroplasty (TKA) precise and accurate bone cuts are critical for proper ligament balance in the reconstructed knee. It is also critical for proper fit of implant components. This reduces wear and increases the life of the implants while decreasing the complications of misalignment and component loosening and consequently the need for another operation. Besides TKA, robots are predominantly used for total hip arthroplasty (THA), revision THA, and unicondylar knee arthroplasty (UKA).

Robotics can enhance surgery through improved precision, stability, and dexterity. With the incorporation of computed tomography, magnetic resonance images and new algorithms operative plans and the execution of these plans are becoming more precise and accurate. The ultimate goal of this technology is to reduce patient morbidity, mortality, and improve functional outcomes. The advantages of robots include good geometric accuracy, stable and untiring, can be designed for a wide range of scales, may be sterilized, and resistant to radiation and infection. Its limitations include poor judgment, limited dexterity and hand-eye coordination, limited to simple procedures, expensive, and difficult to construct and debug. Human strengths include strong hand-eye coordination, dexterous, flexible and adaptable, can integrate extensive and diverse information, able to use qualitative information, good judgment, and easy to instruct and debrief. Human limitations include limited dexterity outside natural scale, prone to tremor and fatigue, limited geometric accuracy, limited ability to use quantitative information, limited sterility and susceptibility to radiation and infection [20].

Minimally invasive surgery (MIS) is not a new concept but one that robotics is applying and capitalizing on in ways unimagined when the procedures began. MIS is beneficial in that it reduces patient discomfort, improves cosmesis, and reduces hospital cost and time away from productive work. Traditional laparoscopic technique requires three to five small incisions to allow long-handled instruments access to the body cavity. This technique places severe limitations on dexterity and creates a fulcrum effect in that the surgeon must move his hand in the opposite direction of his intended target. There is also no tactile information which severely handicaps the surgeon. Robotic systems make up for these limitations by providing dexterity similar to that of the hand and produce images in three dimensions giving the surgeon the perception that he is operating on a patient in front of him while he moves controls at a console away from the patient [20].

There are several types of robotic systems used in orthopedic surgery that can be classified under three main headings. There are active, passive, and semi-active systems. Passive systems are those that have no active role in the procedure. They are utilized as position holders for drills or cutting guides. Active systems as their name suggests play a more aggressive role in the procedure. Though both of these systems have a role in the operating room, the current direction is to combine qualities from both groups to create the semi-active system. A semi-active system allows the surgeon to be an active part of the operation with the added benefit of precision and accuracy afforded by robotic assistance.

27.5 Active Systems

27.5.1 Robodoc

Robodoc (Integrated Surgical Systems, Davis, CA), was the first active robotic system used for an orthopedic procedure. It was developed at the University of California Davis from 1986 to 1992. The need for a better method of THA was realized in the mid 1980s when cementless femoral components were first introduced. There was a significant problem with postoperative thigh pain, intraoperative fracture, and failure of bony ingrowth [21] Many early cementless femoral components were failing secondary to poor fit and stability. The surgical instruments used at the time were carryovers from the era when only cemented components were used. They created gaps at the implant to bone interface that could lead to instability, decreased bony ingrowth, and possible debris pathways for osteolysis [21]. Howard A. Paul, DVM and

William L. Bargar, MD conceived of Robodoc in 1985. Robodoc automatically mills the bone cavity for the femoral component. The design was based on a commercially available Japanese industrial robot. A large clamping arm connects the base of the robot to the patient's leg and a position monitor connected to the robot ensures the procedure is halted if leg movement is detected. A computer assisted orthopedic planning system called Orthodoc was developed for Robodoc. It required the patient to undergo a separate procedure before the THA to place locator pins in the affected femur. A CT scan was then taken of the affected femur and uploaded onto the Orthodoc workstation. An operative plan would then be constructed and used intraoperatively by Robodoc. Before the THA procedure Robodoc registered the patient's anatomy with the preoperative imaging data [22]. A feasibility study of the concept was performed in 1986–1987 at the IBM Thomas Watson Research Center in Yorktown Heights, NY. A canine clinical study started in 1989 and by 1992 the system was adapted for human application. At this point Robodoc was licensed by Integrated Surgical Systems, Inc. The first human feasibility study was conducted on ten patients at Sutter General Hospital in Sacramento, CA. A randomized multicenter study was then conducted beginning in 1994 through 1998 to perform cementless primary hip arthroplasty on 136 consecutive hips in 119 patients at three centers. The results showed statistically improved fit, fill, and alignment when compared with manual THA, and intraoperative fractures were eliminated in the Robodoc group [23]. Though these results were positive there were still issues to reconcile. Surgery times were longer due to the learning curve and the robot shutting down because of leg movement. There was also a small increase in blood loss as compared to controls, which was attributed to the increased surgical time. In 2001, the FDA authorized a second multicenter US trial of the Robodoc system. The aim of the study was to improve the surgical time and blood loss and to show the efficacy of pinless technology. With this technology there was no longer a need for a preliminary surgery to place tracking pins. The anatomy was registered intraoperatively by scanning various points of the patient's boney anatomical landmarks. Through the first 80 cases the average blood loss decreased to 471.54 cc from 1189 cc and the surgical time decreased from 258 to 121.92 min. Outside the US, Robodoc has been used in revision THA. According to Bargar [23] robotic milling reduces the incidence of intraoperative fractures in the decompensated bone of revision THA's. In August 1994, the first Robodoc system in Europe was installed at the Berufsgenossenschaftliche Unfallklinik Clinic in Frankfurt, Germany [21]. Over the next 4 years, there was rapid growth of the Robodoc in Germany with over 30 systems installed. As of 2007 there were 20 Robodoc systems between Austria, France, Switzerland, Japan, Korea, and India [23].

27.5.2 Caspar

The success of Robodoc in Germany fueled competition that came in the form of the CASPAR (Computer Assisted Surgical Planning and Robotics) system produced by Orto Maquet, a subsidiary of the German company Maquet [19]. The CASPAR was based on the Swiss industrial RX 90 6 axes clean-room robot, developed by Stäubli (Horgen, Switzerland) [22]. It was an active robot originally used for THA in 1997 and was later applied to TKR in 1999. In the first 70 TKR patients the mean difference between preoperatively planned and postoperatively achieved tibiofemoral alignment was 0.8° (0–4.1°) which fared better than the control group 2.6° (0–7°) [24]. CASPAR was also used for ACL tunnel placement. In a study conducted by Burkart et al. [25] at the University of Pittsburgh the precision of the CASPAR was compared to four orthopedic surgeons with various levels of experience. The robot and each surgeon drilled tunnels in ten plastic knees. The results showed that the robotic system had the most consistent tunnel directions. Their data suggested that the robotic system had the same precision as the most experienced surgeons. The German company URS Ortho GmbH and Company KG purchased the rights to CASPAR from Orto MAQUET in 2000. URS went into liquidation in 2004 despite good clinical results and sales in Europe [22].

27.5.3 Romeo

Within the last decade a project titled ROMEO (Robotic minimally invasive Endoprosthetics) conducted by Helmut-Schmidt-University/Hamburg and the Department of Trauma and Orthopaedics of the BG Trauma Hospital Hamburg has a focus on smaller incisions and soft tissue protection, an issue that confronted Robodoc and CASPAR. The goal of the ROMEO project is to create a system using minimally invasive surgical approaches and 3D guided navigation to conserv e soft tissue. The project is split in several phases starting with determination of technical and surgical aspects of a minimally invasive approach as well as technological parameters of the robotic process such as bone milling, drilling, and reaming. A milling device with the capability to work around a corner at an angle between 35° and 70° is a product of these efforts [26].

27.5.4 Arthrobot

In an attempt to simplify robot assisted surgery, the Arthrobot (Kaist, Korea) system registration process uses a reamer-attached gauge and a femoral frame to determine the relative position and orientation of the femur in relation to the robot. The reamer-shaped gauge is inserted into the patient's femur as in a manual surgery then the manual broaching process is replaced by the robot. The reported advantage is a less complicated registration process and limiting the reaming process to the metaphyseal region leaving the diaphyseal hard bone untouched. In an experiment with composite and pig bones 93% of the gaps between the bone and implant surface were under the critical .25 mm [27].

27.5.5 Mbars

The Mini Bone-Attached Robotic System (MBARS) is a bone mounted robot designed to assist with patellofemoral arthroplasty. It was developed at the Institute for Computer Assisted Orthopaedic Surgery at Western Pennsylvania Hospital and the Robotics Institute at Carnegie-Mellon University. It has two rigid platforms, an upper and a lower. The lower platform is attached to the femur with three Steinman pins while the upper platform supports a rotary cutting tool and moves relative to the lower platform to machine the trochlea. With regards to preoperative planning, a 3D image is used in which the surgeon marks the patellar tracking line on the preoperative image, the software then positions the artificial component, then the cutting paths are calculated. In an experimental study the process of preparing the cavity required about 2 min [19].

27.5.6 PiGalileo

PiGalileo (Smith and Nephew, Memphis, TN) is a total knee replacement system that utilizes computer navigation and a robotic arm to hold cutting guides. The cutting guides are adjustable and allow precise and individual positioning of the resections without accessing the intramedullary canal. In a small study of 34 TKR surgeries comparing conventional technique to PiGalileo the resection accuracy was greater with the computer assisted device. The gap between implant and bone ranged from 0 to 0.3 mm for the PiGalileo and 0–0.6 mm for the conventional method. The system was able to achieve these results at the cost of only ten additional minutes of OR time.

27.5.7 Romed

In a project titled "Robots and Manipulators for Medical Applications – RoMed" four Fraunhofer institutes are developing a robot system for use in spinal column surgery; more specifically, posterior lumbar interbody fusion, a procedure in which two or more vertebrae are linked with rods. It is a difficult operation that requires great precision and accuracy when drilling and inserting bolts into the vertebrae. Currently, 20–25% of cases involve incorrectly placed hardware. The robot uses six telescopic legs with six degrees of freedom to place pedicel bolts within one tenth of a millimeter. A CT scan is taken preoperatively to make a 3D model the surgeon can use to plan the surgery while an ultrasound-based navigation system provides intraoperative data. The surgeon controls the robot from a monitor located near the robot.

27.6 Semi-Active Systems

27.6.1 Acrobot

The Robodoc was not the only robot in 1991 undergoing human clinical trials. Though the Puma 560 had been restricted from human interaction in the clinical setting it was still being utilized for laboratory studies conducted by Brian Davies of Imperial College of Science, London. He was studying its utility in the transurethral resection of the prostate (TURP) for benign prostatic hyperplasia. The robot was programmed to actively remove soft tissue unlike its original passive role in the neurosurgical biopsies. Based on these laboratory studies, a robotic motorized system was developed specifically for the task of soft tissue resection. It was first used clinically in April of 1991. This active robot eventually evolved into what it is known as Probot.

As Davies began his clinical experience with medical robotic systems in 1991, he soon realized that the surgeon control of the procedure was secondary to that of the robot programmer. The Probot was an autonomous robot that when positioned correctly was turned on and allowed to perform its programmed task with little interaction from the surgeon except for control of an emergency-off button. At first, surgeons thought this autonomous feature was desirable but it became apparent that it was not in their nature to stand back and be a passive participant in the procedure. This led Davies and his Mechantronics in Medicine Group at Imperial College London to develop a hands-on robot in which the surgeon and robot participated in the procedure as a unit. The robot was to be treated as an intelligent tool under the surgeon's control. Davies called this robot Acrobot, (for Active Constraint ROBOT), which actively constrained the surgeons movements through haptic feedback to keep them within a safe predetermined region in the operative field. It was developed for orthopedic surgery, more specifically minimally invasive unicondylar knee replacement. Unlike the da Vinci in which the surgeon sits at a console away from the patient the Acrobot allows the surgeon to remain in contact with the patient and move the robotic arm with his hand while controlling the cutting tool [28]. Cobb et al. compared the results of robotically assisted with conventionally UKAs in a randomized prospective study [29] and found that all patients treated with robotic bone preparation had coronal plane tibiofemoral alignment within 2° of the planned position compared to only 40% in the conventional group.

27.6.2 Mako

The Acrobot technology has been improved upon and made more intuitive with the MAKO Surgical Corp RIO (Robotic Arm Interactive Orthopedic System). With computer assistance the RIO offers the surgeon visual, auditory, and tactile feedback in minimally invasive surgery for medial and lateral UKA components as well as patellofemoral arthroplasty. The RIO system allows the surgeon to manipulate the bone-removing burr within the intended volume but restricts motion beyond that predetermined volume. A preoperative plan is created from a 3D reconstruction of a preoperative CT scan of the patient's leg and computer assisted design (CAD) models of the implanted components. Standard surgical navigation markers are placed in the femur and the tibia and are also mounted on the robotic arm. The virtual modeling of the patient's knee and intra-operative tracking allows real time adjustments to obtain correct knee kinematics and soft tissue balancing. The end of the arm is equipped with a burr that is used to resect the bone. While inside the volume of bone to be resected, the robotic arm operates without offering any resistance [30, 31]. As the burr approaches the boundary, the robotic arm resists that motion and keeps the burr only within the accepted volume [30]. Thus the robotic arm effectively acts as a three-dimensional virtual instrument set that precisely executes the pre-operative plan.

The RIO system has produced initial positive outcomes in the realm of minimally invasive UKA. Sinha [32] evaluated six early-term peer-reviewed scientific presentations for four outcome parameters: accuracy of bone preparation and implant placement; robot system failures; complications, including those both attributable and not attributable to the robot; and patient-specific measures of outcome. With regards to bone preparation and implant placement Roche et al. obtained 344 radiographs for independent review. Three (<1%) femoral components were considered outliers (slight anterior medial overhang, placement of one component too distally). Sinha and colleagues reported one failure of the TGS in tibial registration among their first 37 patients. Three centers contributed 223 cases in which there were six reoperations-two for infection, one for femoral shaft fracture through a navigation pin track, one for arthrofibrotic band release, one for arthrotomy dehiscence, and one for unexplained pain. No implant loosening was reported. Coon and colleagues compared 45 minimally invasive UKAs performed manually with 36 UKAs performed with Tactile Guidance System (TGS) technology. There were no significant differences in mean Knee Society Score (KSS), change in KSS, or Marmor rating between the two groups at any postoperative time. From these early-term results, Sinha concluded that TGS-UKA results are no worse than those of conventional UKA with respect to complications, patient function, and surgeon learning curve. Bone preparation is extremely accurate relative to the preoperative plan and the computer-guided robotic system is very reliable. A recent publication comparing UKA implantation with robotic assistance to manual instrumentation found that final alignment of the components was significantly more precise and less variable when using the robot [33].

27.6.3 BlueBelt

BlueBelt began as the precision freehand sculptor and was developed by the Robotics Institute at Carnegie Mellon University and the Institute for Computer

Assisted Orthopaedic Surgery at the Western Pennsylvania Hospital in Pittsburgh. It consists of a handheld tool, optical tracking system, control computer, and a display monitor. Two prototypes have been developed. The first prototype had a fully exposed spherical blade that featured a clutch that engaged and disengaged the blade from the drive shaft. It did not include a brake so the blade would spin up very fast but took several seconds to spin down. The second prototype features a cylindrical rotating blade which extends and retracts behind a guard so the transition from cutting to not cutting is much faster. It is similar to the Acrobot and MAKO robots in that a tracking system monitors the position of the cutting tool and the computer enables the cutter only when it is held in the preprogrammed region of resection. The difference, as its original name suggests, is that this is a handheld device free from a robotic arm attached to a base. Also, the leg does not need to be fixed in place. There are no clinical trials to report but an experiment on artificial femurs showed an average error of less than 0.2 mm for all cases except for one outlier at 0.27 [19, 34].

27.7 Passive Systems

27.7.1 *Praxiteles*

One thing that all the previous robots have in common, aside from the BlueBelt, is size. For the most part they consist of large bases to which mechanical arms are attached. Praxim-Medivision (LaTronche, France) in collaboration with the Université Joseph Fourier (LaTronche, France) and the department of Mechanical Engineering at the University of British Columbia (Vancouver, Canada) have developed a compact bone-mounted robot, named Praxiteles, to position bone-cutting guides in the appropriate planes surrounding the knee. The surgeon then performs the planar cuts manually using the guides. The robot is comprised of two motorized degrees of freedom whose axes of rotation are aligned in parallel. Two prototypes have been developed, one for TKA surgery and a new version for MIS TKA which mounts on the side of the knee [35].

27.7.2 *Brigit*

The BRIGIT (Bone Resection Instrument Guidance by Intelligent Telemanipulator) system was designed with the goal of easy installation and operation in mind. It was developed by MedTech SA, a small French engineering company, to assist in osteotomies and TKR's. It is a compact robot mounted onto a wheeled trolly along with its control software. It works without preoperative imaging or a navigation system and is utilized as a positioner of tool-guides providing a mechanical support during bone sawing or drilling. First, the bone is rigidly attached to the operating table with a fixation device. The end of the robotic arm is then fitted with a pointing device that collects a series of anatomical landmarks. After the anatomy has been registered the pointing device is replaced by a cutting guide. In the case of a tibial osteotomy the angle of the desired cut is included in the geometrical calculations of the surgical planning parameters and the landmarks. The robot is then positioned and locked in place while the surgeon uses the free hand saw blade. It has a reported average error of 0.7° [36].

27.8 User Experience

The senior author (SK) has been using MAKO since September 2007. 88 surgeries have been performed, which include 74 medial UKA's, 1 lateral UKA's and 13 bicompartmental arthroplasties (medial UKA and patellofemoral replacement). The author's experience with the MAKOplasty has lead to a very high patient satisfaction. The excellent clinical outcome has resulted in a decreased hospital length of stay and has enabled patients to rehabilitate faster than conventional partial knee replacement. From a surgeon's viewpoint, MAKOplasty has a very quick learning curve, particularly if the surgeon is already familiar with computer navigation. Even though this is a highly complex and new technology, there has not been a single case where the surgery had to be aborted due to any malfunctions with the robot. The author believes that this technology allows us to rethink how we approach orthopedic problems, how we design implants, and how we will perform surgery in the future.

27.9 The Future of Robotics in Orthopedics

The future of robotics in orthopedics is promising. The ultimate goals of their utilization are improved functional outcomes, decreased morbidity, and earlier return to previous activity levels. Though the technology is relatively new and long term follow up studies are lacking early results are positive and reflect the basic principles that this technology is based on. The precision and accuracy of robotically assisted bone cuts allow for better fitting prosthetics which lead to less wear, pain, and need for future surgery. The cost of this new technology is one key drawback to its wide acceptance along with the training of surgeons and operating room staff in the use, maintenance, and incorporation of a robotic system in the busy operating room schedule. However, the learning curves for surgeon and staff are smaller than expected and the benefits to the patients more than justify the use of robotic systems. The cost up front is daunting but is warranted because of less patient morbidity and decreased hospital stays. With the advent of semi-active robotic systems surgeons are more likely to accept this technology for it does not replace them but refines them. Patients will also be more likely to accept a technology where the surgeon has complete control of the operation and is using tools that make him more accurate and precise. For these reasons robotically assisted surgery is a technology that will grow to become a permanent part of the orthopedic surgeon's repertoire.

References

1. Merriam-Webster Dictionary. Edited (2009)
2. Gourin, G.C., Terris, J.D.: History of robotic surgery. In: Faust, R.A. (ed.) Robotics in Surgery: History, Current and Future Applications, pp. 3–12. Nova Science Publishers, Inc., New York (2007)
3. RobotWorx: The History and Benefits of Industrial Robots. Edited (2009)
4. RobotWorx: Robot Timeline – Robotic History. Edited (2009)
5. Davies, B.: A review of robotics in surgery. Proc. Inst. Mech. Eng. H **214**(1), 129–140 (2000)
6. Lanfranco, A.R., Castellanos, A.E., Desai, J.P., Meyers, W.C.: Robotic surgery: a current perspective. Ann. Surg. **239**(1), 14–21 (2004)
7. Guidarelli, M.: Robotic surgery. The Next Generation **2**(4) (2006)
8. Marescaux, J., Smith, M.K., Folscher, D., Jamali, F., Malassagne, B., Leroy, J.: Telerobotic laparoscopic cholecystectomy: initial clinical experience with 25 patients. Ann. Surg. **234**(1), 1–7 (2001)
9. Giulianotti, P.C., Coratti, A., Angelini, M., Sbrana, F., Cecconi, S., Balestracci, T., Caravaglios, G.: Robotics in general surgery: personal experience in a large community hospital. Arch. Surg. **138**(7), 777–784 (2003)
10. Sutherland, G.R., Latour, I., Greer, A.D., Fielding, T., Feil, G., Newhook, P.: An image-guided magnetic resonance-compatible surgical robot. Neurosurgery **62**(2), 286–292; discussion 292–293 (2008)
11. CyberKnife Robotic Radiosurgery System. Edited (2009)
12. Clinics, S.H.a.: History of Stanford CyberKnife. Edited, Stanford University (2009)
13. Cho, C.B., Park, H.K., Joo, W.I., Chough, C.K., Lee, K.J., Rha, H.K.: Stereotactic radiosurgery with the CyberKnife for pituitary adenomas. J. Korean Neurosurg. Soc. **45**(3), 157–163 (2009)
14. Subramanian, V.A., Patel, N.U., Patel, N.C., Loulmet, D.F.: Robotic assisted multivessel minimally invasive direct coronary artery bypass with port-access stabilization and cardiac positioning: paving the way for outpatient coronary surgery? Ann. Thorac. Surg. **79**(5), 1590–1596; discussion 1590–1596 (2005)
15. Grossi, E.A. et al.: Minimally invasive versus sternotomy approaches for mitral reconstruction: comparison of intermediate-term results. J. Thorac. Cardiovasc. Surg. **121**(4), 708–713 (2001)
16. Murphy, D., Challacombe, B., Khan, M.S., Dasgupta, P.: Robotic technology in urology. Postgrad. Med. J. **82**(973), 743–747 (2006)
17. Ficarra, V., Cavalleri, S., Novara, G., Aragona, M., Artibani, W.: Evidence from robot-assisted laparoscopic radical prostatectomy: a systematic review. Eur. Urol. **51**(1), 45–55; discussion 56 (2007)
18. Advincula, A.P., Song, A.: The role of robotic surgery in gynecology. Curr. Opin. Obstet. Gynecol. **19**(4), 331–336 (2007)
19. Kazanzides, P.: Robots for orthopaedic joint reconstruction. In: Faust, R.A. (ed.) Robotics in Surgery: History, Current and Future Applications, pp. 61–94. Nova Science Publishers, Inc., New York (2007)
20. Howe, R.D., Matsuoka, Y.: Robotics for surgery. Annu. Rev. Biomed. Eng. **1**, 211–240 (1999)

21. Bargar, W.L.; Bauer, A., Borner, M.: Primary and revision total hip replacement using the Robodoc system. Clin. Orthop. Relat. Res. (354), 82–91 (1998)
22. Davies, B.L., Rodriguez y Baena, F.M., Barrett, A.R., Gomes, M.P., Harris, S.J., Jakopec, M., Cobb, J.P.: Robotic control in knee joint replacement surgery. Proc. Inst. Mech. Eng. H **221** (1), 71–80 (2007)
23. Bargar, W.L.: Robots in orthopaedic surgery: past, present, and future. Clin. Orthop. Relat. Res. **463**, 31–36 (2007)
24. Siebert, W., Mai, S., Kober, R., Heeckt, P.F.: Technique and first clinical results of robot-assisted total knee replacement. Knee **9**(3), 173–180 (2002)
25. Burkart, A., Debski, R.E., McMahon, P.J., Rudy, T., Fu, F.H., Musahl, V., van Scyoc, A., Woo, S.L.: Precision of ACL tunnel placement using traditional and robotic techniques. Comput. Aided Surg. **6**(5), 270–278 (2001)
26. Mantwill, F., Schulz, A.P., Faber, A., Hollstein, D., Kammal, M., Fay, A., Jurgens, C.: Robotic systems in total hip arthroplasty – is the time ripe for a new approach? Int. J. Med. Robot. **1**(4), 8–19 (2005)
27. Chung, J.-H., Ko, S.-Y., Kwon, D.-S., Lee, J.-J., Yoon, Y.-S., Won, C.-H.: Robot-assisted femoral stem implantation using an intramedulla gauge. IEEE Trans. Robot. Autom. **19**(5), 885–892 (2003)
28. Davies, B.: Robotic surgery: from autonomous systems to intelligent tools. In: The Smith and Nephew Annual Lecture. Edited
29. Cobb, J., Henckel, J., Gomes, P., Harris, S., Jakopec, M., Rodriguez, F., Barrett, A., Davies, B.: Hands-on robotic unicompartmental knee replacement: a prospective, randomised controlled study of the acrobot system. J. Bone Joint Surg. Br. **88**(2): 188–197 (2006)
30. Conditt, M.A., Roche, M.W.: Minimally invasive robotic-arm-guided unicompartmental knee arthroplasty. J. Bone Joint Surg. Am. **91** (Suppl 1), 63–68 (2009)
31. Lonner, J.H.: Modular bicompartmental knee arthroplasty with robotic arm assistance. Am. J. Orthop. **38**(2 Suppl), 28–31 (2009)
32. Sinha, R.K.: Outcomes of robotic arm-assisted unicompartmental knee arthroplasty. Am. J. Orthop. **38**(2 Suppl), 20–22 (2009)
33. Lonner, J.H., John, T.K., Conditt, M.A.: Robotic arm-assisted UKA improves tibial component alignment: a pilot study. Clin. Orthop. Relat. Res. **468**(1), 141–146
34. Brisson, G.K.T., Digioia, A., Jaramaz, B.: Precision freehand sculpting of bone. In: International Conference on Medical Image Computing and Computer-Assisted Intervention, pp. 105–112. Edited. St. Malo, France, Springer (2004)
35. Plaskos, C., Cinquin, P., Lavallee, S., Hodgson, A.J.: Praxiteles: a miniature bone-mounted robot for minimal access total knee arthroplasty. Int. J. Med. Robot. **1**(4), 67–79 (2005)
36. Maillet, P., Nahum, B., Blondel, L., Poignet, P., Dombre, E.: BRIGIT, a robotized tool guide for orthopedic surgery. In: Proceedings of the 2005 IEEE International Conference on Robotics and Automation, Barcelona, Spain, pp. 211–216 (2005)

Chapter 28
Robotic-Assisted Urologic Applications

Thomas S. Lendvay and Ryan S. Hsi

Abstract The clinical use of robotic-assisted laparoscopic surgery has been most prevalent in urologic care. The robotic prostatectomy is *the* procedure that has highlighted the potential of robotics. Because this procedure has become so rapidly embraced, urologists have readily adapted the robotic platform to other procedures such as radical cystectomy, partial nephrectomy, pediatric reconstructive procedures, and now female urologic and fertility procedures. Data on comparative effectiveness between the robotic, laparoscopic, and open versions of urologic procedures is still sparse, yet public pressure has forced a demand for access to robotic urologic care. It remains to be seen if adequate robotic training can keep up with the exploding need to provide robotic surgery options, but simulation training is ideally situated to offer a solution to novice trainees and experienced surgeons who wish to embark on robotic urologic surgery.

Keywords Robotic Prostatectomy · Robotic Partial Nephrectomy · Robotic Cystectomy · Robotic Pyeloplasty · Ureteropelvic junction repair · Ureteral reimplantation · Hydronephrosis · Urinary reflux · Pediatric urology · Ultrasound · Assist port

28.1 Background

Robotic surgery has seen the greatest penetration in the field of urology. The advantages robotic techniques afford over laparoscopy coupled with the types of procedures done intracorporeally in urology have provided the perfect setting for clinicians to embrace this modality of surgery. Urologic procedures, whether they are in adults or in children, revolve heavily around anatomic reconstructions requiring a great deal of suturing. Basic laparoscopy was embraced in the 1980s and 1990s

T.S. Lendvay (✉)
Seattle Children's Hospital, 4800 Sand Point Way Northeast, Seattle, WA, USA
e-mail: thomas.lendvay@seattlechildrens.org

J. Rosen et al. (eds.), *Surgical Robotics: Systems Applications and Visions*,
DOI 10.1007/978-1-4419-1126-1_28,

because extirpative procedures such as the removal of cancer laden urological organs were more attractive to patients and clinicians for their purported minimal invasiveness. This set the stage for surgeons in this decade to jump to robotics. Although standard extirpative procedures have been performed robotically, once clinicians realized the value of the robot for intracorporeal suturing – a skill difficult to master with pure laparoscopy – robotic-assistance boomed.

The ability of the only commercially available surgical robot, Intuitive Surgical Inc. da Vinci® (Sunnyvale, CA), to approximate the movements of the human wrist while dampening inherent human hand tremor and magnifying the field of view under 3-D visualization has enabled clinicians to reproduce open surgical techniques intracorporeally. These technological advances have benefitted both patients and clinicians [1,2] Patients are able to avoid large morbid incisions and severe post-operative pain, while experiencing surgical outcomes analogous or better than their open surgical counterparts for some high volume procedures such as prostatectomy and cystectomy (removal of the prostate and bladder) and pyeloplasty (repair of obstructed kidney) [3,4,5].

Adult urological applications of the surgical robot include prostate, bladder, renal, and adrenal cancer surgery, infertility, female incontinence, and neuropathic urinary tract maladies. The most commonly performed urologic procedure is the prostatectomy and in 2009, it is expected that over half of all prostatectomies in the U.S. will be performed robotically [6] which is quite noteworthy since the first robotic prostatectomy was just performed in this decade [7]. Clinicians have expanded the role of robotics into the realm of bladder cancer surgery, whereby the entire bladder and prostate are removed and a new urinary reservoir made out of intestinal segments is reconstructed. Again, a large amount of suturing is required for this procedure. Initially robotic-assistance was shown to be efficacious in radical nephrectomy surgery [8,9], but more recently, clinicians have applied robotics to the more challenging renal-sparing partial nephrectomy in patients with renal masses that allow for sparing of the remainder of the good kidney [10,11,12]. Urologists and general surgeons both perform adrenalectomies robotically and this application is rapidly expanding [13,14]. A few medical centers have begun to describe the use of robotic surgery in male fertility surgery including ligation of the varicoceles (an engorgement of the testicular veins leading to altered sperm production and function) [15,16]. Prior to the FDA approval of the use of the robot for gynecological procedures in 2007, urologists had already been describing applications in female pelvic floor surgery. Sacrocolpopexies (the securing of the vagina to the pelvis for improved pelvic floor support) are now being performed robotically because of the ease of suturing of the support materials to the vagina and the pelvic floor [17].

Just as in laparoscopy, the adoption of robotics in pediatric urology has initially lagged, but is now rapidly being embraced. The major benefits that minimally invasive techniques (MIS) offer such as the smaller incisions, less post-operative pain, and shorter convalescence have not been as dramatically demonstrated in children as in adults, especially pre-adolescents. Few were initially using the robot in children because of the large instrumentation and the perceived limitations of the bulky equipment in small children. Initially, the robotic instruments were 8 mm in diameter and then 5 mm instruments were designed, but the technology to decrease the diameter of the instruments sacrificed the short turning radius of the instruments. This handicap is realized in small children where the distance from the end of the trocar to the target surgical field may be only a few centimeters. In addition, the types of end effectors offered in the 5 mm diameter are limited. Intuitive Surgical relied on a 2D 5 mm scope, and this was also an impediment since one of the key advantages of the robotic platform was the stereoscope rendering 3D visualization. This has since been remedied by the introduction of an 8.5 mm stereoscope avoiding the large 12 mm camera port requirement. Once pediatric urologists developed methods for dealing with space limitations, clinicians realized that the high proportion of birth defect reconstruction procedures in children makes robotic surgery an attractive adjunct to laparoscopy.

The most commonly performed pediatric urological procedure is the pyeloplasty followed by the repair of urinary reflux (ureteral reimplantation) [18,19]. Additional robotic renal applications such as the anastomosing of the upper and lower kidney pole ureters to one another for duplex kidney systems with ectopic ureters and the removal of poorly functioning kidney poles have gained traction [20]. Just as in adults, male adolescents with symptomatic (painful) or testicular stunting varicoceles are being repaired with robotic techniques [16]. New applications of the robot include bladder augmentation (adding segments of the intestine onto dysfunctional bladders to increase capacity and reduce bladder pressures), reconstructions of the bladder neck for incontinence, and the creation of catheterizable channels for children to empty their bladders or add irrigation to flush out their colon in children with neuropathic bladder and bowels from spinal dysraphisms or spinal cord injuries [21].

28.2 Adult Applications

28.2.1 Robotic-Assisted Laparoscopic Radical Prostatectomy (RALRP)

Prostate cancer is the second most common malignancy among men in the United States, with an estimated 192,000 new cases diagnosed and 27,000 deaths in 2009 [22]. The surgical treatment for localized prostate cancer has traditionally been the open radical retropubic prostatectomy (RRP). Long-term outcomes in mature series of patients who underwent RRP have shown the progression-free survival at 68–75% at 10 years and cancer-specific survival rate approaching 97% at 10 years [23,24]. More recently, the robotic-assisted laparoscopic approach has become widely adopted across the United States (Figs. 28.1 and 28.2). A Medicare-based study has shown the utilization of RALRP quickly increased from 12.2% in 2003 to 31.4% in 2005 [25]. RALRP has benefited from a highly successful marketing campaign in addition to surgeon enthusiasm and patient interest. However, there still is a lack of long-term cancer outcomes in the literature, which has fueled a

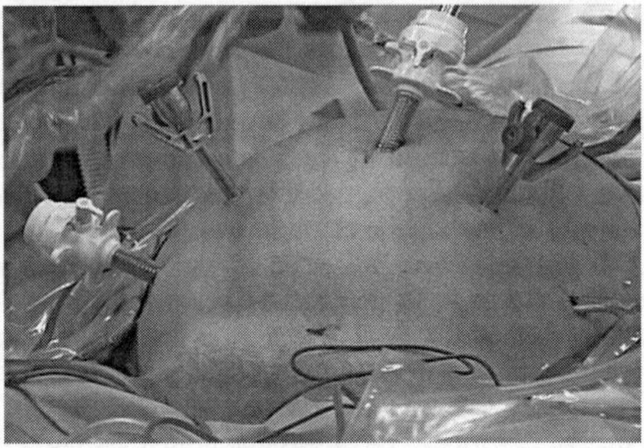

Fig. 28.1 RAL Prostatectomy port placement. (Courtesy Dr. James Porter, Swedish Medical Center, Seattle, WA)

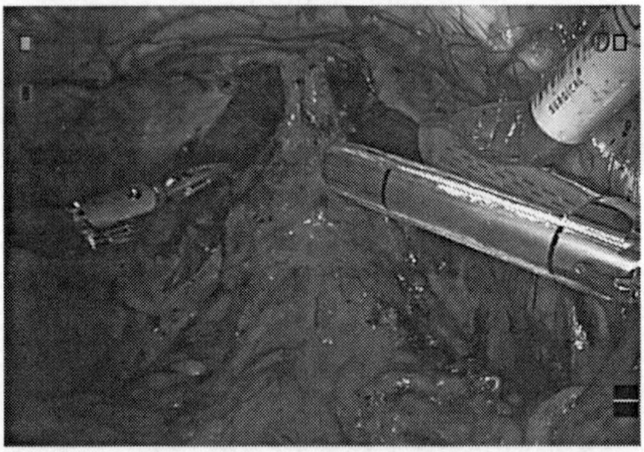

Fig. 28.2 Intracorporeal image looking down the pelvis after stapling of the dorsal venous complex. (Courtesy Dr. James Porter, Swedish Medical Center, Seattle, WA)

healthy debate in the urologic community on which technique is superior. Because of the protracted natural history of prostate cancer, data on disease specific and overall survival for this new treatment has begun to be reported only recently. Badani and colleagues published their outcomes on 2,766 patients who underwent RALRP over 6-year period [26]. With a median follow-up of 22 months [26], 7.3% had PSA recurrence after prostatectomy, and 5-year actuarial biochemical free survival rate as determined by PSA was 84%, which is comparable to the open experience at 5 years [23,24]. Another measure of cancer control is the rate of finding cancer at the edge of the removed tumor by the pathologist during microscopic examination – also known as positive surgical margins – an independent predictor of disease recurrence [27]. Positive surgical margin rates have been reported between 2 and 59% in RALRP [28]. In studies that have compared the positive surgical margin rates between RALRP and RRP, two studies have shown similar positive margin rates [29,30], while one study has shown a higher rate after RRP(23% vs 9% in RALRP) [31]. A separate comparison by Smith and colleagues reported the rate of positive surgical margins at 15% for RALRP and 35% for RRP ($p < 0.001$) [3]. Even when stratified for pathological cancer stage, RALRP had less positive surgical margin rates (RALRP 9.4% vs RRP 24.1 for pT2 [meaning organ confined disease]; $p < 0.001$). For both groups the location of positive margins was most commonly found at the prostatic apex (located at the bottom of the prostate gland) (52% RALRP vs 37% RALRP). In terms of perioperative morbidity and mortality, complication rates have been reported in 6.6% of the robotic series compared with 10.3% in the open series [32,33]. In addition, a prospective, validated, quality of life instrument has shown that patients after RALRP have a faster recovery to their baseline functioning status compared to patients after RRP [34]. Anatomically, there is an intrinsic urinary sphincter at the bladder neck which may become deficient after the prostate is removed and the bladder is reconnected to the urethra, leading to urinary incontinence. There are also two sets of neurovascular structures immediately posterolateral to the prostate responsible for erectile function. Therefore, important measures of functional outcomes after prostatectomy have included continence rates and potency data. Return of continence for both techniques appear to be comparable if not slightly favoring RALRP. Continence rates for RALRP at 3 months are 73–91% and 82–97%, while continence rates for RRP at 3 months are 54–71% at 3 months at 39–87% at 6 months [32,33]. Twelve-month potency data for RALRP is also at least as good as that of the open series. For unilateral nerve-sparing surgery, potency rates at 12 months are 14–61% for RALRP and 17–53% for RRP, while at 24 months they are 24–97 for RALRP and 37–86% for RRP [32,33]. While an increasing number of surgeons are performing RALRP, there also remains a difficult learning curve of more than 150 procedures needed before results comparable with RRP [35]. Overall, the early data on RALRP have shown that it is a promising new technique that offers a minimally invasive, surgical option for prostate cancer.

Meanwhile, skeptics of RALRP have long argued that this approach may not be equivalent for patients with high risk disease, where tactile feedback may be important to delineate more suspicious areas of tumor. In addition, a study of Medicare beneficiaries showed that while patients undergoing minimally invasive radical prostatectomy – including pure laparoscopic and RALRP techniques – were less likely to have perioperative complications (29.8% vs 36.4%; $p = 0.002$) and shorter hospital stays (1.4 vs 4.4 days; $p < 0.01$), they were more likely to receive a secondary salvage therapy (such as radiation therapy) (2.8% vs 9.1%; $p < 0.001$) and have a higher risk for anastomotic stricture where the bladder is reconnected to the urethra (OR 1.40; 95% CI 1.04–1.87) [25]. Proponents of the open approach have also contended that hospital stays for RALRP and RRP can be similar in length depending on postoperative care pathways [36]. Finally, one interesting study from Duke University has examined patient satisfaction after RRP and RALRP [37]. Patients who underwent RALRP were three times more likely to regret choosing that particular technique compared to patients who underwent RRP (OR 3.02, 95% CI 1.50–6.07). The authors hypothesized that this was due to the expectations on a newer, more innovative procedure so it is critical that clinicians prepare their patients for possible morbidity irrespective of the type of procedure utilized.

28.2.2 *Robotic-Assisted Laparoscopic Radical Cystectomy (RALRC)*

Bladder cancer remains the fifth most common malignancy in the United States, with an estimated 71,000 new cases diagnosed and 14,000 deaths in 2009 [22]. Although most patients with bladder cancer have superficial tumors, 20–30% of patients that present with muscle invasive disease [38,39], which has a poorer prognosis. Radical cystectomy has been shown to provide a long-term survival benefit in patients with muscle-invasive bladder cancer [40]. Despite advances in minimally invasive techniques there have been few published series of laparoscopic radical cystectomy since the first laparoscopic radical cystectomy was reported in 1992 [41], primarily due to the significant technical challenges of the procedure. More recently the robotic-assisted laparoscopic approach has gained increasing acceptance for prostatectomy, and with the development of these techniques, surgeons have begun to also perform RALRC [42–47]. The learning curve for RALRC has been shown to be approximately 20 cases, with higher initial operative times and blood loss [48]. In most cases, the extirpative portions of the procedure – cystoprostatectomy and extended or traditional pelvic lymph node dissection – are performed using a robotic-assisted laparoscopic approach, while the urinary diversion portion is performed extracorporeally (the bowel is reconstructed into a urinary diversion outside of the abdomen). Ng and colleagues have reported on 104 open and 83 RALRC from a single surgeon. Compared to the open group, the robotic group had a similar mean operative time (6.25 h vs 5.95 h), less blood loss (460 vs 1,172 mL; $p < 0.001$), similar median lymph node yield (16 vs 15), shorter median length of stay (5.5 vs 8.0 days; $p < 0.0001$), lower major complication rate at 30 days (10% vs 30%; $p = 0.007$), and similar rates of positive margins (7.2% vs 8.7%; $p = 0.77$) [49]. Pruthi and colleagues reported on a series of 50 patients who underwent RALRC, with 0 cases with positive margins and a median lymph node yield of 19. Sixty-six percent of patients had pathologically organ-confined disease (pT2 or less), 14% had local extension of disease (pT3), and 20% had disease found in the lymph nodes. After a mean follow-up of 13.2 months, 14% of patients had evidence of recurrence [50]. These and other preliminary data suggest that RALRC can be performed safely by experienced surgeons with less blood loss, shorter time to return of bowel function, and shorter hospitalization.

28.2.3 *Robotic-Assisted Laparoscopic Partial Nephrectomy (RALPN)*

In 2009 there will be an estimated 57,000 new cases of kidney and renal pelvis cancer with almost 13,000 deaths [22]. While the incidence of renal cancer is increasing in the United States, tumor size is decreasing, which has allowed patients to pursue open partial nephrectomy as an option to preserve renal function [51].

Compared to the open approach, laparoscopic partial nephrectomy (LPN) has recently been shown to have comparable oncological and renal functional outcomes [52]. However, LPN remains technically challenging due to intracorporeal suturing of the kidney bed and difficult hemostasis. Minimizing warm ischemia time to less than the traditional 30 min after clamping of the renal vessels may also rush the surgeon to dissect and remove the tumor. A robotic-assisted laparoscopic approach has been thought to be more accessible to surgeons by allowing greater precision to a procedure that involves a small operative field and extensive knot-tying. Initial small, single institution series consisting of 8–13 patients undergoing RALPN have shown this technique is feasible and safe, with reported mean operative times of 155–279 min, mean blood loss of 92–249 mL, mean tumor size 1.8–3.6 cm, mean warm ischemia time of 21–32 min, and a mean hospital stay of 1.5–4.3 days [53–58] (Figs. 28.3–28.6).

The feasibility and safety of RALPN for nonfunctioning kidneys has also been shown in the pediatric population [59]. Wang and Bhayani have recently published data on a retrospective single-surgeon experience with 40 RALPNs and 62 laparoscopic partial nephrectomies (LPN) [12]. They reported no significant difference between blood loss (136 vs. 173 mL) and tumor size (2.5 vs. 2.4 cm) between RALPN and LPN, respectively. Mean operative times (140 vs 156 min; $p = 0.04$), warm ischemia time (19 vs 25 min; $p = 0.03$), and hospital stay (2.5 vs 2.9 days; $p = 0.03$) favored the RALPN technique. Each group had one patient with positive margins. A multi-institutional study from three minimally invasive experienced surgeons has also retrospectively evaluated 118 LPNs and 129 RALPNs [60]. There was no difference in mean operative times (189 vs 174 min), tumor size (2.8 vs 2.5 cm), and positive margin rate (3.9 and 1%) for RALPN and LPN, respectively. RALPN had significantly less blood loss (155 vs 196 mL; $p = 0.03$), shorter warm ischemia times (19.7 vs 28.4 min; $p < 0.0001$), and shorter hospital stay (2.4 vs 2.7 days; $p < 0.0001$) compared to LPN. There were no intraoperative

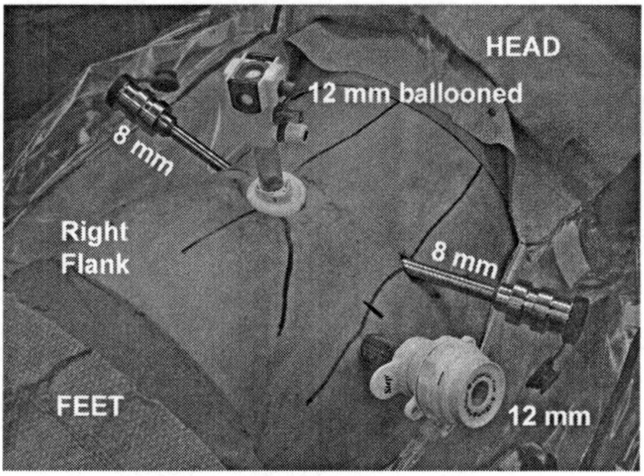

Fig. 28.3 Retroperitoneoscopic RAL right partial nephrectomy port placement. (Courtesy Dr. James Porter, Swedish Medical Center, Seattle, WA)

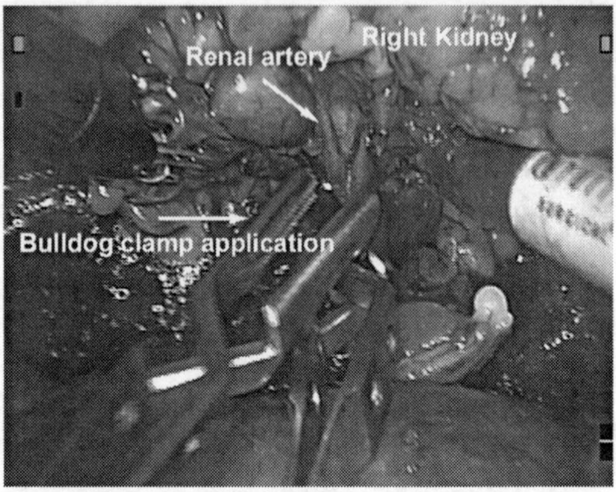

Fig. 28.4 Application of bulldog clamp onto renal artery. (Courtesy Dr. James Porter, Swedish Medical Center, Seattle, WA)

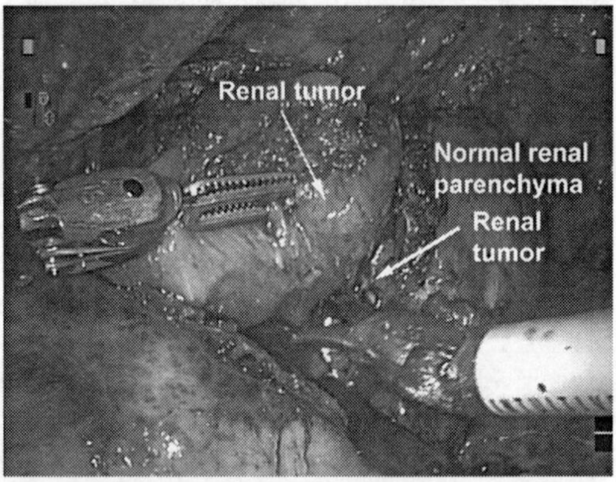

Fig. 28.5 Resection of tumor maintaining Gerota's fat cap. (Courtesy Dr. James Porter, Swedish Medical Center, Seattle, WA)

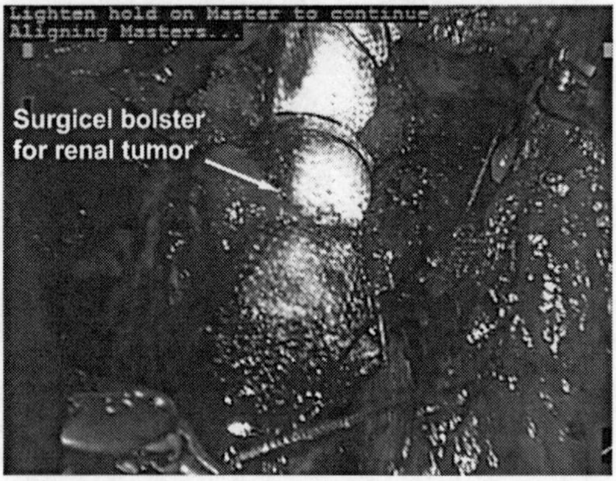

Fig. 28.6 Demonstration of renal tumor bed bolster placement. (Courtesy Dr. James Porter, Swedish Medical Center, Seattle, WA)

complications during RALPN, while there was one case of adrenal injury requiring adrenalectomy for the LPN group. Postoperative complication rates were similar (8.6% for RALPN and 10.2% for LPN). RALPN has also been shown to be feasible in renal hilar tumors that are in physical contact with the renal artery and/or vein [61]. Overall, these data suggest that RALPN is a comparable to LPN in terms of operative and pathological outcomes. RALPN may offer shorter warm ischemia times, less intraoperative blood loss, and shorter hospital stays. Disadvantages of a robotic-assisted approach include the financial costs of using the robot system and the need for an experienced laparoscopic assistant [62]. Long-term data on oncological outcomes and renal function are pending.

28.3 Pediatric Applications

Data is limited in children as to whether there are obvious advantages using robotic-assistance over open surgical techniques. Much like in laparoscopy, instrumentation is generally designed for adult surgery and it is up to the pediatric urologist to develop strategies to apply new technology to small spaces. Because of the increased end-effector dexterity in robotics over pure laparoscopy, the suturing required for many of the reconstructive congenital defect repairs has made robotics popular in pediatric urology. Suturing tends to be performed with finer suture material 5–0, 6–0, and 7–0 sizes enabled by the 12× magnification, 3D visualization, and motion scaling. Replication of open surgical maneuvers in a laparoscopic environment is paramount because these repairs must withstand the majority of a patient's life. Minimal tissue handling, which arguably minimizes tissue ischemia, and precise suture placement are key principles to which every pediatric surgeon tries to adhere, and the robot's degrees of freedom enable the surgeon to maintain these principles. The applications for robotics in pediatric urology have expanded over the last 5 years. Repair of the congenital obstructed kidney or ureteropelvic junction obstruction is the most common procedure followed by repairs for urinary reflux (ureteral reimplantation), duplex kidneys with poorly functioning of obstructed segments (heminephrectomy, uretero-ureterostomy, ureteropyelostomy), abnormal reproductive structures in children with disorders of sexual differentiation, varicoceles, and for more complex reconstructions such as appendicovesicostomy, antegrade continent enema, bladder augmentation, and bladder neck reconstruction. Since far less literature exists about the comparison of robotics and open surgical techniques in children, this section will focus on techniques for some of the more common pediatric urologic procedures.

28.3.1 Pyeloplasty

The second most common fetal abnormality picked up on antenatal ultrasounds is hydronephrosis, the dilation of the kidney pelvis where urine first collects before draining down the ureters. The incidence of hydronephrosis is 1% of all prenatal ultrasounds; within this group, almost half is due to obstruction at the junction between the kidney pelvis and the ureter. The repair of the obstruction involves a reconstructive technique called a pyeloplasty most commonly performed by dividing the narrowed segment and broadening the cross-sectional area of each end and then suturing the ends back together [63]. The first robotic pyeloplasty in children was described by Peters in 2003 [64]. In centers where pediatric urologists have access to the surgical robot, this is the most widely accepted application. Either a transperitoneal or retroperitoneal approach may be employed depending on the surgeon's experience and the patient's size since the retroperitoneoscopic approach affords a much more limited working space. Yueng et al. performed an elegant study looking at the space limitations of the standard robot with 8 mm end effectors and found that the minimal working volume using in vitro cubes was a 5 cm sided cube or 125 cm^3 [65]. This translates into a space that is 130 mL in volume (the average size of a 2 year old child's bladder).

Approach

Standard laparoscopic renal surgery port placement is carried out with the camera port in the umbilicus (we use an 8.5 mm 3D scope and access the abdomen through open laparoscopy to minimize vascular injury). We recommend initial bladder decompression with an indwelling urethral catheter to avoid bladder injury upon initial access and we have frequently observed a natural fascial defect at the umbilicus in many children facilitating ease of open laparoscopic port placement. We then place the patient in a low flank position (~45°) and for children less than 20–30 kg, we prefer placing them on a padded stage of blankets and foam so that the robotic arms do not collide with the table.

The robot is positioned over the ipsilateral shoulder. For left-sided repairs, we use two working ports and because of the limited distances and the limited instrumentation options with the 5 mm instruments, we choose to use the 8 mm instruments. Left-sided repairs may also be amenable to a transmesocolic approach which involves creating a small mesenteric window through the left colon mesentery directly onto the ureteropelvic junction. This avoids mobilizing the colon which can lead to a longer return of bowel function. On the right side, we place ports in the same orientation as the left side and occasionally need to add a left lateral subcostal 5 mm assist port to retract the liver as it sometimes is draped over the area of interest.

We use 8 mm instruments because we use 6–0 monofilament suture material for our anastomosis and have found that the 5 mm needle drivers alter the integrity of the suture, therefore we use the da Vinci Diamond Tip® needle drivers for fine suturing. We have not observed large residual port site scars, even in infants, and only one patient to date developed a port site hernia. We close all port sites at the level of the fascia with absorbable braided polyglycolic acid suture. The working ports are placed in the midline below the xyphoid and on the midclavicular line and on the

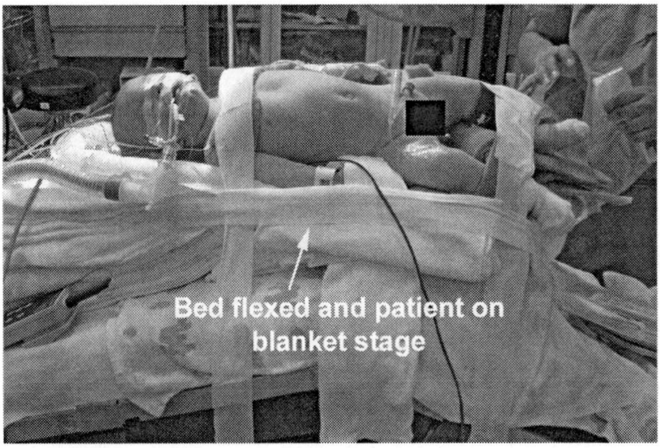

Fig. 28.7 Infant positioning and set-up for left pyeloplasty

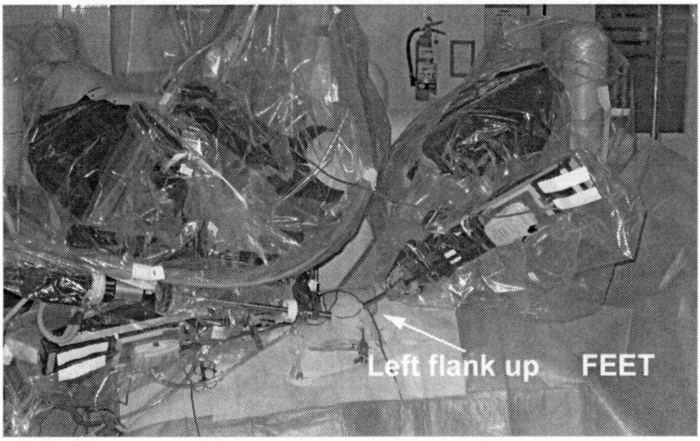

Fig. 28.8 Robot docked for left infant pyeloplasty

ipsilateral side below the umbilicus (Figs. 28.7–28.10). The standard distance that must be maintained between the ports has been advised by Intuitive to be 8 cm which is a distance frequently unattainable especially in infants. Due to the elasticity of the young child's abdomen upon insufflation and the minimal travel required of the instruments once the robot is docked (finite surgical field), we have been able to bring the ports to within 4–5 cm of each other without arm collisions.

A percutaneous hitch stitch is placed through the renal pelvis proximal to the anastomotic site and sometimes through the distal ureter to elevate the anastomosis out of the leaking urine and stabilize the two ends for facilitating the closure. A standard Anderson-Hynes dismembered pyeloplasty technique [66] is employed

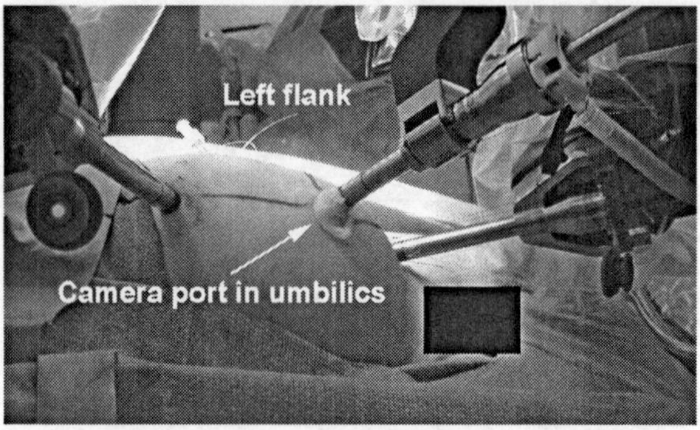

Fig. 28.9 Close-up of camera and two working ports for infant pyeloplasty

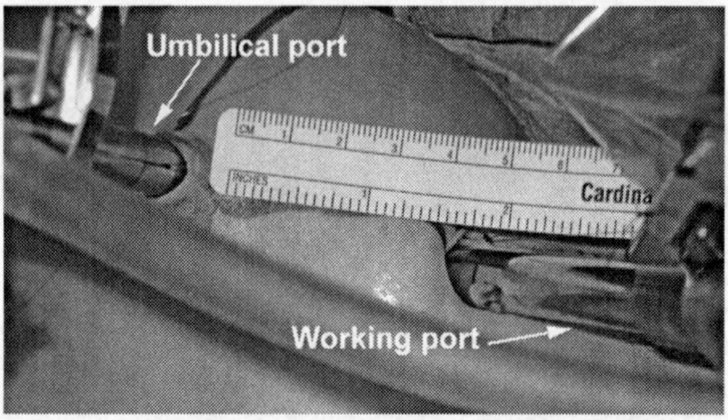

Fig. 28.10 Distance between lower abdominal working port and camera port

and in our practice, we place an internal JJ ureteral stent in an antegrade fashion through a 14 G angiocatheter placed through the abdominal wall [67]. Ureteral stents can also be placed retrograde during cystoscopy, but we have found that antegrade insertion facilitates ease of anastomosis and we use intraoperative bedside ultrasound to confirm adequate distal stent position [68]. The angiocatheter also allows us to avoid an additional assist port as we use off-the-shelf cystoscopic graspers and shears for tissue retraction and suture cutting (Figs. 28.11 and 28.12). The urethral catheter typically remains overnight and is removed on the first post-operative day and discharge is contingent on adequate oral intake and pain control. The stent is removed during a brief general anesthetic 3–6 weeks later via cystoscopy. Kim et al. described their experience with 84 child pyeloplasties and found that children older than 6 months of age could be safely repaired robotically [69]. When looking at post-operative pain control, Lee et al. compared their robotic experience to an age-matched open cohort and found that narcotic requirements were less in the robotic group [70]. What is lacking is a prospective study examining the true benefits of robotic pyeloplasty in children.

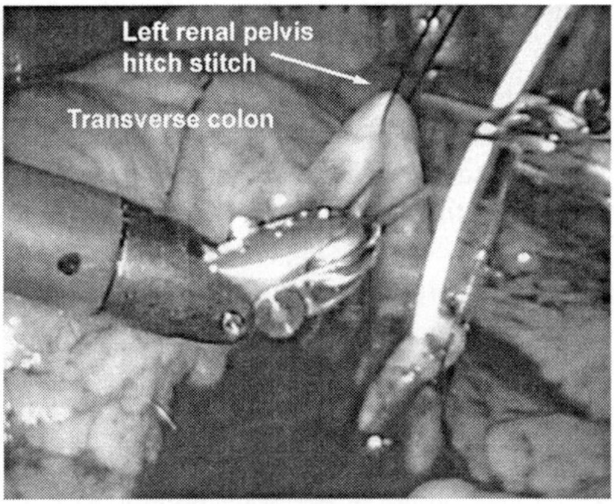

Fig. 28.11 Demonstration of percutaneous hitch stitch with 2–0 monofilament suture and passage of antegrade JJ stent through percutaneous angiocatheter and down ureter

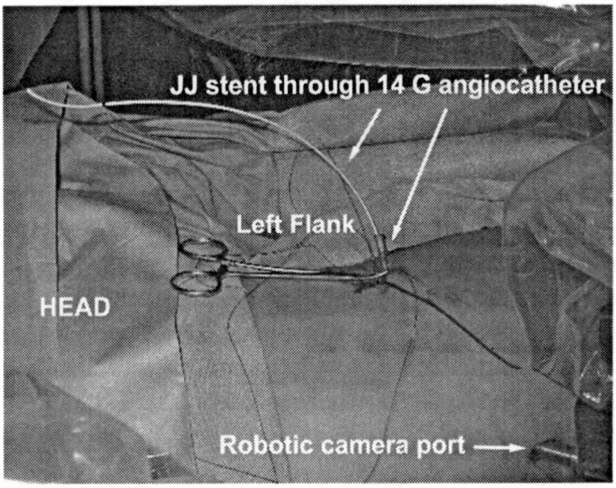

Fig. 28.12 External view of 14 gauge angiocatheter with JJ stent being passed over 0.025″ hydrophilic wire

28.3.2 Ureteral Reimplant

For children with urinary reflux where the urine goes from the bladder backwards up into the kidneys, frequent urinary tract infections and potential permanent renal damage may occur. There are open and endoscopic techniques to reconstruct the ureters as they course through the bladder wall to prevent reflux. The first laparoscopic repair for vesicoureteral reflux disease was described by Atala et al. in minipigs in 1993 followed in humans by Ehrlich et al. in 1994 [71, 72]. Since then pediatric centers have only partially embraced both the laparoscopic extravesical or intravesical cross-trigonal approaches due to the complexity of fine suturing in the small spaces. Success rates have been comparable to open surgical techniques and in 2004, Peters first described his experience using the surgical robot as an adjunct to both transvesical and extravesical repairs [73]. Lendvay and Casale et al. have described their experiences performing the transperitoneal extravesical approach to the robotic reimplant using the Lich technique [74]. The success rates range from 85 to 100% [75,76]. Due to the rapid learning curve with robotic surgery, surgeons are able to utilize the same techniques and suture size as would be used in open surgery without the long learning curve required for pure laparoscopy.

Approach

As with all robotic surgeries in children, appropriate patient positioning is critical to the efficient progression and success of the case. Since it is our practice to perform cystourethroscopy prior to ureteral detrusorraphy surgery, we place the patient in a low lithotomy position and prep the patient for both cystoscopy and laparoscopic access at the same time. We angle the patient in 10° Trendelenberg to encourage the bowel to fall out of the pelvis. For bilateral repairs, we choose to place indwelling stents if the child has a history of a trabeculated thickened bladder due to voiding dysfunction as we have observed post-operative edema at the neo-transmural tunnel causing transient obstruction. For the majority of cases, we typically will place external ureteral catheters attached to a urethral catheter to help guide ureteral dissection during the procedure. These are removed at the end of the surgery.

The camera port is placed in the umbilicus except in children less than 15 kg in which we place it sub-xyphoid. Two subsequent 8 mm ports are placed on either side of the umbilicus in the paramedian lines and for umbilically located camera ports, we lower the level of the working ports below the umbilicus slightly (Figs. 28.13 and 28.14). In children less than 15 kg, we have tended to place the working ports at the level of the umbilicus to ensure a good distance to the target site. For bilateral cases, the robot is situated at the patient's feet in the midline; however, for unilateral repairs we position the robot at the ipsilateral foot. In addition, the ipsilateral working port is placed slightly higher than the contralateral working port. In small infants, we place the camera port sub-xyphoid, to ensure a good working distance from the camera to the target site.

28.4 Intra-Operative Tricks

There are certain maneuvers which are unique to laparoscopic/RAL surgery which assist in expediting the surgeries and allow for the minimum number of ports to be placed. A sharp criticism of minimally invasive surgery in children, especially small children has been that open surgery incisions are not as morbid as in adults and that the additive incisional length of minimally invasive surgeries may equal

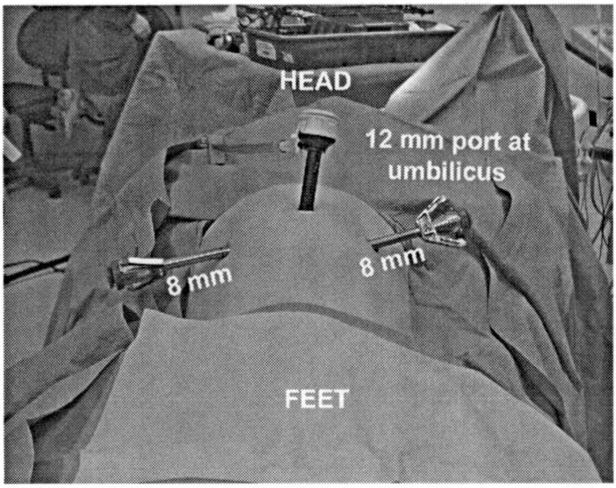

Fig. 28.13 Patient positioning and port placement for bilateral extravesical robotic ureteral reimplantation. (Courtesy Dr. Pasquale Casale, Children's Hospital of Philadelphia, Philadelphia, PA)

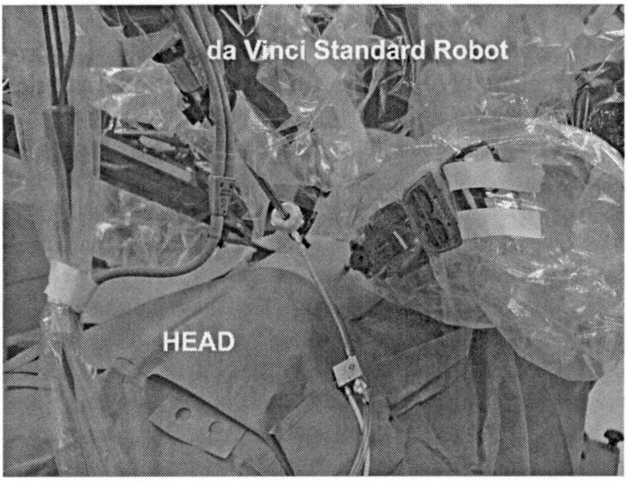

Fig. 28.14 Robot docked for bilateral ureteral reimplantation. (Courtesy Dr. Pasquale Casale, Children's Hospital of Philadelphia, Philadelphia, PA)

and sometimes exceed the total length of a single open surgical incision thereby theoretically causing more post-operative pain. This argument is flawed because Blinman has demonstrated that the sum tensions of port incisions do not equal the whole incisional tensile burden as conjectured by some open surgeons [77]. We believe that the smallest and fewest possible ports should be used to safely and effectively perform MIS surgery therefore we employ the use of hitch-stitches to assist in organ retraction throughout our cases. During an extravesical ureteral

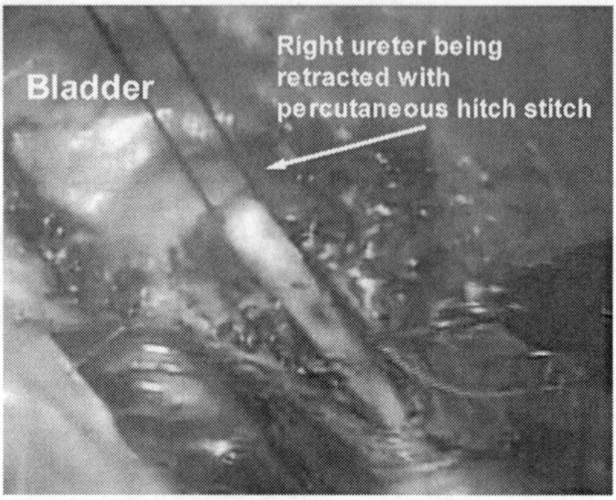

Fig. 28.15 Percutaneous 2–0 monofilament suture for hitching ureter for retraction during extravesical ureteral reimplantation

reimplant, we routinely use monofilament suture placed through the lower abdominal wall to aid in retraction of ureters and the bladder (Fig. 28.15).

During creation and closure of the detrusor bladder flaps, we find that a percutaneously placed hitch stitch to help elongate the bladder anteriorly ensures proper length and straightening of the tunnel. In addition, we use anteriorly retracted stitches around the ureters to assist in laying the ureters down in the detrusor tunnels. When no longer needed, these sutures are removed leaving only behind small needle puncture marks on the suprapubic skin. For children with more subcutaneous fat, we lengthen the hitch stitch needles by partially flattening them (skiing).

Throughout the creation of the detrusor tunnel and the detrusorotomy, we intermittently insufflate the bladder through the indwelling urethral catheter with a second insufflation unit to ensure appropriate position of the ureter as described by Yeung et al. [78]. We have used both manual fluid bladder instillation for distention and gas insufflation and have found the gas to be more rapid in raising and dropping the bladder and in the event of a small mucosotomy which would require oversewing, the liquid distention tends to make for a more tedious closure.

28.5 Future Perspectives

Robotic surgery has been embraced by both adult and pediatric urologists and will become the preferred modality for intracorporeal surgery. The current iteration of the robot is bulky and costly, but ongoing research around the US and developed countries promises to provide robotic surgery platforms that will break the cost barrier. Data on surgical outcomes for adult surgery is just now emerging on long-term outcomes. It remains to be seen if pediatric surgery can replicate the obvious benefits that have been demonstrated in adult surgery. The ability to perform precise fine suture maneuvers lends robotics ideally to pediatric surgery, but instrumentation and overall platform size must decrease before widespread adoption.

In adult surgery, more extensive appraisal of prospective multi-center outcomes need to be performed and in pediatric surgery, more head-to-head-to-head open vs. laparoscopic vs. robotic surgery comparison studies need to be initiated. The advantages of robotic surgery in children are harder to demonstrate than in adults since metrics used in the adult literature to show advantages do not always apply to children such as return to work and wages lost. Metrics such as lost parental work days or days of school missed for the children need to be assessed. Pain score assessments between open and MIS surgeries have not been rigorously tested as randomized trials looking at open vs. robotic urologic surgeries in children are nonexistent.

The advantages for advancing robotics in urology will be greatest in surgical education and patient-specific surgical simulation (Figs. 28.16 and 28.17). With the aid of pre-operative imaging and robotic simulators [79], a surgeon will be able to perform the surgery in a virtual reality arena prior to performing the surgery on the actual patient. In addition, the use of VR robotic "warm-up" prior to surgery is currently being explored and we believe that in the era of surgical simulation training, robotic surgery will allow residents and novice or expert roboticists to acquire technical competence in procedure performance more rapidly with the assistance or patient-specific simulation and warm-up. The development of robotic surgery curricula will be necessary to achieve the highest level of patient outcomes.

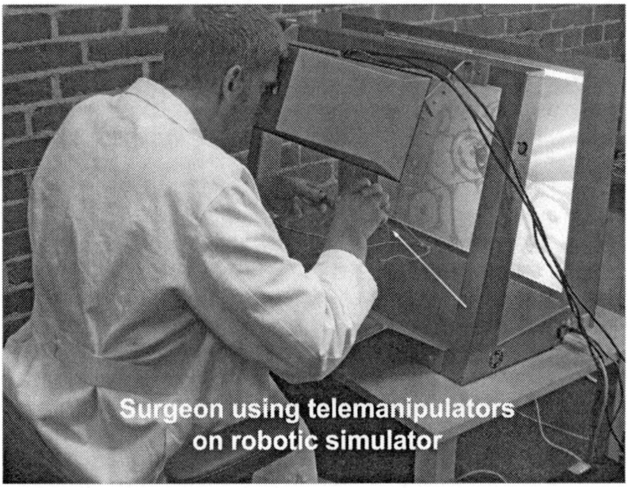

Fig. 28.16 Surgeon at virtual reality robotic console

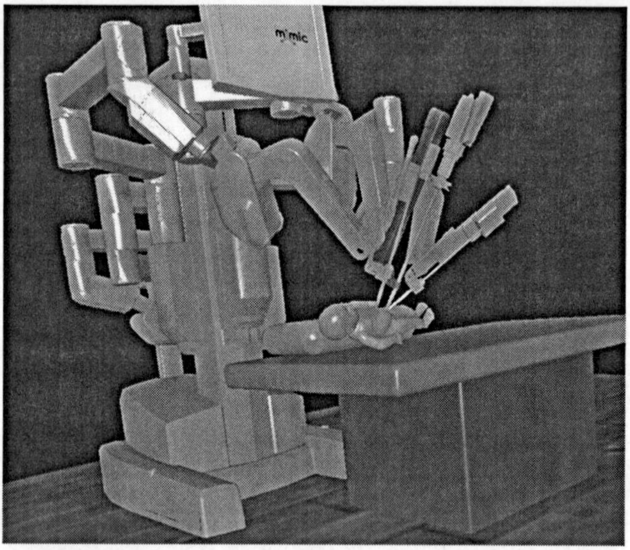

Fig. 28.17 Patient-specific infant surgery simulation model for port placement

Acknowledgments We would like to thank Dr. James Porter, Swedish Medical Center, Seattle, WA for RAL prostatectomy and partial nephrectomy pictures, Dr. Pasquale Casale, Children's Hospital of Philadelphia, Philadelphia, PA for external RAL ureteral reimplantation set-up pictures, and Dr. Jeff Berkley, CEO, MIMIC Technologies, Inc., Seattle, WA for the patient specific infant simulation picture.

References

1. Bagrodia, A., Raman, J.D.: Ergonomics considerations of radical prostatectomy: physician perspective of open, laparoscopic, and robot-assisted techniques. J. Endourol. **23**(4), 627–633 (2009)
2. Khan, M.S., Shah, S.S., Hemel, A., Rimington, P., Dasgupta, P: Robotic-assisted radical cystectomy. Int. J. Med. Robot. Comput. Assist. Surg. MRCAS **4**(3), 197–201 (2008)
3. Smith, J.A., Jr., Chan, R.C., Chang, S.S., Herrell, S.D., Clark, P.E., Baumgartner, R., Cookson, M.S.: A comparison of the incidence and location of positive surgical margins in robotic assisted laparoscopic radical prostatectomy and open retropubic radical prostatectomy. J. Urol. **178**(6), 2385–2389 (2007)
4. Guru, K.A., Wilding, G.E., Piacente, P., Thompson, J., Deng, W., Kim, H.L., Mohler, J., O'Leary, K.: Robot-assisted radical cystectomy versus open radical cystectomy: assessment of postoperative pain. Can. J. Urol. **14**(6), 3753–3756 (2007)
5. Mufarrij, P.W., Woods, M., Shah, O.D., Palese, M.A., Berger, A.D., Thomas, R., Stifelman, M.D.: Robotic dismembered pyeloplasty: a 6-year, multi-institutional experience. J. Urol. **180**(4), 1391–1396 (2008)
6. Pruthi, R.S., Wallen, E.M.: Current status of robotic prostatectomy: promises fulfilled. J. Urol. **181**(6), 2420–2421 (2009)

7. Binder, J., Kramer, W.: Robotically-assisted laparoscopic radical prostatectomy. BJU Int. **87**(4), 408–410 (2001)
8. Guillonneau, B., Jayet, C., Tewari, A., Vallancien, G.: Robot assisted laparoscopic nephrectomy. J. Urol. **166**(1), 200–201 (2001)
9. Rogers, C., Laungani, R., Krane, L.S., Bhandari, A., Bhandari, M., Menon, M.: Robotic nephrectomy for the treatment of benign and malignant disease. BJU Int. **102**(11), 1660–1665 (2008)
10. Rogers, C.G., Singh, A., Blatt, A.M., Linehan, W.M., Pinto, P.A.: Robotic partial nephrectomy for complex renal tumors: surgical technique. Eur. Urol. **53**(3), 514–521 (2008)
11. Patel, M.N., Bhandari, M., Menon, M., Rogers, C.G.: Robotic-assisted partial nephrectomy. BJU Int. **103**(9), 1296–1311 (2009)
12. Wang, A.J., Bhayani, S.B.: Robotic partial nephrectomy versus laparoscopic partial nephrectomy for renal cell carcinoma: single-surgeon analysis of >100 consecutive procedures. Urology **73**(2), 306–310 (2009)
13. Brunaud, L., Ayav, A., Zarnegar, R., Rouers, A., Klein, M., Boissel, P., Bresler, L: Prospective evaluation of 100 robotic-assisted unilateral adrenalectomies. Surgery **144**(6), 995–1001 (2008)
14. Hyams, E.S., Stifelman, M.D.: The role of robotics for adrenal pathology. Curr. Opin. Urol. **19**(1), 89–96 (2009)
15. Shu, T., Taghechian, S., Wang, R.: Initial experience with robot-assisted varicocelectomy. Asian J. Androl. **10**(1), 146–148 (2008)
16. Hidalgo-Tamola, J., Sorenson, M.D., Bice, J.B., Lendvay, T.S.: Pediatric robot-assisted laparoscopic varicocelectomy. J. Endourol. **23**(8), 1297–1300 (2009)
17. Kramer, B.A., Whelan, C.M., Powell, T.M., Schwartz, B.F.: Robot-assisted laparoscopic sacrocolpopexy as management for pelvic organ prolapse. J. Endourol. **23**(4), 655–658 (2009)
18. Casale, P.: Robotic pyeloplasty in the pediatric population. Curr. Opin. Urol. **19**(1), 97–101 (2009)
19. Casale, P.: Robotic pediatric urology. Exp. Rev. Med. Dev. **5**(1), 59–64 (2008)
20. Kutikov, A., Nguyen, M., Guzzo, T., Canter, D., Casale, P.: Laparoscopic and robotic complex upper-tract reconstruction in children with a duplex collecting system. J. Endourol. **21**(6), 621–624 (2007)
21. Lendvay, T.S., Shnorhavorian, M., Grady, R.W.: Robotic-assisted laparoscopic Mitrofanoff appendicovesicostomy and antegrade continent enema colon tube creation in a pediatric spina bifida patient. J. Laparoendosc. Adv. Surg. Tech. A **18**(2), 310–312 (2008)
22. Jemal, A., Siegel, R., Ward, E., et al.: Cancer statistics, 2009. CA Cancer J. Clin. **59**, 225–249 (2009)
23. Roehl, K.A., Han, M., Ramos, C. G., et al.: Cancer progression and survival rates following anatomical radical retropubic prostatectomy in 3,478 consecutive patients: long-term results. J. Urol. **172**, 910–914 (2004)
24. Hull, G.W., Rabbani, F., Abbas, F., et al.: Cancer control with radical prostatectomy alone in 1,000 consecutive patients. J. Urol. **167**, 528–534 (2002)
25. Hu, J.C., Wang, Q., Pashos, C. L., et al.: Utilization and outcomes of minimally invasive radical prostatectomy. J. Clin. Oncol. **26**, 2278–2284 (2008)
26. Badani, K.K., Kaul, S., Menon, M.: Evolution of robotic radical prostatectomy: assessment after 2766 procedures. Cancer **110**, 1951–1958 (2007)
27. Grossfeld, G.D., Chang, J. J., Broering, J. M., et al.: Impact of positive surgical margins on prostate cancer recurrence and the use of secondary cancer treatment: data from the CaPSURE database. J. Urol. **163**, 1171–1177 (2000); quiz 1295
28. Ficarra, V., Cavalleri, S., Novara, G., et al.: Evidence from robot-assisted laparoscopic radical prostatectomy: a systematic review. Eur. Urol. **51**, 45–55 (2007)
29. Ahlering, T.E., Woo, D., Eichel, L., et al.: Robot-assisted versus open radical prostatectomy: a comparison of one surgeon's outcomes. Urology **63**, 819–822 (2004)

30. Touijer, K., Kuroiwa, K., Eastham, J. A., et al.: Risk-adjusted analysis of positive surgical margins following laparoscopic and retropubic radical prostatectomy. Eur. Urol. **52**, 1090–1096 (2007)
31. Tewari, A., Srivasatava, A., Menon, M.: A prospective comparison of radical retropubic and robot-assisted prostatectomy: experience in one institution. BJU Int. **92**, 205–210 (2003)
32. Berryhill, R., Jr., Jhaveri, J., Yadav, R., et al.: Robotic prostatectomy: a review of outcomes compared with laparoscopic and open approaches. Urology **72**, 15–23 (2008)
33. Ficarra, V., Novara, G., Artibani, W., et al.: Retropubic, laparoscopic, and robot-assisted radical prostatectomy: a systematic review and cumulative analysis of comparative studies. Eur. Urol. (2009)
34. Miller, J., Smith, A., Kouba, E., et al.: Prospective evaluation of short-term impact and recovery of health related quality of life in men undergoing robotic assisted laparoscopic radical prostatectomy versus open radical prostatectomy. J. Urol. **178**, 854–858 (2007); discussion 859
35. Herrell, S.D., Smith, J. A., Jr.: Robotic-assisted laparoscopic prostatectomy: what is the learning curve? Urology **66**, 105–107 (2005)
36. Nelson, B., Kaufman, M., Broughton, G., et al.: Comparison of length of hospital stay between radical retropubic prostatectomy and robotic assisted laparoscopic prostatectomy. J. Urol. **177**, 929–31 (2007)
37. Schroeck, F.R., Krupski, T. L., Sun, L., et al.: Satisfaction and regret after open retropubic or robot-assisted laparoscopic radical prostatectomy. Eur. Urol. **54**, 785–93 (2008)
38. Konety, B.R., Joslyn, S. A., O'Donnell, M. A.: Extent of pelvic lymphadenectomy and its impact on outcome in patients diagnosed with bladder cancer: analysis of data from the surveillance, epidemiology and end results program data base. J. Urol. **169**, 946–950 (2003)
39. Snyder, C., Harlan, L., Knopf, K., et al.: Patterns of care for the treatment of bladder cancer. J. Urol. **169**, 1697–1701 (2003)
40. Stein, J.P., Lieskovsky, G., Cote, R., et al.: Radical cystectomy in the treatment of invasive bladder cancer: long-term results in 1,054 patients. J. Clin. Oncol. **19**, 666–675 (2001)
41. Parra, R.O., Andrus, C. H., Jones, J. P., et al.: Laparoscopic cystectomy: initial report on a new treatment for the retained bladder. J. Urol. **148**, 1140–1144 (1992)
42. Hubert, J., Chammas, M., Larre, S., et al.: Initial experience with successful totally robotic laparoscopic cystoprostatectomy and ileal conduit construction in tetraplegic patients: report of two cases. J. Endourol. **20**, 139–143 (2006)
43. Balaji, K.C., Yohannes, P., McBride, C. L., et al.: Feasibility of robot-assisted totally intracorporeal laparoscopic ileal conduit urinary diversion: initial results of a single institutional pilot study. Urology **63**, 51–55 (2004)
44. Sala, L.G., Matsunaga, G. S., Corica, F. A., et al.: Robot-assisted laparoscopic radical cystoprostatectomy and totally intracorporeal ileal neobladder. J. Endourol. **20**, 233–235 (2006); discussion 236
45. Menon, M., Hemal, A. K., Tewari, A., et al.: Nerve-sparing robot-assisted radical cystoprostatectomy and urinary diversion. BJU Int. **92**, 232–236 (2003)
46. Guru, K.A., Kim, H. L., Piacente, P. M., et al.: Robot-assisted radical cystectomy and pelvic lymph node dissection: initial experience at Roswell Park Cancer Institute. Urology **69**, 469–474 (2007)
47. Rhee, J.J., Lebeau, S., Smolkin, M., et al.: Radical cystectomy with ileal conduit diversion: early prospective evaluation of the impact of robotic assistance. BJU Int. **98**, 1059–1063 (2006)
48. Pruthi, R.S., Smith, A., Wallen, E. M.: Evaluating the learning curve for robot-assisted laparoscopic radical cystectomy. J. Endourol. **22**, 2469–2474 (2008)
49. Ng, C.K., Kauffman, E. C., Lee, M. M., et al.: A comparison of postoperative complications in open versus robotic cystectomy. Eur. Urol. (2009)

50. Pruthi, R.S., Wallen, E. M.: Is robotic radical cystectomy an appropriate treatment for bladder cancer? Short-term oncologic and clinical follow-up in 50 consecutive patients. Urology **72**, 617–620 (2008); discussion 620–622
51. Nguyen, M.M., Gill, I. S., Ellison, L. M.: The evolving presentation of renal carcinoma in the United States: trends from the surveillance, epidemiology, and end results program. J. Urol. **176**, 2397–2400 (2006); discussion 2400
52. Lane, B.R., Gill, I. S.: 5-Year outcomes of laparoscopic partial nephrectomy. J. Urol. **177**, 70–74 (2007); discussion 74
53. Phillips, C.K., Taneja, S. S., Stifelman, M. D.: Robot-assisted laparoscopic partial nephrectomy: the NYU technique. J. Endourol. **19**, 441–445 (2005); discussion 445
54. Gettman, M.T., Blute, M. L., Chow, G. K., et al.: Robotic-assisted laparoscopic partial nephrectomy: technique and initial clinical experience with DaVinci robotic system. Urology **64**, 914–918 (2004)
55. Kaul, S., Laungani, R., Sarle, R., et al.: da Vinci-assisted robotic partial nephrectomy: technique and results at a mean of 15 months of follow-up. Eur. Urol. **51**, 186–191 (2007); discussion 191–192
56. Rogers, C.G., Singh, A., Blatt, A. M., et al.: Robotic partial nephrectomy for complex renal tumors: surgical technique. Eur. Urol. **53**, 514–521 (2008)
57. Caruso, R.P., Phillips, C. K., Kau, E., et al.: Robot assisted laparoscopic partial nephrectomy: initial experience. J. Urol. **176**, 36–39 (2006)
58. Deane, L.A., Lee, H. J., Box, G. N., et al.: Robotic versus standard laparoscopic partial/wedge nephrectomy: a comparison of intraoperative and perioperative results from a single institution. J. Endourol. **22**, 947–952 (2008)
59. Lee, R.S., Sethi, A. S., Passerotti, C. C., et al.: Robot assisted laparoscopic partial nephrectomy: a viable and safe option in children. J. Urol. **181**, 823–828 (2009); discussion 828–829
60. Benway, B.M., Bhayani, S. B., Rogers, C. G., et al.: Robot assisted partial nephrectomy versus laparoscopic partial nephrectomy for renal tumors: a multi-institutional analysis of perioperative outcomes. J. Urol. (2009)
61. Rogers, C.G., Metwalli, A., Blatt, A. M., et al.: Robotic partial nephrectomy for renal hilar tumors: a multi-institutional analysis. J. Urol. **180**, 2353–2356 (2008); discussion 2356
62. Gautam, G., Benway, B. M., Bhayani, S. B., et al.: Robot-assisted partial nephrectomy: current perspectives and future prospects. Urology (2009)
63. Carr, M.C., El-Ghoneimi, A.: Anomalies and Surgery of the Ureteropelvic Junction in Children. Campbell-Walsh Urology, 9th edn., pp. 3359–3382. Saunders Elsevier, Philadelphia, PA (2007)
64. Peters, C.A.: Robotic assisted surgery in pediatric urology. Pediatr. Endosurg. Innov. Tech. **7**, 403–413 (2003)
65. Thakre, A.A., Bailly, Y., Sun, L.W., Van Meer, F. Yeung, C.K.: Is smaller workspace a limitation for robot performance in laparoscopy? J. Urol. **179**(3), 1138–1142 (2008)
66. Anderson, J.C., Hynes, W.: Retrocaval ureter. A case diagnosed preoperatively and treated successfully by a plastic operation. *Br. J. Urol.* **21**, 209–212 (1949)
67. Hotaling, J.M., Shear, S., Lendvay, T. S.: 14-Gauge angiocatheter: the assist port. J. Laparoendosc. Adv. Tech. **19**(5) (2009)
68. Ginger, V., Lendvay, T. S.: Intraoperative ultrasound: application in pediatric pyeloplasty. Urology **73**(2), 377–379 (2008)
69. Kim, S., Canter, D., Leone, N., Patel, R., Casale, P.: A comparative study between laparoscopic and robotically assisted pyeloplasty in the pediatric population. In: Abstract #1037, American Urological Association Meeting, Anaheim, CA, 17–22 May 2007
70. Lee, R.S., Retik, A. B., Borer, J. G., Peters, C. A.: Pediatric robot assisted laparoscopic dismembered pyeloplasty: comparison with a cohort of open surgery. J. Urol. **175**(2), 683–687 (2006)
71. Atala, A., Kavoussi, L.R., Goldstein, D.S., Retik, A.B., Peters, C.A.: Laparoscopic correction of vesicoureteral reflux. J. Urol. **150**(*2* Pt 2), 748–751 (1993)

72. Ehrlich, R.M., Gershman, A., Fuchs, G.: Laparoscopic vesicoureteroplasty in children: initial case reports. Urology **43**(2), 255–261 (1994)
73. Peters, C.A.: Robotically assisted surgery in pediatric urology. Urol. Clin. N. Am. **31**(4), 743–752 (2004)
74. Lich, R., Jr., Howerton, L.W., Davis, L.A.: Childhood urosepsis. J. Kentucky Med. Assoc. **59**, 1177 (1961)
75. Lendvay, T.: Robotic-assisted laparoscopic management of vesicoureteral reflux. Adv. Urol. 732942 (2008) (online)
76. Casale, P., Patel, R. P., Kolon, T.F.: Nerve sparing robotic extravesical ureteral reimplantation. J. Urol. **179**(5), 1987–1989 (2008)
77. Blinman, T.: Trocar incision tensions do not sum. In: International Pediatric Endosurgical Group Abstract S009, Buenos Aires, Sept 6–7, 2007.
78. Yeung, C.K., Sihoe, J.D.Y., Borzi, P.A., Endoscopic cross-trigonal ureteral reimplantation under carbon dioxide bladder insufflation: a novel technique. J. Endourol. **19**, 295–299 (2005)
79. Lendvay, T., Casale, P., Sweet, R., Peters, C.: Initial validation of a virtual-reality robotic simulator. J. Robot. Surg. **2**, 145–149 (2008).

Chapter 29
Applications of Surgical Robotics in Cardiac Surgery

E.J. Lehr, E. Rodriguez, and W. Rodolph Chitwood

Abstract Minimally invasive surgery has revolutionized many fields of surgery over the last two decades. Robotic assisted surgery is the latest iteration towards less invasive techniques. Cardiac surgeons have slowly adapted minimally invasive and robotic techniques into their armamentarium. In particular, minimally invasive mitral valve surgery has evolved over the last decade and become the preferred method of mitral valve repair and replacement at certain specialized centers worldwide because of excellent results. We have developed a robotic mitral valve surgery program which utilizes the da Vinci® telemanipulation system allowing the surgeon to perform complex mitral valve repairs through 5 mm port sites rather than a traditional median sternotomy. In this rapidly evolving field, we review the evolution and clinical results of robotically-assisted mitral valve surgery and review other cardiac surgical procedures for which da Vinci® is currently being used.

Keywords Surgical Procedures · Minimally Invasive · Thoracic Surgery · Video-assisted · Robotics · Telemedicine/Instrumentation · Mitral Valve Insufficiency · TECAB

29.1 Introduction

Improvements in endoscopic technology and techniques during the past decade have resulted in a substantial increase in the number of minimally invasive surgical procedures being performed in specialties such as general surgery and urology.

W.R. Chitwood (✉)
East Carolina Heart Institute, Department of Cardiovascular Sciences,
East Carolina University, Greenville, NC 27834, USA
e-mail: chitwoodw@ecu.edu

J. Rosen et al. (eds.), *Surgical Robotics: Systems Applications and Visions*,
DOI 10.1007/978-1-4419-1126-1_29,

Cardiac surgeons were initially reluctant to adopt these new techniques as they limited access to and exposure of the heart during complex and difficult cardiovascular procedures but, in the mid-1990s, they began to recognize the significant advantages of minimizing surgical trauma by reducing incision size.

Minimally invasive cardiac surgery has evolved through graded levels of difficulty each with progressively less exposure and an increasing reliance on video assistance. At East Carolina University, we have moved serially from using direct vision limited incisions for mitral operations (Level I) to assistant-held video-assistance (Level II), to video-directed voice-activated robotic techniques (Level III) and, finally, to robotic mitral valve surgery using the da Vinci® system (Intuitive Surgical Inc., Mountain View, CA)(Level IV). Standard endoscopic instrumentation was used for levels I to III but with only 4 degrees of freedom, dexterity was limited. Additionally, when working through a fixed entry point the surgeon had to reverse hand movements (fulcrum effect) and higher operator handle strain due to instrument-shaft shear forces led to muscle fatigue. These limitations resulted in deteriorating motor skills and disconnection between optimal visual-motor synchrony, commonly associated with endoscopic surgery. Computer-assisted robotic telemanipulation systems have provided a solution to these constraints.

Da Vinci® is the most widely used system in cardiac surgery and consists of three components: a surgeon console, an instrument cart and a visioning platform. The operative console allows the surgeon to immerse himself in a 3-dimensional high-definition videoscopic image. Finger and wrist movements are registered through sensors and then translated into scaled tremor-free movements with 7 degrees of freedom. Wrist-like articulations at the end of microinstruments bring the pivoting action of the instrument to the plane of the operative field improving dexterity in tight spaces and allowing ambidextrous dissection and suture placement. Telesurgery systems have facilitated totally endoscopic robotic cardiac surgery. In this article we review the development and current state of robotic cardiac surgery.

29.2 History

Our surgical forefathers set very high standards by performing magnificent cardiac operations with excellent long-lasting results. Contemporary cardiac surgeons attempt to uphold those high standards while simultaneously trying to adapt to modern pressures in an ever-evolving field. Demand for reduced complication rates, accelerated recovery process, and improved patient satisfaction as well as competition from competing minimally invasive procedures have been the driving forces requiring cardiac surgeons to reproduce the same procedures and results previously accomplished using less invasive techniques. Demand for minimally invasive surgery first influenced other surgical specialties and now has spread to cardiac surgery. Every surgical field has in one way or another introduced minimally invasive techniques, the natural progression of which is robotic surgery.

Cohn, Cosgrove, and Mohr introduced minimally invasive techniques to cardiac surgery in the mid-1990s by modifying cardiopulmonary bypass (CPB) methods and reducing incision sizes to enable safe and effective minimally invasive valve surgery [1–3]. Concurrently, Port-access™ methods, using either an endoaortic balloon occluder or a transthoracic aortic cross clamp, were developed and have rapidly become popular [4, 5]. Assisted vision and advanced instrumentation were paramount advancements for performing cardiac valve operations in restricted spaces through tiny incisions. Minimally invasive techniques began as operations performed under direct vision, through a limited sternotomy or thoracotomy incision [1–3, 6–15]. The evolvement of these techniques to either total or partial videoscopic guidance, provided superior visualization and subsequently allowed access incisions to be reduced in size [16–22]. Long shafted-instruments made it possible to perform procedures at a distance; however, many maneuvers remained ergonomically encumbered by innate tremor, inverted hand motion, and limited access. The da Vinci® Surgical System (Intuitive Surgical, Inc., Sunnyvale, CA) was first applied to cardiac operations in 1998 and has overcome many of the limitations by incorporating true three dimensional (3D) visualization with articulating wrist instrumentation that provides 7 degrees of freedom at the instrument tips. Natural wrist and finger movements are emulated accurately, making delicate intracorporeal surgery possible in the most confined spaces.

Carpentier and colleagues performed the first truly endoscopic mitral valve repair using a prototype of the da Vinci® Surgical System articulated "wrist" device in May 1998 [16]. A week later, Mohr performed the first coronary anastomosis and repaired five mitral valves with the device [23]. Grossi et al. of New York University partially repaired a mitral valve using the Zeus™ system (Computer Motion Inc.) but no annuloplasty ring was inserted. Four days later, in May 2000, our group performed the first complete da Vinci® mitral repair in North America [24]. Lange and colleagues in Munich were the first to perform a totally endoscopic mitral valve repair using only 1 cm ports and the da Vinci® System [25].

Subsequently we performed 20 other mitral repairs as part of a Phase 1 Food and Drug Administration (FDA) trial designed to determine the safety and efficacy of da Vinci®[26]. These initial results were encouraging and prompted a phase II multicenter FDA trial which was completed in 2002 [27]. A total of 112 patients were enrolled at 10 different institutions and all types of repair were performed. At 1 month after surgery, 92% had either no or grade 1 mitral regurgitation, 8% had grade 2 or higher and 5% subsequently required reoperation. Although the reoperative number was high for the number of patients, failures were distributed evenly among centers with some centers having performed fewer than 10 procedures. There were no deaths, strokes or device-related complications. These results prompted FDA approval of the da Vinci® system for mitral valve surgery in November 2002. Based on the success of robotic mitral surgery, surgical telemanipulation has expanded to other cardiac procedures, including coronary revascularization, arrhythmia operations, left ventricular lead placements, congenital heart surgery, and aortic valve replacements.

29.3 Anatomical and Physiological Considerations

The heart lies centrally within the mediastinum, protected by the fibrous pericardium and a rigid superstructure that includes the sternum, ribs, and vertebral column. Standard surgical access to the heart has been through a median sternotomy providing safe access to all cardiac chambers, the great vessels, and the epicardial coronary arteries. However, this approach can be associated with increased pain, risk of wound infection, and/or mediastinitis, and requires prolonged time for complete healing and patient mobility. Moreover, in women, the sternotomy usually does not provide a satisfactory cosmetic result. The advent of coronary artery bypass (CABG) surgery in the late 1960s secured the median sternotomy incision as the preferred exposure of the heart. Interestingly, in the late 1940s, cardiac surgeons used either a large left or right thoracotomy to perform closed mitral commisurotomies and atrial septal defect repairs. This method of exposure continues to be used for reoperations in many centers. Contemporary, minimally invasive, cardiac incisions try to minimize collateral injury by exposing only that which is necessary to perform the operation. The left atrium and mitral valve are located posteriorly and are easily exposed through a right lateral fourth or fifth intercostal space thoracotomy. The right atrium and tricuspid valve can also be exposed easily through this same incision. An alternative approach to the mitral and tricuspid valves is through a lower midline hemisternotomy. Aortic valve and ascending aorta procedures can be approached through an upper hemisternotomy or right second intercostal space thoracic incision. For minimally invasive CABG operations, the left and right internal thoracic arteries (LIMA & RIMA) can be harvested from the left chest using either thorascopic or robotic techniques. Subsequently, the left anterior descending artery (LAD) is exposed via a fifth or sixth intercostal space, left anterior mini-thoracotomy (5 cm) and the LIMA–LAD anastomosis constructed under direct vision. The circumflex artery and its branches can be exposed through a more lateral left thoracotomy using cardiac positioning devices to mobilize coronary artery graft target sites into the incision. Total endoscopic techniques using Da Vinci® further minimize the required incisions for CABG.

Minimally invasive cardiac surgery is performed utilizing the same general principles as for open cardiac surgery. These parameters include myocardial protection, cardiac drainage, maintenance of systemic perfusion pressure, and physiologic maintenance of myocardial and corporeal oxygen supply-demand ratios. Cardiopulmonary bypass is used for all intracardiac procedures, including valvular repair and reconstruction; however, "off-pump" techniques may be employed for CABG operations. Cardiopulmonary bypass is most commonly initiated by peripheral arterial and venous cannulation. However, transthoracic aortic cannulation can be performed, especially in patients with peripheral atherosclerosis. Aortic occlusion techniques, including the transthoracic aortic cross clamp and endoaortic balloon occlusion, are used to induce cardiac arrest. Hypothermic ventricular fibrillation can be used when aortic access is limited, such as in reoperations.

29.4 Anesthetic Considerations

While the underlying anesthetic principles for open cardiac surgery procedures also apply to robotic cardiac surgery cases, some modifications are key to ensuring the success and safety of robotic cases. It is important to consider that most robotic patients can be fast-tracked when planning the anesthetic technique.

Large bore intravenous access is required and usually accomplished by peripheral intravenous catheters and a right internal jugular venous cordis. Often a 17 French right internal jugular venous cannula is used to augment venous drainage. A Swan-Ganz catheter may be required for monitoring. In certain cases, and in particular, when the PORTACCESS platform is employed, a pulmonary artery vent and coronary sinus cardioplegia catheter are also secured by the anesthesiologist via the right internal jugular vein that may prevent the use of a right internal jugular venous cannula. If an endoaortic occlusion device is to be used, bilateral radial arterial catheters are required to monitor correct balloon placement. We also utilize bispectral analysis to monitor depth of anesthesia to assist in fast-tracking patients, aiming for a target of less than 60. Near infrared spectral analysis aids in ensuring adequate brain protection with alternative cannulation strageties. Defibrillator pads are required as internal paddles cannot be applied through limited incisions.

Single lung ventilation is usually preferred for robotic cardiac surgery and may be achieved by either a double lumen tube or occasionally a single lumen tube with a bronchial blocker. Single lung ventilation may reduce cardiac output, and result in hypoxia and hypercapnia thereby increasing pulmonary artery resistance. During robotic IMA harvesting, carbon dioxide insufflation is required which may exacerbate hypercapnia and impede venous return and cardiac filling. The double lumen tube may be exchanged over a tube changer to a single lumen tube prior to transfer to the intensive care unit. If a patient does not tolerate single lung ventilation, CPB may be established prior to entering the chest. Echocardiography has become mandatory for mitral valve repair and forms the roadmap for advanced repair techniques.

29.5 Vascular Access and Perfusion Techniques

29.5.1 Clamp

We expose the anterior surface of the femoral vessels through a 2 cm horizontal groin incision. To facilitate closure of the arteriotomy and venotomy without stenosis, 4-0 polypropylene longitudinal pursestring sutures are secured prior to cannulation. Without encircling the vessles, we routinely cannulate the right femoral artery (17F to 19F) and vein (21F) using Bio-Medicus thin-wall cannulas using a modified Seldinger guide wire technique under transesophageal echocardiography

(TEE) guidance, A long antegrade cardioplegia/aortic root vent needle (Medtronic, Minneapolis, MN) is placed in the ascending aorta after CPB has been established and is secured by a 2-0 Goretex pursestring suture. Alternatively, it can be placed through the lateral chest wall via a 12 mm port to provide a less crowded working area. Kinetic assisted venous drainage is used in every case. To assist with deairing, the surgical field is flooded with CO_2 via a 14F angiocatheter, introduced through the right chest wall. The transthoracic Chitwood clamp (Scanlan International, Minneapolis, MN) is used to directly clamp the aorta and is applied across the aorta in the transverse sinus, taking care not to injure the pulmonary artery or the left atrial appendage [28]. Visualization (direct or videoscopic) of the transthoracic clamp tip and position is imperative (Fig. 29.1). Generally, we place the clamp tip pointing cephalad to provide better aortic occlusion and avoid left atrial appendage injury. The clamping method has achieved standardization in most minimally invasive cardiac surgery centers worldwide.

29.5.2 Endoballoon

Alternatively, some surgeons prefer the Port-Access endovascular balloon occlusion system. This platform includes a pulmonary artery vent and retrograde coronary sinus cardioplegia cannula as well as an endoaortic balloon occlusion catheter with an antegrade cardioplegia perfusion port. The endoballoon is advanced through the side arm of a 21 F or 23 F femoral arterial perfusion catheter. Under TEE and/or fluoroscopic guidance, the balloon occlusion device is advanced to 1 cm above the sinotubular junction. Balloon pressure is gradually increased to between 250 and 340 mmHg to occlude the aorta and the heart is arrested with antegrade cardioplegia. Delivery pressure of antegrade cardioplegia must not exceed the aortic perfusion pressure so as not to displace the balloon clamp. Proper placement of the endoballoon is assessed throughout the case by monitoring bilateral radial artery pressures. Reichenspurner et al. demonstrated an increase in morbidity and cost, as well as total operative and cross clamp times when the balloon technique was used compared with the transthoracic clamp [29].

Vascular injuries from femoral cannulation include arterial occlusions, localized arterial injuries, and aortic dissections. Major aortic dissection is rare but devastating, and occurs in 1 to 2% of patients [27, 30–32]. To avoid these complications, preoperative screening for peripheral vascular disease may include, in addition to a history and physical examination, noninvasive plethsmography, computed tomography, or selective angiography. Intraoperative arterial inspection and palpation give one a good idea of the relative arterial condition. Both venous and arterial cannula deployments should be done using the Seldinger guide wire technique. Guide wire position must be confirmed by TEE prior to dilator or cannula insertion. Any resistance encountered during cannulation should prompt selection of an alternative site. In this circumstance we cannulate the contralateral femoral artery, the aortic arch directly through the chest wall, or the axillary artery. Mobile atheromatous plaque in the descending aorta and

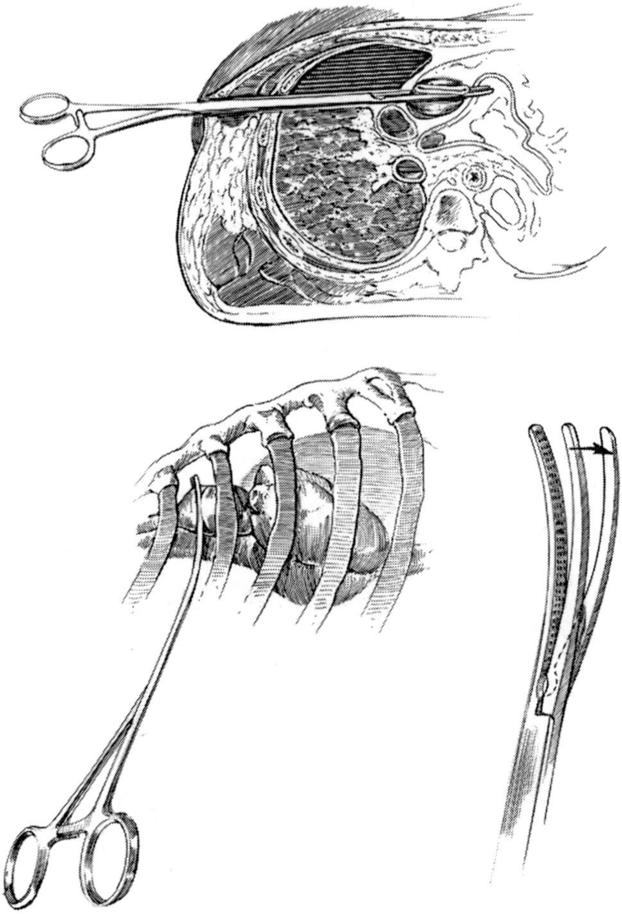

Fig. 29.1 Application of the Chitwood aortic crossclamp. A cross clamp is applied across the aorta in the transverse sinus through a separate stab incision in the third intercostal space in the posterior axillary line. The clamp is applied under visualization to avoid injury to the pulmonary artery and left atrial appendage

peripheral arterial disease are relative contraindications to peripheral cannulation. Retrograde arterial perfusion potentially increases the risk of emboli and strokes in the presence of detachable atheroma. Aortic dissection is more common with the endoaortic balloon technique [38]. We have had no aortic injuries from the transthoracic clamp in over 1,000 minimally invasive patients. In the event of severe mobile atheroma being noticed in the descending or abdominal aorta or in the presence on an abdominal aortic aneurysm, antegrade perfusion techniques should be considered. In redo cardiac surgery or in cases when the ascending aorta is unsuitable for clamping, fibrillatory arrest is an effective myocardial protective strategy to facilitate robotic cardiac surgery.

29.6 Robotic Setup and Surgical Technique

29.6.1 Mitral Valve and Atrial Fibrillation Surgery

Our standard setup for mitral valve, tricuspid valve, maze and other atrial procedures is shown in Fig. 29.2. A double-lumen endotracheal intubation is preferred to allow deflation of the right lung. After placement of a Swan Ganz catheter, a 17F Bio-Medicus venous return cannula (Medtronic Bio-Medicus, Eden Prairie, MN) is introduced via the right internal jugular vein using the Seldinger technique. Transesophageal echocardiography is used to ensure proper positioning of the venous cannula at the superior vena caval–right atrial junction. Patients are placed in a slight left lateral decubitus position (30°). Positioning the right arm along the side with gentle elbow flexion has reduced conflict with robotic arms and decreased risks of a brachial plexopathy. A 3 to 4 cm access incision is made in the right inframammary crease, and the chest is entered through the fourth intercostal space. Rib spreading is minimized by deploying a soft tissue retractor (Cardiovations Ethicon Inc., Somerville, NJ). The pericardium is opened 3 to 4 cm anterior to

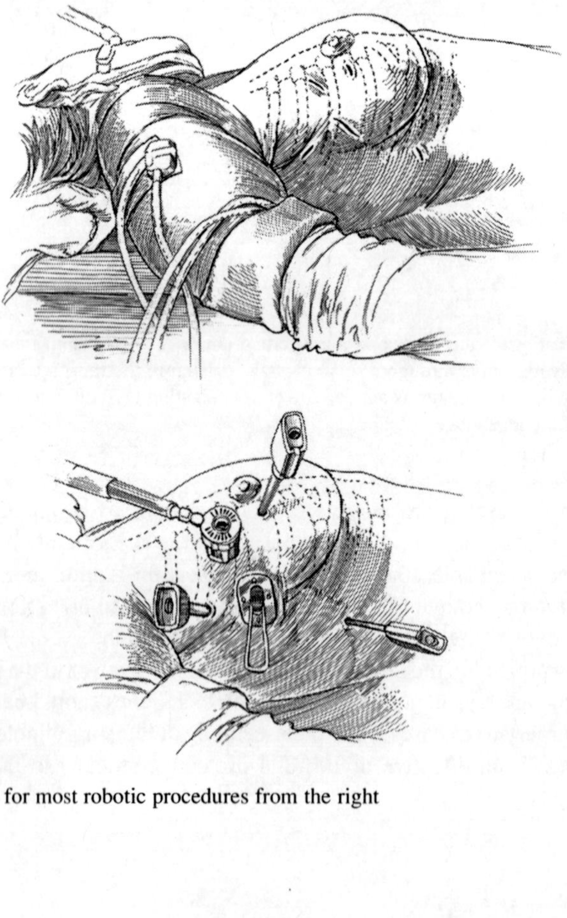

Fig. 29.2 Setup for most robotic procedures from the right

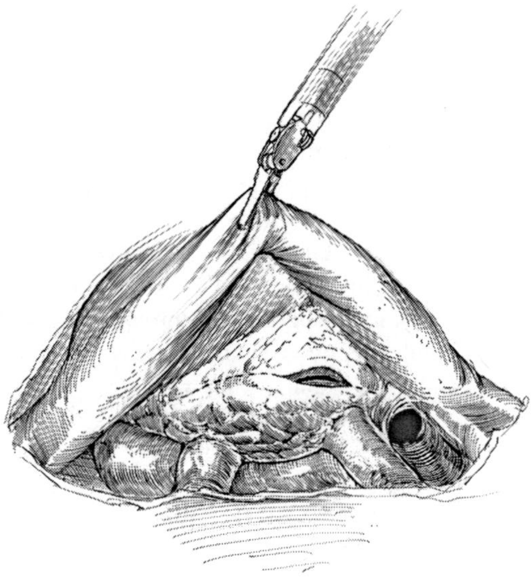

Fig. 29.3 Access to the mitral valve, left atrial appendage and pulmonary veins is achieved by an incision in Sondergaard's groove

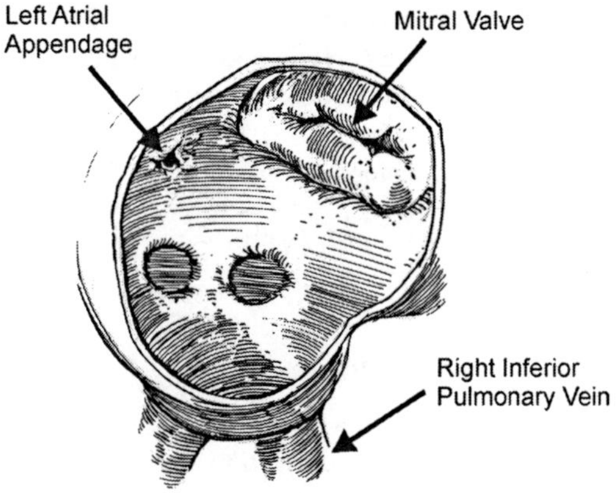

Fig. 29.4 Excellent visualization of all left atrial structures is provided by the right-sided robotic approach

the phrenic nerve to prevent transient phrenic nerve traction injuries. After establishing CPB and cardiac arrest, a left atriotomy is made in Sondergaard's groove (Fig. 29.3).

The four-armed da Vinci® S™ and Si™ Systems, enable deployment of the robotic EndoWrist® left atrial retractor, allowing excellent mitral valve visualization and enhancing exposure of the anterior annulus and both trigones (Fig. 29.4). Releasing the left atrial retractor minimizes displacement of the aortic root which limits aortic insufficiency during administration of cardioplegia and entrainment of aortic root air. Generally, we use conventional mitral valve repair techniques, including posterior leaflet resections, sliding plasties, chordal transfers, and neochord placement. However, our focus has been on limited resections and leaflet conservation. Most recently we are using methods that reduce posterior leaflet width and still provide leaflet-annular continuity. In these circumstances, less annular compression is required and sliding-plasties are performed less frequently. Robotic visualization is particularly advantageous when working deep within the left ventricle as during neochordal replacements or papillary muscle reconstructions. Repairs are supported with an annuloplasty band or ring. We prefer to secure the Cosgrove annuloplasty bands (Edwards Life Science, Irvine, CA) with interrupted mattress 2-0 Cardioflon sutures (Peters, Inc., Paris). ATS annulplasty bands are secure with running 2-0 Cardioflon suture (Peters, Inc., Paris). Each part of the mitral operation is performed through robotic telemanipulation, including leaflet and chordal resections, suturing, and knot tying. The left atriotomy is closed from the operating console with a running 3-0 Gore-Tex® suture. A temporary right ventricular pacing wire should be placed robotically before the transthoracic aortic cross clamp is removed. At this point, the robotic arms are removed, the da Vinci® system secured, and the patient weaned from CPB. After hemodynamic stabilization, TEE is performed to assess repair quality and left ventricular function. Reoperation for bleeding is infrequent, but generally is related to transthoracic instrumentation and usually arises from the chest wall. Videoscopic inspection of the chest wall with a 30 degree endoscope prior to closure is critical. Occasionally, a dental mirror is useful to inspect some areas of the chest wall close to the incision.

Full Cox-Maze III right and left lesion sets can be made with robotic assistance that significantly enhances visualization of the lesions to minimize the risk of gaps which may result in failures. Generally, we perform the right sided lesions on CPB. The left sided lesions are performed after opening the left atrium. We previously published our cryolesion set [33].

29.6.2 Aortic Valve Surgery

Experience with robotic aortic valve surgery is limited to a number of case reports and still requires significant portions of the procedure to be performed without robotic assistance. After instituting CPB as described above, a limited anterior thoracotomy is made in the third or fourth interspace as determined by preoperative gated CT scan. The 30° endoscope is inserted through the thoracotomy and two additional ports for the robot arms are established in the second and either the fourth or fifth intercostals space in the mid axillary line. After applying the transthoracic aortic crossclamp, the heart is arrested with antegrade cardioplegia and additional myocardial protection is achieved by circulating cold saline in the pericardium. An aortotomy is made and the aorta is retracted with commissural stitches. The valve is excised and 2/0 pledgeted polyester sutures are placed around the annulus using the da Vinci system. Sutures are brought through the sewing cuff of the valve which is brought into position after irrigating the aortic root. Sutures are tied using a knot pusher and the aortomy is closed after deairing the heart [34].

29.6.3 Coronary Revascularization

Robotic coronary surgery exists as a spectrum from robotic left internal mammary artery (LIMA) harvesting followed by full sternotomy and standard on-pump CABG techniques to robot-assisted total endoscopic, off pump multivessel revascularization. Patient setup varies according to the exact procedure and vessels that require bypass. Herein we describe the setup for arrested heart robotic assisted total endoscopic LIMA to left anterior descending artery (LAD) grafting.

Following the establishment of general anesthesia with double lumen endotracheal intubation, central venous access, bilateral radial arterial lines, defibrillator pads and TEE probe, the patient is placed in a 30 degree right lateral decubitus position. Dissection is carried out in the left groin to expose the femoral vessels. After administering 10,000 units of heparin, the femoral artery and vein are cannulated and the endoballoon is positioned under echocardiographic guidance in the ascending aorta. With the da Vinci® patient cart positioned on the patient's right side, the camera port in placed in the fourth or fifth left intercostal space in the anterior axillary line and carbon dioxide is insufllated to approximately 10 mmHg. Under visualization with the camera, ports for the left and right robot arms are inserted two interspaces caudal and cranial to the camera port respectively, about 2 cm medial to the anterior axillary line and the robot is deployed (Fig. 29.5).

Fig. 29.5 Port placement for robotic total endoscopic coronary artery bypass grafting. Left (sixth intercostal space) and right (second intercostal space) arm ports are on either side of and slightly anterior to the camera port (fourth intercostal space). The stabilizer port is placed subcostal in the left mid-axillary line. An assistance port in the fourth intercostal space just lateral to the sternum facilitates bringing materials in and out of the chest. For beating heart , we routinely cannulate the left axillary artery with an 8 mm Dacron tube graft with percutaneous femoral venous cannulation in the event that cardiopulmonary bypass assistance is required during the procedure

LIMA harvesting is performed using a skeletonized technique to provide enhanced visualization of the artery, to increase length of the graft and to facilitate multi-arterial Y grafting and flow measurements. After identifying the LIMA, the endothoracic fascia is removed along its entire length. Starting in the midpoint of the artery, the artery is teased from the chest wall using a combination of cautery and clips to occlude the side branches. Dissection is carried out first proximally and then distally. Following systemic heparinization (300 units/kg) a bulldog is placed proximally on the graft and two clips are applied distally. The graft is spatulated and flow assessed prior to fully dividing the artery which is then clipped to the chest wall in preparation for the anastomosis.

Excision of the pericardial fat pad and incision of the pericardium follows. Careful inspection of the epicardial coronary arteries is undertaken to identify the LAD. Two additional ports are then placed, one in the fourth left intercostals space for the assistant and on in the mid-clavicular line at the costal margin for the endostablizer which is then inserted and deployed. The endoballoon is then inflated monitoring bilateral radial artery pressures and the heart is arrested with antegrade blood cardioplegia.

Exposure of the LAD is achieved with the endostablizer and silastic tapes and the robotic I-knife is used to make an arteriotomy which is lengthened with Potts scissors. The anastomosis is completed with a 7 cm 7-0 Pronova (Ethicon) (Fig. 29.6) and hemostasis and patency is assessed using transit-time flow measurements.

The endoballoon is deflated and a stable rhythm is established. Cardiopulmonary bypass is weaned, protamine administered and decannulation ensues. Two chest tubes are secured through port sites and pacing wire are generally not utilized [35].

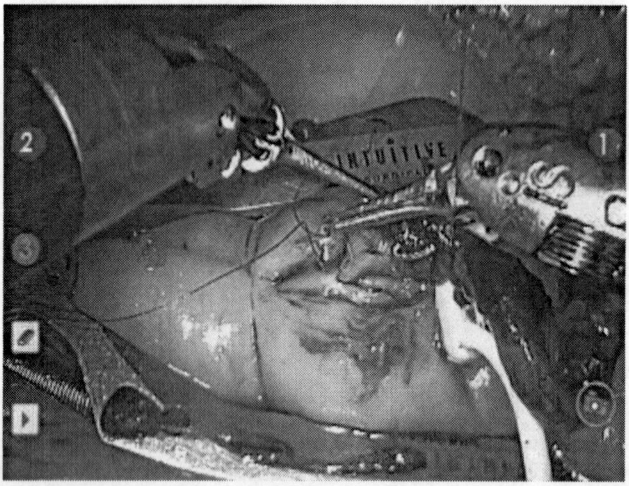

Fig. 29.6 Suturing internal mammary artery anastomoses to the coronary arteries is completed with 6-O Poly (hexafluoropropylene-VDF) in an end-to-side fashion. The back wall is completed counterclockwise sewing outside-in on the coronary artery around the heel of the anastomosis. Needles are exchanged and the toe of the anastomosis is sutured inside-out on the coronary artery

29.7 Surgical Results

29.7.1 Mitral Valve

29.7.1.1 East Carolina Heart Institute Experience

Our experience included 300 patients undergoing robotic mitral valve repair between May 2000 and November 2006 having echocardiographic and survival follow-up in 93 and 100% of patients, respectively [36]. There were 2 (0.7%) 30-day mortalities and 6 (2.0%) late mortalities. No sternotomy conversions or mitral valve replacements were required. Immediate post repair echocardiograms showed the following degrees of MR: none/trivial, 294 (98%); mild, 3 (1.0%); moderate, 3 (1.0%); and severe, 0 (0.0%). Complications included 2 (0.7%) strokes, 2 transient ischemic attacks, 3 (1.0%) myocardial infarctions, and 7 (2.3%) reoperations for bleeding. The mean hospital stay was 5.2 ± 4.2 (standard deviation) days. Sixteen (5.3%) patients required a reoperation. Echocardiographic follow-up demonstrated the following degrees of MR: none/trivial, 192 (68.8%); mild, 66 (23.6%); moderate, 15 (5.4%); and severe, 6 (2.2%).

29.7.1.2 Additional Experience

Other early robotic assisted mitral valve surgery reports included:

1. Tatooles et al. reported their experience with 25 patients and demonstrated excellent results with no mortality, device-related complications, strokes or reoperations for bleeding. One patient had a transient ischemic attack 7 days after surgery. The cardiopulmonary bypass and aortic crossclamp times were 126.6 ± 25.7 min and 87.7 ± 20.9 min, respectively. Eighty-four percent were extubated in the operating room, 8 were discharged home within 24 h and the mean hospital stay was 2.7 days. However, there was a 28% rate of readmission using this aggressive discharge policy and two patients required interval mitral valve replacement [34].
2. Jones et al. reported their series of 32 patients at a community hospital [6]. They performed concomitant procedures in five patients (tricuspid valve repair $n = 3$ and MAZE procedure $n = 2$). There were two deaths in this series where neither was reported as a device-related complication. Complications included reoperations for repair failure ($n = 3$), stroke ($n = 1$), groin lymphocoele ($n = 1$) and pulmonary embolism ($n = 1$) [37].
3. In a nonrandomized single surgeon experience from the University of Pennsylvania, Woo et al. demonstrated that robotic surgery patients had a significant reduction in blood transfusion and length of stay compared to sternotomy patients [38].

4. Folliguet et al. compared patients undergoing robotically-assisted mitral valve repair to a matched cohort undergoing sternotomy ($n = 25$ each). The robotic group had a shorter hospital stay (7 days *vs.* 9 days, $p = 0.05$), but there were no other differences between the two groups [39].

More recent and larger series include:

1. Murphy et al. reported their experience in 127 patients undergoing robotic MV surgery of which five were converted to median sternotomy and one to thoracotomy; seven patients had valve replacement and 114 had repair. There was 1 in-hospital death, 1 late death, 2 strokes and 22 patients developed new onset of atrial fibrillation. Blood product transfusion was required in 31% of patients and two (1.7%) patients required reoperation. Post-discharge echocardiograms were available in 98 patients at a mean follow-up of 8.4 months with no more than 1+ residual MR in 96.2%[40].

Taken together, these series validate the results of previous reports demonstrating that robotic mitral valve surgery is safe and has excellent early- and mid-term results. The introduction of newer robotic instrumentation such as the dynamic left atrial retractor and simpler mitral valve repair techniques including the "Haircut Posterior Leaflet-Plasty" [41] and the "American Correction" [42] will facilitate the use of robotic mitral valve techniques by a larger number of cardiac surgeons.

29.7.2 Atrial Fibrillation Surgery

There have been a few case reports of patients undergoing combined robotic mitral valve and atrial fibrillation (MV/AF) surgery demonstrating that these procedures are safe [43–46]. One small ($n = 16$) series of patients undergoing robotic MV/AF surgery using the Flex-10 microwave catheter (Guidant, Indianapolis, IN) from our own institution has been reported [47]. The ablative procedure added 42 ± 16 min to the mitral valve repair and 1.3 days to hospitalization. At 6 months follow-up, 73% were in sinus rhythm, 20% were paced and 7% were in atrial fibrillation.

Roberts et al. recently reported their robotic endoscopic Cox-Cryomaze technique. Using a warm beating heart strategy they are able to perform a full set of left atrial argon-based cryolesions and closure of the left atrial appendage [48].

29.7.3 Coronary Revascularization

The range of robot-assisted coronary operations ranges from IMA harvest with a hand-sewn anastomosis, performed either on or off-pump through a minithoracotomy or median sternotomy, to totally endoscopic coronary artery bypass grafting (TECAB). Early reports demonstrated the feasibility and safety of harvesting the IMA with the da Vinci® system with harvest times <0 min achievable once the learning curve had been negotiated [49–51].

In 1998 Loulmet et al. demonstrated the feasibility of TECAB on an arrested heart by using da Vinci® to harvest the LIMA and to perform a LIMA to LAD coronary anastomosis in two patients [52]. In 2000, Falk et al. reported TECAB on 22 patients of which 4 were converted to minithoracotomy for anastomotic bleeding or graft issues [53]. In the remaining 18 patients, grafts were widely patent at 3 months with no major complications. The same group subsequently reported the first off-pump TECAB using an endoscopic stabilizing device [54]. Dogan et al. reported 45 arrested heart TECAB procedures in 2002, of which eight patients underwent double vessel revascularization with both IMAs [55]. The initial conversion rate of 22% dropped to 5% in the last 20 patients. The procedural time for single-vessel TECAB was 4.2 $\pm$ 0.4 h, CPB time was 136 +/– 11 minutes and aortic crossclamp time was 61 $\pm$ 5 min.

Subramanian et al. achieved multivessel revascularization (mean number of grafts, 2.6) in 30 patients using robotically-harvested IMAs [56]. Depending on the specific target, either a minithoracotomy or transabdominal approach was employed. Twenty-nine (97%) patients were extubated on the operating table, 77% were discharged within 48 h and only two patients needed readmission. In addition, only one patient needed conversion to sternotomy and there was no mortality. However, the largest single institution series comes from Srivastava et al. with 150 patients undergoing robotic-assisted bilateral IMA harvesting and off-pump CABG via minithoracotomy [57]. Two patients presented with chest pain after discharge secondary to graft occlusion; in both cases, treatment using percutaneous intervention was successful. In 55 patients undergoing computed tomography angiography at 3 months, all 136 grafts were patent.

A multicentre Investigational Device Exemption trial was reported by Argenziano et al. in 2006 [58]. Ninety-eight patients requiring single-vessel LAD revascularization were enrolled at 12 centers; 13 patients (13%) were excluded intraoperatively (e.g., failed femoral cannulation, inadequate working space). In the remaining 85 patients who underwent TECAB, CPB time was 117 $\pm$ 44 min, aortic crossclamp time was 71 $\pm$ 26 min and hospital length of stay was 5.1 $\pm$ 3.4 days. There were five (6%) conversions to open techniques. There were no deaths or strokes, one early reintervention and one myocardial infarction. Three-month angiography was performed in 76 patients, revealing significant anastomotic stenoses (>50%) or occlusions in six patients. Overall freedom from reintervention or angiographic failure was 91% at 3 months. United States FDA approval of use of da Vinci® for coronary revascularization was largely based on this study.

Reports of robotically-assisted coronary surgery have mostly involved highly-selected patient populations which require limited revascularization, usually of the anterior wall [59]. In these circumstances, surgeons have been able to achieve totally endoscopic LIMA-LAD grafting with high success rates after the initial learning curve. Combining robotic TECAB of the LAD with stenting of a second coronary target, so-called *hybrid revascularization*, effectively combines a minimally invasive approach with the proven long-term benefits of LIMA-LAD grafting and is likely to increase particularly with advances in robotic instrumentation. Recent work by Katz et al. has demonstrated that this approach can be accomplished with no mortality, low perioperative morbidity and excellent 3-month angiographic LIMA patency (96.3%) [48]. More recently, these results have been corroborated by Gao et al. in their series of 42 patients undergoing hybrid revascularization [60]. Irrespective of the method, long-term follow-up of these grafts is needed to determine if they have the same excellent patency (>90% at 10 years) as those performed through a median sternotomy.

29.7.4 Left Ventricular Lead Placement

Numerous prospective studies have demonstrated that cardiac resynchronization therapy with or without implantable cardioverter-defibrillator capability improves ventricular function, exercise capacity and quality of life, as well as reducing mortality and heart failure hospitalizations in patients with symptomatic heart failure and delayed intraventricular conduction despite optimal medical therapy [61]. Left ventricular lead placement is usually accomplished percutaneously through coronary sinus cannulation, advancing the lead into a major cardiac vein. This technique is associated with long fluoroscopy times and is not applicable to all patients because of anatomical limitations in coronary venous anatomy. Early and late failures occur in approximately 12% and 10% of procedures, respectively [62]. Surgical epicardial lead placement is often a rescue therapy for these patients.

Early reports by DeRose et al. demonstrated the efficacy of robot-assisted left ventricular lead implantation [63]. They reported results for 13 patients, 6 of whom had previous CABG, with no complications or technical failures. Navia's series of minithoracotomy or robotic/endoscopic left ventricular lead placement included 41 patients without mortality, intraoperative complications or implantation failures [64]. A minimally invasive surgical approach is very attractive as it allows surgeons to determine the best epicardial site for implantation by mapped stimulation and may therefore entail greater success rates than transvenous implantation. A randomized study comparing both techniques is in progress.

29.7.5 Intracardiac Tumor Resection

Cardiac tumors, although relatively uncommon and mostly benign, almost always should be resected to prevent thromboembolic complications. Murphy et al. recently reported endoscopic excision of three left atrial myxomas using either a left atriotomy or right atriotomy with trans-septal approach. Autologous pericardial patches were used to repair septal defects following excision [65]. The mean CPB and aortic crossclamp times were 103 $\pm$ 40 min and 64 $\pm$ 2 min, respectively. Impressive results were reported with all patients being discharged on day 4 and resuming normal activity 3 weeks after surgery. Similarly, Woo et al. used robotic techniques to excise an aortic valve papillary fibroelastoma with the patient being discharged on the third postoperative day and back to work within 1 month [66].

29.7.6 Congenital Surgery

A few congenital cardiac conditions in both children and adults lend themselves to a minimally invasive approach. Del Nido's group from the Boston Children's Hospital published their 2-year experience with 15 patients undergoing patent ductus arteriosus (PDA) closure ($n = 9$) or vascular ring repair ($n = 6$) utilizing the da Vinci® system [67]. The patients were aged 3–18 years old and only one was converted to a thoracotomy because of pleural adhesions. The total operative times were a little prolonged at 170 $\pm$ 46 min (PDA) and 167 $\pm$ 48 min (vascular ring). Nevertheless, all were extubated in the operating room and were discharged after a median of 1.5 days. Bonaros et al. showed that the learning curve is steep and associated with a rapid decrease in operative times [68].

In a US FDA Investigational Device Exemption trial, Argenziano et al. demonstrated that atrial septal defects (ASDs) in adults can be closed safely and effectively using totally endoscopic robotic approaches with a median aortic crossclamp time of 32 min [69]. One of 17 patients had a residual shunt across the atrial septum which was repaired via minithoracotomy on post-operative day 5. The reoperative finding was that the atrial septal primary suture line was intact but there was a tear medial to it. This failure was therefore likely related to use of a direct closure technique rather than using a patch repair and therefore not a failure of the robotic technique per se. Morgan et al. subsequently demonstrated that robotic ASD closure hastens postoperative recovery and improves quality of life compared to either a mini-thoracotomy or median sternotomy approach [70].

29.8 Future of Robotic Cardiac Surgery

Robotic applications to cardiac surgery are gaining momentum. Mitral valve repair and cryomaze procedures are well established particularly at specialized centers. Results for these procedures match are equivalent to standard mitral valve repair and cryomaze techniques, but are associated with shorter recovery periods and enhanced patient satisfaction. As imaging modalities continue to improve the future of robotic mitral valve repair may incorporate 3D mathematical modeling to guide leaftlet resection and optimal annuloplasty ring placement.

Limited centers perform coronary revascularizaiton with robotic techniques. Robotic IMA harvesting and LIMA to LAD grafting can be performed safely. As experience grows we can expect that robotic TECAB will continue to become more widespread. Improvement of anastomotic connectors would made this this surgery more expeditious.

Robotic experience in aortic valve surgery is currently limited. A few case reports are described in the literature but with limited space around the aortic root, the procedures are technically challenging with current technology. The development of rapid deployment valves may make robotic aortic valve replacement more attractive.

Training is the connerstone of the successful future of robotic cardiac surgery. Learning curves for robotic procedures are steeper than for open techniques and consequently require dedication from surgeons and their teams. As the collective experience grows and the technology develops robotic procedures will become more efficient.

29.9 Conclusion

Although robotic cardiac surgery is in a state of evolution, the early results are encouraging with evidence demonstrating fewer blood transfusions, shorter hospital stay, faster return to preoperative function levels and improved quality of life compared to those having a sternotomy. This translates into improved utilization of limited healthcare resources. It is clear that the continued evolution of totally endoscopic cardiac surgery depends on the development of new adjunctive technology, such as retraction systems, perfusion catheters and sutureless anastomotic devices. Thus, although the surgical robot allows unprecedented closed chest surgical access to the heart, it is only one of many new tools that are prerequisite for successful minimally invasive cardiac surgery. It will require a combined effort of physicians with our industry partners to fill in these technological gaps that are present in our current armamentarium of minimally invasive tools. Surgical scientists must continue to critically evaluate this technology and despite enthusiasm, caution cannot be overemphasized. Traditional cardiac operations still enjoy proven long-term success and ever-decreasing morbidity and mortality and remain our measure for comparison. To determine if robotic techniques could become the new standard in cardiac surgery, long-term results are needed.

References

1. Cohn, L.H., Adams, D.H., Couper, G.S., Bichell, D.P.: Minimally invasive aortic valve replacement. Semin. Thorac. Cardiovasc. Surg. 9, 331-336 (1997)
2. Cosgrove, D.M., III, Sabik, J.F.: Minimally invasive approach for aortic valve operations. Ann. Thorac. Surg. 62, 596-597 (1996)

3. Cosgrove, D.M., III, Sabik, J.F., Navia, J.L.: Minimally invasive valve operations. Ann. Thorac. Surg. 65, 1535-1538 (1998)
4. Pompili, M.F., Stevens, J.H., Burdon, T.A., Siegel, L.C., Peters, W.S., Ribakove, G.H., Reitz, B.A.: Port-access mitral valve replacement in dogs. J. Thorac. Cardiovasc. Surg. 112, 1268-1274 (1996)
5. Falk, V., Walther, T., Diegeler, A., Wendler, R., Autschbach, R., van Son, J.A., Siegel, L.C., Pompilli, M.F., Mohr, F.W.: Echocardiographic monitoring of minimally invasive mitral valve surgery using an endoaortic clamp. J. Heart Valve Dis. 5, 630-637 (1996)
6. Arom, K.V., Emery, R.W.: Minimally invasive mitral operations. Ann. Thorac. Surg. 63, 1219-1220 (1997)
7. Arom, K.V., Emery, R.W., Kshettry, V.R., Janey, P.A.: Comparison between port-access and less invasive valve surgery. Ann. Thorac. Surg. 68, 1525-1528 (1999)
8. Cohn, L.H., Adams, D.H., Couper, G.S., Bichell, D.P., Rosborough, D.M., Sears, S.P., Aranki, S.F.: Minimally invasive cardiac valve surgery improves patient satisfaction while reducing costs of cardiac valve replacement and repair. Ann. Surg. 226, 421-426 (1997)
9. Fann, J.I., Pompili, M.F., Burdon, T.A., Stevens, J.H., St Goar, F.G., Reitz, B.A.: Minimally invasive mitral valve surgery. Semin. Thorac. Cardiovasc. Surg. 9, 320-330 (1997)
10. Fann, J.I., Pompili, M.F., Stevens, J.H., Siegel, L.C., St Goar, F.G., Burdon, T.A., Reitz, B.A.: Port-access cardiac operations with cardioplegic arrest. Ann. Thorac. Surg. 63, S35-S39 (1997)
11. Gundry, S.R., Shattuck, O.H., Razzouk, A.J., del Rio, M.J., Sardari, F.F., Bailey, L.L.: Facile minimally invasive cardiac surgery via ministernotomy. Ann. Thorac. Surg. 65, 1100-1104 (1998)
12. Konertz, W., Waldenberger, F., Schmutzler, M., Ritter, J., Liu, J.: Minimal access valve surgery through superior partial sternotomy: a preliminary study. J. Heart Valve Dis. 5, 638-640 (1996)
13. Mohr, F.W., Falk, V., Diegeler, A., Walther, T., van Son, J.A., Autschbach, R.: Minimally invasive port-access mitral valve surgery. J. Thorac. Cardiovasc. Surg. 115, 567-574 (1998)
14. Navia, J.L., Cosgrove, D.M., III.: Minimally invasive mitral valve operations. Ann. Thorac. Surg. 62, 1542-1544 (1996)
15. Spencer, F.C., Galloway, A.C., Grossi, E.A., Ribakove, G.H., Delianides, J., Baumann, F.G., Colvin, S.B.: Recent developments and evolving techniques of mitral valve reconstruction. Ann. Thorac. Surg. 65, 307-313 (1998)
16. Carpentier, A., Loulmet, D., Carpentier, A., Le Bret, E., Haugades, B., Dassier, P., Guibourt, P.: Open heart operation under videosurgery and minithoracotomy. First case (mitral valvuloplasty) operated with success. C. R. Acad. Sci. III 319, 219-223 (1996)
17. Chitwood, W.R., Jr., Elbeery, J.R., Chapman, W.H., Moran, J.M., Lust, R.L., Wooden, W.A., Deaton, D.H.: Video-assisted minimally invasive mitral valve surgery: the "micro-mitral" operation. J. Thorac. Cardiovasc. Surg. 113, 413-414 (1997)
18. Chitwood, W.R., Jr., Elbeery, J.R., Moran, J.F.: Minimally invasive mitral valve repair using transthoracic aortic occlusion. Ann. Thorac. Surg. 63, 1477-1479 (1997)
19. Chitwood, W.R., Jr., Wixon, C.L., Elbeery, J.R., Moran, J.F., Chapman, W.H., Lust, R.M.: Video-assisted minimally invasive mitral valve surgery. J. Thorac. Cardiovasc. Surg. 114, 773-780 (1997)
20. Falk, V., Walther, T., Autschbach, R., Diegeler, A., Battellini, R., Mohr, F.W.: Robot-assisted minimally invasive solo mitral valve operation. J. Thorac. Cardiovasc. Surg. 115, 470-471 (1998)
21. Felger, J.E., Nifong, L.W., Chitwood, W.R., Jr. The evolution of and early experience with robot-assisted mitral valve surgery. Surg. Laparosc. Endosc. Percutan. Tech. 12, 58-63 (2002)
22. Loulmet, D.F., Carpentier, A., Cho, P.W., Berrebi, A., d'Attellis, N., Austin, C.B., Couetil, J.P., Lajos, P.: Less invasive techniques for mitral valve surgery. J. Thorac. Cardiovasc. Surg. 115, 772-779 (1998)
23. Mohr, F.W., Falk, V., Diegeler, A., Autschback, R.: Computer-enhanced coronary artery bypass surgery. J. Thorac. Cardiovasc. Surg. 117, 1212-1214 (1999)

24. Chitwood, W.R. Jr., Nifong, L.W., Elbeery, J.E., Chapman, W.H., Albrecht, R., Kim, V., Young, J.A.: Robotic mitral valve repair: trapezoidal resection and prosthetic annuloplasty with the da vinci surgical system. J. Thorac. Cardiovasc. Surg. 120, 1171-1172 (2000)
25. Mehmanesh, H., Henze, R., Lange, R.: Totally endoscopic mitral valve repair. J. Thorac. Cardiovasc. Surg. 123, 96-97 (2002)
26. Nifong, L.W., Chu, V.F., Bailey, B.M., Maziarz, D.M., Sorrell, V.L., Holbert, D., Chitwood, W.R., Jr. Rsobotic mitral valve repair: experience with the da Vinci system. Ann. Thorac. Surg. 75, 438-442 (2003)
27. Nifong, L.W., Chitwood, W.R., Pappas, P.S., Smith, C.R., Argenziano, M., Starnes, V.A., Shah, P.M.: Robotic mitral valve surgery: a United States multicenter trial. J. Thorac. Cardiovasc. Surg. 129, 1395-1404 (2005)
28. Nifong, L.W., Chitwood, W.R., Jr. Challenges for the anesthesiologist: robotics? Anesth. Analg. 96, 1-2 (2003)
29. Reichenspurner, H., Detter, C., Deuse, T., Boehm, D.H., Treede, H., Reichart, B.: Video and robotic-assisted minimally invasive mitral valve surgery: a comparison of the Port-Access and transthoracic clamp techniques. Ann. Thorac. Surg. 79, 485-490 (2005)
30. Casselman, F.P., Van Slycke, S., Wellens, F., De Geest, R., Degrieck, I., Van Praet, F., Vermeulen, Y., Vanermen, H.: Mitral valve surgery can now routinely be performed endoscopically. Circulation 108(Suppl 1), II48-II54 (2003)
31. Greelish, J.P., Cohn, L.H., Leacche, M., Mitchell, M., Karavas, A., Fox, J., Byrne, J.G., Aranki, S.F., Couper, G.S.: Minimally invasive mitral valve repair suggests earlier operations for mitral valve disease. J. Thorac. Cardiovasc. Surg. 126, 365-371 (2003)
32. Muhs, B.E., Galloway, A.C., Lombino, M., Silberstein, M., Grossi, E.A., Colvin, S.B., Lamparello, P., Jacobowitz, G., Adelman, M.A., Rockman, C., Gagne, P.J.: Arterial injuries from femoral artery cannulation with port access cardiac surgery. Vasc. Endovasc. Surg. 39, 153-158 (2005).
33. Rodriguez, E., Cook, R.C., Chu, M.W., Chitwood, W.R.: Minimally invasive bi-atrial cryomaze operation for atrial fibrillation. Optechstcvs. 14, 208-223 (2010)
34. Tatooles, A.J., Pappas, P.S., Gordon, P.J., Slaughter, M.S.: Minimally invasive mitral valve repair using the da Vinci robotic system. Ann. Thorac. Surg. 77, 1978-1982 (2004)
35. Bonatti, J., Schachner, T., Bonaros, N., Ohlinger, A., Danzmayr, M., Jonetzko, P., Friedrich, G., Kolbitsch, C., Mair, P., Laufer, G.: Technical challenges in totally endoscopic robotic coronary artery bypass grafting. J. Thorac. Cardiovasc. Surg. 131, 146-153 (2006)
36. Chitwood, W.R., Jr., Rodriguez, E., Chu, M.W., Hassan, A., Ferguson, T.B., Vos, P.W., Nifong, L.W.: Robotic mitral valve repairs in 300 patients: a single-center experience. J. Thorac. Cardiovasc. Surg. 136, 436-441 (2008)
37. Jones, B.A., Krueger, S., Howell, D., Meinecke, B., Dunn, S.: Robotic mitral valve repair: a community hospital experience. Tex. Heart Inst. J. 32, 143-146 (2005)
38. Woo, Y.J., Nacke, E.A.: Robotic minimally invasive mitral valve reconstruction yields less blood product transfusion and shorter length of stay. Surgery 140, 263-267 (2006)
39. Folliguet, T., Vanhuyse, F., Constantino, X., Realli, M., Laborde, F.: Mitral valve repair robotic versus sternotomy. Eur. J. Cardiothorac. Surg. 29, 362-366 (2006)
40. Murphy, D.A., Miller, J.S., Langford, D.A., Snyder, A.B.: Endoscopic robotic mitral valve surgery. J. Thorac. Cardiovasc. Surg. 132, 776-781 (2006)
41. Chu, M.W., Gersch, K.A., Rodriguez, E., Nifong, L.W., Chitwood, W.R., Jr. Robotic "haircut" mitral valve repair: posterior leaflet-plasty. Ann. Thorac. Surg. 85, 1460-1462 (2008)
42. Lawrie, G.M.: Mitral valve: toward complete repairability. Surg. Technol. Int. 15, 189-197 (2006)
43. Bolotin, G., Kypson, A.P., Nifong, L.W., Chitwood, W.R., Jr.: Robotically-assisted left atrial fibrillation ablation and mitral valve repair through a right mini-thoracotomy. Ann. Thorac. Surg. 78, e63-e64 (2004)
44. Loulmet, D.F., Patel, N.C., Patel, N.U., Frumkin, W.I., Santoni-Rugiu, F., Langan, M.N., Subramanian, V.A.: First robotic endoscopic epicardial isolation of the pulmonary veins with

microwave energy in a patient in chronic atrial fibrillation. Ann. Thorac. Surg. 78, e24-e25 (2004)

45. Akpinar, B., Guden, M., Sagbas, E., Sanisoglu, I., Caynak, B., Bayramoglu, Z.: Robotic-enhanced totally endoscopic mitral valve repair and ablative therapy. Ann. Thorac. Surg. 81, 1095-1098 (2006)
46. Pruitt, J.C., Lazzara, R.R., Dworkin, G.H., Badhwar, V., Kuma, C., Ebra, G.: Totally endoscopic ablation of lone atrial fibrillation: initial clinical experience. Ann. Thorac. Surg. 81, 1325-1330 (2006)
47. Reade, C.C., Johnson, J.O., Bolotin, G., Freund, W.L., Jr., Jenkins, N.L., Bower, C.E., Masroor, S., Kypson, A.P., Nifong, L.W., Chitwood, W.R., Jr.: Combining robotic mitral valve repair and microwave atrial fibrillation ablation: techniques and initial results. Ann. Thorac. Surg. 79, 480-484 (2005)
48. Katz, M.R., Van Praet, F., de Canniere, D., Murphy, D., Siwek, L., Seshadri-Kreaden, U., Friedrich, G., Bonatti, J.: Integrated coronary revascularization: percutaneous coronary intervention plus robotic totally endoscopic coronary artery bypass. Circulation 114, I473-I476 (2006)
49. Falk, V., Jacobs, S., Gummert, J., Walther, T.: Robotic coronary artery bypass grafting (CABG)–the Leipzig experience. Surg. Clin. North Am. 83, 1381-1386, ix (2003)
50. Vassiliades, T.A., Jr.: Technical aids to performing thoracoscopic robotically-assisted internal mammary artery harvesting. Heart Surg. Forum 5, 119-124 (2002)
51. Kappert, U., Cichon, R., Gulielmos, V., Schneider, J., Schramm, I., Nicolai, J., Tugtekin, S.M., Schueler, S.: Robotic-enhanced Dresden technique for minimally invasive bilateral internal mammary artery grafting. Heart Surg. Forum 3, 319-321 (2000)
52. Loulmet, D., Carpentier, A., d'Attellis, N., Berrebi, A., Cardon, C., Ponzio, O., Aupecle, B., Relland, J.Y.: Endoscopic coronary artery bypass grafting with the aid of robotic assisted instruments. J. Thorac. Cardiovasc. Surg. 118, 4-10 (1999)
53. Falk, V., Diegeler, A., Walther, T., Banusch, J., Brucerius, J., Raumans, J., Autschbach, R., Mohr, F.W.: Total endoscopic computer enhanced coronary artery bypass grafting. Eur. J. Cardiothorac. Surg. 17, 38-45 (2000)
54. Falk, V., Diegeler, A., Walther, T., Jacobs, S., Raumans, J., Mohr, F.W.: Total endoscopic off-pump coronary artery bypass grafting. Heart Surg. Forum 3, 29-31 (2000)
55. Dogan, S., Aybek, T., Andressen, E., Byhahn, C., Mierdl, S., Westphal, K., Matheis, G., Moritz, A., Wimmer-Greinecker, G.: Totally endoscopic coronary artery bypass grafting on cardiopulmonary bypass with robotically enhanced telemanipulation: report of forty-five cases. J. Thorac. Cardiovasc. Surg. 123, 1125-1131 (2002)
56. Subramanian, V.A., Patel, N.U., Patel, N.C., Loulmet, D.F.: Robotic assisted multivessel minimally invasive direct coronary artery bypass with port-access stabilization and cardiac positioning: paving the way for outpatient coronary surgery? Ann. Thorac. Surg. 79, 1590-1596 (2005).
57. Srivastava, S., Gadasalli, S., Agusala, M., Kolluru, R., Naidu, J., Shroff, M., Barrera, R., Quismundo, S., Srivastava, V.: Use of bilateral internal thoracic arteries in CABG through lateral thoracotomy with robotic assistance in 150 patients. Ann. Thorac. Surg. 81, 800-806 (2006)
58. Argenziano, M., Katz, M., Bonatti, J., Srivastava, S., Murphy, D., Poirier, R., Loulmet, D., Siwek, L., Kreaden, U., Ligon, D.: Results of the prospective multicenter trial of robotically assisted totally endoscopic coronary artery bypass grafting. Ann. Thorac. Surg. 81, 1666-1674 (2006)
59. Anderson, C.A., Rodriguez, E., Chitwood, W.R., Jr.: Robotically assisted coronary surgery: what is the future? Curr. Opin. Cardiol. 22, 541-544 (2007)
60. Gao, C., Yang, M., Wu, Y., Wang, G., Xiao, C., Liu, H., Lu, C.: Hybrid coronary revascularization by endoscopic robotic coronary artery bypass grafting on beating heart and stent placement. Ann. Thorac. Surg. 87, 737-741 (2009)

61. McAlister, F.A., Ezekowitz, J., Hooton, N., Vandermeer, B., Spooner, C., Dryden, D.M., Page, R.L., Hlatky, M.A., Rowe, B.H.: Cardiac resynchronization therapy for patients with left ventricular systolic dysfunction: a systematic review. JAMA. 297, 2502-2514 (2007)
62. Alonso, C., Leclercq, C., d'Allonnes, F.R., Pavin, D., Victor, F., Mabo, P., Daubert, J.C.: Six year experience of transvenous left ventricular lead implantation for permanent biventricular pacing in patients with advanced heart failure: technical aspects. Heart 86, 405-410 (2001)
63. Derose, J.J., Jr., Belsley, S., Swistel, D.G., Shaw, R., Ashton, R.C., Jr.: Robotically assisted left ventricular epicardial lead implantation for biventricular pacing: the posterior approach. Ann. Thorac. Surg. 77, 1472-1474 (2004)
64. Navia, J.L., Atik, F.A., Grimm, R.A., Garcia, M., Vega, P.R., Myhre, U., Starling, R.C., Wilkoff, B.L., Martin, D., Houghtaling, P.L., Blackstone, E.H., Cosgrove, D.M.: Minimally invasive left ventricular epicardial lead placement: surgical techniques for heart failure resynchronization therapy. Ann. Thorac. Surg. 79, 1536-1544 (2005)
65. Murphy, D.A., Miller, J.S., Langford, D.A.: Robot-assisted endoscopic excision of left atrial myxomas. J. Thorac. Cardiovasc. Surg. 130, 596-597 (2005)
66. Woo, Y.J., Grand, T.J., Weiss, S.J.: Robotic resection of an aortic valve papillary fibroelastoma. Ann. Thorac. Surg. 80, 1100-1102 (2005)
67. Suematsu, Y., Mora, B.N., Mihaljevic, T., del Nido, P.J.: Totally endoscopic robotic-assisted repair of patent ductus arteriosus and vascular ring in children. Ann. Thorac. Surg. 80, 2309-2313 (2005)
68. Bonaros, N., Schachner, T., Oehlinger, A., Ruetzler, E., Kolbitsch, C., Dichtl, W., Mueller, S., Laufer, G., Bonatti, J.: Robotically assisted totally endoscopic atrial septal defect repair: insights from operative times, learning curves, and clinical outcome. Ann. Thorac. Surg. 82, 687-693 (2006)
69. Argenziano, M., Oz, M.C., Kohmoto, T., Morgan, J., Dimitui, J., Mongero, L., Beck, J., Smith, C.R.: Totally endoscopic atrial septal defect repair with robotic assistance. Circulation 108(Suppl 1), II191-II194 (2003)
70. Morgan, J.A., Peacock, J.C., Kohmoto, T., Garrido, M.J., Schanzer, B.M., Kherani, A.R., Vigilance, D.W., Cheema, F.H., Kaplan, S., Smith, C.R., Oz, M.C., Argenziano, M.: Robotic techniques improve quality of life in patients undergoing atrial septal defect repair. Ann. Thorac. Surg. 77, 1328-1333 (2004)

Chapter 30
Robotics in Neurosurgery

L.N. Sekhar, D. Ramanathan, J. Rosen, L.J. Kim, D. Friedman, D. Glozman, K. Moe, T. Lendvay, and B. Hannaford

"Concern for man and his fate must always form the chief interest of all technical endeavors. Never forget this in the midst of your diagrams and equations"

– Albert Einstein.

Use of robots in surgery, especially in neurosurgery, has been a fascinating idea since the development of industrial robots. Using the advantages of a robot to complement human limitations could potentially enhance surgical possibilities, other than making it easier and safer. Over the last few decades, much progress has been made in this direction across various disciplines of neurosurgery such as cranial surgery, spinal surgery and radiation therapy. This chapter details the necessity, principles and the future directions of robotics in neurosurgery. Also, the concept of curvilinear robotic surgery and associated instrumentation is discussed.

The idea of using robots in surgery has fascinated surgeons since the making of the first robots for industrial and military use. The first robots were developed in the late fifties for use in industry mainly as transfer machines, used for transporting objects across a few feet. Further design modifications with articulated multi axial arms helped in the making of robots such as Stanford Arm and Programmable Universal Machines for Assembly (PUMA), which were used for automation of manufacturing processes.

Robotics in surgery has made giant strides in recent years with its increasing use in certain specialties like urology and gynecology. Use of the robot da Vinci (Intuitive Surgical, Sunnyvale, CA) for surgeries such as prostatectomy and hysterectomy, has come a long way from hype to hope, creating new benchmarks for surgical care [1]. Robots are also being researched and developed for use in other specialties like neurosurgery, cardiothoracic surgery, etc. The first instance of use of a robot in neurosurgery was in 1985 for stereotaxy, where an industrial robot (PUMA) was used for holding and orienting a biopsy needle (Kwoh et al. [20]).

L.N. Sekhar (✉)
Department of Neurological Surgery, University of Washington,
325, 9th Avenue, Seattle, WA 98104, USA
e-mail: lsekhar@u.washington.edu

J. Rosen et al. (eds.), *Surgical Robotics: Systems Applications and Visions*,
DOI 10.1007/978-1-4419-1126-1_30,

Since then, robotic applications have developed in safety and functionality. They have been tested and some practiced in neurosurgical procedures such as brain irradiation (using the CyberKnife), pedicle screw placement, navigation in neuroendoscopy, robotic frameless stereotaxy and even robotic or robot assisted microsurgery [2–4]. However, there still are a few large chasms that need to be bridged, for this giant technological leap to be seen as a standard of patient care in neurosurgery. This chapter focuses on the current state of robotic applications in neurosurgery, its current limitations, challenges in development and their future.

30.1 What is a Robot?

Robot is a programmable computer device with mechanical abilities to perform tasks, generally by interacting with the environment. As defined by the Robotic Institute of America, it is "a reprogrammable, multifunctional manipulator designed to move materials, parts, tools, or other specialized devices though various programmed motions for the performance of a variety of tasks." Generally robots used in medicine are made of multi-jointed links, which are controlled by a computer device. The end-piece or the end link of such a construct is called an "end-effector," to which attaches various instruments for performance of any desired activity. The end effector can have many degrees of freedom, which translates to the degrees of dexterity of the device.

Robots are indefatigable, accurate and have the ability to process a large amount of data simultaneously. They have the advantage of having near absolute 3-dimensional geometric accuracy apart from being able to be fast in performing their tasks with minimal or no tremor. Robots can reduce tremor of the surgeon's hand, from approximately 40 μm of the human hand to around 4 μm or less by dexterity enhancement techniques [5]. They can also be tele-controlled, thereby giving the advantage of remote operation. The disadvantages include lack of judgment and decision-making capacity, inability to spontaneously react to new situations, and poor spatial coordination, which are attributes of human performance.

30.2 Classification

There are many classifications of robots used in medicine. Broadly based on their usage Taylor classified them into (1) intern replacements (2) telesurgical systems (3) precise path systems (e.g. navigational systems) (4) precise positioning systems (e.g. stereotaxy system). Based on the type of the control system it is broadly divided into active and passive systems, though many robots would fit somewhere in the middle of this broad dichotomy. "Active" refers to the motion of the robotic device directed by a non-human device usually aided by a computer. The robot performs a part or whole of the surgical procedure autonomously. For example, ROBODOC (Integrated Surgical Systems Inc, Fremont, CA), used for hip replacement surgery, is an active system.

In a passive robotic system, the surgeon usually provides the input to move and control the device. A master–slave robotic system is an example of a passive system where the robot performs by constantly responding to the instructions of the surgeon.

Robots can also be semi-active, meaning they can provide guidance to the operator to provide the input for motion. For instance, navigational devices could help guide the surgeon in performing a stereotactic procedure. When using the NeuroMate, for example, the surgeon has complete control of the stereotactic procedure, but is aided by the guidance of the robotic device. Most of the robots used in neurosurgery are of this type wherein there is a "shared control system." The surgeon performs the procedure with the guidance of the robot.

30.3 Robots in Neurosurgery: What For?

Neurosurgery is a specialty which involves operating under a microscope for high precision and careful tissue handling. The brain is a 3-dimensional structure enclosed within the skull by rigid bone and easily damaged by even minor excursions of surgical instruments [6]. Human limits to safe tissue handling are a few hundred microns under the best of conditions, which is much more than the range of visual recognition with new microscopes. Such a discrepancy is due to the physical limitations of the human hand. Addressing this discrepancy with appropriate technological breakthroughs and innovation would help perform better surgeries.

Currently, numerous studies have been reported on the use of robots for specific surgical procedures, including robotic assisted pedicle screw placement, epilepsy surgery, robot assisted stereotactic procedures, and robotic brain irradiation.

30.4 Construction of a Robot

The construction of a robot essentially involves sensors and an operator console for acquiring information, a computer control system for processing information and the manipulator (base, links, actuators and end effectors) for task performance (Fig. 30.1).

The operator console is the interface between the robot and the input from the surgeon. It can vary from joystick or a finger glove to a voice operated system depending on the use of the robot and preference of the surgeon. Movements performed by the surgeon on the console can be scaled and reproduced in the end effector of the robot. By downscaling certain hand movements, the robotic arms can essentially eliminate tremor, thereby delivering only purposeful, intended motion.

Sensors are the other source for information input to the robot. Sensors may be vision or non-vision types [7, 9]. Vision sensors may be from optical fiber cameras mounted at desired locations; they may be mobile or fixed to a certain part of the

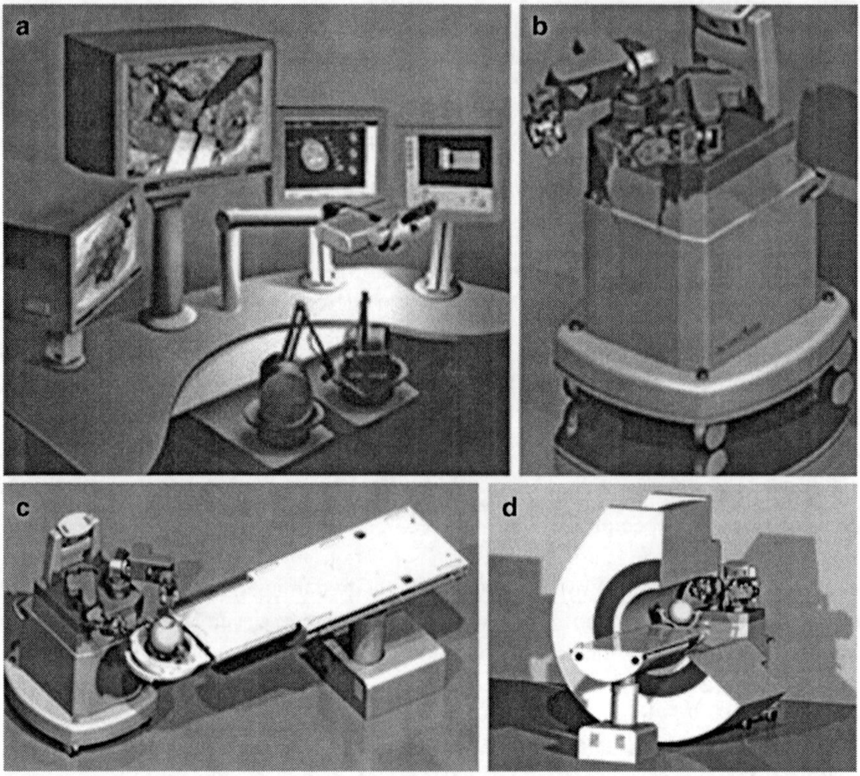

Fig. 30.1 (**a**) Picture showing the console of NeuroArm with video screen shots of microscopic view, external view of the head, and radiology images. (**b**) Structural design of NeuroArm with a base, joints, links and end-effector. (**c**) NeuroArm in position for performing microsurgery. Surgical microscope can be positioned adjacent to the robot's base. (**d**) NeuroArm attached to MRI machine for performing stereotactic procedures. (Pictures from Sutherland et. al. [7, 8])

manipulator. When fixed to the finger (end-effector) or wrist joint they can function as "eye-in-hand" devices. Non-vision sensors can process touch, pressure, temperature and object proximity. They also can provide information about the 3D positioning of the manipulator, thereby providing a feedback mechanism for the function of actuators. Haptic systems (from hapto in Greek, meaning "to touch") are sensors attached to actuators that provide force feedback from environment or virtual situations, thereby providing a real immersive physical feeling to the operator.

The computer system receives information from the sensors and operator interface (console) and processes it to direct the manipulator to perform the appropriate action. Often this computer interacts with multiple other computers, a mechanism that also allows for redundancy in the system in case of malfunction [7]. The computer system's ability to process vast amounts of data contributes to the ability of the robot to be precise in its actions. The software design for processing the information is a critical component in the efficiency of the robot. Basing the operating system of the robot on commercially available software packages may be an easy and attractive solution, and is done with most medical robotics projects. However, development of an original software tool based on the functional design of the robot and surgeons' need can also be a productive measure [10].

The manipulator is the mechanical component of a robot that consists of a base, links, end-effector(s), and the actuators. The end-effector is the final distal link where the action is performed. Actuators convert the signaling from the computer output into mechanical movements to position and orient the links of the manipulator. In image-guided surgeries like stereotactic procedures, the process of registration provides geometric inputs for the actuator (after being processed by the computer control system). The base helps in positioning the robot in a required place. The links are connected by joints, which in turn connect to the robot. The joints connecting the links can be either prismatic (meaning translation between joints possible) or revolute (able to rotate but not translate) [8]. Each joint denotes a degree of freedom. There could be numerous (up to six) joints in the design of robot. In such cases, the proximal three joints are usually the major joints, which determine the 3-D workspace (called work envelope) and the position of the end effector in space. The distal three joints determine the orientation of the object in space. The orientation is regulated by the junction of pitch, roll and yaw at the wrist (penultimate joint) [8].

30.5 Current Trends with Robots

Since the advent of medical robotics, robots have passed through a few stages of technological innovations. The first use of robots was for retraction purposes in surgery. This was followed by the use of robot named NeuroMate in surgical planning and for performing stereotactic procedures. However, these robots relied on preoperative images for positioning and lacked proper safety mechanisms. The first system to use real time guidance system was Minerva (University of Lausanne, Switzerland) which had an inbuilt CT scanner in its robotic arm. Following this, efforts to incorporate MRI robotic image guidance resulted in three different groups, from Harvard University, University of Tokyo, and the University of Calgary to develop them independently.

The development of individual robots has been targeted mostly to address specific kinds of procedures. The majority of the initial robots developed were for stereotactic surgeries, helping in positional 3D access accuracy. These include NeuroMate, Minerva and IMARL for precise needle insertion and biopsy, instrument holding and moving motion. Robots to help in open neurosurgery were developed later including the Robot Assisted Microsurgery Systems (RAMS) and the Steady hand system (Johns Hopkins University). RAMS was a master slave robotic arm with six degrees of freedom and equipped with tremor reduction technology such as motion scaling and tremor filters. Experiments to perform microanastomosis with this robot were performed in rats; the main disadvantage noted was it took twice the time compared to performance with hands. Robots also have been developed for radiosurgery for accurate delivery of radiation with out frame fixation, such as the CyberKnife for tumor resection endoscopic neurosurgery. Recent development of NeuroArm is a significant milestone in combining the abilities of stereotactic surgery and microsurgery in a single system with intraoperative real-time MRI navigation.

30.6 Robots for Position or Stereotaxy Based Procedures

Stereotactic procedures employ robotic systems for their near perfect accuracy in 3-dimensional space. The robot is used for the process of registration with CT/MRI images and trajectory planning to position a mechanical guide. Through the mechanical guide a surgical tool such as an electrode probe can be passed. NeuroMate is a standard robot used in stereotactic procedures that can reduce human error and save time in performing biopsies. This is a passive robotic system that guides the surgeon on the trajectory. It has five degrees of freedom and can hold tools such as electrodes or needles. The main disadvantage of this device is that it is bulky and occupies too much space in the operating room.

In patients with medically refractory epilepsy, surgical treatment with robots has been experimented and found to be a technically safe, feasible and an efficient procedure [11, 12]. For example, using SurgiScope, a handheld probe was jointly used with a stereotactic guide to accurately place subdural monitoring electrodes while the patients were undergoing craniotomy. Such accurate placement of the electrodes for recording the epileptic focus in the brain reduces the necessity to remove the frame or reposition the patient for further attempts [12]. Another robot called PathFinder (Armstrong Healthcare Ltd, High Wycombe, UK) was used in epilepsy surgery to locate the temporal horn and epileptic focus of the brain accurately. The device had a proximal link rotating in a horizontal axis and two links rotating in a vertical axis. An instrument holder that can rotate 180° is attached to the end of the arm (Fig. 30.2a). The system is registered to an MRI scan superimposed onto a CT scan with fiducials, and then attached to the Mayfield head holder [11]. After craniotomy, electrodes are passed into the hippocampus by the robotic device and a catheter is introduced into the temporal horn under image guidance from the robot. This system was found to be more accurate and less time consuming when compared to using a navigation system alone for such procedures [11].

A robotic stereotactic gamma radiation system named CyberKnife (Accuray, Sunnyvale, CA) has been used for precise irradiation of some brain and spinal pathologies such as tumors and arteriovenous malformation. This system, with the MRI registration of the patients head, avoids the frame usage in conventional gamma knife radiation techniques.

Robot assisted spine surgery studies for placement of pedicle screws (including trans-laminar facet screws, kyphoplasty and vertebroplasty) have been described [13, 14]. A commercially available system called SpineAssist (Mazor Surgical

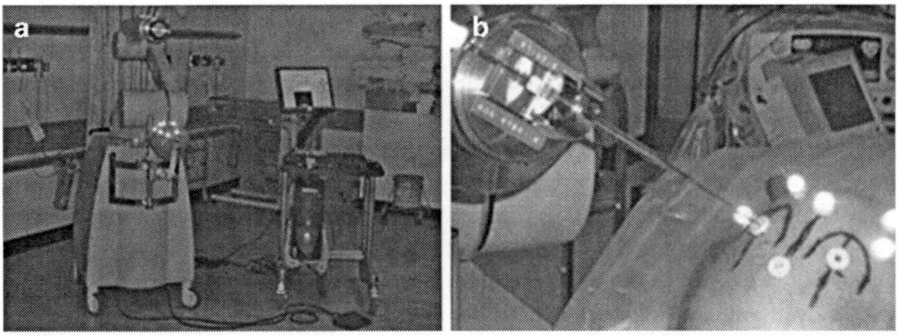

Fig. 30.2 (**a**, **b**) PathFinder and the instrument holder attachment inserting the electrode (adopted from Eljamel et al. [11])

Technologies, Caesarea, Israel) was used for these procedures. This is a miniature robot that mounts to the bony anatomy or to the patient's spine. After the mounting of the robot, pre-operative CT scan images are merged with intraoperative fluoroscopy images and registered to the operating field, with which the robot guides and assists the surgeon to execute a pre-planned procedure. Numerous cohort studies using this robot for minimally invasive spine surgeries have been reported with excellent results on safety and accuracy. Consensus of these experiments is that the robot is "helpful but not a *conditio sine qua non*" for performing these surgeries (Hardenbrook and Dominique et al. [14]). Controlled, head to head studies comparing the use of robots and freehand/fluoronavigation procedures by the surgeons, for efficacy and cost might help to clarify the relative benefit of a robot as compared to human operators. Having established its accuracy and safety, some design related modifications for better planning of surgical windows, graphical representation of virtual anatomy, and better connections of the end-effector to the bony anatomy are being advocated for further improvement of this system [14].

30.7 Robots for Microsurgery

Developing robotic devices for microsurgery is more challenging than for stereotactic procedures, as there are more functional parameters to be considered for design and construct of such a device. Microsurgical robots can be endoscopic robots, which can perform through a keyhole, or open microsurgical robots, which can operate by an open, larger incision and craniotomy.

The endoscopic tools for the brain have been useful for observing and performing minor operative actions like biting, penetrating, or dilating a hole with a balloon (for ventriculostomies). Angled rigid and flexible endoscopes especially help in observing around critical structures [15]. However, due to non-availability of working channels in a rigid endoscope and just one working channel in flexible endoscopes, much of any necessary surgical procedure might not be possible to be performed. NeuRobot, a telecontrolled micromanipulator system was developed to address these inadequacies [9]. This essentially consists of a manipulator with diameter of around a centimeter, which houses a 3D endoscope and three micromanipulators (each 1 mm in diameter) (see Fig. 30.3). This setup is mounted on a manipulator-supporting device, which has six degrees of freedom, and each micromanipulator has three degrees of freedom (up and down, rotational, flexion from 0 to 90°). Basic surgical procedures like dissecting, cutting, coagulating, stitching and tying sutures can all be performed by the surgeon, with visual feedback provided via 3D monitors. Haptic feedback is also provided to help with movement.

This device was used in cadaver experiments to perform surgery through endoscopic and a larger regular incision (pterional approach). This device is reportedly able to reach out to structures around a point to a limited extent.

Robot assisted surgical planning for tumor resection, craniotomy and reconstruction have been performed. The reconstruction of the bony part can be performed after the primary surgery for tumor resection by computer-aided design and planning of the implant size and shape that would be needed for a reconstructive surgery. This helps to avoid the time delay to design an implant and schedule a second cranioplasty, as is done currently in most cases [16].

NeuroArm is a comprehensive robotic system developed at the University of Calgary (Sutherland et al.) with intraoperative MRI ability and the ability to perform both image-guided procedures (stereotaxy) and motion scaled fine open micro-neurosurgery. This is a master slave robotic system, which consists of a robot, a controller, and a workstation or console. The robot's design is adaptable to the kind of procedure performed and based on surgeons' dual arm (ambidextrous) design. The robot has two arms, each with seven degrees of freedom and one degree of freedom for tool actuation, attached to each end effector (see Table 30.1). This, along with the intraoperative imaging, is considered a crucial design feat that can benefit in bringing dexterity and accuracy to the procedures performed. The tools attached to the arms can be either standard tools such as bipolar forceps, needle drivers and dissectors, or stereotactic instruments such as electrodes. The end effectors have a haptic feedback mechanism in place that helps in precise controlled movements by the operator.

Real time MRI is an important addition to NeuroArm over previous generation of robots. It helps in navigation of the tools with improved tool positioning and adequate tissue sampling during stereotaxy. For the microsurgery, MRI hasn't been clearly examined, nevertheless it is supposed that having constant intraoperative MRI would help monitor the position of the tool tips and help avoid a "no-entry" zone before and during the surgery, adding a safety mechanism [7].

The workstation is designed to provide an immersive environment for the surgeon. It has tactile, audio and visual feedback with binocular display providing three-dimensional vision of the operative site (see Table 30.2). Other than this there are desk-mounted displays of MRI, a robot operative parameters display and multidirectional surgical site views. The tools, attached to end effectors, can be superimposed on the MRI to provide navigation to the surgeon.

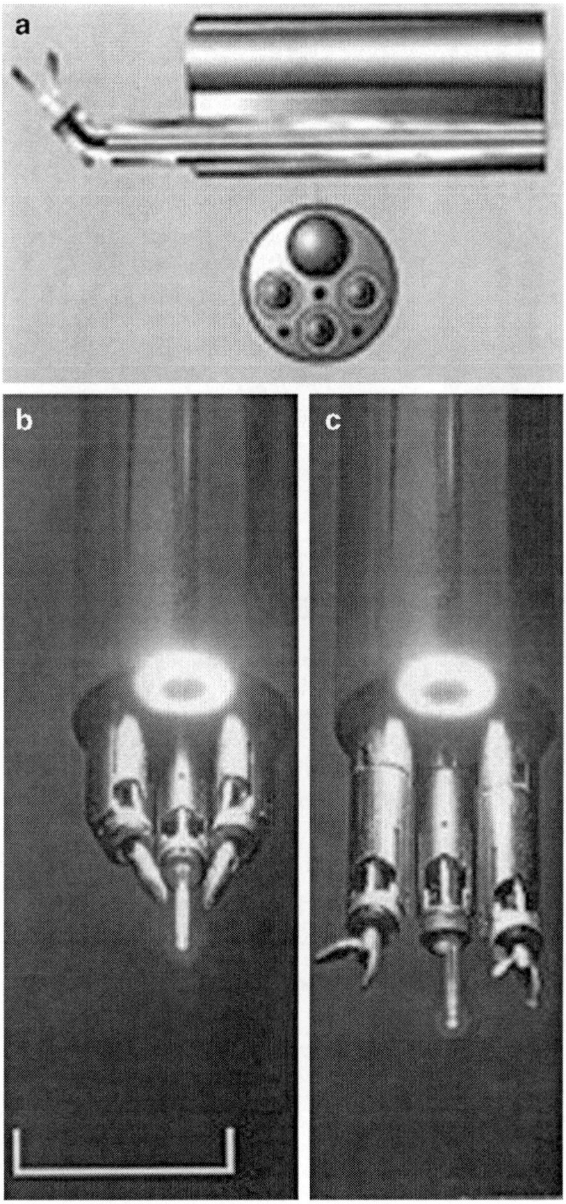

Fig. 30.3 (**a**, **b**, **c**) Design of the NeuRobot manipulator and associated instruments – endoscope, micromanipulators and a laser source (from Hongo et al. [9])

Other than performing surgery with NeuroArm, image processing and integration with the robot helps by providing simulations of surgery before the actual surgery. These virtual surgery trials could possibly help neurosurgeons practice,

Table 30.1 NeuroArm mechanical specifications

Parameters	Specification
Degrees of freedom	8 (including tool actuation) for each arm; 16 total
Payload	0.5 kg
Force (static)	10 N
Tool tip speed	Surgery: 0.5–50 mm/s Tool change: 200 mm/s
Positional accuracy	
Payload $< 100\ g$	± 1 mm absolute 100 μm resolution
Payload $> 100\ g$	± 2 mm absolute 1 mm resolution
Optical force sensors	Sensitivity: 0.02–5 N Dynamic range: 450:1
Continuous operation time	> 10 h

Table 30.2 NeuroArm workstation specifications*

Parameters	Specification
Hand controller	6-DOF position sensing 3-DOF translational force feedback using direct current motors Workspace (tool tip) $x \times y \times z$ (ellipsoid) 40 × 25 × 50 cm Pitch, ± 130 degrees, × 150 degrees, roll, × 168 degrees
Microscope	Counterbalanced microscope equipped with motorized and high-quality optics Beam splitter with two high-resolution IVC camera High-definition format
Visual display	Binoculars using miniature display technology XGA resolution
Voice communication	Simultaneous talk/listen voice communication Wireless digital headset

*DOF, degrees of freedom, XGA, extended *graphic array.*

compare and analyze difficult techniques to arrive at an optimal solution for complex problems. By combining image processing with brain biophysical property modeling and with data on tool–tissue interactions, realistic projections of hemorrhage will help eliminate the gap between virtuality and reality.

Robots have also been used to enhance surgeon presence in neurocritical care units. With camera and video screen mounted on a remotely controlled mobile robot called the RP-6 (In Touch Health Inc, Santa Barbara, CA), the surgeon is able to be virtually present near the patient to observe and verbally respond [17].

30.8 Surgical Robotics Research at the University of Washington: Perspective of a Research Group

Raven is a surgical robot developed at the University of Washington. The main advantages of Raven is its relatively smaller size and design features for being operated remotely. The other advantage is a spherical design of the effectors that limits the range of motion at the surgical port location. This mechanical safety design is fail safe with respect to a software-based control in other robotic systems. This robot was initially developed for general surgical and urologic procedures. Later, it was adapted to perform suctioning in micro anastomotic procedures with a surgical suction tool attached to the end effector. It was used to experiment in micro anastomotic procedures in chicken wings. The Raven is a robot with master slave control system, with the movements of the surgical assistant on a console being downscaled and reproduced in the surgical field.

30.9 Roboscope in Neurosurgery: Minimal Access Curvilinear Surgery in the Brain

Currently minimal access surgery in neurosurgery is in its initial stages. A few endoscopic procedures like endoscopic ventriculostomies and transnasal transphenoidal procedures to the median anterior skull base have recently been introduced to mainstream neurosurgery, albeit with reservations. Minimal access surgery in its current form is performed with instruments that can work only along a straight line of access. The ability to work along a curved line will confer better surgical range and more applications for endoscopic procedures. Nevertheless, a whole new array of surgical tools will be required to operate along a curved access pathways. Robots in neurosurgery can aid in performing surgery through minimal access curvilinear approach similar to that being performed in Natural Orifice Transluminal Endoscopic Surgery (NOTES). Pre-operative planning, instrument navigation and advancement can all benefit with the superior geometric accuracy of a robotic device (Figs. 30.4 and 30.5).

Our team is pursuing the idea of minimal access neurosurgery with the design and development of a flexible robotic sheath called Roboscope (in collaboration with SPI Surgical, Seattle, WA, USA). This robotic device is a multi-jointed flexible tube with multiple degrees of freedom through which various operating instruments like dissectors, suction tubes, scopes, etc. pass through. The advantage of a flexible design is to take a curvilinear approach to the site of surgery, along the path of safe entry zones. This robotic device is computer guided, which helps direct through the required turns at specific anatomic points (Fig. 30.6). A CT or MRI guidance can be used for this purpose. Such image guided flexible robotic systems may provide endoscopic surgical options for conditions that are currently treated with open microsurgery.

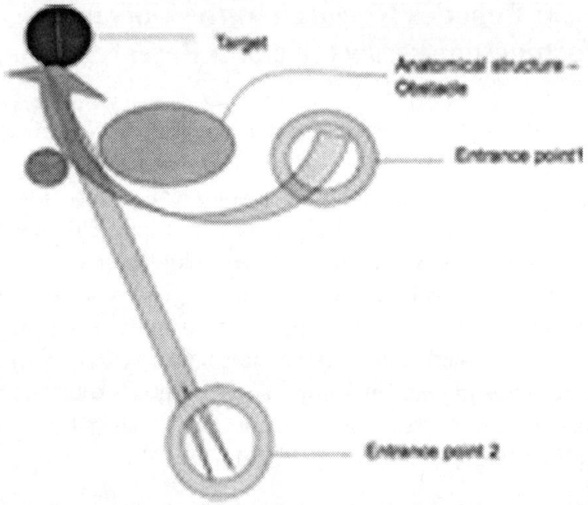

Fig. 30.4 This conceptual figure shows the need for a curvilinear pathway to negotiate obstacles when the entrance site must be in one area (entrance point 1) vs another (entrance point 2)

This device can be compositely used with other technologies being developed for minimal access surgery. Nanotechnology based tumor treatments, cryoablation, and high frequency ultrasound for tumors can all be performed though this device.

30.10 Design of Roboscope's Main Flexible Access Port System

The access port system is a flexible construct, with multi jointed links connected in a serial fashion. This serves the purpose of a **maneuverable** channel though which surgical instruments can pass to the site of surgery. The jointed links are connected and mechanically operated through cables or wires running along the circumference, at certain points in association with wheels (pulley wheel mechanism). Depending on the design of two cables or four cables, the device can bend in one plane or have biplanar bending ability (Fig. 30.6). This movement is controlled through the external robotic device. The movement of the cables provides up to two degrees of freedom for the movement of the robotic scope (Fig. 30.6). The cables at the instrument end however, have articulations with a sphere, which provides additional rotational degrees of freedom along with axial movement (Fig. 30.7). Depending on the nature of the design employed at the end articulation of instrument, it can have different degrees of freedom and movement at the working end (Figs. 30.9–30.12).

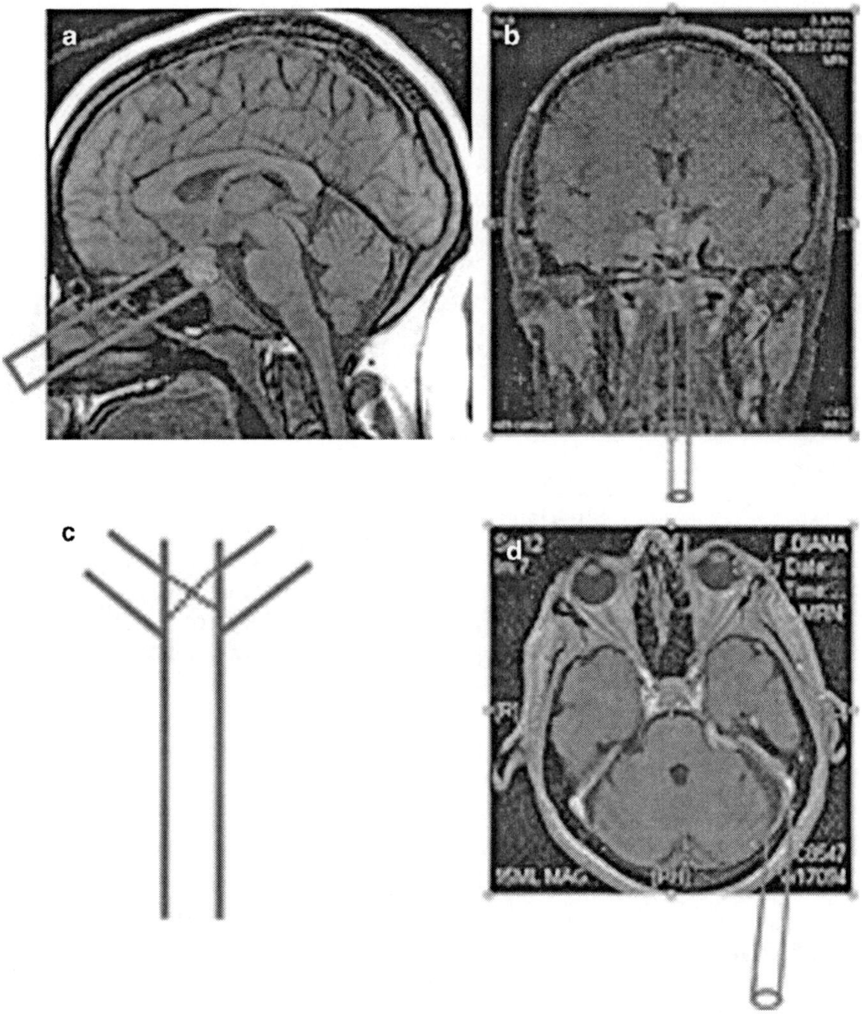

Fig. 30.5 (**a**, **b** and **c**) MRI image showing a pituitary tumor extending laterally that cannot be operated via a transnasal endoscopic approach. A curvilinear approach with the Roboscope would make such an approach possible. (**d**) Posterior fossa meningioma which can be approached and removed by a standard retrosigmoid craniotomy. In order to access and remove this tumor, however, a minimal access approach through the retrosigmoid area will require curvilinear instrumentation. This is a case for performing an endoscopic surgery where open microsurgery would normally be done

For the advancement of Roboscope through the brain, the Roboscope at the functional end has two movable curved plaves which oppose each other to form a pointed surface. The pointed surface can help pierce through planes, by separating tissues on either side, thereby making a plane for advancement of robotic scope in a planned trajectory.

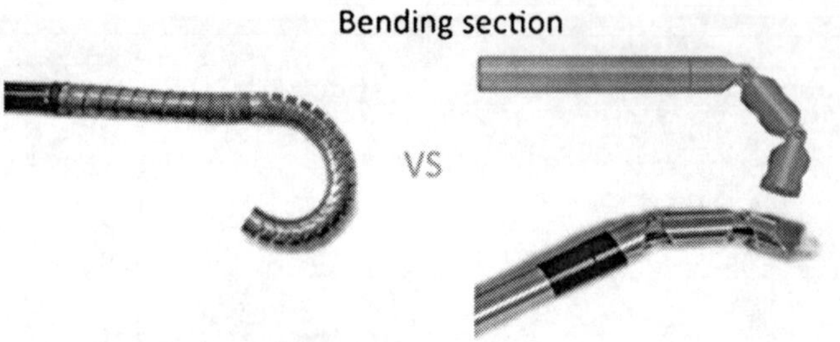

2 cables	2 way articulation (Left/Right)
4 cables	4 way spatial / 2 way planar "S" shape

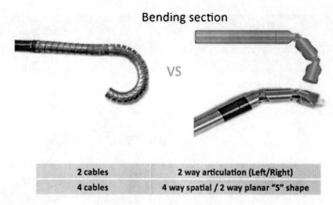

Fig. 30.6 Illustrating the bendable design of the roboscope

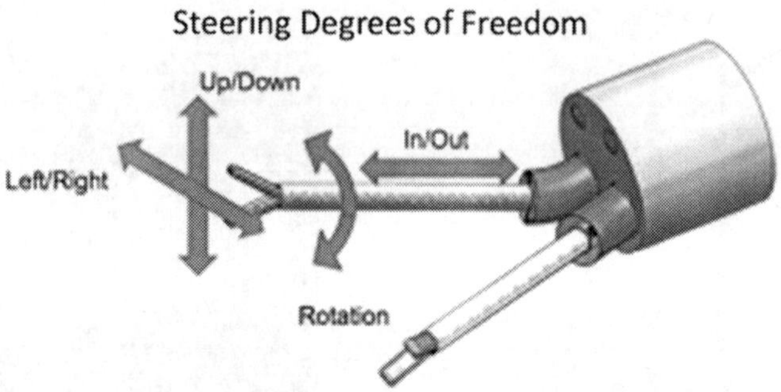

Fig. 30.7 Illustration showing the degrees of freedom needed at the end of operating tools

30.11 Instruments Passing Through the Roboscope

This robotic port system will house atleast two working channels through which surgical instruments can pass, including a modified bipolar instrument (Fig. 30.8). The flexible scope can have a cross section of a circle or oval design. The instruments inside the Roboscope are actually held in a sheath, which can be either fixed or freely movable within the scope. The sheaths, depending on the design, have an axial or rotational movement capability (Figs. 30.9–30.12). The scope also houses two camera heads providing for binocular vision. A suction device is located radially, which could also be maneuvered directionally. This device also has the ability to spray clean the endoscopic camera heads. A flexible CO_2 laser tube can also be used through this port, which holds the suction device.

The two working channels used in the robotic device will be designed to accommodate cryoablation or thermo ablation devices, or future nanotechnology instruments helping in advanced imaging or drug delivery.

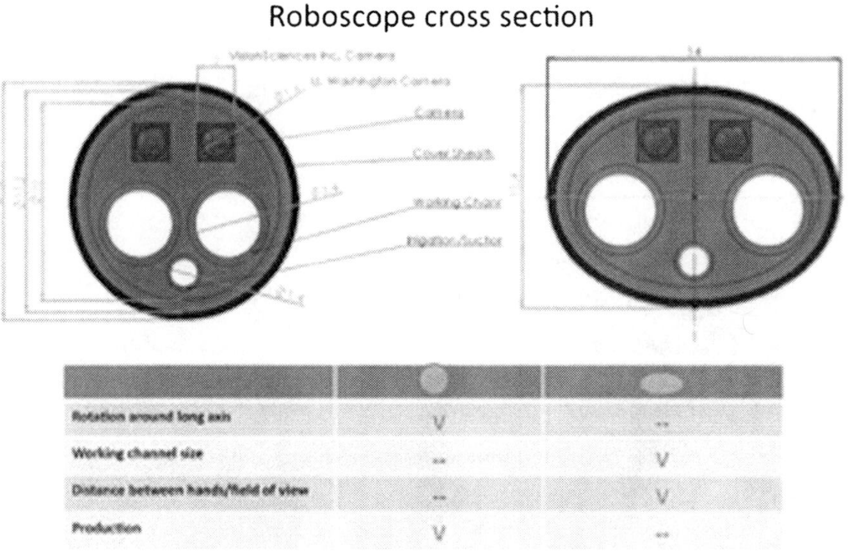

Fig. 30.8 Cross section of the Roboscope showing the arrangement of ports for intruments and the degrees of freedom. Description of degrees of freedom: (1) Cable up/down, (2) Cable left/right, (3) Tool in/out, and (4) Tools/rotation

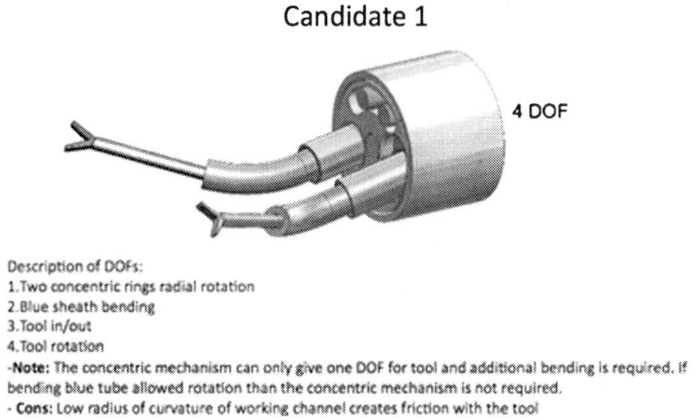
Candidate 1
4 DOF
Description of DOFs:
1.Two concentric rings radial rotation
2.Blue sheath bending
3.Tool in/out
4.Tool rotation
-Note: The concentric mechanism can only give one DOF for tool and additional bending is required. If bending blue tube allowed rotation than the concentric mechanism is not required.
- Cons: Low radius of curvature of working channel creates friction with the tool

Fig. 30.9 Illustration showing the possible movements and degrees of freedom of the end operating instruments

30.12 Future of Robotics in Neurosurgery

The future of robotics in neurosurgery will be based on the need for precision in smaller operative spaces.

Robotic master slave robots take more time than a standard microsurgical operation performed by a surgeon. This may limit the use of robots in long surgical

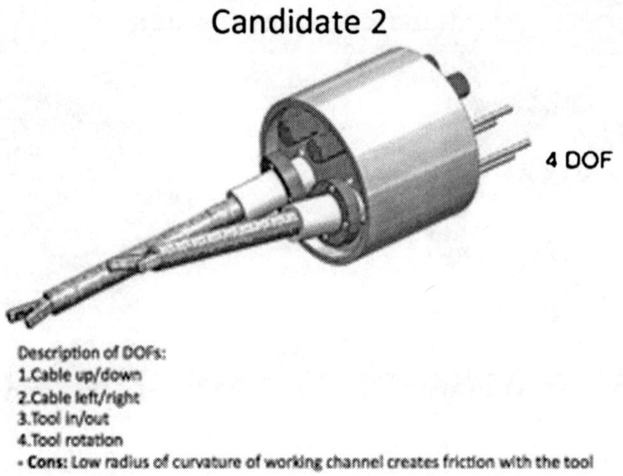
Candidate 2
4 DOF
Description of DOFs:
1.Cable up/down
2.Cable left/right
3.Tool in/out
4.Tool rotation
• Cons: Low radius of curvature of working channel creates friction with the tool

Fig. 30.10 Alternative design, with narrower field of access and radius of curvature

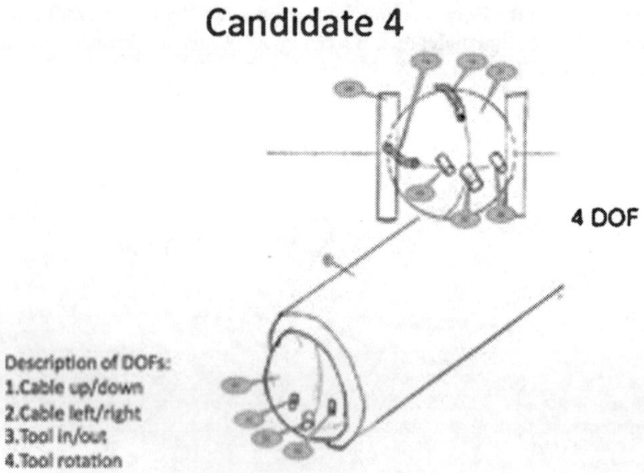
Candidate 4
4 DOF
Description of DOFs:
1.Cable up/down
2.Cable left/right
3.Tool in/out
4.Tool rotation

Fig. 30.11 Mechanism of " wire and sphere" movements and description of the degrees of freedom

procedures. However, the operative time while using a robot can be decreased with operator training and design optimization of the robot (as experienced in many of our robotic experiments).

Incorporation of various preoperative images such as functional MRI, diffusion tensor imaging, three dimensional angiography, and intelligent solutions to help get the surgeon to the target site without the destruction of normal tissues will be key reasons for using robotic devices. Smaller sized robotic machines (possibly micro or nano) may help to create robots which can self assemble, and disassemble after performing the required task. Such robots can find a place in endovascular surgery, intraventricular surgery, and tumor surgery through a small space.

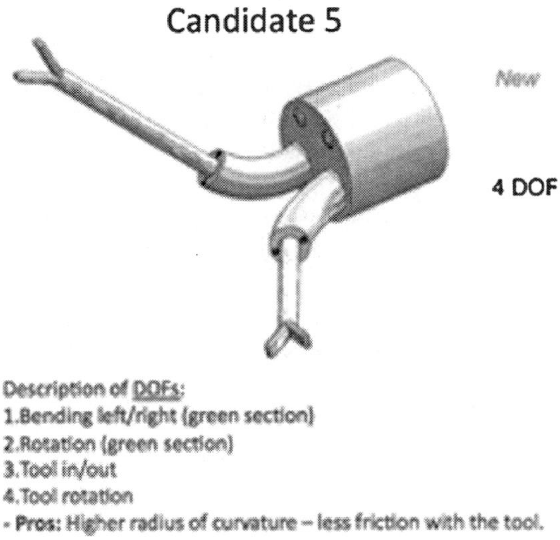

Fig. 30.12 This figure illustrates the wide range of movement of the surgical devices at the end which can be useful in dissecting larger tumors/lesions without much movement of the robotic scope. Most of the range of movement is conferred by the flexibility of the sheath that houses the surgical instrument

Robots such as NeuroArm, despite being commercially available for more than a year now, are not widely used in operative microsurgery. A system that confers on us the ability to perform surgeries that can otherwise not be done with existing technology will be readily tested and adopted. Curvilinear, minimal access surgery is one such technology, which can help neurosurgeons reach beyond the current frontiers.

30.13 Teamwork

Creation of medical robots require a multidisciplinary team with close collaboration between surgeons and engineers. The research group at the University of Calgary that developed NeuroArm may be a good example of a successful team effort. Our research group at the University of Washington, working on Robo scope is based on similar lines with collaborative effort involving surgeons, engineers and business associates from industry. A typical team would consist of lead surgeons with expertise in surgical specialties such as neurosurgery and ENT detailing the requirements of a proposed device to a team of engineers and business associates. The team of engineers involves professors and graduate students in electrical and mechanical engineering and nanotechnology to translate the surgical requirements into a manufacturable design. Business representatives help with financial plans and timelines for these processes.

30.14 Discussions

Despite the advances in the use of robots in neurosurgery, there are some downsides, some of which might be generalized to all surgical specialties. Robots, like any other machine, would have the risk of technical failure. As seen with few examples, initial stages of their development and use would have more of such risk and failures, which may be corrected in course of time as with any new complex technological application [3,18].

The issue of safety will be the primary concern that any new device needs to address first. There are numerous mechanisms developed in robotics which prevent the robot from a lock down, such as dual mechanisms or feedback loops for critical steps. These would help avoid unexpected errors that could be potentially harmful. Such safety features are also a necessity to help meet the standards of complex regulations in place for medical device industry.

Secondly, robots, being bulky by design, could occupy a lot of operative space, making it difficult for surgeons to operate with them. This is especially true in neurosurgery with smaller operative exposures and deep location of actual field of surgery. The issue of sterilization before use also needs to addressed in the design of the robot.

Finally, the quality of work performed with robots needs to be superior or at least equal to that of good surgeons, within reasonably similar cost brackets. Fulfilling this criterion would be an absolute necessity for a robot to be embraced by the surgeons. Currently, open micro-neurosurgery as performed by experienced neurosurgeons meets or exceeds the expected standards for the procedure. Use of a robot to substitute surgeons in this situation might be not be essential in performing or improving the surgery, though it might help make performing the procedure easier for the surgeon. Such redundancy can be an important determinant in wide acceptance or usage of robots among surgeons. The da Vinci is a good example of this [1]. Though initially designed for performing cardiac bypass surgeries, it is not being used for its intended purpose, as cardiac surgeons are able to perform the procedures equally well or better without the robot. However, it has found its place in bettering prostatic and gynecologic surgeries, where surgeons were not traditionally microsurgery trained. On this note, a definite case where robots can help neurosurgeons by providing valuable tools is curvilinear endoscopic surgery. Such surgeries are beyond the scope of neurosurgery in its current form.

After the stage of the acceptance of robotic surgery as a standard of care, the overall benefit to the population would largely depend on adequate training of surgeons and complication avoidance. This brings the need to have a quantitative evaluation system for assessing surgical skills in utilizing such technology, as with several studies now being performed for the evaluation of minimally invasive surgery techniques in general surgery [19].

One of the other major hurdles for such ambitious ventures is in getting research funding [10]. The road from design and manufacture of the robot in the lab to the operating room is a very long and tedious one. It is a difficult task to sustain the economic means to pursue such endeavors, more so with the ongoing debate and downsizing of federal health spending. Increasing government control of the health care in many countries can impose limitations on the development and adoption of new medical technology, especially in the neurosurgical market as it is smaller campared to other specialties.

30.15 Conclusion

For robots to be embraced in neurosurgery, it would need a fine complement of human strengths such as judgment and ability to react to situations, with the advantages of the robot. Reaching this fine balance is a function of advancing technology and appropriate design. Design of robots that can contribute accuracy, indefatigability and zero tremor to a surgeon's judgement could possibly help push the limits of human performance in microsurgery. Neurosurgery in the future, especially with the minimal access techniques, requiring superlative technical and fine motor skills would benefit from such a system. Much remains to be seen whether the heights of engineering can appropriately complement the finesse of the fingers, which has evolved over millions of years.

References

1. Guru, K.A., Hussain, A., Chandrasekhar, R., Piacente, P., Bienko, M., Glasgow, M., Underwood, W., Wilding, G., Mohler, J.L., Menon, M., Peabody, J.O.: Current status of robot-assisted surgery in urology: a multi-national survey of 297 urologic surgeons. Can. J. Urol. **16**, 4736–4741 (2009); discussion 4741
2. Eljamel, M.S.: Robotic neurological surgery applications: accuracy and consistency or pure fantasy? Stereotact. Funct. Neurosurg. **87**, 88–93 (2009)
3. Zimmermann, M., Krishnan, R., Raabe, A., Seifert, V.: Robot-assisted navigated neuroendoscopy. Neurosurgery **51**, 1446–1451 (2002); discussion 1451–1442
4. Zimmermann, M., Krishnan, R., Raabe, A., Seifert, V.: Robot-assisted navigated endoscopic ventriculostomy: implementation of a new technology and first clinical results. Acta Neurochir. (Wien) **146**, 697–704 (2004)
5. Nathoo, N., Cavusoglu, M.C., Vogelbaum, M.A., Barnett, G.H.: In touch with robotics: neurosurgery for the future. Neurosurgery **56**, 421–433 (2005); discussion 421–433
6. Buckingham, R.A., Buckingham, R.O.: Robots in operating theatres. Br. Med. J. **311**, 1479–1482 (1995)
7. Louw, D.F., Fielding, T., McBeth, P.B., Gregoris, D., Newhook, P., Sutherland, G.R.: Surgical robotics: a review and neurosurgical prototype development. Neurosurgery **54**, 525–536 (2004); discussion 536–527
8. McBeth, P.B., Louw, D.F., Rizun, P.R., Sutherland, G.R.: Robotics in neurosurgery. Am. J. Surg. **188**, 68S–75S (2004)
9. Hongo, K., Kobayashi, S., Kakizawa, Y., Koyama, J., Goto, T., Okudera, H., Kan, K., Fujie, M.G., Iseki, H., Takakura, K.: NeuRobot: telecontrolled micromanipulator system for minimally invasive microneurosurgery-preliminary results. Neurosurgery **51**, 985–988 (2002); discussion 988
10. Zamorano, L., Li, Q., Jain, S., Kaur, G.: Robotics in neurosurgery: state of the art and future technological challenges. Int. J. Med. Robot. **1**, 7–22 (2004)
11. Eljamel, M.S.: Robotic application in epilepsy surgery. Int. J. Med. Robot. **2**, 233–237 (2006)
12. Spire, W.J., Jobst, B.C., Thadani, V.M., Williamson, P.D., Darcey, T.M., Roberts, D.W.: Robotic image-guided depth electrode implantation in the evaluation of medically intractable epilepsy. Neurosurg. Focus **25**, E19 (2008)
13. Barzilay, Y., Kaplan, L., Libergall, M.: Robotic assisted spine surgery – a breakthrough or a surgical toy? Int. J. Med. Robot. **4**, 195–196 (2008)
14. Pechlivanis, I., Kiriyanthan, G., Engelhardt, M., Scholz, M., Lucke, S., Harders, A., Schmieder, K.: Percutaneous placement of pedicle screws in the lumbar spine using a bone mounted

miniature robotic system: first experiences and accuracy of screw placement. Spine (Phila Pa 1976) **34**, 392–398 (2009)

15. Fries, G., Perneczky, A.: Endoscope-assisted brain surgery: part 2 – analysis of 380 procedures. Neurosurgery **42**, 226–231 (1998); discussion 231–222
16. Bast, P., Popovic, A., Wu, T., Heger, S., Engelhardt, M., Lauer, W., Radermacher, K., Schmieder, K.: Robot- and computer-assisted craniotomy: resection planning, implant modelling and robot safety. Int. J. Med. Robot. **2**, 168–178 (2006)
17. Vespa, P.M.: Multimodality monitoring and telemonitoring in neurocritical care: from microdialysis to robotic telepresence. Curr. Opin. Crit. Care **11**, 133–138 (2005)
18. Pandya, S., Motkoski, J.W., Serrano-Almeida, C., Greer, A.D., Latour, I., Sutherland, G.R.: Advancing neurosurgery with image-guided robotics. J. Neurosurg. (2009)
19. Winckel, C.P., Reznick, R.K., Cohen, R., Taylor, B.: Reliability and construct validity of a structured technical skills assessment form. Am. J. Surg. **167**, 423–427 (1994)
20. Kwoh, Y.S., Hou, J., Jonckheere, E.A., Hayati, S.: A robot with improved absolute positioning accuracy for CT guided stereotactic brain surgery. IEEE trans. Biomed. Eng. **35**(2), 153–160 (1988)

Chapter 31
Applications of Surgical Robotics in Pediatric General Surgery

John Meehan

31.1 Introduction

Robotic technology poses some distinct challenges in pediatric general surgery. The biggest problem is simply a matter of size. The current robot is huge when compared to a neonate and the instruments were not designed with small patients in mind. In this chapter, we will present the areas where robotic surgery can excel while also discussing the issues and problems with the current technology as it pertains to the huge variety of congenital anomalies and patient sizes that a pediatric general surgeon encounters.

31.2 Hurdles and Challenges

While pediatric urologists have rapidly expanded the use of robotic surgery, pediatric general surgeons have been slow to embrace robotic technology. The reasons are multifactorial and are related to cost, instrumentation size, equipment size, and a level of comfort that has developed in the last 10–15 years with standard hand held laparoscopic instrumentation. When the robotic surgical technology first emerged, several pioneering pediatric centers selected the Zeus surgical system manufactured by Computer Motion. The pediatric surgeons at these institutions gravitated towards the Zeus because of the smaller instrumentation which, at the time, was not available in the Da Vinci. Most of the initial assessments were performed on animal models – such as piglets – a model which simulated the small size of infants and neonates [1, 2]. As a result, the Zeus system gained mild popularity with pediatric general surgeons from 2000 to 2003 with several pediatric hospitals betting on this system as the future of robotic surgery. But the system was

J. Meehan
Department of Surgery, Seattle Children's Hospital, University of Washington,
Seattle Washington, USA
e-mail: john.meehan@seattlechildrens.org

J. Rosen et al. (eds.), *Surgical Robotics: Systems Applications and Visions*,
DOI 10.1007/978-1-4419-1126-1_31, © Springer Science+Business Media, LLC 2011

ergonomically more like a video game joystick and not all that similar to open surgery. Meanwhile, the Da Vinci rapidly gained popularity in adult surgical settings where instrument size was not such an issue. Computer Motion soon found itself with significant financial difficulties and was eventually taken over by Intuitive Surgical. The Zeus system was quickly abandoned by Intuitive with equipment support and maintenance stopped almost immediately. Financially, this hurt many pediatric hospitals that had purchased the Zeus and now had no support for their investment. The abandonment of the Zeus left many children's hospitals and their surgeons bitter in regards to robotic surgery and this may have tarnished its popularity amongst pediatric general surgeons. However, a few centers began using robotic surgery on a regular basis. Small series and case reports began to emerge over the next 2 years [3–5]. Finally, the first large pediatric general surgery series consisting of 100 cases was presented at SAGES in 2006 [6]. This retrospective series highlighted the diversity of procedures possible with the Da Vinci robot in both abdominal and thoracic procedures. Several procedure specific series followed reporting outcomes and challenges using this new technology in children [7–12].

Instrumentation design and manufacturing is based on necessity and frequency of utilization. Since the dawn of modern surgery, development of pediatric specific instrumentation has tremendously lagged behind the advances seen in adult surgery. Much of this is due to the standard axiom that it's a matter of supply and demand. First, there are far fewer pediatric patients undergoing surgical procedures than adults. Additionally, the huge variety of acquired and congenital conditions in pediatric surgery mandate a variety of highly specialized instrumentation which may have excellent utility for one procedure but little use in another. Therefore, the business of making pediatric surgical instrumentation is not seen as a high return investment for many equipment manufacturers. Therefore, pediatric surgeons have often needed to settle on using instrumentation and equipment that was intended for much larger patients. Robotic technology is certainly no exception to this ongoing challenge.

31.3 Patient Size and General MIS Considerations

The varying size of pediatric patients requires relentless innovation from pediatric surgeons as their patients are not simply "little adults." The physical dimensions of the abdominal cavity of a child cannot be calculated by a simple one to one conversion factor from that of an adult. For example, the toddler's abdomen is much wider left to right proportionally than an adult whereas an adult abdomen is usually longer from cranial to caudal. This is one of the main reasons why pediatric surgeons make transverse incisions for an open case while adult surgeons generally select vertical incisions. Beyond this simple illustration, every organ system is different too with different characteristics. A good example of this phenomena is the neonatal liver. While the liver is a fairly sturdy regenerative organ in the adult, the neonatal liver, particularly in a premature infant, is extremely fragile and the premature baby can bleed to exsanguination with the slightest bit of trauma to the liver capsule. Significant care must be taken by the pediatric surgeon regardless to the method of approach. In dealing with MIS procedures, the size of the patients small abdomen or chest become problematic for instruments that are either too wide or have long articulating lengths. Moreover, the insulating subcutaneous fatty layer common in adults can be negligible in the child. This layer helps maintain a pneumoperitoneum as the fat collapses around the trocar. In kids with nearly no fatty abdominal wall, the surgeon needs to make the trocar incision as small as possible or significant pneumoperitoneal CO_2 loss can occur. This loss is further amplified with larger ports. A small amount of CO_2 leakage can make the case very frustrating as the neonatal abdomen deflates completely even with small to moderate CO_2 escape.

31.3.1 Robot Overview

The Da Vinci Standard, The Da Vinci S, and the new Da Vinci SI are the three robots currently in use today. Although the robot has been streamlined somewhat over the progression of newer versions, the large robot cart still poses an intimidating challenge for the pediatric surgeon. The robot weights over 500 kg and stands nearly 6 feet in height. Placed over a small pediatric patient such as a neonate, the robot appears daunting. However, there are simple maneuvers a surgeon can perform in the positioning of the patient and the placement of the trocars in order to overcome these hurdles, particularly for small children [13].

31.3.2 Patient Positioning

Patient positioning needs to be adjusted for pediatric patients. In a general sense, the robotic instrument arms external to the patient must have enough room to move without colliding with each other or with the OR table. These collision problems are amplified as the patients get smaller. This can be solved by raising the smaller pediatric patients above the OR table on foam padding (Fig. 31.1). This simple maneuver helps on a number of levels. First, it allows for more lateral trocar placement. With the instrument trocars more lateral, collisions between the instrument arms and the camera arm are less likely and also allows the instruments to come down to a near horizontal level without colliding with the table. Additionally, the assistant can access the patient more easily. In general, our practice was to place children less than 10 kg on two large egg crate foam pads and on one if they were between 10 and 20 kg.

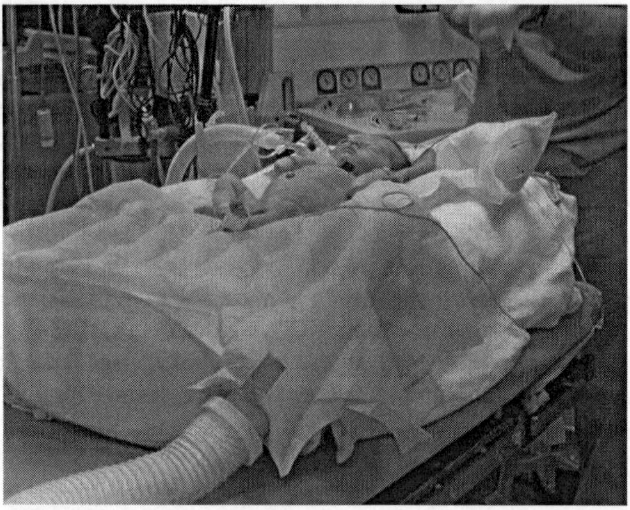

Fig. 31.1 A 2.7 kg neonate positioned for a fundoplication. Foam padding elevates the patient off of the table allowing better access to the patient by the bedside assistant and also reducing the likelihood of robot arm interference with the OR table

31.3.3 Robot Cart Location and the OR Table Positioning

A number of issues must be considered in patient positioning. First, access to the patient by both anesthesiologist as well as the surgical team is hindered by the presence of the large Da Vinci robot hovering over the small patient. Careful planning of lines, monitoring electrodes, and ventilator tubing are important parameters for the anesthesiologist to consider. Pediatric general surgery cases can vary tremendously in regards to the anatomic location a procedure must occur. For example, a fundoplication is performed in the upper abdomen requiring the robot to come in over the patients head. An ovarian teratoma requires a pelvic approach with the robot wheeled in to the patient from the feet. In our first few cases, the patient and the OR table were kept in the same standard location no matter what type of case we were doing. This required us to move the robotic console, the robot, the video tower, back table, and all the support equipment such as cautery machine, suction setup and many other items to different locations for each procedure. Every day was a new floor plan depending on the type of case. Moreover, the anesthesiologist would also have to move their equipment. This became too taxing to orchestrate every day. Our solution was relatively simple: instead of rearranging every piece of equipment in the room except the patient, why not just move the patient and essentially leave everything else in its usual location? The robot was left in one corner of the room for every case. Then, after anesthetic induction, intubation, and all their lines were set, we would unlock the OR table rotating the OR table and the patient such that the anatomy of interest was in alignment with the robot. For example, the feet would be rotated toward the robot for a case in the pelvis and the head would be rotated toward the robot for a case in the upper abdomen. Once the trocars were placed, the robot was moved the final few feet up to the OR table with minimal steering needed. Minimal movement of the rest of the OR equipment was needed with this strategy and made the setup significantly simplified for our nursing staff. Our anesthesiologist and the nursing team became very comfortable with this strategy and our case efficiency improved dramatically. This was adopted for all of our patients regardless of size, and – interestingly – was also adopted by our adult robotic colleagues in urology, gynecology, and cardiothoracic surgery.

31.3.4 Trocar Positioning

Ideal robotic trocar locations may not necessarily correlate with ideal laparoscopic trocar locations. In robotic surgery, a wider distance between trocar sites helps avoid external arm collisions where such lateral positioning may prove ergonomically awkward for the laparoscopic surgeon holding the hand-held instruments. For example, Fig. 31.2 demonstrates trocar placement for a robotic fundoplication for an infant. Even travelling a short 1 cm laterally from the midline camera will increase the angle between the camera and the external portion of the instrument arm a significant amount on the convex shape of the rotund insufflated neonatal abdomen. However, placing the trocar site too far laterally may lead to collisions between the instrument arm and the operating table. Careful planning will help the surgeon individualize the best trocar location for each patient based on size and type of procedure performed.

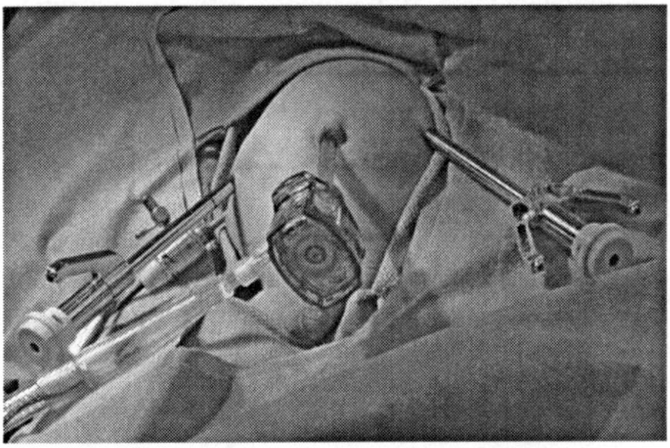

Fig. 31.2 Robotic port placement for a fundoplication in a 2.7 kg infant. Notice the lateral trocar placement which is significantly more lateral than the typical laparoscopic trocar location. This adjustment avoids external collisions between the robotic instrument arms and the robot camera arm

31.3.5 Trocar Depth

One disadvantage of the current robotic instruments is the long length of the articulations of the instrument when compared to the shorter length of the pivoting jaws of standard hand held MIS instruments. If the entire articulating robotic length is not inserted passed the end of the trocar, the Da Vinci software will not allow that instrument to function. This means that a minimum length of each instrument must be external to the trocar in order for the instrument to function. Next, a minimum length of the trocar must be inside the patient simply to avoid trocar displacement out the patient. Each robotic trocar has a thick band intended to aid the surgeon in placing the remote center at the ideal location (Fig. 31.3). The remote center is the location in 3 dimensional space that the robotic arm pivots around. The manufacturer recommends that the trocar's remote center, demarcated by the thick black, be placed just inside the patient. The remote center is a set length from the end of the trocar and is measured at 2.90 cm. This means a trocar length of 2.90 cm is internal before any instrument is even placed. If we select even the shortest 5 mm instrument, the Da Vinci needle driver with an articulating length of 2.71 cm, a total of 2.71 + 2.90 cm, or 5.61 cm must be internal to the patient before the instrument can even function. This requirement may use up all of the available space in the smallest patients and these size limitations have been previously noted [13, 14].

Although there is nothing we can do about changing the articulating length of the instrument, we can certainly adjust the trocar depth. What if we were to "cheat" the trocar just outside the relatively thin neonatal body wall instead of just inside (Fig. 31.4)? This simple maneuver can buy the surgeon an extra 1.0–1.5 cm of maneuverability which we have found to make a significant difference. Although this may sound like a small length, it can have dramatic results when attempting to maneuver the robotic instruments in the small working space of a neonate. This adjustment appears to place no untoward torque on the entry point yet has allowed us to gain a significant advantage with available working domain.

31.3.6 Camera and Instrumentation

One obstacle to the growth of pediatric robotic surgery has been the relatively large camera and instrument diameters. After being accustomed to 3 and 5 mm diameter hand held laparoscopic instruments, pediatric surgeons viewed the Da Vinci 8 mm

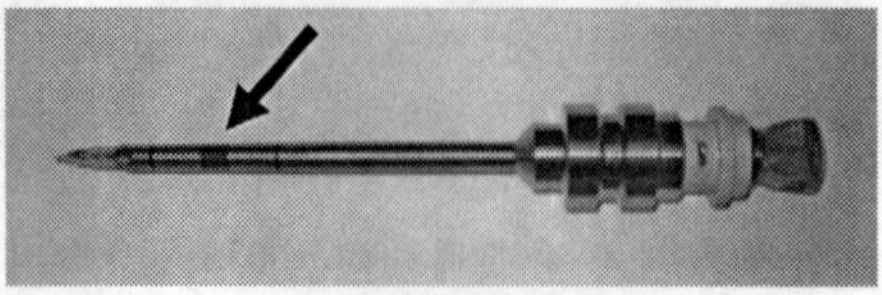

Fig. 31.3 Robot trocar. The *thick black band* denotes the remote center of the robot (*arrow*), the pivot point at which the robotic arm rotates in space

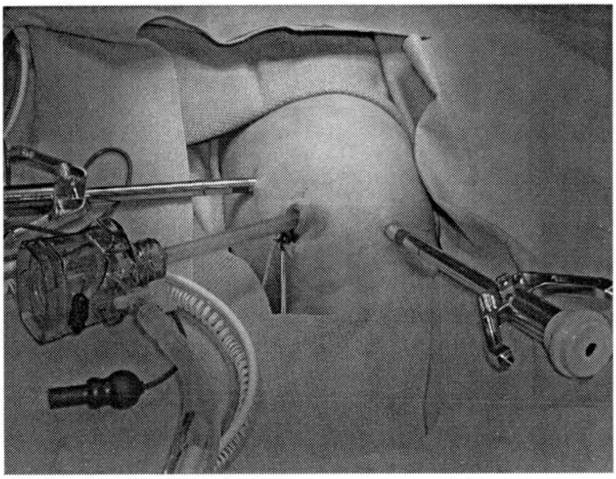

Fig. 31.4 Robot trocar adjustment. By "cheating" the trocar a cm or so further outside the patient, significant room can be achieved which will allow more maneuverability of articulating robotic arm

instruments and 12 mm camera as simply too large for many procedures in small children. Between 2000 and 2003, these large diameter devices were the only hardware available. Fortunately, a number of equipment improvements have helped open the door for robotic surgery in smaller patients. The first advancement was the development of the 5 mm instruments initially released in December 2003. A recall occurred shortly thereafter and they were re-released by the summer of 2004. The smaller diameter advantage of the 5 mm instrument is somewhat offset by the longer articulating length when compared to the 8 mm pitch-roll-yaw movement. Meanwhile, engineers at Intuitive were also working on an alternative for the large 12 mm 3-D camera. Pediatric surgeons suggested that sacrificing the 3-D vision in a 12 mm diameter for a 2-D standard 5 mm laparoscope was an acceptable trade. In early 2004, the 5 mm 2-D camera specific for the Da Vinci Surgical System was introduced. The initial design of the Da Vinci robot camera arm anticipated a specific size and weight for the camera. The 5 mm 2-D scope was markedly lighter than the 12 mm 3-D camera making the camera robot arm movement unbalanced. Therefore, a weight adaptation in the form of a docking collar was added (Fig. 31.5). These two changes in 2004 made robotic surgery using the Da Vinci more practical in small children, and robotic surgery was finally available for the smallest patients. The robotic experience at the University of Iowa Children's Hospital increased dramatically immediately after these advances with 50% of their entire robotic patient list weighing less than 20 kg and 25% weighing less than 10 kg. In 2008, a new 3-D 8.5 mm camera was released which allowed pediatric surgeons to perform robotic procedures with 3-D vision in many more small patients. However, this diameter is still too big for thoracic procedures in children less than 7 or 8 kg because the camera will not fit into the narrow rib space of smaller infants and neonates. Therefore, the 5 mm 2-D camera still has utility in pediatric robotic surgery. Unfortunately, Intuitive Surgical decided to end support

Fig. 31.5 The docking collar for the 5 mm robotic camera. The weighted collar allowed for the robotic camera arm to be properly balanced in the first version of the Da Vinci robot

of the 5 mm 2-D camera in the fall of 2009 and this abandonment will significantly limit the ability of pediatric surgeons to perform robotic procedures in the smaller neonates.

31.4 Specific Procedures

31.4.1 Thoracic

One of the best applications for robotic surgery in children is thoracic surgery, particularly the resection of mediastinal tumors. A variety of tumors, benign and malignant, have been successfully removed using robotic technology in children [10]. Consider the concept of removing such a rigid mediastinal mass trying to use standard hand held instruments: the surgeon is trying to circumnavigate around a rigid solid mass using rigid non-articulating instruments through the rigid chest wall. Although small masses can be easily excised using this type of equipment, larger tumors become problematic as the angles to safely excise such a mass become difficult to approach. Articulating instrumentation can easily overcome this obstacle while providing a 3-dimensional view of the surrounding mediastinal structures. The general principles behind patient positioning and robot cart location are fairly simple. With the patient in a decubitus position, camera port location should be selected first and ideally placed in the most lateral position of the hemithorax. Once the camera port is placed, a quick view with the camera to confirm the location of the mass will help select the other port locations keeping in mind that the Da Vinci robot needs to come in from the direction of the tumor (Fig. 31.6). For a mass that is anterior and superior, the robot cart needs to be placed anteriorly and superiorly. If a tumor is posterior and inferior, the robot cart should come in from a posterior and inferior location. Some of the benign masses and anomalies that have been resected robotically include bronchogenic cyst, esophageal duplication, ganglioneuroma, cystic hygroma, and teratoma (Fig. 31.7). Intermediate lesions such as ganglioneuroblastoma along with higher grade malignancies such as germ cell tumor and neuroblastoma have also been resected (Fig. 31.8).

A number of congenital conditions and anomalies have been repaired robotically including tracheoesophageal fistula, congenital diaphragmatic hernia, diaphragm eventration, and partial and complete pulmonary lobectomies for congenital cystic adenomatoid malformation and pulmonary sequestration [11].

31.4.2 Abdominal Procedures

The variety of general intra-abdominal procedures which have been accomplished in children robotically is extensive (table). This is largely due to the wide range of pediatric congenital anomalies and acquired diseases that exist in the general population. A review of the literature suggests that a fundoplication is probably the most commonly performed robotic procedure [4, 6, 8, 12]. Some papers suggest that this procedure is not cost effective since it can be done laparoscopically with similar results [12]. However, this claim should be also interpreted in terms of the much longer learning curve of a laparoscopic fundoplication. Fundoplications performed laparoscopically have historically taken the pediatric surgeon at least 25 cases before proficiency can be expected [15, 16]. Robotically performed fundoplications may have a learning curve of less than five cases [8]. There has been no financial analysis published in pediatric fundoplications regarding the cost in regards to the amount of time required for learning a laparoscopic technique

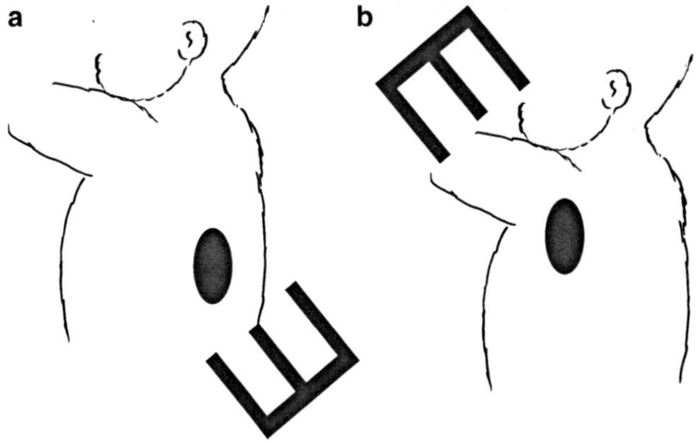

Fig. 31.6 Ideal robot cart location and direction (E) for a mediastinal mass resection is dependent on the location of the mass. (**a**) Posterior inferior tumor. (**b**) Anterior superior tumor.

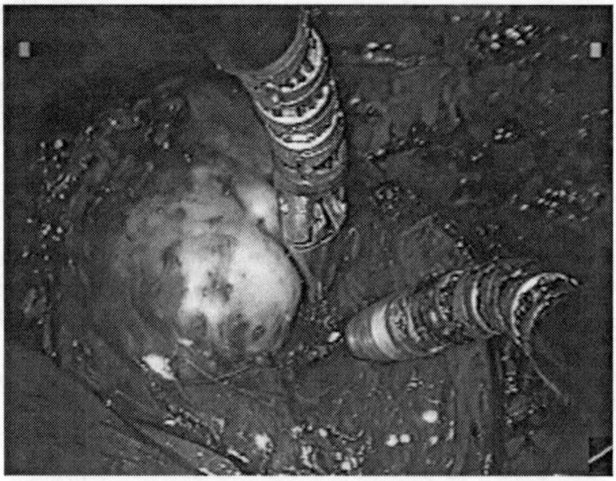

Fig. 31.7 Resection of a mediastinal teratoma

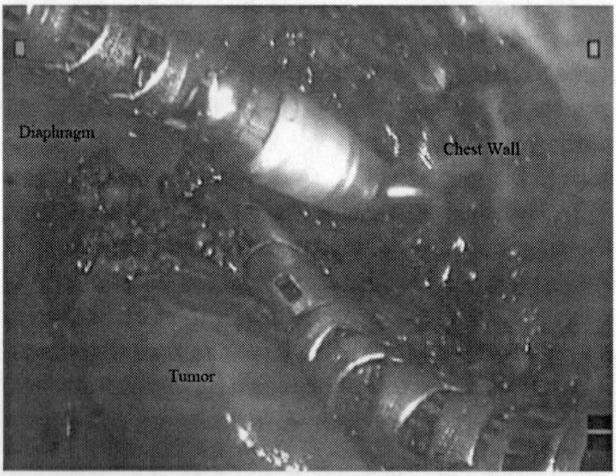

Fig. 31.8 Resection of a neuroblastoma in an infant

when compared to learning the robotic equivalent. Interestingly, resident teaching has also been subjectively faster and more efficient with the robot [8]. These attributes have left us to conclude that the pediatric fundoplication is an ideal introductory teaching case when the inexperienced robotic surgeon is familiarizing themselves with the robotic equipment. The concept of using the fundoplication as the introductory robotic teaching case is further enhanced by the relatively frequency of fundoplications with the opportunity for repetitive experience. There are other advantages of using the robot for a fundoplication as well. Many pediatric surgeons advocate the use of five ports for a standard laparoscopic fundoplication [17]. It is this authors experience in over 100 pediatric fundoplications that only four ports are required for any fundoplication which includes one camera port, one port for a liver retractor, and two instrument ports for the robotic arms. Moreover, we have never been in a situation where we needed to takedown a pre-existing gastrostomy tube in any robotic fundoplication. Many patients referred for fundoplication already have a gastrostomy tube tacking the stomach to the anterior abdominal wall. This can be problematic for the pediatric laparoscopist using standard rigid laparoscopic instruments as they have to steer around the g-tube site which is in close alignment with the camera line-of-sight from the umbilicus to the esophageal hiatus. The articulating robotic instruments make it quite easy to steer around this problem. In over 50 cases performed robotically with a g-tube already in place, we have never had to take the g-tube site down and have accomplished fundoplications with minimal discernable change in operative time from those cases without a previously placed G-tube.

Another commonly performed introductory case is the robotic cholecystectomy. There is little advantage to performing a cholecystectomy robotically except that it is helpful for training new personnel and residents, but it does provide the novice robotic surgeon with a relatively easy and familiar minimally invasive case in order to determine the nuances of the robot. Likewise, this is also an excellent case to introduce the junior resident to the robot. However, additional procedures which may be relatively common such as performing a cholangiogram or common duct exploration pose certain challenges with the large robot hovering over the patient. The possibility of needing these additional procedures as part of the cholecystectomy need to be taken into consideration when planning a robot case. For example, a surgeon may need to temporarily undock the robot and back it out of the way in order to bring in ancillary equipment such as fluoroscopy for a cholangiogram. The robot can be redocked after the cholangiogram is accomplished.

Other common MIS procedures include the splenectomy and adrenalectomy [6, 18]. The addition of the robotic Gyrus PK (Gyrus ACMI, Minneapolis, MN) to the instrument list for the Da Vinci has allowed a greater variety of common abdominal procedures as well as bowel resections to be performed more readily. The Gyrus PK is a type of bipolar technology that will thermally seal vessels up to 7 mm in size. This can greatly facilitate taking down mesenteric vessels in bowel resections and short gastric or even splenic hilar vessels that are within the size parameters for this instrument in splenectomies [19]. The Gyrus PK has also been used in pediatric thoracic procedures such as pulmonary resections for congenital cystic adenomatoid malformation (CCAM) and pulmonary sequestration [11]. An intraoperative photograph from a sequestration resection is shown in Fig. 31.9.

The great advantage of the robot is certainly in the most complicated pediatric minimally invasive procedures requiring the high precision and fine suturing capabilities afforded by robotic surgery. These procedures include complex hepatobiliary procedures such as a Kasai portoenterostomy and the resection of the choledochal cyst [7, 9]. The creation of a Kasai portoenterostomy is a highly detailed and precise operation. At the annual IPEG scientific meeting in 2007, the world's top leaders in minimally invasive pediatric surgery recommended a moratorium on minimally invasive Kasai because of port results when compared to open surgery. This is may not be as unexpected as some would like to believe. In reality,

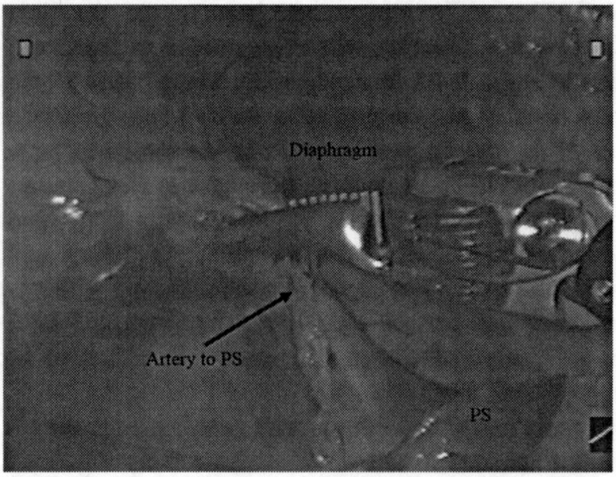

Fig. 31.9 Sealing of the artery to a pulmonary sequestration (PS) using the Gyrus PK in an infant

the backwards fulcrum effect hand movements of standard non-articulating laparoscopic instruments are less precise. Coupled with 2-dimensional imaging, this operation may not be well suited for laparoscopy. Interestingly, the Kasai's we performed robotically at our institution worked well and we felt we saw the anatomy far better robotically then we ever saw in an open procedure since it was magnified 12 times in 3 dimensional viewing (Fig. 31.10). However, we had very limited experience with this procedure as biliary atresia is somewhat rare. In fact, we only had two patients with biliary atresia before the moratorium was put into effect. Both of the robotic Kasai's we performed worked well. One patient still had a working Kasai at the time of this writing. The other underwent transplantation for primary liver disease reasons unrelated to the Kasai which, incidentally, was working well at the time of transplantation. However, we have agreed to the moratorium in principle until a better understanding of the laparoscopic failure can be investigated.

Complex procedures such as a choledochal cyst resection (Fig. 31.11) and the lateral pancreaticojejunostomy (Fig. 31.12) are also more easily accomplished robotically over the laparoscopic counterpart. We performed the first Puestow robotically in 2008 and the case report is still pending publication. Although a handful of pediatric surgeons have performed these procedures laparoscopically, the grand majority of our colleagues are unlikely to attempt these operations laparoscopically due to the challenges posed by rigid non-articulating instruments.

A number of other procedures involving congenital anomalies have been reported. Among these reports are a few congenital diaphragmatic hernia (CDH) repairs in both the Morgagni and Bochdalek varieties. The use of the robot for repairing Bochdalek hernias may be of particular value as early reports of the standard thoracoscopic results were sub-optimal [20]. Success for standard rigid hand held instruments came at a price of a long learning curve of 20–40 cases [21,

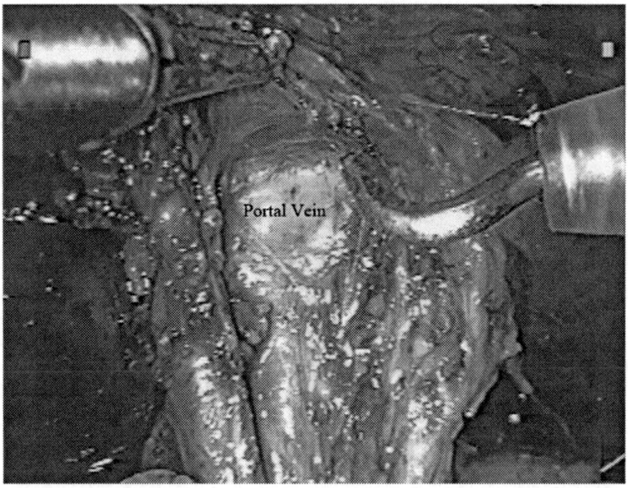

Fig. 31.10 Dissecting the portal vein in a Kasai portoentersotomy

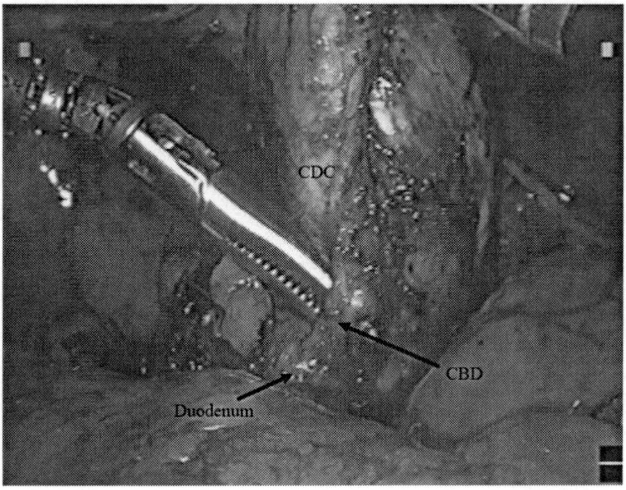

Fig. 31.11 Dissection of the common bile duct (CBD) as it enters the duodenum during a choledochal cyst (CDC) resection

22]. The early robotic success for these repairs is documented in a number of case reports and may prove to have a substantially shorter learning curve [14, 23–25]. Moreover, the benefit of the articulating robotic instruments may be of particular benefit for reaching the most difficult to reach area of the repair, the posterior and lateral aspect of the Bochdalek CDH (Fig. 31.13), an area where no diaphragm is present on the lateral wall. This is the portion of the closure that poses significant

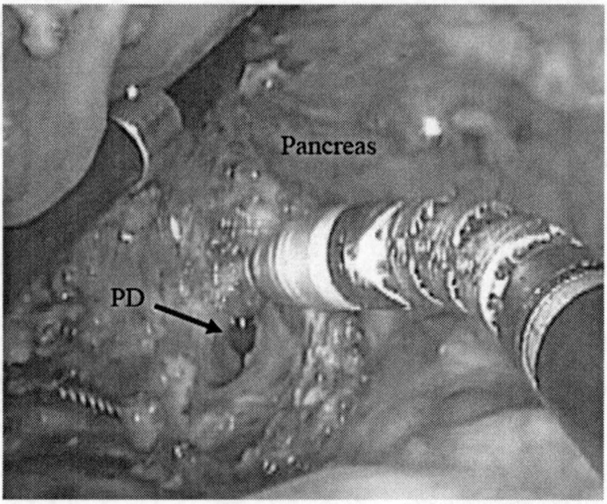

Fig. 31.12 Opening the chronically dilated pancreatic duct (PD) in preparation to construct a lateral pancreaticojejunostomy (Puestow)

challenges for the standard laparoscopic instruments. The rarity of these procedures, however, will make it difficult to ascertain outcomes in a scientific fashion. Long term outcomes studies, even with open techniques, are small and add little insight into the issues and problems these patients encounter. Other intraabdominal congenital anomalies we have dealt with robotically include resections of congenital lymphangiomas and intestinal duplications and repair of a duodenal atresia [26]. The duodenal atresia repair is the youngest patient on record to ever undergo robotic surgery and was less than 1 day of age with a weight of 2.4 kg at the time of surgery. The smallest patient by weight was 6 days old but weighed only 2.2 kg in size and had a Bochdalek CDH repair.

Size is an issue but not as limiting as we had originally imagined. In a review of our first 5 years of robotic surgery, we discovered that 50% of our patients were under 20 kg and 25% were under 10 kg. However, we also discovered that the Da Vinci robot's range of motion and operative utility becomes increasingly limited when the patient's size is less than 3 kg for an abdominal procedure and less than 4 kg for a thoracic procedure [13].

31.5 Recommendations and the Future

It is helpful to utilize the small tricks which make the large Da Vinci robot adapt to small children. Careful attention to patient positioning, robot cart location, and trocar placement will help pediatric surgeons perform even very complex MIS procedures with relative ease. However, repetition is one of the hallmarks of

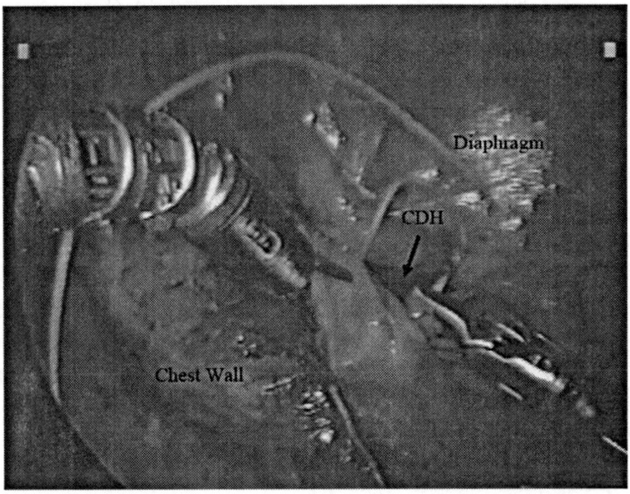

Fig. 31.13 Closing the lateral most aspect of a foramen of Bochdalek CDH with interrupted horizontal mattress stitches

successful surgery and robotics is no exception. Reserving the robot for only the most complex procedures is a common pitfall that will doom any pediatric robotic program. It is critical to maintain familiarity with the nuances of the robot by keeping the OR team involved on a weekly basis. This includes performing routine cases such as cholecystectomies and fundoplications using the robot on a regular basis in order to make things go smoothly when the rare choledochal cyst, mediastinal mass, or diaphragmatic hernia suddenly come in.

The biggest problem facing children's hospitals is the cost. When laparoscopy emerged in pediatric surgery in the mid 1990s, the equipment cost – while increased over open surgery – was within reach of most children's hospitals. Many initial papers condemned laparoscopic surgery due to the cost and increased operative time. But persistent pediatric surgeons thwarted these early criticisms and laparoscopy is the standard of care for many basic pediatric operations. The shorter hospital stays, decreased pain, and improved cosmesis over an open procedure eventually won out, even without definitive prospective randomized proof that laparoscopy was either safer or provided better outcomes. Soon, the increased operating room cost was justified by the lower cost of shorter hospital length of stay and faster return to full activity. A new pediatric surgical society – the International Pediatric Endoscopic Group or IPEG – was formed. As the years passed, MIS became mainstream and has replaced open surgery for many procedures, once again without any proof that laparoscopy was better. Interestingly, the surgeons who made their careers by pushing the envelope in laparoscopic surgery, and were so heavily criticized during those early days, are now the loudest critics of pediatric robotic surgery.

But the technology is making progress at an amazing rate. With the rigid and fulcrum/opposite movement of laparoscopic instruments coupled with 2-dimensional one-eyed view, it's hard to imagine how such limitations can be justified over more precise filtered articulating real time hand movements that mimic the exact motion of the wrist and hand, all utilizing 3-D vision with depth perception. As laparoscopy has been around for 20 years, it can be compared to the experienced golfer who took just a few lessons and has a wicked slice. This golfer has played the same course so many times, and he knows his own wicked slice really well, that he can still get it on the fairway from time to time even though the swing isn't pretty. Robotic surgery puts the precision and accuracy back into pediatric MIS. But the exorbitant cost of the Da Vinci robot has clearly overshadowed its opportunity for growth in general pediatric surgery. The cost of one Da Vinci robot can nearly wipe out the surgical capital budget of any children's hospital for an entire year. This makes purchasing such a device extremely problematic for most pediatric hospitals.

References

1. Hollands, C.M., Dixey, L.N., Torma, M.J.: Technical assessment of porcine enteroenterostomy performed with ZEUS robotic technology. J. Pediatr. Surg. **36**(8), 1231–1233 (2001)
2. Lorincz, A., Langenburg, S., Klein, M.D.: Robotics and the pediatric surgeon. Curr. Opin. Pediatr. **15**(3), 262–266 (2003) Review
3. Gutt, C.N., Markus, B., Kim, Z.G., Meininger, D., Brinkmann, L., Heller, K.: Early experiences of robotic surgery in children. Surg. Endosc. **16**(7), 1083–1086 (2002)
4. Knight, C.G., Lorincz, A., Gidell, K.M., Lelli, J., Klein, M.D., Langenburg, S.E.: Computer-assisted robot-enhanced laparoscopic fundoplication in children. J. Pediatr. Surg. **39**(6), 864–866 (2004); discussion 864–866. Review
5. Lehnert, M., Richter, B., Beyer, P.A., Heller, K.: A prospective study comparing operative time in conventional laparoscopic and robotically assisted Thal semifundoplication in children. J. Pediatr. Surg. **41**(8), 1392–1396 (2006)
6. Meehan, J.J., Sandler, A.: Pediatric robotic surgery: a single-institutional review of the first 100 consecutive cases. Surg. Endosc. **22**(1), 177–182 (2008)
7. Dutta, S., Woo, R., Albanese, C.T.: Minimal access portoenterostomy: advantages and disadvantages of standard laparoscopic and robotic techniques. J. Laparoendosc. Adv. Surg. Tech. A **17**(2), 258–264 (2007)
8. Meehan, J.J., Meehan, T.D., Sandler, A.: Robotic fundoplication in children: resident teaching and a single institutional review of our first 50 patients. J. Pediatr. Surg. **42**(12), 2022–2025 (2007)
9. Meehan, J.J., Elliott, S., Sandler, A.: The robotic approach to complex hepatobiliary anomalies in children: preliminary report. J. Pediatr. Surg. **42**(12), 2110–2114 (2007)
10. Meehan, J.J., Sandler, A.D.: Robotic resection of mediastinal masses in children. J. Laparoendosc. Adv. Surg. Tech. A **18**(1), 114–119 (2008)
11. Meehan, J.J., Phearman, L., Sandler, A.: Robotic pulmonary resections in children: series report and introduction of a new robotic instrument. J. Laparoendosc. Adv. Surg. Tech. A **18**(2), 293–295 (2008)
12. Albassam, A.A., Mallick, M.S., Gado, A., Shoukry, M.: Nissen fundoplication, robotic-assisted versus laparoscopic procedure: a comparative study in children. Eur. J. Pediatr. Surg. **19**, 316–319 (2009)
13. Meehan, J.J.: Robotic surgery in small children: is there room for this? J. Laparoendosc. Adv. Surg. Tech. A **19**, 707–712 (2009)

14. Meehan, J.J., Sandler, A.: Robotic repair of a Bochdalek congenital diaphragmatic hernia in a small neonate: robotic advantages and limitations. J. Pediatr. Surg. **42**(10), 1757–1760 (2007)
15. Rothenberg, S.S.: Experience with 220 consecutive laparoscopic Nissen fundoplications in infants and children. J. Pediatr. Surg. **33**(2), 274–278 (1998)
16. Meehan, J.J., Georgeson, K.E.: The learning curve associated with laparoscopic antireflux surgery in infants and children. J. Pediatr. Surg. **32**(3), 426–429 (1997)
17. Allal, H., Captier, G., Lopez, M., Forgues, D., Galifer, R.B.: Evaluation of 142 consecutive laparoscopic fundoplications in children: effects of the learning curve and technical choice. J. Pediatr. Surg. **36**(6), 921–926 (2001)
18. Rogers, C.G., Blatt, A.M., Miles, G.E., Linehan, W.M., Pinto, P.A.: Concurrent robotic partial adrenalectomy and extra-adrenal pheochromocytoma resection in a pediatric patient with von Hippel-Lindau disease. J. Endourol. **22**(7), 1501–1503 (2008)
19. Meehan, J.J.: The robotic Gyrus PK: a new articulating thermal sealing device and a preliminary series report. J. Laparoendosc. Adv. Surg. Tech. A **18**(1), 183–185 (2008)
20. Arca, M.J., Barnhart, D.C., Lelli, J.L., Jr, Greenfeld, J., Harmon, C.M., Hirschl, R.B., Teitelbaum, D.H.: Early experience with minimally invasive repair of congenital diaphragmatic hernias: results and lessons learned. J. Pediatr. Surg. **38**(11), 1563–1568 (2003) Review
21. Kim, A.C., Bryner, B.S., Akay, B., Geiger, J.D., Hirschl, R.B., Mychaliska, G.B.: Thoracoscopic repair of congenital diaphragmatic hernia in neonates: lessons learned. J. Laparoendosc. Adv. Surg. Tech. A **19**(4), 575–580 (2009)
22. Nguyen, T.L., Le, A.D.: Thoracoscopic repair for congenital diaphragmatic hernia: lessons from 45 cases. J. Pediatr. Surg. **41**(10), 1713–1715 (2006)
23. Meehan, J.J. Robotic repair of a Morgagni congenital diaphragmatic hernia in an infant. J. Robot. Surg. **2**(2), 97–99 (2008)
24. Anderberg, M., Kockum, C.C., Arnbjornsson, E.: Morgagni hernia repair in a small child using da Vinci robotic instruments – a case report. Eur. J. Pediatr. Surg. **19**(2), 110–112 (2009)
25. Slater, B.J., Meehan, J.J.: Robotic repair of congenital diaphragmatic anomalies. J. Laparoendosc. Adv. Surg. Tech. **19**(s1), s123–s127 (2009)
26. Meehan, J.J.: Robotic repair of congenital duodenal atresia: a case report. J. Pediatr. Surg. **42**(7), E31–E33 (2007)

Chapter 32
Applications of Surgical Robotics in Gynecologic Surgery

Rabbie K. Hanna and John F. Boggess

32.1 Introduction

Surgeons strive to minimize surgical complications and new procedures are developed with this goal in mind. It is critical when evaluating new innovations that their be no compromise in overall surgical technique and treatment goals. The emergence of minimal invasive surgery (MIS) has led to a significant reduction in perioperative morbidity, mortality and length of hospital stay as compared to traditional laparotomy. However, current conventional laparoscopy has seen limited application in many complex pelvic procedures due to the pelvis's limited space and complex anatomy. The introduction of robotic assisted MIS has overcome many of these limitations by providing superior dexterity, intuitive movement, 3-D vision, ergonomics and autonomy. The use of the da Vinci surgical system has now become an integral surgical tool in gynecologic surgeries. This chapter will review the current robotic platform's development and use in the field of gynecology and gynecologic oncology since 2005 demonstrating its feasibility and safety.

32.2 Applications in Gynecologic Oncology

Approximately 78,490 women were diagnosed and 28,490 women died from their disease in the United States in 2008 with gynecologic malignancies [1]. The three most common cancers affecting these women are uterine, ovarian and cervical cancers. Treatment often includes major abdominal surgery to remove the primary cancer, usually a hysterectomy and/or bilateral salpingo-oophorectomy, combined with surgical and pathologic assessment of regional lymph nodes to exclude the presence of metastatic disease. Adjuvant treatment with chemotherapy and

J.F. Boggess (✉)
The Division of Gynecologic Oncology, Department of Obstetrics and Gynecology, University of North Carolina, Campus Box 7572, Chapel Hill, NC 27599-7572, USA
e-mail: jboggess@med.unc.edu

J. Rosen et al. (eds.), *Surgical Robotics: Systems Applications and Visions*,
DOI 10.1007/978-1-4419-1126-1_32, © Springer Science+Business Media, LLC 2011

radiation therapy is then prescribed based upon surgical staging. Gynecologic oncology patients are often elderly and have medical co-morbidities that increase their surgical risk. Efforts to reduce surgical morbidity are needed to improve outcomes in this patient population [2].

Minimal invasive surgery (MIS), specifically laparoscopic surgical procedures have gained wide acceptance in the evaluation and treatment of gynecologic cancer patients over the past 10–15 years. Feasibility studies have shown that MIS allows for a quicker postoperative recovery, less postoperative pain, and lower blood loss for most patients. These benefits translate into potential cost savings for the health care system. The use of MIS in gynecologic oncology procedures does not appear to compromise oncologic outcomes [3]. Laparoscopic endometrial cancer staging, including hysterectomy with pelvic and para-aortic lymphadenectomy [4], laparoscopic radical hysterectomy [5–7], and laparoscopic staging of early ovarian cancer [8], have shown improvements with regards to blood loss and recovery compared to laparotomy [8–10]. There are inherent limitations in conventional laparoscopic tools, however. These limitations include a 2-dimensional imaging format, which lacks depth perception, instruments with limited range of motion, poor ergonomics for the surgical team, counter-intuitive surgical movements, and picture and instrument instability secondary to a hand tremor. These limitations have restricted (MIS) utilization in complex pelvic surgical procedures. The introduction of the da Vinci robotic system (Intuitive Surgical Corporation, Sunnyvale, California, USA) has enhanced laparoscopic surgery by introducing 3-D vision, instrument dexterity and precision due to seven degrees of freedom and tremor reduction. The da Vinci surgical system was approved by the FDA in 2005 for use in gynecologic surgery. The three-dimensional image provided by the camera gives the surgeon a life-like view of the surgical field with the added benefit of magnification. The surgeon manipulates instruments equipped with 'endowrist' movements that duplicate the movements of the hand and computer assistance filters tremor and scales movement at the surgeon's discretion [11].

32.2.1 Endometrial Cancer

One significant application of robotics in gynecologic oncology has been in the management of endometrial cancer, the most common gynecologic malignancy [1]. The foundation of endometrial cancer treatment is surgical staging, which begins with hysterectomy and removal of the ovaries and in most cases includes pelvic and para-aortic node dissection. Appropriate evaluation of the regional lymph nodes directs adjuvant therapy with chemotherapy and/or radiation therapy [12]. Staging endometrial cancer, therefore is both a pelvic and abdominal procedure (Fig. 32.1).

While endometrial cancer staging can be performed with the first generation or "Standard" da Vinci robotic system, limited range of arm motion can preclude appropriate para-aortic lymph node evaluation in some patients. Even with placing the robotic ports very high on the abdomen, proper range of motion to reach high on

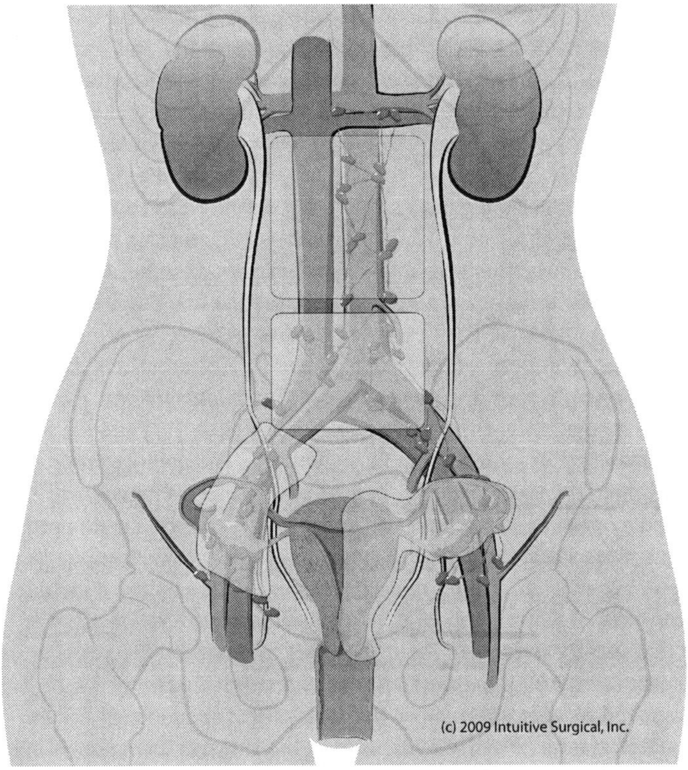

Fig. 32.1 Surgical staging of endometrial cancer involves a total hysterectomy with bilateral salpingo-oophorectomy (**a**), bilateral pelvic (**b**) and paraaortic lymph node dissection (**c**). Copyright 2009 Intuitive surgical, Inc. Used with permission

the aorta and still operate to the pelvic floor can be difficult. Some surgeons have overcome this problem by docking the robot twice, once oriented to pelvic surgery and a second time over the patient's head to complete the high para-aortic node dissection. This method works well, but does add to the complexity of the procedure and adds the number of surgical ports required to complete the operation. The second generation da Vinci robotic system, with its wider range of arm motion and longer instruments improves the range of motion of the platform to sufficiently perform both the pelvic and abdominal components of the surgery with a single docking procedure [13]. The da Vinci S system has allowed a single docking set-up to be utilized for both portions of the staging without re-docking the system. This allows for a simplified surgical technique with fewer surgical ports.

As mentioned, endometrial carcinoma is the most common gynecologic malignancy in the United States and it is estimated that in 2008 that 40,100 women were diagnosed with endometrial cancer and 7,470 died of their disease [1].

Because women with uterine cancer are often obese, diabetic and hypertensive, laparotomy is associated with a high operative morbidity and long postoperative recovery [13, 14]. MIS has decreased complications and shortened recovery for women with endometrial cancer compared with laparotomy [13]. However, conventional laparoscopy has been associated with long operative times and high conversion rates [15]. The role of robotic surgery in the surgical management of endometrial cancer has been gaining a slow but definite acceptance among gynecologic oncology surgeons. Table 32.1 demonstrates the various studies published in the field of robotic surgeries and endometrial cancer. Marchal et al. [16] were the first to publish their initial experience in utilizing the Da Vinci Robotic system in gynecologic malignancies. They reported on five patients who underwent robotic assisted surgical treatment of endometrial cancer. The authors did not provide details on demographics or operative data for the patients treated. This early report was followed by the initial experience of Reynolds et al. [17] and the use of robotic assisted laparoscopy in gynecologic malignancies. The authors reported their data of four endometrial cancer patients, two of whom had a diagnosis of endometrial cancer found by prior hysterectomy and underwent a surgical staging procedure and two patients who underwent total robotic assisted hysterectomies, bilateral salpingo-oophorectomy and lymph node dissections. The authors concluded that the mechanical-wristed instruments of the da Vinci Surgical System allowed improved dexterity that readily overcame the difficulties imposed by the scarring from prior surgical procedures and the complexity of the surgical staging itself. The limitations recognized by this group were the lack of tactile feedback for the operating surgeon, which can be partially overcome as the surgeon learns to recognize force applied by visual cues, such as tissue blanching or deflection of soft tissue structures. Veljovich and colleagues [18] reported a comparison of 25 patients who underwent robotic surgical staging to 131 patients undergoing the procedure though conventional laparotomy. This group found a longer operative time (283 vs 139 min, $p < 0.0001$) significantly less blood loss and shorter hospital stay in favor of robotic surgery. The lymph node count was comparable (17.5 vs. 13.1). They documented the feasibility of robotic surgery in the field of gynecologic oncology as its use in their practice has facilitated a dramatic expansion of their MIS practice. In a study to compare the operative and peri-operative outcomes, complications, adequacy of staging, and cost in hysterectomy and lymphadenectomy completed via robotic assistance, laparotomy, and laparoscopy for endometrial cancer staging, Bell et al. [19] reported on a total of 110 patients who's surgeries were performed by a single surgeon at a single institution. In Bell's series, operative time did not differ between the robotic and laparoscopic approaches (184 vs. 171 min) but was longer than laparotomy (108.6 min). Lymph node retrieval was similar in all three groups (14–17 lymph nodes). Boggess et al. [13] published their experience of robotic assisted endometrial cancer staging as compared to a historical cohort of surgical staging performed via the laparotomy and conventional laparoscopic approaches. In their series, a total of 322 women underwent endometrial cancer staging: 138 by laparotomy; 81 patients by laparoscopy and 103 by utilizing the robotic platform. The robotic surgery cohort had the highest lymph

Table 32.1 The various studies published in the field of robotic surgeries and endometrial cancer

Author	Mode of surgical approach	*N*	Age in years	BMI	Stage	Operative time	LN count	EBL	Transfusion	Hospitalization (days)	Major complications	Conversion to laparotomy
Marchal et al. (2005) [16]	Robotic	5	53** (29–27)	NR*	I	181 (107–300)	11 (4–21)	83 (0–900)	NR	8** (4–33)	17%**	1
Reynolds et al. (2005) [17]	Robotic	4	47.6** (42–68)	27** (22–39.6)	IA–IIIA	257** (174–345)	15 (5–29)	50**	0	2** (1–6)	0%	0
Veljovich et al. (2008) [18]	Laparotomy	131	63 (30–92)	32.2 (16.4–65.8)	NR	139 (69–294)	13.1 (1–42)	197.6 (25–900)	NR	127 (13–576)	20.6%	N/A
	Laparoscopy	4	54 (51–67)	24.6 (22–29)	NR	255 (220–305)	20.3 (7–39)	75 (50–100)	NR	28.8 h (22–47)	NR	0
	Robotic	25	59.5 (36–85)	27.6 (18.7–49.5)	IA–III	283 (171–443)	17.5 (2–32)	66.6 (10–300)	0	40.3 h (17–215)	8%	1
Bell et al. (2008) [19]	Laparotomy	40	72.3 ± 12.5	31.8 ± 7.7	NR	108.6 ± 41.4	14.9 ± 4.8	316.8 ± 282.1	6	4.0 ± 1.5	27.5%	N/A
	Laparoscopy	30	68.4 ± 11.9	31.9 ± 9.8	NR	171.1 ± 36.2	17.1 ± 7.1	253.0 ± 427.7	3	2.0 ± 1.2	20.0%	NR
	Robotic	40	63.0 ± 10.1	33.0 ± 8.5	NR	184.0 ± 41.3	17.0 ± 7.8	166.0 ± 225.9	2	2.3 ± 1.3	7.5%	NR
Boggess et al. (2008) [13]	Laparotomy	138	64 ± 12.8	34.7 ± 9.2	IA–IVB	146.5 ± 48.8	14.9 ± 13.7	266.0 ± 184.5	1.5%	4.4 ± 2.0	21.7%	N/A
	Laparoscopy	81	62 ± 10.8	29.0 ± 6.5	IA–IVB	213.4 ± 34.7	23.1 ± 11.4	145.8 ± 105.6	2.5%	1.2 ± 0.5	8.6%	4
	Robotic	103	61.9 ± 10.6	32.9 ± 7.6	IA–IIIC	191.2 ± 36.0	32.9 ± 26.2	74.5 ± 101.2	1.0%	1.0 ± 0.2	2.9%	3
	Laparotomy	106	62.5 ± 10.8	34.0 ± 9.3	IA–IV	79 ± 17		241 ± 115	8.5%	3.2 ± 1.2	20.8%	N/A

(continued)

Table 32.1 (continued)

Author	Mode of surgical approach	*N*	Age in years	BMI	Stage	Operative time	LN count	EBL	Transfusion	Hospitalization (days)	Major complications	Conversion to laparotomy
DeNardis et al. (2008) [22]	Robotic	56	58.9 ± 10.3	28.5 ± 6.4	IA–IIIC	177 ± 55	18.0 ± 9.6 18.6 ± 12.4	105 ± 77	0%	1.0 ± 0.5	3.6%	3
Seamon et al. (2008) [23]	Robotic	92	59 (34–82)	34 (19–58)	IA–IV	242 ± 50	(9–56) 29	99 ± 83	3	(1–46) 1	13%	12.4% 13/105
Peiretti et al. (2009) [24]	Robotic	80	58.3 (55.7–60.9)	25.2 (23.6–26.7)	1A–IIIC	181.1 (166.7–195.5)	15.5 (3–33)	50 (5–150)	1/80	2.5 (2.2–2.7)	6%	3.7% (3/80)

Numbers in Median (Range) where applicable or Mean ± SD
* *NR* Not reported
** Figures are derived from all the patients in the series

node count, shortest hospital stay and the least blood loss among the three groups. Conversion-to-laparotomy rates were lower in the robotic group (2.9%) than the laparoscopic group (4.9%). Boggess et al. [13] concluded that robotic assistance may allow for an easier and more comprehensive MIS technique than laparoscopy. Additionally, they also echoed the feasibility and safety of the utilization of the robotic platform in the surgical staging of endometrial cancer. In a follow-up study from this group of patients Gerhig et al. [14] analyzed the efficacy of each surgical method in obese patients. The role of obesity in the development of endometrial cancer has been well established and approximately 70–90% of estrogen dependent endometrial cancer patients are obese [20]. Furthermore, endometrial cancer patients with a BMI more then 40 kg/m^2 were found to have a higher relative risk of death at 6.25 [21]. Thus, it would be important to establish a surgical approach that minimizes surgical morbidity in this high-risk population. Gehrig et al. [14] found in a cohort of 36 obese and 13 morbidly obese women who underwent surgery with the da Vinci robotic system compared with 25 obese and seven morbidly obese women who underwent traditional laparoscopy that more women were successfully staged in the robotic cohort (92% vs. 84%). While there was not a significant difference in surgical complications between the two groups, none of the planned robotic procedures as compared to two of the latter were converted to laparotomy. Furthermore, a shorter operative period of 26 min was noted in the robotic group [14]. Based on these findings, it is suggested that robotic assisted surgery is the treatment of choice for obese endometrial cancer patients.

DeNardis et al. compared the operative performance, pathology, and morbidity between robotic assisted laparoscopy and laparotomy for endometrial cancer staging in a retrospective study [22]. The patients selected for the robotic platform were significantly thinner and younger with less co-morbidities than the laparotomy group. The findings of their study were not surprising and echoed the findings of other studies in terms of less blood less, hospital stay and peri-operative complications. The conversion rate was 5.4% and the operative time for the robotic group was 177 vs 79 min of the laparotomy group.

Seamon et al. [23] reported the Ohio State University experience in attempting a robotic assisted comprehensive surgical staging in 105 patients with a conversion rate of 12.4%. Seventy-nine of the 92 patients who had the robotic assisted procedure had a comprehensive staging procedure (total hysterectomy, bilateral salpingo-oophorectomy, pelvic and para-aortic lymph node dissection) [23]. In a study to show the effect of robotic surgery on the surgical approach in patients with endometrial cancer, Peiretti et al. [24], out of Italy, reported on 80 consecutive patients with endometrial cancer. They reported a mean BMI of 25.2, a mean operative time of 181 min, and a mean hospital stay of 2.5 days. The conversion rate in this series was 3.7% (three patients).

The only study to date to evaluate survival was conducted by Mendivil et al. [25]. They reported on a 3 year interim survival analysis of patients who had undergone the surgical staging via robotic assisted laparoscopy. This data was compared to the progression free (PFS) and overall survival (OS) of endometrial cancer cohorts who were surgically staged via laparotomy or laparoscopy cohorts.

One hundred and forty-one women had undergone robotic assisted comprehensive surgical staging during the study period of May 2005 to June 2008. At 40 months of follow up, the PFS for the TRH cohort was 95% and the OS was 94%. This was not significantly different amongst the three groups [25].

32.2.1.1 Recurrent Endometrial Cancer

In a systemic literature review [26], recurrent endometrial cancer occurs in approximately 13% of patients. Recurrences are pelvic in 39% vs. distant in the remainder 61% of patients. The different treatment modalities for recurrent endometrial cancer are surgical resection, radiotherapy or systemic agents consisting of chemotherapy or hormonal therapy. The use of surgical resection is the first choice in patients with local recurrence in an irradiated area [27]. Although no defined patient selection criteria exist for surgical resection, several authors have published their experience in this matter [28–31], none of whom were performed by robotic assisted laparoscopy.

The use of robotic assisted laparoscopic resection for recurrent endometrial cancer has been performed at our institution. An otherwise healthy patient with a localized recurrence of endometrial cancer with a prior incomplete surgical staging 5 years prior to presentation was evaluated and was found to have a left pelvic sidewall isolated recurrence measuring 6×8 cm with no radiologic evidence of distant metastasis, in addition to complaints of left lower limb venous congestion and negative Doppler findings of venous thromboembolism. An optimal cytoreduction was successfully achieved. In this case, partial resection of the external iliac vein was required due to tumor involvement and the venous defect was repaired robotically by the primary surgeon. The patient was discharged 2 days later, and has received three cycles of carboplatin and Taxol and remains without evidence of disease (unpublished data).

32.2.1.2 Patient Evaluation for Robotic Endometrial Surgical Staging

After an appropriate diagnosis of endometrial with an endometrial biopsy and/or a diagnostic dilation and curettage, patients should be evaluated for a comprehensive surgical staging. As many of these patients have associated co-morbidities, an appropriate evaluation in preparation for operative management is emphasized. Otherwise, other modalities of management (primary radiation or chemotherapy) are applied to those who cannot undergo surgical intervention. An anesthesia evaluation for patient tolerability to general anesthesia and surgical positioning in a steep Trendelenberg fashion as both are essential requirements for a successful robotic assisted comprehensive staging for endometrial cancer.

For patients with a diagnosis of a grade 1 endometrial cancer (the most indolent form of this malignancy) only a comprehensive physical examination and a chest radiograph are required [12]. Although some authorities do not recommend preoperative imaging for higher grade endometrial cancer [12], it is our practice to obtain preoperative computer tomography scans to assist in surgical planning or changing the overall plan of management if necessary. When there is confirmed cervical involvement, the plan of surgical management changes from a simple hysterectomy to that of a radical hysterectomy in most of the patients and at times, primary radiation is offered. Preoperative CA-125 levels, in patients with endometrial cancers of high grades, may assist in predicting treatment response or in post-treatment surveillance [12]. In patients who are suspected to have a large uterus on physical examination or those whose examination is limited by their body habitus, a pelvic sonogram to assess for uterine size assists in counseling these patients for a possible mini-laparotomy for specimen removal rather than morcellating it.

In our practice, almost all patients are eligible for surgical management of endometrial cancer are candidates for robotic-assisted surgery; we do not have specific age, weight, or body mass index (BMI) limits. For those who have large uteri due to fibroids, a mini-laparotomy is performed for removal of the uterus after completion of the full surgical staging. This has not affected the overall morbidity and length of hospitalization for these patients.

Operative Technique

A comprehensive endometrial surgical staging includes a hysterectomy, bilateral salpingo-oophorectomy, and para-aortic and pelvic lymph node dissection according to the International Federation of Gynecology and Obstetrics staging system [32]. The senior author has described this surgical procedure [13, 33] and an archived demonstration can be accessed at the National Library of Congress at http://www.nlm.nih.gov/medlineplus/surgeryvideos.html [34]. Other descriptions can be found within feasibility papers of robotic assisted surgeries for uterine cancer [19, 20].

32.2.2 *Cervical Cancer*

Radical hysterectomy (Fig. 32.2) is the gold standard surgical treatment of cervical cancer and is considered the most elegant and complex pelvic procedure by many gynecologic oncology surgeons. The robotic platform has emerged as an essential surgical tool in performing this procedure as it allows excellent surgical maneuverability, especially around the great vessels and ureters in the pelvis.

Cervical cancer is the second most common female malignancy worldwide with 11,070 new cases diagnosed and 3,870 deaths in the USA alone in 2008 [1]. The treatment of early cervical cancer often includes surgical management. The treatment of stage 1A-1 is accomplished by a cone biopsy or simple hysterectomy, while Stage 1A-2 and 1B-1 are treated with radical hysterectomy. Radical hysterectomy differs from simple hysterectomy in that it includes removal of the

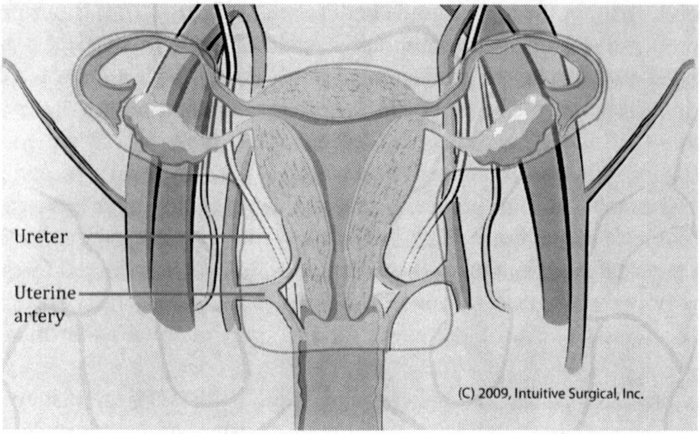

Fig. 32.2 A radical hysterectomy involves removal of the uterus in addition to the parametrial tissue and a larger portion of the upper vagina. The ureters are dissected laterally and the uterine arteries are divided close to their origins. Copyright 2009 Intuitive surgical, Inc. Used with permission

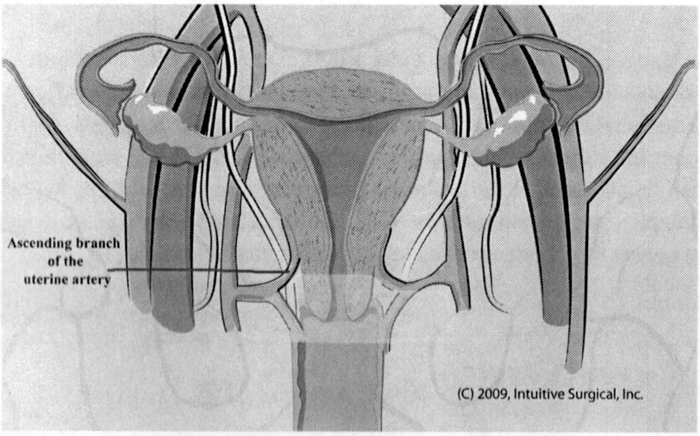

Fig. 32.3 A radical trachelectomy involves removal of the distal portion of the cervix and proximal vagina in addition to preservation of the ascending branch of the uterine artery. Copyright 2009 Intuitive surgical, Inc. Used with permission

parametrial tissues, pelvic lymph nodes and upper third of the vagina. Women who develop cervical cancer during their reproductive years can be treated with radical trachelectomy, a technique that preserves the proximal cervix and uterus in an effort to preserve fertility options (Fig. 32.3) rather than a radical hysterectomy without impairing their survival [35, 36].

32.2.2.1 Radical Hysterectomy

Operative Technique

Various authors have described the surgical technique in performing a robotic assisted radical hysterectomy [37–42]. The senior author demonstrated the current technique performed at The University of North Carolina during a worldwide webcast in February 2006. A full-narrated version can be accessed at http://www.nlm.nih.gov/medlineplus/cervicalcancer.html#cat63 [43].

The development of (MIS) has allowed an appropriate and comparable surgical treatment option for cervical cancer patients as compared to that of laparotomy. Since the earliest publications of laparoscopic radical hysterectomy [44, 45], the feasibility and safety of this procedure have been demonstrated.

The first feasibility study employing the robotic platform in the treatment of stage Ib1 cervical cancer to determine whether type III radical hysterectomy and pelvic node dissection could be performed with acceptable surgical outcomes was presented by the senior author and his colleagues [46] at the 37th annual meeting of the Society of Gynecologic Oncologists in 2006. Seven consecutive modified type III radical hysterectomies with pelvic lymph node dissection for stage Ib1 cervical cancer were performed using the da Vinci Robotic System. An oophoropexy was performed when indicated. The average patient age was 41 years old, and the average BMI was 27.1 (18.6–31.8). The largest lesion was 4 cm. A type III radical hysterectomy with pelvic node dissection was successfully performed in all seven patients without any intra-operative complications. The mean blood loss was 143 (25–300) cc, and operative time was 252 (187–290) min. All patients were discharged on postoperative day 1, none received intravenous pain medication, and none required a blood transfusion. The only postoperative complication was that of a vaginal cuff abscess 2 weeks postoperatively. All patients passed their voiding trial within 1 week. All surgical margins were negative, and the average number of pelvic nodes recovered was 35 (25–55). The authors concluded that robotic modified type III radical hysterectomy with pelvic node dissection is feasible in patients with stage Ib1 cervical cancer. Surgical margins and lymph node retrieval were at least comparable to those of open surgery. When indicated, oophoropexy is facilitated by the robotic suturing advantage. The operative time in this initial series was comparable to published times for open and laparoscopic radical hysterectomy, with significantly lower blood loss and hospital stay.

The first published RRH was that Marchal et al. [16], seven of the 30 patients described had cervical cancer. Details pertinent to the operative outcomes for the RRH are not detailed in their publication. This publication was followed by that of Sert et al. [47], these authors published a case report describing a RRH and bilateral pelvic lymph node dissection for a patient with Stage Ib1 cervical carcinoma concluding that the radical dissection is much more precise than the conventional laparoscopic radical hysterectomy [47]. The same authors evaluated the feasibility and efficacy of RRH and bilateral pelvic lymph node dissection for early cervical carcinoma as compared to LRH [48]. Fifteen patients with early-stage cervical carcinoma were involved in this pilot case-control study. One patient from the conventional laparoscopic cohort was excluded from the analysis due to equipment failure necessitating conversion to laparotomy. There were no conversions observed. The median operative time, blood loss and hospital stay were in favor of the robotic group.

Several authors have published their series of robotic radical hysterectomies and have shown the advantages of robotic assisted laparoscopy as compared to laparotomy and conventional laparoscopy. Table 32.2 represents a summary of the various published studies of robotic radical hysterectomies for cervical cancer. Magrina et al. [49] published their initial cohort of 27 RRH and compared it to two matched groups of patients who underwent a radical hysterectomy via the laparoscopic and laparotomy approaches. A mixture of cervical and endometrial cancer was involved in all three cohorts of Magrina's comparison. Both laparoscopic approaches took longer but were associated with less blood loss and hospital stay as compared to the laparotomy group, there was no difference in the number of lymph node yield nor in the intra- and postoperative complication rates. Kim et al. [38] described ten patients with early stage cervical cancer who had a RRH concluding the feasibility of this surgical approach. Fanning et al. [50] reported a series of RRH procedures in 20 consecutive stage IB1–IIA cervical cancer patients confirming a better anatomic dissection and operative maneuverability. Nezhat et al. [40] compared two cohorts of patients with early cervical cancer. Thirteen RRH were compared to 21 patients who had a conventional laparoscopic radical hysterectomy LRH. This group concluded that both approaches are equivalent with respect to operative time, blood loss, hospital stay and oncologic outcome.

The group of the University of North Carolina published their experience [37] of 51 RRH and compared them to a historical cohort of 49 ARH. In this comparison, significant differences were found in operative times, blood loss, lymph node yield, and hospital stay, all in favor of the RRH group concluding the superiority of RRH over ARH. Ko and colleagues [51] reported a cohort of 48 patients with early stage cervical cancer including 16 patients who underwent a RRH and 32 had an ARH. Their conclusions reinforced those cited above.

A multi-institutional experience of RRH among five different gynecologic oncologists without prior experience of laparoscopic radical hysterectomy was reported by Lowe et al. [52]. They reported on 42 patients with early stage cervical cancer (1A1–1B2). The median BMI was 25.1. The median operative time was 215 min and a median EBL of 50 cc. The median hospital stay was 1 day. From an operative complication standpoint, there was one ureteral injury repaired via the robot and one bladder injury near the trigone requiring conversion to laparotomy for repair. They echoed the feasibility of the robotic platform and added that it represents a tremendous technological advancement offering the potential to redefine the surgical options for the gynecologic oncologists.

Another multi-surgeon experience of three different surgeons was reported by Persson et al. [41] in which 64 patients with cervical cancer patients underwent RRH were included. Sixty-four had cervical cancer and 16 had stage II endometrial cancer. The reported operative time reduced as the number of procedures performed

Table 32.2 The various studies published in the field of robotic surgeries and cervical cancer

Author	Mode of surgical approach	*N*	Age in years	BMI	Stage	Operative time	LN count	EBL	Transfusion	Hospitalization (days)	Major complications	Conversion to laparotomy
Magrina et al. (2008) [49] This series included radical hysterectomies for endometrial cancer and cervical cancer as well	Laparotomy	35	52 (30–65)	27 (16–39)	N/A	166.8 (122–237)	26 (18–46)	350 (50–1200)	3	3 (2–7)	5 (15%)	N/A
	Laparoscopy	31	53 (26–90)	26 (20–43)	N/A	220.4 (165–300)	25 (14–52)	200 (50–520)	0	2 (1–8	3 (9%)	0
	Robotic	27	48 (2686)	26 (20–50)	N/A	189.9 (119–281)	26 (10–36)	100 (50–600)	1	1 (1–4)	2 (7%)	0
Sert et al. (2008) [48]	Laparoscopy	7	45	NR	IA1 & IB1	300 (225–375)	15	160	NR	8	5/7	0
	Robotic	7	41	NR	IA2 & IB1	241 (160–445)	13	71	NR	4	4/7	0
Kim et al. (2008) [38]	Robotic	10	50 (34–75)	(20.3–31.6)	IA2–IB1	207 (120–240)	27.6 (12–52)	355	0	7.9 (5–17)	0%	0
Fanning et al. (2008) [50]	Robotic	20	44 (32–53)	69.9 kg (49.9–79.4)	IB1–IIA	6.5 h (3.5–8.5)	18 (15–35)	300 (100–475)	0	1	10%	0
Nezhat et al. (2008) [40]	Laparoscopic	21	46.8 (29–63)	NR	IA1–IIA	318 (200–464)	31 (10–61)	200 (100–500)	NR	3.8 (2–11)	6 (20%)	0
	Robotic	13	54.8 (39–78)	NR	IA1–IIA	323 (232–453)	25 (11–51)	157 (50–400)	NR	2.7 (1–6)	3 (23%)	0
Boggess et al. (2008) [37]	Laparotomy	49	41.9 ± 11.2	26.1 ± 5.1	IA2–IIA	247.8 ± 48.8	23.3 ± 12.7	416.8 ± 188.1	4	3.2	6 (12.2%)	N/A
	Robotic	51	47.4 ± 12.9	28.6 ± 7.2	IA1–IIA	210.9 ± 45.5	33.8 ± 14.2	96.5 ± 85.5	0	1	2 (3.9%)	0
Ko et al. (2008) [51]	Laparotomy	32	41.7 ± 8.1	26.6 ± 5.9	IA1–IIA	213 (113–308)	17.1	450 (200–3,500)	10	4 (3–8)	7(28.8%)	N/A
	Robotic	16	42.3 ± 7.9	27.6 ± 6.4	IA1–IB1	287 (199–364)	15.6	50 (20–400)	1	1.5 (1–4)	2 (12.5%)	0

(continued)

Table 32.2 (continued)

Author	Mode of surgical approach	*N*	Age in years	BMI	Stage	Operative time	LN count	EBL	Transfusion	Hospitalization (days)	Major complications	Conversion to laparotomy
Lowe et al. (2009) [52]	Robotic	42	41	25.1	IA1–IB2	215 (120–606)	25	50 (25–150)	0	1	7 (16.6%)	1
Persson et al. (2009) [41]	Robotic	80	48 (23–86)	24.4 (17.5–39.0)	IA1–IIA	(132–475) 262	26 (15–55)	150 (25–1,300)	0	3 (1–9)	34 (42.5%)	4
Estape et al. (2009) [42]	Laparotomy	14	42.0 ± 12.0	29.5 ± 6.4	1B1–1B2	114 ± 36	25.7 ± 11.5	621.4 ± 294.0	5	4.0 ± 1.7	4 (28.6%)	N/A
	Laparoscopy	17	52.8 ± 14.2	28.1 ± 4.8	1A2–1B2	132 ± 42	18.6 ± 5.3	209.4 ± 169.9	0	2.3 ± 1.4	4 (23.5%)	NR
	Robotic	32	55.0 ± 12.7	29.7 ± 3.2	1B1–1B2	144 ± 119.4	32.4 ± 10.0	130.0 ± 119.4	1	2.6 ± 2.1	6 (18.7%)	NR

Numbers in Median (Range) where applicable or Mean ± SD
*NR Not reported
** Figures are derived from all the patients in the series ± 36.2

reached 176 and 132 min after 9 and 34 procedures respectively. Additionally, the rate of preoperative complications reduced significantly as more procedures were performed. Their findings were consistent with other reports in terms of feasibility of this procedure in addition to recommending nerve-sparing techniques to be developed. In a study to add to the growing body of literature of robotics feasibility in gynecologic oncology, Estape et al. [42] reported 32 consecutive patients with RRH for cervical cancer and compared them to a historical cohort of 17 patients with LRH and 14 patients with RAH. In this series, the laparoscopic intra-operative complications were significantly decreased with the introduction of robotic assisted laparoscopy from 11.8 to 3.1%. Additionally, they favored the robotic platform over conventional laparoscopy even during the surgeon's learning phase.

32.2.2.2 Radical Trachelectomy (Fig. 32.3)

In patients with early cervical cancer, preservation of fertility is an option for which a radical trachelectomy is offered. This procedure consists of removal of the uterine cervix with its associated parametria in addition to an upper vaginectomy and a regional lymphadenectomy.

The patient must meet the following criteria before offering a fertility preserving radical procedure [53]:

1. Desire to preserve fertility
2. No clinical evidence of impaired fertility
3. Stage IA2 to IB (International Federation of Gynecology and Obstetrics)
4. Lesion size <2.0 cm
5. Limited endocervical involvement at colposcopic evaluation
6. No evidence of pelvic node metastasis

This procedure has been approached through traditional laparotomy [54], vaginally [55] as well as laparoscopically [53, 56, 57]. Most procedures are started with a laparoscopic lymph node dissection and the radical trachelectomy is completed through the vaginal route, a pure laparoscopic approach is not favored secondary to the complexity of the procedure. The first reported cases of robotic radical trachelectomy were by Persson et al. [58]. Two patients with early stage cervical cancer (1A1 and a small 1B1) were reported. The operative times were 359 and 387 min with an estimated blood loss (EBL) was 100–150. The prolonged operative time was attributed to performing a sentinel lymph node biopsy and to the time waiting for frozen section results. No peri-operative complications were reported. They concluded that the robotic approach for this surgical procedure is a safe and feasible alternative to its combined laparoscopic and vaginal counterparts [59].

Another report of a radical trachelectomy for a cervical adenosarcoma via robotic assistance was reported by Geisler et al. [59]. The operative time was 172 min with an EBL of 100 ml. Chuang et al. [60] reported a case report of a robotic assisted radical trachelectomy with a detailed description of the surgical procedure. The operative time was 345 min, an EBL of 200 ml and a lymph node yield of 43.

To date, we have performed a total of six robotic assisted fertility-sparing surgeries at The University of North Carolina (unpublished data). The median age was 27.3 yo (19.6–33.4 yo), and a median BMI of 23.49 kg/m^2 (16.61–31.73). All the patients had stage 1B-1 cervical cancer except for one who had a 3 mm II-A lesion. The ability to preserve the ascending branch of the uterine artery was successful in nine out of ten arteries among the last five patients in whom the original plan was uterine artery preservation. The median console time was 193 min (range, 177–277) with an improvement of 70–90 min after the first two cases. The median estimated blood loss was 63 ml (range 25–100) and the mean hospitalization was 1 day (range 1–2). There were no laparotomy conversions, nor any blood transfusions. Two minor complications were observed; slight left thigh numbness and a vaginal apex cellulitis treated with antibiotics. One patient presented with pelvic pain and radiologic imaging suspicious for a large lymphocyst which was managed via laparoscopic drainage. Experience performing robotic assisted radical trachelectomy is early, but it appears that the advantages of an abdominal approach, such as a greater parametrial dissection combined with the patient advantages of a laparoscopic/vaginal approach may make this an attractive procedure for fertility preservation in young women with cervical cancer.

32.2.2.3 Radical Parametrectomy

An occasional diagnosis of cervical cancer is made after a simple hysterectomy for benign indications; such patients are offered a radical parametrectomy that includes an upper vaginectomy, removal of the parametria and a regional lymph node dissection or adjuvant radiation therapy. Daniel and Brunschwig were the first to describe this procedure [61]. It was routinely performed via a laparotomy approach but with the development of laparoscopic techniques, several authors have described this procedure utilizing conventional laparoscopy [62–65].

Ramirez and colleagues [66] reported the first series of robotic assisted parametrectomy and pelvic lymphadenectomy. Five patients with squamous cell cervical carcinoma were described, with no conversions to laparotomy. The median operative time was 365 min, an estimated blood loss of 100 ml and a hospital stay of 1 day. There was one intra-operative complication, a cystotomy. Two postoperative complications were reported which occurred in the same patient; a vesicovaginal fistula and a lymphocyst. They concluded that a robotic radical parametrectomy is a feasible and safe procedure.

The senior author has performed a series of radical parametrectomy, upper vaginectomy with bilateral pelvic lymph node dissection utilizing the robotic platform on five patients with incidental 1B-1 cervical carcinomas from prior simple hysterectomies (unpublished data). The route of prior hysterectomies: two total vaginal hysterectomies and bilateral salpingo-oophorectomy (BSO), two via a laparotomy and one via a robotic hysterectomy and BSO. The mean console time was 208 min (170–270), EBL 75 (50–250 ml) and length of stay was 1 day (1–2). The mean lymph node count was 29 (18–41). Two intra-operative cystotomies were encountered and repaired with any residual dysfunction. There was one mild right genitofemoral nerve dysfunction. Robotic assistance for this complex procedure allows for better surgical technique in the face of adhesions from the preceding hysterectomy and may become the procedure of choice in women who choose surgery over radiation when an occult cervical cancer is diagnosed post hysterectomy.

32.2.2.4 Retroperitoneal Lymph Node Dissection

In patients with advanced cervical cancer (IB2–IIIB) the incidence of para-aortic lymph node metastasis despite a negative FDG-PET/CT imaging is 8–11% [67, 68]. Thus many authors have reported on surgically staging advanced cervical cancer with retro-peritoneal para-aortic lymphadenectomy to help design the radiation fields. To eliminate or decrease the amount of postoperative intestinal adhesions that occur after a transperitoneal approach, a retroperitoneal (extraperitoneal) approach has been developed [69–75]. Because the robotic platform provides a significant advantage over conventional laparoscopy, authors have developed and reported their techniques in robotic assisted extraperitoneal para-aortic lymphadenectomy [76, 77]. Vergote et al. [77] from Belgium reported the first description of this technique as it was applied to five patients (four with IIB and one with IIIB squamous cervical carcinoma) after having negative preoperative staging Positron Emission Tomography – Computed Tomography (PET-CT) scans for para-aortic lymph node involvement. There was one operative complication: damage to the right ureter that was repaired robotically with a stent placement and appropriate suturing. There were no conversions to laparotomy. An EBL of less than 50 ml was reported. The operative times decreased as each procedure was performed from 139 to 60 min. The authors concluded that the robotic platform offers advantages due to the steady 3-D visualization, instrumentation with articulating tips that allow for 7 degrees of movement surpassing the human hands mobility, and in addition if needed a downscaling of the surgeons movements (without tremor) increasing the accuracy and precision. They emphasize certain technical aspects of the procedure that can be found in their publication [77].

Magrina et al. [76] of the Mayo Clinic described a case report of a 32 yo female with IB2 cervical with CT scan findings of a few enlarged aortic lymph nodes. They performed a retro-peritoneal robotic assisted node dissection with an operative time of 155 min and an EBL of 30 ml. Both groups concluded the needs for further evaluation of the techniques as the robotic platform offers significant advantages over conventional laparoscopy.

Narducci et al. [78] reported their robotic experience performing extraperitoneal para-aortic lymphadenectomy in six patients (a male with germ of the testicle, four women with locally advance cervical cancer and another woman with bulky cancer of the vaginal cuff). A complete para-aortic lymphadenectomy was performed in one patient complicated by a lymphocele with moderate hematoma managed by drainage via CT scan guidance. A left para-aortic lymphadenectomy was performed in the other five patients without complications. They reported an operative time of

130–240 min and an EBL of <50–200 cc. The lymph node yield was 3–20.They concluded that a para-aortic lymphadenectomy up to the left renal vein is feasible with the robotic platform.

32.2.3 *Ovarian and Fallopian Tube Cancer*

Ovarian cancer was the seventh most common malignancy affecting women in the USA in 2008 with an estimated 21,650 new cases and 15,520 deaths [1]. Survival is favorable for early stage ovarian cancer; i.e. stage Ia–Ic granted that an appropriate surgical staging is performed. A comprehensive surgical staging procedure for epithelial ovarian and fallopian tube cancers includes a total abdominal hysterectomy, bilateral salpingo-oophorectomy, cytologic washings, biopsies of adhesions and peritoneal surfaces, omentectomy, and retroperitoneal lymph node sampling from the pelvic and para-aortic regions [79]. Patients with a suspected early stage ovarian or fallopian tube cancer can be comprehensively staged via laparoscopy when performed by experienced gynecologic oncologists [80]. Laparoscopy offers many advantages over laparotomy including avoidance of an abdominal incision, shorter hospital stay, and a more rapid recovery time.

32.2.3.1 Initial Surgical Staging Procedure

As with other laparoscopic surgical procedures in gynecologic oncology, surgeons have proven the feasibility of robotics in performing a comprehensive ovarian or fallopian cancer staging. Three patients included in the series reported by Reynolds et al. [17] in describing their initial experience with the robotic platform in gynecologic oncology had ovarian and fallopian tube cancers. They were all comprehensively staged.

Although no published studies or reports exist for the role of robotics in the surgical management of advanced ovarian cancer, Bandera and Magrina point out in their review of robotic surgery in gynecologic oncology [11] their unpublished experience in 12 primary robotic debulking and five secondary debulking procedures for ovarian cancer. Included in these procedures were a rectosigmoid resection, small bowel resection, diaphragmatic resection and hepatic tumor resection. The upper abdominal procedures were achieved by undocking the robot and rotating the table to enable re-docking the robot at the patient's head [11].

Debulking advanced ovarian cancer is not feasible, due to the uncertainty of the tumor's extent in the abdomen and the complexity of the surgical procedures required for an effective debulking surgery. The limitations of the current robotic platform in exploring the upper abdomen thoroughly prohibit an adequate surgical management of advanced ovarian cancer.

32.2.3.2 Recurrent Ovarian Cancer

Although no published reports exist for either conventional or robotic assisted laparoscopy for secondary cytoreduction in recurrent ovarian cancer setting, Bandera and Magrina [11] have performed five robotic procedures for secondary cytoreduction with plans to publish their data. We advocate for the use of the robotic platform if the recurrence is localized and resectable. In general, any procedure that can be approached with conventional laparoscopy can be, at the least, equally performed robotically.

32.2.4 Pelvic Exenteration

Pelvic exenteration, an en-bloc resection of pelvic organs of multiple subtypes, is offered to patients with persistent or central mobile recurrence in the pelvis (i.e. with no pelvic sidewall extension and resectable disease) in a field of prior radiation. Reports of laparoscopic exenteration have been published in the gynecologic and urologic oncology literatures [81–83]. Randell and Eun presented a live webcast of a robotic assisted anterior exenteration performed for a woman with recurrent cervical cancer in the pelvis with a 20-year history of pelvic radiation as primary treatment (this can be accessed through the following web link: http://www.or-live.com/daVinci/2271/). These surgeons concluded that in experienced hands, such an extensive morbid procedure could be performed robotically safely. No other reports have been published on pelvic exenteration in the gynecologic literature. Pruthi et al. [84] reported a case series of 12 patients who underwent robotic assisted anterior pelvic exenteration and extracorporeal urinary diversion for clinically localized bladder cancer concluding that this approach appears to achieve the clinical and oncologic goals of radical cystectomy in the female patient. There are no reported robotic assisted exenteration procedures in the gynecologic oncology literature. We have performed a robotic assisted total pelvic exenteration on a 70 yo female with recurrent IIB cervical cancer after 4.5 years of remission. The procedure included a total pelvic exenteration with end colostomy, and an incontinent ileal conduit. The total operating room time was approximately 11 h with an EBL of 800 ml. The final pathology demonstrated widely negative margins.

32.3 Applications in Benign Gynecologic Diseases

Following early reports on the use of the da Vinci robotic system in gynecologic procedures by Diaz Arrastia et al. [85] and Marchal et al. [16], Advincula et al. [86], provided feasibility and safety data allowing the food and drug administration to approve the use of da Vinci robotic platform for gynecologic surgical procedures in April 2005.

32.3.1 Simple Hysterectomy with or Without Salpingo-Oophorectomy

Over 600,000 hysterectomies are performed on annual basis in the United States [87] and a small proportion are performed with minimal invasive surgery [88]. Several case series and surgeons' experiences have been published in the literature confirming the advantages of the robotic platform as compared to the conventional platform [89–91].

In a retrospective review of 200 hysterectomies performed before and after implementing a robotic program, Payne et al. [90] reported 100 hysterectomies performed mainly via standard laparoscopy and another 100 procedures performed robotically. Significant decreases in blood loss, conversion rates (9% vs 4% in standard and robotic laparoscopic procedures respectively), and hospital stay were observed. Additionally, it was also noted that 11% of all hysterectomies were performed via traditional laparotomy in the pre-robotic period whereas none existed after implementation of a robotics program [90].

We have demonstrated our robotic experience in performing hysterectomies for benign cases with complex pathology in 152 patients [92]. Obese patients comprised 64.4% (98 patients of the cohort). Uterine weight was 250–500 g in 33 patients (21.7%) and more than 500 g in 29 patients (19.1%). Seventy-one patients had one prior abdominal or pelvic surgery while 23 patients had two or more prior surgeries. The overall operative time of 122.9 min (43–325 min), with the observation that two variables contributed to longer operating times; uteri greater than 250 g (due to morcellation time) and resident/fellow involvement in performing the procedure. There were three intra-operative complications (2.1%) and no perioperative blood transfusions nor any conversions to laparotomy. The robotic platform appears to facilitate MIS for complex hysterectomy [92].

32.3.2 Myomectomy

A myomectomy is offered to women with uterine fibroids who wish to retain their fertility or in whom it is thought that the fibroids are a mechanical hindrance to appropriate transportation of the gametes within the genital tract. This procedure is usually offered through an abdominal approach for subserous and intramural myomas, and occasionally by the utilization of a hysteroscopic approach for submucosal myomas. The use of laparoscopic myomectomy has proven to cause less postoperative pain, a shorter hospital stay, and faster postoperative recovery but does not allow for a standard multi-layer uterine closure [93].

The first case series in the literature was reported by Advincula et al. [86] in which 35 patients who underwent a robotic assisted myomectomy were reported underscoring that this platform is promising and may overcome conventional laparoscopy's surgical limitations. The same group out of the University of

Michigan reported a second case matched series of 58 women treated via robotic assistance or the standard laparotomy approach [94]. This series represents the largest comparative study to date. The results demonstrated less blood loss, hospital stay and a lower complication rate (all of which bore statistical significance) in favor of robotic assistance. The operative time was longer, however, in the robotic group (with a mean of 231.38 vs 154.41 for the laparotomy group, $p < 0.05$). The operative technique was described in detail in this study [94].

32.4 Applications in Reproductive Endocrinology and Infertility

32.4.1 Uterine Anomalies

Malformations of the genitourinary tract are rare with an overall occurrence of 0.16% and may remain silent until incidental detection in evaluation and treatment of other pathologies during adulthood. Laparoscopic management of genital anomalies have been describe in the gynecologic literature [95–97]. Although there are no reported robotic assisted repair of these congenital anomalies, it is our belief that the robotic platform offers an effective minimally invasive tool and increased precision during these complex surgical procedures.

32.4.2 Tubal Anastomosis

Tubal anastomosis is offered to women of reproductive age who desire reversal of their tubal ligature. Tubal anastomosis is a delicate surgical procedure that requires precise suturing and excellent visualization of the tubal lumen. In an attempt to minimize the inferior properties offered by conventional laparoscopy, Falcone et al. [98] reported the first successful case report of bilateral tubal anastomosis in 1999 using the Zeus surgical system. The authors pointed out the significant advantages of a robotic platform for this delicate surgical operation. The tremor that is noted by many authors in conventional laparoscopy is filtered through computer enhancement, the stability of the camera allows scaling movements, and enabling an unwavering view of the surgical site to be maintained [98]. Ergonomically, the surgeon is in a sitting comfortable position rather than moving his body in an anti-ergonomic fashion as noticed in conventional laparoscopic procedures which could affect surgical efficiency [99]. Following their first case report, Falcone et al. [100] performed a pilot study on ten patients with previous tubal ligations. A tubal ligation reversal was performed using a Zeus robotic suturing device. No conversions to laparotomy were required, chromotubation at the end of the procedure showed patency in all tubes anastomosed, and in 17 of 19 tubes when performed 6 weeks postoperatively. There were five pregnancies without any complications at the time of the publication.

The first anastomosis with the da Vinci robotic system was published by Degueldre et al. [101]. The authors of this study proved the feasibility of the current robotic platform for tubal anastomosis in a series of eight women. This was followed by and an evaluation of 28 women by the same group [101] concluding that the robot with its three-dimensional vision, allows ultra-precise manipulations with intra-abdominal articulated instruments while providing the necessary degrees of freedom required to perform the microscopic technique.

32.4.3 Endometriosis

Minimally invasive techniques are of paramount importance in the in the diagnosis, and treatment of endometriosis as these techniques offer a lower rate of adhesions which improve the clinical outcomes in patients with endometriosis. The reports of robotic assisted surgery for endometriosis are limited. Nezhat et al. [102] described their experience of robotic use in various gynecologic indications including the treatment of endometriosis and have demonstrated the feasibility of this platform when excising infiltrating endometriotic lesions in the pelvis.

32.5 Applications in Urogynecology

The rate of pelvic prolapse is steadily increasing as our population is aging. With an approximate 200,000 procedures performed for surgical prolapse annually [103], a minimal invasive procedure will significantly decreased the peri-operative morbidity associated with the procedure as it is performed in elderly patients utilizing the traditional abdominal approach in most instances.

32.5.1 Pelvic Organ Prolapse

The most effective surgical correction of vaginal vault prolapse is an abdominal sacrocolpopexy [103]. Laparoscopic sacrocolpopexy has not gained widespread acceptance secondary to the difficulty suturing. The robotic platform has made this surgical procedure feasible as it allows for better maneuvering in the narrow pelvic cavity and the ability to suture.

The first experience published was that of Elliott et al. of the Mayo clinic [104] in which 30 patients with post-hysterectomy vaginal vault prolapse underwent robotic assisted laparoscopic sacrocolpopexy with a mean follow-up of 24 months and a minimum of 12 months after surgery. Patients had a mean age of 67 years

(47–83). The mean operative time was 3.1 h (range 2.15–4.75). All patients were discharged on postoperative day 1 and the other on postoperative day 2. Regarding the outcomes, there was one grade 3 rectocele recurrence, one recurrent vault prolapse and two with vaginal extrusion of mesh. All patients were satisfied with outcome. These authors concluded that the robotic platform combines the advantages of open sacrocolpopexy with the decreased morbidity of laparoscopy. Daneshgari et al. [105] reported feasibility and short-term outcome of 15 cases of robotic abdominal sacrocolpopexy and sacrouteropexy for advanced pelvic organ prolapse, concluding that the technique is safe and its outcomes compare favorably with open or laparoscopic abdominal sacrocolpopexy. One patient was converted to a laparoscopic assisted sacrocolpopexy, another one to laparotomy and a third to transvaginal repair. The mean (range) patient age was 64 (50–79) years. The mean (range) estimated blood loss was 81 (50–150) ml. The mean hospital stay was 2.4 (1–7) days.

The largest comparative study of robotic vs. abdominal sacrocolpopexy, for vaginal vault prolapse, was that published by Geller at al [103] in which the short term outcome of these two modalities were compared. There were 73 patients in the robotic cohort and 105 in the abdominal cohort. The two groups were similar from a demographic standpoint. Robotic sacrocolpopexy was associated with less blood loss (103 ± 96 ml compared with 255 ± 155 ml), longer total operative time (328 ± 55 compared with 225 ± 61 min), shorter length of stay (1.3 ± 0.8 compared with 2.7 ± 1.4 days), and a higher incidence of postoperative fever (4.1% compared with 0%). Robotic sacrocolpopexy demonstrated similar short-term vaginal vault support compared with abdominal sacrocolpopexy, with less blood loss and shorter length of stay. These authors suggested the long-term data are needed to assess the durability of this new minimally invasive procedure.

32.6 Advantages and Disadvantages

As with any other technological innovation, the advantages the robotic platform offers are faced with some disadvantages that are offset by its superiority over conventional laparoscopy. The da Vinci robotic system offers: (1) a better and stable 3-D operative visual picture with the ability of digital zooming, (2) greater degrees of freedom of articulation and improved dexterity with the elimination of the fulcrum effect, (3) better ergonomics, (4) motion scaling and filtration of physiologic tremor [106].

32.7 Conclusion

The utilization of the robotic platform for laparoscopic procedures has augmented the surgeon's superiority during any of the abovementioned surgical procedures without sacrificing surgical technique. Additionally, robotic assisted procedures have made MIS surgery more generalizable for more procedures and diseases. The learning curves for robotic assisted procedures are faster when compared to those of conventional laparoscopy, which is reflected in the rapid adoption of this tool set by gynecologists and gynecologic oncologists. While there has been great success with the current robotic platform, there is a desire for future systems. In addition, better level data is necessary to ensure that the outcomes reported are generalizable to all surgeons and that clinical long-term outcomes are satisfactory.

References

1. Jemal, A., Siegel, R., Ward, E., Hao, Y., Xu, J., Murray, T., Thun, M.J.: Cancer statistics, 2008. CA. Cancer J. Clin. **58**, 71–96 (2008)
2. Boggess, J.: Robotic surgery in gynecologic onology: evolution of a new surgical paradigm. J. Robotic. Surg. **1**, 31–37 (2007)
3. Magrina, J.F., Weaver, A.L.: Laparoscopic treatment of endometrial cancer: five-year recurrence and survival rates. Eur. J. Gynaecol. Oncol. **25**, 439–441 (2004)
4. Childers, J.M., Brzechffa, P.R., Hatch, K.D., Surwit, E.A.: Laparoscopically assisted surgical staging (LASS) of endometrial cancer. Gynecol. Oncol. **51**, 33–38 (1993)
5. Spirtos, N.M., Eisenkop, S.M., Schlaerth, J.B., Ballon, S.C.: Laparoscopic radical hysterectomy (type III) with aortic and pelvic lymphadenectomy in patients with stage I cervical cancer: surgical morbidity and intermediate follow-up. Am. J. Obstet. Gynecol. **187**, 340–348 (2002)
6. Ramirez, P.T., Slomovitz, B.M., Soliman, P.T., Coleman, R.L., Levenback, C.: Total laparoscopic radical hysterectomy and lymphadenectomy: the M. D. Anderson Cancer Center experience. Gynecol. Oncol. **102**, 252–255 (2006)
7. Pomel, C., Atallah, D., Le Bouedec, G., Rouzier, R., Morice, P., Castaigne, D., Dauplat, J.: Laparoscopic radical hysterectomy for invasive cervical cancer: 8-year experience of a pilot study. Gynecol. Oncol. **91**, 534–539 (2003)
8. Abu-Rustum, N.R., Gemignani, M.L., Moore, K., Sonoda, Y., Venkatraman, E., Brown, C., Poynor, E., Chi, D.S., Barakat, R.R.: Total laparoscopic radical hysterectomy with pelvic lymphadenectomy using the argon-beam coagulator: pilot data and comparison to laparotomy. Gynecol. Oncol. **91**, 402–409 (2003)
9. Malur, S., Possover, M., Michels, W., Schneider, A.: Laparoscopic-assisted vaginal versus abdominal surgery in patients with endometrial cancer – a prospective randomized trial. Gynecol. Oncol. **80**, 239–244 (2001)
10. Zapico, A., Fuentes, P., Grassa, A., Arnanz, F., Otazua, J., Cortes-Prieto, J.: Laparoscopic-assisted vaginal hysterectomy versus abdominal hysterectomy in stages I and II endometrial cancer. Operating data, follow up and survival. Gynecol. Oncol. **98**, 222–227 (2005)
11. Bandera, C.A., Magrina, J.F.: Robotic surgery in gynecologic oncology. Curr. Opin. Obstet. Gynecol. **21**, 25–30 (2009)
12. ACOG Practice Bulletin: Clinical management guidelines for obstetrician-gynecologists, number 65, August 2005: management of endometrial cancer. Obstet. Gynecol. **106**, 413–425 (2005)
13. Boggess, J.F., Gehrig, P.A, Cantrell, L., Shafer, A., Ridgway, M., Skinner, E.N., Fowler, W.C.: A comparative study of 3 surgical methods for hysterectomy with staging for endometrial cancer: robotic assistance, laparoscopy, laparotomy. Am. J. Obstet. Gynecol. **199**(360), e1–e9 (2008)

14. Gehrig, P.A., Cantrell, L.A., Shafer, A., Abaid, L.N., Mendivil, A., Boggess, J.F.: What is the optimal minimally invasive surgical procedure for endometrial cancer staging in the obese and morbidly obese woman? Gynecol. Oncol. **111**, 41–45 (2008)
15. Walker, J., Mannel, R., Piedmonte, M., Schlaerth, J., Spirtos, N., Spiegel, G.: Phase III trial of laparoscopy versus laparotomy for surgical resection and comprehensive surgical staging of uterine cancer: a gynecologic oncology group study funded by the National Cancer Institute. Gynecol. Oncol. **101**, S11–S12 (2006)
16. Marchal, F., Rauch, P., Vandromme, J., Laurent, I., Lobontiu, A., Ahcel, B., Verhaeghe, J.L., Meistelman, C., Degueldre, M., Villemot, J.P., Guillemin, F.: Telerobotic-assisted laparoscopic hysterectomy for benign and oncologic pathologies: initial clinical experience with 30 patients. Surg. Endosc. **19**, 826–831 (2005)
17. Reynolds, R.K., Burke, W.M., Advincula, A.P.: Preliminary experience with robot-assisted laparoscopic staging of gynecologic malignancies. JSLS **9**, 149–158 (2005)
18. Veljovich, D.S., Paley, P.J., Drescher, C.W., Everett, E.N., Shah, C., Peters, W.A., 3rd.: Robotic surgery in gynecologic oncology: program initiation and outcomes after the first year with comparison with laparotomy for endometrial cancer staging. Am. J. Obstet. Gynecol. **198**, 679 e1–e9 (2008); discussion 679 e9–e10
19. Bell, M.C., Torgerson, J., Seshadri-Kreaden, U., Suttle, A.W., Hunt, S.: Comparison of outcomes and cost for endometrial cancer staging via traditional laparotomy, standard laparoscopy and robotic techniques. Gynecol. Oncol. **111**, 407–411 (2008)
20. Fader, A.N., Arriba, L.N., Frasure, H.E., von Gruenigen, V.E.: Endometrial cancer and obesity: epidemiology, biomarkers, prevention and survivorship. Gynecol. Oncol. **114**, 121–127 (2009)
21. Calle, E.E., Rodriguez, C., Walker-Thurmond, K., Thun, M.J.: Overweight, obesity, and mortality from cancer in a prospectively studied cohort of U.S. adults. N. Engl. J. Med. **348**, 1625–1638 (2003)
22. DeNardis, S.A., Holloway, R.W., Bigsby, G.E.T., Pikaart, D.P., Ahmad, S., Finkler, N.J.: Robotically assisted laparoscopic hysterectomy versus total abdominal hysterectomy and lymphadenectomy for endometrial cancer. Gynecol. Oncol. **111**, 412–417 (2008)
23. Seamon, L.G., Cohn, D.E., Richardson, D.L., Valmadre, S., Carlson, M.J., Phillips, G.S., Fowler, J.M.: Robotic hysterectomy and pelvic-aortic lymphadenectomy for endometrial cancer. Obstet. Gynecol. **112**, 1207–1213 (2008)
24. Peiretti, M., Zanagnolo, V., Bocciolone, L., Landoni, F., Colombo, N., Minig, L., Sanguineti, F., Maggioni, A.: Robotic surgery: changing the surgical approach for endometrial cancer in a referral Cancer Center. J. Minim. Invasive Gynecol. **16**, 427–431 (2009)
25. Mendivil, A.C.L., Gehrig, P.A., Boggess, J.F.: Survival outcomes for women undergoing robotic surgery for endometrial cancer: a three-year experience. Gynecol. Oncol. **112**, S170 (2008)
26. Fung-Kee-Fung, M., Dodge, J., Elit, L., Lukka, H., Chambers, A., Oliver, T.: Follow-up after primary therapy for endometrial cancer: a systematic review. Gynecol. Oncol. **101**, 520–529 (2006)
27. van Wijk, F.H., van der Burg, M.E., Burger, C.W., Vergote, I., van Doorn, H.C.: Management of recurrent endometrioid endometrial carcinoma: an overview. Int. J. Gynecol. Cancer **19**, 314–320 (2009)
28. Awtrey, C.S., Cadungog, M.G., Leitao, M.M., Alektiar, K.M., Aghajanian, C., Hummer, A.J., Barakat, R.R., Chi, D.S.: Surgical resection of recurrent endometrial carcinoma. Gynecol. Oncol. **102**, 480–488 (2006)
29. Bristow, R.E., Santillan, A., Zahurak, M.L., Gardner, G.J., Giuntoli, R.L., 2nd, Armstrong, D.K.: Salvage cytoreductive surgery for recurrent endometrial cancer. Gynecol. Oncol. **103**, 281–287 (2006)
30. Campagnutta, E., Giorda, G., De Piero, G., Sopracordevole, F., Visentin, M.C., Martella, L., Scarabelli, C.: Surgical treatment of recurrent endometrial carcinoma. Cancer **100**, 89–96 (2004)
31. Scarabelli, C., Campagnutta, E., Giorda, G., DePiero, G., Sopracordevole, F., Quaranta, M., DeMarco, L.: Maximal cytoreductive surgery as a reasonable therapeutic alternative for recurrent endometrial carcinoma. Gynecol. Oncol. **70**, 90–93 (1998)

32. Mikuta, J.J.: International Federation of Gynecology and Obstetrics staging of endometrial cancer 1988. Cancer **71**, 1460–1463 (1993)
33. Shafer, A., Boggess, J.F.: Robotic-assisted endometrial cancer staging and radical hysterectomy with the da Vinci surgical system. Gynecol. Oncol. **111**, S18–S23 (2008)
34. Boggess, J.: Robotic-assisted hysterectomy for endometrial cancer. National Library of Medicine Archives. Available at: http://www.nlm.nih.gov/medlineplus/surgeryvideos.html (2007)
35. Beiner, M.E., Hauspy, J., Rosen, B., Murphy, J., Laframboise, S., Nofech-Mozes, S., Ismiil, N., Rasty, G., Khalifa, M.A., Covens, A.: Radical vaginal trachelectomy vs. radical hysterectomy for small early stage cervical cancer: a matched case-control study. Gynecol. Oncol. **110**, 168–171 (2008)
36. Mathevet, P., Laszlo de Kaszon, E., Dargent, D. [Fertility preservation in early cervical cancer]. Gynecol. Obstet. Fertil. **31**, 706–712 (2003)
37. Boggess, J.F., Gehrig, P.A., Cantrell, L., Shafer, A., Ridgway, M., Skinner, E.N., Fowler, W. C.: A case-control study of robot-assisted type III radical hysterectomy with pelvic lymph node dissection compared with open radical hysterectomy. Am. J. Obstet. Gynecol. **199**(357), e1–e7 (2008)
38. Kim, Y.T., Kim, S.W., Hyung, W.J., Lee, S.J., Nam, E.J., Lee, W.J.: Robotic radical hysterectomy with pelvic lymphadenectomy for cervical carcinoma: a pilot study. Gynecol. Oncol. **108**, 312–316 (2008)
39. Magrina, J.F., Kho, R., Magtibay, P.M.: Robotic radical hysterectomy: technical aspects. Gynecol. Oncol. **113**, 28–31 (2009)
40. Nezhat, F.R., Datta, M.S., Liu, C., Chuang, L., Zakashansky, K.: Robotic radical hysterectomy versus total laparoscopic radical hysterectomy with pelvic lymphadenectomy for treatment of early cervical cancer. JSLS **12**, 227–237 (2008)
41. Persson, J., Reynisson, P., Borgfeldt, C., Kannisto, P., Lindahl, B., Bossmar, T.: Robot assisted laparoscopic radical hysterectomy and pelvic lymphadenectomy with short and long term morbidity data. Gynecol. Oncol. **113**, 185–190 (2009)
42. Estape, R., Lambrou, N., Diaz, R., Estape, E., Dunkin, N., Rivera, A.: A case matched analysis of robotic radical hysterectomy with lymphadenectomy compared with laparoscopy and laparotomy. Gynecol. Oncol. **113**, 357–361 (2009)
43. Boggess, J.: Radical hysterectomy procedure for cervical cancer. National Library of Medicine Archives. Available at: http://www.nlm.nih.gov/medlineplus/cervicalcancer.html#cat63 (2007)
44. Canis, M., Mage, G., Pouly, J.L., Pomel, C., Wattiez, A., Glowaczover, E., Bruhat, M.A.: Laparoscopic radical hysterectomy for cervical cancer. Baillieres Clin. Obstet. Gynaecol. **9**, 675–689 (1995)
45. Nezhat, C., Nezhat, F., Burrell, M.O., Benigno, B., Welander, C.E.: Laparoscopic radical hysterectomy with paraaortic and pelvic node dissection. Am. J. Obstet. Gynecol. **170**, 699 (1994)
46. Boggess, J.G.P., Rutledge, T., Ridgeway, M., Skinner, E., Bae-Jump, V., Fowler, W.: Robotic type III radical hysterectomy with pelvic lymph node dissection: description of a novel technique for treating stage Ib1 cervical cancer. Gynecol. Oncol. **103**, S18–S19 (2006)
47. Sert, B.M., Abeler, V.M.: Robotic-assisted laparoscopic radical hysterectomy (Piver type III) with pelvic node dissection – case report. Eur. J. Gynaecol. Oncol. **27**, 531–533 (2006)
48. Sert, B., Abeler, V.: Robotic radical hysterectomy in early-stage cervical carcinoma patients, comparing results with total laparoscopic radical hysterectomy cases. The future is now? Int. J. Med. Robot. **3**, 224–228 (2007)
49. Magrina, J.F., Kho, R.M., Weaver, A.L., Montero, R.P., Magtibay, P.M.: Robotic radical hysterectomy: comparison with laparoscopy and laparotomy. Gynecol. Oncol. **109**, 86–91 (2008)
50. Fanning, J., Fenton, B., Purohit, M.: Robotic radical hysterectomy. Am. J. Obstet. Gynecol. **198**(649), e1–e4 (2008)

51. Ko, E.M., Muto, M.G., Berkowitz, R.S., Feltmate, C.M.: Robotic versus open radical hysterectomy: a comparative study at a single institution. Gynecol. Oncol. **111**, 425–430 (2008)
52. Lowe, M.P., Chamberlain, D.H., Kamelle, S.A., Johnson, P.R., Tillmanns, T.D.: A multi-institutional experience with robotic-assisted radical hysterectomy for early stage cervical cancer. Gynecol. Oncol. **113**, 191–194 (2009)
53. Roy, M., Plante, M.: Pregnancies after radical vaginal trachelectomy for early-stage cervical cancer. Am. J. Obstet. Gynecol. **179**, 1491–1496 (1998)
54. Cibula, D., Slama, J., Fischerova, D.: Update on abdominal radical trachelectomy. Gynecol. Oncol. **111**, S111–S115 (2008)
55. Dargent, D.: [Radical trachelectomy: an operation that preserves the fertility of young women with invasive cervical cancer]. Bull. Acad. Natl. Med. **185**, 1295–1304 (2001); discussion 1305–1306
56. Shepherd, J.H., Spencer, C., Herod, J., Ind, T.E.: Radical vaginal trachelectomy as a fertility-sparing procedure in women with early-stage cervical cancer-cumulative pregnancy rate in a series of 123 women. BJOG **113**, 719–724 (2006)
57. Diaz, J.P., Sonoda, Y., Leitao, M.M., Zivanovic, O., Brown, C.L., Chi, D.S., Barakat, R.R., Abu-Rustum, N.R.: Oncologic outcome of fertility-sparing radical trachelectomy versus radical hysterectomy for stage IB1 cervical carcinoma. Gynecol. Oncol. **111**, 255–260 (2008)
58. Persson, J., Kannisto, P., Bossmar, T.: Robot-assisted abdominal laparoscopic radical trachelectomy. Gynecol. Oncol. **111**, 564–567 (2008)
59. Geisler, J.P., Orr, C.J., Manahan, K.J.: Robotically assisted total laparoscopic radical trachelectomy for fertility sparing in stage IB1 adenosarcoma of the cervix. J. Laparoendosc. Adv. Surg. Tech. A **18**, 727–729 (2008)
60. Chuang, L.T., Lerner, D.L., Liu, C.S., Nezhat, F.R.: Fertility-sparing robotic-assisted radical trachelectomy and bilateral pelvic lymphadenectomy in early-stage cervical cancer. J. Minim. Invasive Gynecol. **15**, 767–770 (2008)
61. Daniel, W.W., Brunschwig, A.: The management of recurrent carcinoma of the cervix following simple total hysterectomy. Cancer **14**, 582–586 (1961)
62. Nezhat, F., Prasad Hayes, M., Peiretti, M., Rahaman, J.: Laparoscopic radical parametrectomy and partial vaginectomy for recurrent endometrial cancer. Gynecol. Oncol. **104**, 494–496 (2007)
63. Magrina, J.F., Walter, A.J., Schild, S.E.: Laparoscopic radical parametrectomy and pelvic and aortic lymphadenectomy for vaginal carcinoma: A case report. Gynecol. Oncol. **75**, 514–516 (1999)
64. Lee, Y.S., Lee, T.H., Koo, T.B., Cho, Y.L., Park, I.S.: Laparoscopic-assisted radical parametrectomy including pelvic and/or paraaortic lymphadenectomy in women after prior hysterectomy-three cases. Gynecol. Oncol. **91**, 619–622 (2003)
65. Fleisch, M.C., Hatch, K.D.: Laparoscopic assisted parametrectomy/upper vaginectomy (LPUV)-technique, applications and results. Gynecol. Oncol. **98**, 420–426 (2005)
66. Ramirez, P.T., Schmeler, K.M., Wolf, J.K., Brown, J., Soliman, P.T.: Robotic radical parametrectomy and pelvic lymphadenectomy in patients with invasive cervical cancer. Gynecol. Oncol. **111**, 18–21 (2008)
67. Boughanim, M., Leboulleux, S., Rey, A., Pham, C.T., Zafrani, Y., Duvillard, P., Lumbroso, J., Haie-Meder, C., Schlumberger, M., Morice, P.: Histologic results of para-aortic lymphadenectomy in patients treated for stage IB2/II cervical cancer with negative [18F] fluorodeoxyglucose positron emission tomography scans in the para-aortic area. J. Clin. Oncol. **26**, 2558–2561 (2008)
68. Mortier, D.G., Stroobants, S., Amant, F., Neven, P., Van Limbergen, E., Vergote, I.: Laparoscopic para-aortic lymphadenectomy and positron emission tomography scan as staging procedures in patients with cervical carcinoma stage IB2–IIIB. Int. J. Gynecol. Cancer **18**, 723–729 (2008)

69. Burnett, A.F., O'Meara, A.T., Bahador, A., Roman, L.D., Morrow, C.P.: Extraperitoneal laparoscopic lymph node staging: the University of Southern California experience. Gynecol. Oncol. **95**, 189–192 (2004)
70. Gil-Moreno, A., Diaz-Feijoo, B., Perez-Benavente, A., del Campo, J.M., Xercavins, J., Martinez-Palones, J.M.: Impact of extraperitoneal lymphadenectomy on treatment and survival in patients with locally advanced cervical cancer. Gynecol. Oncol. **110**, S33–S35 (2008)
71. Gil-Moreno, A., Franco-Camps, S., Diaz-Feijoo, B., Perez-Benavente, A., Martinez-Palones, J. M., Del Campo, J.M., Parera, M., Verges, R., Castellvi, J., Xercavins, J.: Usefulness of extraperitoneal laparoscopic paraaortic lymphadenectomy for lymph node recurrence in gynecologic malignancy. Acta Obstet. Gynecol. Scand **87**, 723–730 (2008)
72. Lowe, M.P., Bahador, A., Muderspach, L.I., Burnett, A., Santos, L., Caffrey, A., Roman, L.D., Morrow, C.P.: Feasibility of laparoscopic extraperitoneal surgical staging for locally advanced cervical carcinoma in a gynecologic oncology fellowship training program. J. Minim. Invasive Gynecol. **13**, 391–397 (2006)
73. Lowe, M.P., Tillmanns, T.: Outpatient laparoscopic extraperitoneal aortic nodal dissection for locally advanced cervical carcinoma. Gynecol. Oncol. **111**, S24–S28 (2008)
74. Mehra, G., Weekes, A.R., Jacobs, I.J., Visvanathan, D., Menon, U., Jeyarajah, A.R.: Laparoscopic extraperitoneal paraaortic lymphadenectomy: a study of its applications in gynecological malignancies. Gynecol. Oncol. **93**, 189–193 (2004)
75. Nagao, S., Fujiwara, K., Kagawa, R., Kozuka, Y., Oda, T., Maehata, K., Ishikawa, H., Koike, H., Kohno, I.: Feasibility of extraperitoneal laparoscopic para-aortic and common iliac lymphadenectomy. Gynecol. Oncol. **103**, 732–735 (2006)
76. Magrina, J.F., Kho, R., Montero, R.P., Magtibay, P.M., Pawlina, W.: Robotic extraperitoneal aortic lymphadenectomy: development of a technique. Gynecol. Oncol. **113**, 32–35 (2009)
77. Vergote, I., Pouseele, B., Van Gorp, T., Vanacker, B., Leunen, K., Cadron, I., Neven, P., Amant, F.: Robotic retroperitoneal lower para-aortic lymphadenectomy in cervical carcinoma: first report on the technique used in 5 patients. Acta Obstet. Gynecol. Scand **87**, 783–787 (2008)
78. Narducci, F., Lambaudie, E., Houvenaeghel, G., Collinet, P., Leblanc, E.: Early experience of robotic-assisted laparoscopy for extraperitoneal para-aortic lymphadenectomy up to the left renal vein. Gynecol. Oncol. **115**, 172–174 (2009)
79. Trimbos, J.B., Bolis, G.: Guidelines for surgical staging of ovarian cancer. Obstet. Gynecol. Surv **49**, 814–816 (1994)
80. Chi, D.S., Abu-Rustum, N.R., Sonoda, Y., Ivy, J., Rhee, E., Moore, K., Levine, D.A., Barakat, R. R.: The safety and efficacy of laparoscopic surgical staging of apparent stage I ovarian and fallopian tube cancers. Am. J. Obstet. Gynecol. **192**, 1614–1619 (2005)
81. Pomel, C., Rouzier, R., Pocard, M., Thoury, A., Sideris, L., Morice, P., Duvillard, P., Bourgain, J.L., Castaigne, D.: Laparoscopic total pelvic exenteration for cervical cancer relapse. Gynecol. Oncol. **91**, 616–618 (2003)
82. Iavazzo, C., Vorgias, G., Akrivos, T.: Laparoscopic pelvic exenteration: a new option in the surgical treatment of locally advanced and recurrent cervical carcinoma. Bratisl. Lek. Listy. **109**, 467–469 (2008)
83. Pulliam, S.J., Berkowitz, L.R.: Smaller Pieces of the Hysterectomy Pie: Current Challenges in Resident Surgical Education. Obstet. Gynecol. **113**, 395–398 (2009)
84. Pruthi, R.S., Stefaniak, H., Hubbard, J.S., Wallen, E.M.: Robot-assisted laparoscopic anterior pelvic exenteration for bladder cancer in the female patient. J. Endourol. **22**, 2397–2402 (2008); discussion 2402
85. Diaz-Arrastia, C., Jurnalov, C., Gomez, G., Townsend, C., Jr.: Laparoscopic hysterectomy using a computer-enhanced surgical robot. Surg. Endosc. **16**, 1271–1273 (2002)
86. Advincula, A.P., Song, A., Burke, W., Reynolds, R.K.: Preliminary experience with robot-assisted laparoscopic myomectomy. J. Am. Assoc. Gynecol. Laparosc. **11**, 511–518 (2004)

87. Learman, L.A., Kuppermann, M., Gates, E., Gregorich, S.E., Lewis, J., Washington, A.E.: Predictors of hysterectomy in women with common pelvic problems: a uterine survival analysis. J. Am. Coll. Surg. **204**, 633–6341 (2007)
88. Advincula, A.P., Wang, K.: Evolving role and current state of robotics in minimally invasive gynecologic surgery. J. Minim. Invasive Gynecol. **16**, 291–301 (2009)
89. Lenihan, J.P., Jr., Kovanda, C., Seshadri-Kreaden, U.: What is the learning curve for robotic assisted gynecologic surgery? J. Minim. Invasive Gynecol. **15**, 589–594 (2008)
90. Payne, T.N., Dauterive, F.R.: A comparison of total laparoscopic hysterectomy to robotically assisted hysterectomy: surgical outcomes in a community practice. J. Minim. Invasive Gynecol. **15**, 286–291 (2008)
91. Reynolds, R.K., Advincula, A.P.: Robot-assisted laparoscopic hysterectomy: technique and initial experience. Am. J. Surg. **191**, 555–560 (2006)
92. Boggess, J.F., Gehrig, P.A., Cantrell, L., Shafer, A., Mendivil, A., Rossi, E., Hanna, R.: Perioperative outcomes of robotically assisted hysterectomy for benign cases with complex pathology. Obstet. Gynecol. **114**, 585–593 (2009)
93. Mais, V., Ajossa, S., Guerriero, S., Mascia, M., Solla, E., Melis, G.B.: Laparoscopic versus abdominal myomectomy: a prospective, randomized trial to evaluate benefits in early outcome. Am. J. Obstet. Gynecol. **174**, 654–568 (1996)
94. Advincula, A.P., Xu, X., Goudeau, St., Ransom, S.B.: Robot-assisted laparoscopic myomectomy versus abdominal myomectomy: a comparison of short-term surgical outcomes and immediate costs. J. Minim. Invasive Gynecol. **14**, 698–705 (2007)
95. Bailez, M.M.: Laparoscopy in uterovaginal anomalies. Semin. Pediatr. Surg. **16**, 278–287 (2007)
96. Patterson, D., Mueller, C., Strubel, N., Rivera, R., Ginsburg, H.B., Nadler, E.P.: Laparoscopic neo-os creation in an adolescent with uterus didelphys and obstructed hemivagina. J. Pediatr. Surg. **41**, E19–E22 (2006)
97. Boudhraa, K., Barbarino, A., Gara, M.F.: Laparoscopic hemi-hysterectomy in treatment of a didelphic uterus with a hypoplastic cervix and obstructed hemi-vagina. Tunis Med. **86**, 1008–1010 (2008)
98. Falcone, T., Goldberg, J., Garcia-Ruiz, A., Margossian, H., Stevens, L.: Full robotic assistance for laparoscopic tubal anastomosis: a case report. J. Laparoendosc. Adv. Surg. Tech. A **9**, 107–113 (1999)
99. Berguer, R., Forkey, D.L., Smith, W.D.: Ergonomic problems associated with laparoscopic surgery. Surg. Endosc. **13**, 466–468 (1999)
100. Falcone, T., Goldberg, J.M., Margossian, H., Stevens, L.: Robotic-assisted laparoscopic microsurgical tubal anastomosis: a human pilot study. Fertil. Steril. **73**, 1040–1042 (2000)
101. Degueldre, M., Vandromme, J., Huong, P.T., Cadiere, G.B.: Robotically assisted laparoscopic microsurgical tubal reanastomosis: a feasibility study. Fertil. Steril. **74**, 1020–1023 (2000)
102. Nezhat, C., Saberi, N.S., Shahmohamady, B., Nezhat, F.: Robotic-assisted laparoscopy in gynecological surgery. JSLS **10**, 317–320 (2006)
103. Geller, E.J., Siddiqui, N.Y., Wu, J.M., Visco, A.G.: Short-term outcomes of robotic sacrocolpopexy compared with abdominal sacrocolpopexy. Obstet. Gynecol. **112**, 1201–1206 (2008)
104. Elliott, D.S., Krambeck, A.E., Chow, G.K.: Long-term results of robotic assisted laparoscopic sacrocolpopexy for the treatment of high grade vaginal vault prolapse. J. Urol. **176**, 655–659 (2006)
105. Daneshgari, F., Kefer, J.C., Moore, C., Kaouk, J.: Robotic abdominal sacrocolpopexy/sacrouteropexy repair of advanced female pelvic organ prolaspe (POP): utilizing POP-quantification-based staging and outcomes. BJU Int. **100**, 875–879 (2007)
106. Schreuder, H.W., Verheijen, R.H.: Robotic surgery. BJOG **116**, 198–213 (2009)

Chapter 33
Applications of Surgical Robotics in General Surgery

Ozanan Meireles and Santiago Horgan

33.1 Introduction

From the first laparoscopic cholecystectomy performed in 1985 to the introduction of robotic surgical telemanipulators in general surgery in the early 2000s, the field of general surgery has changed tremendously, to the point that isn't general anymore; indeed this concept barely exists as a surgical discipline outside rural areas. It has however evolved into the last subspecialty of what was once known as general surgery, and it called Gastrointestinal or Alimentary Tract surgery. The factor the influenced the most on this change was the adoption of Minimally Invasive Surgery in the late 1980s, which selected the General Surgeons that were interested on the abdominal cavity and the gastrointestinal tract. Furthermore, the success and wide spread acceptance of the Laparoscopic Cholecystectomy, brought this field to the spotlight. Nowadays laparoscopic anti-reflux surgical procedures, esophagectomies, colectomies and bariatric surgery are done routinely in major centers across the United States, Europe and many parts of the world.

Minimally Invasive Surgery has benefited patients with small incisions, less traumatic operations, and improved cosmesis [1]. However these procedures requires long instruments that are introduced into the abdominal cavity, and use the abdominal wall as a fulcrum at point of insertion requiring to execute hand movements that are directed to the opposite direction to the instruments tip, furthermore they amplify motion tremors, impair tactile sensation, and limit the degrees of freedom; additionally the TV monitors used in laparoscopy imposes loss of stereoscopic view [2, 3].

Robotic assisted surgery debuted in 2000, aiming to return to the Minimally Invasive Surgical field the benefits of open surgery, such as three-dimensional vision, and wrist action with seven degrees of freedom at the surgical site [4]. Robotic surgical systems are capable to generate duplication of the surgeon's hand movements [5]; moreover, they add unique advantages during laparoscopy, by

S. Horgan (✉)
Department of Surgery, University of California, San Diego, San Diago, CA 92103, USA
e-mail: shorgan@ucsd.edu

J. Rosen et al. (eds.), *Surgical Robotics: Systems Applications and Visions*,
DOI 10.1007/978-1-4419-1126-1_33, © Springer Science+Business Media, LLC 2011

eliminating tremor, enhancing precision, and reducing fatigue due to comfortable ergonomics [6]. However still some drawbacks with the current generation of robots, which include the operational cost and the absolutely lack of haptics [7, 8].

The introduction of robots in general surgery for human application started in 1993 with the use of automated camera positioning robots to maneuver the laparoscope with a hands-free approach, and therefore eliminate the need for the assistant to hold the camera; these systems were the EndoAssist (Armstrong Health care, High Wycombe, UK) and the AESOP (Computer Motion, Goleta, California). The EndoAssist used infrared sensors to detect the movements of the surgeon's head in order to maneuver the camera, and the AESOP which was initially maneuvered by a foot-pedal interface, and later by a voice-controlled [9].

From camera holders, the surgical robotic systems evolved to offer surgeons the ability to perform tele-manipulation with active and total control of the robotic arms equipped with surgical tools, such as graspers, scissors, dissectors and needle holders in a master–slave fashion, where the robotic arms act only when the surgeon at the master control performs an action [10].

Currently there are two available robotic systems with tele-manipulation capabilities adopted for gastrointestinal surgery and approved by the FDA for clinical use. They are the ZEUS robot (Computer Motion, Goleta, California) and the da Vinci Surgical System (Intuitive Surgical, Sunny Valle, California). Intuitive Surgical now owns both systems, and since the acquisition of Computer Motion by Intuitive Surgical, only the da Vinci system still commercially available; yet both systems continue to be used in clinical settings. Both robots are designed to function in a master–slave fashion with several robotic arms managed by a surgeon from a computerized console remotely located from the surgical table.

The Zeus robotic system has a control console where the surgeon visualizes the operative field via high resolution monitors and operates the master controls, which communicate with three robotic arms mounted to the surgical table. The central robotic arm is an AESOP, which allows the surgeon to manage the laparoscopic camera using voice-controlled commands; peripherally, two tele-manipulator robotic arms are located to the right and left of the central robotic arm and are controlled by the master controls [11, 12].

Feasibility studies for the use of the Zeus robotic system for abdominal surgeries were initially conducted with experimental models using swines, and in 2001, the FDA granted clinical approval for limited application of the Zeus system for abdominal operations in the USA. However, in Europe, Mareschaux et al. were already performing robot-assisted laparoscopic cholecystectomies, and performed the first transatlantic cholecystectomy, with the patient located in Strasbourg, France and the robot's console in New York, USA [13]. The Zeus' application in remote telesurgery was then further explored in rural Canada by Anvari [14].

The da Vinci Surgical System has a large console where the surgeon receives binocular image from the surgical field and operates the robotic arms through a computerized system using master controls. In this system, depending on the model, there are three or four robotic arms assembled on a movable single unit, with the central arm supporting two parallel cameras and the lateral arms serving as instrument arms. The image are acquired and delivered simultaneously to the console's two monitors who are observed in a parallel binocular fashion by the surgeon's eyes, allowing stereoscopic view. The tips of the instruments have seven degrees of freedom and wrist action that are remarkably similar to the mechanics of open surgery, allowing more complex and delicate tasks than standard laparoscopic instruments [15].

Since the Zeus was discontinued, the great majority of scientific research and clinical data using robotic systems have been done with the Da Vinci. The main advantages of surgical robotic systems over standard laparoscopy are the elimination of the counter intuitive motion of standard laparoscopy and the alignment of the surgeon's eyes and hands over the area of interest, while positioned on a comfortable console, therefore improving ergonomics. They mimic open surgery due to the wristed action movements, which add more degrees of freedom than laparoscopic instruments and allow more precision in the surgical field. Moreover, those systems filter and eliminate the surgeon's natural tremors and are capable to scale the movements to millimetric precision, making them ideal to work in small working spaces such as the male pelvis and hiatus.

By offering three-dimensional stereoscopic vision with its dual camera technology the da Vinci system returns to the surgeon the depth of perception, which was lost with the use of standard laparoscopic monitors. Furthermore, with the addition of the fourth arm, surgeons are now able to perform certain operations without the need of an assistant. In bariatric surgery the use of surgical robotic systems also helps to decrease the surgeon's fatigue caused by the abdominal wall torque in morbidly obese patients. Lastly, those platforms offer unique and astonishing capability of performing remote telesurgery. However there are some disadvantages of the current commercially available platforms, which are the high operational cost, the suboptimal application for multi-quadrant operations and the absence of tactile feedback requiring the surgeon to depend on visual cues instead of tactile cues.

33.2 Specific Applications

33.2.1 Esophagectomy

Nowadays the major indication for esophagectomy is the Carcinoma of the Esophagus, with has an incidence of 5 per 100,000 people and is responsible for about 12,000 related deaths per year in the United States [16]. Currently, adenocarcinoma accounts for about 50% of all new cases of esophageal cancer in the US. It is usually located in the lower esophagus or gastroesophageal junction, and the most important risk factor associated with esophageal adenocarcinoma is gastroesophageal reflux disease (GERD). Approximately 15% of patients with GERD will develop intestinal metaplasia and 1% of these will develop esophageal cancer. Consequently, when the diagnosis of high-grade dysplasia (HGD) is confirmed, esophagectomy should be considered as the first line of treatment, since up to 60% of high-grade dysplastic mucosa harbors unsuspected adenocarcinoma [17].

Esophagectomy is one of the most complex and traumatic operations in gastrointestinal surgery and is associated with significant postoperative morbidity and mortality [18]. The esophagus spans from the neck to the abdominal cavity, passing through the posterior mediastinum in close proximity to the great vessels. The first successful esophagectomy with anastomosis was performed through the left chest in 1933 [19]. Since then several surgical approaches have been developed including the Ivor Lewis procedure with esophageal resection through separate laparotomy and right thoracotomy with an intra-thoracic anastomosis. In 1972, McKeown described a modified Ivor–Lewis procedure with esophagogastric anastomosis in the neck through a separate cervical incision. Furthermore in 1978, Orringer proposed the transhiatal transcervical esophagectomy technique without a thoracotomy.

With the introduction of Minimally Invasive Surgery and in an effort to lessen the invasiveness of the open approach, laparoscopic and thoracoscopic minimally invasive techniques have been proposed and validated for esophageal resection [20, 21]. In the mid 1990s De Paula [22] and Swanstrom [23] reported laparoscopic techniques where through a trans-hiatal approach, the stomach and esophagus were mobilized with excellent visualization up to the level of the inferior pulmonary veins, with the indisputable advantages of a minimized blood loss, shorter operative times and hospital stay [24].

Nevertheless, the shortcoming of laparoscopy, such as lack of stereoscopic view, unstable camera platform, straight and short laparoscopic instruments with limited degrees of freedom, and poor ergonomics, made it difficult to perform an adequate middle and upper esophageal and mediastinal nodal dissection [25]. For this reason, some surgeons advocated performing a combined laparoscopic and thoracoscopic approach; consequently, imposing the need of one lung ventilation, patient repositioning during surgery, and the associated perioperative morbidities of violating the thoracic cavity [26]. Hence, to date the most beneficial approach has not been established yet and no surgical technique has prevailed over the others [27]. Robotic technology has the promise of changing this paradigm by offering to the surgeon the benefits that are lacking in standard laparoscopic surgery.

The robotic system allows the surgeon to work in the narrow space of the mediastinum, therefore overcoming spatial limitations experienced during standard laparoscopy. It offers stereoscopic view and utilizes instruments that are 7.5 cm longer than standard laparoscopic instruments, allowing more proximal mobilization, sometimes reaching beyond the level of the carina. Also, the dissection in the vicinity of the pulmonary veins, aorta, and parietal pleura can be accomplished safely due to the articulated instruments tips, three dimensional view and the magnification of the operative field. It has also been shown that robotic THE offers a potential added benefit of maintain the oncological principles without the need of concomitant thoracoscopic approach when compared with standard laparoscopic THE [28–30].

In 2003, we described the very first robotic assisted trans-hiatal esophagectomy [31], which was followed by several reports that confirmed the feasibility and safety of this technique. The goal of this operation is to resect the specimen and perform lymphadenectomy when indicated, create a gastric conduit and perform a cervical anastomosis. The steps are the following: after proper positioning the patient and obtaining peritoneum access, the exposure of the Crura and mobilization of the Stomach is accomplished with standard laparoscopic instrumentation, then the robot is docked to perform the mediastinal esophageal dissection, followed by open cervical dissection where the cervical and transhiatal dissection are connected; the gastric tube formation is then performed with standard laparoscopic instruments and the gastric tube is pulled up into the mediastinum and out through the cervical incision. And finally the anastomosis between the esophagus and the stomach is performed at the cervical level.

When compared with standard laparoscopy the daVinci robotic system enhances dexterity by nearly 50% [32]. The abovementioned benefits of the daVinci robotic system allow the surgeon to minimize intraoperative complications during critical portions of the esophageal dissection, especially near the pulmonary veins, aorta, and parietal pleura, resulting in minimal cardiac and pulmonary complications, as well as significant decrease in blood loss [33]. Although still controversial whether the application of minimally invasive techniques for the treatment of esophageal malignancy, the three-dimensional visualization and the magnification of the operative field associated with the articulated wristed instruments allow precise dissection with optimal proximal and distal resection margins and a mean number of lymph nodes harvested comparable to series of thoracoscopic esophageal mobilization [34–36].

In conclusion, the use of robotic surgical systems during THE, enhance the surgeon's ability to perform fine and precise dissection in a very narrow field with the benefits of Minimally Invasive Surgery. Still, the high cost of the equipment and the necessity of a highly trained Minimally Invasive esophageal surgeon to operate a machine that lacks in tactile feedback, an important drawback for this application. And certainly, long-term survival and regional recurrence need to be determined to assess if robotic THE has comparable long-term outcomes with open transhiatal or transthoracic techniques.

33.2.2 Re-do Nissen Fundoplication

Gastroesophageal reflux disease (GERD) has a high prevalence in the western world; and it is estimated that approximately 20% of the population in the United States experience characteristic symptoms of this disease once a week [11]. GERD is a multifactorial disease, where the size, location and tonicity of the lower esophageal sphincter (LES), and the presence of a hiatal hernia are implicated on its pathophysiology. If not treated, this disease can lead to complications such as esophagitis, strictures, intestinal metaplasia, and adenocarcinoma.

Before the advent the minimally invasive surgery and a better understanding of the pathophysiology of GERD, surgery was only indicated for patient who failed medical management. However after laparoscopic fundoplications were proven feasible, safe and with comparable results of open surgery, the ideal candidate for surgery became the patient who responds well to PPIs, presents with typical symptoms and has a positive 24-h pH study [12]. Additionally, patients with GERD that were diagnosed early in their life, patients that are not willing to receive long-term medical treatment, patients at high risk of developing complications or with Barrett's esophagus should also be consider as candidates for laparoscopic fundoplication.

Since Dallemagne and colleagues [37] described the first laparoscopic Nissen fundoplication in 1991, the technique became well accepted and has revolutionized antireflux surgery. From 1993 to 2002 the numbers of fundoplications performed in the United States increased from 22,000 to 41,000 [38]. The success rate of laparoscopic fundoplication is around 90–95%, with 2.8% of reoperations due to recurrence in specialized centers [39]. Laparoscopic revisional fundoplication procedures although feasible and with comparable results to open surgery, are extremely challenging cases [40–42]; mostly due to the well-known disadvantages and limitations of standard laparoscopic surgery. These limitations however, can be overcome with the robotic assisted surgery, which allows surgeons to perform these complex procedures with more accuracy.

Laparoscopic revisions of Nissen Fundoplication consist on the identification of the wrap and the esophageal hiatus, the assessment of the intactness of the fundoplication and crurorrhaphy and as well their relation. In the majority of the times the wrap needs to be undone to identify the structures; moreover the reduction of a herniated stomach needs to be performed in at least half of the cases [43, 44]. When trying to accomplish those tasks the surgeon may injures the liver, spleen, gastric wall, pleura, esophagus, inferior vena cava or aorta, due to the fibrotic adhesions created around the wrap and at the previous site of the hiatal hernia repair. In this hostile environment standard laparoscopic surgery exposes its shortcomings, due to the loss of stereoscopic view and the limitation of the rigid instruments movements in the narrow operative field around the thoracic esophagus, consequently limiting the surgeon's ability to perform these procedures in the safest possible manner.

Robotic surgery technology has the potential to overcome some pitfalls of standard laparoscopic surgery. In 1999 Candire performed the first robotic assisted fundoplication [45], and since then, clinical trials comparing robotic-assisted Nissen fundoplication vs. laparoscopic Nissen fundoplication have been reported in the literature [46–48]. These investigators not only demonstrated that robotic approaches were feasible and safe; but also noted that they added the benefits of three-dimensional view and precise dissection at the narrow mediastinum space, therefore avoiding the drawbacks associated with standard laparoscopic approaches [49–51].

Although robust supportive literature is available for primary Robotic Nissen fundoplication, to date there is no published scientific data that compares robotic assisted redo fundoplication with purely laparoscopic redo fundoplication, or even a compilation of case series on the former. There is, however, significant anecdotal and extrapolated data from the overall robotic Nissen literature and from personal experience of esophageal surgeons who perform these procedures with assistance of the robot.

On these premises, experienced surgeons have observed and reported that the employment of robotic assistance in revisional fundoplication has some clear advantages over standard laparoscopy. Among those advantages is the utilization of 3D visualization, which allows a better identification of the Vagus nerve and the differentiation of muscle fibers from fibrosis, avoiding inadvertent gastric or esophageal perforation. Overall, the robotic design augments the surgeon's capability to maneuver instruments in the narrow space of the hiatus and allows fine and precise dissection of the warp and esophagus from vital structures such as the aorta, vena cava, pleura and pericardium; therefore providing a safer operation in skillful hands.

Conversely, from the same source of unpublished experience, there are reports of the noticeable increased operation room time and cost. It was also noted that the lack of tactile feedback can too predispose to iatrogenic injuries to the surrounding structures in inexperienced hands.

Robotic assisted redo fundoplication procedures are certainly a promising alternative for laparoscopic treatment of failed anti-reflux surgery, and despites the increased operative time and cost, they have the promise of improving individual surgical outcomes when wisely employed. Nevertheless prospective randomized data still needed before we can justify a more liberal utilization of this device in the routine surgical practice.

33.2.3 Heller Myotomy

Achalasia is the most common dysmotility of the esophagus, with annual incidence of approximately one in 100,000 individuals in North America. Its etiology is unknown and the treatment is mainly palliative, intended to decrease the intraluminal pressure at the level of the lower esophageal sphincter (LES). With the advent of less invasive techniques [52, 53] and their well established long-term results [54], the initial treatment pathway shifted toward operative procedures instead of endoscopic dilatation or Botulinum toxin injections.

Laparoscopic Heller myotomy (LHM) has become the procedure of choice for the treatment of Achalasia, with satisfactory results in 85–90% of patients [55]. LHM can be safely performed on hands of skillful laparoscopic surgeons; however the average incidence of esophageal perforation reported is about 5–8% [56–58]. The goal of this operation is to perform complete myotomy of the esophagus at 12 o'clock position at least 6 cm proximal to the gastroesophageal junction (GEJ) and 2 cm distally into the stomach, exposing the submucosal plane and preserving the Vagus nerve. During standard laparoscopy, surgeons rely on monoscopy view, instruments with limited degrees of freedom and use the abdominal wall as a swivel, making the movements contra intuitive and augmenting the surgeon's natural tremors. This could justify limited or incomplete myotomies and iatrogenic mucosal perforations.

Heller Myotomy, is probably the gastrointestinal procedure where robotic surgery finds one of its best application. The first case of robotically assisted Heller myotomy (RAHM) was performed by our group in 2001, with was compiled with eight other cases and published on the same year. Our experience demonstrated that the robotic system allowed to complete the myotomy in about half the time spent with the standard laparoscopic approach (17 vs 33 min), with greater comfort and precision for the operating surgeon [59]. By adding stereoscopic view, instruments with seven degrees of freedom, elimination of inherited tremors and fine motion scaling, RAHM has established itself as a safer alternative to LHM [60]

In a multi-institution trial assessing efficacy and safety of RAHM compared with LHM for treatment of achalasia, a total of 121 patients underwent (RAHM) or (LHM), and where followed at 18 and 22 months, with 92% and 90% of patients having persistent relief of their dysphagia respectively. There was no difference between the two approaches regarding alleviation of symptoms or postoperative heartburn, therefore demonstrating the comparable efficacy of RAHM to LHM. Conversely, there were more intra-operative complications with LHM, resulting in esophageal perforation (16%), whereas RAHM had zero complication (0%). There was however, a shorter operative time for LHM (141 ± 49 vs 122 ± 44 min, $P < 0.05$), which was observed only in the first half of the trial. As the surgical team learning curve achieved its pinnacle, there was no more difference between the operative times between the groups.

Although RAHM is more expensive and may add extra operating room time in less experienced hands, when compared with LHM; studies suggest that RAHM has at least the same efficacy and is safer than LHM, because it achieves comparable results and sharply decreases the incidence of esophageal perforation to 0%, even on patients who had previous endoscopic treatment.

33.2.4 Robotic Assisted Gastric Bypass

Obesity has become an endemic problem in the United States. In 2007 was estimated that 34% of U.S. adults aged 20 and over, were obese and 13% of the entire population were morbidly obese. Long-term weight loss with diet and exercise are in general unsatisfactory. For about two decades the National Institute of Health (NIH) have been pointing obesity as an epidemic problem, and procedures such the vertical-banded gastroplasty (VBG) back on 1990s and gastric bypass were considered suitable treatments for long term results [61]. The number of bariatric procedures performed in the USA in 2007 was estimated in 180,000 cases. Initially those procedures were performed in an open fashion, indisputably adding short and long term morbidity due to the laparotomy, which included lengthen hospital stay, and diseases associated to the short term immobility such as pneumonia and pulmonary embolism; additionally incision related complications such as wound infections and high incidence of incisional hernias were quite frequently observed [62]. With introduction of minimally invasive bariatric surgery, and its well proven decreased morbidity compared with open surgery, the number of laparoscopic bariatric operations has grown tremendously in the last decade.

Laparoscopic adjustable gastric band placement and laparoscopic Roux-en-Y gastric bypass (LRYGBP) are the most common bariatric procedures performed these days. Adjustable gastric band procedures can be safely done laparoscopically, as they don't require major advanced laparoscopic skills and have a short learning curve; on the other hand LRYGBP is one of the most complex and challenging minimally invasive procedures in gastrointestinal surgery. LRYGBP requires very advanced laparoscopic surgical skills, such as suturing, intra-corporeal knot tying, stapling, two-handed tissue manipulation, and the ability to operate in multiple quadrants of the abdomen; consequently the learning curve is very steep [63, 64]. Moreover, some additional challenges may be present during Laparoscopic RYGBP, particularly on super morbidly obese patients, where the abdominal wall torque induces surgeon's fatigue and decreases the ability of fine tissue manipulation. Those impeding factors led some surgeons not to adopt this technique, which is now mainly confined to Minimally Invasive Surgery fellowship trained bariatric surgeons [65].

The first report series of robotic-assisted cases concluded that the use of robots is most beneficial when fine manipulation is required, such as when working at the hiatus [66]. As discussed on this chapter, the first generation of the da Vinci's design is best suited to work in single quadrant operations and confined spaces, and is capable to perform fine tissue manipulation and precise suturing tasks. Therefore for procedures such as LRYGBP, that requires multiquadrant manipulation, the robotic equipment is either used for only a portion of the entire procedure or requires extensively repositioning throughout the case if more than one quadrant needs to be accessed with the robot, which adds considerable extra time to the operation. Consequently, most of the bariatric applications involve the utilization of the robot for only one step of the procedure; usually involving the gastrojejunal anastomosis. Some centers however were able to use the robot for the entire operation by using a customized port placement with median operative time of 140 min (80–312 min) [67].

33.2.4.1 Adjustable Gastric Banding

The laparoscopic adjustable gastric banding is a purely restrictive procedure, which allows a creation of a gastric pouch by applying extraluminal restriction to the stomach. It has become increasingly popular due to its simplicity, lower morbidity and potential reversibility. Nowadays, laparoscopic adjustable gastric banding accounts for a large number of the bariatric cases performed every year [68].

Cadiere and colleagues performed the first robotically assisted laparoscopic gastric banding procedure in 1998, using the Mona robotic surgical system (Intuitive Surgical, Mountain View, CA), with an operating time of 90 min [69]. Following this case, others have reported that robotic assisted gastric banding has comparable clinical outcomes to traditional laparoscopic procedures, though the operative times were significantly longer in the robotic procedures [70]. The experience drawn from those studies demonstrated that the main advantage of the robotic approach was noted in patients with BMIs greater than 60 kg/m^2, where abdominal wall torque was surpassed by the robotic system's mechanics, furthermore the ability of performing delicate and precise movements was helpful in this tight space were a large liver and excess of intra-abdominal fat limit the surgeons range of motion [71].

Still, many surgeons agree that robotically assisted adjustable gastric banding has no major benefit due to the simplicity of the procedure, the increased setup time and cost, and therefore does not justify its use in routine bases [72], restricting its application to large BMI (>60 kg/m^2), where it would benefit the surgeons ergonomics and potentially save operating room time due to the enhanced dexterity.

33.2.4.2 Roux-en-Y Gastric Bypass

Laparoscopic Roux-en-Y gastric bypass is one of the most complex and challenging minimally invasive procedure in gastrointestinal surgery; it is an ergonomically demanding procedure, and requires performing precise anastomosis in confined spaces using extra long instruments that traverse a thick abdominal wall, which increases the instrument torque and augments the surgeon's natural tremors and therefore limiting dexterity. By acting as an interface between the surgeon and the patient, the robotic system resolves several of those issues; as the surgeon can sit at a comfortable and ergonomically designed console, where there is no noticeable abdominal wall torque or tremor, and it makes the suturing tasks as natural as during open surgery.

The laparoscopic Roux-en-Y gastric bypass was first described by Wittgrove and colleagues in 1994 [73], and is a restrictive and malabsorptive procedure, where certain crucial surgical steps must be properly accomplished. They include transabdominal liver retraction to expose the surgical field, serial transection of the stomach using a linear cutting stapler angling toward the gastroesophageal fat pad, creation of a residual gastric pouch distal to the gastroesophageal junction; transection of the small bowel 50–100 cm distal to the ligament of Treitz, advancement of the Roux limb toward the gastric pouch with a length of 100–150 cm measured from the point where the small bowel was transected, creation of the side-to-side jejunojejunostomy, closure of the mesenteric defect, and creation of the gastrojejunostomy by anastomosing the gastric pouch and the Roux limb.

Our group, performed the first robotically assisted LRYGBP in 2000, using the da Vinci robotic surgical system (Intuitive Surgical, Sunnyvale, CA); this along with six others cases, later on published in 2001, were performed in a hybrid fashion where the initial part of the gastric mobilization was performed laparoscopically and then the robot was docked and used to perform the gastro-jejunostomy [74]. Following this sentinel case the literature on robotic assisted Roux-en-Y gastric bypass has been populated with cases ranging from hybrid laparoscopic and robotic procedures to purely robotic procedures; however in the great majority, it still limited to performing a robotically sewn gastrojejunostomy.

During RYGBP, the robotic system greatly improves ergonomics for the surgeon, avoiding fatigue and related orthopedics problems. It also provides a steady robotic camera controlled by the surgeon, eliminating the assistant tremor and sometimes inattention which can occur in long operations [75]. Multiple case series have demonstrated the safety and efficacy of robotic gastric bypass [76, 77], but some authors consider that well trained laparoscopic bariatric surgeons can accomplish this procedure with standard laparoscopic techniques without requiring robotic assistance [78]. Still the robotic system offers important advantages, especially during surgery on super obese patients (BMI greater than 60 kg/m^2), where it facilitates the creation of the gastrojejunal anastomosis, by eliminating the abdominal wall torque and operator's tremor, which invariably occurs during standard laparoscopy in this subset of patients. It has been reported that a robotically assisted hand-sewn gastrojejunostomy is superior to other currently available minimally invasive anastomotic techniques. Additionally, when a total robotic LRYGBP was performed, the surgeons perceived that was technically easier to suture both the gastrojejunostomy and the jejunojejunostomy robotically when compared to standard laparoscopy.

Although fellowship training is highly recommended to safely perform laparoscopic or robotic LRYGBP [79], it was observed that Robotic RYGBP has an apparent shorter learning curve when compared with standard LRYGBP [80]. In some training Institutions that have been using the daVinci Surgical System for cases such as Nissen and cholecystectomy, the learning curve on the system was about 20 operations; which was noticed when the operating room times using the robotic system were matched with those of the equivalent laparoscopic cases. Additionally it was demonstrated that inexperienced laparoscopic surgeons were able to perform Robotic LRYGBP an average of 29.6 min faster than traditional LRYGB for patients with larger body mass indexes [81].

There are, however, some disadvantages of the robotic surgical systems in bariatric surgery, including the system's bulky size, the suboptimal multi-quadrant range, the increase in operative time due to the set up procedures, and the increased operative cost which is becoming a considerable issue in times where health care expenditures are under scrutiny. Additionally, the absolute lack of tactile feedback makes the surgeon to rely on visual cues to assess tissue and suture tension. Among those issues, some are being overcome by the resourcefulness of talented operating room staff, as was demonstrated that setup times decrease rapidly with experience

[82]; others by the miniaturization of those robotic units and their adequacy for multiquadrant operation, such promises the da Vinci S unit.

33.2.5 Colorectal

Since the first laparoscopic right hemicolectomy performed in 1991 by Jacobs, Minimally Invasive Surgery has beneficially impacted colorectal surgery [83]. When compared with open colorectal surgery, laparoscopic colectomy has shown an improvement in postoperative pain, better cosmesis, earlier return to normal activities, resumption of bowel function, and decreased length of hospital stay [84–89]. Among the COST [90] and CLASICC [91] trials, laparoscopic colorectal operations have also shown to have superior postoperative outcomes and comparable lymph node harvesting, surgical margins, local recurrence and survival rates compared with open surgery. However, the limitations of laparoscopy in colorectal surgery are exposed when addressing the extraperitoneum Rectum. The major limitations for standard laparoscopic proctetomies reside on the anatomic characteristics of the pelvis and the technological restraints of the laparoscopic instruments. Although proven feasible, laparoscopic total mesorectal excision (TME) for rectal malignancies is still technically challenging and has a very steep learning curve. On this environment, robotic surgery offers a stable platform, with stereoscopic visualization, and is capable of fine and precise movements in the narrow surgical fields, with tremor elimination, motion scaling, and increased dexterity, which are tremendous advantages in narrow and limited working spaces, such as the pelvic basin [92, 93].

Since its introduction, robotic assisted surgery has been shown safe and feasible for colon surgery, however has not gain much popularity for hemicolectomies due to their simplicity and ease accomplishment by skilled laparoscopic surgeons and neither for total colectomies due to its multiquadrant dissection, which requires frequent repositioning of the robotic platform during the operation, making it cumbersome and adding valuable time to the operation. In the other hand Robotic assisted Surgery techniques are especially suited for dissection in confined spaces, where precise movements and fine tissue dissection offer great advantages.

33.2.5.1 Proctectomy

Early experiences with robotically assisted lower anterior resection (LAR), and abdominal-perineal resection (APR) for malignancy including TME technique have been demonstrated in the recent literature [94–96]. Those reports showed some advantages over laparoscopic surgery and open surgery.

The fundamental oncologic principals for rectal cancer must be followed during robotic proctetomies, therefore proper lymph node dissection and macroscopic completeness of the resected rectal specimen are crucial, and total mesorectal excision (TME) is the surgical treatment of choice [97]. TME is a technically demanding procedure and requires dissection of the avascular plane between the presacral fascia and the fascia propria of the rectum without injuring the latter. The working space for TME is very limited due to the narrow pelvis and the need of macroscopic completeness of the resected specimen, especially in obese male. On this environment, robotic surgery facilitates the preservation of the pelvic autonomic nerve and ensures the completeness of resection.

In general for both LAR and APR the mobilization of the splenic flexure is performed laparoscopically and the robot is then docked for the TME portion. The reason is that when utilizing the first-generation da Vinci system for proctectomy, which requires access to two quadrants of the abdomen, (the left upper quadrant and the pelvis) it is necessary to either reposition the robot for each quadrant or perform a laparoscopic robotic hybrid procedure for the splenic flexure mobilization [98]. It has been reported that with the utilization of the da Vinci S system, which has smaller profile arms and increased instrument length allowing greater intracorporeal range of motion, a total robotic LAR including the mobilization of the splenic flexure can be accomplished in non obese patients, without the need of redocking the robot [99].

Similarly to prostatectomies, rectal surgery does benefits from robotic surgery by enhancing pelvic autonomic nerve sparing, which is associated with preservation of postoperative sexual and voiding functions, without compromising adequate surgical margins and lymph node harvesting [100].

33.2.5.2 Rectopexy

Colorectal surgery for benign disease confined to the pelvis can also benefit from robotic surgery, such as the surgical treatment of rectal prolapsed with abdominal rectopexy.

Nowadays, several different techniques are available for the treatment of rectal procidentia, and they need to be costumed to each patient needs and general health. The trans-abdominal procedure generally has a greater success rate compared to perineal procedures and should be considered in healthy patients [101–103]. Laparoscopic repair seems to be as effective as open surgery with potential advantages such as faster recovery, less blood loss, lower medical costs, and less postoperative pain and many authors have advocated this approach as the preferred technique [104–108].

Likewise for intra-abdominal procedures for rectal prolapsed, rectopecxy combined or not with sigmoidectomy can be performed laparoscopically and are well suited for robotic assistance due to the narrow pelvis. Minimal anterior mobilization of the rectum is usually necessary and the placement of sutures to fix the rectum or any prosthetic material to the sacrum is straightforward with the da Vinci's enhanced dexterity, when compared with standard laparoscopic instrumentation.

To date, there have been several reports of rectopexies using the da Vinci Surgical System. A study comparing conventional laparoscopic, robot-assisted rectopexy, and open rectopexy showed that open surgery seems to lead to fewer recurrences [109]. The author speculated that the use of different fixation instruments or techniques during standard laparoscopy, may had played a role on the less encouraging results, although this was also observed in the robot-assisted rectopexy group. Nevertheless it is valid to mention that the variation in results could be due to the possibility that open surgery leads to more adhesions, therefore creating a better fixation of the rectum to the promontory and as a result, less recurrences. Additionally the author also commented that if recurrence rates were statistically stratified by age, the differences in recurrence rate for the various operative techniques would become statistically non-significant.

Robotic assistance in laparoscopic rectopexy is a safe and feasible procedure, and even though demonstrating similar functional results compared with open procedures, such as constipation and incontinence, still may be associated with similar recurrence rates, lengthier operating time and higher operational costs compared to conventional laparoscopy [110, 111].

33.2.6 Stricturoplasty

Stricturoplasty in gastrointestinal surgery is a surgical procedure intended to alleviate stenotic areas of the intestinal wall which are caused by ischemia, infection, previous surgery or by chronic inflammatory bowel pathologies such as Crohn's disease. The procedure consists in transversely incising the involved segment of bowel at the anti-mesenteric portion, and then suturing the edges longitudinally, causing a shortening and widening the bowel segment thus resolving the stricture. This procedure can be preferable in those who have already had bowel resections and are at risk for short bowel syndrome.

Standard laparoscopic surgery has a strong reliance on surgical staplers to perform bowel anostomosis and resections. However one of the limitations of surgical staplers is when the target tissue has different thickness; on those situations the stapler devices tend to adjust the height of the staples to the thickness of the thickest portion of the entire tissue where staples are being applied, therefore leaving the thinner portions of tissue at the staple line with staples heights wider than the tissue thickness, predisposing staple line leaks [112].

Pathologies such as Crohn's disease that require stricturoplaties as part of the surgical treatment invariable present with irregular tissue thickness at the site of the stricture due to the inflammatory nature of the disease, therefore making suboptimal the utilization of surgical staplers to accomplish stricturoplasty. On those instances where stapler devices cannot be utilized safely, the surgeons has to either perform a laborious laparoscopic instrument-sewn stricturoplasty with all the inherited limitation of standard laparoscopy or to exteriorize the segment of bowel that needs to be treated to then perform open hand-sewn stricturoplasty.

Robotic surgical systems can also be employed on tasks where surgical staplers would reach their limitations. It has been previously discussed on this chapter that robotic hand-sewn anastomosis for gastrojejunostomies and jejunojejunostomies during LRYGBP are superior to other currently available minimally invasive anastomotic techniques and are also technically easier to be accomplished. Therefore robotic assisted hand-sewn stricturoplasty can be utilized for tissues with different thickness, since it mimics open surgery hand-sewn techniques which tailor every individual stitch to the local tissue thickness.

It has been some reports on the scientific literature, although not particularly focused on this subject, which demonstrated the feasibility and safety of robotic stricturoplasty [113]. With the robotic system stereoscopic visualization and the enhanced dexterity, stricturoplasty can be easily performed; nevertheless specialized training is necessary since these systems lack in tactile feedback and the surgeon has to rely on visual cues to avoid bowel injuries, especially when dealing with inflamed tissue.

33.3 Future of Robotics in General Surgery

Some of the new advancements in robotic surgery that will likely be implemented in the next decades and would benefit gastrointestinal surgery are the miniaturization the units, utilization of multiple individual robotic arms, utilization of multiple consoles, introduction of computer-interfaced proprioception for haptics and a greater utilization of remote robotic telesurgery with telecollaboration.

33.3.1 Miniaturization

Some of the limitations of the current robotic system are the large size and the wide rotational radius of the arms; those factors limit the number of robotic arms that can be introduced into a specific body cavity. With miniaturization, the placement of multiple robotic arms in restricted spaces would be reasonably accomplished.

33.3.2 Multiple Consoles

The utilization of dual or multiple consoles would make possible for the surgeon and the assistant to manipulate multiple robotic arms at the same time. This would also allow the incorporation of a "teaching console" and collaborative remote telemanipulation.

33.3.3 Haptics

Another very important limitation of the current robotic systems is the absolute lack of tactile feedback. These systems lack in haptic feedback due to its reliance on computer interface between the master console and the slave's arms. In some academic institutions this is a field of major research where tactile surrogates are being added to the system, either by visual or auditory cues, or by computer-interfaced proprioceptive instrument feedback.

33.3.4 Flexible Endoluminal Robotic Arms

Even though on its infancy, Natural Orifice Transluminal Endoscopic Surgery (NOTES) has been employed on transgastric and transvaginal intraperitoneal procedures, however stills some major limitations due to its rudimentary instrumentation [114]. With the creation of small flexible robotic arms operated by telemanipulation that can be adopted for NOTES or endoluminal surgery, more complex surgical tasks can be accomplished making possible to advance those fields.

33.4 Conclusions

Although the future of robotics for gastrointestinal surgery seams pragmatically structured on a famous quote from Jules Verne "Anything one man can imagine, other men can make real", it actually relies on collected scientific data, patient demand, surgeon's necessity and healthcare cost. Yet, like Verne, academic institutions should think outside the box and push the envelope to create and identify new technologies, and test them before proven beneficial for physicians and patients.

To date robotic surgery has been responsible for a major revolution in areas such as urology and cardiac surgery, however its adoption in general surgery has been a very slow a tedious process, mainly due to the nature of the gastrointestinal operations. Furthermore the highly advanced laparoscopic skill set possessed by minimally invasive surgeons, make them less dependent on robots than other specialties.

The correct implementation of robotic assistance in laparoscopic surgery should be based the on added benefits for surgeons and patients, and can be stratified on necessity and feasibility. For the abdominal cavity, more simple operations can be easily performed with standard laparoscopic instrumentation, therefore the use of robots isn't necessary; on the other hand more complex operations would benefit the most form the use of robotic systems. Additionally, the robotic system is designed for single or dual quadrant application; thus complex cases that would benefit from those systems but at the same time require multiquadrant manipulation are not suited for the solo use of robots due to system limitations. Therefore, following these premises, complex surgeries for the hiatus, mediastinum and pelvic basin, plus necessity of precise hand-sewn anastomosis and need of stereoscopic visualization are the procedures where the robots should be wisely employed.

The healthcare cost has limited the utilization of gastrointestinal robotic surgery to academic institution. Perhaps the popularization and redistribution of old units to regional hospitals and production cost reduction, would positively impact the number of robotic cases performed for less complex operations. Lastly, it is very important to emphasize that a major limitation of those systems still the absolute lack of tactile feedback, which should be address with future prototypes.

Robotic surgical systems are in their infancy. For the next decades, they will undergo evolutionary changes, with units miniaturization, incorporation of haptics, implementation of integrated independent arms capable to perform multi-quadrant surgery, deployable intra-abdominal multitask toolboxes, utilization of augmented reality with the integration of computerized tomography and intra-operative ultrasound, and further utilization of remote telemanipulation. Hopefully all those improvements will come with cost reduction and in this near future those changes will make robotic surgical systems be part of the daily routine of the gastrointestinal surgeons.

References

1. Jones, D.B., Soper, N.J.: Laparoscopic general surgery: current status and future potential. AJR Am. J. Roentgenol. **163**(6), 1295–1301 (1994)
2. Gallagher, A.G., McClure, N., McGuigan, J., Ritchie, K., Sheehy, N.P.: An ergonomic analysis of the fulcrum effect in the acquisition of endoscopic skills. Endoscopy **30**(7), 617–620 (1988)
3. Hanuschik, M.: The technology of robotic surgery, Chapter 2 in Robotic Surgery, Gharagozloo, F. and Najam, F. (eds.), McGraw Hill (2008)
4. Satava, R.M.: Future applications of robotics. Prob. Gen. Surg. **20**(2), 79–85 (2003)
5. Sung, G.T., Gill, I.S.: Robotic laparoscopic surgery: a comparison of the da Vinci and ZEUS systems. Urology **58**(6), 893–898 (2001)
6. Stylopoulos, N., Rattner, D.: Robotics and ergonomics. Surg. Clin. N. Am. **83**, 1321–1337 (2003)
7. MacFarlane, M., Rosen, J., Hannaford, B., et al.: Force feedback grasper helps restore the sense of touch in minimal invasive surgery. J. Gastrointest. Surg. **3**, 278–285 (1999)
8. Panait, L., Rafiq, A., Mohammed, A., Mora, F., Merrell, R.C.: Robotic assistant for laparoscopy. J. Laparoendosc. Adv. Surg. Tech. A **16**(2), 88–93 (2006)
9. Finlay, P.A., Ornstein, M.H.: Controlling the movement of a surgical laparoscope. IEEE Eng. Med. Biol. Mag. **14**, 289–291 (1995)
10. Gill, I.S., Sung, G.T., Hsu, T.H., Meraney, A.M.: Robotic remote laparoscopic nephrectomy and adrenalectomy: the initial experience. J. Urol. **164**, 2082–2085 (2000)
11. Locke, G.R., III, Talley, N.J., Fett, S.L., et al.: Prevalence and clinical spectrum of gastroesophageal reflux: a population-based study in Olmsted County, Minnesota. Gastroenterology **112**(5), 1448–1456 (1997)

12. Campos, G.M., Peters, J.H., DeMeester, T.R., et al.: Multivariate analysis of factors predicting outcome after laparoscopic Nissen fundoplication. J. Gastrointest. Surg. **3**(3), 292–300 (1999)
13. Mareschaux, J., Smith, M.K., Folscher, D., Jamali, F., Malassagne, B., Leroy, J.: Telerobotic laparoscopic cholecystectomy: initial clinical experience with 25 patients. Ann. Surg. **234**, 1–7 (2001)
14. Anvari, M., McKinley, C., Stein, H.: Establishment of the world's first telerobotic remote surgical service: for provision of advanced laparoscopic surgery in a rural community. Ann. Surg. **241**(3), 460–464 (2005)
15. Kim, V.B., Chapman, W.H., Albrecht, R.J.: Early experience with telemanipulative robot-assisted laparoscopic cholecystectomy using Da Vinci. Surg. Laparosc. Endosc. Percutan. Tech. **12**, 34–40
16. Ries, L.A.G., Eisner, M.P., Kosary, C.L., et al.: SEER Cancer Statistics Review, 1975–2001. National Cancer Institute, Bethesda, MD (2004)
17. Korst, R.J., Altorki, N.K.: High grade dysplasia: surveillance, mucosal ablation, or resection? World J. Surg. **27**, 1030–1034 (2003)
18. Birkmeyer, J.D., Siewers, A.E., Finlayson, E.V., et al.: Hospital volume and surgical mortality in the United States. N. Engl. J. Med. **346**, 1128–1137 (2002)
19. Ohsawa, T.: Esophageal surgery. J. Jpn. Surg. Soc. **34**, 1318–1950 (1933)
20. Nguyen, N. T., Roberts, P., Follette, D.M., Rivers, R., Wolfe, B.M.: Thoracoscopic and laparoscopic esophagectomy for benign and malignant disease: lessons learned from 46 consecutive procedures. J. Am. Coll. Surg. **197**, 902–913 (2003)
21. Luketich, J.D., Schauer, P.R., Christie, N.A., Weigel, T.L., Raja, S., Fernando, H.C., Keenan, R.J., Nguyen, N.T.: Minimally invasive esophagectomy. Ann. Thorac. Surg. **70**, 906–911 (2000); discussion 911–902
22. De Paula, A.L., Hashiba, K., Ferreira, E.A., de Paula, R.A., Grecco, E.: Laparoscopic transhiatal esophagectomy with esophagogastroplasty. Surg. Laparosc. Endosc. **5**, 1–5 (1995)
23. Swanstrom, L.L., Hansen, P.: Laparoscopic total esophagectomy. Arch. Surg. **132**, 943–947 (1995); discussion 947–949
24. Nguyen, N.T., Roberts, P., Follette, D.M., Rivers, R., Wolfe, B.M.: Thoracoscopic and laparoscopic esophagectomy for benign and malignant disease: lessons learned from 46 consecutive procedures. J. Am. Coll. Surg. **197**, 902–913 (2003)
25. Espat, N.J., Jacobsen, G., Horgan, S., Donahue, P.: Minimally invasive treatment of esophageal cancer: laparoscopic staging to robotic esophagectomy. Cancer J. **11**, 10–17 (2005)
26. Law, S., Fok, M., Chu, K.M., Wong, J.: Thoracoscopic esophagectomy for esophageal cancer. Surgery **122**, 8–14 (1997)
27. Oelschlager, B.K., Pellegrini, C.A.: Role of laparoscopy and thoracoscopy in the treatment of esophageal adenocarcinoma. Dis. Esophagus. **14**, 91–94 (2001)
28. Luketich, J.D., Alvelo-Rivera, M., Buenaventura, P.O., Christie, N.A., McCaughan, J.S., Litle, V.R., Schauer, P.R., Close, J.M., Fernando, H.C.: Minimally invasive esophagectomy: outcomes in 222 patients. Ann. Surg. **238**, 486–494 (2003); discussion 494–485
29. Kawahara, K., Maekawa, T., Okabayashi, K., Hideshima, T., Shiraishi, T., Yoshinaga, Y., Shirakusa, T.: Video-assisted thoracoscopic esophagectomy for esophageal cancer. Surg. Endosc. **13**, 218–223 (1999)
30. Law, S., Wong, J.: Use of minimally invasive oesophagectomy for cancer of the oesophagus. Lancet. Oncol. **3**, 215–222 (2002)
31. Horgan, S., Berger, R.A., Elli, E.F., Espat, N.J.: Robotic-assisted minimally invasive transhiatal esophagectomy. Am. Surg. **69**(7), 624–626 (2003)
32. Moorthy, K., Munz, Y., Dosis, A., Hernandez, J., Martin, S., Bello, F., Rockall, T., Darzi, A.: Dexterity enhancement with robotic surgery. Surg. Endosc. **18**(5), 790–795 (2004)

33. Galvani, C.A., Gorodner, M. V., Moser, F., Jacobsen, G., Chretien, C., Espat, N. J., Donahue, P., Horgan, S.: Robotically assisted laparoscopic transhiatal esophagectomy. Surg. Endosc. **22**(1), 188–195 (2008)
34. Nguyen, N. T., Follette, D. M., Wolfe, B. M., Schneider, P. D., Roberts, P., Goodnight, J. E., Jr.: Comparison of minimally invasive esophagectomy with transthoracic and transhiatal esophagectomy. Arch. Surg. **135**, 920–925 (2000)
35. Okushiba, S., Ohno, K., Itoh, K., Ohkashiwa, H., Omi, M., Satou, K., Kawarada, Y., Morikawa, T., Kondo, S., Katoh, H.: Handassisted endoscopic esophagectomy for esophageal cancer. Surg. Today **33**, 158–161 (2003)
36. Bodner, J., Wykypiel, H., Wetscher, G., Schmid, T.: First experiences with the da Vinci operating robot in thoracic surgery. Eur. J. Cardiothorac. Surg. **25**, 844–851 (2004)
37. Dallemagne, B., Weerts, J.M., Jehaes, C., Markiewicz, S., Lombard, R.: Laparoscopic Nissen fundoplication: preliminary report. Surg. Laparosc. Endosc. **1**(3), 138–143 (1991)
38. Detailed Diagnoses and Procedures. National Hospital Discharge Survey (years 1993, 1998, and 2002). Atlanta, GA: Centers for Disease Control and Prevention. Available at: www.cdc.gov/nchs/about/major/hdasd/nhds.htm
39. Carlson, M.A., Frantzides, C.T.: Complications and results of primary minimally invasive antireflux procedures: a review of 10,735 reported cases. J. Am. Coll. Surg. **193**, 428–439 (2001)
40. Curet, M. J., Josloff, R.K., Schoeb, O., Zucker, K.A.: Laparoscopic reoperation for failed antireflux procedures. Arch. Surg. **134**, 559–563 (1999)
41. DePaula, A.L., Hashiba, K., Bafutto, M., Machado, C.A.: Laparoscopic reoperations after failed and complicated antireflux operations. Surg. Endosc. **9**, 681–686 (1995)
42. Frantzides, C.T., Carlson, M.A.: Laparoscopic redo Nissen fundoplication. J. Laparoendosc. Adv. Surg. Tech. A **7**, 235–239 (1997)
43. Coelho, J.C., Goncalves, C.G., Claus, C.M., Andrigueto, P.C., Ribeiro, M.N.: Late laparoscopic reoperation of failed antireflux procedures. Surg. Laparosc. Endosc. Percutan. Tech. **14**, 113–117 (2004)
44. Granderath, F.A., Kamolz, T., Schweiger, U.M., Pointer, R.: Laparoscopic refundoplication with prosthetic hiatal closure for recurrent hiatal hernia after primary failed antireflux surgery. Arch. Surg. **138**, 902–907 (2003)
45. Cadiere, G.B., Himpens, J., Vertruyen, M., Bruyns, J., Fourtanier, G.: Nissen fundoplication done by remotely controlled robotic technique. Ann. Chir. **53**(2), 137–141 (1999)
46. Cadiere, G.B., Himpens, J., Vertruyen, M., et al.: Evaluation of telesurgical (robotic) NISSEN fundoplication. Surg. Endosc. **15**(9), 918–923 (2001)
47. Heemskerk, J., van Gemert, W.G., Greve, J.W., Bouvy, N.D.: Robot-assisted versus conventional laparoscopic Nissen fundoplication: a comparative retrospective study on costs and time consumption. Surg. Laparosc. Endosc. Percutan. Tech. **17**(1), 1–4 (2007)
48. Ceccarelli, G., Patriti, A., Biancafarina, A., Spaziani, A., Bartoli, A., Bellochi, R., Casciola, L.: Intraoperative and postoperative outcome of robot-assisted and traditional laparoscopic Nissen Fundoplication. Eur. Surg. Res. **43**, 198–203 (2009)
49. Melvin, W.S., et al.: Computer-assisted robotic antireflux surgery. J. Gastrointest. Surg. **6**(1), 11–15 (2002); discussion 15–16. [PubMed: 11986012]
50. Talamini, M.A., Chapman, S., Horgan, S., Melvin, W.S.: The academic robotics group. A prospective analysis of 211 robotic-assisted surgical procedures. Surg. Endosc. **17**(10), 1521–1524 (2003). [PubMed: 12915974]
51. Horgan, S., Vanuno, D.: Robots in laparoscopic surgery. J. Laparoendosc. Adv. Surg. Tech. A. **11**(6), 415–419 (2001). [PubMed: 11814134]
52. Shimi, S., Nathanson, L. K., Cuschieri, A.: Laparoscopic cardiomyotomy for achalasia. J.R. Coll. Surg. Edinb. **36**(3), 152–154 (1991)
53. Pellegrini, C.A., Wetter, L.A., Pellegrini, C., et al.: Initial experience with a new approach for the treatment of achalasia. Ann. Surg. **216**, 291–296 (1992). [PubMed: 1417178]

54. Spiess, A.E., Kahrilas, P.J.: Treating achalasia: from whalebone to laparoscope. JAMA **280** (7), 638–642 (1998). [PubMed: 9718057]
55. Ellis, F.H., Jr.: Oesophagomyotomy for achalasia: a 22-year experience. Br. J. Surg. **80**(7), 882–885 (1993). [PubMed: 8369925]
56. Finley, R.J., Clifton, J.C., Stewart, K.C., et al.: Laparoscopic Heller myotomy improves esophageal emptying and the symptoms of achalasia. Arch. Surg. **136**, 892–896 (2001)
57. Zaninotto, G., Costantini, M., Molena, D., et al.: Minimally invasive surgery for esophageal achalasia. J. Laparoendosc. Adv. Surg. Tech. A **11**, 351–359 (2001)
58. Bloomston, M., Serafini, F., Rosemurgy, A.S.: Videoscopic Heller myotomy as first-line therapy for severe achalasia. Am. Surg. **67**, 1105–1109 (2001)
59. Horgan, S., Vanuno, D.: Robots in laparoscopic surgery. J. Laparoendosc. Adv. Surg. Tech. A **11**(6), 415–419 (2001)
60. Horgan, S., Galvani, C., Gorodner, M.V., Omelanczuck, P., Elli, F., Moser, F., et al.: Robotic-assisted Heller myotomy versus laparoscopic Heller myotomy for the treatment of esophageal achalasia: multicenter study. J. Gastrointest. Surg. **9**, 1020–1029 (2005); discussion 1029–1030
61. NIH Conference. Gastrointestinal surgery for severe obesity. Ann. Intern. Med. **115**, 956–961 (1991)
62. Nguyen, N. T., Goldman, C., Rosenquist, C. J., Arango, A., Cole, C. J., Lee, S. J., Wolfe, B. M.: Laparoscopic versus open gastric bypass: a randomized study of outcomes, quality of life, and costs. Ann. Surg. **234**(3), 279–89 (2001); discussion 289–291
63. Schauer, P., Ikramuddin, S., Hamad, G., Gourash, W.: The learning curve for laparoscopic Roux-en-Y gastric bypass is 100 cases. Surg. Endosc. **17**(2), 212–215 (2003)
64. Oliak, D., Owens, H., Schmidt, H. J.: Laparoscopic Roux-en-Y gastric bypass: Defining the learning curve. Surg. Endosc. **17**(3), 405–408 (2003)
65. Jacobson, G., Berger, R., Horgan, S.: The role of robotic surgery in morbid obesity. J. Laparoendosc. Adv. Surg. Tech. 13, 279–284 (2003)
66. Cadiere, G.B., Himpens, J., Germay, O., et al.: Feasibility of robotic laparoscopic surgery: 146 cases. World J. Surg. **25**(11), 1467–1477 (2001)
67. Mohr, C.J., Nadzam, G.S., Alami, R.S., Sanchez, B.R., Curet, M.J.: Totally robotic laparoscopic Roux-en-Y Gastric bypass: results from 75 patients. Obes. Surg. **16**(6), 690–696 (2006)
68. Favretti, F., Ashton, D., Busetto, L., Segato, G., De Luca, M.: The gastric band: first-choice procedure for obesity surgery. World J. Surg. **33**(10), 2039–2048 (2009)
69. Cadiere, G., Himpens, J., Vertruyen, M., Favretti, F.: The world's first obesity surgery performed by a surgeon at a distance. Obes. Surg. **9**, 206–209 (1999)
70. Muhlmann, G., Klaus, A., Werner, K., Wykypiel, G.: DaVinci robotic-assisted laparoscopic bariatric surgery: is it justified in a routine setting? Obes. Surg. **13**, 848–854 (2003)
71. Galvani, C., Horgan, S.: Robots in general surgery: present and future. Cir. Esp. **78**(3), 138–147 (2005)
72. Mühlmann, G., Klaus, A., Kirchmayr, W., et al.: DaVinci robotic-assisted laparoscopic bariatric surgery: is it justified in a routine setting? Obes. Surg. **13**(6), 848–854 (2003)
73. Wittgrove, A.C., Clark, G.W., Tremblay, L.J.: Laparoscopic gastric bypass, Roux-en-Y: preliminary report of five cases. Obes. Surg. **4**, 353–357 (1994)
74. Horgan, S., Vanuno, D.: Robots in laparoscopic surgery. J. Laparoendosc. Adv. Surg. Tech. A **11**(6), 415–419 (2001)
75. Cadière, G.B., Himpens, J., Germay, O. et al.: Feasibility of robotic laparoscopic surgery: 146 cases. World J. Surg. **25**, 1467–1477 (2001)
76. Sanchez, B.R., Mohr, C.J., Morton, J.M., et al.: Comparison of totally robotic laparoscopic Roux-en-Y gastric bypass and traditional laparoscopic Roux-en-Y gastric bypass. Surg. Obes. Relat. Dis. **1**(6), 549–554 (2005)
77. Ali, M.R., Bhaskerrao, B., Wolfe, B.M.: Robot-assisted laparoscopic Roux-en-Y gastric bypass. Surg. Endosc. **19**(4), 468–472 (2005)

78. Hubens, G., Balliu, L., Ruppert, M., et al.: Roux-en-Y gastric bypass procedure performed with the da Vinci robot system: is it worth it? Surg. Endosc. **22**(7), 1690–1696 (2008)
79. Oliak, D., Owens, M., Schmidt, H.J.: Impact of fellowship training on the learning curve for laparoscopic gastric bypass. Obes. Surg. **14**, 197–200 (2004)
80. Yu, S.C., Clapp, B.L., Lee, M.J., et al.: Robotic assistance provides excellent outcomes during the learning curve for laparoscopic Roux-en-Y gastric bypass: results from 100 robotic-assisted gastric bypasses. Am. J. Surg. **192**(6), 746–749 (2006)
81. Sanchez, B.R., Mohr, C.J., Morton, J.M., Safadi, B.Y., Alami, R.S., Curet, M.J.: Comparison of totally robotic laparoscopic Roux-en-Y gastric bypass and traditional laparoscopic Roux-en-Y gastric bypass. Surg. Obes. Relat. Disord. **1**, 549–554 (2005)
82. Jacobsen, G., Berger, R., Horgan, S.: The role of robotic surgery in morbid obesity. J. Laparoendosc. Adv. Surg. Tech. A **13**, 229–283 (2003)
83. Jacobs, M., Verdeja, J.C., Goldstein, H.S.: Minimally invasive colon resection (laparoscopic colectomy). Surg. Laparosc. Endosc. **1**(3), 144–150 (1991)
84. Lacy, A.M., Garcia-Valdecasas, J.C., Pique, J.M., et al.: Short-term outcome analysis of a randomized study comparing laparoscopic vs. open colectomy for colon cancer. Surg. Endosc. **9**(10), 1101–1105 (1995)
85. Schwenk. W., Böhm, B., Haase, O., Junghans, T., Muller, J.M.: Laparoscopic versus conventional colorectal resection: a prospective randomised study of postoperative ileus and early postoperative feeding. Langenbecks Arch. Surg. **383**(1), 49–55 (1998)
86. Young-Fadok, T.M., Radice, E., Nelson, H., Harmsen, W.S.: Benefits of laparoscopic-assisted colectomy for colon polyps: a case-matched series. Mayo Clin. Proc. **75**(4), 344–348 (2000)
87. Chen, H.H., Wexner, S.D., Iroatulam, A.J., et al.: Laparoscopic colectomy compares favorably with colectomy by laparotomy for reduction of postoperative ileus. Dis. Colon Rectum **43**(1), 61–65 (2000)
88. Duepree, H.J., Senagore, A.J., Delaney, C.P., Fazio, V.W.: Does means of access affect the incidence of small bowel obstruction and ventral hernia after bowel resection? Laparoscopy versus laparotomy. J. Am. Coll. Surg. **197**(2), 177–181 (2003)
89. Chen, H.H., Wexner, S.D., Iroatulam, A.J., et al.: Laparoscopic colectomy compares favorably with colectomy by laparotomy for reduction of postoperative ileus. Surg. Laparosc. Endosc. Percut. Tech. **12**, 52–57 (2002)
90. Weeks, J.C., Nelson, H., Gelber, S., Sargent, D., Schroeder, G.: Clinical outcomes of surgical therapy (COST) study group. Short-term quality-of-life outcomes following laparoscopic-assisted colectomy vs. open colectomy for colon cancer: a randomized trial. JAMA **287**(3), 321–328 (2002)
91. Ng, S.S., Leung, K.L., Lee, J.F., Yiu, R.Y., Li, J.C.: MRC CLASICC trial. Lancet **366**(9487), 713 (2005)
92. Lanfranco, A.R., Castellanos, A.E., Desai, J.P., Meyers, W.C.: Robotic surgery: a current perspective. Ann. Surg. **239**, 14–21 (2004)
93. Ballantyne, G.H., Moll, F.: The da Vinci telerobotic surgical system: the virtual operative field and telepresence surgery. Surg. Clin. N. Am. **83**, 1293–1304 (2003)
94. Hellan, M., Anderson, C., Ellenhorn, J.D., Paz, B., Pigazzi, A.: Short-term outcomes after robotic-assisted total mesorectal excision for rectal cancer. Ann. Surg. Oncol. **14**, 3168–3173 (2007)
95. DeNoto, G., Rubach, E., Ravikumar, T.S.: A standardized technique for robotically performed sigmoid colectomy. J. Laparoendosc. Adv. Surg. Tech. A **16**, 551–556 (2006)
96. Anvari, M., Birch, D.W., Bamehriz, F., Gryfe, R., Chapman, T.: Robotic-assisted laparoscopic colorectal surgery. Surg. Laparosc. Endosc. Percutan. Tech. **14**, 311–315 (2004)
97. Enker, W.E., Thaler, H.T., Cranor, M.L., Polyak, T.: Total mesorectal excision in the operative treatment of carcinoma of the rectum. J. Am. Coll. Surg. **181**, 335–346 (1995)

98. Pigazzi, A., Ellenhorn, J. D., Ballantyne, G. H., Paz, I. B.: Robotic-assisted laparoscopic low anterior resection with total mesorectal excision for rectal cancer. Surg. Endosc. **20**(10), 1521–1525 (2006)
99. Luca, F., Cenciarelli, S., Valvo, M., Pozzi, S., Faso, FL., Ravizza, D., Zampino, G., Sonzogni, A., Biffi, R.: Full robotic left colon and rectal cancer resection: technique and early outcome. Ann. Surg. Oncol. **16**(5), 1274–1278 (2009)
100. Patel, V.R., Chammas, M.F., Jr., Shah, S.: Robotic assisted laparoscopic radical prostatectomy: a review of the current state of affairs. Int. J. Clin. Pract. **61**, 309–314 (2007)
101. Altemeier, W.A., Giuseffi, J., Hoxworth, P.: Treatment of extensive prolapse of the rectum in aged or debilitated patients. AMA Arch. Surg. **65**(1), 72–80 (1952)
102. Chow, P.K., Ho, Y.H.: Abdominal resection rectopexy versus Delorme's procedure for rectal prolapse: comparison of clinical and physiological outcomes. Int. J. Colorectal Dis. **11**(4), 201–202 (1996)
103. Rose, S. M.: Classic articles in colonic and rectal surgery. Edmond Delorme 1847–1929. Dis. Colon Rectum **28**(7), 544–553 (1985)
104. D'Hoore, A., Cadoni, R., Penninckx, F.: Long-term outcome of laparoscopic ventral rectopexy for total rectal prolapse. Br. J. Surg. **91**(11), 1500–1505 (2004)
105. Rose, J., Schneider, C., Scheidbach, H., Yildirim, C., Bruch, H.P., Konradt, J., et al.: Laparoscopic treatment of rectal prolapse: experience gained in a prospective multicenter study. Langenbecks Arch. Surg. **387**(3–4), 130–137 (2002)
106. Solomon, M.J., Young, C.J., Eyers, A.A., Roberts, R.A.: Randomized clinical trial of laparoscopic versus open abdominal rectopexy for rectal prolapse. Br. J. Surg. **89**(1), 35–39 (2002)
107. Kairaluoma, M.V., Viljakka, M.T., Kellokumpu, I.H.: Open vs. laparoscopic surgery for rectal prolapse: a case-controlled study assessing short-term outcome. Dis. Colon Rectum **46** (3), 353–360 (2003)
108. Kariv, Y., Delaney, C.P., Casillas, S., Hammel, J., Nocero, J., Bast, J., et al.: Long-term outcome after laparoscopic and open surgery for rectal prolapse: a case-control study. Surg. Endosc. **20**(1), 35–42 (2006)
109. de Hoog, D. E., Heemskerk, J., Nieman, F. H., van Gemert, W. G., Baeten, C. G., Bouvy ND recurrence and functional results after open versus conventional laparoscopic versus robot-assisted laparoscopic rectopexy for rectal prolapse: a case-control study. Int. J. Colorectal Dis. **24**(10), 1201–1206 (2009)
110. Heemskerk, J., de Hoog, D.E., van Gemert, W.G., Baeten, C.G., Greve, J.W., Bouvy, N.D.: Robot-assisted vs. conventional laparoscopic rectopexy for rectal prolapse: a comparative study on costs and time. Dis. Colon Rectum **50**(11), 1825–1830 (2007)
111. Delaney, C.P., Lynch, A.C., Senagore, A.J., Fazio, V.W.: Comparison of robotically performed and traditional laparoscopic colorectal surgery. Dis. Colon Rectum **46**, 1633–1639 (2003)
112. Baker, R.S., Foote, J., Kemmeter, P., et al.: The science of stapling and leaks. Obes. Surg. **14** (10), 1290–1298 (2004)
113. Hanly, E.J., Talamini, M.A.: Robotic abdominal surgery. Am. J. Surg. **188**(4A Suppl), 19S–26S (2004)
114. Horgan, S., Cullen, J.P., Talamini, M.A., Mintz, Y., Ferreres, A., Jacobsen, G.R., Sandler, B., Bosia, J., Savides, T., Easter, D.W., Savu, M.K., Ramamoorthy, S.L., Whitcomb, E., Agarwal, S., Lukacz, E., Dominguez, G., Ferraina, P.: Natural orifice surgery: initial clinical experience. Surg. Endosc. **23**(7), 1512–1518 (2009)

Index

CPSIA information can be obtained at www.ICGtesting.com
Printed in the USA
LVOW070048180113

316216LV00003B/38/P

9 781441 911278